18 0118562 3

UNIVERSITY COLLEGE OF WALES
LIBRARY
ABERYSTWYTH

AF606614

ORGANIC REACTION MECHANISMS · 1974

ORGANIC REACTION MECHANISMS · 1974

An annual survey covering the literature dated December 1973 through November 1974

Edited by

A. R. BUTLER, University of St. Andrews

M. J. PERKINS, Chelsea College, University of London

An Interscience® Publication

JOHN WILEY & SONS

London · New York · Sydney · Toronto

An Interscience® Publication

Library of Congress Catalog Card Number 66-23143

ISBN 0 471 12693 4

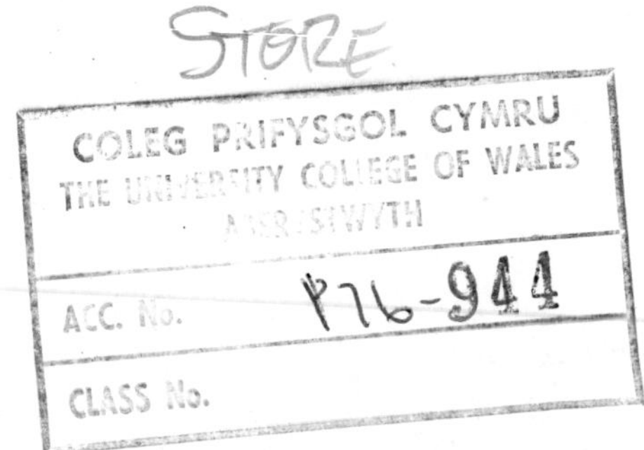

Printed in Great Britain by
William Clowes & Sons Limited
London, Colchester and Beccles

Contributors

M. S. BAIRD	Department of Organic Chemistry, University of Newcastle-upon-Tyne
R. BAKER	Department of Chemistry, The University, Southampton
T. W. BENTLEY	Department of Chemistry, University College of Swansea
R. B. BOAR	Department of Chemistry, Chelsea College, University of London
A. R. BUTLER	Department of Chemistry, The University, St. Andrews
B. CAPON	Department of Chemistry, Glasgow University
M. R. CRAMPTON	Department of Chemistry, Durham University
A. C. KNIPE	Department of Chemistry, The New University of Ulster
G. V. MEEHAN	Department of Chemistry and Biochemistry, James Cook University of North Queensland, Australia
D. C. NONHEBEL	Department of Pure and Applied Chemistry, University of Strathclyde
M. I. PAGE	Department of Chemistry, The Polytechnic, Huddersfield
M. J. PERKINS	Department of Chemistry, Chelsea College, University of London
B. V. SMITH	Department of Chemistry, Chelsea College, University of London
I. D. R. STEVENS	Department of Chemistry, The University, Southampton
J. C. WALTON	Department of Chemistry, St. Salvator's College, University of St. Andrews
W. E. WATTS	School of Physical Sciences, The New University of Ulster

Preface

This volume, continuing the general pattern now established for the series, covers literature actually dated December 1973 to November 1974. The principal aim is a general summary of the progress of work on organic reaction mechanism, rather than a detailed evaluation of selected topics; therefore, some sections are inevitably somewhat fragmentary, but all are concise, although our contributors are encouraged to emphasise results which appear at the time to represent major advances.

Since this is the tenth volume in the series, we have incorporated a second quinquennial subject index (1970–74), and would like to thank Terry Jenkins and Elizabeth Leitch for their major contributions to the demanding task of preparing this.

Unfortunately, owing to circumstances beyond his control, Dr. Meehan was unable to complete the Chapter on Molecular Rearrangements by our normal deadline; however, with his help we have been able to compile a survey which we hope affords adequate coverage of this important topic. In doing this we have been fortunate to have the sympathetic and expert co-operation of the British office of John Wiley and Sons, who also have once again done everything possible to expedite the rapid publication of this book.

August 1975

A.R.B.
M.J.P.

Contents

CHAPTER 1

Reactions of Aldehydes and Ketones and their Derivatives

B. CAPON

Chemistry Department, Glasgow University

Formation and Reactions of Acetals and Ketals[1–3]

Craze and Kirby[4] have studied the hydrolysis of a series of 4- and 5-substituted 2-(methoxymethoxy)benzoic acids (**1**) and analysed the results by Jaffé's method to yield values of $\rho_{\text{phenol}} = 0.89$ and $\rho_{\text{carboxy}} = 0.02$ for the spontaneous hydrolysis of the

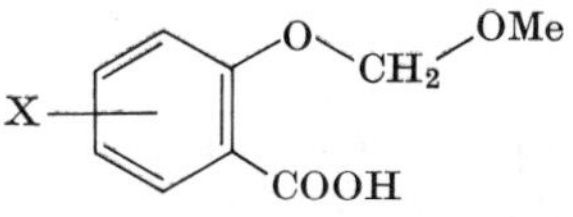

(**1**)

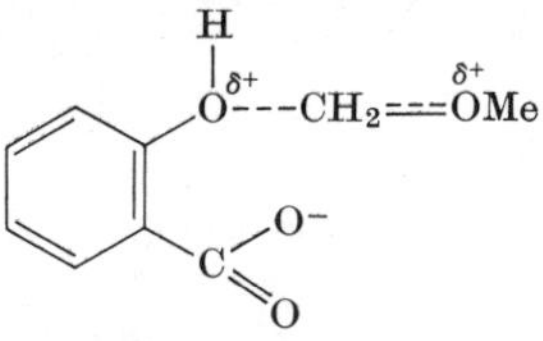

(**2**)

un-ionized form. The small value of $\rho_{carboxy}$ suggests that there is little proton transfer in the transition state which is not consistent with a mechanism (**2**) in which the ionized form undergoes a rapid specific acid-catalysed hydrolysis owing to an electrostatic effect[5] as here the proton should be completely transferred in the transition state. A mechanism involving intramolecular general-acid catalysis with a small amount of proton transfer seems more reasonable.[5]

It has been suggested that the hydrolysis of *o*-hydroxybenzaldehyde diethyl acetal involves intramolecular catalysis.[6] Unfortunately, the rate constant was not compared with that for the corresponding *para*-compound and so it is difficult to judge this claim. The corresponding 2-(hydroxyphenyl)-1,3-dioxans were studied in 1963 by Bender and Silver[7] who concluded that there was no intramolecular catalysis in the hydrolysis of the *ortho*-compound as it and the *para*-analogue reacted at very similar rates. The anion of *o*-hydroxybenzaldehyde diethyl acetal also undergoes measurable hydrolysis with $k_D/k_H = 0.469$; spontaneous ionization with a transition state (**3**) is a reasonable mechanism.[6]

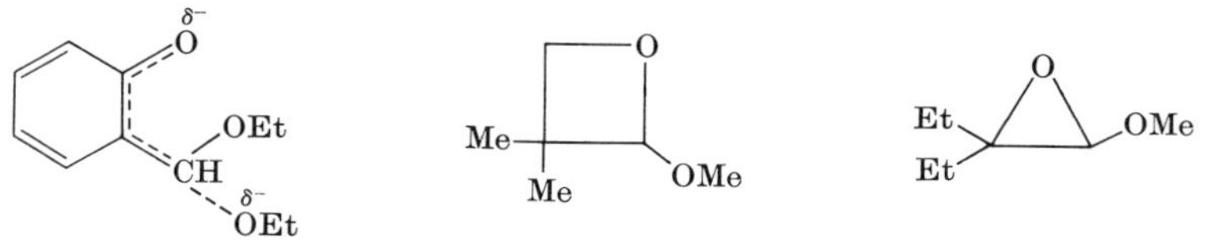

(**3**) (**4**) (**5**)

It has been shown that hydrolysis of the four-membered cyclic acetal (**4**), like that of similar three-membered ones,[8] is general-acid catalysed.[9] Presumably ring opening is such an easy process that it starts on only partial transfer of a proton. The hydronium-ion-catalysed hydrolysis of the three-membered compound (**5**) is about ten times faster than that of (**4**), and the solvent isotope effects are $k(D_3O^+)/k(H_3O^+) = 2.02$ and $1.8(\pm 0.9)$, respectively. At high pH, dehydration of the hemiacetal is the rate-limiting step (see p. 8).

General-acid catalysis has been detected in the conversion of the spiro-Meisenheimer complex (**6**) into (**7**).[10] This reaction stands in a logical sequence since it has been shown that increasing the stability of the intermediate carbonium ion formed from an acetal causes the reaction to become general-acid catalysed.[11] Since (**6**) is negatively charged

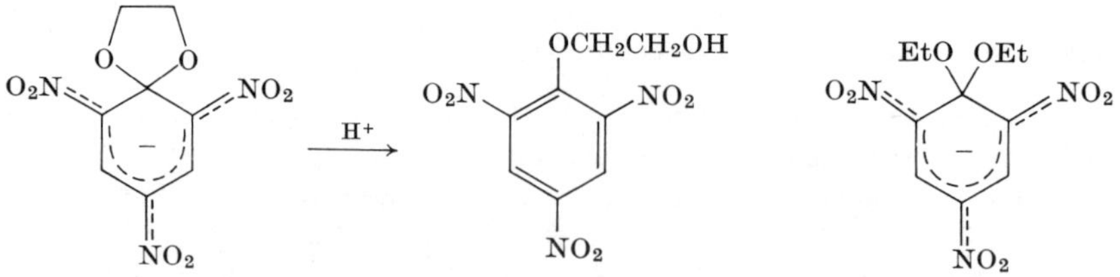

(**6**) (**7**) (**8**)

the fragment which corresponds to a carbonium ion is uncharged and so the detection of general-acid catalysis in the conversion of (**6**) into (**7**) is not unexpected. It is interesting that decomposition of the 1,1-diethoxy-complex (**8**) is not general-acid catalysed.

The hydrolysis of tetrahydro-2-(*p*-tolylthio)pyran is not general-acid catalysed. The plot of $\log k_{obs}$ against H_0 is a straight line with slope –1.10 in hydrochloric acid and –1.19 in perchloric acid. The solvent isotope effect is $k(D_3O^+)/k(H_3O^+) = 1.3$, the entropy of activation +7.4 cal mol^{-1} K^{-1}, and the ρ-value for hydrolysis of a series of tetrahydro-2-(*p*-substituted phenylthio)pyrans is –0.96. An *A*-1 mechanism which involves rate-limiting carbon—sulphur bond fission was proposed. The mercuric-ion-promoted reaction was also studied.[12a]

Acid hydrolysis of 2,2-diphenyl-4-(2-piperidyl)-1,3-dioxolan is about 50 times slower than that of 2,2-diphenyl-1,3-dioxolan, presumably because of the electrostatic effect of the protonated piperidyl residue.[12b] The hydrolyses of 4,4-dimethyl-1,3-dioxan in aqueous dioxan,[12c] of diethyl acetal catalysed by ferric ions,[13a] and of the acetals formed from substituted diphenylacetaldehydes and ethylene glycol[13b] have been studied.

It has been shown that the hydronium-catalysed hydrolyses of substituted acetophenone dimethyl acetals do not yield a linear Hammett plot when σ^+ constants are used.[13c] This behaviour is similar to that reported for the hydrolysis of substituted benzaldehyde diethyl acetals in 50% aqueous dioxan[14] but differs from that reported for their hydrolysis in water[15] (see also p. 10).

If the group R of dioxolans (**9**) is changed from H to Ph there is only a four-fold increase in the rate of hydrolysis. The rate constants for the hydrolysis of (**9**) with R = substituted phenyl are not correlated by the σ^+ or σ constants. Application of the

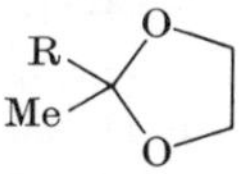

(**9**) 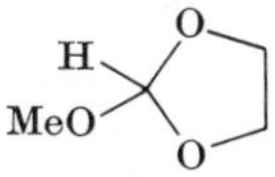 (**10**) 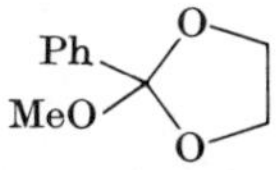 (**11**)

Yukawa–Tsuno equation leads to a ρ-value of –2.9 which is only slightly less than that for the hydrolysis of benzaldehyde diethyl acetals ($\rho = -3.35$). This suggests that the small effect of introducing a phenyl group does not arise mainly because there is steric inhibition to conjugation of this group with the developing carbonium ion in the transition state, but from some other effect. It was suggested that this was steric interaction of the 2-methyl group with the dioxolan ring.[16]

General-acid catalysis has been detected in the hydrolysis of the orthoesters (**10**) and (**11**). The catalytic constants for hydrolysis of (**11**) are about forty times greater than for (**10**). This contrasts with the results obtained with the acyclic orthoesters where the phenyl-substituted compound reacts more slowly than the unsubstituted one. It was suggested that the latter result arose from steric inhibition of resonance because of steric interaction, absent in compound (**11**), between the methoxyl and the phenyl groups.[17] It would be interesting if the ρ-value for hydrolysis of (**11**) were compared with that for the hydrolysis of trimethyl orthobenzoate.

The heats of hydrolysis of some orthoesters have been measured and their heats of formation calculated. There is a stabilizing interaction of 6–7 kcal mol^{-1} when three alkoxy-groups are attached to a single carbon atom.[18]

The hydrolysis of ortho-thio-esters has been studied.[19]

Hydrolysis of the acylals (**12**), (**13**) and (**14**) has been studied. The reactions are specific-acid and general-base catalysed and also show a pH-independent reaction. The entropy of activation and the variation of rate-constant with acid concentration suggest that the mechanism of the acid-catalysed hydrolysis of (**12**) is *A*2, presumably with

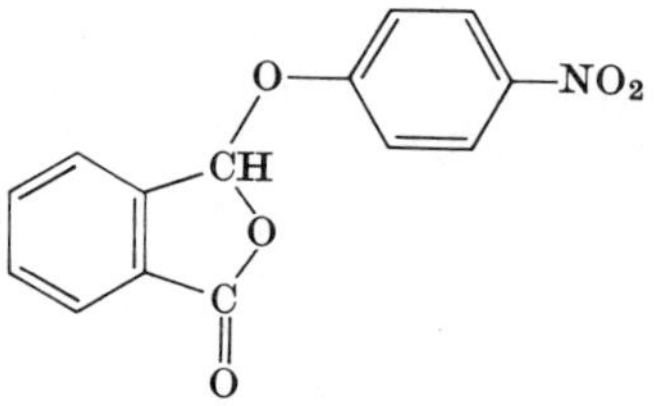

(12) 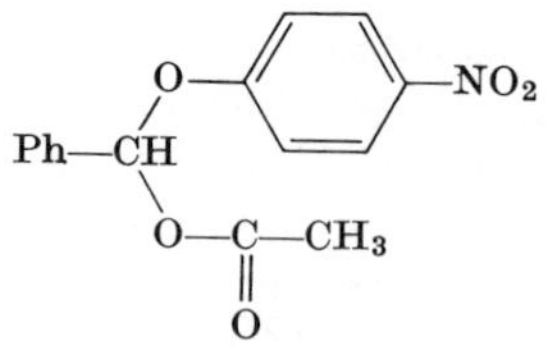 (13)

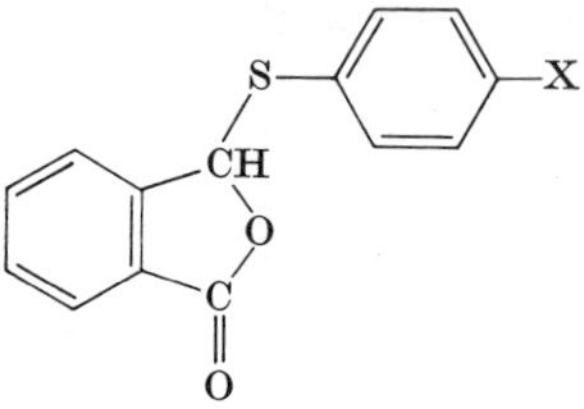

(14) 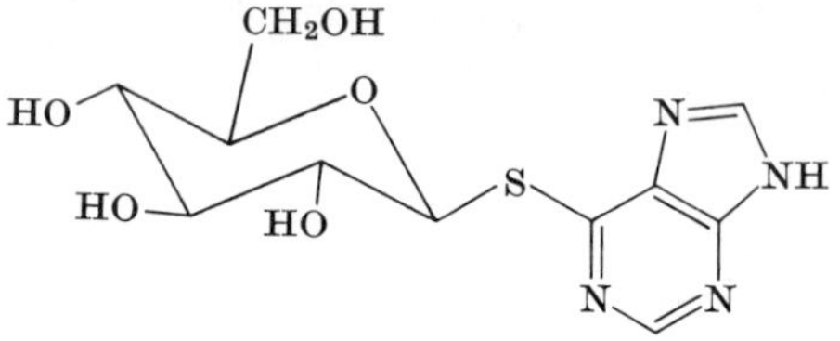 (15)

acyl—oxygen fission. It was also thought that the pH-independent reaction of (**13**) proceeded with attack of water on the acyl-carbon atom. The acid-catalysed reaction of (**12**) occurs about 1000 times slower than that of 3-methoxyphthalide, which suggests that the mechanism involving a cyclic carbonium ion previously proposed[20] is incorrect. It is possible but not certain that the acid-catalysed hydrolysis of the acyclic acetal-acylal (**13**) also proceeds by an *A*2 mechanism, with attack of water on the acyl-carbon atom. The base-catalysed reactions of (**12**) and its thio-analogues (**14**) occur with attack at the acyl-carbon atom (see p. 29).[21]

Hydrolysis of 1-ethoxyethyl *o*-acetoxybenzoate[22] and of acylals of substituted benzoquinones[23a] has been studied. Trifluoroacetic anhydride reacts with aldehydes to form *gem*-bis(trifluoroacetates). The ρ-value for the reaction of substituted benzaldehydes is -2.5.[23b]

Ion cyclotron resonance has been used to study the reactions of aldehydes with alcohols to form alkoxycarbonium ions.[24] The synthesis of 1,3-dioxolans using a catalytic dehydrator[25a] and the hydrolysis of dioxolans[25b] have been studied.

There have been investigations of the formation of acetals from crotonaldehyde catalysed by $Rh(CO)_4Cl_2$,[26a] from aldehydes and amide acetals,[26b] from 4-substituted indanediones,[27a] and from butyraldehyde and galactitol.[27b]

There have been numerous investigations of the conformations of cyclic acetals. The classes of compounds studied include 1,3-dioxans,[28,29] 1,3-dioxolans,[29] 1,3-oxathians,[30] 1,3-oxathiolans,[31] and 1,3-dithians.[29,32]

Hydrolysis and Formation of Glycosides[33]

Non-enzymic Reactions

No dimethyl acetals were detected in the Fischer glycoside synthesis with D-fructose and L-sorbose[34a] but they were present in that of D-galacturonic acid.[34b] The Fischer glycosidation of D-glucose has also been studied.[35a]

The rate of hydrolysis of 6-purinyl β-D-glucothiopyranoside (**15**) shows a first- and second-order dependence on the concentration of hydrogen ions, which suggests that reaction occurs via a mono- and a di-protonated conjugate acid. The reaction occurs about 10^6 times faster than that of phenyl β-D-thioglucoside. The solvent deuterium isotope effects for the reactions of mono- and di-protonated forms are $k_D/k_H = 2.5$ and 1.9, respectively. The hydrolysis catalysed by the β-glucosidase from almond emulsin was also studied.[35b]

The acid-catalysed hydrolysis of substituted phenyl α-D-galactopyranosides,[36] maltose and cellobiose 1-phenylflavazoles,[37] the alkaline hydrolysis of *p*-nitrophenyl β-D-xylopyranoside,[38] and the acetolysis of disaccharides[39] have been studied.

There has been a quantum-mechanical study of the acid-catalysed hydrolysis of alkyl and aryl hexopyranosides.[40]

Enzymic Reactions

(a) *Glucosidases.* A rapid-initial burst experiment[41] which indicates the formation of a glucosyl enzyme has been performed with almond-emulsin β-glucosidase and *p*-nitrophenyl β-D-glucoside in 50% aqueous dimethyl sulphoxide below −15°.

The α-glucosidase from *Saccharomyces oviformis* and methyl [^{14}C]-α-D-glucoside have been allowed to react in a rapid-mixing and -quenching apparatus which denatures the enzyme with phenol. The incorporation of radioactivity into the denatured enzyme suggested that 27% of it was present as a covalent glucosyl enzyme.[42]

Another investigation of the β-glucosidase from almond emulsin is described in reference 43; and the β-glucosidase from *Actinoplanes* sp. No. 1700[44] and the steroid β-glucosidase from rabbit muscle[45] have also been studied.

(b) *Lysozymes.*[46] The value of k_{cat}/k_m for hydrolysis of the 3,4-dinitrophenyl glycoside of $(NAG)_3$ catalysed by lysozyme from hen's egg white is at least 1400 times greater than that for the hydrolysis of the 3,4-dinitrophenyl glycoside of $(NAG)_2NAX$ (NAG = 1,4-β-linked *N*-acetyl-D-glucosamine, NAX = 1,4-β-linked *N*-acetyl-D-xylosamine). This is consistent with distortion of the conformation of the residue bound in the *D*-site owing to non-bonding interactions between the CH_2OH group and the enzyme and with contribution from this distortion to the driving force for enzymic catalysis. If this explanation is correct this driving force is at least 4.4 kcal mol^{-1}.[47]

Phillips and his co-workers have studied the binding of the δ-lactone derived from $(NAG)_4$ to lysozyme by X-ray crystallography and concluded that the lactone ring occupies the *D*-site and exists in a sofa conformation (**16**). It also seems that $(NAG)_4$

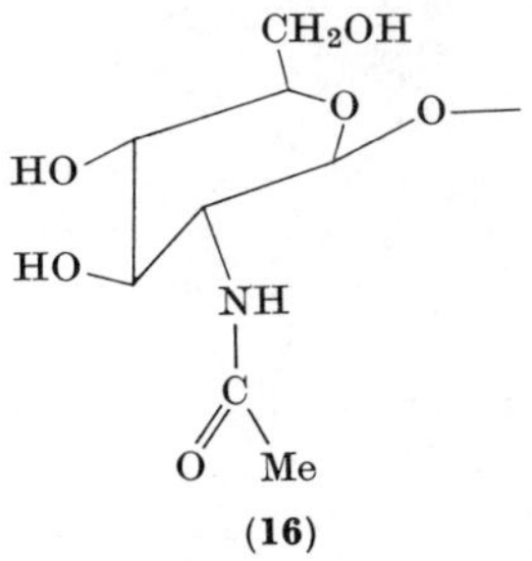

(**16**)

itself binds similarly and with one residue in the *D*-site in a sofa conformation.[48] It is interesting that this conformation, besides being a favourable conformation for formation of a carbonium ion at C-1, would probably also be more favourable for participation by the 2-acetamido-group than the original chair conformation.

There is only a factor of 2–3 between the binding constants of $(NAG)_3$ and $(NAG)_4$ to lysozyme and the complex of lysozyme with Gd^{3+}, which is consistent with the major mode of binding of these compounds being in the *A*, *B*, and *C* sites since it is thought that the Gd^{3+} ion binds at the junction of the *D* and the *E* sites. In contrast, the lactone from $(NAG)_4$ is bound 70 times less strongly to the Gd^{3+} complex than to lysozyme itself, which is consistent with this being bound with one residue in the *D*-site.[49] Biebrich Scarlet has also been used to block the *D*-site.[50]

There have been several investigations of the binding of sugars to lysozyme.[51]

Complexing of 3-carbamoyl-1-methylpyridinium chloride with lysozyme has been studied: the indole ring of tryptophan 62 and the phenol ring of tyrosine 23 probably act as donors.[52]

There have been other investigations of hen's egg-white lysozyme,[53] human lysozyme,[54,55] turkey lysozyme[55] and rat lysozyme.[56] There have also been further refinements of the three-dimensional structure of lysozyme from hen's egg white,[57] and an attempt has been made to separate the factors responsible for catalysis by lysozyme.[58]

(c) *Galactosidases.* β-D-Galactosylpyridinium salts are well-behaved substrates for the β-galactosidase of *E. coli*. The plot of $\log k_{cat}$ against pK_a of the pyridine residue has slope –0.93. The hydrolysis of these compounds presumably proceeds without electrostatic catalysis by the enzyme. The pyridinium salts are hydrolysed 10^4–10^5 times more slowly than aryl glycosides with leaving groups having the same pK_a.[59] Galactosylpyridinium salts are bound to the β-galactosidase 2–3 times less strongly than analogous *C*-glycosides, which suggests a small electrostatic repulsion by the active site. The effect is about 10 times larger with the corresponding β-D-glucosyl compounds and the β-glucosidase from almond emulsin, suggesting that there is a greater electrostatic repulsion here.[60]

The binding of D-galactal to the β-galactosidase from *E. coli* has been studied. The D-galactal probably reacts with the enzyme to form a covalent 2-deoxy-D-galactosyl–enzyme complex.[61]

The α-D-galactosidase from *Lens esculanta*,[62a] the α- and β-D-galactosidases from *Cajanus indicus*,[62b] and the galactosyl transferase from human milk[63] have been studied.

(d) *Amylases.*[64] The effect of chloride ion on the α-amylase from hog pancreas has been studied. It increases k_{cat} for starch and for *p*-nitrophenyl maltoside thirty-fold without changing K_m. Bromide, iodide, nitrite, nitrate, perchlorate, thiocyanate, azide and cyanate also increase k_{cat} but the effect decreases with increasing size of the ionic radius. Chloride ions also suppress the exchange of 26 protons and cause a 240-fold increase in the binding of calcium ions. It was thought that a conformational change occurs when chloride is bound to the enzyme but this was too small to be detected by circular dichroism, protein fluorescence or specific probes attached to the enzyme.[65]

From studies of hydrolysis of malto-oligosaccharides catalysed by saccharifying α-amylase from *B. subtilis* it has been concluded that the active site of the enzyme has 4 sub-sites.[66]

There have also been investigations on liquefying α-amylase from *B. subtilis*,[67,68] Taka amylase[68,69] and β-amylases.[70]

(e) *Other Glycosidases.* There have been investigations on the α-mannosidase from *Medicayo sativa L.*,[71] α-fucosidases from marine gastropods,[72] the 1,3-β-D-gluconase from *Rhizopus arrhizus*,[73] glucamylases,[74] Taka *N*-acetyl-β-D-glucosamidase,[75a], β-D-glucosaminidase from bull epididymis,[75b] β-*N*-glycylglucosaminidase from *Chaetopterus variopedatus*,[76] human β-D-*N*-acetylhexoseaminidases[77] and glucuronidases.[78]

Hydration of Aldehydes and Ketones and Related Reactions

The kinetics of mutarotation of series of 5-substituted (**17**)[79] and 2-substituted (**18**)[79,80] aldoses have been measured. When the substituent X in (**17**) is electron-withdrawing, the rate of the base-catalysed reaction is increased and the rates of the hydronium-ion catalysed and the spontaneous mutarotation are decreased.[79] When the substituent Y in (**18**) is electron-withdrawing the rates of the acid- and base-catalysed and the spontaneous mutarotations are decreased.[79,80] It was suggested that the hydronium-ion and base-catalysed mutarotations proceeded by step-wise pathways (eqns. 1 and 2) but that the spontaneous mutarotation proceeded by a concerted mechanism.[79] Intramolecular catalysis in the mutarotation of the sugars (**17**) with X = CH_2COOH and *o*-hydroxyphenoxymethyl was also studied.[79,81]

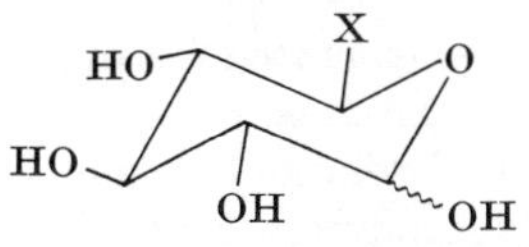

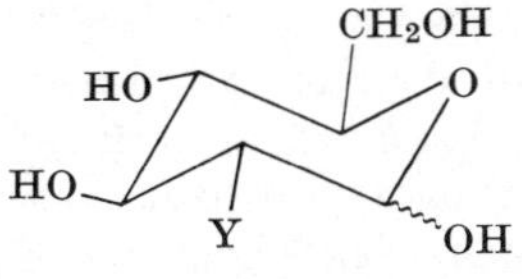

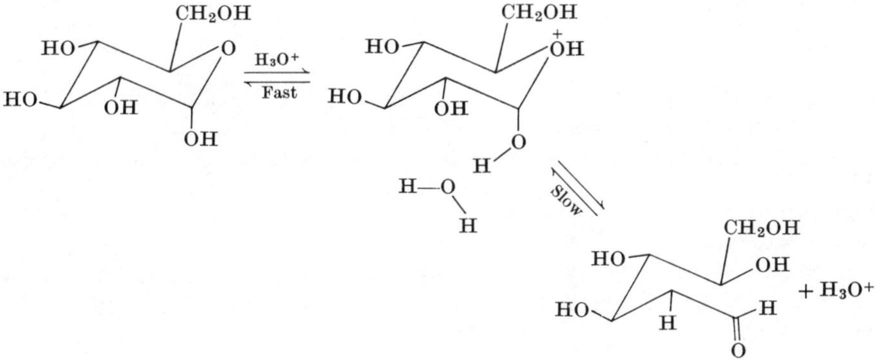

(**17**) (**18**)

(1)

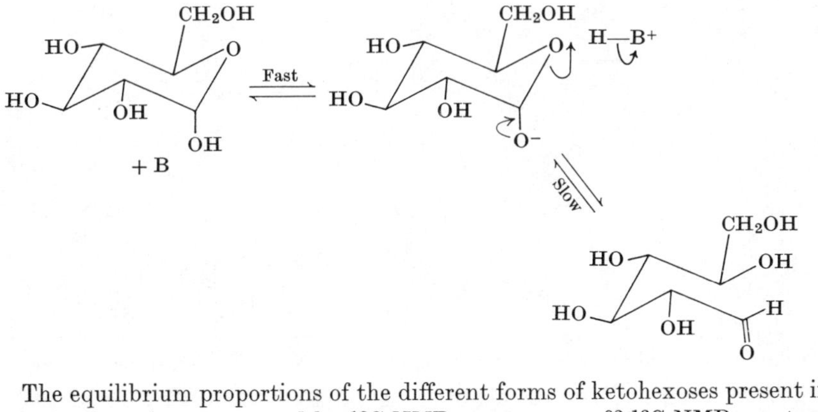

(2)

The equilibrium proportions of the different forms of ketohexoses present in aqueous solution have been measured by ^{13}C-NMR spectroscopy.[82] ^{13}C-NMR spectroscopy was also used to show that glutaraldehyde exists mainly in the cyclic hemiacetal form in

25% aqueous solution.[83] 8-Formylriboflavine contains a hemiacetal linkage between the formyl group and one of the hydroxyl groups (5′?) of the ribitol residue.[84]

The anomerization of D-glucose 6-phosphate catalysed by mutarotases[85] and the hydration of alkyl pyruvates catalysed by carbonic anhydrase[86] have been studied.

The decomposition of the methyl hemiacetal of 3-hydroxy-2,2-dimethylpropionaldehyde, which is an intermediate in the hydrolysis of the four-membered cyclic acetal (**4**), has been found to be general-acid and general-base catalysed[87a] (see p. 2).

The dehydration of hydrates of aromatic aldehydes generated by reduction of the corresponding acids at a rotating-disc electrode has been studied.[87b]

The following topics have also been investigated: formation of hemiacetals from aromatic aldehydes,[88a] hydration of glyoxylic acid,[88b] formation of internal hemiacetals from hydroxyalkylquinones,[89a] reaction of *p*-nitrobenzaldehyde with hydroxide ion,[89b] equilibration of tetracyclic hemiacetals,[90] hydration of *N*,*N*,*N*-triethyl-*N*-(β-formylvinyl)ammonium perchlorate,[91] addition of water to quinazoline,[92a] reaction of formaldehyde in water–methanol mixtures,[92b] and ring–chain tautomerism of 1-aryl-1-hydroxy-3,3-dimethylphthalans.[92c]

Ring–chain tautomerism has been reviewed.[93]

Reactions with Nitrogen Bases

Schiff Bases

Calculations which support the suggestion that the angle of attack of a nucleophile on a carbonyl group is ca. 105° have been reported.[94]

The proportion of quinonoid tautomer present in a solution of *N*-salicylideneisopropylamine has been measured; the neutral hydrolysis is thought to proceed through this tautomer.[95]

The exchange reaction of *N*-(*p*-nitrobenzylidene)aniline with *N*-benzylidene-*p*-anisidine to give *N*-(*p*-nitrobenzylidene)-*p*-anisidine and *N*-benzylideneaniline is of the first order in each Schiff base and also in hydrogen ions (solvent unspecified). The mechanism of equation (3) was proposed.[96]

$$\mathrm{Ar^1{-}CH{=}\overset{+}{N}H{-}R^1 \;+\; R^2{-}N{=}CH{-}Ar^2 \;\rightleftharpoons\; Ar^1{-}CH(\overset{+}{C}H{-}Ar^2)(N{-}R^2)\;\;NH{-}R^1 \;\rightleftharpoons\; \text{ring }[Ar^1CH{-}\overset{+}{N}H(R^1){-}CH(Ar^2){-}N(R^2)]}$$

$$\text{ring }[Ar^1CH{-}\overset{+}{N}H(R^1){-}CH(Ar^2){-}N(R^2)] \;\rightleftharpoons\; \mathrm{Ar^1{-}\overset{+}{C}H{-}N(R^2){-}CH(Ar^2){-}NH{-}R^1}$$

$$\mathrm{Ar^1{-}\overset{+}{C}H{-}N(R^2){-}CH(Ar^2){-}NH{-}R^1 \;\rightleftharpoons\; Ar^1{-}CH{=}N{-}R^2 \;+\; R^1{-}\overset{+}{N}H{=}CH{-}Ar^2} \qquad (3)$$

Amine exchange of substituted benzylideneanilines with *n*-butylamine has been studied: those compounds with an *ortho*-hydroxyl group were reported to react rapidly owing to hydrogen bonding.[97]

The kinetics of formation of isobutylideneanilines from isobutyraldehyde and substituted anilines in methanol have been measured. Formation of the carbinolamine is rapid and dehydration of this to the Schiff base is the slow step. Both steps are general-acid catalysed with the rates increased by *p*-methoxy-substituents and decreased by *p*-bromo-substituents ($\rho = 1.6 \pm 0.2$ and 3.4 ± 0.4).[98a]

N-Isopropylideneanilines undergo three kinds of reaction in CD_3OD: (1) Exchange of hydrogen, presumably via an enamine, the rate of which is increased by electron-releasing substituents and decreased by electron-withdrawing substituents. (2) Addition of methanol; this was detected only for compounds with electron-withdrawing substituents in the benzene ring; the rate decreases with increasing electron-withdrawing power of the substituent, but the extent of conversion at equilibrium increases. (3) Formation of acetone dimethyl ketal; this was also only observed for compounds with electron-withdrawing substituents.[98b]

The equilibrium constants for the base-catalysed addition of methanol to substituted benzylideneanilines in 90:10 v/v methanol–acetonitrile have a ρ-value +0.86 at 25° and the addition to benzylidene(substituted anilines) has $\rho = +1.58$. The rate constants show $\rho = 1.7$ for the reaction benzylidene(substituted anilines).[98c]

The enthalpy of addition of dimethylamine to formaldehyde has been measured.[98d]

The acid-catalysed hydrolysis of (α-cyanobenzylidene)anilines has been studied. The plot of rate constant against H_0 and pH has a maximum and it was suggested that hydration of the protonated form is rate-limiting on the less acidic side and that decomposition is rate-limiting on the more acidic side.[99a] The hydrolysis of (*p*-nitrobenzylidene)aniline has also been studied.[99b]

The cyanohydrin is formed rapidly and reversibly in aqueous solutions of acetone, cyanide and ammonia in the pH-range 8–12, but the α-amino-nitrile is formed via the imine.[100]

Formation of Schiff bases,[101] cyclization of *N*-(benzoylmethyl)-*o*-phenylenediamine,[102] and formation of the carbinolamine from pyridine-4-carboxaldehyde and sarcosine[103] have been studied. The equilibrium proportions of the *cis*- and *trans*-forms of Schiff bases have been measured.[104]

Transamination has been investigated.[105, 106a]

An intermediate Schiff base has been detected in the dehydroquinase-catalysed dehydrogenation of dehydroquinic acid to dehydroshikmic acid.[106b]

Enamines

Hydrolysis of the enamine (**19**) and incorporation of deuterium in D_2O have been measured by NMR and UV spectroscopy. At high acetate-buffer concentration (0.278 M), exchange is ten times faster than hydrolysis and this ratio decreases with decreasing buffer concentration.[107]

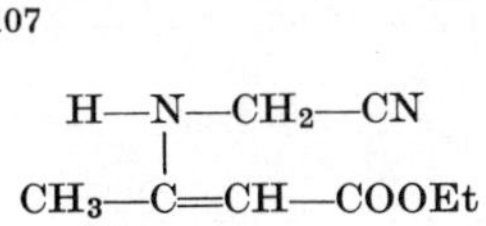

$$\begin{array}{c} H—N—CH_2—CN \\ | \\ CH_3—C{=}CH—COOEt \end{array}$$

(**19**)

Some kinetic studies on the hydrolysis of enaminones have been reported and a mechanism which involves protonation on oxygen was proposed.[108]

An enamine has been implicated as an intermediate in the conversion of β-amino-α-aryl-alcohols into β-arylcarbonyl compounds.[109]

There have been investigations of the stereochemistry of hydrolysis of enamines,[110] the kinetics of hydrolysis of enamines[111] and ynamines,[112] and the reaction of dienamines with electrophiles.[113]

Nucleosides and Glycosylamines

It has been shown that thymidine (**20**; X = Me) and deoxyuridine (**20**: X = H) undergo anomerization and ring expansion to be six-membered cyclic forms concurrently with hydrolysis.[114] These rearrangements must proceed via an acyclic form such as (**21**). At

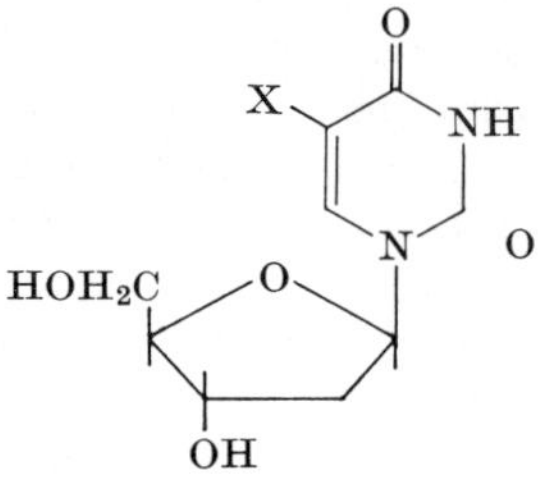

(**20**) 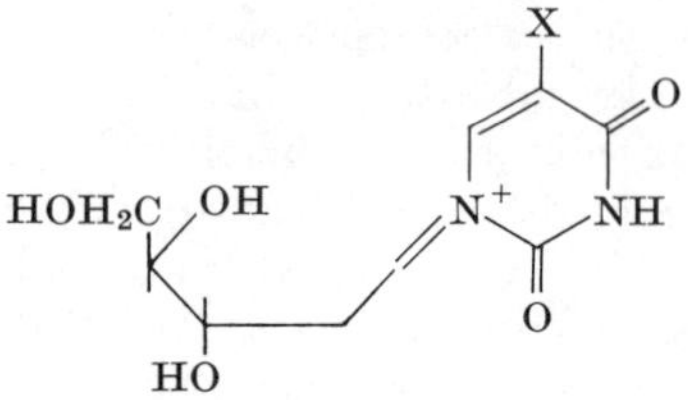 (**21**)

present it is difficult or impossible to decide whether this is also an intermediate in the hydrolysis or whether hydrolysis proceed via a cyclic ion. However, the demonstration that species resembling (**21**) are present under the hydrolysis conditions must reopen the question as to whether they are intermediates.[115] Isomerization of purine nucleosides in dilute alkali[116a] and their hydrolysis on ion-exchange resins[116b] have also been studied.

The kinetics of the Amadori rearrangement of D-glucosylamines[116c] have been measured, and the catalytic effect of alkali-metal chlorides on the rate of mutarotation of β-chloro-*N*-D-glucosylaniline has been studied.[116d]

Hydrazones, Oximes, Semicarbazones and Related Compounds

The kinetically formed ratio of *cis* to *trans* oximes from (trimethylammonio)acetaldehyde and hydroxylamine is 54:46 whereas the equilibrium ratio is 25:75.[117a]

Details of the determination of the α-deuterium isotope effects by Cordes and his co-workers for the reactions of semicarbazide and hydroxylamine with substituted benzaldehydes have been reported.[117b]

Thio-oximes have been prepared.[118]

Other reactions studied include hydrolysis of steroid oximes,[119a] hydrolysis of hydrazones,[119b] conversion of tosylhydrazones into olefins by organolithium reagents,[120] reaction of 1-alkyl-*cis*-3,5-diphenylpiperidin-4-ones with hydroxylamine,[121] formation of osazones,[122a] formation of dihydropyridazino[4,5-*b*]indoles from the benzylidene-hydrazides of indole-2-carboxylic acids,[122b] and oxidative hydrolysis of oximes by metal ions.[122c]

Hydrolysis of Enol Ethers

The rate constants for the acetic-acid catalysed and hydronium ion catalysed hydrolysis of α-methoxystyrenes are not correlated by the Hammett equation using σ^+ constants. The results were analysed by the Yukawa–Tsuno equation and were considered to arise from perturbation of LUMO of the α-methylbenzyl cation by the α-methoxy substituent[123] (see also p. 3).

The conversion of *o*-(1-methoxyvinyl)benzoic acid (**22**) into 2-*o*-acetylbenzoic acid (**24**) occurs partly via 3-methoxy-3-methylphthalide (**23**). There appears to be no intramolecular catalysis by the un-ionized carboxyl group of (**22**), and the ionized

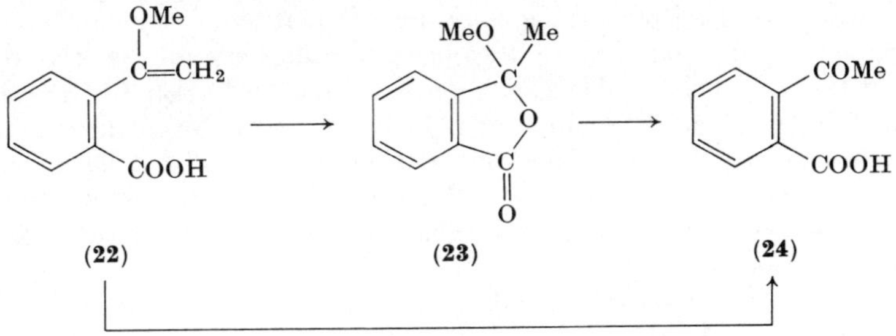

(22) **(23)** **(24)**

carboxyl group can provide only very little electrostatic or nucleophilic catalysis since the effect of ionization of the carboxyl group on the rate is only slightly greater than that for the *para*-isomer.[124]

The *cis*-forms of alkyl propenyl ethers undergo acid-catalysed hydrolysis 4 times more rapidly than the *trans*-forms. The methyl propenyl sulphides react about 800 times more slowly than the corresponding *O*-ethers.[125]

Hydrolysis of the vinyl ethers $CH_2{=}CR{-}CR'{=}CR''{-}OEt$ and $R''CH{=}CR'{-}C(OEt){=}CHR$ have been studied. The former compounds were shown by reaction in D_2O to react with deuteriation (and protonation) at the terminal CH_2 group.[126]

2-Methoxyfuran (**25**) yields 4-hydroxycrotonic acid lactone (**26**), methyl *cis*-4-hydroxycrotonate (**27**), and methyl succinaldehydate (**28**) in acidic DMSO–H_2O. The first two products were thought to arise from protonation at the 5-position and the last from protonation at the 3-position.[127]

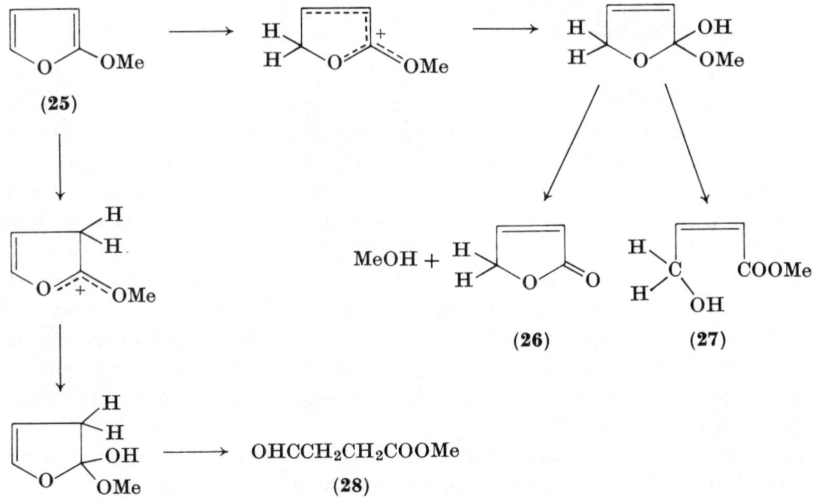

(25)

MeOH +

(26) **(27)**

$OHCCH_2CH_2COOMe$
(28)

The hydrolysis of divinyl acetals[128a] and of 1-alkoxybut-1-ene-3-ynes[128b] have been studied.

Enolization and Related Reactions

In 10% sulphuric acid the ratio of 3:1 chlorination (i.e. chlorination at position 3 to chlorination at position 1) of butan-2-one is 2.1:1, which is almost the same as that of

3:1 bromination (2.4:1). Under these conditions formation of the enols are the rate-determining steps and the ratio of 3:1 halogenation depends on the relative rates of formation of the two enols. At high acid concentration (96% H_2SO_4) the ratio of 3:1 chlorination is 9:1 and of 3:1 bromination is 11.5:1. Under these conditions enolization must have become a fast equilibrium step, so that the ratio of 3:1 halogenation now depends on the equilibrium proportions of the two enols and their relative rates of reaction with halogen. It was estimated that the enols react 6.2 times faster with bromine than with chlorine.[129]

The rate constant for the acetate-catalysed iodination of 4-acetylpyridine is 4.6 times greater than for 2-acetylpyridine. Protonation of the pyridine-nitrogen atom increases these constants by 20- and 200-fold, respectively, which suggests that there may be a small electrostatic stabilization of the transition state for reaction of the 2-isomer. The reaction of 2-acetylpyridine is strongly catalysed by Zn^{++}, Ni^{++} and Cu^{++}, the catalytic efficiency increasing in this order. A transition state (**29**) was proposed. The deuterium isotope effect was ca. 7 for the acetate-catalysed reaction of the conjugate acid and of the metal-ion complexes of 2-acetylpyridine.[130]

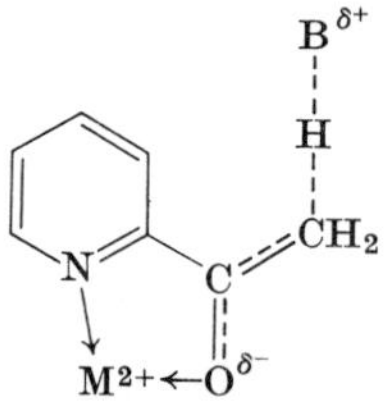

(**29**) 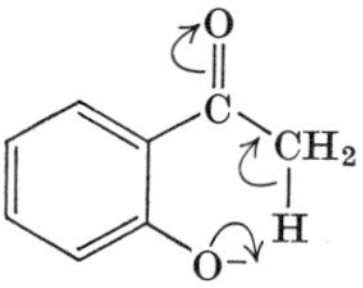 (**30**)

Iodination of the anion of *o*-hydroxyacetophenone occurs more rapidly than is normally found for aliphatic ketones, which suggests that the rate-limiting proton abstraction is intramolecularly catalysed, probably as shown in (**30**). Comparison with an analogous intermolecular process indicates that the effective concentration of the internal phenoxide group is quite small, ca. 1.3 mol l^{-1}. In contrast, the intramolecular nucleophilic substitution which follows the halogenation step has an effective concentration of ca. 400 mol l^{-1}. It was suggested that this difference arose because the bending frequencies in the transition state for the proton-transfer reaction are lower than in the transition state for the nucleophilic substitution. Hence the entropy loss on formation of the transition state should be less for the proton transfer than for the nucleophilic substitution, thus leading to a smaller difference between the rates for the inter- and intra-molecular reactions.[131]

The base-catalysed and spontaneous iodination of 2-oxocyclohexane-carboxylic acid are 500–1000-fold faster than with the methyl ester, which suggests that the carboxyl group provides intramolecular catalysis.[132]

The kinetic of the deprotonation of β-dicarbonyl compounds by OH^- have been measured by fast-reaction techniques. The deprotonation of the enol is fast with $k = 2.5 \times 10^8$ l mol^{-1} s^{-1} for the enol of ethyl acetoacetate and 2.4×10^8 l mol^{-1} s^{-1} for ethyl 2-hydroxycyclohex-2-enecarboxylate. As these rate constants are less than the rate constant for diffusion it was suggested that the enols were intramolecularly hydrogen-bonded. The rate constants for deprotonation of the keto forms are much lower, 10^1–10^3 l mol^{-1} s^{-1}.[133]

Details of Toullec and Dubois' work on the isotope effects for ketonization of enols has been published.[134]

Catalysis of dedeuteriation of [$^2H_6$]acetone by diamines, which occurs by intramolecular deuteron abstraction in an intermediate Schiff base, has been demonstrated. With acyclic amines the most effective catalyst was 3-(dimethylamino)propylamine which would react via an eight-membered ring including the deuteron. This is presumably the result of a combination of several factors, including the requirement for a linear transition state, eclipsing of the CH bonds and loss of rotational entropy. Enhanced catalytic activity was found with *cis*- and *trans*-2-(dimethylamino)cyclopentylamine: with these compounds there are no new eclipsing interactions on going to the transition state.[135] Dedeuteriation of [2-^{2}H]isobutyraldehyde catalysed by polyethyleneamines[136] and octakis-*O*-(3-aminopropyl)sucrose[137] has also been studied.[138]

An investigation of the iodination of glycolaldehyde dihydrogenphosphate catalysed by ethylenediamine has been reported.[139]

Solvent effects in proton transfer from carbon acids (including ketones) have been studied.[140]

The "principle of least motion" has been applied to enolization and homoenolization.[141]

Other topics that have been studied include stereochemistry of halogenation of exocyclic cyclohexyl enols,[142a] detritiation of ketones with trifluoromethyl-groups,[142b] iodination of 2,3-dimethyl-1-phenylpyrazolin-5-one and 3-methyl-1-phenylpyrazolin-5-one,[143] reaction of ketones with iodine monochloride in methanol,[144] zinc-catalysed enolization of oxaloacetate which precedes decarboxylation,[145] hydration of $\alpha\beta$-

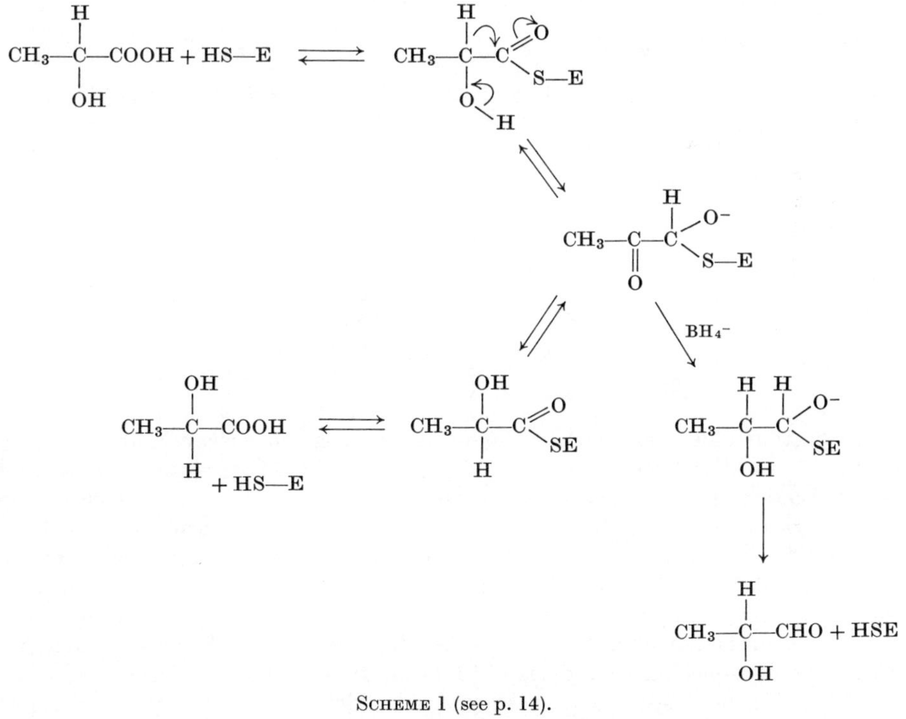

SCHEME 1 (see p. 14).

unsaturated ketones,[146] base-catalysed tautomerization of anthrone to anthrol,[147] conversion of ketones into their potassium derivatives by treatment with potassium hydride in tetrahydrofuran,[148] iodine exchange between molecular iodine and iodomethyl-carbonyl compounds,[149] and alkylation of enolates.[150, 151]

There have been several investigations of keto–enol equilibria.[152, 153]

Dihydroxyacetone phosphate undergoes stereoselective reduction by borohydride when bound to triose phosphate isomerase from chicken muscle. The reaction occurs about seven times faster than in free solution, which is consistent with mild electrophilic catalysis by the enzyme for the reduction.[154]

An α-carbonyl intermediate has been trapped by borohydride in the racemization of lactic acid catalysed by lactate racemase from *Clostridium butylicium*. When the reaction was carried out in the presence of borohydride, lactaldehyde was isolated from the reaction mixture. A mechanism as shown in Scheme 1 (p. 13) was proposed.[155]

Other work on triose phosphate isomerase[156] and enolase[157] has also been reported.

Homoenolization

Homoenolization of adamantanone in KOBu^t/Bu^{t}OD has been studied by ^{2}H-NMR spectroscopy. Proton exchange at the *exo*-β-position occurs about 15 times faster than at the *endo*-β-position. α-Exchange occurs at about half the rate of *exo*-β-exchange. The preferred stereochemistry is the same as that found with camphenilone.[158]

It has been shown that homoketonization of 1-acetoxynortricyclane to yield 2-norbornanone occurs with 94% inversion of configuration in D_2SO_4–CH_3COOD whereas in CH_3COOD–D_2O–D_2SO_4 it occurs with retention of configuration.[159]

Aldol and Related Reactions

Catalysis by primary amines for the retroaldol reaction of diacetone alcohol has been measured. The catalytic activity of $Me_2NCH_2CH_2NH_2$ is about four times greater than would be expected from its pK_a and the Brönsted plot for other amines, and it was suggested that the reaction proceeded via a cyclic transition state (**31**).[160] The retroaldol

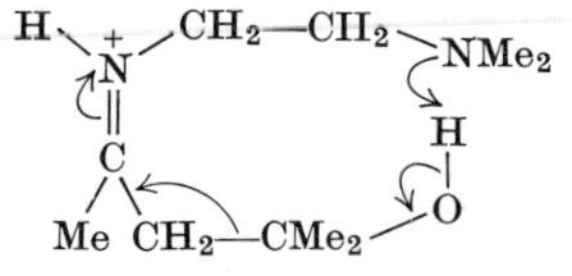

H, $\overset{+}{N}$, CH_2—CH_2, NMe_2, H, O, C, Me, CH_2—CMe_2

(**31**)

reaction of crotonaldehyde to yield acetaldehyde,[161] the alkaline cleavage of ketones,[162a] and the base-catalysed decomposition of diacetone alcohol[162b] have also been studied.

Aldol reactions investigated include condensation of methyl acetoacetate dianion with ketones and aldehydes to yield δ-hydroxy-β-oxo-esters,[163] cyclization of alkane-2,5-diones,[164a] self-condensation of glyceraldehyde[164b] and of aliphatic aldehydes catalysed by phenoxymagnesium bromide,[165] the aldol reactions of methylglyoxal,[166a] cyclohexanone[166b] and β-methyl branched ketones,[166c] and the condensation of the following reactants: aldehydes with formaldehyde,[167a] *p*-(dimethylamino)benzaldehyde with *N*-methyl-4-picolinium iodide,[167b] 1,4-diacetylpiperazine-2,5-dione with aldehydes,[168] acetone with acetaldehyde,[169a] aromatic aldehydes with acetone catalysed by cupric ion,[169b] *n*-heptaldehyde with phosphatidyl ethanolamine,[170] *n*-valeraldehyde with cyclopentanone,[171] and arylpyruvic acids with rhodanine.[172]

The conversion of [2-^{18}O]-fructose 1,6-diphosphate into dihydroxyacetone phosphate catalysed by rabbit-muscle aldolase occurs with complete loss of the label and when catalysed by yeast aldolase with retention of the label. This is consistent with the former reaction proceeding via a Schiff base and the latter via a metal complex.[173]

Ferricyanide is reduced by a mixture of aldolase and dihydroxyacetone phosphate. This was thought to involve reaction of the carbanion of the latter formed by proton abstraction by the aldolase.[174] It appears that the enzyme utilizes the keto- and hydrated forms of dihydroxyacetone phosphate.[175]

N-(Bromoacetyl)ethanolamine is an active-site directed, irreversible inhibitor for rabbit- and rat-muscle aldolase.[176a] The amino-acid sequence of rabbit-muscle aldolase has been determined.[176b]

There have been other investigations on aldolase from rabbit muscle[177] and *Lactobacillus casei*.[178]

Other Reactions

The cyclization of 1,3,5-triphenylpent-2-ene-1,5-dione into the 2,4,6-triphenylpyrylium ion has been studied. The reaction probably proceeds via the enol which cyclizes to the hemiacetal as shown in Scheme 2.[179]

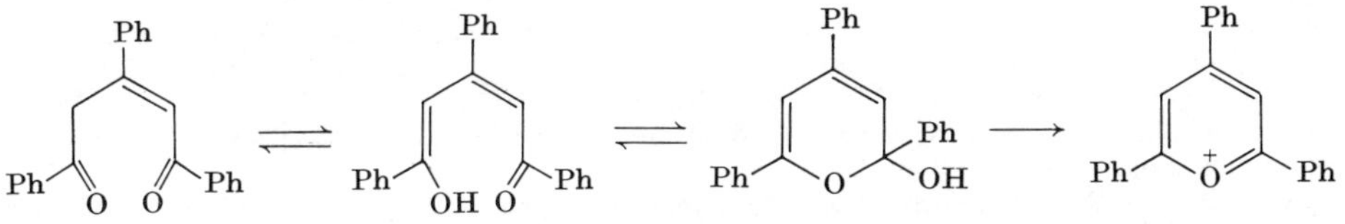

Scheme 2.

Proton exchange of the conjugate acid of *p*-chlorobenzaldehyde in SbF_5–FSO_3H has been studied by measuring the variation of the proton and ^{13}C-NMR spectra under conditions where rotation in the deprotonated form was fast and deprotonation was the rate-limiting step in the interconversion of the different rotamers.[180]

The protonation of ketones[181] and the relative stability of *O*-methylated ketones[182a] have been studied.

The oxidation of several 4-deoxy-analogues of methyl β-D-galactopyranoside by D-galactose oxidase follows the sequence OH > NH_2 > F $\gg$ Cl > H. It was concluded that hydrogen-bonding between the 4-hydoxyl group of the substrate and the enzyme was not important for catalytic action.[182b]

There have also been investigations of the following reactions: condensation of formaldehyde with diphenyl sulphide,[183] acyloin condensation of benzaldehyde catalysed by thiazolium salts,[184] Leuckart–Wallach reaction of formaldehyde with dimethylformamide,[185] acid-catalysed disproportionation of formaldehyde,[186] condensation of phenol with formaldehyde,[187,188] the Schmidt reaction,[189] the Kiliani reaction of galactose and arabinose with HCN,[190a] acid-catalysed isomerization of D-glucose into D-fructose.[190b] Wittig reaction,[191] nucleophilic alkylidene transfer by sulphonium and oxosulphonium ylides,[192] the Biginelli reaction,[193] Cannizzaro reaction of anisaldehyde[194] and *m*-nitrobenzaldehyde,[195] Voyes–Proskauer reaction,[196] Knoevenagel reaction,[197a] Baeyer–Villiger reaction,[197b] stereochemistry of addition of trimethylaluminium to decalones,[198] reaction of aldehydes and ketones with chloro- and fluoro-sulphonyl isocyanate,[199] and the Grignard reaction.[200]

The mechanism of the Grignard reagent addition to ketones[201] and the acid-catalysed rearrangements of β,γ-unsaturated ketones[202] have been reviewed.

References

1 E. H. Cordes and H. G. Bull, *Chem. Rev.*, **74**, 581 (1974).
2 B. M. Dunn, *Int. J. Chem. Kinet.*, **6**, 143 (1974); *Chem. Abs.*, **80**, 94861 (1974).
3 R. M. Gandour, G. M. Maggiora and R. L. Schowen, *J. Am. Chem. Soc.*, **96**, 6967 (1974).
4 G.-A. Craze and A. J. Kirby, *J. C. S. Perkin II*, **1974**, 61.
5 See *Org. Reaction Mech.*, **1971**, 394; **1972**, 390.
6 A. Kankaanperä, L. Oinonen and M. Lahti, *Acta Chem. Scand. A*, **28**, 442 (1974).
7 M. L. Bender and M. L. Silver, *J. Am. Chem. Soc.*, **85**, 3006 (1963).
8 See *Org. Reaction Mech.*, **1972**, 389.
9 R. F. Atkinson and T. C. Bruice, *J. Am. Chem. Soc.*, **96**, 819 (1974).
10 M. Crampton and M. J. Willison, *J. C. S. Perkin II*, **1974**, 1686.
11 See *Org. Reaction Mech.*, **1969**, 399–400; **1971**, 393.
12a L. R. Fedor and B. S. R. Murty, *J. Am. Chem. Soc.*, **95**, 8407 (1973).
12b L. W. Brown and A. A. Forist, *J. Pharm. Sci.*, **62**, 1365 (1973).
12c V. M. Zakoshanki, G. S. Idlis and S. K. Ogorodnikoc, *Zh. Org. Khim.*, **10**, 1967 (1974).
13a G. Wada and M. Sakamoto, *Bull. Chem. Soc. Japan*, **46**, 3378 (1973).
13b B. L. Jensen and R. E. Counsell, *Can. J. Chem.*, **51**, 3820 (1973).
13c G. M. Loudon and C. Berke, *J. Am. Chem. Soc.*, **86**, 4508 (1974).
14 T. H. Fife and L. K. Jao, *J. Org. Chem.*, **30**, 1492 (1965).
15 R. B. Dunlap, G. A. Ghanim and E. H. Cordes, *J. Phys. Chem.*, **73**, 1898 (1969).
16 V. P. Vitullo, R. M. Pollack, W. C. Faith and M. L. Keiser, *J. Am. Chem. Soc.*, **96**, 6682 (1974).
17 Y. Chiang, A. J. Kresge, P. Salomaa and C. I. Young, *J. Am. Chem. Soc.*, **96**, 4494 (1974).
18 J. Hine and A. W. Klueppel, *J. Am. Chem. Soc.*, **96**, 2924 (1974).
19 R. A. Ellison, W. D. Woessner and C. C. Williams, *J. Org. Chem.*, **39**, 1430 (1974).
20 See *Org. Reaction Mech.*, **1973**, 2.
21 T. H. Fife and N. C. De, *J. Am. Chem. Soc.*, **96**, 6158 (1974).
22 A. Hussain, M. Yamasaki and J. E. Truelove, *J. Pharm. Sci.*, **63**, 627 (1974).
23a R. T. Borchardt and L. A. Cohen, *J. Am. Chem. Soc.*, **95**, 8319 (1973).
23b A. L. Ternay, D. Deavenport and G. Bledsoe, *J. Org. Chem.*, **39**, 3268 (1974).
24 J. K. Pau, J. K. Kim and M. C. Caserio, *Chem. Comm.*, **1974**, 120.
25a V. I. Sternberg and D. A. Kubik, *J. Org. Chem.*, **39**, 2815 (1974).
25b J. Y. Conan, A. Natat, F. Guinot and G. Lamaty, *Bull. Soc. chim. France*, 1405 (1974); J. Y. Conan, F. Guinot, G. Lamaty and A. Natat, *Tetrahedron Letters*, **1974**, 1667.
26a R. V. Hoffman, *Tetrahedron Letters*, **1974**, 2415.
26b H. Vorbrüggen, *Annalen*, **1974**, 821.
27a K. Sakane, Y. Otsuji and E. Imoto, *Bull. Chem. Soc. Japan*, **47**, 2410 (1974).
27b T. G. Bonner, E. J. Bourne, D. Lewis and L. Yuceer, *Carbohydrate Res.*, **33**, 1 (1974).
28 E. L. Eliel and O. Hofer, *J. Am. Chem. Soc.*, **95**, 8041 (1973); E. L. Eliel and F. Alcudia, *ibid.*, **96**, 1939 (1974); W. F. Bailey and E. L. Eliel, *ibid.*, p. 1798; E. L. Eliel, J. R. Powers and F. W. Nader, *Tetrahedron Letters*, **30**, 513 (1974); G. Binsch, E. L. Eliel and S. Mager, *J. Org. Chem.*, **38**, 4079 (1973); G. Borgen and J. Dale, *Chem. Comm.*, **1974**, 484; S. Mager and E. L. Eliel, *Rev. Roum. Chim.*, **18**, 2097 (1973); M. C. Vertut, J. P. Fayet, G. Chassaing and L. Cazaux, *Compt. Rend. Ser. C.*, **277**, 635 (1973); M. Anteunis and M. Coryn, *Bull. Soc. Chem. Belges*, **83**, 133 (1974); E. Bernaert, M. Anteunis and D. Tavernier, *ibid.*, p. 357.
29 P. E. Stevenson, G. Bhat, C. H. Bushweller and W. G. Anderson, *J. Am. Chem. Soc.*, **96**, 1067 (1974); D. Chambenois and G. Mousset, *Compt. rend., Ser. C.*, **278**, 1231 (1974).
30 K. Pihlaja and P. Pasanen, *J. Org. Chem.*, **39**, 1948 (1974).
31 G. E. Wilson, *J. Am. Chem. Soc.*, **96**, 2426 (1974); R. Keskinen, A. Nikkila, K. Pihlaja and F. G. Riddell, *J. C. S. Perkin II*, **1974**, 466.
32 K. Pihlaja, *J. C. S. Perkin II*, **1974**, 890.
33 R. V. Lemieux and S. Koto, "The Conformational Properties of Glycosidic Linkages", *Tetrahedron*, **30**, 1933 (1974).
34a G. S. Bethell and R. J. Ferrier, *Carbohydrate Res.*, **31**, 69 (1973).

[34b] K. Larsson and G. Petersson, *Carbohydrate Res.*, **34**, 323 (1974).
[35a] D. F. Mowery, Jr., *J. Phys. Chem.*, **78**, 1918 (1974).
[35b] L. R. Fedor and B. S. R. Murty, *J. Am. Chem. Soc.*, **95**, 8410 (1973).
[36] C. K. De Bruyne and H. Carchon, *Carbohydrate Res.*, **33**, 117 (1974).
[37] J. N. BeMiller, D. R. Smith, V. K. Ghanta, T. O. Lyles and E. R. Doyle, *Carbohydrate Res.*, **35** 255 (1974).
[38] C. K. De Bruyne, F. Van Wijnedaele and H. Carchon, *Carbohydrate Res.*, **33**, 75 (1974).
[39] L. Rosenfeld and C. E. Ballou, *Carbohydrate Res.*, **32**, 287 (1974).
[40] I. Farkas and Z. Dinya, *Acta Chim. Acad. Sci. Hung.*, **82**, 75 (1974).
[41] A. L. Fink and N. E. Good, *Biochem. Biophys. Res. Comm.*, **58**, 126 (1974).
[42] H.-Y. L. Lai, L. G. Butler and B. Axelrod, *Biochem. Biophys. Res. Comm.*, **60**, 635 (1974).
[43] G. Legler and F. Witassek, *Z. Physiol. Chem.*, **355**, 617 (1974).
[44] C. J. Michalski and A. Domnas, *Carbohydrate Res.*, **33**, 153 (1974).
[45] J. D. Mellor and D. S. Layne, *J. Biol. Chem.*, **249**, 361 (1974).
[46] K. Hayaski, in *Molecular Mechanisms of Enzyme Action*, Y. Ogura, Y. Tougmura and T. Nakamura, eds., University Park Press, Baltimore, 1972, p. 217.
[47] B. Capon and W. M. Dearie, *Chem. Comm.*, **1974**, 370.
[48] L. O. Ford, L. N. Johnson, P. A. Machin, D. C. Phillips and R. Tjian, *J. Mol. Biol.*, **88**, 349 (1974).
[49] I. I. Secemski and G. E. Lienhard, *J. Biol. Chem.*, **249**, 2932 (1974); see also K. Schray and J. P. Klinman, *Biochem. Biophys. Res. Comm.*, **57**, 641 (1974).
[50] E. Holler, J. A. Rupley and G. P. Hess, *FEBS Letters*, **40**, 25 (1974).
[51] S. K. Banerjee, G. E. Vandenhoff and J. A. Rupley, *J. Biol. Chem.*, **249**, 1439 (1974); A. Cooper, *Biochemistry*, **13**, 2853 (1974).
[52] L. M. Hinman, C. R. Coan and D. A. Deranleau, *J. Am. Chem. Soc.*, **96**, 7067 (1974).
[53] R. S. Norton and J. H. Bradbury, *Chem. Comm.*, **1974**, 870; I. D. Campbell, C. H. Dobson and R. J. P. Williams, *ibid.*, p. 888; J. V. von Schütz, J. Zuclich and A. H. Maki, *J. Am. Chem. Soc.*, **96**, 714 (1974); N.-T. Yu, *ibid.*, p. 4664; M. Z. Atassi and M. C. Rosemblatt, *J. Biol. Chem.*, **249**, 4802 (1974); R. E. Greene and C. N. Pace, *ibid.*, p. 5388; T. Imoto, M. Fujimoto and K. Yagishita, *J. Biochem.* (*Tokyo*), **76**, 745 (1974); M. Z. Atassi, M. C. Rosemblatt and A. F. S. A. Habeeb, *Immunochemistry*, **11**, 495 (1974); R. R. Wickett, G. J. Ide and A. Rosenberg, *Biochemistry*, **13**, 3273 (1974); R. Jones, R. A. Dwek and S. Forsén, *Eur. J. Biochem.*, **47**, 271 (1974); T. V. Chentsova and N. A. Kravchenko, *Biokhimiya*, **39**, 491 (1974); K. W. Snape, R. Tjian, C. C. F. Blake and D. E. Koshland, *Nature* (*London*), **250**, 295 (1974); R. N. Sharma and C. C. Bigelow, *J. Mol. Biol.*, **88**, 247 (1974).
[54] R. S. Mulvey and S. Beychok, *Biochemistry*, **13**, 2980 (1974); N. V. Pokidova, S. S. Babyan and T. P. Shuvaleva, *Antiobiotiki* (*Moscow*), **19**, 721 (1974).
[55] S. Kuramitsu, K. Ikeda, K. Hamaguchi, H. Fujio, T. Amano, S. Miwa and T. Nishina, *J. Biochem.* (*Tokyo*), **76**, 671 (1974).
[56] R. S. Mulvey, R. J. Gualtieri and S. Beychok, *Biochemistry*, **13**, 782 (1974).
[57] R. Diamond, *J. Mol. Biol.*, **82**, 371 (1974); P. K. Warme and H. A. Scheraga, *Biochemistry*, **13**, 757 (1974).
[58] J. A. Thoma, *J. Theor. Biol.*, **44**, 305 (1974).
[59] M. L. Sinnott and S. G. Withers, *Biochem. J.*, **143**, 751 (1974).
[60] R. S. T. Loeffler, M. L. Sinnott and M. C. Whiting, *Chem. Comm.*, **1974**, 984.
[61] D. F. Wentworth and R. Wolfenden, *Biochemistry*, **13**, 4715 (1974).
[62a] P. M. Dey and K. Wallenfels, *Eur. J. Biochem.*, **50**, 107 (1974).
[62b] P. D. Dey and M. Dixon, *Biochem. Biophys. Acta*, **370**, 269 (1974).
[63] B. S. Khatra, D. G. Herries and K. Brew, *Eur. J. Biochem.*, **44**, 537 (1974).
[64] K. Hiromi, in *Molecular Mechanisms of Enzyme Action*, Y. Ogura, T. Tonomura and T. Nakamura, eds., University Park Press, Baltimore, 1972, p. 241.
[65] A. Levitzki and M. L. Steer, *Eur. J. Biochem.*, **41**, 171 (1974).
[66] T. Shibaoka, K. Miyano and T. Watanabe, *J. Biochem.* (*Tokyo*), **76**, 475 (1974).
[67] B. Iwasa, H. Aoshima, K. Hiromi and H. Hatano, *J. Biochem.* (*Tokyo*), **75**, 969 (1974); H. Fujimori, M. Ohnishi and K. Hiromi, *ibid.*, p. 767.
[68] M. Ohnishi, T. Sugamima, H. Fujimori and K. Hiromi, *J. Biochem.* (*Tokyo*), **74**, 1271 (1973).
[69] N. Suetsuga, S. Koyama, K. Takeo and T. Kuge, *J. Biochem.* (*Tokyo*), **76**, 57 (1974).
[70] M. Kato, K. Hiromi and Y. Morita, *J. Biochem.* (*Tokyo*), **75**, 563 (1974); J. A. Thoma, *Eur. J. Biochem.*, **44**, 139 (1974).
[71] J. De Prijcker, A. Vervoort and C. De Bruyne, *Eur. J. Biochem.*, **47**, 561 (1974).
[72] M. Nishigaki, T. Muramatsu, A. Kobata and K. Maeyama, *J. Biochem.* (*Tokyo*), **75**, 509 (1974).

[73] J. J. Marshall, *Carbohydrate Res.*, **34**, 289 (1974).
[74] K. Hiromi, M. Kawai, N. Suetsuga, Y. Nitta, T. Hosotani, A. Nagao, T. Nakajima and S. Ono, *J. Biochem.* (*Tokyo*), **74**, 935 (1973).
[75a] K. Yamamoto, *J. Biochem.* (*Tokyo*), **75**, 659, 931 (1974); **76**, 385 (1974).
[75b] M. Pokorny, E. Zissis, H. G. Fletcher and N. Pravdić, *Carbohydrate Res.*, **37**, 321 (1974).
[76] N. V. Molodtsov and M. G. Vafina, *Biochem. Biophys. Acta*, **364**, 296 (1974).
[77] S. K. Srivastava, A. Yoshida, Y. C. Awasthi and E. Beutler, *J. Biol. Chem.*, **249**, 2049 (1974); S. K. Srivastava and E. Beutler, *ibid.*, p. 2054.
[78] V. A. Nesmeyanov, S. E. Zurabyan, I. E. Sosinskaya and A. Y. Khorlin, *Biokhimiya*, **39**, 116 (1974).
[79] B. Capon and R. B. Walker, *J. C. S. Perkin II*, **1974**, 1600.
[80] D. M. L. Morgan and A. Neuberger, *Proc. Roy. Soc.*, *A*, **337**, 317 (1974).
[81] See *Org. Reaction Mech.*, **1971**, 403.
[82] L. Que and G. R. Gray, *Biochemistry*, **13**, 146 (1974).
[83] E. B. Whipple and M. Ruta, *J. Org. Chem.*, **39**, 1666 (1974).
[84] D. E. Edmondson, *Biochemistry*, **13**, 2817 (1974).
[85] B. Wurster and B. Hess, *FEBS Letters*, **38**, 33 (1973); *Z. Physiol. Chem.*, **355**, 255 (1974); B. Wurster and B. Hess, *FEBS Letters*, **40**, S112 (1974).
[86] Y. Pocker, J. E. Meany and B. C. Davis, *Biochemistry*, **13**, 1411 (1974).
[87a] R. F. Atkinson and T. C. Bruice, *J. Am. Chem. Soc.*, **96**, 819 (1974); cf. *Org. Reaction Mech.*, **1972**, 398.
[87b] R. G. Barradas, O. Kutowy and D. W. Shoesmith, *Can. J. Chem.*, **52**, 1635 (1974).
[88a] R. Fuchs, T. M. Young and R. F. Rodewald, *Can. J. Chem.*, **51**, 4122 (1973).
[88b] P. E. Sørensen, K. Bruhn and F. Lindeløv, *Acta Chem. Scand.*, A, **28**, 162 (1974).
[89a] R. T. Borchardt and L. A. Cohen, *J. Am. Chem. Soc.*, **95**, 8308, 8318, 8319 (1973).
[89b] M. I. Vinnik and I. S. Kislina, *Kinet. Katal.*, **15**, 311 (1974); *Chem. Abs.*, **81**, 12735 (1974).
[90] A. Winston, W. D. Rightler, F. G. Bollinger and R. F. Bargiband, *J. Org. Chem.*, **38**, 4249 (1973).
[91] K. Sommermeyer, *Z. Naturforsch.*, **29**b, 254 (1974).
[92a] M. J. Cho and I. H. Pitman, *J. Am. Chem. Soc.*, **96**, 1843 (1974).
[92b] Z. Fiala and M. Navratil, *Coll. Czech. Chem. Comm.*, **39**, 2200 (1974).
[92c] V. S. Sorokina, I. G. Savich and L. A. Pavlova, *Zh. Org. Khim.*, **9**, 1967 (1973); *Chem. Abs.*, **79**, 145541 (1973).
[93] R. Escale and J. Verducci, *Bull. Chem. Soc. France*, **1974**, 1203.
[94] H. B. Bürgi, J. M. Lehn and G. Wipff, *J. Am. Chem. Soc.*, **96**, 1956 (1974); H. B. Bürgi, J. D. Dunitz, J. M. Lehn and G. Wipff, *Tetrahedron*, **30**, 1563 (1974); see *Org. Reaction Mech.*, **1973**, 6.
[95] R. Herscovitch, J. J. Charette and E. de Hoffmann, *J. Am. Chem. Soc.*, **96**, 4954 (1974); see also *idem*, *ibid.*, **95**, 5135 (1973).
[96] G. Tóth, I. Pinter and A. Messmer, *Tetrahedron Letters*, **1974**, 735.
[97] P. Nagy, *Magyar Kem. Foly.*, **80**, 92 (1974); *Chem. Abs.*, **80**, 120001 (1974).
[98a] A. Bartecek, J. Kaválek, V. Macháček and V. Štěrba, *Coll. Czech. Chem. Comm.*, **39**, 1722 (1974).
[98b] H. Iwamura, M. Tsuchimoto and S. Nishimura, *Bull. Chem. Soc. Japan*, **47**, 193 (1974).
[98c] Y. Ogata and A. Kawasaki, *J. Org. Chem.*, **39**, 1058 (1974).
[98d] F. E. Rogers and R. J. Rapiejko, *J. Phys. Chem.*, **78**, 599 (1974).
[99a] M. Masui, H. Ohmori, C. Ueda and M. Yamauchi, *J. C. S. Perkin II*, **1974**, 1448.
[99b] D. Pitea, G. Favini and D. Grasso, *J. C. S. Perkin II*, **1974**, 1595.
[100] J. Taillades and A. Commeyras, *Tetrahedron*, **30**, 2493, 3407 (1974); see also J. S. Walia, S. N. Bannore, A. S. Walia and L. Cruillot, *Chem. Letters* (*Tokyo*), **1974**, 1005.
[101] R. Toyoda, T. Nakagawa and T. Uno, *Bunseki Kagaku*, **22**, 914 (1973); *Chem. Abs.*, **80**, 59136 (1974); I. Wiesner and L. Wiesnerova, *Chem. Prum.*, **24**, 304 (1974); *Chem. Abs.*, **81**, 62722 (1974),
[102] D. M. Aleksandrova, L. J. Kolotova and L. Ya. Kheifets, *Zh. Org. Khim.*, **9**, 2107 (1973); *Chem. Abs.*, **80**, 36469 (1974).
[103] R. N. F. Thorneley and H. Diebler, *J. Am. Chem. Soc.*, **96**, 1072 (1974).
[104] J. Bjørgo, D. R. Boyd, C. G. Watson and W. B. Jennings, *J. C. S. Perkin II*, **1974**, 757; J. Bjørgo, D. R. Boyd, C. G. Watson, W. B. Jennings and D. M. Jerina, *ibid.*, p. 1081.
[105] U. Gebert and B. von Kerékjartó, *Annalen*, **1974**, 644; S. Matsumoto and Y. Matsushima, *J. Am. Chem. Soc.*, **96**, 5228 (1974); *Kagaku* (*Kyoto*), *Zokan*, **1974**, 49; *Chem. Abs.*, **81**, 62550 (1974).
[106a] M. D. Broadhurst and D. J. Cram, *J. Am. Chem. Soc.*, **96**, 581 (1974).
[106b] J. R. Butler, W. A. Alworth and M. J. Nugent, *J. Am. Chem. Soc.*, **96**, 1617 (1974).
[107] F. Jordan, *J. Am. Chem. Soc.*, **96**, 825 (1974); see also J. P. Guthrie and F. Jordan, *ibid.*, **94**, 9132, 9136 (1972).
[108] K. Dixon and J. V. Greenhill, *J. C. S. Perkin II*, **1974**, 164.

[109] S. A. Fine and J. Shreiner, *J. Org. Chem.*, **39**, 1009 (1974).
[110] F. Johnson, L. G. Duquette, A. Whitehead and L. C. Dorman, *Tetrahedron*, **30**, 3241 (1974).
[111] W. F. Verhelst and W. Drenth, *J. Am. Chem. Soc.*, **94**, 6692 (1974).
[112] A. B. Grigor'ev, M. K. Polievktov, Yu. V. Tsukanov, V. G. Smirnova, V. G. Granik and R. G. Glushkov, *Zh. Org. Khim.*, **10**, 855 (1974); *Chem. Abs.*, **81**, 24670 (1974).
[113] F. Weisbuck and G. Dana, *Tetrahedron*, **30**, 2873 (1974).
[114] J. Cadet and R. Teoule, *J. Am. Chem. Soc.*, **96**, 6517 (1974).
[115] See *Org. Reaction Mech.*, **1970**, 435; **1972**, 404.
[116a] Y. Suzuki and S. Yatabe, *Bull. Chem. Soc. Japan*, **47**, 2353 (1974).
[116b] Y. Suzuki, *Bull. Chem. Soc. Japan*, **47**, 2077 (1974).
[116c] S. Kolka and J. Sokolowski, *Rocz. Chem.*, **48**, 439 (1974).
[116d] K. Smiataczowa, G. Szukay and T. Jasinski, *Rocz. Chem.*, **48**, 1059 (1974).
[117a] I. N. Somin, Yu. A. Ignat'ev and V. A. Gindin, *Zh. Org. Khim.*, **9**, 1756 (1973); *Chem. Abs.*, **79**, 115009 (1973).
[117b] L. do Amaral, M. P. Bastos, H. B. Bull and E. H. Cordes, *J. Am. Chem. Soc.*, **95**, 7369 (1973); see *Org. Reaction Mech.*, **1972**, 405.
[118] D. H. R. Barton, P. D. Magnus and S. I. Pennaen, *Chem. Comm.*, **1974**, 1007; C. Brown, B. T. Grayson and R. F. Hudson, *ibid.*, p. 1007.
[119a] R. E. Huettemann and A. P. Shroff, *J. Pharm. Sci.*, **63**, 74 (1974).
[119b] A. B. Dekelbaum, B. V. Passet and G. F. Fiodorov, *Organic Reactivity* (*Tartu*), **10**, 645 (1973).
[120] R. H. Shapiro and E. C. Hornaman, *J. Org. Chem.*, **39**, 2302 (1974).
[121] R. Andvisano, A. S. Angeloni and G. Gottarelli, *Tetrahedron*, **30**, 3827 (1974).
[122a] T. Okuda, S. Saito and K. Uobe, *Tetrahedron*, **30**, 1187 (1974).
[122b] N. A. Kogan, A. I. Balagurova and M. I. Vlassova, *Organic Reactivity* (*Tartu*), **10**, 1093 (1973).
[122c] A. L. Meenakshi, P. Maruthamuthu, K. V. Seshadri and M. Santappa, *Indian J. Chem.*, **11**, 608 (1973); *Chem. Abs.*, **79**, 114751 (1973).
[123] G. M. Loudon and C. Berke, *J. Am. Chem. Soc.*, **96**, 4508 (1974).
[124] G. M. Loudon, C. K. Smith and S. E. Zimmerman, *J. Am. Chem. Soc.*, **96**, 465 (1974).
[125] T. Okuyama and T. Fueno, *J. Org. Chem.*, **39**, 3156 (1974).
[126] J.-P. Gouesnard and M. Blain, *Bull. Soc. Chim. France*, **1974**, 338.
[127] J. E. Garst and G. L. Schmir, *J. Org. Chem.*, **39**, 2920 (1974); see *Org. Reaction Mech.*, **1972**, 407.
[128a] G. C. Balezina, N. M. Deriglazov, S. F. Zanina and S. M. Shostakovskii, *Khim. Vysokomol. Soedin. Neftekhim.*, **1973**, 33; *Chem. Abs.*, **80**, 132373 (1974).
[128b] A. N. Khudyakova, A. N. Volkov and B. A. Trofimov, *Organic Reactivity* (*Tartu*), **10**, 987 (1973).
[129] N. C. Deno and R. Fishbein, *J. Am. Chem. Soc.*, **95**, 7445 (1973).
[130] B. G. Cox, *J. Am. Chem. Soc.*, **96**, 6823 (1974).
[131] R. P. Bell, D. W. Earls and B. A. Timimi, *J.C.S. Perkin II*, **1974**, 811; D. B. Cook and J. McKenna, *ibid.*, p. 1223.
[132] B. G. Cox and R. E. J. Hutchinson, *J. C. S. Perkin II*, **1974**, 613.
[133] R. Brouillard and J.-E. Dubois, *J. Org. Chem.*, **39**, 1137 (1974).
[134] J. Toullec and J.-E. Dubois, *J. Am. Chem. Soc.*, **96**, 3524 (1974); see *Organic Reaction Mech.*, **1969**, 421.
[135] J. Hine, M. S. Cholod and R. A. King, *J. Am. Chem. Soc.*, **96**, 835 (1974).
[136] J. Hine and R. L. Flachskam, *J. Org. Chem.*, **39**, 863 (1974).
[137] J. Hine and S. S. Ulrey, *J. Org. Chem.*, **39**, 3231 (1974).
[138] Cf. *Org. Reaction Mech.*, **1971**, 409–410; **1973**, 10.
[139] L. S. Mushketik and N. V. Volkova, *Ukr. Khim. Zh.* (*Russ. Ed.*), **39**, 962 (1973); *Chem. Abs.*, **79**, 145592 (1973).
[140] B. G. Cox and A. Gibson, *Chem. Comm.*, **1974**, 638.
[141] O. S. Tee, J. A. Altmann and K. Yates, *J. Am. Chem. Soc.*, **96**, 3141 (1974).
[142a] M. Bettahar and M. Charpentier-Morize, *Tetrahedron*, **30**, 1373 (1974).
[142b] J. R. Jones and S. R. Patel, *J. Am. Chem. Soc.*, **96**, 574 (1974).
[143] M. Orban, E. Koros and K. Krudy-Sandor, *Acta Chim.* (*Budapest*), **79**, 155 (1973); *Chem. Abs.*, **80**, 14212 (1974).
[144] J. M. Cachaza, A. Castro and M. A. Herraez, *An. Quím.*, **69**, 847 (1973); *Chem. Abs.*, **80**, 69952 (1974).
[145] W. D. Covey and D. L. Leussing, *J. Am. Chem. Soc.*, **96**, 3860 (1974); *Chem. Comm.*, **1974**, 31.
[146] J. L. Jensen and D. J. Carré, *J. Org. Chem.*, **39**, 2103 (1974).
[147] F. M. Menger and R. F. Williams, *J. Org. Chem.*, **39**, 2131 (1974).
[148] C. A. Brown, *J. Org. Chem.*, **39**, 1324 (1974).

[149] E. Koros, M. Orban, M. Burger, K. Voros and E. Fussy, *Acta Chim.* (*Budapest*), **74**, 387 (1972); *Chem. Abs.*, **78**, 57541 (1973).
[150] G. H. Posner, C. E. Whitten, J. J. Sterling and D. J. Brunelle, *Tetrahedron Letters*, **1974**, 2591.
[151] A. de Groot and B. J. M. Jansen, *Rec. Trav. chim.*, **93**, 153 (1974).
[152] M. Dreyfus and F. Garnier, *Tetrahedron*, **30**, 133 (1974); V. A. Gindin, B. A. Ershov and A. I. Kol'tsov, *Zh. Org. Khim.*, **9**, 1985 (1973); *Chem. Abs.*, **79**, 145800 (1973); A. Kettrup and M. Grote, *Z. Anal. Chem.*, **269**, 118 (1974); *Chem. Abs.*, **81**, 12863 (1974); L. Cazaux, G. Chassaing and P. Maroni, *Compt. rend.*, *Ser. C.*, **278**, 1509 (1974).
[153] J. Terpinski and T. Kozluk, *Rocz. Chem.*, **48**, 29 (1974).
[154] M. R. Webb and J. R. Knowles, *Biochem. J.*, **141**, 589 (1974).
[155] A. Cantwell and D. Dennis, *Biochemistry*, **13**, 287 (1974).
[156] C. A. Browne and S. G. Waley, *Biochem. J.*, **141**, 753 (1974).
[157] R. H. Lane and J. K. Hurst, *Biochemistry*, **13**, 3292 (1974).
[158] J. B. Stothers and C. T. Tan, *Chem. Comm.*, **1974**, 738; see also ref. 141 and *Org. Reaction Mech.* **1966**, 326–327.
[159] A. Nickon, J. J. Frank, D. F. Covey and Y. Lin, *J. Am. Chem. Soc.*, **96**, 7574 (1974); see *Org. Reaction Mech.*, **1966**, 326–327.
[160] J. Hine and W. H. Sachs, *J. Org. Chem.*, **39**, 1937 (1974).
[161] J. P. Guthrie, *Can. J. Chem.*, **52**, 2037 (1974).
[162a] V. I. Stanko, T. V. Klimova and I. P. Beletskaya, *Dokl. Akad. Nauk SSSR*, **216**, 329 (1974); *Chem. Abs.*, **81**, 104244 (1974).
[162b] C. Kalidas and N. Chattanathan, *J. Indian Chem. Soc.*, **51**, 479 (1974).
[163] S. N. Huckin and L. Weiler, *Can. J. Chem.*, **52**, 2157 (1974).
[164a] P. M. McCurry and R. K. Singh, *J. Org. Chem.*, **39**, 2316, 2317 (1974).
[164b] Yu. K. Glotova, V. I. Irzhak and N. S. Enikolopyan, *Kinet. Katal.*, **14**, 1419 (1973); *Chem. Abs.*, **80**, 94859 (1974).
[165] G. Casnati, A. Pochini, G. Salerno and R. Ungaro, *Tetrahedron Letters*, **1974**, 959.
[166a] J. Konigstein and M. Fedoronko, *Coll. Czech. Chem. Comm.*, **38**, 3801 (1973).
[166b] V. P. Baeva, A. V. Iogansen, G. A. Kurkchi and V. M. Furman, *Zh. Org. Khim.*, **10**, 1446 (1974); *Chem. Abs.*, **81**, 104249 (1974).
[166c] V. A. Rybakov, V. Yu. Gankin and I. S. Fuks, *Zh. Prikl. Khim.* (*Leningrad*), **46**, 2741 (1973); *Chem. Abs.*, **80**, 59138 (1974).
[167a] J.-E. Vik, *Acta Chem. Scand.*, **B28**, 325, 509 (1974).
[167b] D. N. Kramer, L. P. Bisauta, R. Bato and B. L. Muir, *J. Org. Chem.*, **39**, 3132 (1974).
[168] C. Gallina and A. Liberatori, *Tetrahedron*, **30**, 667 (1974).
[169a] M. I. Arkhipov, N. I. Borodkina, G. I. Tsar'kova and V. D. Zhestkora, *Tr. Ivanov. Khim-Tekhnol. Inst.*, **1972**, 146; *Chem. Abs.*, **79**, 125460 (1973).
[169b] M. Iwata and S. Emoto, *Chem. Letters* (*Tokyo*), **1974**, 959.
[170] T. Nakanishi and K. Suyama, *Nippon Nogei Kagaku Kaishi*, **47**, 313 (1973); *Chem. Abs.*, **79**, 125669 (1973).
[171] S. S. Baba-Zade, A. G. Gasanov and O. A. Narimanbekov, *Dokl. Akad. Nauk Azerb. SSR*, **29**, 17 (1973); *Chem. Abs.*, **80**, 119869 (1974).
[172] B. A. Alekseyenko and S. N. Baranov, *Organic Reactivity* (*Tartu*), **9**, 144 (1972).
[173] E. J. Heron and R. M. Caprioli, *Biochemistry*, **13**, 4371 (1974).
[174] E. Grazi, *Biochem. Biophys. Res. Comm.*, **56**, 106 (1974).
[175] E. Grazi, *Biochem. Biophys. Res. Comm.*, **59**, 450 (1974).
[176a] F. C. Hartman, B. Suh, M. H. Welch and R. Barker, *J. Biol. Chem.*, **248**, 8233 (1973).
[176b] C. Y. Lai, N. Nakai and D. Chang, *Science*, **183**, 1205 (1974).
[177] I. Gibbons and R. N. Perham, *Biochem. J.*, **139**, 331 (1974); P. J. Anderson and H. Kaplan, *ibid.* **137**, 181 (1974); E. Kuhn and K. Brand, *Biochemistry*, **12**, 5217 (1974).
[178] G. S. Kaklij and G. B. Nadkarni, *Arch. Biochem. Biophys.*, **160**, 4752 (1974).
[179] R. Hubaut and J. Landais, *Compt. rend.*, *Ser. C*, **278**, 1247 (1974); Z. Csuros, P. Sallay and G. Deak, *Acta Chim.* (*Budapest*), **79**, 345, (1973); *Chem. Abs.*, **80**, 59143 (1974).
[180] T. Drakenberg, S. Forsén and J. M. Sommer, *J. C. S. Perkin II*, **1974**, 520.
[181] A. Levi, G. Modena and G. Scorrano, *J. Am. Chem. Soc.*, **96**, 6585 (1974).
[182a] R. P. Quirk and C. R. Gambill, *Chem. Comm.*, **1974**, 503.
[182b] A. Maradulfu and A. S. Perlin, *Carbohydrate Res.*, **32**, 93 (1974).
[183] N. Kunieda, A. Suzuki and M. Kinoshita, *Makromol. Chem.*, **170**, 243 (1973); *Chem. Abs.*, **79** 145637 (1973).
[184] A. A. Yasnikov and A. F. Babicheva, *Ukr. Khim. Zh.*, **40**, 52 (1974); *Chem. Abs.*, **80**, 145006 (1974).

[185] G. V. Mihailova and Ya. J. Tur'yan, *Sb. Nauch. Tr., Yaroslav. Tekhnol. Inst.*, **22**, 105 (1972); *Chem. Abs.*, **80**, 59168 (1974).
[186] L. M. Pugach, N. D. Gil'chenok, G. S. Idlis and S. K. Ogorodnikov, *Zh. Org. Khim.*, **10**, 925 (1974); *Chem. Abs.*, **81**, 49048 (1974).
[187] B. Ya. Aksel'rod and M. I. Siling, *Zh. Fiz. Khim.*, **48**, 323 (1974); *Chem. Abs.*, **80**, 132587 (1974).
[188] R. Natesan and L. M. Yeddanapalli, *Indian J. Chem.*, **12**, 691 (1974).
[189] T. M. Koshtaleva, V. A. Ostrovski, G. I. Koldobskii and B. V. Gidaspov, *Zh. Org. Khim.*, **10**, 546 (1974); *Chem. Abs.*, **80**, 132381 (1974).
[190a] J.-S. Wang, *Bull. Inst. Chem., Acad. Sinica*, **1972**, 109; *Chem. Abs.*, **79**, 136199 (1973).
[190b] D. W. Harris and M. S. Feather, *Carbohydrate Res.*, **30**, 359 (1973).
[191] G. Aksnes and F. Y. Khalil, *Phosphorus*, **3**, 79, 109 (1973); *Chem. Abs.*, **79**, 145749 (1973); **80** 107556 (1974).
[192] C. R. Johnson, C. W. Schroeck and J. R. Shanklin, *J. Am. Chem. Soc.*, **95**, 7424 (1973).
[193] F. Sweet and J. D. Fissekes, *J. Am. Chem. Soc.*, **96**, 8741 (1973).
[194] I. S. Kislina and M. I. Vinnik, *Kinet. Katal.*, **14**, 1134 (1973); *Chem. Abs.*, **80**, 26494 (1974).
[195] I. S. Kislina and M. I. Vinnik, *Kinet. Katal.*, **14**, 860 (1973); *Chem. Abs.*, **79**, 136110 (1973).
[196] K. Kijima and T. Sakaguchi, *Yakugaku Zasshi*, **93**, 831 (1973); *Chem. Abs.*, **80**, 2818 (1974).
[197a] J. L. van der Baan and F. Bickelhaupt, *Tetrahedron*, **30**, 2447 (1974).
[197b] M. A. Winnik, V. Stoute and P. Fitzgerald, *J. Am. Chem. Soc.*, **96**, 1977 (1974).
[198] T. Suzuki, T. Kobayashi, Y. Takegami and Y. Kawasaki, *Bull. Chem. Soc. Japan.* **47**, 1971 (1974).
[199] K. Clauss, H.-J. Friedrich and H. Jensen, *Annalen*, **1974**, 561.
[200] A. Tuulmets, *Organic Reactivity (Tartu)*, **11**, 81 (1974); S. Viirlaid and A. Tuulmets, *ibid.*, p. 65; S. Viirlaid, S. Kurrikoff and A. Tuulmets, *ibid.*, p. 73.
[201] E. C. Ashby, J. Laemmle and H. M. Neumann, *Accounts Chem. Res.*, **7**, 272 (1974).
[202] R. L. Cargill, T. E. Jackson, N. P. Peet and D. M. Pond, *Accounts Chem. Res.*, **7**, 106 (1974).

CHAPTER 2

Reactions of Acids and their Derivatives

M. I. PAGE

Department of Chemistry, The Polytechnic, Huddersfield

CARBOXYLIC ACIDS

Tetrahedral Intermediates

The method used for estimating the free energies of formation of orthoacids from those of orthoesters[1] has been extended to the free energy of formation of orthoacid monoamides, $RC(OH)_2NR'_2$, the tetrahedral intermediates formed in the hydrolysis of amides and also $RC(OH)(OMe)(NR'_2)$, the intermediate produced in the aminolysis of methyl esters. The free-energy profiles for amide hydrolysis and ester aminolysis have been constructed and discussed.[1] For example, the equilibrium constant for the formation of the tetrahedral intermediate (**1**) formed from dimethylamine and methyl acetate is estimated to be 9.3×10^{-11} l mol^{-1}.

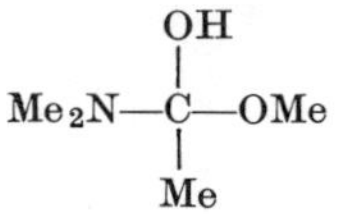

(1) 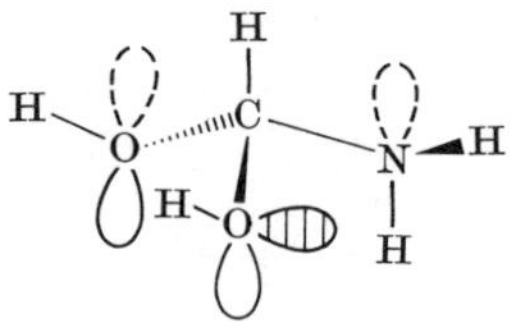 (2) 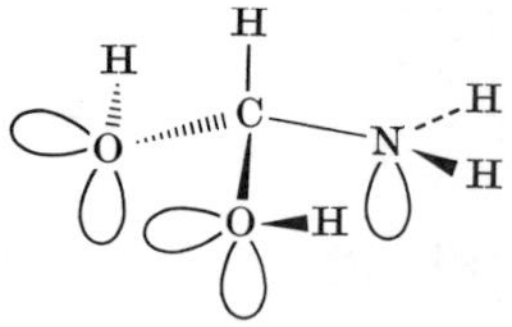 (3)

Ab initio quantum calculations of aminomethanediol indicate a marked stereo-electronic effect on the properties of these species, as suggested by earlier experimental results.[2a] The carbon—nitrogen bond is strongest in conformation (**2**) and weakest in (**3**), which is the least stable conformation. In general, the longest and weakest carbon—X bond is that which has the highest number of antiperiplanar (app) lone pairs and no X-lone pair app to an adjacent polar bond. The thermodynamically preferred conformer may not be kinetically the most labile.[2a]

The total energies, atomic charges, bond lengths and bond orders of the reactants, products and intermediate in the reaction of ammonia with methyl acetate have been calculated by CNDO and MINDO/2. Cleavage of a nitrogen—hydrogen bond in the intermediate is necessary for the ammonolysis to proceed.[2b]

Details of the tetrahedral intermediate isolated in an acetyl-transfer reaction have been reported.[3] Amines attack the carbonyl-carbon rather than the tetrahedral one (**4**).

The formation and breakdown of the tetrahedral intermediate (**6**) have been observed directly during the hydrolysis of *N,O*-trimethylenephthalimidium cation (**5**) in weakly acidic solutions. A two-point Brønsted plot for general-base catalysed hydration gives a β-value of 0.25. The β-value for the general-base catalysed breakdown of the water-addition intermediate to (**7**) is 0.57. The mechanism for this reaction is formulated as the kinetically equivalent general-acid catalysed departure of alcohol from the anion of the tetrahedral intermediate (general-base catalysis of ROH attack in the reverse direction).[4] The breakdown of the amine-addition compound (**8**) shows no buffer catalysis and apparently proceeds by a stepwise mechanism of specific-acid catalysis with rate-limiting expulsion of HNR_2 from the protonated amine (**9**).[5] Rate constants for the

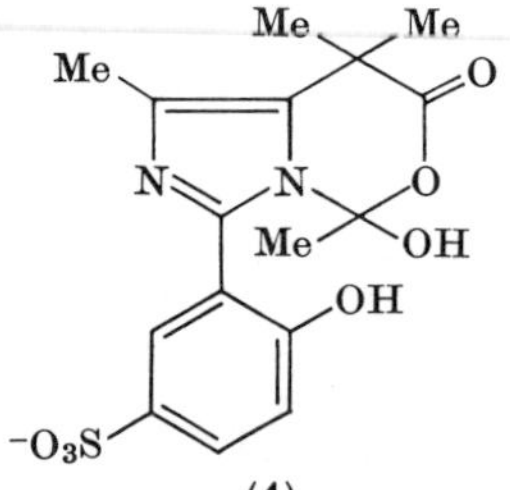

(4) 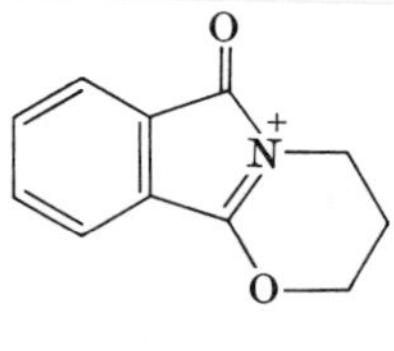 (5) 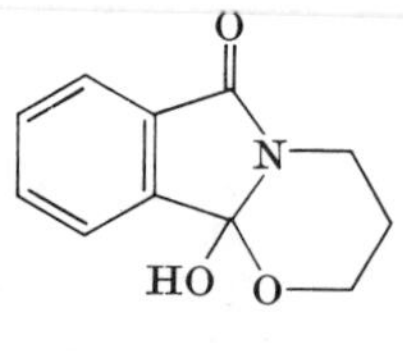 (6)

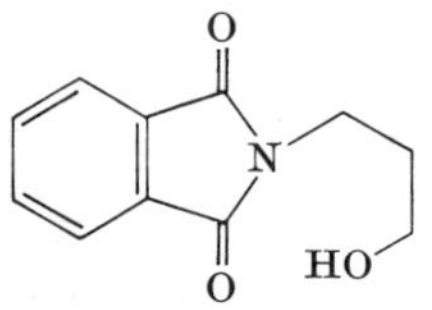

(7) 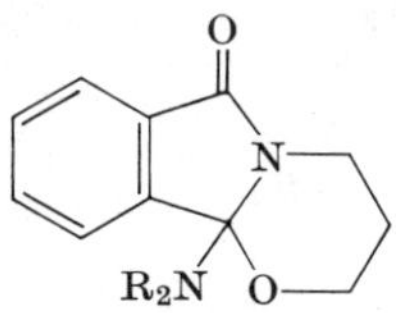 (8) 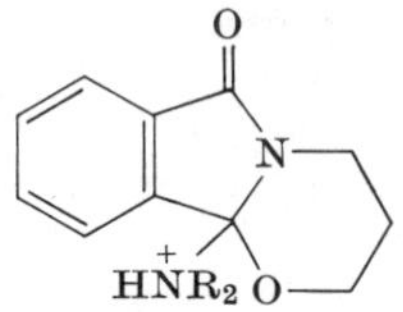 (9)

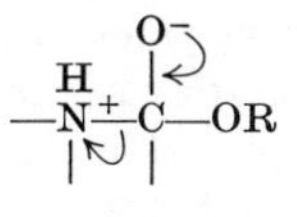

(10) 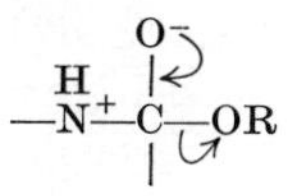 **(11)** 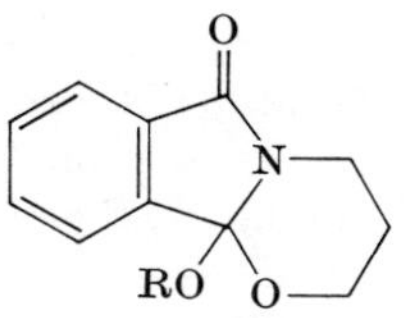 **(12)** 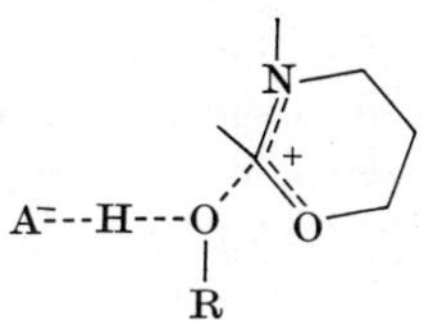 **(13)**

breakdown of (**9**) yield a Brønsted β-value of -0.9, indicating a late transition state. A comparison of the rate constants for expulsion of amines (**10**) with those for alkoxide-ions (**11**) shows that for a given basicity the amines are better leaving groups by a factor of ca. 10^5. This suggests that alkoxide-ion departure will ordinarily be rate-limiting in the uncatalysed aminolysis of esters.[5] However, general-acid and general-base catalysis is observed for the formation and breakdown of the tetrahedral intermediate (**12**) formed from alcohols and (**5**). The Brønsted α-values for the general-acid catalysed expulsion of alcohol *increase* with increasing alcohol basicity. Similarly the β-value for the leaving alcohol *decreases* as the acid catalyst becomes weaker. This is interpreted as requiring that formation and breakdown of the carbon—oxygen bond is concerted with proton transfer (**13**). These observations are opposite to the predictions of the Hammond–Leffler postulate. The rate constants for the attack of alkoxide-ions on (**5**) are ca. 30-fold larger than those for amines of comparable pK.[6]

Brønsted plots for both the general-acid and the general-base catalysed methoxyaminolysis of 1-acetyl-1,2,4-triazole are non-linear[7] with equal limiting rate constants for strongly acidic and strongly basic catalysts.[8] This is interpreted in terms of rate-limiting, diffusion-controlled encounter of a tetrahedral intermediate $T^{\pm}$ with the catalyst for strongly basic or acidic catalysts, and breakdown of the intermediate $T^{\pm}$ or T^0 for weak catalysts (Scheme 1). The change from a mechanism that gives a linear Brønsted plot for general-acid catalysis of the hydrazine-acetylimidazole reaction[7] is suggested to be due to the lower basicity of acetyltriazole.

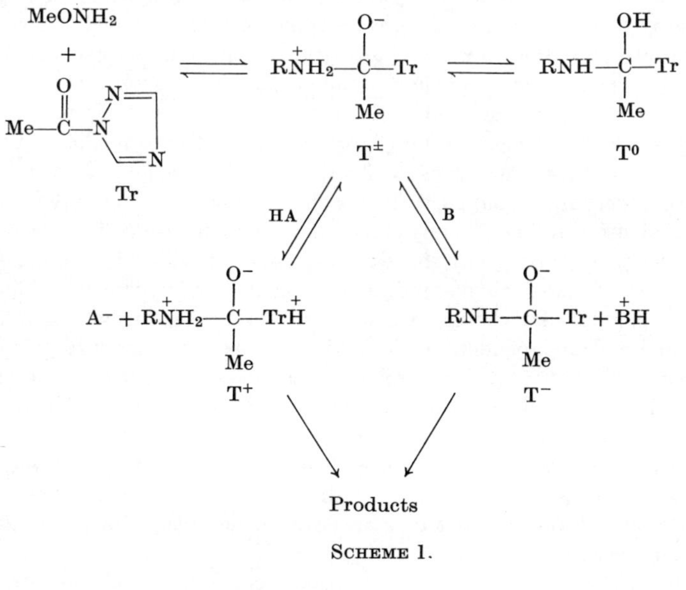

MeONH$_2$ + Tr ⇌ $T^{\pm}$ ⇌ T^0

$T^{\pm}$ —HA→ $A^- + T^+$; $T^{\pm}$ —B→ $T^- + \overset{+}{B}H$

T^+, T^- → Products

SCHEME 1.

The rate-limiting step for the uncatalysed aminolysis of all but the most reactive phenyl esters is phenoxide-ion expulsion ($k_{\pm}$) and for alkyl acetates the formation of the tetrahedral intermediate T^0 from $T^{\pm}$, a proton switch through water rather than, as thought previously, attack of amine on the ester. For very fast reactions there is a change to rate-limiting formation of $T^{\pm}$ accompanied by a sharp decrease in sensitivity to substituent effects (Scheme 2).[9] The general-base catalysed aminolysis of both alkyl and phenyl esters is not concerted but involves a kinetically significant intermediate with proton-removal from $T^{\pm}$ as the rate-limiting step. Proton-donation to give T^+ is rate-limiting for the general-acid catalysed reaction (Scheme 2).[9] Brønsted plots for the

$$RNH_2 + R\text{-}C(=O)\text{-}OR' \rightleftharpoons R\overset{+}{N}H_2\text{-}C(O^-)(R)\text{-}OR'\ (T^{\pm}) \underset{}{\overset{k_s[H_2O]}{\rightleftharpoons}} RNH\text{-}C(OH)(R)\text{-}OR'\ (T^0) \longrightarrow \text{Products}$$

$$T^{\pm} \overset{HA}{\rightleftharpoons} R\overset{+}{N}H_2\text{-}C(OH)(R)\text{-}OR'\ (T^+) \overset{H_2O}{\longleftarrow} RH\overset{+}{N}{=}C(R)(OR') \overset{^-OH}{\longrightarrow} T^0$$

$$T^{\pm} \overset{k_{\pm}}{\longrightarrow} \text{Products}; \qquad T^{\pm} \overset{B}{\rightleftharpoons} RNH\text{-}C(O^-)(R)\text{-}OR'\ (T^-) \longrightarrow \text{Products}$$

SCHEME 2.

general-acid catalysed formation of esters on hydrolysis of *p*-tolyl imidates are curved, which is consistent with rate-limiting proton-transfer in the product-determining step. Different intermediates, that are not in protonic equilibrium, are generated in imidate hydrolysis and in the aminolysis of the corresponding phenyl ester, so that the products formed from imidate hydrolysis do not necessarily show which step is rate-limiting in ester aminolysis.[10] For phenyl esters, the tetrahedral intermediate $T^{\pm}$ does not equilibrate with its other ionic forms in the absence of buffers, i.e. $k_{\pm} > k_s$ (Scheme 2). The following observations support this path: (i) the pH- and buffer-independent product ratio of ester/amide remains constant on substitution of electron-attracting groups into the phenyl ring, (ii) the decrease in ester yield with increasing pH is correlated with the fraction of the imidate that reacts with hydroxide ion, and (iii) the observed rate constants for the general-acid catalysed ester formation remain almost constant as the buffer-independent ester yield decreases with increasing pH.[10]

The Brønsted plot for the general-base catalysed methoxyaminolysis of the formamidine (**14**) is non-linear, approaching ca. 1.0 for weakly basic catalysts and ca. 0 for more basic catalysts. This suggests the formation of an intermediate and kinetically significant transport processes. The rate-limiting step is thought to be a diffusion-controlled encounter of the intermediate (**15**) and the general-base for strongly basic catalysts, and diffusion apart or a reorganization within the encounter complex (**16**) for weaker bases.[11]

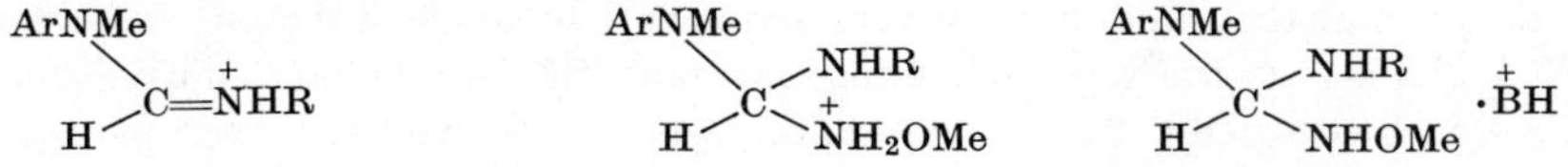

(14) **(15)** **(16)**

If the tetrahedral intermediate (**18**), obtained in an acid-catalysed imidate (**17**) hydrolysis, can expel alcohol to give amide, the analogous intermediate for an acid-catalysed amide hydrolysis (R = H) could expel water, a process which, on the assumption of rapid proton-transfer, would result in ^{18}O exchange.[12] An alternative explanation is that imidates hydrolyse to give amide without involving a tetrahedral intermediate. Unlike basic hydrolysis of ^{18}O-labelled imidate esters, which involves acyl-carbon—oxygen fission, the acid-catalysed hydrolysis occurs by alkyl-carbon—oxygen fission. Thus a tetrahedral intermediate is *not* formed during imidate hydrolysis to amide and a comparison of imidate and amide hydrolysis in concentrated acid solution is not valid. The mechanism suggested for this imidate hydrolysis is an S_N2-type of displacement (**19**).[13]

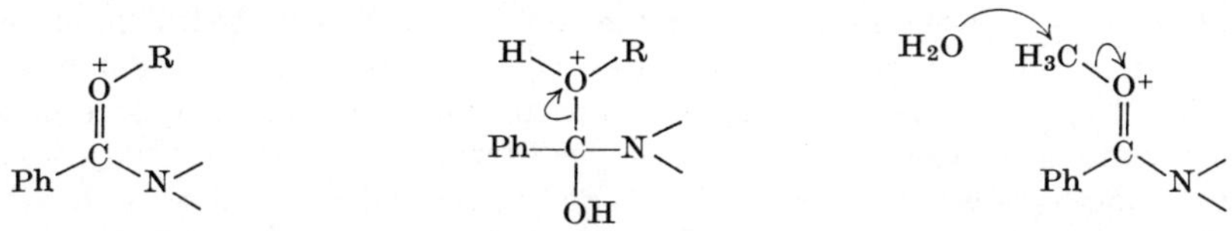

(17) **(18)** **(19)**

The solvent isotope effect k_{H_2O}/k_{D_2O} is 2.17 for the hydrolysis of phenyl *N*-methylacetamidate, but for the attack of nucleophiles it is ca. 1. The mechanism is thought to involve nucleophilic attack upon the conjugate acid of the imido-ester.[14]

Only uncatalysed aminolysis and hydrolysis are observed in the reaction of amines with methyl chloroformate which shows decreasing sensitivity to the basicity of the attacking amine with increasing p*K*. This curvature of the Brønsted plot is attributed either to a change in rate-limiting step from breakdown to formation of the tetrahedral intermediate (**20**), or to a change in geometry of the transition state in a concerted reaction.[15]

(20) **(21)**

The mechanism for the acetate- and pyridine-catalysed hydrolysis of phenyl chloroformates (**21**) is thought to involve nucleophilic catalysis with rate-limiting expulsion of chloride ion. This is compatible with the small Hammett ρ-value of 0.47.[16]

The hydrolysis of ring-substituted trifluoroacetanilides is catalysed by imidazole acting as a nucleophile. It is claimed that there is spectrophotometric evidence for the formation of a tetrahedral intermediate (**22**). The solvent isotope effect k_{H_2O}/k_{D_2O} of 4.3 and the non-linear Hammett plot are interpreted in terms of the relative importance of proton-donation to the anilide-nitrogen compared with breaking of the carbon—nitrogen bond.[17] A claim that the tetrahedral intermediate (**23**) accumulates during alkaline hydrolysis of substituted trifluoroacetanilides[18] has been shown to be incompatible with the kinetic analysis scheme used; furthermore, NMR studies show that such an intermediate must be present to a maximum extent of less than 0.5%.[19]

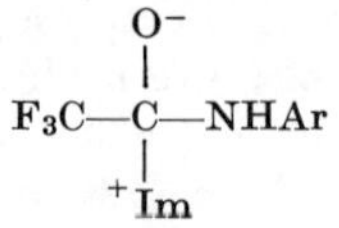

$$F_3C-C(O^-)(^+Im)-NHAr \quad \textbf{(22)} \qquad F_3C-C(O^-)(OH)-NHAr \quad \textbf{(23)}$$

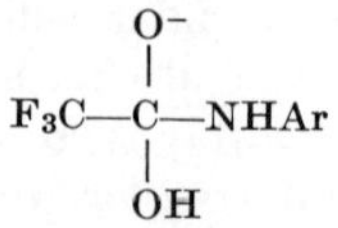

The effect of raising the mole fraction of MeOH in aqueous base-catalysed hydrolysis of aryl-substituted *N*-methylacetanilides depends upon the substituent;[12] there is a rate acceleration for good leaving groups and a rate deceleration for poor leaving groups. This is interpreted in terms of a change in rate-limiting step from formation to breakdown of the tetrahedral intermediate.[20] However, the effect of added dimethyl sulphoxide (DMSO) on the alkaline hydrolysis of *p*-nitroacetanilide, under different conditions where either formation or breakdown of the tetrahedral intermediate is rate-limiting, indicates that this cannot be used as a probe for the rate-determining step in anilide hydrolysis. The results are compatible with a transition state of different geometry for rate-limiting breakdown of the intermediate.[12, 21] The rate constants for the basic methanolysis of a series of *N*-methylbenzanilides, unlike those for the corresponding acetanilides, give a linear Hammett plot with a ρ-value of 2.5. This indicates rate-limiting breakdown of the tetrahedral intermediate through a transition state involving solvent-assisted carbon—nitrogen bond cleavage (**24**). The Hammett ρ-value for the basic methanolysis of *N*-arylazetidinones (**25**), which is thought to involve rate-limiting formation of the tetrahedral intermediate, is 1.7.[22]

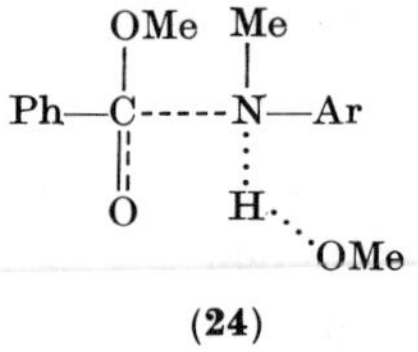

OMe Me
Ph—C-----N—Ar
O H···OMe

(24)

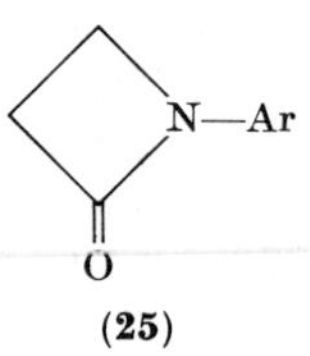

N—Ar
O

(25)

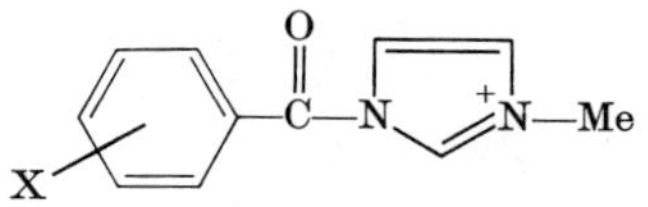

X–C₆H₄–C(=O)–N(imidazolium)$^+$N—Me

(26)

Ar—C(OH)(SO$_3$H)—O—C(=O)—Ar

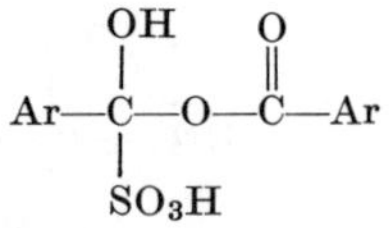

(27)

"Uncatalysed" hydrolyses of substituted *N*-benzoylimidazoles and *N*-benzoyl-*N'*-methylimidazolium ions (**26**) show solvent isotope effects k_{H_2O}/k_{D_2O} of ca. 2.5 which are essentially independent of the substituent. It is suggested that general-base catalysed hydrolysis of the benzoylimidazoles consists of rate-limiting nucleophilic attack by water with one proton partly transferred to the general base but not being transferred as part of the reaction co-ordinate motion. Coupling of heavy-atom motions with proton transfer is avoided by two consecutive transition states.[23]

A tetrahedral intermediate has been isolated from the reaction of sulphuric acid with carboxylic acid anhydrides (**27**). This may be the kinetically important intermediate in the reaction forming an anhydride mixture.[24]

The hydrolysis of *N*-methyldiacetamide is specific-acid catalysed. The buffer-catalysed hydrolysis shows a non-linear dependence on buffer concentration, which is interpreted in terms of a tetrahedral intermediate and a change in the rate-limiting step.[25]

Volumes of activation for the hydroxide-ion- and "Tris"-catalysed hydrolyses of *p*-nitrophenyl esters have been measured and compared with those for alkyl ester hydrolysis and with the volume changes for ionization of the corresponding carboxylic acids and hydration of the homologous aldehydes.[26]

High-pressure kinetics in solution have been reviewed.[26a]

Intermolecular Catalysis and Reactions

Reactions in Hydroxylic Solvents

The hydrolysis of methylated glucono-δ-lactones is general-acid and general-base catalysed with Brønsted exponents α and β of ca. 0.5 and ca. 0.35, respectively. There is no observed general-acid catalysis in the hydrolysis of tetra-*O*-methylmannono-δ-lactone (**28**). The susceptibility to general-acid catalysis is interpreted in terms of enhanced reactivity of δ-lactones due to changes in strain energy. There is no ^{18}O-exchange with solvent during the hydrolysis of D-glucono-δ-lactone.[27]

Second-order rate constants for the hydroxide-ion catalysed hydrolysis of lactones and esters decrease in the order methyl formate, δ-valerolactone, β-propiolactone, γ-butyrolactone, ϵ-caprolactone, methyl acetate. The differences in reactivity are interpreted in terms of the changes in the observed activation parameters. Pyridine and 1-methylimidazole react only with the β-lactone and proceed mainly with alkyl—oxygen fission to form betaines. This is attributed to the greater bond-angle strain in β-propiolactone.[28]

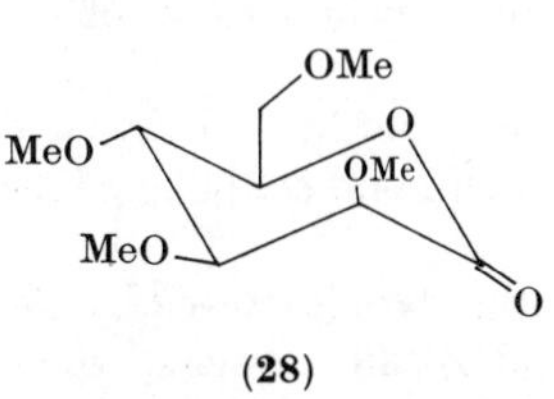

MeO, MeO, OMe, OMe, O, O

(**28**)

H, OAr, C, O, C, O, N, N—H, :N, NH

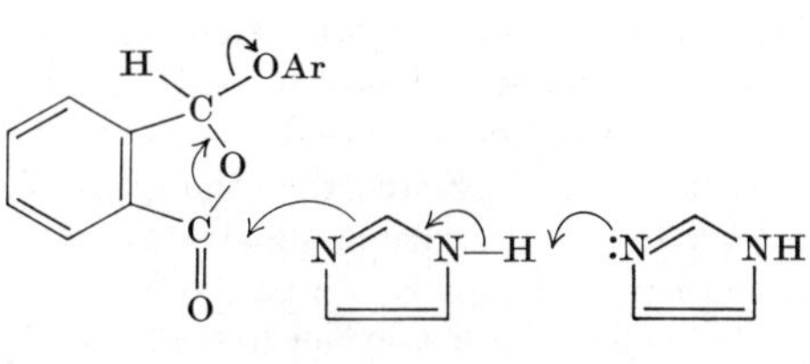

(**29**)

The buffer-catalysed methanolysis of *p*-nitrophenyl benzoate proceeds along both a general-base and a nucleophilic pathway. It is concluded that nucleophilic catalysis of acyl-transfer to alcoholic functions is preferred to protolytic catalysis when compared with acyl-transfer to water.[29]

The pH-independent hydrolysis of 3-(*p*-nitrophenoxy)phthalide has an entropy of activation of -47 cal deg^{-1} mol^{-1}. There is both a first- and a second-order dependence on the base component of the buffer, suggesting a pathway involving general-base catalysed nucleophilic attack (**29**). Similar ring-opening is suggested to take place on the specific-acid catalysed pathway, so that these acylals are hydrolysing as esters and not as acetals. The solvent isotope effect k_{H_2O}/k_{D_2O} for the ethanolamine general-base catalysed hydrolysis of 3-(*p*-nitrophenyl)phthalide is 1.8, the Hammett ρ-value is 0.55 and a mechanism analogous to (**29**) is suggested for the hydrolysis of these compounds.[30]

The rate of hydrolysis of 1-ethoxyethyl 2-acetoxybenzoate is pH-independent and very sensitive to solvent polarity and shows a solvent isotope effect k_{H_2O}/k_{D_2O} near unity. The mechanism of hydrolysis of this acylal is thought to be of the classical S_N1 type.[30a]

Phenyl-substitution at the pro-acyl-carbon atom of the cyclic orthoester 2-methoxy-1,3-dioxolane increases its rate of hydrolysis by a factor of 40. This is in contrast to the retardation normally found in acyclic systems, indicating steric inhibition of resonance. General-acid catalysed hydrolysis is observed, and since exocyclic carbon—oxygen bond fission occurs in the transition state this suggests a concerted mechanism (**30**) or spectator catalysis (**31**).[31]

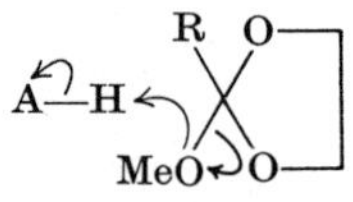

(**30**) 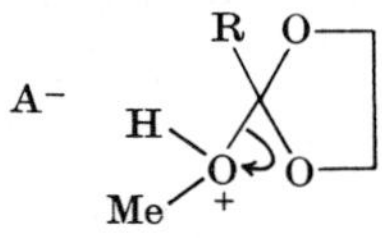 (**31**)

The enthalpies of formation of orthoesters have been determined, and the stabilization that accompanies attachment of several alkoxy-groups to the same saturated carbon atom has been discussed.[32]

The hydrolysis and alcoholysis of orthothioesters have also been investigated.[33]

The reactivity-selectivity principle has again[34] been criticized on the basis of data obtained from acid-catalysed hydrogen-exchange in acetophenones.[35] It has been suggested that a dual-parameter treatment should be used in the Hammett $\sigma\rho$-relationship.[36]

The use of the Hammett acidity function concept has been criticized.[37] The relationship of Bunnett–Olsen ϕ parameters to reaction mechanisms has been discussed.[38] Other aspects of acidity functions are reported in ref. 39.

The fractionation factor for L_3O^+ in H_2O–D_2O mixtures should continue to be 0.69,[40] and should not be increased as suggested earlier.[41]

It has been suggested that the rates of reaction of 1,4-diazabicyclo[2.2.2]octane and quinuclidine with various substrates may be explained by through-bond coupling of the lone pairs in the diaza-compound.[42]

The non-linear first-order hydrolysis of *p*-tolyl *N*-methylacetimidate is caused by the different rates of hydrolysis of slowly equilibrating *E* (**32**) and *Z* (**33**) isomers, which may be identified by NMR spectroscopy. The conjugate acid of the *E* form is 3–10 times more reactive than its isomer and has a pK_a of 6.55 compared with 8.0 for the *Z*-isomer. Since the conjugate acid is the reactive species even at high pH, the *Z*-isomer reacts more rapidly at higher pH values.[43]

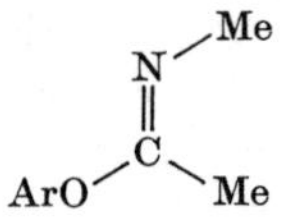

(**32**) 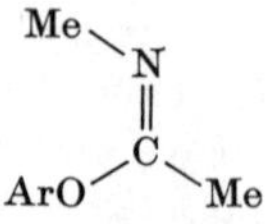 (**33**)

E–Z-Isomerization in *N*-arylmaleisoimides and *N*-aryl-2,2-dimethylsuccinoisoimides is thought to occur by the inversion mechanism.[44]

Kinetic *versus* thermodynamic control for the mechanisms of folate-catalysed transfer reactions have been discussed with reference to the reaction of amines with formamidinium tetrahydroquinoxaline analogues.[45]

The Hammett ρ-value for alkaline hydrolysis of aryl *N,N*-dimethylcarbamates is 1.0, compared with 0.98 for the same reaction of aryl acetates. The reaction order for the methanolysis of *N,N*-dimethylcarbamates decreases with increasing methoxide-ion concentration and it is claimed that this is due to a change from rate-limiting formation to rate-limiting breakdown of the tetrahedral intermediate. The rate of the acid-catalysed hydrolysis of *p*-nitrophenyl *N,N*-dimethylcarbamate does not reach a maximum value with increasing acid concentration, which is in contrast to that of the *N*-monomethyl derivative.[46]

The rate of hydrolysis of *N*-(phenylcarbamoyl)imidazole (**34**) is a complex function of pH, reaching two maxima at pH ca. 4.5 and 10.0. The zwitterionic tautomer (**35**) is thought to be the reactive species over the entire pH range, although arguments used to justify this depend partly on assigning a p*K* of 7 to the leaving imidazole when it is known that protonated imidazole behaves like a protonated amine of p*K* ca. 11.[5] The phenyl isocyanate intermediate may be trapped with imidazole. The hydrolysis of (**34**) is general-acid and general-base catalysed between pH 4 and 10, and the observed rate constants have a non-linear dependence on buffer concentration. Similar results are obtained for the hydrolysis of 1-(phenylcarbamoyl)benzimidazole and the values of the observed rate constants at "infinite" buffer concentration are independent of the nature of the buffer. These observations are rationalized in terms of changes in pH and buffer concentration causing changes in the rate-limiting step from decomposition of the zwitterion (**35**) to its formation by a diffusion-controlled proton transfer either from the conjugate acid (**36**) or to the conjugate base (**37**).[47]

PhNH—C(=O)—N(imidazole) (**34**)

$Ph\bar{N}$—C(=O)—$\overset{+}{N}$(imidazole)NH (**35**)

PhNH—C(=O)—$\overset{+}{Im}H$ (**36**)

Ph—$\bar{N}$—C(=O)—Im (**37**)

Ar—C(=O)—$\bar{N}$—C(=NH)—N(3,5-dimethylpyrazole, Me, Me) (**38**)

(1-C(=O)NHC_4H_9)benzimidazol-2-yl—NHCOOMe (**39**)

The hydrolysis of 1-(*N′*-aroylamidino)-3,5-dimethylpyrazoles occurs by a specific-base catalysed elimination reaction (**38**) at pH > 11 to give *N*-cyanoarenecarboxamides and 3,5-dimethylpyrazole. The Hammett ρ-value for this reaction with *N*-aroyl substituents is 0.45. In acid solution, specific-acid catalysed hydrolysis occurs to give aroylureas, probably by a mechanism involving attack by water on the conjugate acid of the substrate.[48]

The hydrolysis of the (*N*-butylcarbamoyl)benzimidazol-1-ylcarbamate (**39**) occurs by nucleophilic attack on the butylcarbamoyl-carbonyl group in both acid and alkaline solution. A second-order dependence upon hydroxide-ion concentration is interpreted in terms of the tetrahedral intermediate dianion.[49]

The pathways in the base-catalysed decomposition of cyclic *N*-nitrosocarbamates have been examined.[50]

An observed favourable enthalpy of activation, but an unfavourable entropy of activation, in the imidazole-catalysed hydrolysis of *N*,*O*-diacetylserinamide has led to the claim that a properly oriented imidazole side-chain in chymotrypsin may increase the rate of reaction 10^5-fold by general-base catalysis.[51]

The lower reactivity of *p*-nitroacetanilide compared with its *N*-alkyl derivatives in their base-catalysed methanolysis is attributed to ground-state effects.[52]

The rate constants for reversible hydrolysis of *o*- and *p*-nitroformanilides show a maximum as a function of formic acid concentration.[53] The perchloric acid-catalysed hydrolysis of 2,4-dinitroacetanilide has been described.[53a]

The p*K*'s of 1-acylpyrrolidines (**40a**) and 1-acylpiperidines (**40b**) are similar. Since *N*-protonation would involve a change in hybridization from sp^2 to sp^3, which should be more favourable for a six-membered ring, the results indicate *O*-protonation. The acid-catalysed hydrolysis of (**40b**) occurs only 5 times faster than that of (**40a**) under conditions of complete substrate protonation. This is consistent with rate-limiting formation of a tetrahedral intermediate from the *O*-protonated cation.[54]

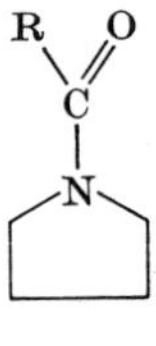

(**40a**) 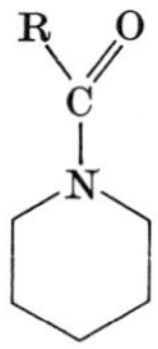 (**40b**) 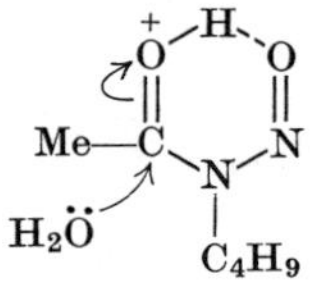 (**41**) 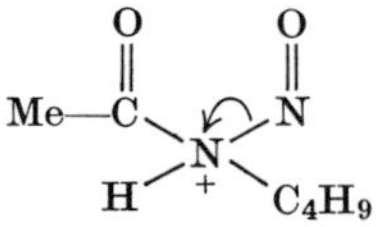 (**42**) 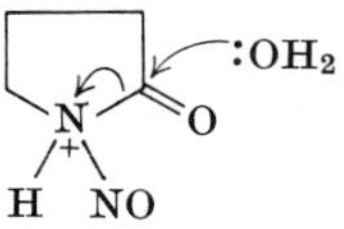 (**43**)

Decomposition of *N*-*n*-butyl-*N*-nitrosoacetamide in aqueous acid involves both deamination and denitrosation. Acidity dependences, activation parameters, and solvent deuterium isotope effects all indicate that deamination involves rate-limiting attack by water on an *O*-conjugate acid (**41**). Protonation of the amide-nitrogen (**42**) is considered to be the rate-limiting step for denitrosation.[55] It is suggested that acid-catalysed deamination and denitrosation of 1-nitroso-2-pyrrolidone occurs by a common rate-limiting step, which is thought to be formation of the *N*-conjugate acid with deamination occurring by direct S_N2 displacement by water (**43**).[56]

Acid-catalysed hydrolysis of thiobenzamide leads initially to both benzamide and thiobenzoic acid, which is interpreted in terms of attack by water upon the *S*-conjugate acid.[57]

The reaction profile of the $A_{AC}1$ hydrolysis of formamide has been calculated *ab initio*, including the orbital energies for *N*- and *O*-protonated formamide.[58]

Ultraviolet spectra of benzamide have been interpreted in terms of a tautomeric change from *N*-protonation in 60% sulphuric acid to *O*-protonation in 100% sulphuric acid.[59]

Differential NMR broadening indicates that the mechanism of acid-catalysed hydrogen exchange in amides does not involve *O*-protonation followed by N–H proton abstraction but rather *N*-protonation followed by proton-removal from a species in which there is restricted rotation about the carbon—nitrogen bond.[60] Other work on hydrogen exchange in amides is reported in ref. 61.

The rate of alkaline hydrolysis of phenylurea shows a less than first-order dependence upon hydroxide-ion concentration, and this is interpreted in terms of an unreactive conjugate base.[62]

Kinetic independence of hydroxide-ion concentration above 0.1 M, activation parameters and a Hammett ρ-value of 0.10 for the alkaline hydrolysis of *N*-aroylthioureas have been interpreted in terms of hydroxide-ion attack on the un-ionized substrate.[63]

Other studies include: hydrolysis of semicarbazide,[64] of peptides containing glutaminyl or asparaginyl residues,[65] of *N*-(5-carboxypentyl)perfluorocyclohexanamide[66] and of other amides;[67] acid-catalysed hydrolysis of γ-butyrolactam,[68] *cis*- and *trans*-*N*-styrylacetamides[69] and azodicarboxamide;[70] cyzlization of 2-chloroacetamides to 2,5-dioxopiperazines and hydrolysis of these heterocyclic compounds,[71] and the protonation of very weak bases in acetonitrile.[71a]

Reactions in Aprotic Solvents

The deuterium kinetic isotope effect in the acetic acid-catalysed reaction of benzoyl fluoride with arylamines in toluene significantly exceeds that for the corresponding reaction of benzoyl chloride. An eight-membered cyclic transition state is proposed for both reactions.[72]

The rate of the 5-bromo-2-hydroxypyridine-catalysed benzoylation of *m*-chloroaniline by benzoyl halides in benzene increases in the order F < Cl < Br; the suggested mechanism involves bifunctional catalysis.[73]

O-Nucleophilic catalysis has been suggested for the benzoylation of arylamines by benzoyl chloride catalysed by organophosphorus acid amides.[74, 75]

Phosphinic acids are extremely effective catalysts for the acylation of arylamines by benzoyl fluoride, the third-order rate constants being up to 10^7 mol^{-1} greater than the uncatalysed second-order ones. The mechanism is thought to involve bifunctional catalysis.[76] Similar results are obtained on using benzoyl chloride as the acylating agent.[77]

The Brønsted β-value for the benzoylation of substituted imidazoles by benzoyl chloride in benzene is 1.9. A stepwise mechanism with ion-pair formation is discussed.[78]

It has been suggested, on the basis of activation parameters and Hammett plots, that the mechanism for reaction of aniline with 2- and 3-furoyl chloride in benzene is similar to that for benzoyl chloride.[79]

Other work on the acylation of amines by acyl halides is listed in ref. 80.

The rate of esterification of substituted aliphatic alcohols with *p*-nitrobenzoyl chloride decreases as the electron-withdrawing ability and the volume of substituents in the alcohol increase. The Taft ρ^*-value of -2.5 and δ-value of 0.46 are compatible with rate-limiting nucleophilic addition of the alcohol to the acid halide.[81]

The selective esterification of 1,6-anhydrohexopyranoses with acid chlorides or acid anhydrides in pyridine is attributed to intramolecular hydrogen bonding.[82]

The esterification of phenols with acyl chlorides in inert solvents is catalysed by trifluoromethanesulphonic acid.[83]

The Hammett ρ-value for the reaction of substituted anilines with thiophene-3-

sulphonyl chloride in methanol is −2.25. An addition–elimination mechanism is proposed in which the rate-limiting step is addition.[84]

The effect of solvent polarity and the catalytic effect of pyridine derivatives and amines have been examined for sulphonamide formation from arylamines and arenesulphonyl halides.[85]

The rates of *N*-acylation of arylamines with arenesulphonic anhydrides in nitrobenzene are more sensitive to the nature of substituents in the anhydride than are the corresponding reactions with sulphonyl halides. Since arenesulphonate is a better leaving group than halide it is claimed that this supports a stepwise mechanism.[86]

Unlike acylation with acyl halides, the uncatalysed reaction of arylamines with acetic anhydride in toluene shows a primary isotope effect.[87]

Other studies on acylation with acid anhydrides are cited in ref. 88.

It is claimed that the catalysed and the uncatalysed amidation of *N*-(benzyloxycarbonyl)-L-phenylalanine *p*-nitrophenyl ester by the enantiomers of $PhCHMeNH_2$ in benzene are stereospecific.[89]

The rate of aminolysis of *p*-nitrophenyl *N*-phenylcarbamate in toluene occurs at least 10^4 times faster than that of the corresponding *N*-methyl derivative. This indicates an elimination mechanism giving the isocyanate as an intermediate, the formation of which, it is suggested, occurs through an ionic mechanism, *E*1*cB* (**44**) or *E*2 (**45**), rather than by a concerted cyclic six-centred process generating little or no charge. This is supported by the finding that tertiary amines are as effective catalysts as secondary ones.[90]

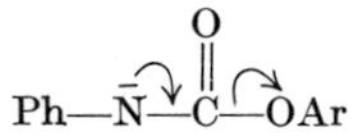

Ph—N̄—C(=O)—OAr

(**44**)

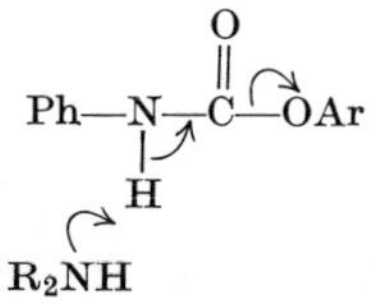

Ph—N(H)—C(=O)—OAr R_2NH

(**45**)

Me—C(O^-)($H_2\overset{+}{N}R$)—OAr

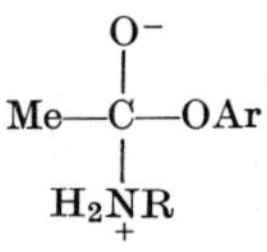

(**46**)

The rate-constants for the general-base catalysed *n*-butylaminolysis of *p*-nitrophenyl acetate in chlorobenzene correlate with the catalyst's ability to hydrogen-bond to *p*-fluorophenol. This is interpreted in terms of little proton-transfer in the rate-limiting decomposition of the tetrahedral intermediate (**46**). Amides are effective catalysts in this reaction and it is suggested that they could act as general-base catalysts in enzymes with hydrophobic active sites.[91]

Other studies on acylation reactions with esters are cited in ref. 92.

It is suggested that carboxylic acid-catalysed ethanolysis of *p*-chlorophenyl isocyanate in diethyl ether proceeds through a cyclic intermediate adduct of acid and ethanol which then attacks the isocyanate in a slow step.[93]

Reports on other aspects of acylation reactions in aprotic solvents are collected in ref. 94.

In contrast to the behaviour observed in solution, esters appear to protonate at the ether-oxygen in the gas phase.[95] Ion-molecule esterification reactions with carboxylic acids and alcohols occur in the gas phase when the reaction is exothermic and the proton affinity of the acid is higher than that of the alcohol.[96]

The Hammett ρ-value for the ionization of substituted benzoic acids in the gas phase is ca. 10, showing the large attenuation brought about in aqueous solution.[97]

Intramolecular Catalysis and Neighbouring-group Participation

As suggested previously,[98] the changes in strain energy of analogous hydrocarbons may be used to rationalize the relative rates of intramolecular reactions.[99] "Orbital steering" has received further criticism.[100, 101]

It has been suggested that intramolecular proton-transfer occurs most readily when the ring involved is eight-membered. The occurrence of such reactions involving six-membered rings is questioned and it is suggested that they probably involve transfer through a solvent bridge increasing the ring size to eight-membered.[102]

The hydrolysis of dialkylmaleamic acids, but not of the unsubstituted derivatives, is subject to general-acid catalysis. On the basis of an apparently non-linear Brønsted plot the rate-limiting step for strongly acidic catalysts is claimed to be diffusive separation of the general-acid and the zwitterion intermediate (**47**). There is a non-linear dependence of the rate on the buffer concentration, and at high buffer concentrations the rate-limiting step is thought to be fission of the carbon—nitrogen bond in (**47**).[103] When the

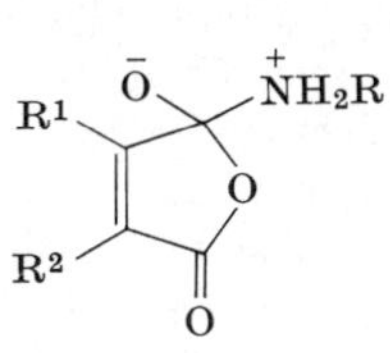

(47) 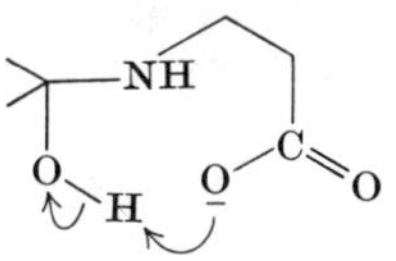 **(48)** 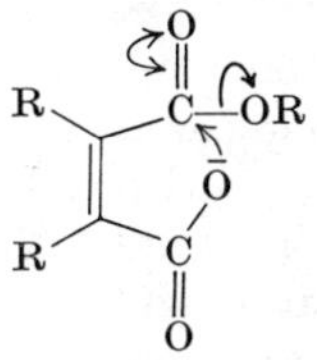 **(49)**

displaced alkylamine contains a carboxyl group there is no or little intramolecular general-acid catalysis, but when this carboxyl group is ionized a small rate enhancement is observed which is attributed to intramolecular general-base catalysed interconversion of neutral and zwitterionic forms of the tetrahedral intermediate (**48**).[104]

The kinetics and mechanism of the hydrolysis of alkylmonoesters of dialkylmaleic acid are similar to those reported for phthalic acid monoesters.[105] The Brønsted β-value for the leaving alkoxide anion in the reaction of the monoanion (**49**) is 1.43.[106]

Details of the intramolecular nucleophilic catalysis by carboxylate group in the hydrolysis of carbamate esters have been reported. The Hammett ρ-value for the leaving group is 2.0, whereas in nucleophilic attack by hydroxide ion on carbamates the rate is relatively independent of the nature of the leaving group. This is interpreted in terms of rate-limiting breakdown of the tetrahedral intermediate (**50**) in the intramolecular reaction.[107]

The hydrolysis of *ortho*-(*N*-phenylcarbamoyl)benzoic acid proceeds with formation of phthalic anhydride as an intermediate from pH 4 to 13 M-sulphuric acid. The two terms in the rate law refer to the neutral amide and its conjugate acid.[108]

The rate constant for carboxylate anion nucleophilic catalysis is only three-fold greater for the monosuccinate ester (**51**; R = H) of hexachlorophene than for the corresponding methylated derivative (**51**; R = Me). There is thus no experimental basis to support the earlier suggestion of a concerted nucleophilic general-acid catalysed reaction to explain the bell-shaped pH–rate profile for (**51**; R = H).[109] The nucleophilic attack of the carboxyl group and the aminolysis of other hexachlorophene esters involve rate-limiting breakdown of the tetrahedral intermediate but show a small dependence upon the basicity of the leaving group with a Brønsted α-value of 0.2. Conversely, the hydroxide-ion catalysed hydrolysis of these esters, which involves rate-limiting formation of the tetrahedral intermediate, shows a larger dependence on the basicity of the leaving group with a Brønsted α-value of 1.0. It is suggested that this dependence is controlled mainly by the ground-state conformation of the ester.[110]

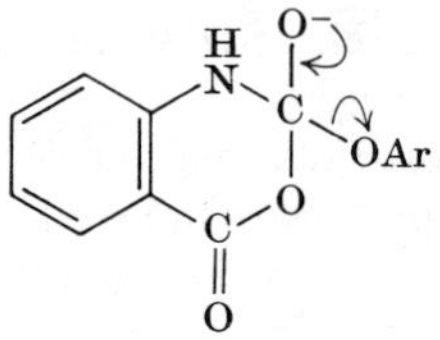

(50) 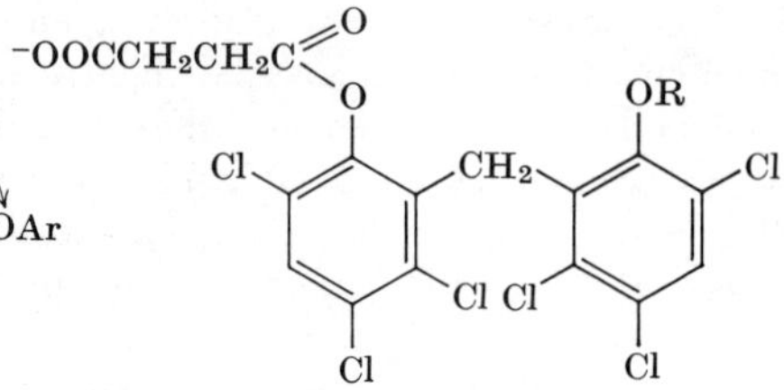 **(51)** 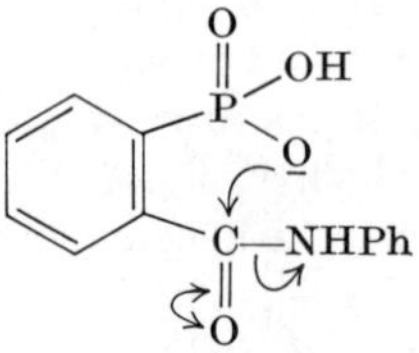 **(52)**

The effect of substituents and acid concentration on the rates of solvolysis of some acylamino-acids in acetic acid, 0.25 M in water, is interpreted in terms of neighbouring-group participation by the terminal carboxyl group to form an azlactone.[111]

The pH-rate profile for the hydrolysis of 2′-phosphonobenzanilide indicates that the undissociated acid and its monoanion are the reactive species. The monoanion is hydrolysed 6×10^5-fold faster than the extrapolated value of k_{H^+} for benzanilide at pH 6. There is no detectable intermolecular phosphate-catalysis of benzanilide hydrolysis giving a lower limit to the effective concentration of 10^3 M. It is suggested that the neighbouring phosphonate group acts as a nucleophilic catalyst forming a cyclic acyl phosphonate intermediate (**52**) which is rapidly hydrolysed to products.[111a]

The pH–rate profiles for lactonization of methyl-substituted coumaric acids vary with the position of the methyl substituents on the ring and on the side-chain. The pH–rate profile for 4,6,8-trimethylcoumaric acid is consistent with the existence of three kinetically significant intermediates and it is suggested that the additional one[112] is the anionic tetrahedral intermediate (**53**). Non-linear dependence of the rates of lactonization on buffer concentration is also taken as evidence of intermediates in these reactions.[113]

As reported earlier,[114] phenyl *N*-(*o*-hydroxyphenyl)carbamate cyclizes to benzoxazolinone by the ionized phenoxy-group acting as an intramolecular nucleophile. Even poor leaving groups such as methoxide ion may be displaced (**54**).[107]

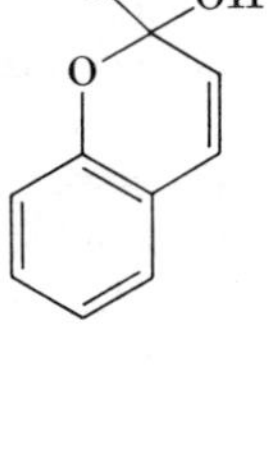

(53) 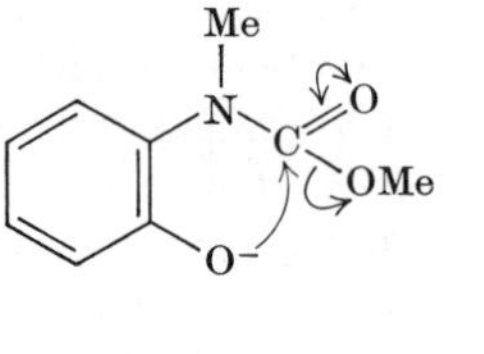 **(54)** 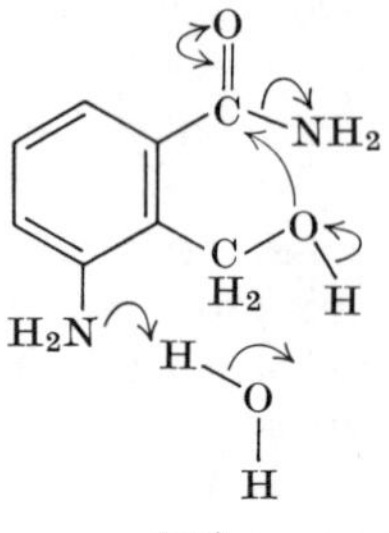 **(55)** 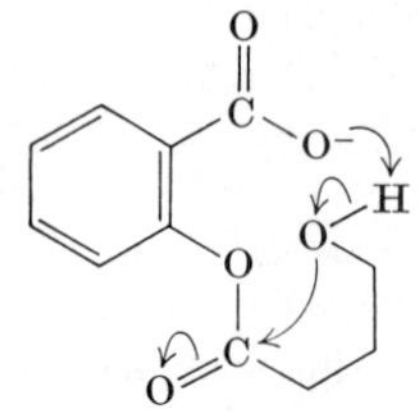 **(56)**

The kinetics of intramolecular attack of a carbonyl group on a nitrile catalysed by ethoxide ion have been described.[115]

Cyclization of 3-amino-2-(hydroxymethyl)benzamide to 4-aminophthalide is pH-independent from pH 3.5 to 10 and occurs at a rate ca. 10^3 times that for the 4-, 5- and 6-amino- or unsubstituted derivatives. Together with a solvent isotope effect k_{H_2O}/k_{D_2O} of 2.8 and an entropy of activation of -23 J K^{-1} mol^{-1}, these observations are taken to indicate intramolecular general-base catalysis through one or more solvent molecules (**55**).[116] Similarly, the rate of hydrolysis of the anion of *o*-carboxyphenyl 4-hydroxy-butyrate is ca. 500-fold greater than the spontaneous hydrolysis of phenyl 4-hydroxy-butyrate. This is analogous to the previously reported intermolecularly catalysed reaction,[117] the effective concentration of the intramolecular carboxyl group being 14 M. The mechanism is thought to involve concerted intramolecular general-base nucleophilic catalysis by the carboxylate and hydroxy-groups, respectively (**56**).[117,118]

There is no rate increase in the catalysed hydrolysis of *p*-nitrophenyl acetate by 2-(mercaptoalkyl)benzimidazoles compared with that by simple thiols.[119]

The rate constants for uncatalysed hydrolysis of the anion of 3,5-dinitro-*O*-acetyl-salicylic acid have a non-linear dependence upon the atom fraction of deuterium in mixtures of deuterium and protium oxide. This is interpreted in terms of two parallel mechanisms for the hydrolysis of the salicylic acetic anhydride intermediate: first, an uncatalysed or electrostatically stabilized attack by water at the benzoyl-carbonyl group of the anhydride intermediate catalysed intermolecularly by another water molecule (**57**); secondly, intramolecular general-base catalysed attack by water at the acetyl-carbonyl group (**58**). It is suggested that formal general-base catalysis by the neighbouring phenoxide is unlikely because of the non-linear transition state involved for proton transfer (**59**).[120]

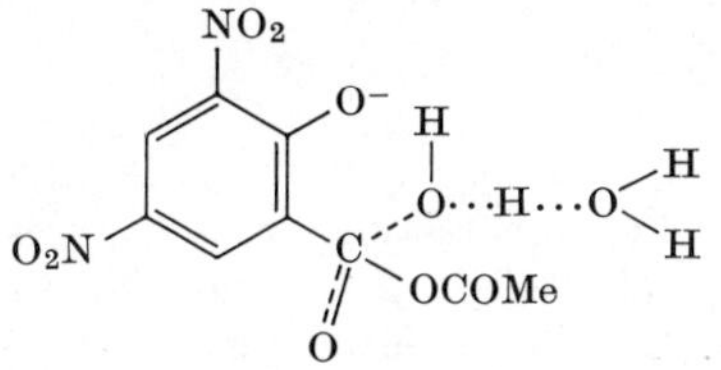

(57)

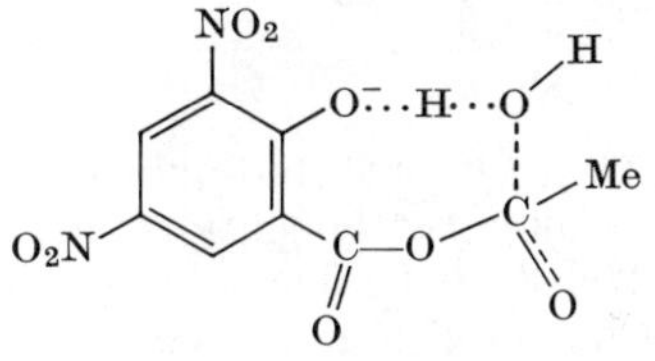

(58)

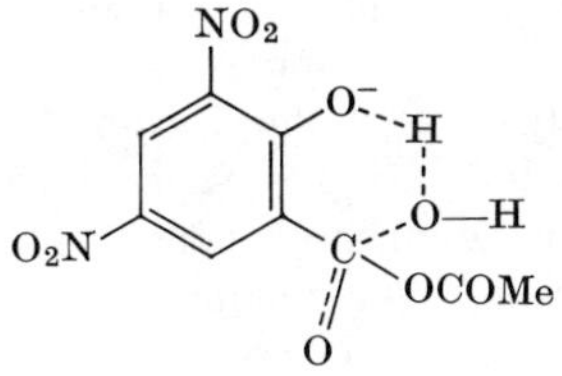

(59)

The first step in the alkaline hydrolysis of methyl salicylate is the formation of methyl salicylate anion, which is then hydrolysed to products by a mechanism involving intramolecular general-base catalysis.[121]

The increase in the rates of alkaline hydrolysis of diol monoesters compared with those of suitable model compounds is solvent-dependent and is interpreted in terms of intramolecular hydrogen-bonding in the transition state.[122]

The hydrolysis of the ion (**60**) at low pH is pH-independent and shows a rate enhancement of 25-fold. This is attributed to intramolecular general-acid catalysed attack by water (**60**). At neutrality a second plateau is observed and is attributed to general-base catalysed attack by water (**61**). At pH 9–11 additional plateaus in the pH–rate profile are seen. Intramolecular nucleophilic catalysis by the imidazole anion (**62**) is believed to occur in this region and in one case the acetylimidazole intermediate was observed (**63**), being subsequently hydrolysed by specific-base catalysis.[123] Incorporation of a

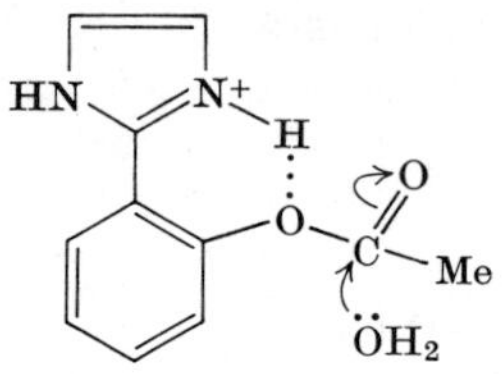

(**60**) 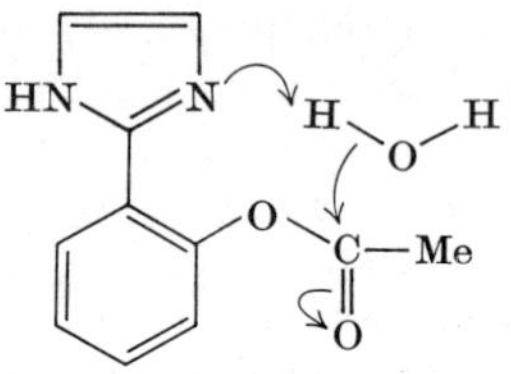 (**61**) 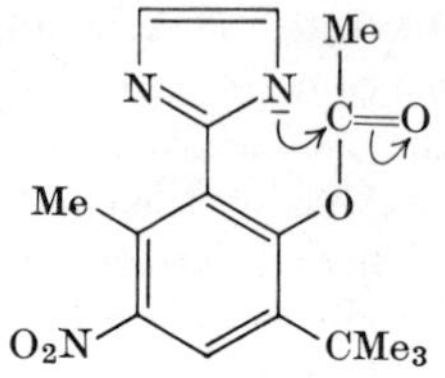 (**62**)

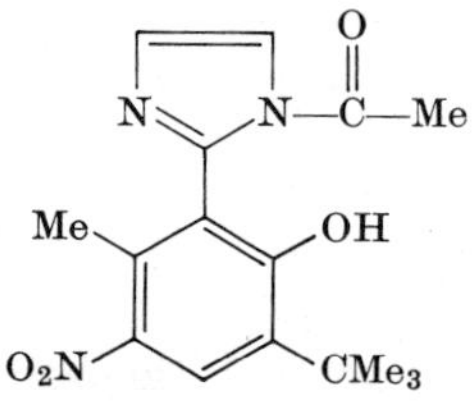

(**63**) 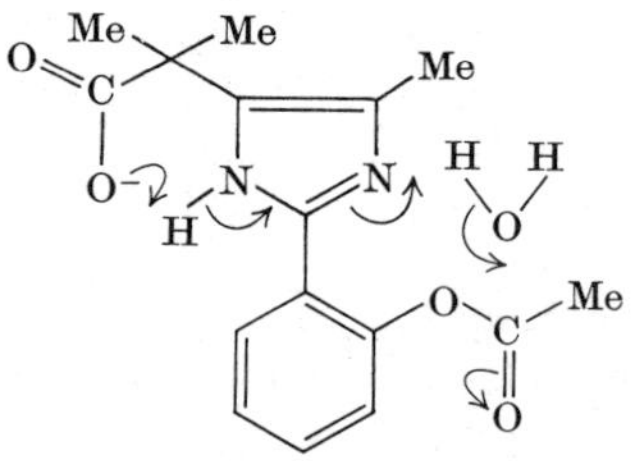 (**64**)

carboxylate function juxtaposed to the imidazolyl NH provides no appreciable assistance, as may be thought by analogy to the "charge relay system" suggested for chymotrypsin (**64**). There is no observable catalysis in 6:94 water–acetonitrile. In solvents of low dielectric constant there is no inversion of the carboxylate and imidazolyl base strength.[124,125] The catalytic triad of amino-acid residues at the active site of the serine esterases is discussed.[126]

The rate of water-catalysed hydrolysis of aryl quinoline-8-carboxylates proceeds an estimated 10^4 times faster than the corresponding 6-isomer. This is attributed to intramolecular general-base catalysis and the Hammett ρ-value of 0.97 for aryl-substituted esters is assumed to indicate a concerted (**65**) rather than a stepwise mechanism. Brønsted β-values, deuterium solvent kinetic isotope effects, and Hammett ρ-values indicate that with catalysis by a tertiary amine hydrolysis proceeds by a general-base catalysed mechanism, except for the 2,4-dinitrophenyl ester for which the amine acts as a nucleophilic catalyst.[127] A plot of absorbance against time passes through a minimum during the methylaminolysis of *p*-cyanophenyl quinoline-8-carboxylate; it is suggested that this is due to the formation of an unusually stable tetrahedral intermediate. The Hammett plots for the second-order reaction of primary and secondary amines with the substituted quinoline-8- and -6-carboxylate esters are curved, and this is interpreted in terms of a change in the rate-limiting step with respect to formation and breakdown of the tetrahedral intermediate. The change from rate-limiting breakdown of the tetrahedral intermediate to rate-limiting attack by amine occurs with a poorer leaving group for the 8-isomer than for the 6-isomer, indicating that the quinoline-nitrogen atom promotes collapse of the intermediate to products, presumably by intramolecular general-base catalysis. The efficiency of this catalysis appears to be independent of the basicity of the amine and it is suggested that the function of the quinoline-nitrogen is to transfer a proton from the amine nitrogen to negatively charged oxygen in the tetrahedral intermediate (**66**), although it is difficult to see how this leads to an overall more favourable pathway.[128]

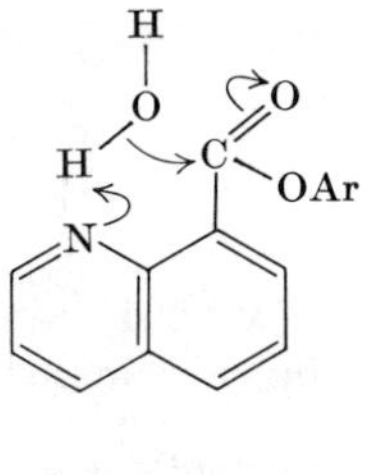

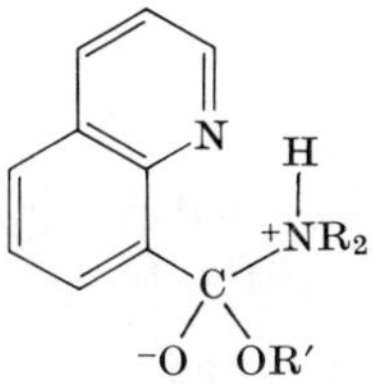

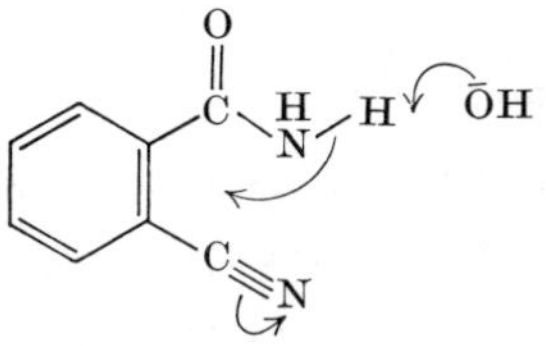

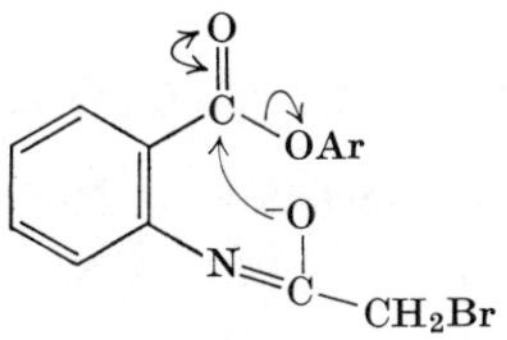

(65) **(66)** **(67)** **(68)**

Neighbouring-group participation by piperidine- and pyrrolidine-nitrogen in the solvolysis of esters and thioesters has been described.[129]

The rate of cyclization of *o*-cyanobenzamide to 3-iminoisoindolin-1-one is first order in hydroxide-ion and shows a solvent isotope effect k_{H_2O}/k_{D_2O} of 0.60. The mechanism is thought to be that depicted in (**67**).[130]

The second-order rate constant for the alkaline hydrolysis of *p*-nitrophenyl *N*-(bromoacetyl)anthranilate is 3.8×10^6 l mol^{-1} s^{-1} and is ca. 5×10^5 greater than that for comparable model compounds. A cyclic intermediate is detected and nucleophilic attack probably occurs through the conjugate base of the amide (**68**).[131]

The isocyanate intermediate formed in the *E*1*cB* hydrolysis of phenyl *N*-(*o*-carbamoylphenyl)carbamate is trapped by the neighbouring amide group. Nucleophilic attack by the amide anion occurs (**69**) when the isocyanate pathway is blocked by *N*,*N*-disubstitution.[132] An *o*-amide function in the leaving phenoxide ion of salicylanilide carbamates increases the rate of hydrolysis by 2500 compared with that for the unsubstituted compound but by only 25 compared with that for the corresponding *para*-isomer. This is explicable in terms of electronic and steric effects[132] without involving participation of the amide group as was suggested earlier.[133] Activation parameters for the hydrolysis of the carbamoyl group in salicylanilide *N*-methylcarbamate have been reported and were still interpreted in terms of participation by the carboxanilide group (**70**).[134]

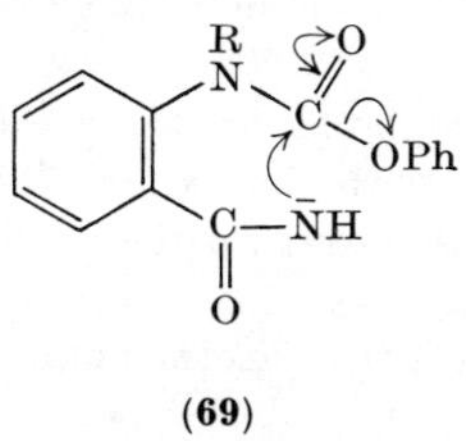

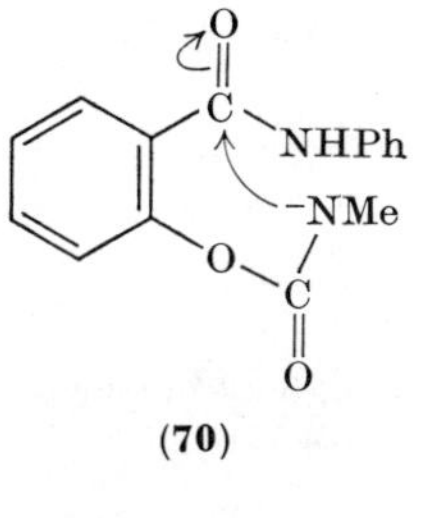

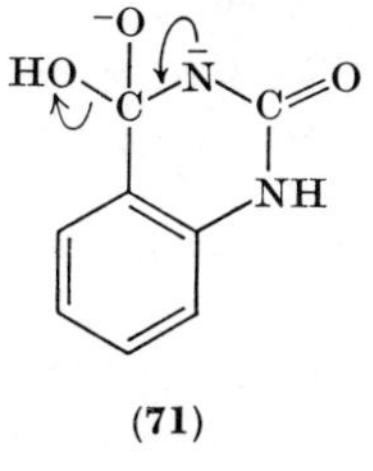

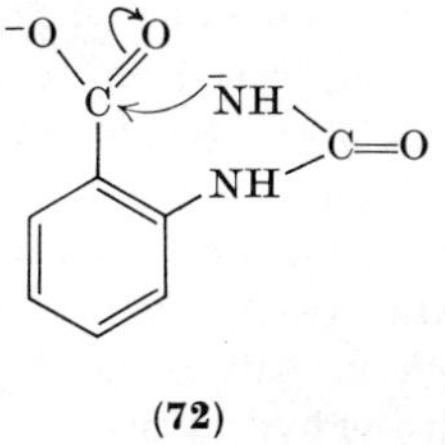

(69) **(70)** **(71)** **(72)**

Intermolecular nucleophilic attack of the ureido-nitrogen anion on the β-lactam carbonyl-carbon of penicillin derivatives has been described.[135]

The first-order dependence on hydroxide ion of the cyclization of *N*-(*o*-carboxyphenyl)urea in strongly basic solution could be explained by a rate-limiting breakdown of the tetrahedral intermediate (**71**) rather than by the originally suggested[136] anionic attack upon a carboxylate anion as in (**72**). However, cyclization of the *N*-methyl derivative is twice as fast as that of the corresponding non-methylated urea. Mechanism (**71**) therefore does not occur, and intramolecular anionic nucleophilic attack upon a carboxylate anion (**72**) can take place.[137]

Other reactions involving neighbouring-group participation are listed in ref. 138.

Association-prefaced Catalysis[139]

The acylation of *N*-methyldodecanohydroxamic acid, $CH_3(CH_2)_{10}CON(Me)OH$, by *p*-nitrophenyl acetate in the presence of hexadecyltrimethylammonium bromide, $CH_3(CH_2)_{15}\overset{+}{N}Me_3Br^-$, shows Michaelis–Menten kinetics and proceeds 333-fold faster than in the absence of the micelle. This is attributed to increased nucleophilicity, because of desolvation of the hydroxamate anion.[140] The acylation of *p*-nitrophenyl chloroacetate (**73**) catalysed by this mixed micelle has a second-order rate constant of ca. 10^6 l mol^{-1} s^{-1} which is comparable with that for the α-chymotrypsin-catalysed reaction. The rates of deacylation by the two catalysts are also similar.[141]

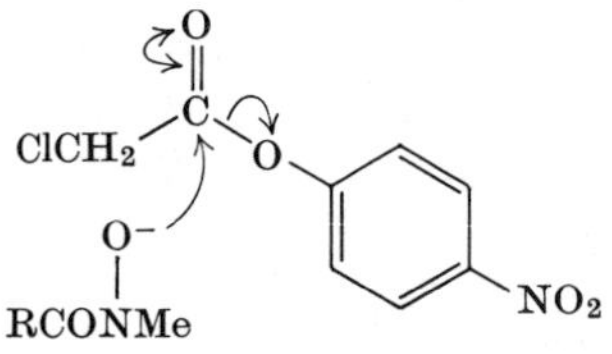

ClCH2–C(=O)–O–C6H4–NO2 ; O⁻–N(Me)COR

(**73**)

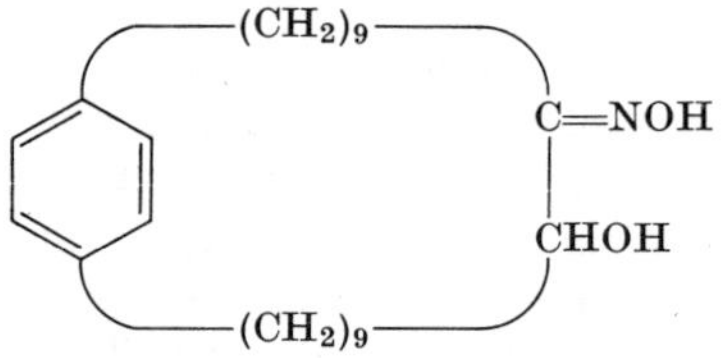

(CH2)9 ; C=NOH ; CHOH ; (CH2)9

(**74**)

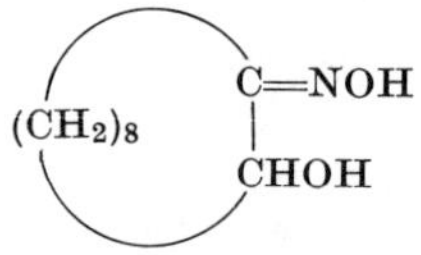

(CH2)8 ; C=NOH ; CHOH

(**75**)

Water-soluble copolymers containing hydroxamic acid groups react more slowly with *p*-nitrophenyl acetate than do the corresponding monomeric compounds. Deacylation is rate-limiting, but the catalytic efficiency of the hydroxamic acid can be enhanced by the introduction of an intramolecular imidazole unit which assists deacylation of the acetohydroxamate intermediates.[142]

Cyclohexa-amylose-*N*-methylacetohydroxamic acid reacts 20- and 70-fold faster with *p*-nitrophenyl acetate and 2-hydroxy-5-nitrotoluene-α-sulphonic acid, respectively, than does *N*-methylmethoxyacetohydroxamic acid. It also shows, in contrast to cyclohexa-amylose itself, a marked kinetic stereospecificity for *p*-nitrophenyl compared with *m*-nitrophenyl acetate. There is a reversible binding step between substrate and cycloamylose prior to nucleophilic attack.[143]

Only *p*-nitrophenyl esters of long-chain fatty acids are deacylated by the paracyclophane oxime (**74**), and none of the esters examined reacts with the oxime (**75**). It is suggested that reaction occurs by incorporation of the substrates into the cyclic oxime cavity.[144]

A bile salt containing an oxime group is a slightly better catalyst than acetone oxime for the hydrolysis of *p*-nitrophenyl dodecanoate.[145]

Poly-5-vinylbenzimidazole is a better catalyst than benzimidazole for solvolysis of negatively, but not positively, charged esters of various aliphatic chain lengths. Its activity is controlled by the degree of ionization and the polymer conformation. It is suggested that bifunctional catalysis by the polymer, and by copolymers with acrylic acid, facilitates the reaction.[146]

The water-soluble steroid (**76**) catalyses the hydrolysis of aryl acetates and, compared with the imidazole-catalysed reaction of esters with anionic substituents on the phenolic residue, shows rate enhancements of up to 15-fold. A small rate retardation is observed with esters containing cationic substituents. These results are attributed to electrostatic catalysis and inhibition.[147]

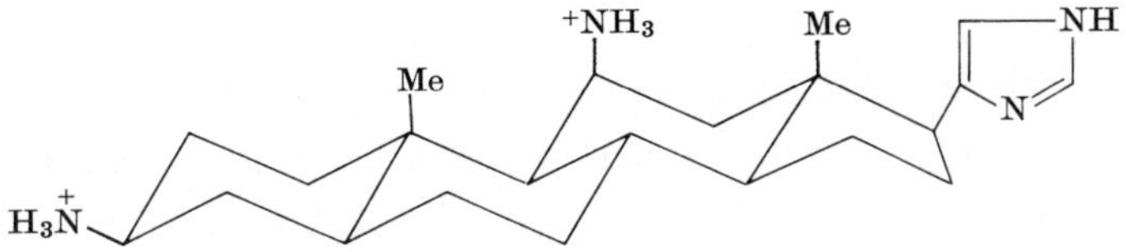

(**76**)

Alkaline hydrolysis of *p*-nitrophenyl hexanoate is catalysed by poly-soaps formed by alkylation of poly(vinylpyridine) with ethyl and dodecyl bromide. Catalysis increases with increasing ratio of dodecyl to ethyl groups, indicating the importance of the hydrophobic properties of the polymer. The poly-soap also catalyses nucleophilic attack of the azide ion on the ester.[148]

Decomposition of acetylsalicylic acid in aqueous solutions of poly(ethylenimine), a weak cationic polyelectrolyte containing free amino-groups, is 10^3 faster at pH 8 than in the absence of polyelectrolyte. At pH 11 the two rates are similar. This is attributed to action of the amino-groups on the polyelectrolyte as nucleophiles attacking the substrate. At high pH there is a large concentration of free amino-groups but the concentration of charged ester near the now neutral polymer is decreased.[149]

The heats of transfer of non-aromatic compounds from water to hexadecyltrimethylammonium bromide micelles are small (0 ± 2 kcal mol^{-1}), but for aromatic molecules the reactions are exothermic and their heats of transfer are as large as 9 kcal mol^{-1}, and different from heats of solution in *n*-hexane. This indicates that aromatic compounds are not solubilized in the hydrocarbon-like interior of the micelle.[150] Methyl salicylate has a small negative enthalpy of transfer from water to the micelle, but salicylic acid (one of its hydrolysis products) interacts very favourably with the micelle. However, the presence of the micelle has no effect on the rates of hydrolysis of salicylate esters, indicating that this favourable interaction with salicylic acid is not felt in the rate-limiting step. It is suggested that cationic micelles enhance the base-catalysed hydrolysis of esters with good leaving groups but not of those with poor leaving groups because of the different rate-limiting steps involved.[151]

Ca^{2+} ions enhance the rate of the octylamine-catalysed methanolysis of choline phosphoglyceride. This is attributed to lowering of the pK of methanol by the metal ion, thus facilitating proton-transfer to the amine.[151a]

The reactions of side-chain groups of angiotensin II, an octapeptide, with *p*-nitrophenyl acetate have been investigated.[151b]

Hydrophobic and electrostatic interactions affect the acidity of carboxylic acids incorporated into micelles of hexadecyltrimethylammonium bromide.[152]

The micelle-catalysed hydrolysis of ethyl benzoates,[153] ethyl *p*-aminobenzoates[154] and *p*-nitrophenyl decanoate,[155] reaction of anions with esters,[156] and acylation of benzimidazole derivatives[157] have been studied.

The previous results that led to the claim[158] for stereochemical differentiation for micelle-catalysed hydrolysis of some ester are not reproducible.[159] The hydrolysis of *p*-nitrophenyl (+)-, (−)-, and (±)-α-methoxyphenylacetates in the presence of surfactants containing one centre of chirality show micellar catalysis but no stereoselectivity.[160]

The rate constants for the *cis–trans*-isomerization of bis(oxalato)diaquochromate(III) anion in the polar cavities of reversed alkylammonium carboxylate micelles in benzene are up to 63-fold faster than that in pure water.[161]

A kinetic study of micelle-formation has been reported.[162]

Metal-ion Catalysis[163]

Hydrolysis of a dipeptide co-ordinated to Co^{3+} (**77**) proceeds 10^8 times faster than that of glycylglycine at pH 7, and still faster on addition of bases. It is suggested that a stepwise mechanism occurs involving rate-limiting diffusion-controlled proton-transfer from the tetrahedral intermediate (**78**) or (**79**).[164]

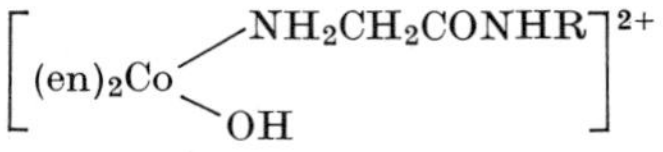

$[(en)_2Co(OH)(NH_2CH_2CONHR)]^{2+}$

(**77**)

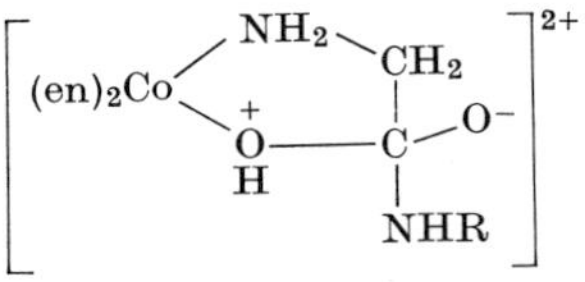

(**78**)

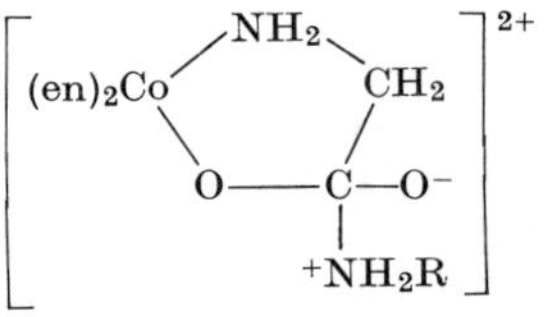

(**79**)

There are two reaction paths in the base-catalysed hydrolysis of oxygen-bonded dimethylformamide in $[Co(NH_3)_5DMF]^{3+}$. One involves hydroxide-ion attack at the carbonyl centre and the other probably proceeds through the conjugate base of the amide. Compared with the hydrolysis of the unco-ordinated species, amide cleavage is accelerated by greater than ca. 10^4. It is suggested that only single co-ordination of the carbonyl-oxygen atom is required to activate such centres towards nucleophiles.[165]

The inert metal-complex (**80**) forms the imine (**81**) in aqueous solution. The dependence of the rate upon hydroxide-ion concentration is not simple but the reaction is assumed to occur through the deprotonated ammonia *cis* to the carboxylate ligand.[166]

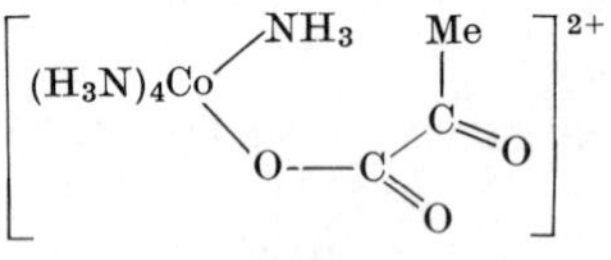

(**80**)

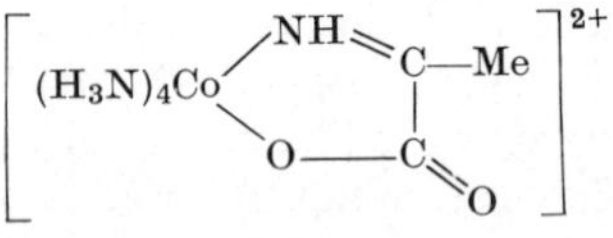

(**81**)

The structure of the diastereoisomeric forms of β_1-glycinatotriethylenetetraminecobalt(III) ions, determined by X-ray analysis of their iodides, may be reproduced by calculated strain-energy minimization. The mutarotation of the two species occurs 10^6 more slowly than proton exchange of the pertinent asymmetric nitrogen centres.[167]

Hydrolysis of glycine methyl ester catalysed by a $Cu(L)^{2+}$ complex proceeds by pre-equilibrium formation of the metal-co-ordinated ester. Weakly donating L ligands produce the most active catalysts, presumably by enhancing the Lewis acidity of the Cu(II), which promotes binding of the ester and also facilitates hydroxide-ion attack on the co-ordinated ester.[168]

Hydroxide-ion-catalysed hydrolysis of a nitrile is subject to large rate enhancement if the nitrile is co-ordinated to a metal ion such that only external hydroxide-ion attack can take place:[169]

$$(H_3N)_5Co^{3+}(NCR) + {}^{-}OH \rightarrow (H_3N)_5Co^{2+}(NHCOR)$$

Both nitrile and ester hydrolysis of ethyl cyanoformate occurs readily in $[(H_3N)_5Ru(NCCOOEt)][PF_6]_2$.[170]

In contrast to the uncatalysed reaction, structural variation in the alkyl group of the acid chloride in its Cu(I)-catalysed reaction with a Grignard reagent is independent of the steric parameter and yields a Taft ρ^*-value of 2.96. These results are consistent with a cyclic transition state in which chloride is displaced with no change in hybridization of the carbonyl group by either a mixed cuprate(I) intermediate $[RCuX]^{-}MgX^{+}$ (**82**) or a loose $RCu \cdots\cdots MgX_2$ complex (**83**).[171]

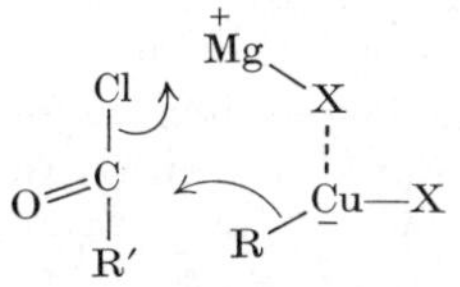

(**82**) 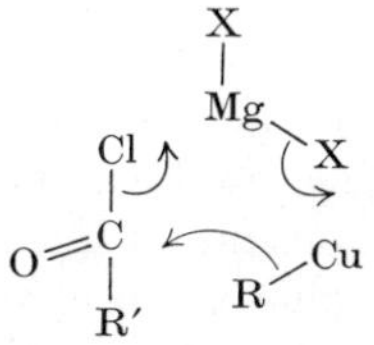 (**83**)

Other systems studied include peptide hydrolysis catalysed by metal ions,[172] oxidative decarboxylation of Cu(II) complexes of aminomalonic acid[173] and chromic acid oxidation of oxalic acid.[174]

Enzymic Catalysis[175]

It has been suggested that there is a lowering of vibational frequencies on going from the Michaelis complex to the transition state and that this lowers the free energy of activation.[176]

Intermediates accumulating after the initial Michaelis complex are undesirable, and enzymes whose function is to optimize reaction rates should have evolved to exhibit K_M-values above those of accessible substrate concentrations.[177]

Serine Proteinases

The mechanism for the chymotrypsin-catalysed hydrolysis of amides is suggested to be analogous to that for general-base catalysed aminolysis of esters in aqueous solution.[178] The diffusion process in solution is replaced by rate-limiting movement of the imidazole group relative to the substrate (k_2 in Scheme 3).[9]

The "charge-relay" system in chymotrypsin, the Asp-His-Ser catalytic triad, is negatively charged above pH 7 and either neutral or zwitterionic below this pH. On the basis of coupling of surface carboxyl groups of the protein by a carbodi-imide with glycinamide or semicarbazide, followed by titration of the proton uptake on denaturation, it is thought that the Asp-carboxyl group is ionized at pH-values greater than 4.2.[179] NMR studies of the histidine enriched with ^{13}C at C-2 in the Asp-His-Ser catalytic

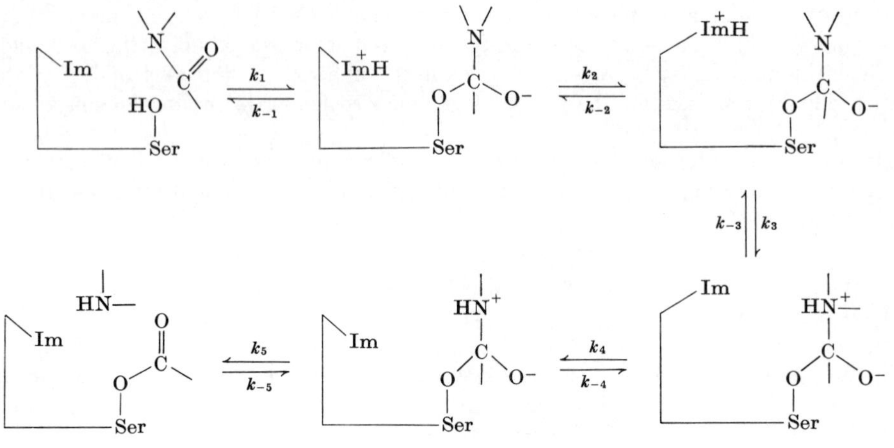

SCHEME 3.

triad of α-lytic protease have led to the suggestion that only below pH 4 does the imidazole residue become protonated and only above pH 6.7 does the aspartic acid residue become ionized. This would mean that over the pH range 4–6.7 the catalytic triad consists of neutral histidine and aspartic acid and not the ionized forms as usually assumed. It is claimed that the group in the enzyme known to ionize with pK 6.7 is the aspartic acid residue and that the function of this residue is to act as a general-base-avoiding charge separation (**84**).[180,181] High-resolution proton NMR has located the

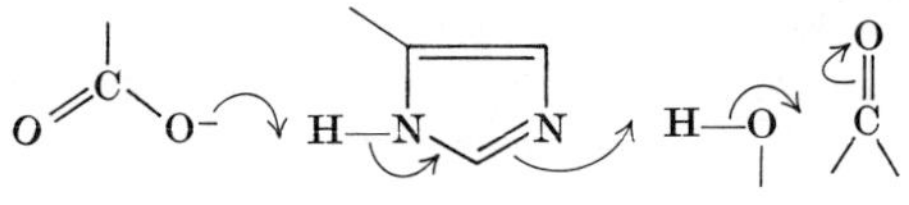

(**84**)

proton in a hydrogen bond between His-57 and Asp-102 of chymotrypsin A_δ and chymotrypsinogen A. The pK for ionization of the imidazolium ion is 7.5. Chemical-shift differences between the enzyme and a stable acyl-enzyme analogue are interpreted in terms of a hydrogen bond between Ser-195 and His-57 in the active enzyme and zymogen. The influence of substrate analogues and competitive inhibitors indicates that there is direct interaction between His-57 and the atom occupying the leaving-group position of the substrate.[182]

It has been suggested that fission of the hydrogen bond between His-57 and Asp-102 does not occur easily in chymotrypsin and that this step is essential in the catalytic process.[183]

The pH-dependence of $k_{\text{cat}}/K_{\text{M}}$ for the α-chymotrypsin-catalysed hydrolysis of several substrates indicates that the pK_{a} of the active site is 6.80. Measurement of proton release on binding hydrazide substrates shows that this pK_{a} is lowered in the enzyme substrate complex, and this is explained by a hydrogen bond between the hydrazide leaving group and the oxygen of Ser-195. It is concluded that there is no evidence for accumulation of a tetrahedral intermediate in the chymotrypsin-catalysed hydrolysis of small substrates. Log $k_{\text{cat}}/K_{\text{M}}$ for the hydrolysis of acetyl-L-phenylalanine p-nitrophenyl ester catalysed by δ-chymotrypsin decreases linearly with pH from pH 5 to pH 2 with unit slope when the conformational equilibria in the enzyme are taken into account. This indicates that the first ionization of the "charge-relay" system

Asp-His$\overset{+}{H}$-Ser occurs below pH 2.[184]

There is little or no effect of polar substituents on the values of k_{cat} or K_M for the α-chymotrypsin-catalysed hydrolysis of substituted anilides. These results are not consistent with the pretransition-state hypothesis and probably not with accumulation of a tetrahedral intermediate, i.e. the latter is thermodynamically less stable than the Michaelis complex.[185]

The second-order rate constants for the acylation of α-chymotrypsin by *p*-nitrophenyl acetate and *N*-acetyl-L-tryptophan methyl ester decrease with increasing concentration of added organic solvent. Solubility measurements of the two esters indicate that the ratio of the activity coefficients of the two transition states is nearly independent of solvent composition. It is concluded that a large part of the difference between the two esters is a ground-state effect and that the substrate undergoes significant desolvation in the activation process.[186]

The rate constants for the acylation step of the α-chymotrypsin-catalysed hydrolysis of *O*- and analogous *S*-esters of *N*-acetyl-L-tryptophan are nearly identical. Since a thiol is a better leaving group than an alcoholic hydroxyl group by at least 100-fold, this indicates that formation of the tetrahedral intermediate is the rate-limiting step.[187]

Volumes of activation for α-chymotrypsin-catalysed hydrolysis of *p*-nitrophenyl esters are thought to correspond to the volume changes for deacylation of the corresponding acyl-enzymes, which are characterized by a small decrease in volume.[26]

A comparative study of the effect of substrate α-acetamido-groups on the α-chymotrypsin- and subtilisin-catalysed hydrolysis of some specific and non-specific esters indicates that the maximum effect of the α-acetamido-group on deacylation rates is ca. 100 for the α-chymotrypsin-catalysed hydrolysis but that there is no rate enhancement for the subtilisin-catalysed reaction.[188]

Transient-phase kinetic studies of α-chymotrypsin have been described.[189]

The pH–rate profile for the hydrolysis of what is thought to be α-benzamido-*trans*-cinnamoyl-α-chymotrypsin has been reported.[190]

A comparison of α-chymotrypsin-catalysed hydrolysis of different substrates must be made by means of the second-order rate constant k_{cat}/K_M since the rate-limiting step changes with the nature of the leaving group.[191] The use of *p*-nitroanilides as substrates for proteolytic enzymes has been discussed.[192]

Peptide condensation catalysed by α-chymotrypsin has also been described.[193]

The rate of deacylation, but not of acylation, and values of K_M for hydrolysis reactions catalysed by insolubilized α-chymotrypsin bound to porous glass gel are dependent on the concentration of dioxan.[194] The chymotrypsin-catalysed hydrolysis of sepharose-bound L-phenylalanine *p*-nitroanilide has been described.[195]

When *N*-acetyl-L-tyrosine semicarbazide is bound to chymotrypsin A_δ NMR studies indicate that it is not in a strained state.[196]

The binding of boronic acid derivatives to α-chymotrypsin indicates that the structure of the boronic acid–enzyme complex resembles the transition state for the reaction of the corresponding substrates.[197]

The individual subsite-binding free energies for a series of *N*-peptidylphenethylamine competitive inhibitors with α-chymotrypsin have been determined and are not additive.[198]

Kinetic, binding and optical rotatory dispersion data for a group of anionic and cationic inhibitors indicate that inhibition of α-chymotrypsin is a slow, uncompetitive deactivation process that is controlled from a hydrophobic site separate from the active site. Hydrophobic interaction at two sites may account, in general, for the parabolic structure–reactivity relationship that commonly occurs for a homologous series of substrates.[199]

Other investigations of α-chymotrypsin include several inhibition studies,[200] the kinetics of inactivation with substituted benzenesulphonyl fluorides,[201] modification studies,[202] photo-affinity labelling,[203] the interaction with D-amino-acid residues,[204] the effect of dimethylsulphoxide at subzero temperatures on its catalysed hydrolysis of *N*-acetyl-L-phenylalanine methyl ester,[205] changes in conformation with changing pH,[206] reactivation of phenylmethanesulphonyl-chymotrypsin with hydrogen peroxide,[207] its racemase activity[208] and its potential for the resolution of alcohols.[209]

It has been suggested that the inferior catalytic properties of the zymogens of the pancreatic serine proteases are primarily due to ineffective substrate binding and only secondarily to a less efficient catalytic process.[210]

Other studies with zymogens are cited in ref. 211.

On the basis of values of k_{cat}/K_M, bovine A- and B-trypsin display similar catalytic efficiency towards some substrates; differences are explained by greater molecular flexibility of α-trypsin.[212]

X-Ray structural analysis of the trypsin–trypsin inhibitor complex indicates that the preferred conformation of the imidazolium cation, with a hydrogen bond to the serine-oxygen atom, is different from the optimum position for proton-transfer to or from the nitrogen atom of the addition intermediate.[213]

The effect of subzero temperatures and dimethyl sulphoxide on the trypsin-catalysed hydrolysis of *N*-benzoylcarbonyl-L-lysine *p*-nitrophenyl ester has been investigated.[214]

The largest known values of k_{cat}/K_M for the subtilisin-catalysed hydrolysis of peptide methyl esters have been determined and approximate to the rates observed for the α-chymotrypsin-catalysed hydrolysis of *p*-nitrophenyl esters. pH–rate profiles were obtained and rate-limiting binding is thought not to occur.[215]

The acylation of porcine pancreatic elastase with acetic anhydride indicates that a free amino-group of Val-16 is required for full enzyme activity.[216]

Inhibition studies of thermolysin by ethoxyformic anhydride suggest that a histidine residue is essential for enzyme activity.[217]

Dioxan enhances the esterase activity of thrombin and increases both the apparent K_M and the k_{cat} values.[218]

Studies with choline-esterases are reported in ref. 219.

Thiol Proteinases

Papain-catalysed hydrolysis of *N*-benzoyl-L-arginamide shows nitrogen isotope effects k^{14}/k^{15} of 1.021, 1.024 and 1.023 at pH's 8.0, 6.0 and 4.0, respectively. This indicates nitrogen bond-fission occurs in the rate-limiting step which is thought to be breakdown of the tetrahedral intermediate to the acyl-enzyme.[220]

Modification of tryptophan residues of papain with a 2-hydroxy-5-nitrobenzyl group increases k_{cat}/k_M for the enzyme-catalysed hydrolysis of benzoyl-L-arginyl substrates. This is attributed to an increase in the rate of the acylation step.[221]

Studies of the ionization of papain that is alkylated at Cys-25 with 2-(bromoacetamido)-4-nitrophenol indicate the existence of an area of high negative charge-density in the region of the active site.[222]

Amino-, hydroxy- and carbonyl derivatives of *n*-butane cause a time-dependent irreversible inactivation of papain. From the effect of substrate on the inactivation process it is concluded that papain undergoes a conformational change during the catalysed hydrolysis of ester substrates.[223]

Other studies with papain include its catalysed hydrolysis of *N*-benzoyl-L-serine methyl ester[224] and its inhibition by *N*-ethylmaleimide.[225]

Ficin-catalysed hydrolysis of a series of hippuric esters shows identical values of k_{cat}, suggesting rate-limiting deacylation of a hippuryl-ficin intermediate.[226]

Acid Proteinases

The active site of pepsin has been reviewed.[227]

Modification of pepsin by reaction with a diazocarbonyl compound destroys its peptidase activity, although catalytic activity for the hydrolysis of sulphite esters is retained. pH–k_{cat}/K_M profiles indicate that a carboxylate group at only a single active site is necessary for sulphite esterase activity.[228]

Metallo-proteinases

Possible mechanisms, intermediates, pH-effects and conformational changes have been discussed for the peptidase activity of carboxypeptidase A.[229]

The carboxypeptidase A-catalysed hydrolysis of a series of esters of *O*-acyl-2-hydroxy-butanoic acid shows large variations in values of K_M for relatively minor structural variations in the acid residue.[230] The Hammett ρ-value for k_{cat} for the carboxypeptidase A-catalysed hydrolysis of substituted *O*-benzoyl-2-hydroxybutanoic acids is 1.7 and that for K_M is –0.53. The observed differences in the pH–rate profiles for k_{cat}/K_M and k_{cat} for the enzyme-catalysed hydrolysis of *O*-(*p*-nitrobenzoyl)mandelic acid and *O*-hippuryl-L-3-phenyl-lactic acid are suggested as a criterion for distinction between specific and non-specific substrates.[231]

Kinetic studies of carboxypeptidase A_α and A_γ crystals and solutions indicate that the physical state of the enzyme is important and the kinetic parameters may be used to probe the conformational differences of the enzyme in these two physical states.[232]

The binding of alkylmalonic acid dianions to carboxypeptidase A is specific, and the inhibition constants correlate with the Hansch π-parameter.[233]

Other studies of metallo-proteinases are cited in ref. 234.

Esterases

Work on esterases is reported in ref. 235.

Other Enzymes

Bovine carbonic anhydrase-catalysed hydrolysis of methyl *p*-nitrophenyl carbonate produces an intermediate monoester, methyl hydrogen carbonate. The rate of release of *p*-nitrophenol increases with increasing pH, but the rate of the enzyme-catalysed decarboxylation of methyl hydrogen carbonate decreases with increasing pH. The second-order rate constant for the enzyme-catalysed dehydration of bicarbonate is 10^3-fold greater than that for the decomposition of methyl carbonate which is discussed in terms of the proton-transfer mechanisms available to bicarbonate.[236]

The kinetics of the carbonic anhydrase-catalysed hydration of alkyl pyruvates to their corresponding 2,2-dihydroxy derivatives have been described.[237]

The Mn(II) complex of bovine carbonic anhydrase B is ca. 10^7-fold less stable than the corresponding Zn(II) complex.[238]

Esterification of carboxyl groups in carbonic anhydrase B has been reported.[239]

Carbonyl phosphate binds to aspartate transcarbamylase before aspartic acid, and carbamoyl phosphate is released before phosphate ion.[240]

The role of thiol groups at the active site of aspartate transcarbamylase has been studied.[241]

Other studies of aspartate transcarbamylase are reported in ref. 247.

Other systems studied include the hydrolysis and aminolysis of thiol esters catalysed by fibrinoligase,[243] the transglutaminase-catalysed hydrolysis and aminolysis of esters,[244] ribulose biphosphate carboxylase,[245] an ether carbon—oxygen-cleaving lyase,[246] an enzyme catalysing the reversible sulphitolysis of glutathione disulphide,[247] affinity labelling of L-aspartate-β-decarboxylase[248] and inhibition of acetoacetate decarboxylase.[249]

Decarboxylation

The pH–rate profile for the decarboxylation of α-phenyl-α-aminomalonic acid is bell-shaped, the reactive species being the neutral zwitterionic form, $^{+}H_3NC(C_6H_5)$-(COOH)COO^- which decarboxylates ca. 5×10^6-fold faster than malonic acid monoanion.[250]

The decarboxylation of substituted phenylacetic acids catalysed by ceric ammonium nitrate in aqueous acetonitrile, which gives carbon dioxide, the corresponding benzyl alcohol, benzaldehyde and benzyl nitrate, exhibits a ρ^+-value of 2.91 and proceeds through benzyl radicals which may be trapped with oxygen. It is suggested that rate-limiting decomposition of carboxylic acid–cerium(IV) complex to a benzyl radical and CO_2 occurs through a very polar transition state.[251]

The two relaxations in the stopped-flow analysis of Zn(II)-catalysed decarboxylation of oxaloacetic acid are identified as initial metal-ion promoted keto–enol tautomerism and catalysed decarboxylation.[252]

Other studies include the decarboxylation of 1,3-dimethylorotic acid,[253] the effect of chloride ion on the decarboxylation of esters in wet dimethyl sulphoxide,[254] the alkylative decarboxylation of *N*-(alkoxycarbonyl)pyrazoles,[255] the lead tetra-acetate-catalysed oxidative decarboxylation of bridged bicyclic carboxylic acids[256] and the photodecarboxylation of esters.[257]

Other Reactions[258]

Hydrolysis of *p*-nitrophenyl acetate at pH 6 to pH 9 does not go to completion in 3.3% v/v dioxan–water with 0.1 M-phosphate or -tris as buffer. This is attributed to the establishment of an equilibrium between reactants and products, probably including *p*-nitrophenyl phosphate.[259]

There have been investigations of the base-catalysed hydrolysis of the following compounds: benzoic acid esters of amino-alcohols,[260] vinyl esters of a series of furan-carboxylic acids,[261] di- and mono-esters of dicarboxylic acids,[262] "di-isooctyl" adipate and phthalate,[263] triesters of isocyanuric acids,[264] haloalkyl esters[265] and ethyl β-aryl-α-cyanoacrylates.[266]

The relative rates of the base-catalysed hydrolysis of *n*-alkyl acetates, their ω-tert-butyl, ω-(trimethylsilyl) and ω-(trimethylgermyl) derivatives,[267] of sterically hindered and unhindered aryl acetates[268] and of thioacetates and methanethiol esters[269] have been interpreted in terms of steric effects.

On the evidence of dipole-moment studies, the relative rates of hydrolysis of esters of trifluoroacetic and trifluorothiolacetic acids have been interpreted in terms of *d*-orbital conjugation of sulphur.[270]

Rates and equilibrium constants for the formation of thiol esters by reaction of choline selenol esters with thiols have been determined.[271]

Theoretical predictions have been compared with the observed effect of ionic strength on the alkaline hydrolysis of ethyl succinate and glutarate.[272]

Enthalpies of solution of ethyl acetate and sodium hydroxide in aqueous dimethyl sulphoxide solutions and the observed enthalpy of activation for the hydroxide-ion catalysed hydrolysis of ethyl acetate, in conjunction with extrathermodynamic assumptions, suggest that the increasing reaction rates observed with increasing sulphoxide concentration do not result from the large enthalpy of desolvation of hydroxide ion, which is compensated by desolvation of the transition state, but rather by an entropy effect.[273] The relative increase in the rate of hydrolysis of esters in aqueous dimethyl sulphoxide compared with that in aqueous ethanol depends on the structure of the ester; solvation of the transition state is thought to be important as well as hydroxide-ion desolvation; differences in solvation of the esters in the two solvent systems is claimed not to be important.[274] Neither properties of the solvent alone, nor generalizations involving the structure of water, predict the relative rates of hydrolysis of acetic anhydride and the reaction of *p*-nitrophenyl acetate with imidazole in aqueous-organic solvent mixtures.[275] As is often found for other reactions, the neutral hydrolysis of *p*-nitrophenyl dichloroacetate in highly aqueous *tert*-butyl alcohol exhibits extrema in its thermodynamic parameters as a function of the mole fraction of water.[276] The logarithms of the rates of the alkaline hydrolysis of ethyl acetate in water–dioxan mixtures are linearly related to the reciprocal of the solvent dielectric constant.[277]

The rate of the alkaline hydrolysis of lactones in aqueous-organic solvent mixtures is greater than that of the open-chain esters. The rate differences are greater in aqueous ethanol than in aqueous dimethyl sulphoxide or aqueous acetone. This is interpreted in terms of the minor importance of the electrostatic repulsion between hydroxide ion and the "lone-pair dipole" of oxygen, the absence of which in the case of lactones is sometimes thought to cause their higher reactivity in $B_{AC}2$ reactions.[278]

The rate of the alkaline hydrolysis of the lactone ring in coumarin decreases with increasing amount of organic solvent in water–organic solvent mixtures.[279]

Ring-opening reactions of lactones with chloroethyldistannoxanes have been described.[280]

The hydrolysis of aroyl chlorides in water–acetone mixtures is bimolecular at low concentration of water, but the S_N1 pathway predominates at higher water concentration.[281]

Papers on other reactions of acid halides are listed in ref. 282.

Changes from linear to non-linear Hammett plots and less negative entropies of activation with increasing acid concentration in the hydrolysis of substituted benzoic anhydrides in dioxan–water are taken to indicate a change in mechanism from *A*2 to *A*1.[283]

Acetaldehyde-inhibited hydrolysis of acetic anhydride has been described.[284]

The pH–rate profile for the hydrolysis of *n*-butyl xanthate fits the following scheme:

$$ROCS_2^- + H^+ \rightleftharpoons ROCS_2H \rightarrow \text{Products}$$

The reaction of *tert*-butyl xanthate gives largely isobutene, demonstrating alkyl—oxygen fission. At pH <0 the rate of decomposition of *n*-butyl xanthate decreases because of the formation of unreactive protonated xanthic acid but protonation of *tert*-butyl xanthate increases the rate.[285]

Acid-catalysed hydrolysis of a series of *ortho*-substituted benzohydroxamic acids, ArCONHOH, shows a Taft ρ^*-value of -0.87 and a δ-value of 0.76.[286]

The rate of the acid-catalysed hydrolysis of phenylacetohydroxamic acid in aqueous sulpholane shows a linear relationship between $\log k/(H_2O)$ and the mole fraction of sulpholane.[287]

The following reactions have been studied: alkaline hydrolysis of 1,2-diaryl-1,4,5,6-tetrahydropyrimidines[288] and of *N*-alkoxy-*N*-aryl-carboxamides to azo-compounds,[289] dissociation of arylcarbamoyl-imidazolines, ArNHCOIm, to aryl isocyanates and imidazole,[290] hydrolysis of *N*-acyl derivatives of thiobenzamides,[291] of thiobarbital[292] and of allobarbital,[293] reaction of 1,3-*tert*-alkylaziridin-2-ones with *tert*-butyl-lithium to give α-hydroxy-imines,[294] and acid-catalysed condensation of an aldehyde, a keto-ester and urea to form pyrimidones.[295]

Solvent effects on the acid-catalysed hydrolysis of esters have been described.[296]

The acid-catalysed hydrolysis of $Ph_3CSC(O)Ar$ proceeds by the $A_{AL}1$ mechanism with protonation on sulphur.[297]

In contrast to previous series studied, α-methyl-substitution decreases the rate more than β-methyl-substitution in the acid-catalysed esterification of the isomeric pair of acids (**85**) and (**86**).[298]

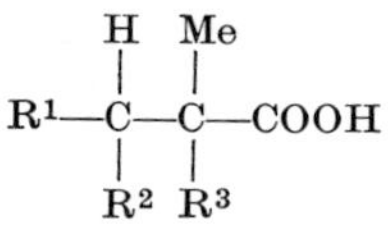

```
      H  Me                          Me H
      |  |                           |  |
R¹—C—C—COOH                 R¹—C—C—COOH
      |  |                           |  |
      R² R³                          R² R³
      (85)                           (86)
```

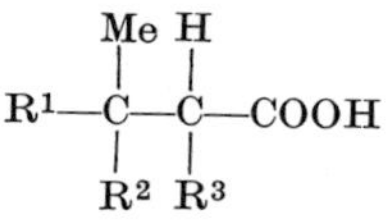

The kinetics of various esterification reactions catalysed by acids,[299] cation-exchange resins[300] and Lewis acids[301] have been studied.

There have been several reports of transesterification reactions;[302] and acetyl-transfer reactions with 3-acetoxy-1-acetyl-5-methylpyrazole,[303] and the imidazole-catalysed deacylation of thioesters,[304] have been described.

Acyl migrations between formally neutral carbon and nitrogen in 2-(2-mercapto-benzimidazolyl)1-phenylbutane-1,3-dione in benzene have been reported,[305] and the effects of substituents in position 1 of 2-formylimidazoles in their transamination reactions with alanine have been discussed.[306]

The Bunnett–Olsen ϕ-parameter for the acid-catalysed hydrolysis of *o*-tolunitrile is 0.61 but no mechanistic conclusions have been deduced.[307] There have also been other reports of the reactions of nitriles.[308]

The *N*-chloramide group of (*N*-chlorocarbamoyl)ethyl-starch reacts with amines in alkaline solution through the formation of an isocyanate group. The reactivities of the amino-groups with the isocyanate are comparable with that of hydroxide-ion.[309]

In the reaction of carboxylic acids with phenyl isocyanate to give the corresponding acid anhydride, CO_2 and $(PhNH)_2CO$, the CO_2 originates from the isocyanate.[310] There have been several other reports of the reactions of isocyanates.[311]

The volume of activation for the thermal decomposition of *tert*-butyl peroxypivalate in cumene ranges from ca. 0.3 to 1.9 ml mol^{-1}, over a 5000-atm. pressure range, while that for *tert*-butyl peroxyisobutyrate changes from ca. 1.6 to 3.2. These observations are taken to indicate a polar contribution to the transition state for decomposition.[312]

The reaction of phenyl salicylate with perbenzoic acid in aqueous ethanol at $pH > 9$ gives some *o*-ethoxyphenol and catechol. This novel alkoxylation probably proceeds through an aryl cation intermediate.[313]

Other reactions of peracids are reported in ref. 314.

Yet other studies relate to decomposition of *N*-monosubstituted dithiocarbamic acids[315] and tertiary alkyl peroxides,[316] synthesis of urea from ammonia and carbon dioxide in the liquid phase,[317] imidazole-catalysed formation of acid chlorides from carboxylic acids and phosgene,[318] various electrolytic reactions.[319] chloromercuriolactonization[320] and iodolactonization[321] of unsaturated acids, dissociation constants of non-*ortho*-substituted benzoic acids interpreted in terms of electrostatic, solvation and steric effects,[322] the basicity of substituted 2-acetylthiophens.[323] the stereochemistry of reactions of heterocyclic *N*-oxides with acetic anhydride and acetyl chloride,[324] the hydrolysis of cyclic carbonates,[325] acidic hydrolysis of esters catalysed by oxidized carbon,[326] the von Braun amide degradation,[327] cleavage of acetoxysilanes to siloxanes,[328] pyrolysis of *N*-(trimethylsilyl)amides,[329] reaction of *p*-bromoaniline with methyl hydrazone-*N*-dithiocarboxylates,[330] thermal decomposition of 1,1′-azobisformamide,[331] reaction of xenon difluoride with trifluoroacetic acid,[332] and application of I.R. spectroscopy to kinetic investigations of peptide coupling.[333]

Sodium cyanide in hexamethylphosphoric triamide cleaves methyl esters selectively in the presence of ethyl esters. It is thought that the reaction occurs by nucleophilic displacement of the carboxylate ion by cyanide-ion attack on the alcohol carbon, a $B_{AL}2$ mechanism.[334]

The following solid-phase reactions have been studied: acetamide with hydroxylamine,[335] amide formation from carboxylic acids and amines catalysed by *NN*′-dicyclohexylcarbodi-imide,[336] and peptide synthesis.[337]

Formic anhydride adopts the *EZ*-configuration in solution with a low barrier (4.3 kcal mol^{-1}) for topomerization.[338] The conformations of various arenecarbothioamides in benzene have been determined.[339]

NON-CARBOXYLIC ACIDS

Phosphorus-containing Acids

Non-enzymic Reactions

Hydrolysis of the diphenyl phosphate esters of *cis*-3-*exo*-hydroxynorbornane-2-carboxylic acid (**87**) and the corresponding *cis*-*endo*-isomer proceeds with phosphorus—phenoxy bond fission but that of *exo*- and *endo*-2norbornan-2-ol occurs with alkyl—oxygen fission. It is estimated that at pH 5 the neighbouring carboxyl groups increase the rate of hydrolysis by up to 10^7-fold compared wih the calculated value for phosphorus—phenoxy bond fission in the unsubstituted compound. It is claimed that intramolecular nucleophilic catalysis by carboxylate occurs although there is a very high rate of solvolysis at low pH.[340]

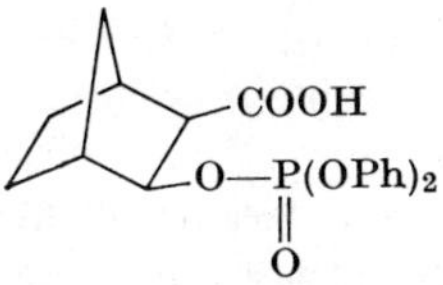

COOH; O—P(OPh)₂ (=O)

(**87**)

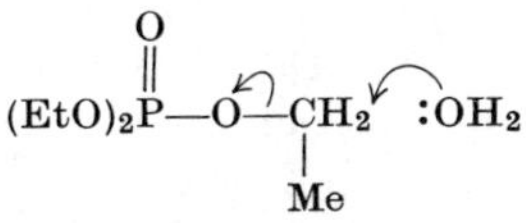

$(EtO)_2P(=O)—O—CH(Me)—CH_2$:OH_2

(**88**)

$(EtO)_2P(=\overset{+}{O}H)—O—CH(Me)—CH_2$:OH_2

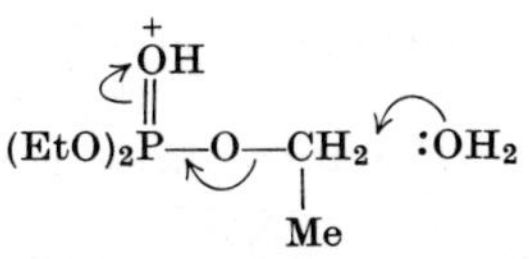

(**89**)

The hydrolysis of triethyl phosphate in water shows a solvent isotope effect k_{H_2O}/k_{D_2O} of 1.3. Carbon—oxygen bond fission occurs and there is no catalysis by 0.5 M-sulphuric acid. These results are consistent with the $B_{AL}2$ mechanism (**88**). The perchloric acid-catalysed hydrolysis in dioxan–water shows a solvent isotope effect k_{H_2O}/k_{D_2O} of 0.6 and also occurs with carbon—oxygen bond fission, indicating the $A_{AL}2$ pathway (**89**).[341]

The pH–rate profile and activation parameters have been obtained for the hydrolysis of adenosine triphosphate.[342]

Other reactions studied include that of amines with tris-(2-chloroethyl) phosphate,[343] synthesis and reactions of the first five-membered cyclic acyl phosphate,[344] thermodynamics of complex formation between Ni^{2+} and adenosine 5′-monophosphate,[345] and the kinetics of phosphorylation of *m*-cresol with $POCl_3$ in the presence of magnesium chloride.[346]

Nucleophilic substitution at phosphorus in *cis*- and *trans*-isomers of 2-substituted 5-chloromethyl-5-methyl-2-thio-1,3,2-dioxaphosphorinanes proceeds with a greater degree of retention of configuration as the basicity of the nucleophile increases, but the amount of inversion increases with added salts. Separate mechanisms for inversion and retention are thought to be involved.[347]

A monomeric metaphosphate has been produced for the first time by gas-phase pyrolysis of methyl but-2-enylphostonate and trapped by *N*-methylaniline.[348]

The reactions of hexafluorobiacetyl with mixed anhydrides of phosphorus and phosphoric acid give stable compounds which are oxyphosphorane models (**90**) of the hypothetical intermediates derived by addition of nucleophiles to the phosphorus of phosphate esters.[349]

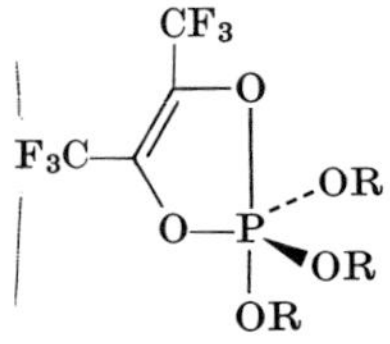

(**90**)

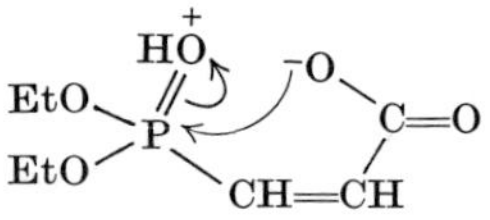

(**91**)

Reversible protonation of the phosphorus in trineopentyl phosphite by hydrogen chloride occurs in dioxan and is followed by nucleophilic displacement of the alkyl group by a second molecule of hydrogen chloride.[350]

Other reactions of phosphites are reported in ref. 351.

The hydrolysis of diethyl *cis*-(2-carboxyvinyl)phosphonate occurs by a stepwise release of the ethyl ester groups with phosphorus—oxygen bond fission. The reactions are ca. 10^6-fold faster than that for the release of one ethyl group of the corresponding *trans*-isomer. It is proposed that the mechanism involves intramolecular nucleophilic attack by the carboxyl group (cf. **91**).[352]

Nucleophilic attack by methoxide-ion on *O*-menthyl *S*-methyl phenylphosphonothiolate displaces the methylthio-group with inversion of configuration at phosphorus. Racemization, presumably through an achiral intermediate, is competitive with displacement of the alkylthio-ligand, $k_{-1} > k_{-3'}$ (Scheme 4). A surprising feature of the kinetic pathway in Scheme 4 is that attack of methoxide-ion opposite to the methylthio-ligand is preferred by a factor of ca. 5 over attack opposite the methoxy-ligand.[353]

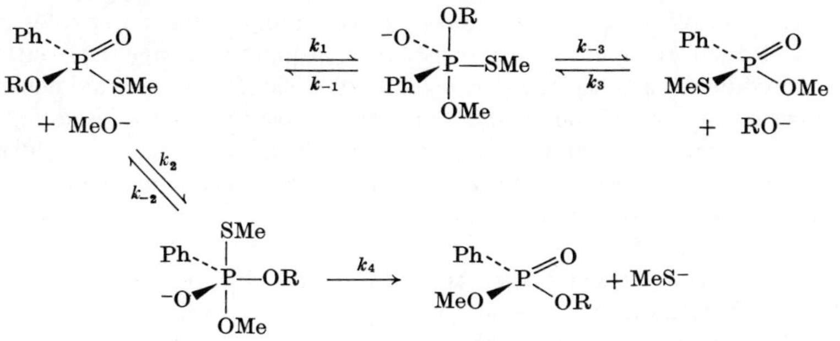

SCHEME 4.

Hydrolysis of phosphorus acid chlorides, RR′POCl, in acetone–water mixtures is second-order in water. This order declines with increasing size of the substituents, presumably because of a contribution from the S_N1 mechanism.[354]

The effect of the structure of the alkyl halide on its reaction with ethyl-(or phenyl-)-dithiophosphonous acid esters has been reported.[355]

The alkaline hydrolysis of diallylphosphinic acid ester occurs by the S_N2 mechanism.[356]

The energy difference between a five-membered pentacovalent phosphorane with the ring occupying apical-equatorial (**92**) rather than diequatorial positions (**93**) depends, not only on the nature of the heteroatom which moves from an axial to an equatorial position, but also on the nature of the atom which remains equatorial.[357]

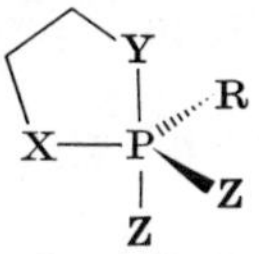

(**92**) 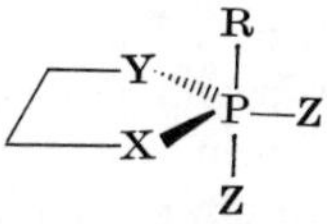 (**93**)

Bunnett plots and pH–rate profiles for oxygen-18 exchange in phosphine oxides have been used as criteria for distinguishing between mechanisms proceeding through dihydroxyphosphoranes in which both oxo-ligands are apical and those in which they are situated apical-equatorial. The mechanism operating depends on the structure of the phosphine oxide.[358]

Reaction of aryl disulphides with triphenylphosphine in water to give the substituted thiophenol and triphenylphosphine oxide is thought to occur by a two-step mechanism.[359]

The kinetics of esterification of substituted phosphorus monothio-acids by diazomethane have been described.[360]

The thermodynamic parameters for hydrolysis of pyrophosphate ion to orthophosphate ion at pH 7.4 as functions of the concentrations of Mg^{2+} and K^+ and of the ionic strength have been determined from equilibrium studies of the reaction.[361]

Finally, the role of electron-deficient sulphides as intermediates in oxidative phosphorylation has been examined.[362]

Enzymic Reactions

Binding of the substrate to alkaline phosphatase is independent of the binding of Mg^{2+} and *vice versa*. It is therefore suggested that Mg^{2+} must play a part in the substrate decomposition.[363]

Two isozymes of *E. coli* alkaline phosphatase show similar initial-burst rates in the enzyme-catalysed hydrolysis of 2,4-dinitrophenyl phosphate at pH 5.5, but their steady-state rates differ by a factor of 2. However, hydrolysis of *p*-nitrophenyl phosphate at pH 8.0 shows no significant burst and the steady-state rate is the same for both isozymes. These observations support the previous suggestion that the molecular process that determines the steady-state rate at pH 8.0 is responsible for the pre-steady-state burst at pH 5.5. The steady-state process at pH 5.5 must be a subsequent step in the catalytic mechanism.[364]

Alkaline phosphatase-catalysed hydrolysis of substrates with different net negative charges indicates that the active site of the enzyme is positively charged but that hydrophobic effects are also important.[365]

A kinetic study of intestinal alkaline phosphatase has been reported.[366]

The elongation factor G-catalysed hydrolysis of guanosine triphosphate occurs with bond fission between γ-phosphorus and oxygen bridging β- and γ-phosphorus.[367]

The site of reaction of pyridoxal phosphate with fructose-1,6-diphosphatase is a lysyl residue at or near the 6-phosphate substrate-binding site.[368]

Fructose-1,6-diphosphatase catalyses the hydrolysis of naphthyl phosphates.[369]

Acetylimidazole produces a pseudo-first-order loss of fructose diphosphatase activity with fructose diphosphate as substrate, but with *p*-nitrophenyl phosphate as substrate the rate is first enhanced and then slowly inhibited. The results agree with the previous suggestion that hydrolysis of *p*-nitrophenyl phosphate is catalysed at or near the regulatory site of this enzyme.[370]

The alkaline form of fructose-1,6-diphosphatase utilizes the β-form 5- to 10-fold less rapidly than the α-form of fructose diphosphate.[371] Rabbit-muscle 6-phosphofructokinase uses only the β-furanose anomer.[372]

The order of addition of substrates to 6-phosphofructokinase is random.[373] Other studies of this enzyme are reported in ref. 374.

The hydrolysis of uridine 2′,3′-phosphate catalysed by ribonuclease is competitively inhibited by uridine, 2′-deoxyuridine and oxovanadium(IV) ion. The strong binding of the complexes to ribonuclease is interpreted in terms of their being transition-state analogues.[375]

The Lys-41 amino-group of ribonuclease *A* has a p*K* of 9.03, and the p*K* values of kinetically important amino-groups in the reaction of the enzyme with trinitrobenzene-sulphonic acid have also been determined.[376] The amino-acid sequence of pancreatic ribonuclease A has been reported.[377] The microenvironment of tris-12 ribonuclease has been studied by using ^{13}C-NMR spectroscopy.[378]

Bovine pancreatic deoxyribonuclease A is inactivated by reaction with methane-sulphonyl chloride; at pH 5 a serine residue is sulphonated. Calcium ions protect six to eight carboxyl groups from reacting with a carbodi-imide reagent.[379]

There is ordered binding to uridine phosphorylase; phosphate binds to the enzyme before uridine, and uracil leaves the enzyme before ribose 1-phosphate. The postulated mechanism involves S_N2 displacement of uridine by phosphate dianion.[380]

Sulphur-containing Acids[381]

It has been suggested that Cu(II) catalyses the hydrolysis and alcoholysis of 8-quinolyl sulphate through metal-ion co-ordination to oxygen and nitrogen atoms of the substrate.[382]

Rate maxima as a function of acid concentration are observed for the acid-catalysed hydrolysis of sodium *N*-1-naphthylsulphamate; Bunnett ω^*-plots, solvent isotope effects, and salt effects suggest rate-limiting attack of water on the conjugate acid, but it is argued that entropies of activation of –5 to –17 J mol^{-1} K^{-1} indicate considerable nitrogen—sulphur bond-breaking and little bond-formation with the nucleophile.[383]

The rate-limiting step in the solvolysis of sulphamic acid, H_2NSO_3H, in dimethylformamide is thought to be formation of the complex DMF·SO_3 which then reacts to form H_3N·DMF·SO_3 and $HO_3SNHSO_3NH_4$.[384]

The general stability of tetrasubstituted sulphamides has been compared with the reactivity of other substituted sulphamides.[385]

Carboxylate-ion catalysed hydrolyses of diaryl sulphites show a Brønsted β-value of 0.85 and a solvent isotope effect $k^{H_2O}{}_{AcO^-}/k^{D_2O}{}_{AcO^-}$ of 1.32. Together with the trapping of the intermediate mixed anhydride with hydroxylamine this indicates nucleophilic catalysis (**94**).[386]

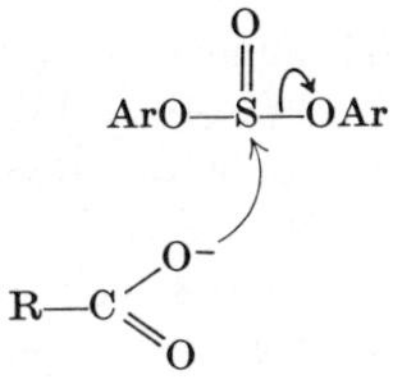

ArO—S(=O)—OAr ← $^-$O—C(=O)—R

(**94**)

The influence of specific solvation effects on the rate of solvolysis of arenesulphonyl halides has been investigated.[387]

The Hammett ρ-value for the hydroxide-ion catalysed hydrolysis of aryl toluene-α-sulphonates is 4.84, indicating considerable phenoxide-ion character in the transition state of the rate-limiting step. Amine buffers catalyse the release of *p*-nitrophenol from the corresponding ester, k_B in Scheme 5 being rate-limiting, but they have no effect at high buffer concentration where k_2 is rate-limiting and the sulphonamide is the sole product. The observations are consistent with a stepwise elimination–addition mechanism, *E*1*cB*, involving a sulphene intermediate.[388]

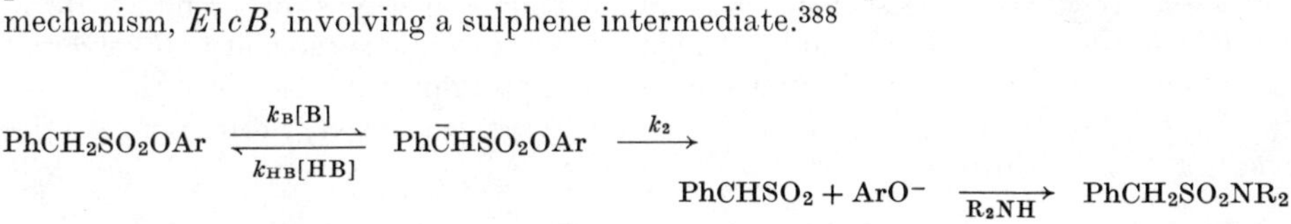

$$PhCH_2SO_2OAr \underset{k_{HB}[HB]}{\overset{k_B[B]}{\rightleftharpoons}} Ph\bar{C}HSO_2OAr \xrightarrow{k_2} PhCHSO_2 + ArO^- \xrightarrow[R_2NH]{} PhCH_2SO_2NR_2$$

SCHEME 5.

Details of the *E*1*cB* mechanism for the hydrolysis of aryl (*N*-methylamino)sulphonates have been reported.[389]

*E*1*cB* hydrolyses of arenesulphonylmethyl perchlorates, $ArSO_2CH_2OClO_3$, have been further investigated.[390]

The rate of hydrolysis of the sulphonamide (**95**) is apparently pH-independent in the region 0.1–1 M-hydrochloric acid and is orders of magnitude greater than that of the analogous *p*-carboxy-derivative. It is suggested that this is due to intramolecular nucleophilic catalysis by the neighbouring carboxylate group as indicated.[391]

The acid-catalysed decomposition of *o*-nitrophenylsulphonyldiazomethane (**96**) is thought to involve neighbouring-group participation of the nitro-group.[392]

The hydrolysis of *p*-nitrophenyl *p*-toluenesulphonate is catalysed by 1-ethylpyrrolidine but not by triethylamine. A solvent isotope effect k_{H_2O}/k_{D_2O} of 1.45 is taken to indicate a nucleophilic mechanism for the catalysed reaction (**97**).[393]

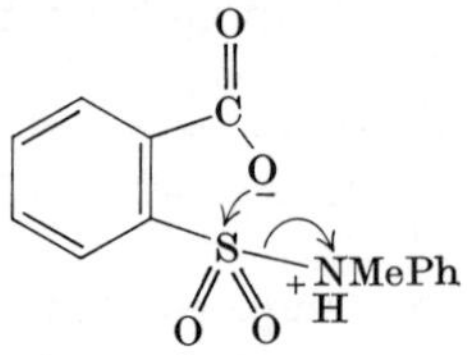

(**95**) 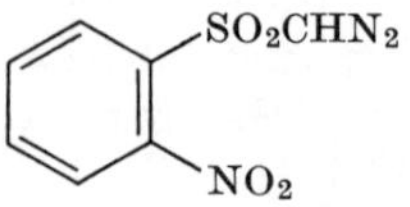 (**96**) 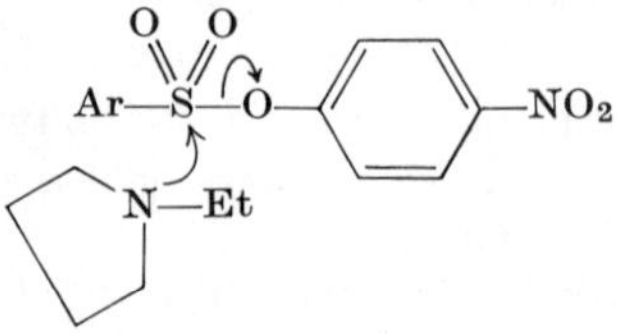 (**97**)

The influence of substituents, steric hindrance and intramolecular co-ordination on the exchange rates and equilibria in arenesulphonamides and their *N*-arylmercury derivatives have been described.[394]

Other studies of sulphonamides are noted in ref. 395.

The second-order rate constant for attack by hydrogen peroxide anion on *n*-butyl methanesulphonate is 55-fold greater than that for hydroxide ion, and the reaction occurs solely by alkyl—oxygen fission.[396]

Other investigations of sulphonic acids and their esters are reported in ref. 397.

Alkaline hydrolysis of *para*-substituted benzenesulphonyl fluorides yields a large Hammett ρ-value of 2.79. This is taken to support an addition–elimination mechanism with a large amount of charge development in the transition state.[398] The alkaline hydrolysis of 2,4,6-trimethylbenzenesulphonyl chloride is, contrary to a previous report, of the first order in both substrate and hydroxide ion. There is thus now no unambiguous evidence of an S_N1 mechanism in nucleophilic substitution at sulphur.[399]

The aminolysis of benzenesulphonyl chloride by *N*-methylaniline in chloroform and acetone is accelerated by amine salts, which is held to be due to general-base catalysis by the ion-pair of the salt rather than to a salt-effect.[400]

Pyrolysis of dialkyl thiolsulphinates gives alkanesulphenic and alkanethiosulphoxylic acids, and the reactions of these compounds were also studied.[401]

Base-catalysed hydrolysis of ethyl 2-nitro-4-(trifluoromethyl)benzenesulphenate is of the first order in both ester and hydroxide ion. The reaction is thought to involve the formation of a sulphenic acid intermediate, the disappearance of which is of the second order in sulphenate anion. The rate of the latter process is inversely proportional to the concentration of hydroxide ion:[402]

$$\mathrm{ArSX + HO^- \rightarrow ArSOH + X^-}$$
$$\mathrm{2ArSO^- + H_2O \rightarrow ArSOSAr + 2HO^-}$$
$$\mathrm{ArSOSAr + 2HO^- \rightarrow ArSO_2^- + ArS^- + H_2O}$$

Such a mechanistic scheme is supported by detecting the various intermediates by the ^{19}F-NMR chemical shift of the trifluoromethyl group.[403]

An oxoazetidinesulphenic acid has been isolated, supporting its intermediacy in the penicillin–cephalosporin rearrangement.[404]

The reaction of tertiary amines with 2,4-dinitrobenzenesulphenyl chloride gives an enaminic product.[405]

The reactions of sulphoxides in moderately concentrated acid have been reviewed,[406] and the importance of catalysis arising from strong acid formation *in situ* in dimethyl sulphoxide has been discussed.[407]

Other Acids

Both alkyl–oxygen and nitro–oxygen bond fission occurs in the acid-catalysed hydrolysis of 2-methyl-2-nitratopropionic acid.[408]

Other reactions studied include acid-catalysed S_N1 hydrolysis of α-nitrato-carboxylic acids[409] and acid-catalysed hydrolysis of glyceryl nitrates,[410] a new decomposition route of nitrate esters,[411] reaction of diazodiphenylmethane with pyridinecarboxylic acids giving a Hammett ρ-value of 1.08,[412] and the kinetics of its reaction with monomeric and dimeric benzoic acids in donor aprotic solvents,[413] the conversion of a carbonyl group of an electronegatively acyl-substituted ester into an oxiran ring with diazoethane,[414] and the reactivities of nucleophiles towards nitrosyl chloride.[415]

References

1 J. P. Guthrie, *J. Am. Chem. Soc.*, **96**, 3608 (1974); see *Org. Reaction Mech.*, **1973**, 22.
2a J. M. Lehn and G. Wipff, *J. Am. Chem. Soc.*, **96**, 4048 (1974); see *Org. Reaction Mech.*, **1973**, 23.
2b K. Ya. Brushtein and Yu. I. Khurgin, *Izv. Akad. Nauk SSSR, Ser. Khim*, **1974**, 579; *Chem. Abs.*, **80**, 145211 (1974).
3 G A. Rogers and T. C. Bruice, *J. Am. Chem. Soc.*, **96**, 2481 (1974); see *Org. Reaction Mech.*, **1973**, 21; see also p. 38.
4 N. Gravitz and W. P. Jencks, *J. Am. Chem. Soc.*, **96**, 489 (1974).
5 N. Gravitz and W. P. Jencks, *J. Am. Chem. Soc.*, **96**, 499 (1974).
6 N. Gravitz and W. P. Jencks, *J. Am. Chem. Soc.*, **96**, 507 (1974).
7 Cf. *Org. Reaction Mech.*, **1973**, 23.
8 J. P. Fox and W. P. Jencks, *J. Am. Chem. Soc.*, **96**, 1436 (1974).
9 A. C. Satterthwait and W. P. Jencks, *J. Am. Chem. Soc.*, **96**, 7018 (1974).
10 A. C. Satterthwait and W. P. Jencks, *J. Am. Chem. Soc.*, **96**, 7031 (1974).
11 W. P. Bullard, L. J. Farina, P. R. Farina and S. J. Benkovic, *J. Am. Chem. Soc.*, **96**, 7295 (1974).
12 See *Org. Reaction Mech.*, **1973**, 26.
13 R. A. McClelland, *J. Am. Chem. Soc.*, **96**, 3690 (1974).
14 Y. Pocker, M. W. Beng and K. L. Stephens, *J. Am. Chem. Soc.*, **96**, 174 (1974).
15 E. A. Castro and R. B. Moodie, *J.C.S. Perkin II*, **1974**, 658.
16 A. R. Butler, I. H. Robertson and R. Bacaloglu, *J.C.S. Perkin II*, **1974**, 1733.
17 C. E. Stauffer, *J. Am. Chem. Soc.*, **96**, 2489 (1974).
18 See *Org. Reaction Mech.*, **1972**, 424.
19 J. P. Guthrie, *J. Am. Chem. Soc.*, **96**, 588 (1974).
20 V. Gani and P. Viout, *Tetrahedron Letters*, **1974**, 3663.
21 R. M. Pollack, *J. Org. Chem.*, **39**, 2108 (1974).
22 T. J. Broxton and L. W. Deady, *J. Org. Chem.*, **39**, 2767 (1974).
23 M. Choi and E. R. Thornton, *J. Am. Chem. Soc.*, **96**, 1428 (1974).
24 H. Collet, A. Germain and A. Commeyras, *Bull. Soc. Chim. France*, **1974**, 279.
25 E. Laurent and N. Pellissier, *Bull. Soc. Chim. France*, **1974**, 1904.
26 G. D. Lockyer, Jr., D. Owen, D. Crew and R. C. Neuman, Jr., *J. Am. Chem. Soc.*, **96**, 7303 (1974); see also p. 45.
26a C. A. Eckert, "High Pressure Kinetics in Solution" in *Ann. Rev. Phys. Chem.*, **23**, 239 (1972).
27 Y. Pocker and E. Green, *J. Am. Chem. Soc.*, **96**, 166 (1974).
28 G. M. Blackburn and H. L. H. Dodds, *J.C.S. Perkin II*, **1974**, 377.
29 A. E. Williams, J. K. Lee and R. L. Schowen, *J. Org. Chem.*, **38**, 4053 (1973).
30 T. H. Fife and N. C. De, *J. Am. Chem. Soc.*, **96**, 6158 (1974).
30a A. Hussain, M. Yamasaki and J. E. Truelove, *J. Pharm. Sci.*, **63**, 627 (1974).
31 Y. Chiang, A. J. Kresge, P. Salomaa and C. I. Young, *J. Am. Chem. Soc.*, **96**, 4494 (1974).
32 J. Hine and A. W. Klueppel, *J. Am. Chem. Soc.*, **96**, 2924 (1974).
33 R. A. Ellison, W. D. Woessner and C. C. Williams, *J. Org. Chem.*, **39**, 1430 (1974).
34 See *Org. Reaction Mech.*, **1973**, 27.
35 T. J. Gilbert and C. D. Johnson, *J. Am. Chem. Soc.*, **96**, 5846 (1974).
36 R. T. C. Brownlee and R. D. Topsom, *Tetrahedron Letters*, **1973**, 5187.

[37] M. M. Kreevoy and E. H. Baughman, *J. Am. Chem. Soc.*, **95**, 8178 (1973).
[38] J. F. Bunnett, R. L. McDonald and F. P. Olsen, *J. Am. Chem. Soc.*, **96**, 2855 (1974).
[39] R. F. Cookson, *Chem. Rev.*, **74**, 5 (1974); E. H. Baughman and M. M. Kreevoy, *J. Phys. Chem.*, **78**, 421 (1974); D. G. Lee and M. H. Sadar, *J. Am. Chem. Soc.*, **96**, 2862 (1974).
[40] A. J. Kresge and W. J. Albery, *J.C.S. Chem. Comm.*, **1974**, 507.
[41] B. D. Batts and J. Kilford, *J.C.S. Faraday I*, **69**, 1033 (1973).
[42] P. Frøyen and R. F. Hudson, *Acta Chem. Scand. B*, **27**, 4001 (1973).
[43] A. C. Satterthwait and W. P. Jencks, *J. Am. Chem. Soc.*, **96**, 7045 (1974).
[44] C. B. Sauers and H. M. Relles, *J. Am. Chem. Soc.*, **95**, 7731 (1973).
[45] S. J. Benkovic, T. H. Barrows and P. R. Fariva, *J. Am. Chem. Soc.*, **95**, 8414 (1973).
[46] T. Vontor, V. Drobilič, J. Socha and M. Večeřa, *Coll. Czech. Chem. Comm.*, **39**, 281 (1974).
[47] A. F. Hegarty, C. N. Hegarty and F. L. Scott, *J.C.S. Perkin II*, **1974**, 1258.
[48] A. F. Hegarty, C. N. Hegarty and F. L. Scott, *J.C.S. Perkin II*, **1973**, 2054.
[49] J.-P. Calmon and D. Sayag, *Compt. Rend. Ser. C*, **277**, 875 (1973).
[50] A. Hassner and R. H. Reuss, *J. Org. Chem.*, **39**, 553 (1974).
[51] M. J. Boland, M. J. Hardman and I. D. Watson, *Bioorg. Chem.*, **3**, 213 (1974).
[52] T. J. Broxton, L. W. Deady and P. R. A. Williamson, *Austral. J. Chem.*, **27**, 1053 (1974).
[53] I. M. Medvetskaya and M. I. Vinnik, *Zh. Fiz. Khim.*, **47**, 1223 (1973); *Chem. Abs.*, **79**, 136233 (1973).
[53a] M. I. Vinnik and L. R. Andreeva, *Zh. Org. Khim.*, **9**, 2532 (1973); *Chem. Abs.*, **80**, 59161 (1974).
[54] A. J. Kresge, P. H. Fitzgerald and Y. Chiang, *Am. J. Chem. Soc.*, **96**, 4698 (1974).
[55] C. N. Berry and B. C. Challis, *J.C.S. Perkin II*,, **1974**, 1638.
[56] B. C. Challis and S. P. Jones, *J.C.S. Chem. Comm.*, **1974**, 748.
[57] A. J. Hall and D. P. N. Satchell, *J.C.S. Perkin II*, **1974**, 1077; *Chem. Ind.* (*London*), **1974**, 527.
[58] A. C. Hopkinson and I. G. Csizmadia, *Theor. Chim. Acta*, **31**, 83 (1973); *Chem. Abs.*, **80**, 26645 (1974).
[59] M. Liler, *J.C.S. Perkin II*, **1974**, 71.
[60] C. L. Perrin, *J. Am. Chem. Soc.*, **96**, 5628, 5631 (1974).
[61] J. F. Whidby and W. R. Morgan, *J. Phys. Chem.*, **77**, 2999 (1973); Y. Kakuda and D. D. Mueller, *Bioorg. Chem.*, **3**, 311 (1974).
[62] J.-P. Calmon and C. Doux, *Compt. Rend. Ser. C*, **277**, 699 (1973).
[63] W. I. Congdon and J. T. Edward, *Can. J. Chem.*, **52**, 697 (1974).
[64] Ya. Yu. Rekshinskii and G. M. Strongin, *Tr. Khim. Khim. Tekhnol.*, **1973**, 146; *Chem. Abs.*, **80**, 26487 (1974).
[65] A. B. Robinson, J. W. Scotchler and J. H. McKerrow, *J. Am. Chem. Soc.*, **95**, 8156 (1973).
[66] M. Napoli and G. Gambaretto, *Atti Ist. Veneto Sci., Lett. Arti. Cl. Sci. Mat. Natur.*, **130**, 291 (1972); *Chem. Abs.*, **80**, 107610 (1974).
[67] S. Goto, N. Murao and S. Iguchi, *Yakuzaigaku*, **33**, 139 (1973); *Chem. Abs.*, **81**, 104274 (1974); V. G. Smirnova, A. B. Crigor'ev, M. K. Polievktov, V. G. Granik, N. P. Kostyuchenko, I. V. Persianova and R. G. Glushkov, *Khim. Geterotsikl. Soedin.*, **1974**, 530; *Chem. Abs.*, **81**, 49085 (1974).
[68] M. M. Mhala and M. H. Jagdale, *J. Shivaji Univ.*, **5**, 23 (1972); *Chem. Abs.*, **80**, 69995 (1974).
[69] S. Goszcynski and W. Zielinski, *Zh. Org. Khim.*, **9**, 2103 (1973); *Chem. Abs.*, **80**, 36456 (1974).
[70] Ya. Yu. Rekshinskii and G. M. Strongin, *Tr. Khim. Khim. Tekhnol.*, **1973**, 142; *Chem. Abs.*, **80**, 26488 (1974).
[71a] Yu. V. Svetkin, I. B. Abdrakhmanov and M. Augustin, *Wiss. Z. Martin-Luther-Univ., Halle-Wittenberg, Math.-Naturwiss, Reihe*, **22**, 7 (1973); *Chem. Abs.*, **80**, 107430 (1974).
[71a] I. M. Kolthoff and M. K. Chantooni, Jr., *J. Am. Chem. Soc.*, **95**, 8539 (1973).
[72] N. M. Oleinik, L. M. Litvinenko and M. N. Sorokin, *Zh. Org. Khim.*, **9**, 1693 (1973); *Chem. Abs.*, **79**, 125477 (1973).
[73] L. M. Litvinenko, V. A. Savelova and A. V. Skripka, *Dokl. Akad. Nauk SSSR*, **216**, 1327 (1974); *Chem. Abs.*, **81**, 104278 (1974).
[74] L. M. Litvinenko, G. D. Titskii, O. P. Stepko and N. F. Kirpenko, *Zh. Obshch. Khim.*, **43**, 1974 (1973); *Chem. Abs.*, **79**, 145789 (1973).
[75] See *Org. Reaction Mech.*, **1970**, 463.
[76] L. M. Litvinenko, G. D. Titskii and O. P. Stepko, *Organic Reactivity* (*Tartu*), **10**, 1069 (1973).
[77] G. D. Titskii, O. P. Stepko and L. M. Litvinenko, *Zh. Org. Khim.*, **9**, 2341 (1973); *Chem. Abs.*, **80**, 36468 (1974).
[78] Yu. S. Simanenko, L. M. Litvinenko and V. A. Dadali, *Zh. Org. Khim.*, **10**, 1308 (1974); *Chem. Abs.*, **81**, 77300 (1974).
[79] A. Arcoria, S. Fisichella, G. Scarlata and D. Sciotto, *J. Org. Chem.*, **39**, 3025 (1974); A. Arcoria

and S. Fisichella, *Gazz. Chim. Ital.*, **103**, 813 (1973); *Chem. Abs.*, **81**, 62989 (1974); see *Org. Reaction Mech.*, **1973**, 30.

80 N. K. Vorob'ev, E. A. Chizhova and L. A. Malysheva, *Izv. Vyssh. Ucheb. Zaved., Khim. Khim. Tekhnol.*, **16**, 1366 (1973); *Chem. Abs.*, **80**, 47101 (1974); P. A. Wegner, D. M. Adams, F. J. Callabretta, L. T. Spada and R. G. Unger, *J. Am. Chem. Soc.*, **95**, 7513 (1973); I. V. Shpanko, G. D. Titskii, L. M. Litvinenko and M. A. Yeremeyev, *Organic Reactivity (Tartu)*, **10**, 203 (1973); L. V. Kuritsyn, *Izv. Vyssh. Ucheb. Zaved., Khim. Khim. Tekhnol.*, **17**, 40, 521, 685 (1974); *Chem. Abs.*, **80**, 94836 (1974); **81**, 24629, 62909 (1974).

81 L. G. Babaeva, S. V. Bogatkov, R. I. Kruglikova and B. V. Uhkovskii, *Zh. Org. Khim.*, **10**, 250 (1974); *Chem. Abs.*, **80**, 107773 (1974).

82 J. M. McLeod, L. R. Schroeder and P. A. Seib, *Carbohydrate Res.*, **30**, 337 (1973).

83 F. Effenberger and H. Klenk, *Chem. Ber.*, **107**, 175 (1974).

84 A. Arcoria, E. Maccarone, G. Musumarra and G. A. Tomaselli, *J. Org. Chem.*, **39**, 1689 (1974).

85 T. N. Solomoichenku, V. A. Savelova and L. M. Litvinenko, *Zh. Org. Khim.*, **10**, 534 (1974); *Chem. Abs.*, **80**, 132379 (1974); V. A. Savelova, V. A. Shatskaya, L. M. Litvinenko and N. I. Nikishina, *Zh. Obshch. Khim.*, **44**, 1124 (1974); *Chem. Abs.*, **81**, 48990 (1974); L. V. Kuritsyn, *Tr. Jvanov. Khim.-Tekhnol. Inst.*, **1970**, 69; *Chem. Abs.*, **79**, 125666 (1973).

86 L. M. Litvinenko, N. T. Maleeva, V. A. Savelova and O. I. Butko, *Zh. Org. Khim.*, **9**, 2123 (1973); *Chem. Abs.*, **80**, 36578 (1974).

87 L. M. Litvinenko, N. M. Oleinik and M. N. Sorokin, *Zh. Org. Khim.*, **10**, 770 (1974); *Chem. Abs.*, **81**, 12739 (1974).

88 Z. Szponar and T. Jasinski, *Zcsz. Nauk. Wydz. Mat., Fiz. Chem., Uniw. Gdanski, Chem.*, **2**, 35 (1972); *Chem. Abs.*, **81**, 104259 (1974); A. W. Spassov and A. Alexiev, *Z. Chem.*, **14**, 58 (1974); A. F. Marton, T. Komives and F. Dutka, *Radiochem. Radioanal. Lett.*, **17**, 211 (1974); *Chem. Abs.*, **81**, 62802 (1974); M. Dror and M. Levy, *J.C.S. Perkin II*, **1974**, 1425; T. R. Radhakrishnan, M. Balasubramanian and V. Baliah, *Indian J. Chem.*, **11**, 562 (1973); *Chem. Abs.*, **79**, 114737 (1973); E. A. Artemova, M. K. Shchennikova, N. B. Tret'yakova and N. E. Falaleeva, *Tr. Khim. Khim. Tekhnol.*, **1973**, 94; *Chem. Abs.*, **80**, 81649 (1974).

89 L. I. Dereza, A. F. Popov, L. M. Litvinenko and A. V. Anikeev, *Zh. Org. Khim.*, **10**, 522 (1974); *Chem. Abs.*, **80**, 132669 (1974).

90 F. M. Menger and L. E. Glass, *J. Org. Chem.*, **39**, 2469 (1974).

91 C. Su and J. H. Watson, *J. Am. Chem. Soc.*, **96**, 1854 (1974).

92 W. Konig and R. Geiger, *Chem. Ber.*, **106**, 3626 (1973); N. M. Ollinik, L. M. Litvinenko, Yu. S. Sadovskii, B. G. Igolkina, A. A. Popkova and S. E. Terekhova, *Dokl. Akad. Nauk SSSR*, **213**, 390 (1973); *Chem. Abs.*, **80**, 59140 (1974); H. Tanaka, T. Endo and M. Okawara, *Nippon Kagaku Kaishi*, **1973**, 1780; *Chem. Abs.*, **79**, 145632 (1973).

93 S. A. Lammiman and R. S. Satchell, *J.C.S. Perkin II*, **1974**, 877.

94 A. P. Grekov, S. A. Sukhorukova and G. V. Otroshko, *Zh. Org. Khim.*, **10**, 526 (1974); *Chem. Abs.*, **80**, 132377 (1974); A. P. Grekov and G. V. Otroshko, *Zh. Org. Khim.*, **10**, 530, 783 (1974); *Chem. Abs.*, **80**, 132378 (1974); **81**, 12962 (1974); O. Mauz, *Annalen*, **1974**, 345; N. K. Vorob'ev and O. K. Shebanova, *Izv. Vyssh. Ucheb. Zaved., Khim. Khim. Tekhnol.*, **17**, 688 (1974); *Chem. Abs.*, **81**, 62973 (1974); A. Uejima and H. Munakata, *Nippon Kagaku Kaishi*, **1973**, 1496; *Chem. Abs.*, **79**, 114744 (1973); N. Yamazaki, F. Higashi and M. Niwano, *Tetrahedron*, **30**, 1319 (1974); J. F. Wolfe, D. E. Portlock and D. J. Feuerbach, *J. Org. Chem.*, **39**, 2006 (1974).

95 C. V. Pesheck and S. E. Buttrill, Jr., *J. Am. Chem. Soc.*, **96**, 6027 (1974).

96 P. W. Tiedemann and J. M. Riveros, *J. Am. Chem. Soc.*, **96**, 185 (1974).

97 R. Yamdagni, T. B. McMahon and P. Kebarle, *J. Am. Chem. Soc.*, **96**, 4035 (1974).

98 M. I. Page, *Chem. Soc. Rev.*, **2**, 295 (1973).

99 D. F. DeTar, *J. Am. Chem. Soc.*, **96**, 1254, 1255 (1974).

100 W. P. Jencks and M. I. Page, *Biochem. Biophys. Res. Comm.*, **57**, 887 (1974).

101 L. L. Ingraham, *Biochem. Biophys. Acta*, **279**, 8 (1972); C. S. Kim and L. L. Ingraham, *ibid.*, **297**, 220 (1973).

102 R. D. Gandour, *Tetrahedron Letters*, **1974**, 295.

103 M. F. Aldersley, A. J. Kirby, P. W. Lancaster, R. S. McDonald and C. R. Smith, *J.C.S. Perkin II*, **1974**, 1487.

104 A. J. Kirby, R. S. McDonald and C. R. Smith, *J.C.S. Perkin II*, **1974**, 1445.

105 See *Org. Reaction Mech.*, **1966**, 342.

106 M. F. Aldersley, A. J. Kirby and P. W. Lancaster, *J.C.S. Perkin II*, **1974**, 1504.

107 A. F. Hegarty, L. N. Frost and D. Cremin, *J.C.S. Perkin II*, **1974**, 1249; see also *Org. Reaction Mech.*, **1973**, 31.

[108] P. P. Nechaev, Y. V. Moiseev, Y. S. Vygodskii and G. E. Zaikov, *Int. J. Chem. Kinet.*, **6**, 245 (1974).
[109] See *Org. Reaction Mech.*, **1971**, 433.
[110] T. C. Bruice and I. Oka, *J. Am. Chem. Soc.*, **96**, 4500 (1974).
[111] R. J. L. Martin, C. H. Skovron and D. L. H. Yiu, *J.C.S. Perkin II*, **1974**, 125; R. J. L. Martin and S. L. Tan, *ibid.*, p. 129.
[111a] R. Kluger and J. L. W. Chan, *J. Am. Chem. Soc.*, **96**, 5337 (1974).
[112] See *Org. Reaction Mech.*, **1973**, 32.
[113] R. Hershfield and G. L. Schmir, *J. Am. Chem. Soc.*, **95**, 8032 (1973).
[114] See *Org. Reaction Mech.*, **1973**, 32.
[115] S. N. Semenova, N. F. Bondar and T. I. Temnikova, *Zh. Org. Khim.*, **9**, 2111 (1973); *Chem. Abs.*, **80**, 36461 (1974).
[116] T. H. Fife and B. M. Benjamin, *J.C.S. Chem. Comm.*, **1974**, 525.
[117] See *Org. Reaction Mech.*, **1973**, 32.
[118] A. J. Kirby and G. J. Lloyd, *J.C.S. Perkin II*, **1974**, 637.
[119] J. Schoenleber and P. Lochon, *Compt. Rend. Ser. C*, **278**, 1293 (1974).
[120] R. D. Gandour and R. L. Schowen, *J. Am. Chem. Soc.*, **96**, 2231 (1974).
[121] S. K. Pal, S. C. Lahini and S. Aditya, *Z. Phys. Chem. (Leipzig)*, **255**, 226 (1974).
[122] M. Balakrishnan, G. V. Rao and N. Venkatasubramanian, *Int. J. Chem. Kinet.*, **6**, 103 (1974).
[123] G. A. Rogers and T. C. Bruice, *J. Am. Chem. Soc.*, **96**, 2463 (1974).
[124] Cf. p. 43.
[125] See ref. 180.
[126] G. A. Rogers and T. C. Bruice, *J. Am. Chem. Soc.*, **96**, 2473 (1974).
[127] P. Y. Bruice and T. C. Bruice, *J. Am. Chem. Soc.*, **96**, 5523 (1974).
[128] P. Y. Bruice and T. C. Bruice, *J. Am. Chem. Soc.*, **96**, 5533 (1974).
[129] S. Ikegami, K. Uoji and S. Akaboshi, *Tetrahedron*, **30**, 2077 (1974); S. Ikegami, T. Asai, K. Tsuneoko, S. Matsumura and S. Akaboshi, *ibid.*, p. 2087.
[130] A. R. Butler, *J.C.S. Perkin II*, **1974**, 1239.
[131] J.-J. Béchet, R. Alazard, A. Dupaix and C. Roucous, *Bioorg. Chem.*, **3**, 55 (1974).
[132] A. F. Hegarty, L. N. Frost and J. H. Coy, *J. Org. Chem.*, **39**, 1089 (1974).
[133] See *Org. Reaction Mech.*, **1972**, 432.
[134] L. W. Brown, R. S. P. Hsi and A. A. Forist, *J. Pharm. Sci.*, **63**, 459 (1974).
[135] M. Bundgaard, *Acta Pharm. Suec.*, **10**, 309 (1973); *Chem. Abs.*, **80**, 14211 (1974).
[136] See *Org. Reaction Mech.*, **1969**, 449.
[137] F. M. Menger, T. E. Thanos and J. L. Lynn, *J. Org. Chem.*, **39**, 1771 (1974).
[138] J. L. van der Baan and F. Bickerlhaupt, *Tetrahedron*, **30**, 2447 (1974); M. Bodanszky, M. L. Fink, K. W. Funk, M. Kondo, C. Y. Lin and A. Bodanszky, *J. Am. Chem. Soc.*, **96**, 2234 (1974); A. Tai, Y. Yokoyama, K. Shindo and O. Fujii, *Yuki Gosei Kagaku Kyokai Shi*, **31**, 410 (1973); *Chem. Abs.*, **80**, 81673 (1974).
[139] H. Morawetz, *Pure Appl. Chem.*, **38**, 267 (1974); Y. Murakami and J. Sunamoto, *Kagaku No Ryoiki*, **27**, 1144 (1973); *Chem. Abs.*, **80**, 132229 (1974).
[140] I. Tabushi, Y. Kuroda and S. Kita, *Tetrahedron Letters*, **1974**, 643.
[141] I. Tabushi and Y. Kuroda, *Tetrahedron Letters*, **1974**, 3613.
[142] T. Kunitake, Y. Okahata and R. Ando, *Bull. Chem. Soc. Japan*, **47**, 1509 (1974); *Macromolecules*, **7**, 140 (1974); T. Kunitake and Y. Okahata, *Chem. Letters (Tokyo)*, **1974**, 1057.
[143] W. B. Gruhn and M. L. Bender, *Bioorg. Chem.*, **3**, 324 (1974).
[144] Y. Murakami, J. Sunamoto and K. Kano, *Bull. Chem. Soc. Japan*, **47**, 1238 (1974).
[145] F. M. Menger and M. J. McCreery, *J. Am. Chem. Soc.*, **96**, 121 (1974).
[146] C. G. Overberger and C. J. Podsiadly, *Bioorg. Chem.*, **3**, 16, 35 (1974).
[147] J. P. Guthrie and Y. Ueda, *J.C.S. Chem. Comm.*, **1974**, 111.
[148] T. Rodulfo, J. A. Hamilton and E. H. Cordes, *J. Org. Chem.*, **39**, 2281 (1974).
[149] R. Fernandez-Prini and E. Baumgartner, *J. Am. Chem. Soc.*, **96**, 4489 (1974).
[150] J. W. Larsen and L. J. Magid, *J. Phys. Chem.*, **78**, 834 (1974).
[151] L. J. Magid and J. W. Larsen, *J. Org. Chem.*, **39**, 3142 (1974).
[151a] M. A. Wells, *Biochemistry*, **13**, 2258 (1974).
[151b] L. Juliano, P. Boschov and A. C. M. Paiva, *Biochemistry*, **13**, 4263 (1974).
[152] C. A. Bunton and M. J. Minch, *J. Phys. Chem.*, **78**, 1490 (1974).
[153] I. K. Winterborn, B. J. Meakin and D. J. G. Davies, *J. Pharm. Sci.*, **63**, 64 (1974).
[154] G. G. Smith, D. R. Kennedy and J. G. Nairn, *J. Pharm. Sci.*, **63**, 712 (1974).
[155] T. Okonogi, Y. Yano and W. Tagaki, *Bull. Chem. Soc. Japan*, **47**, 771 (1974).
[156] C. Lapinte and P. Viout, *Tetrahedron Letters*, **1974**, 2401.

[157] A. P. Osipov, K. Martinek, A. K. Yatsimirskii and I. V. Berezin, *Dokl. Akad. Nauk SSSR*, **215**, 914 (1974); *Chem. Abs.*, **81**, 12758 (1974).
[158] See *Org. Reaction Mech.*, **1971**, 442.
[159] D. Hindman and J. Jacobus, *Tetrahedron Letters*, **1974**, 1619.
[160] R. A. Moss and W. L. Sunshine, *J. Org. Chem.*, **39**, 1083 (1974).
[161] C. J. O'Connor, E. J. Fendler and J. H. Fendler, *J. Am. Chem. Soc.*, **96**, 370 (1974).
[162] N. Tatsumoto, K. Takeda, S. Isshiki and T. Yasunaga, *Bull. Chem. Soc. Japan*, **47**, 289 (1974).
[163] B. Jezowska-Trzebiatowska, *Pure Appl. Chem.*, **38**, 367 (1974).
[164] D. A. Buckingham, F. R. Keene and A. M. Sargeson, *J. Am. Chem. Soc.*, **96**, 4981 (1974).
[165] D. A. Buckingham, J. M. Harrowfield and A. M. Sargeson, *J. Am. Chem. Soc.*, **96**, 1726 (1974).
[166] J. M. Harrowfield and A. M. Sargeson, *J. Am. Chem. Soc.*, **96**, 2634 (1974).
[167] D. A. Buckingham, P. J. Cresswell, R. J. Dellaca, M. Dwyer, C. J. Gainsford, L. G. Marzilli, I. E. Maxwell, W. T. Robinson, A. M. Sargeson and K. R. Turnbull, *J. Am. Chem. Soc.*, **96**, 1713 (1973).
[168] R. Nakon, P. R. Rechani and R. J. Angelici, *J. Am. Chem. Soc.*, **96**, 2117 (1974).
[169] R. J. Balahura, P. Cock and W. L. Purcell, *J. Am. Chem. Soc.*, **96**, 2739 (1974).
[170] S. E. Diamond and H. Taube, *J.C.S. Chem. Comm.*, **1974**, 622.
[171] J. A. McPhee, M. Boussu and J. E. Dubois, *J.C.S. Perkin II*, **1974**, 1525.
[172] E. Kimura, *Inorg. Chem.*, **13**, 951 (1974); S. K. Oh and C. B. Storm, *Biochemistry*, **13**, 3250 (1974).
[173] J. H. Fitzpatrick, Jr., and D. Hopgood, *Inorg. Chem.*, **13**, 568 (1974).
[174] F. Hasan and J. Roček, *Tetrahedron*, **30**, 21 (1974).
[175] A. S. Mildvan, *Ann. Rev. Biochem.*, **43**, 850 (1974); R. J. P. Williams, *Pure Appl. Chem.*, **38**, 249 (1974); G. L. Kenyon and J. A. Fee, *Progr. Phys. Org. Chem.*, **10**, 381 (1974); G. E. Means, D. S. Ryan and R. E. Feeney, *Accounts Chem. Res.*, **7**, 315 (1974); H. Tschesche, *Angew. Chem. Int. Ed.*, **13**, 10 (1974); H. Brockerhoff, *Bioorg. Chem.*, **3**, 176 (1974); S. G. Waley, *Biochem. J.*, **139**, 789 (1974); D. J. Bates and C. Friedman, *J. Biol. Chem.*, **248**, 7878 (1973).
[176] D. B. Cook and J. McKenna, *J.C.S. Perkin II*, **1974**, 1223.
[177] A. R. Fersht, *Proc. Roy. Soc. (London)*, *B*, **187**, 397 (1974).
[178] See p. 26.
[179] A. R. Fersht and J. Sperling, *J. Mol. Biol.*, **74**, 137 (1973).
[180] M. W. Hunkapiller, S. H. Smallcombe, D. R. Whitaker and J. H. Richards, *Biochemistry*, **12**, 4732 (1973); *J. Biol. Chem.*, **248**, 8306 (1973).
[181] See also p. 38.
[182] G. Robillard and R. G. Shulman, *J. Mol. Biol.*, **86**, 519, 541 (1974).
[183] W. H. Cruickshank and H. Kaplan, *J. Mol. Biol.*, **83**, 267 (1974).
[184] A. R. Fersht and M. Renard, *Biochemistry*, **13**, 1416 (1974).
[185] M. Philipp, R. M. Pollack and M. L. Bender, *Proc. Nat. Acad. Sci. U.S.*, **70**, 517 (1973).
[186] R. P. Bell, J. E. Critchlow and M. I. Page, *J.C.S. Perkin II*, **1974**, 66.
[187] H. Hirohara, M. L. Bender and R. S. Stark, *Proc. Nat. Acad. Sci. U.S.*, **71**, 1643, 3317 (1974).
[188] M. S. Matta and D. D. Staley, *J. Biol. Chem.*, **249**, 732 (1974).
[189] R. J. Maguire, N. H. Hijazi and K. J. Laidler, *Biochem. Biophys. Acta*, **341**, 1 (1974).
[190] K. Brocklehurst and K. Williamson, *Tetrahedron*, **30**, 351 (1974).
[191] A. Dupaix and J.-J. Becket, *FEBS Letters*, **34**, 185 (1973).
[192] R. M. Pollack and T. C. Dumsha, *FEBS Letters*, **38**, 292 (1974).
[193] M. Yamashita, S. Arai, S.-Y. Tanimoto and M. Fujimaki, *Biochem. Biophys. Acta*, **358**, 105 (1974).
[194] K. Tanizawa and M. L. Bender, *J. Biol. Chem.*, **249**, 2130 (1974).
[195] H.-D. Jakubke and L. Lange, *FEBS Letters*, **43**, 281 (1974).
[196] G. Robillard, E. Shaw and R. G. Shulman, *Proc. Nat. Acad. Sci. U.S.*, **71**, 2623 (1974).
[197] J. D. Rawn and G. E. Lienhard, *Biochemistry*, **13**, 3124 (1974).
[198] J. C. Powers, B. L. Baker, J. Brown and B. K. Chelm, *J. Am. Chem. Soc.*, **96**, 238 (1974).
[199] R. N. Smith and C. Hansch, *Biochemistry*, **12**, 4924 (1973).
[200] U. Quast, J. Engel, H. Heumann, G. Krause and E. Steffen, *Biochemistry*, **13**, 2512 (1974); A. B. Cohen, *J. Biol. Chem.*, **248**, 7055 (1973); P. Rogers and G. C. K. Roberts, *FEBS Letters*, **36**, 330 (1973); I. V. Berezin, S. D. Varfolomeyer, A. M. Klibanov and K. Martinek, *ibid.*, **39**, 329 (1974); A. L. Fink, *Biochemistry*, **13**, 277 (1974); H. R. Bosshard and A. Berger, *ibid.*, p. 266.
[201] J. T. Gerig and D. C. Roe, *J. Am. Chem. Soc.*, **96**, 233 (1974); L. J. Berliner and S. S. Wong, *J. Biol. Chem.*, **249**, 1668 (1974).
[202] B. H. Havsteen, *Acta Chem. Scand.*, *B*, **27**, 3769 (1973).
[203] H. Wakayama and Y. Kanaoka, *FEBS Letters*, **37**, 200 (1974).
[204] H. R Bosshard, *FEBS Letters*, **38**, 139 (1974)
[205] A. L. Fink and E. Wildi, *J. Biol. Chem.*, **249**, 6087 (1974).

[206] J.-R. Garel, S. Epely and B. Labouesse, *Biochemistry*, **13**, 3117 (1974); A. Mavridis, A. Tulinsky and M. N. Liebman, *ibid.*, p. 3661.
[207] N. A. Radomsky, M. J. Gibian and D. L. Elliot, *J. Am. Chem. Soc.*, **95**, 8717 (1973).
[208] M.-A. Coletti-Previero, F. Kraicsovits and A. Previero, *FEBS Letters*, **37**, 93 (1974).
[209] Y. Y. Lin, D. N. Palmer and J. B. Jones, *Can. J. Chem.*, **52**, 469 (1974).
[210] A. Gertler, K. A. Walsh and H. Neurath, *Biochemistry*, **13**, 1302 (1974); *FEBS Letters*, **38**, 157 (1974); P. H. Morgan, A. Walsh and H. Neurath, *ibid.*, **41**, 108 (1974).
[211] E. J. East and C. G. Trowbridge, *J. Biol. Chem.*, **248**, 7552 (1973); H. T. Wright, *J. Mol. Biol.*, **79**, 1 (1973); J. E. Gomez, E. R. Birnbaum and D. W. Darnall, *Biochemistry*, **13**, 3745 (1974).
[212] D. V. Roberts and D. T. Elmore, *Biochem. J.*, **141**, 545 (1974).
[213] A. Rühlmann, D. Kukla, P. Schwager, K. Bartels and R. Huber, *J. Mol. Biol.*, **77**, 417 (1973).
[214] A. L. Fink, *J. Biol. Chem.*, **249**, 5027 (1974).
[215] M. Philipp and M. L. Bender, *FEBS Letters*, **42**, 282 (1974).
[216] D. Karibian, C. Jones, A. Gertler, K. J. Dorrington and T. Hofmann, *Biochemistry*, **13**, 2891 (1974).
[217] Y. Burstein, K. A. Walsh and H. Neurath, *Biochemistry*, **13**, 205 (1974).
[218] G. B. Villanevva, C. W. Batt and W. Brunner, *J.C.S. Chem. Comm.*, **1974**, 766.
[219] L. M. Chan, C. M. Himel and A. R. Main, *Biochemistry*, **13**, 86 (1974); G. M. Steinberg and M. L. Mednick, *Bioorg. Chem.*, **3**, 81 (1974).
[220] M. H. O'Leary, M. Urberg and A. P. Young, *Biochemistry*, **13**, 2077 (1974).
[221] J. E. Mole and H. R. Horton, *Biochemistry*, **12**, 5285 (1973).
[222] S. D. Lewis and J. A. Shafer, *Biochemistry*, **13**, 690 (1974).
[223] A. L. Fink and C. Gwyn, *Biochemistry*, **13**, 1190 (1974).
[224] C. W. Wharton, A. Cornish-Bowden, K. Brocklehurst and E. M. Crook, *Biochem. J.*, **141**, 365 (1974).
[225] L. J. Brubacher and B. R. Glick, *Biochemistry*, **13**, 915 (1974).
[226] A. A. Kortt, J. A. Hinds and B. Zerner, *Biochemistry*, **13**, 2029 (1974).
[227] J. S. Fruton, *Accounts Chem. Res.*, **7**, 241 (1974).
[228] H. J. Chen and E. T. Kaiser, *J. Am. Chem. Soc.*, **96**, 625 (1974).
[229] W. N. Lipscomb, *Tetrahedron*, **30**, 1725 (1974).
[230] J. W. Bunting and J. Murphy, *Can. J. Chem.*, **52**, 2640 (1974).
[231] J. W. Bunting, J. Murphy, C. D. Myers and G. G. Cross, *Can. J. Chem.*, **52**, 2648 (1974).
[232] C. A. Spilburg, J. L. Bethune and B. L. Valle, *Proc. Nat. Acad. Sci. U.S.*, **71**, 3922 (1974).
[233] J. W. Bunting and C. D. Myers, *Can. J. Chem.*, **52**, 2053 (1974).
[234] J.-M. Ghuysen, P. E. Reynolds, H. R. Perkins, J.-M. Frère and R. Moreno, *Biochemistry*, **13**, 2539 (1974); C. W. Garner, Jr., and F. J. Behal, *ibid.*, **18**, 3227 (1974).
[235] E. Simonianová and M. Petáková, *Coll. Czech. Chem. Comm.*, **39**, 1925 (1974); M. Salmona, C. Saronio, R. Bianchi, F. Marcucci and E. Mussini, *J. Pharm. Sci.*, **53**, 222 (1974).
[236] Y. Pocker and L. J. Guilbert, *Biochemistry*, **13**, 70 (1974).
[237] Y. Pocker, J. E. Meany and B. C. Davis, *Biochemistry*, **13**, 1411 (1974).
[238] R. G. Wilkins and K. R. Williams, *J. Am. Chem. Soc.*, **96**, 2241 (1974).
[239] G. Fölsch, *Chem. Scripta*, **6**, 32 (1974).
[240] F. C. Wedler and F. J. Gasser, *Arch. Biochem. Biophys.*, **163**, 57 (1974).
[241] G. R. Jacobson and G. R. Stark, *J. Biol. Chem.*, **248**, 8003 (1973).
[242] D. R. Evans, S. C. Pastra-Landis and W. N. Lipscomb, *Proc. Nat. Acad. Sci. U.S.*, **71**, 1351 (1974); N. J. Moogenraad, *Arch. Biochem. Biophys.*, **161**, 76 (1974); T.-Y. Chang and M. E. Jones, *Biochemistry*, **13**, 638, 646 (1974); E. Heyde, A. Nagabhushanam and J. F. Morrison, *ibid.*, **12**, 4718 (1973); E. Heyde and J. F. Morrison, *ibid.*, p. 4727.
[243] C. G. Curtis, P. Stenberg, K. L. Brown, A. Baron, K. Chen, A. Gary, I. Simpson and L. Lorand, *Biochemistry*, **13**, 3257 (1974).
[244] M. Gross and J. E. Folk, *J. Biol. Chem.*, **249**, 3021 (1974).
[245] F. C. Hartman, M. H. Welch and I. L. Norton, *Proc. Nat. Acad. Sci. U.S.*, **70**, 3721 (1973).
[246] D. Peterson and J. Llaneza, *Arch. Biochem. Biophys.*, **162**, 135 (1974).
[247] B. Mannervik, G. Persson and S. Eriksson, *Arch. Biochem. Biophys.*, **163**, 283 (1974).
[248] N. M. Relya, S. S. Tate and A. Meister, *J. Biol. Chem.*, **249**, 1519 (1974).
[249] R. Kluger and K. Nakaoka, *Biochemistry*, **13**, 910 (1974).
[250] J. W. Thanassi, *Biochemistry*, **12**, 5109 (1973).
[251] W. S. Trahanovsky, J. Cramer and D. W. Brixius, *J. Am. Chem. Soc.*, **96**, 1077 (1974).
[252] W. D. Covey and D. L. Leussing, *J. Am. Chem. Soc.*, **96**, 3860 (1974).
[253] P. Beak and B. Siegel, *J. Am. Chem. Soc.*, **95**, 7919 (1973).
[254] A. P. Krapcho, E. G. E. Jahngen, Jr., A. J. Lovey and F. W. Short, *Tetrahedron Letters*, **1974**, 1091; C. L. Liotta and F. C. Cook, *Tetrahedron Letters*, **1974**, 1095.

[255] J. J. Wilczynski and H. W. Johnson, Jr., *J. Org. Chem.*, **39**, 1909 (1974).
[256] B. C. C. Cantello, J. M. Mellor and G. Scholes, *J.C.S. Perkin II*, **1974**, 348.
[257] R. S. Givens, B. Matuszewski and C. V. Neywick, *J. Am. Chem. Soc.*, **96**, 5547 (1974).
[258] Y. Murakami and J. Sunamoto, *Kagaku No Ryoiki*, **27**, 1055 (1973); *Chem. Abs.*, **80**, 81521 (1974); R. Oda, *Kagaku (Kyoto)*, **29**, 372 (1974); *Chem. Abs.*, **81**, 119299 (1974).
[259] H. J. Goren and M. Fridkin, *Eur. J. Biochem.*, **41**, 263 (1974) *Chem. Abs.*, **80**, 94900 (1974).
[260] S. V. Bogatkov, A. G. Gaganova, D. A. Kereselidze and E. M. Cherkasova, *Zh. Org. Khim.*, **9**, 2096 (1973); *Chem. Abs.*, **80**, 26649 (1974).
[261] G. G. Skvortsova, Yu. A. Mansurov and N. M. Deriglazov, *Khim. Geterotsikl. Soedin.*, **1974**, 33; *Chem. Abs.*, **80**, 107564 (1974).
[262] K. Watanabe, T. Takahashi and N. Shinagawa, *Nippon Kagaku Kaishi*, **1973**, 1831; *Chem. Abs.*, **80**, 26512 (1974).
[263] P. Tribunescu and P. Velica, *Bul. Stinnt. Tch. Inst. Politch. Timirisoara, Ser. Chim.*, **16**, 187 (1971); *Chem. Abs.*, **80**, 2809 (1974).
[264] T. Koyama, T. Ishikawa and Y. Yamazaki, *Nippon Kagaku Kaishi*, **1973**, 2170; *Chem. Abs.*, **80**, 59157 (1974).
[265] N. J. Cleve and E. K. Euranto, *Finnish Chem. Letters*, **1974**, 1, 82; N. J. Cleve, *ibid.*, p. 78
[266] S. Kondo, M. Otani, N. Yoshiga and K. Tsuda, *Nagoya Kogyo Daigaku Gakuho*, **24**, 429 (1972); *Chem. Abs.*, **80**, 145014 (1974).
[267] J. Pola and V. Chvalovský, *Coll. Czech. Chem. Comm.*, **39**, 2247, 2637 (1974).
[268] J. F. W. McOmie and S. A. Saleh, *Tetrahedron*, **29**, 4003 (1973).
[269] J. P. Idoux, P. T. R. Hwang and C. K. Hancock, *J. Org. Chem.*, **38**, 4239 (1973).
[270] E. Bock, A. Queen, S. Brownlee, T. A. Nour and M. N. Paddon-Row, *Can. J. Chem.*, **52**, 3113 (1974).
[271] A. Makriyannis, W. H. H. Günther and H. G. Mautner, *J. Am. Chem. Soc.*, **95**, 8403 (1973).
[272] V. Holba and M. Rievaj. *Coll. Czech. Chem. Comm.*, **38**, 3283 (1973).
[273] R. Fuchs, C. P. Hagan and R. F. Rodewald, *J. Phys. Chem.*, **78**, 1509 (1974).
[274] M. Balakrishnan, G. V. Rao and N. Venkatsubramanian, *J.C.S. Perkin II*, **1974**, 6; *J. Indian Chem. Soc.*, **51**, 537 (1974).
[275] D. G. Oakenfull, *Austral. J. Chem.*, **27**, 1423 (1974).
[276] J. F. J. Engbersen and J. B. F. N. Engberts, *J. Am. Chem. Soc.*, **96**, 1231 (1974).
[277] N. K. Vorob'ev and O. K. Shebanova, *Tr. Ivanov. Khim.-Tekhnol. Inst.*, **1970**, 35; *Chem. Abs.*, **79**, 125667 (1973).
[278] M. Balakrishnan, G. V. Rao and N. Venkatasubramanian, *J.C.S. Perkin II*, **1974**, 1093; M. Balakrishnan, G. Venkoba and N. Venkatasubramanian, *Current Sci.*, **43**, 308 (1974); *Chem. Abs.*, **81**, 49101 (1974).
[279] Yu. E. Orlov and R. V. Piskareva, *Khim. Prir. Soedin.*, **10**, 87 (1974); *Chem. Abs.*, **80**, 119877 (1974).
[280] S. Sakai, Y. Kiyohara, M. Ogura and Y. Ishii, *J. Organometal. Chem.*, **72**, 93 (1974).
[281] W. K. Kim and I. Lee, *Daehan Mwahak Mwoejee*, **17**, 235 (1973); *Chem. Abs.*, **79**, 136102 (1973).
[282] R. Vîlcu and I. Ciocázanu, *Rev. Roumaine Chim.*, **19**, 785 (1974); R. P. Tiger, L. S. Bekhli, S. P. Bondarenko and S. G. Entelis, *Zh. Org. Khim.*, **9**, 1563 (1973); *Chem. Abs.*, **79**, 114795 (1973); C. Rüchardt and S. Rochlitz, *Annalen*, **1974**, 15; J. K. Stille and M. T. Regan, *J. Am. Chem. Soc.*, **96**, 1508 (1974); J. K. Stille and R. W. Fries, *ibid.*, p. 1514; J. K. Stille, F. Huang and M. T. Regan, *ibid.*, p. 1518; J. W. Munson, *J. Pharm. Sci.*, **63**, 252 (1974); T. Nakabayashi, Y. Tanaka and S. Ito, *Bull. Chem. Soc. Japan*, **46**, 3853 (1973).
[283] G. Calvaruso, F. P. Cavasino and E. D. Dio, *J.C.S. Perkin II*, **1974**, 1108.
[284] Z. P. Prisyazhnyuk, S. S. Levush and A. L. Bel'ferman, *Kinet. Katal.*, **15**, 540 (1974); *Chem. Abs.*, **81**, 24622 (1974).
[285] C. A. Bunton, P. Ng and L. Sepulveda, *J. Org. Chem.* **39**, 1130 (1974).
[286] D. C. Berndt and I. E. Ward, *J. Org. Chem.*, **39**, 841 (1974).
[287] D. C. Berndt, *J. Org. Chem.*, **39**, 840 (1974).
[288] B. Fernández, I. Perillo and S. Lamdan, *J.C.S. Perkin II*, **1974**, 1416.
[289] R. Kreher, A. Bauer and H. Hennige, *Z. Naturforsch.*, **29**, 231 (1974).
[290] A. A. Zalikin, L. P. Nikitenkova and Yu. A. Strepikheev, *Zh. Obshch. Khim.*, **43**, 1766 (1973); *Chem. Abs.*, **79**, 145741 (1973).
[291] J. Mirek, B. Nawalck, *Rocz. Chem.*, **48**, 243 (1974); *Chem. Abs.*, **81**, 24638 (1974).
[292] J. Bojarski, *Rocz. Chem.*, **48**, 619 (1974); *Chem. Abs.*, **81**, 62731 (1974).
[293] T. Hermann, *Pharmazie*, **29**, 453 (1974); *Chem. Abs.*, **81**, 90729 (1974).
[294] E. R. Talaty and C. M. Utermoehlen, *J.C.S. Chem. Comm.*, **1974**, 204.
[295] F. Sweet and J. D. Fissekis, *J. Am. Chem. Soc.*, **95**, 8741 (1973).

296 F. Texier, J. Guenzet and A. Derdour, *Bull. Soc. Chim. France*, **1974**, 495; G. Costeanu, O. Landauer and C. Mateescu, *Bul. Inst. Politk. "Gheorghe Gheorghiu-Dej" Bucuresti*, **35**, 51 (1973); *Chem. Abs.*, **81**, 90750 (1974); T. P. Visvanathan, *Indian J. Chem.*, **11**, 766 (1973); *Chem. Abs.*, **79**, 145628 (1973).

297 A. N. Assad and R. Tewfik, *Egypt. J. Chem.*, **15**, 85 (1972); *Chem. Abs.*, **79**, 125440.

298 P. J. Sniegoski, *J. Org. Chem.*, **39**, 3141 (1974).

299 V. S. Smorodinov, *Sb. Tr. Molodykh. Uch., Tomsk. Politekh. Inst.*, **1973**, 152; *Chem. Abs.*, **81**, 48988 (1974); J. P. Hardy, S. L. Kerrin and S. L. Manatt, *J. Org. Chem.*, **38**, 4196 (1973); V. Sreeramulu and P. B. Rao, *Ind. Eng. Chem., Process Res. Develop.*, **12**, 843 (1973); *Chem. Abs.*, **79**, 125458 (1973); V. Baliah and P. A. Nador, *Indian J. Chem.*, **11**, 886 (1973); *Chem. Abs.*, **80**, 36478 (1974); Yu. E. Nosovskii, S. A. Osintseva, A. J. Kutsenko and A. N. Alleva, *Khim. Prom. (Moscow)*, **1974**, 108; *Chem. Abs.*, **80**, 119883 (1974); Yu. B. Kagan, L. I. Zvedzdkina, A. Ya. Rozovskii, G. M. Pakhomova and A. N. Bashkirov, *Kinet. Katal.*, **15**, 368 (1974); *Chem. Abs.*, **81**, 12768 (1974).

300 O. N. Sharma, G. D. Nageshwar and P. S. Mene, *Indian J. Technol.*, **11**, 360 (1973); *Chem. Abs.*, **80**, 107555 (1974).

301 J. Guenzet and M. Camps, *Tetrahedron*, **30**, 849 (1974); N. N. Lebedev, V. N. Sapunov and H. M. Lehmann, *Izv. Vyssh. Ucheb. Zaved., Khim. Khim. Tekhnol.*, **16**, 1762 (1973); *Chem. Abs.*, **80**, 59137 (1974).

302 G. A. Chubarov, S. M. Danov and R. V. Efremov, *Zh. Prikl. Khim. (Leningrad)*, **47**, 476 (1974); *Chem. Abs.*, **80**, 119865 (1974); L. V. Pol'stichikova, V. P. Savel'yanov and N. N. Lebedeva, *Izv. Vyssh. Ucheb. Zaved., Khim. Khim. Tekhnol.*, **17**, 144 (1974); *Chem. Abs.*, **80**, 94837 (1974); V. O. Reikhsfel'd, N. A. Filippov and G. P. Yavshchits, *Zh. Prikl. Khim. (Leningrad)*, **46**, 2550 (1973); *Chem. Abs.*, **80**, 59141 (1974); F. Dutka, A. F. Marton and I. Sarkady, *Radiochem. Radioanal. Letters*, **15**, 325 (1973); *Chem. Abs.*, **80**, 47151 (1974).

303 K. Arakawa and T. Miyasaka, *Chem. Pharm. Bull.*, **22**, 214 (1974); *Chem. Abs.*, **80**, 107600 (1974).

304 H. Wenck and M. Schallies, *Helv. Chim. Acta*, **56**, 3059 (1973).

305 Y. Akasaki and A. Ohno, *J. Am. Chem. Soc.*, **96**, 1957 (1974).

306 V. Gebert and B. von Kerékjártó, *Annalen*, **1974**, 644.

307 J. F. Bunnett and F. P. Olsen, *J. Org. Chem.*, **39**, 1156 (1974).

308 M. Lasperas, J. Taillades and A. Commeyras, *Bull. Soc. Chim. France*, **1974**, 1322; D. Zavioanu, *An. Univ. Bucuresti, Chim.*, **20**, 183 (1971); *Chem. Abs.*, **80**, 26507 (1974); E. N. Zil'berman, E. N. Mizinov, S. M. Danov, R. V. Efremov and O. P. Drachkova, *Zh. Org. Khim.*, **10**, 197 (1974); *Chem. Abs.*, **80**, 107572 (1974).

309 M. Higuchi, K. Takeshita and R. Senju, *Bull. Chem. Soc. Japan*, **47**, 1451 (1974).

310 M. L. P. De Troparevsky, A. E. A. Mitta and A. Troparevsky, *An. Asoc. Quím. Argent.*, **61**, 227 (1973); *Chem. Abs.*, **80**, 107612 (1974).

311 V. A. Grigor'eva, O. V. Gelonogova, S. M. Baturin and S. G. Entelis, *Izv. Akad. Nauk SSSR, Ser. Khim.*, **1974**, 812; *Chem. Abs.*, **81**, 48973 (1974); R. P. Tiger, L. S. Bekhli and S. G. Entelis, *Kinet. Katal.*, **15**, 586 (1974); *Chem. Abs.*, **81**, 119437 (1974); G. Zinner and B. Geister, *Arch. Pharmazie*, **307**, 39 (1974); A. P. Grekov, S. A. Sukhorukova and G. V. Otroshko, *Dopov. Akad. Nauk Ukr. RSR, Ser. B*, **35**, 832 (1973); *Chem. Abs.*, **79**, 145587 (1973); R. Parthasarathy, J. Ohrt, S. P. Dutta and G. B. Chheda, *J. Am. Chem. Soc.*, **95**, 8141 (1973); M. L. P. DeTroparevsky, A. Troparevsky and A. E. A. Mitta, *Argent., Com. Nac. Energ. At.*, [*Informe*], **354**, 12 (1973); *Chem. Abs.*, **81**, 77081 (1974); Ya. N. Chirkov, O. V. Nesterov and S. G. Entelis, *Kinet. Katal.*, **14**, 916 (1973); *Chem. Abs.*, **79**, 145598 (1973).

312 R. C. Neuman, Jr., and R. P. Pankratz, *J. Am. Chem. Soc.*, **95**, 8372 (1973).

313 Y. Ogata, Y. Sawaki and M. Furuta, *J. Org. Chem.*, **39**, 216 (1974).

314 C. V. Gurumurthy, *J. Appl. Chem. Biotechnol.*, **23**, 775 (1973); *Chem. Abs.*, **80**, 107638 (1974); J. Lilie, E. Heckel and R. C. Lamb, *J. Am. Chem. Soc.*, **96**, 5543 (1974).

315 F. Takami, K. Ikawa, K. Tokuyama, S. Wakahara and T. Maeda, *Chem. Pharm. Bull.*, **22**, 275 (1974); *Chem. Abs.*, **80**, 119975 (1974).

316 W. H. Richardson, M. B. Yelvington, A. H. Andrist, E. W. Ertley, R. S. Smith and T. D. Johnson, *J. Org. Chem.*, **38**, 4219 (1973).

317 V. I. Kucjeryavyi, V. A. Gal'perin, E. M. Moncharzh and A. I. Finkel'shtein, *Zh. Prikl. Khim. (Leningrad)*, **47**, 529 (1974); *Chem. Abs.*, **80**, 145007 (1974).

318 C. F. Hauser and L. F. Theiling, *J. Org. Chem.*, **39**, 1134 (1974).

319 D. W. Shoesmith, O. Kutowy and R. G. Barradas, *J. Electroanal. Chem. Interfacial Electrochem.*, **48**, 167 (1973); *Chem. Abs.*, **80**, 119932 (1974); S. Arita, N. Hirai, Y. Nishimura and K. Takeshita, *Nippon Kagaku Kaishi*, **1973**, 2160; *Chem. Abs.*, **80**, 59122 (1974).

320 O. A. El Seoud, A. T. do Amaral, M. M. Campos and L. do Amaral, *J. Org. Chem.*, **39**, 1915 (1974).

[321] T. A. Degurko and V. I. Staninets, *Dopov. Akad. Nauk Ukr. RSR, Ser. B*, **36**, 255 (1974); *Chem. Abs.*, **80**, 145031 (1974).
[322] M. K. Chantooni, Jr., and I. M. Kolthoff, *J. Phys. Chem.*, **78**, 839 (1974).
[323] L. A. Kutulia, L. P. Pisovarevich, V. G. Gordienko, S. V. Tsukerman and V. F. Lavrushin, *Organic Reactivity* (*Tartu*), **9**, 1053 (1972).
[324] W. E. Parham and P. E. Olson, *J. Org. Chem.*, **39**, 2916 (1974).
[325] B. L. Vorabiov, T. E. Zhesko and A. L. Shapiro, *Organic Reactivity* (*Tartu*), **10**, 1127 (1973).
[326] V. A. Liferova, Z. D. Skripnik and G. F. Yankovskaya, *Adsorbtsiya Adsorbenty*, **2**, 43 (1974); *Chem. Abs.*, **81**, 104246 (1974).
[327] B. A. Phillips, G. Fodor, J. Gal, F. Letourneau and J. J. Ryan, *Tetrahedron*, **29**, 3309 (1973).
[328] J. Pola, M. Jakoubková and V. Chvalovský, *Coll. Czech. Chem. Comm.*, **39**, 1169 (1974).
[329] F. N. Vishnevskii, V. P. Kozyukov, N. A. Kondratova, I. I. Skorokhodov, and N. S. Fedotov, *Zh. Obshch. Khim.*, **43**, 2236 (1973); *Chem. Abs.*, **80**, 59230 (1974).
[330] I. Porvbszky, G. Liptay and J. Korosi, *Acta Chim.* (*Budapest*), **79**, 179 (1973); *Chem. Abs.*, **81**, 3063 (1974).
[331] J. E. Herweh and R. M. Fantazier, *J. Org. Chem.*, **39**, 786 (1974).
[332] E. I. Duman and T. I. Yurasova, *Zh. Obshch. Khim.*, **44**, 1413 (1974); *Chem. Abs.*, **81**, 62745 (1974).
[333] M. Mutter and H. Hagenmaier, *Angew. Chem. Int. Ed.*, **13**, 149 (1974).
[334] P. Müller and B. Siegfried, *Helv. Chem. Acta*, **57**, 987 (1974).
[335] G. B. Sergeev, V. A. Batyuk and B. M. Sergeev, *Kinet. Katal.*, **15**, 326 (1974); *Chem. Abs.*, **81**, 12718 (1974).
[336] J. Rebek and D. Feitler, *J. Am. Chem. Soc.*, **96**, 1606 (1974).
[337] M. Rothe and J. Mazánek, *Annalen*, **1974**, 439.
[338] E. A. Noe and N. Raban, *J.C.S. Chem. Comm.*, **1974**, 479.
[339] G. C. Pappalardo and S. Gruttadauria, *J.C.S. Perkin II*, **1974**, 1441.
[340] S. S. Simons, Jr., *J. Am. Chem. Soc.*, **96**, 6492 (1974).
[341] E. P. Lyznicki, Jr., K. Oyama and T. T. Tidwell, *Can. J. Chem.*, **52**, 1066 (1974).
[342] M. M. T. Khan and M. S. Mohan, *J. Inorg. Nucl. Chem.*, **36**, 707 (1974).
[343] A. A. Fedorov, F. K. Samigullin and S. A. Sokolov, *Plaste Kaut.*, **20**, 909 (1973); *Chem. Abs.*, **81**, 119441 (1974).
[344] F. Ramirez, S. L. Glaser and P. Stern, *Tetrahedron*, **29**, 3741 (1973).
[345] J. L. Banyasz and J. E. Stuehr, *J. Am. Chem. Soc.*, **96**, 6481 (1974).
[346] N. V. Malysheva and N. A. Rus'yanova, *Vop. Tekhnol. Ulavlivoniya Pererab. Prod. Koksovaniya*, **1972**, 106; *Chem. Abs.*, **81**, 62739 (1974).
[347] W. S. Wadsworth, Jr., and Y.-G. Tsay, *J. Org. Chem.*, **39**, 984 (1974).
[348] C. H. Clapp and F. H. Westheimer, *J. Am. Chem. Soc.*, **96**, 6710 (1974).
[349] F. Ramirez, Y. F. Chaw, J. F. Marecek and I. Ugi, *J. Am. Chem. Soc.*, **96**, 2429 (1974).
[350] H. R. Hudson and J. C. Roberts, *J.C.S. Perkin II*, **1974**, 1575.
[351] D. B. Denney and F. A. Wagner, Jr., *Phosphorus*, **3**, 27 (1973); *Chem. Abs.*, **81**, 119786 (1974); A. N. Pudovik, I. V. Konovalova, V. P. Kakurina and V. A. Fomin, *Zh. Obshch. Khim.*, **44**, 263 (1974); *Chem. Abs.*, **80**, 119829 (1974).
[352] J. P. J. van der Holst, C. van Hooidonk and H. Kienhuis, *Rec. Trav. chim.*, **93**, 40 (1974).
[353] K. E. DeBruin and D. M. Johnson, *J. Am. Chem. Soc.*, **95**, 7921 (1973).
[354] A. A. Neimysheva, M. V. Ermolaeva and I. L. Knunyants, *Zh. Obshch. Khim.*, **43**, 2608 (1973); *Chem. Abs.*, **80**, 107552 (1974).
[355] E. A. Krasil'nikova, T. V. Zykova, A. M. Razumov, I. M. Starshov, G. V. Orlova and R. A. Salakhutdinov, *Zh. Obshch. Khim.*, **43**, 1701 (1973); *Chem. Abs.*, **79**, 145593 (1973).
[356] A. I. Razumov, I. A. Krivosheeva, B. G. Liorber, T. A. Tarzivolova, Z. M. Khammatova and V. A. Pavlov, *Zh. Obshch. Khim.*, **44**, 51 (1974); *Chem. Abs.*, **80**, 119807 (1974).
[357] S. A. Bone, S. Trippett, M. W. White and P. J. Whittle, *Tetrahedron Letters*, **1974**, 1795.
[358] R. B. Wetzel and G. L. Kenyon, *J. Am. Chem. Soc.*, **96**, 5199 (1974).
[359] L. E. Overman, D. Matzinger, E. M. O'Connor and J. D. Overman, *J. Am. Chem. Soc.*, **96**, 6081 (1974).
[360] T. A. Mastryukova, M. Orlov, L. S. Butorina, E. I. Matrosov, T. M. Shcherbina and M. I. Kabachnik, *Zh. Obshch. Khim.*, **44**, 1001 (1974); *Chem. Abs.*, **81**, 48991 (1974).
[361] H. Flodgaard and P. Fleron, *J. Biol. Chem.*, **249**, 3465 (1974).
[362] R. S. Glass, E. B. Williams, Jr., and G. S. Wilson, *Biochemistry*, **13**, 2800 (1974).
[363] J. Ahlers, *Biochem. J.*, **141**, 257 (1974).
[364] W. Bloch and M. J. Schlesinger, *J. Biol. Chem.*, **249**, 1760 (1974).
[365] M. Julien and L. Ouellet, *Can. J. Chem.*, **52**, 3087 (1974).

[366] D. Chappelet-Tordo, M. Fosset, M. Iwatsubo, C. Gache and M. Lazdunski, *Biochemistry*, **13**, 1788 (1974).
[367] M. S. Rohrbach, M. E. Dempsey and J. W. Bodley, *J. Biol. Chem.*, **249**, 5094 (1974).
[368] G. Colombo and F. Marcus, *Biochemistry*, **13**, 3085 (1974).
[369] E. Hubert and F. Marcus, *FEBS Letters*, **40**, 37 (1974).
[370] M. E. Kurtley and J. C. Dix, *Biochemistry*, **13**, 4469 (1974).
[371] P. A. Benkovic, W. P. Bullard, M. de Maine, R. Fishbein, K. J. Schray, J. J. Steffens and S. J. Benkovic, *J. Biol. Chem.*, **249**, 930 (1974).
[372] R. Fishbein, P. A. Benkovic, K. J. Schray, I. J. Siewers, J. J. Steffens and S. J. Benkovic, *J. Biol. Chem.*, **249**, 6047 (1974).
[373] R. L. Hanson, F. B. Rudolph and H. A. Lardy, *J. Biol. Chem.*, **248**, 7852 (1973).
[374] G. Sauermann, *Z. physiol. Chem.*, **355**, 1058 (1974).
[375] R. N. Lindquist, J. L. Lynn, Jr., and G. E. Lienhard, *J. Am. Chem. Soc.*, **95**, 8762 (1973).
[376] A. R. Goldfarb, *Bioorg. Chem.*, **3**, 249 (1974); A. R. Goldfarb, D. N. Buchanan and D. E. Sutton, *ibid.*, p. 260.
[377] R. Kobayashi and C. H. W. Hirs, *J. Biol. Chem.*, **248**, 7833 (1973).
[378] I. M. Chaiken, J. S. Cohen and E. A. Sokolski, *J. Am. Chem. Soc.*, **96**, 4703 (1974).
[379] T. L. Poulos and P. A. Price, *J. Biol. Chem.*, **249**, 1453 (1974).
[380] R. Bose and E. W. Yamada, *Biochemistry*, **13**, 2051 (1974).
[381] J. G. Tillet, *Int. J. Sulfur Chem.*, **3**, 289 (1973); *Chem. Abs.*, **79**, 136013 (1973).
[382] K. Nagasawa and H. Yoshidome, *J. Org. Chem.*, **39**, 1681 (1974).
[383] W. J. Spillane, N. Regan and F. L. Scott, *J.C.S. Perkin II*, **1974**, 445.
[384] Yu. B. Kagan, G. M. Pakhomova, N. A. Shimanko, A. Yu. Koshevnik, M. V. Shishkina and R. G. Lokteva, *Kinet. Katal.*, **15**, 23 (1974); *Chem. Abs.*, **80**, 119804 (1974).
[385] W. J. Spillane, J. A. Barry, W. A. Heapy and F. L. Scott, *Z. Naturforsch.*, **29**, 702 (1974).
[386] L. H. King and E. T. Kaiser, *J. Am. Chem. Soc.*, **96**, 1410 (1974).
[387] A. F. Popov and V. I. Tokarev, *Organic Reactivity* (*Tartu*), **9**, 946 (1972).
[388] A. Williams, K. T. Douglas and J. S. Loran, *J.C.S. Chem. Comm.*, **1974**, 689.
[389] A. Williams and K. T. Douglas, *J.C.S. Perkin II*, **1974**, 1727; see *Org. Reaction Mech.*, **1973.**
[390] L. Menninga and J. B. F. N. Engberts, *Tetrahedron Letters*, **1974**, 733; see *Org. Reaction Mech.*, **1973.**
[391] A. Wagenaar, A. J. Kirby and J. B. F. N Engberts, *Tetrahedron Letters*, **1974**, 3735
[392] A Wagenaar, G. Kransen and J. B. F. N. Engberts, *J. Org. Chem.*, **39**, 411 (1974).
[393] J. L. Kice, C. A. Walters and S. B. Burton, *J. Org. Chem.*, **39**, 346 (1974).
[394] A. S. Peregudov, D. N. Kravtsov and L. A. Fedorov, *J. Organometal. Chem.*, **71**, 347 (1974).
[395] K. A. Kovar, G. P. Wojtovicz and H. Auterhoft, *Arch. Pharmazie*, **307**, 657 (1974); M. A. Portnov, M. N. Vaisman, T. A. Dubinina and V. A. Zasosov, *Khim.-Farm. Zh.*, **8**, 53 (1974); *Chem. Abs.*, **81**, 90748 (1974).
[396] N. A. Radomsky and M. J. Gibian, *J. Am. Chem. Soc.*, **95**, 8713 (1973).
[397] S. N. Semenova, L. P. Dementieva and T. I. Temnikova, *Organic Reactivity* (*Tartu*), **10**, 1086 (1973); A. I. Scott and E. M. Gordon, *Tetrahedron Letters*, **1974**, 3507; G. M. Kusho, B. G. Boldyrey and M. E. Yarish, *Khim. Seraorg. Soedin., Soderzhashchikhsya Neft. Nefteprod.*, **9**, 295 (1972); *Chem. Abs.*, **80**, 59163 (1974); P. K. Maarsen, R. Bregman and H. Cerfontain, *Tetrahedron*, **30**, 1211 (1974).
[398] E. Ciuffarin and L. Senatore, *Tetrahedron Letters*, **1974**, 1635.
[399] L. Senatore, L. Sagramora and E. Ciuffarin, *J.C.S. Perkin II*, **1974**, 722.
[400] S. D. Ross, M. Finkelstein and F. S. Dunkl, *J. Org. Chem.*, **39**, 134 (1974).
[401] E. Block and J. O'Connor, *J. Am. Chem. Soc.*, **96**, 3921, 3929 (1974).
[402] D. R. Hogg and J. Stewart, *J.C.S. Perkin II*, **1974**, 43.
[403] D. R. Hogg and J. Stewart, *J.C.S. Perkin II*, **1974**, 436.
[404] T. S. Chou, J. R. Burgtorf, A. L. Ellis, S. R. Lammert and S. P. Kukolja, *J. Am. Chem. Soc.*, **96**, 1609 (1974).
[405] V. J. Traynelis and J. N. Rieck, *J. Org. Chem.*, **38**, 4339 (1973).
[406] G. Scorrano, *Accounts Chem. Res.*, **6**, 132 (1973).
[407] T. M. Santosusso and D. Swern, *Tetrahedron Letters*, **1974**, 4255.
[408] V. N. Kreitsberg, I. Z. Kurostyshevskii, L. A. Kiptianova, M. S. Rusakova and I. P. Gragerov, *Fiz. Khim.*, **1972**, 21; *Chem. Abs.*, **80**, 132437 (1974).
[409] M. S. Rusakova, V. N. Kreitsberg, V. A. Podgornova, B. F. Ustavshchikov and V. V. Voronenkov, *Kinet. Katal.*, **15**, 601 (1974); *Chem. Abs.*, **81**, 104314 (1974).
[410] M. T. Rosseel, M. G. Bogaert and D. DeKeukeleire, *Bull. Soc. Chim. belges*, **83**, 211 (1974).

[411] K. S. Pillay, R. N. Lockhart, T. Tezuka and Y. L. Chow, *J.C.S. Chem. Comm.*, **1974**, 80.
[412] D. M. Dimitrijević, Z. D. Tadič, M. M. Mišićyuković and M. Muškatirović, *J.C.S. Perkin II*, **1974**, 1051.
[413] N. B. Chapman, M. R. J. Dack, D. J. Newman, J. Shorter and R. Wilkinson, *J.C.S. Perkin II*, **1974**, 962, 971.
[414] F. M. Dean and B. K. Park, *J.C.S. Chem. Comm.*, **1974**, 162.
[415] D. L. H. Williams, *J.C.S. Chem. Comm.*, **1974**, 324.

CHAPTER 3

Radical Reactions

D. C. NONHEBEL

Department of Pure and Applied Chemistry, University of Strathclyde

and J. C. WALTON

Department of Chemistry, University of St. Andrews

Introduction

The continued high level of interest in free-radical chemistry has been marked by the publication of three books, one a comprehensive treatise[1] and the other two at undergraduate level.[2,3] Briefer surveys of the chemistry of free radicals[4] and radical anions[5] have also appeared. The whole saga of the quinonoid structure of the triphenylmethyl radical dimer has been reviewed.[6] A quantum-mechanical treatment of various physical properties and the chemical reactivity of radicals has appeared.[7] Chirality in radicals has been discussed.[7a] Included in this year's Report as a separate section are photochemical reactions involving radical intermediates. Autoxidation is now dealt with in the Chapter on Oxidation and Reduction.

Structure, Stereochemistry and Stability

Carbon Radicals

The use of ESR spectroscopy to determine structures of organic radicals has been reviewed.[8] The stability of organic radicals can be dramatically increased by steric shielding of the radical centre by bulky hydrocarbon groups;[9] this is seen in the remarkable stability of the $Bu^t{}_2\dot{C}H$, $Bu^t{}_3C\cdot$ and $(Me_3Si)_3C\cdot$ radicals, which cannot either dimerize readily because of steric hindrance or disproportionate.[10,11] The di-*tert*-butylmethyl radical and related species decay by a first-order process probably involving β-scission of a methyl group; this results in relief of strain: $Bu^t{}_2\dot{C}H \rightarrow Bu^tCH{=}CMe_2 + Me\cdot$. The 2,2,4,6,6-pentamethyl-$\dot{}$ and 2,2,4,4,6,6-hexamethyl-cyclohexyl radicals, which are closely related to $Bu^t{}_2\dot{C}H$, have similar stability.[10] Their ESR spectra are characterized by a doublet splitting with the α-hydrogen and a quartet splitting with the three equivalent protons of the equatorial 4-methyl group.[12] The W-plan arrangement of bonds between the radical centre and this group results in an anomalously large splitting (2.30 and 3.07 G, respectively). 2-Substituted 1,1-di-*tert*-butylethyl radicals, $Bu^t{}_2\dot{C}CH_2R$, are also very stable and like $Bu^t{}_2\dot{C}H$ radicals decay in a unimolecular process with β-scission:[10] $Bu^t{}_2\dot{C}{-}CH_2R \rightarrow Bu^t{}_2C{=}CH_2 + R\cdot$. The order of stability of these radicals is related to the stability of $R\cdot$ and is in the order $Ph \gg OCF_3 \gg CCl_3 > P(O)(OEt)_2 > SiBu_3 \gg SCF_3$. These radicals exist in conformation (**1**) regardless of the nature of the group R.[13]

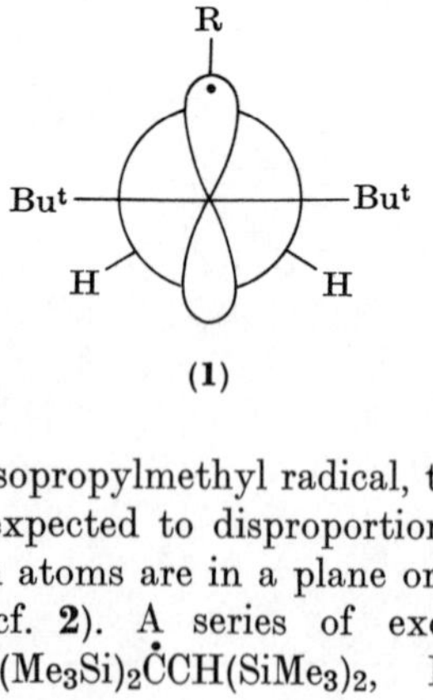

(**1**)

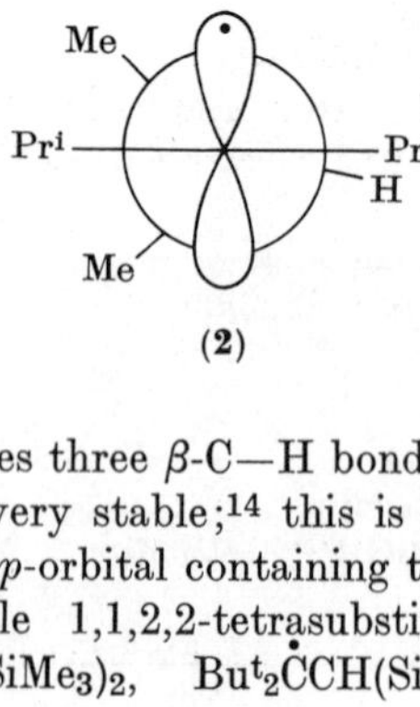

(**2**)

The tri-isopropylmethyl radical, though it possesses three β-C—H bonds and hence might be expected to disproportionate readily, is very stable;[14] this is because the β-hydrogen atoms are in a plane orthogonal to the *p*-orbital containing the unpaired electron (cf. **2**). A series of exceptionally stable 1,1,2,2-tetrasubstituted ethyl radicals, $(Me_3Si)_2\dot{C}CH(SiMe_3)_2$, $Bu^t(Me_3Si)\dot{C}CH(SiMe_3)_2$, $Bu^t{}_2\dot{C}CH(SiMe_3)_2$ and

$Bu^t_2\dot{C}CHBu^t_2$ have been prepared;[15] no coupling with the lone proton is observable in their ESR spectra, indicating that they are locked in the conformation (**3**) as a result of steric factors. The radical $Bu^t_2\dot{C}CHBu^t_2$ is the least stable of this series as it can decay by β-scission of a *tert*-butyl radical. The remaining radicals probably decay by elimination of a methyl group from an α-substituent, such a process occurring more readily from a Me_3Si group than from a Bu^t group.

(**3**) (**4**)

α-Aminoalkyl radicals, $Bu^t_2\dot{C}NR^1R^2$, like $Bu^t_2\dot{C}CH_2R$ radicals, are remarkably long-lived.[16] They appear to be locked in the conformation (**4**) as a result of steric interactions. This is in marked contrast to simple aminoalkyl radicals which prefer a conformation that allows overlap of the half-filled *p*-orbital and the nitrogen lone pair. The hyperconjugative interaction with a group R is about half that found for $Bu^t_2\dot{C}CH_2R$, possibly reflecting the influence of greater electronegativity of nitrogen.

Substitution by fluorine at the radical centre results in some deviation from planarity in α-fluoroalkyl radicals, while the corresponding α,α-difluoroalkyl radicals are substantially pyramidal as indicated by their a_F and $a_{\beta\text{-H}}$ hyperfine splittings.[17,18] The study of the temperature-dependence of line shapes in solution ESR spectra of alkyl radicals has yielded valuable information about the barriers to rotation and about the equilibrium conformation of the radicals.[19] Thus α,α-difluoroalkyl radicals have a barrier to rotation about the $C_\alpha—C_\beta$ bond of 2.2–2.85 kcal mol^{-1}, indicative of a threefold axis of internal rotation.[20] This barrier to rotation is a consequence of bending at the radical centre which decreases hyperconjugative interaction of the unpaired electron with the three β-protons and the three β-fluorines in the $CH_3CF_2\cdot$ and the $CF_3CF_2\cdot$ radical, respectively. No appreciable barrier to rotation is found for the planar $CF_3CH_2\cdot$, $CHF_2CH_2\cdot$ and $CH_2FCH_2\cdot$ radicals.[18] The preferred conformations for the $XCH_2CF_2\cdot$ ($X = H$, CF_3O, Cl, RS), $CF_3CF_2\cdot$ and $XCF_2CF_2\cdot$ ($X = F$, CF_3, C_5H_{11}) radicals are (**5**) and (**6**), respectively.[18,19,22] There may be some evidence for bridging between X when

(**5**) (**6**)

$X = Cl$ or RS as indicated by the smaller values of $a_{\beta-H}$, but the situation is complex owing to the tetrahedral radical centre. The extent of bending at the radical centre is much less for α-chloro- and α-bromo-alkyl radicals.[17,21] The ESR spectrum of

$\cdot CHICONH_2$ is also consistent with a planar structure.[23] *Ab initio* SCF MO calculations for trigonal radicals $\cdot AX_3$ indicate that deviation from planarity increases as X becomes more electronegative, i.e. in the order $\cdot CH_3 < \cdot CCl_3 < \cdot CF_3$.[24] The unpaired electron is delocalized on to bromine to the extent of 19% in α-bromoalkyl radicals.[25]

The eclipsed conformation (**7**) is the preferred conformation of $R_nMCH_2CH_2\cdot$ radicals where M is an element in row 2, 3 or 4, whilst when M is C the staggered conformation (**8**) is preferred.[13, 26] The preference for the eclipsed conformation produced by row 2

(**7**) (**8**)

elements has been attributed to hyperconjugative and/or homoconjugative interaction between the element and the half-filled *p*-orbital.[27] The radicals $Bu^t_2\dot{C}CH_2MR_n$ invariably adopt the eclipsed conformation even when M is C. In these radicals the extent of hyperconjugative interaction is about the same whether M is C or Si.[13] It is thus concluded that homoconjugative interaction involving *pπ–dπ* bonding must be invoked to explain the preference that $\cdot CH_2CH_2MR_n$ radicals have for the eclipsed conformation when M is in row 2. The eclipsed conformation can be induced in $\cdot CHR^1CH_2OR^2$ radicals when the radical centre is secondary ($R^1 \neq H$) and electronegative groups are attached to oxygen ($R^2 = CF_3$, PhCO, Pr^iOCO) but not when $R^2 = H$, CH_3, or Bu^t. In these instances the staggered conformation is preferred.[26] Electronegative groups attached to oxygen enhance homoconjugative interaction between the oxygen and the half-filled *p*-orbital. In contrast to this picture, $\cdot CMe_2CH_2Br$ radicals exist in the staggered conformation even though $\cdot CMe_2CH_2Cl$ prefers the normal eclipsed conformation.[27a] Primary and secondary alkyl radicals normally exist in the staggered conformation, though tertiary radicals prefer the eclipsed conformation as steric interactions force the β-alkyl group away from the nodal plane. An eclipsed conformation (**10**) is also favoured for the 1-aziridylcarbinyl radical (**9**)[28] in contrast to its carbon analogue which prefers the bisected conformation.[29] In the eclipsed con-

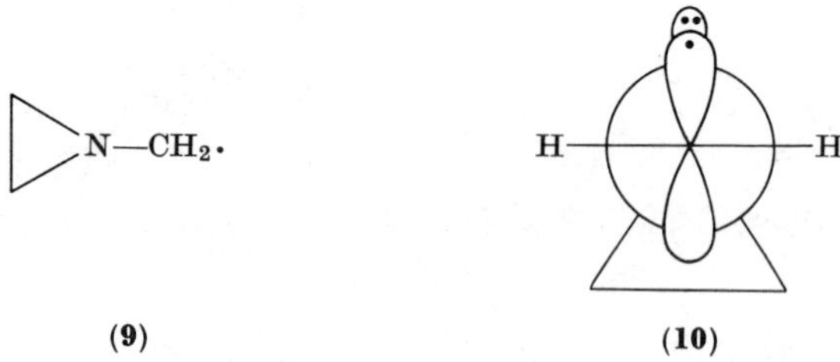

(**9**) (**10**)

formation, stabilization is achieved through interaction between the nitrogen lone-pair and the half-filled *p*-orbital. INDO calculations also indicate that the eclipsed conformation is favoured for (**10**) but not for the cyclopropylamino-radical.

Ab initio MO calculations indicate that the barriers to rotation about the C_α—C_β bond in 2-substituted ethyl radicals with first-row substituents are considerably smaller than in the corresponding cations.[30] Hyperconjugative interaction is smaller in radicals because the 2*p*-orbital is no longer vacant. The ESR spectrum of the propyl radical indicates that it is rapidly reorientating between approximately equivalent conformations,[31] whereas at 4°K it is rigid, as is the allyl radical.[32] The methyl and ethyl radicals are, however, rapidly reorientating about their axes even at 4°K.[32] The hindered internal motions of alkyl radicals in urea clathrates[33] and the conformations that the alkyl chain adopts in the $CH_3\dot{C}HCH_2CH_2\overset{+}{N}H_3$ and related radicals have been studied.[34]

The hydroxyl group plays an important role in determining the preferred conformation of 1-hydroxyalkyl radicals.[35] The 1-hydroxycyclopentyl radical (and the 1-hydroxycyclobutyl radical) is in a rigidly planar conformation in contrast to the cyclopentyl radical and also the five-membered heterocyclic radicals (**11**) and (**12**) in which there is rapid interconversion between two half-chair geometries.[36] The 1-hydroxycyclohexyl

X = $^+NH_2$, ^+NHR

(**11**)

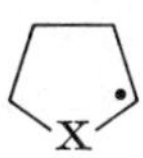

X = O, S

(**12**)

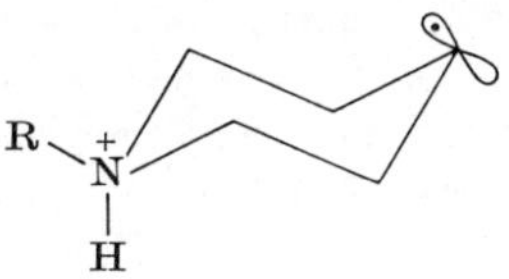

(**13**)

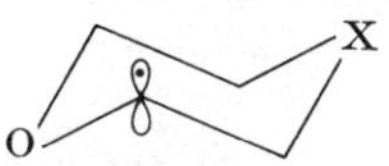

X = ^+NHR, O

(**14**)

radical adopts a twisted chair conformation at low temperatures[35] whereas the cyclohexyl radical and the radicals (**13**) and (**14**) exist as rapidly interconverting chair conformers. The absorption spectra of cyclohexyl and cyclohexenyl radicals have been recorded.[37] The conformations of $HOCH_2\dot{C}HCH_2X$ radicals have also been studied;[38] ε-splitting is observed when X = OCH_2CH_3 as the radical adopts a cyclic conformation in which the half-filled *p*-orbital can interact with the protons of the methyl group.

The stabilization energy of the benzyl radical has been estimated as 11.2 kcal mol^{-1} by the HF^+ infrared chemiluminescence technique.[39] This compares with a value of 13 kcal mol^{-1} from bond-dissociation energies. The angle of twist in $(C_6F_5)_3C\cdot$ is slightly greater than that in $Ph_3C\cdot$ as a result of increased steric interaction.[40] The perchlorobenzyl radical, $C_6Cl_5CCl_2\cdot$, has been reported;[41] it is probably non-planar and its relatively high stability is attributed to steric shielding of the radical centre and steric hindrance to dimerization. The α,α-di-*tert*-butylbenzyl radical (**15**) is a perpendicular radical in which there is no π–π delocalization.[42] There is however some π–σ

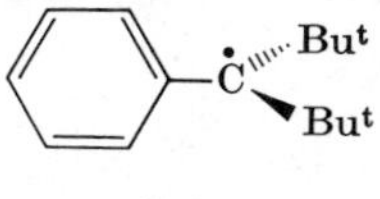

(**15**)

delocalization as indicated from the ESR coupling constants. Several MO calculations have been carried out on the benzyl radical and the isoelectronic phenoxyl and anilino-radicals;[43–45] reasonable correlations of calculated and experimental splitting constants have been obtained. The trend in splitting constants shows that the overall spin density in the ring increases from $PhCH_2\cdot$ to $PhO\cdot$, indicating the increasing importance of an ionic resonance form. The fluorescence lifetimes[46] and optical excitation and emission spectra[47] of benzyl radicals have also been reported.

The resonance energy of allyl radicals has been estimated as 14.0 ± 2.0 kcal mol^{-1} from the rate of racemization of 1,3-dimethyl- and 1,3-di-*tert*-butyl-allene.[48] The inequivalence of the terminal protons in the 1,1-dimethylallyl radical (**16**) at temperatures up to 180° indicates that the activation energy for rotation about either allylic

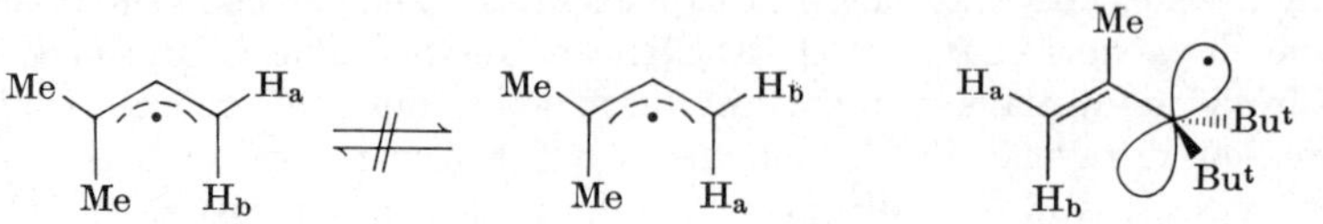

(**16**) (**17**)

bond is at least 14 kcal mol^{-1}.[49] This is effectively a measure of the stabilization energy of the allylic radical. The ESR spectrum of the perpendicular allyl radical (**17**), in which there is no π–π delocalization, indicates considerable coupling to the protons H_a and H_b as a consequence of homohyperconjugation.[50] *Ab initio* MO calculations on the pentadienyl radical have been carried out.[51]

Further details on the structure of 2-oxoalkyl radicals have appeared;[52–54] ESR studies suggest that this class of radical is a resonance hybrid of the two allylic structures $R^1R^2\dot{C}COR^3 \leftrightarrow R^1R^2C{=}C(O\cdot)R$ with the second structure contributing about 15% to the resonance hybrid. The very different contributions of the two canonical structures suggest that the resonance energy of this type of radical is small. Consistent with the allylic nature of the radical, restricted rotation about the C—CO bond is observed (barrier to rotation 9 kcal mol^{-1}) which may be sufficient to control the stereochemistry in reactions involving 2-oxoalkyl radicals. The ESR spectrum of $PhCOCH_2\cdot$ has also been described.[54a]

The ESR spectrum of the cyclopropenyl radical is consistent with rapid interchange between three equivalent structures.[55] MO calculations on the isomeric linear C_3H_3 radical indicate that it corresponds to a π-delocalized propargylic structure (**18**);[56]

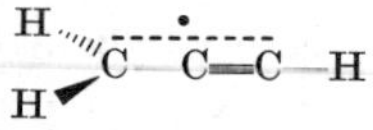

(**18**)

reactions of this radical tend to give predominantly acetylenic products, which is consistent with there being a higher spin density at C-1 than at C-3. The ESR spectra of hydroxycyclohexadienyl radicals are consistent with the radicals being planar.[57]

The 2,4,6-tri-*tert*-butylphenyl radical (**19**) is the first moderately stable phenyl radical;[58] its stability arises from steric shielding of the radical centre; decay is by a

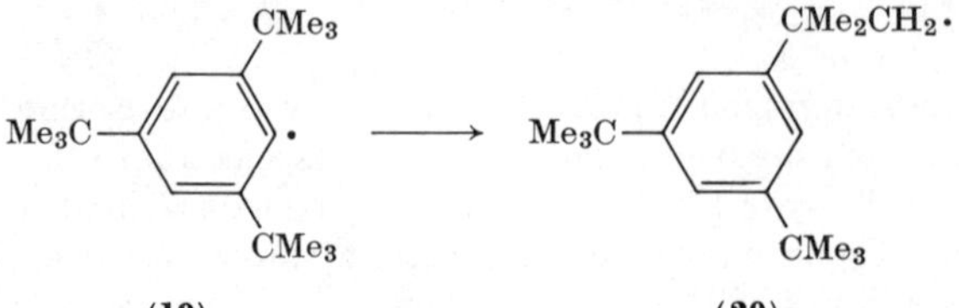

(**19**) (**20**)

unimolecular process to (**20**). The ESR spectra of the *p*-tolyl,[59] and 2-hydroxy- and 2-methoxy-phenyl radicals[60] indicate that they are σ-radicals.

It has been found that, somewhat anomalously, alkyl groups apparently exert a destabilizing influence at a radical centre,[61] whilst a bromomethyl group has a stabilizing effect;[62] these conclusions are indicated from the effect of substituents on the ease of ring opening of 1-substituted 4-chlorobicyclo[2,2,0]hexanes (**21**) (R = H, Me, Et or CH_2Br) which proceed via diradical intermediates (**22**). The stabilizing influence of the bromomethyl group has been attributed to the involvement of a bridged transition state. Cyano-groups similarly stabilize the radical centre.[62a]

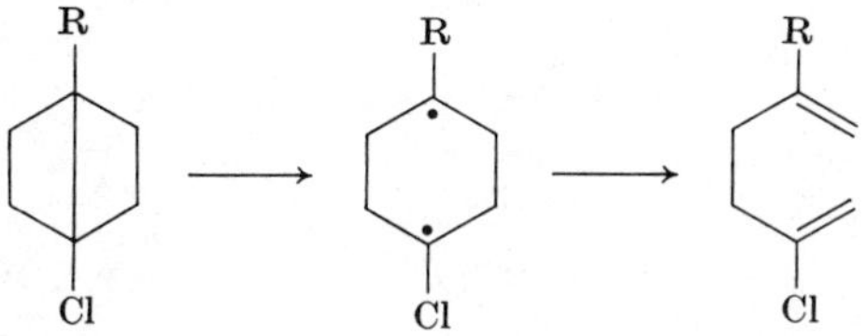

(**21**) (**22**)

The ESR spectrum of the biradical, Schlenk's hydrocarbon, shows a strong triplet pattern due to interaction of the unpaired spins.[63,64] The ESR spectrum of Tschitschibabin's hydrocarbon has also been described.

The fluoroacyl radicals $CF_3\dot{C}O$ and $CF_3CF_2\dot{C}O$ are σ-radicals.[65] The latter shows an unusual doublet splitting in its ESR spectrum, implying that the CF_3 group is locked in a conformation so that only one of the three fluorines can interact with the half-filled *p*-orbital. The microwave spectrum of the formyl radical lends further support to its formulation as a σ-radical.[66]

The electronic configurations of formamido-, and *N*-hydroxy- and *N*-methyl-formamido-radicals have been calculated.[67] The lowest states are predicted to be π with the possible exception of the *N*-methyl compound where the σ state is of very similar energy. The $HCO_2\cdot$ radical is also calculated to be in a π state; this correlates with an excited state of carbon dioxide and hence decarboxylation of the ground state would be energetically forbidden by the non-crossing rule.[68]

Nitrogen Radicals

Radicals may be stabilized by delocalization of the unpaired electron or by steric shielding of the radical site. A third general way which is particularly important for nitrogen radicals is by push-pull effects or merostabilization, i.e. stabilization of radicals by simultaneous conjugation with electron-donor and electron-acceptor groups;[69–74] this results in stabilization of zwitterionic resonance forms. Dimerization of this class of radical does not lead to a great gain in stability. Consideration of these factors led to the preparation of the stable radicals (**23**)–(**27**).[69–74] The stability of (**28**) is similarly explained.[75]

The ESR spectra of simple hydrazyl radicals (**29**) indicate that the unpaired spin is associated with both nitrogens. In 1-alkylhydrazyls the spin density on N-1 is greater than that on N-2, indicating that canonical form (**29a**) is the more important owing to the electron-donating influence of the alkyl group.[77] The same is true for 1,2-dialkyl- and trialkyl-hydrazyls, whereas the canonical form (**29b**) is the major form for 2,2-dialkylhydrazyls and also for $H_2N\dot{N}COOEt$.[75] In a series of bicyclic hydrazyls the

$Ph_2\ddot{N}\dot{N}CN \longleftrightarrow Ph_2\overset{+\bullet}{N}\overset{-}{\ddot{N}}CN \longleftrightarrow Ph_2\overset{+\bullet}{N}N{=}C{=}N^-$

(23)

$Ph_2N\dot{N}SO_2CF_3$ **(24)**

R_2N–C_6H_4–$\dot{N}SO_2Ar$ **(25)**

R_2N–C_6H_4–$\dot{N}COAr$ **(26)**

(27)

(28)

(29a) **(29b)**

near equality of a_N values suggests that both canonical forms contribute significantly to the resonance hybrid.[78] Triorganosilylhydrazyls $(R_3Si)_2N\dot{N}SiR_3$ are much more stable than simple trialkylhydrazyls and are characterized by a much smaller coupling constant to N-2 than to N-1; this is attributed to delocalization of unpaired spin on to the silicon;[79] a second radical, tentatively assigned the triazyl structure $(R_3Si)_2N\dot{N}N(SiR_3)_2$ was also observed.

Steric shielding of the radical centre accounts for the exceptional stability of species (**30**) and (**31**).[80] The less hindered but structurally similar radicals (**32**) and (**33**) are much less stable. Steric shielding also accounts for the high stability of the perchlorodiphenylaminyl radical $(C_6Cl_5)_2N\cdot$.[81]

$Bu^t_2C{=}N{-}\dot{N}Ph$ **(30)** $Bu^t_2C{=}N{-}\dot{N}Bu^t$ **(31)** $Ph_2C{=}N{-}\dot{N}Ph$ **(32)** $Bu^tN{=}C(Me){-}\dot{N}Bu^t$ **(33)**

Contact shifts induced in the NMR spectra of 7-azabenzonorbornene and 7-azabenzonorbornadiene by bis(acetylacetonato)nickel(II) are consistent with a non-planar radical centre in the radical cations (**34**) and (**35**) with unpaired spin in an orbital directed

(34) **(35)**

towards the benzene ring;[82] the positive spin densities observed for the vinylic protons in (**35**) can only be attributed to a homohyperconjugative interaction. The methyl group rotates rapidly about the axis of the C—N bond in the $CH_3NH\cdot$ radical.[83]

The a_N splitting constants for diarylaminyls show no Hammett correlation though there is an approximate correlation for the extent of dissociation of the parent tetra-arylhydrazines, where substituents influence the ground state of the hydrazine;[84,85] *o*-disubstituted aminyls are much more stable as a consequence of steric hindrance to dimerization. The ΔS° for the dissociation of 1,2-di-*tert*-butyl-1,2-bis(3′,5′-di-*tert*-butylphenyl)hydrazine is astonishingly high, reflecting the release of steric strain on dissociation to *N*,3,5-tri-*tert*-butylanilino-radicals.[86] The ΔS^* for dimerization of these radicals is very negative, indicating a very restricted transition state.

The 1-pyrrolyl radical has been designated as a π-radical,[68] though a more detailed calculation in which configurational and geometric problems were considered suggested that it is a σ-radical;[87] the imidazolyl and benzimidazolyl radicals have been calculated to have σ- and π-structures, respectively.[87] The α- and β-aryl groups in tetra-aryl-pyrrolyl radicals make angles of 25° and 60°, respectively, with the pyrrole ring.[88] The influence of substituents on the stability of tetra-arylpyrrolyl[88a] and of 2-(α-naphthyl)-4-phenyl-5-(*p*-substituted phenyl)imidazolyl radicals[89] has been described; the 1-α-cumyl-4-methylurazolyl radical appears to be indefinitely stable in inert solvents.[89a]

Syntheses of several new verdazyls and bisverdazyls have been described;[90,91] the extent of interaction between the two unpaired electrons in the bisverdazyls decreases with increasing separation of the radical centres.[91] Both γ- and δ-splittings have been observed in alkyl substituted verdazyls as a result of W-plan long-range interactions.[92]

The planar nature of the iminyl radical $H_2C{=}N\cdot$ has been confirmed.[93] Its ESR spectrum is characterized by a large hyperconjugative coupling, suggesting that the radical should be relatively stable. The 1,1′-*gem*-biadamantyl grouping effectively shields the radical centre in the $(1\text{-Ad})_2C{=}N\cdot$ radical, thus accounting for its high stability.[94]

Miscellaneous Radicals[95]

MO calculations have indicated that the degree of non-planarity in $\cdot AX_3$ radicals increases with increasing electronegativity difference between A and X, i.e. in the order $\cdot AH_3 < \cdot ACl_3 < \cdot AF_3$ and $\cdot CX_3 < PX_3^{+\cdot} < \cdot SiX_3$.[96] In particular, $\cdot SiH_3$ is essentially tetrahedral, as also are $\cdot SiH_2F$, $\cdot SiHF_2$ and $\cdot SiF_3$. When X is more electronegative than A, the A—X bonds have less *s*-character and the half-filled orbital has more *s*-character than the 25% for a tetrahedral bond orbital; this leads to increased non-planarity. The ESR spectrum of (**36**) is characterized by a small β-hydrogen coupling constant which is ascribed to the pyramidal nature of the bridgehead silyl radical site.[97]

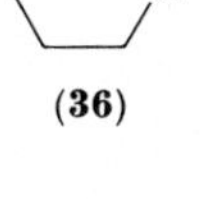

Si·

(36)

The structures and electronic states of the methoxyl and ethoxyl radicals have been determined.[98,99] The $CH_3O\cdot$ radical has geometry very similar to that of methanol.

A review of the structure and behaviour of the stable 2,4,6-tri-*tert*-butylphenoxy-radical has appeared.[100] ESR studies reveal strong intramolecular hydrogen bonding in 2-hydroxy- and 2,3-dihydroxy-phenoxy-radicals, involving the radical centre.[101] The coupling constants in 1- and 2-naphthyloxy-[102] and in *o*-, *m*- and *p*-substituted phenoxy-radicals[103] have been assigned. ESR studies reveal that tropolonyl radicals (**37**) have a spin-density distribution closely resembling that of phenoxy-radicals.[104, 105] MO calculations have been carried out on galvinoxyl to explain its reactivity.[106]

O· O

(**37**)

The ESR spectrum of a ^{17}O-labelled 1-tetralylperoxy-radical indicates that the spin densities on O-1 and O-2 are 0.62 and 0.41, respectively.[107] The *g*-factor for the *tert*-butylsulphinyl radical, $Bu^tSO\cdot$, is smaller than that for the isoelectronic $Bu^tOO\cdot$[108] as a consequence of different geometries; the former is also shorter-lived.

Alkylthiyl radicals RS· are generally not observable by ESR spectroscopy and instead $RS\dot{-}SR_2$ radicals are seen.[109] These are essentially σ^* radicals. Similarly irradiation of trimethyl phosphite gives the σ^* species $[(MeO)_3P{-}P(OMe)_3]^{\dot{+}}$;[110] the unpaired electron occupies a σ^* orbital built from 3*s* and 3*p* orbitals of two phosphorus atoms. Very little delocalization of spin on to phosphorus is observed in $(HO)_2P(O)CH_2\cdot$.[111] There appears to be little delocalization of unpaired spin on to the phenyl groups in $Ph_2P\cdot$[112] and $Ph_4As\cdot$[113] though this is not the case for the radical anions $Ph_3P^{\dot{-}}$, $Ph_3PO^{\dot{-}}$ and $Ph_4Si^{\dot{-}}$.[114] It is estimated for $(EtO)\dot{P}O_2^-$ that 70% of the unpaired spin is in an sp^3-orbital centred on phosphorus.[115]

The 1,1′-*gem*-biadamantyl group is very effective in stabilizing the iminoxy-radical $(1\text{-}Ad)_2C{=}NO\cdot$ because of its bulk.[94] The ESR spectra of 2- and 4-alkyl- and *cis*-3,5-dimethyl-cyclohexanone iminoxy-radicals indicate that there is a strong conformational preference for the alkyl groups to occupy an equatorial position whereas the methyl groups in the *cis*-2,6-dimethylcyclohexanone iminoxy-radical assume the preferred conformation in which the methyl groups are axial to avoid *cis*-methyl–oxygen interaction.[115] The bicyclo[3.3.1]nonan-9-one iminoxy-radical adopts a chair–chair conformation (**38**), as do its *endo*-2-, *endo,endo*-2,4-, and *endo,exo*-2,4-derivatives.[116]

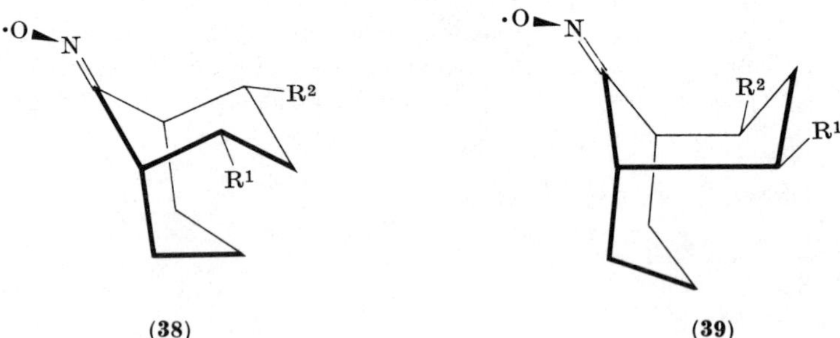

(**38**) (**39**)

Introduction of a bulky *exo*-substituent at the 2-position, however, forces the radical into a chair–boat conformation (**39**). Long-range W-plan couplings in these and other iminoxy-radicals derived from bicyclic ketones are observed.[116, 117] Both isomeric α-sulphonyl iminoxy-radicals Ar(Ar′SO_2)C=NO• are generated by oxidation of the oxime, implying their ready interconversion.[118]

Radical Cations

The ESR spectra of tetra-alkylhydrazine radical cations indicate that they are non-planar with deviation from planarity in an eclipsed sense (**40**).[119]

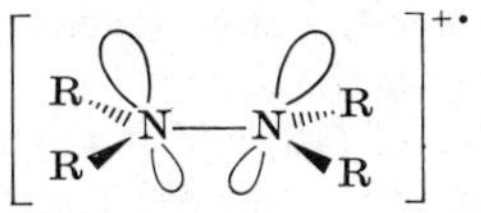

(**40**) 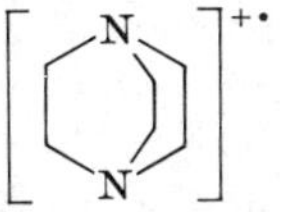 (**41**) 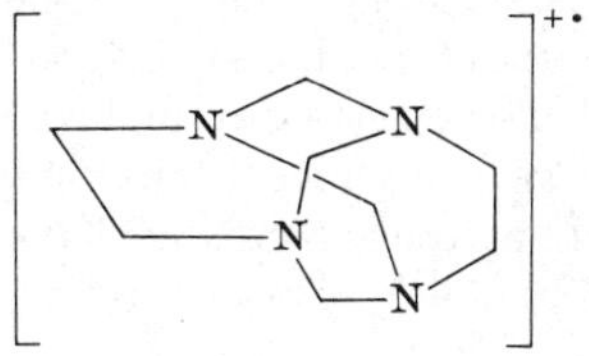 (**42**)

The stabilities of the radical cations (**41**) and (**42**) are much greater than those of the tetra-aza-adamantyl radical cation or other tertiary amine radical cations.[120] This is attributed to the favourable geometry in (**41**$^{+\bullet}$) and (**42**$^{+\bullet}$) for n–σ_{CC} interaction which is not present in the tetra-aza-adamantane radical cation.

Biphenylene radical cations are sufficiently stable to allow their isolation as stable salts.[121] The ESR spectra have been recorded and the spin densities determined of *ortho*-Wurster-type salts,[122] and of the radical cations of 2,7- and 2,8-dihalophenoxanthiin, dibenzo-*p*-dioxin,[123] 1,5-dihydronaphthyridine, di-*N*-methyl-1,5-naphthyridine,[124] tetra-alkylporphyrins,[125] bis(phenazine di-*N*-oxide,[126] indole[127] and pleiadiene.[128] The electronic structure of the 1,5-cyclo-octadiyne radical cation has been calculated.[129]

Radical Anions

Ring strain has a pronounced influence on the spin-density distribution in benzocycloalkenes, the spin density at the α-carbon atom increasing with increased strain due to rehybridization of the orbitals at the ring junction and consequent increase in *s*-character at the α-carbon.[130,131] Ring strain effects are also important in determining the spin-density distribution in the benzocyclobutadienoquinone radical anion.[132] The ESR spectra of the radical anions of benzocyclobutene and indene indicate that there is strong 1,3-*p*–σ overlap, i.e. homohyperconjugation.[133,134] The ESR spectra have been reported, and the hyperfine splitting constants calculated, for the radical anions of azuleno[5,6,7-*cd*]phenalene,[135] xanthene dyes,[136] tetramethoxybiphenyls,[137] spirobifluorene,[138] diphenylmethane,[139] triphenylmethane,[140] *p*-dibenzylbenzene[141] and *p*-di-isopropylbenzene.[142] In the last case the rotational barrier for the isopropyl group was also calculated. Spin-density distributions have been obtained for the phosphabenzene[143] and 2-phosphanaphthalene[144] radical anions. The hyperfine coupling constants of the radical anions of (trimethylsilyl)benzenes indicate the importance of π-electron interaction between silicon and the aromatic system.[145,146] The unpaired electron is delocalized throughout the whole system in the radical anions of 9-sila- and 9,10-disila-9,10-dihydroanthracenes.[147] The ESR spectra of trimethyl- and triphenyllead halides have been reported.[148] The unpaired electron in the 4-nitro-*cis*-stilbene is associated exclusively with the *p*-nitrophenyl moiety.[149] The spin densities in nitroaromatic radical anions are solvent-dependent as a result of hydrogen-bonding effects.[150,151] The radical anion of cyclo-octatrienyne has been observed at −100°.[152] The radical anions of butadiene, isoprene and 2,3-dimethylbutadiene appear to be predominately in the *trans*-configuration.[153] Coupling with the protons of the trimethylsilyl group is observed in the radical anions of 1-phenyl-2-(trimethylsilyl)acetylene and 1,4-bis(trimethylsilyl)buta-1,3-diyne but not with the *tert*-butyl protons of the 1-*tert*-butyl-2-phenylacetylene and 1,4-di-*tert*-butylbuta-1,3-diyne.[154,155] Transmission of spin through silicon to the methyl groups occurs via d_π–p_π interaction. The rate of dimerization of the 1,4-di-*tert*-butylbuta-1,3-diyne radical anion is greater than that of 1,4-bis(trimethylsilyl)buta-1,3-diyne radical anion as in the latter the spin density on the carbon atoms is lower than that on the central carbon atoms of the former.[155]

Long-range couplings have been observed in the ESR spectra of (**43**) and (**44**) with H^a and Me^a, respectively, as a result of homohyperconjugative interaction.[156] There is

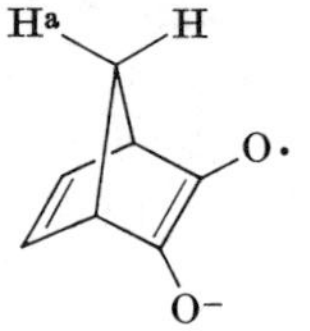

(**43**) 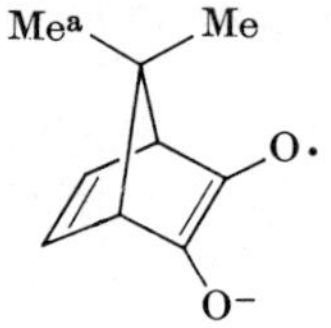 (**44**) 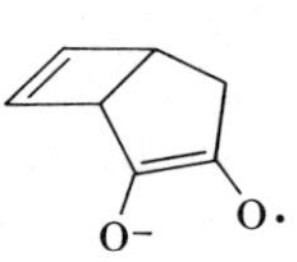 (**45**)

little homoallylic delocalization of electron spin between the two π-systems of (**45**).[157] The electronic absorption spectrum of the benzil radical anion indicates that it adopts the *trans*-configuration.[158]

The di-2-thienyl ketone radical anion exists as a mixture of the *trans,trans*- (**46**) and the *cis-trans*-isomer (**47**). None of the *cis-cis*-isomer (**48**) could be detected, presumably owing to steric hindrance between H-3 and H-3′. The barrier to rotation in ketyl radical

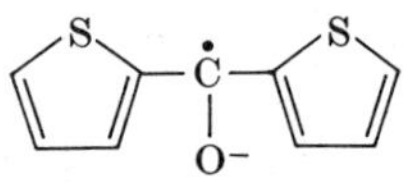

(**46**) 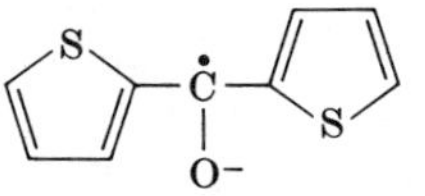 (**47**) 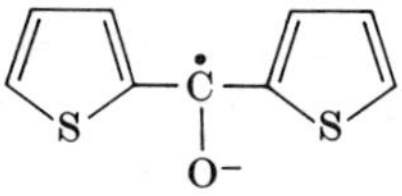 (**48**)

anions, $Ar_2\dot{C}{-}O^-$, increases in the order: phenyl < 2-thienyl < 2-thieno[2,3-*b*]thienyl < 2-thieno[3,2-*b*]thienyl, reflecting the double-bond character of the bond between the aromatic portion and the ketyl group.

The ESR spectra have been recorded, and the spin densities calculated, for the radical anions of a range of tricyclic ketones including anthrones,[160] cyclopentathiophenones[161] and dibenzo[*b*,*e*]thiepinones.[162] Similar studies have been carried out on the radical anion of Michler's ketone,[163] bis(tetracyclone) ketyl biradicals,[164] indane-1,3-diones,[165] and imidazobenzoquinones.[166]

Monothio-oxamide radical anions (**49**) are stable only if they can attain a planar structure.[167] The stabilities of dithio-oxamide radical anions (**50**) are likewise dependent on their attaining a planar structure as a result of increased resonance stabilization and intramolecular hydrogen-bonding.[168] The radical anion of $H_2NC(=S)C(=S)NMe_2$, which is non-planar, is unstable though anomalously the radical anion of $Me_2NC(=S)C(=S)NMe_2$ is stable.

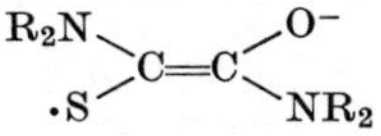

$R_2N(\cdot S)C{=}C(O^-)NR_2$

(49)

$R{-}NH\cdots S^-$ / $C{=}C$ / $\cdot S\cdots HN{-}R$

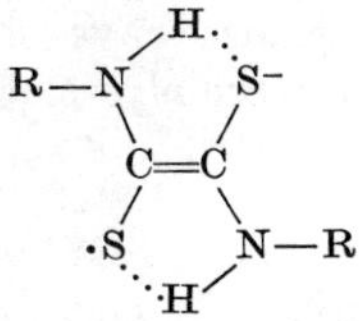

(50)

The radical anions of dihydroxybenzenes and 2- and 4-hydroxypyridines exist in the keto-forms whilst that of 3-hydroxypyridine exists in the enol form.[169] ESR studies of the radical dianions and trianions of galvinoxyl,[170] and of semiquinone radical anions,[171,172] have been reported, as have studies of monocyclic oxocarbon (e.g. squarate)[173] and ascorbic acid radical anions.[174]

There has been considerable interest in ion-pairing phenomena of a variety of radical anions including metal complexes of dithienobenzoquinones,[175] alkylbenzenes,[176] tetracyanoethylene,[177] fluorenone,[178] and perfluorobiacetyl.[179] The radical anion of sodium 9,10-anthraquinone-2-sulphonate is stabilized by binding with micelles.[180]

Azo-compounds

The question whether homolysis of azoalkanes proceeds by a one-bond or a two-bond process continues to attract attention.[181] CIDNP studies confirm that (phenylazo)-triphenylmethane decomposes by a two-step process:[182]

$$PhN{=}NCPh_3 \rightarrow PhN{=}N\cdot + Ph_3C\cdot$$
$$PhN{=}N\cdot \rightarrow Ph\cdot + N_2$$

The rates of decomposition of (arylazo)ethane-1,1-dinitriles, $ArN{=}NCMe(CN)_2$, are sensitive to the nature of the aryl group, being greater the higher the resonance energy of the aryl component.[183] Lack of coplanarity of the aromatic system and the azo-group induced by 2,6-disubstitution also reduces the stability of the azo-compound by reducing the C—N bond strength. These results appear to indicate that there is at least considerable stretching of the Ar—N bond in the decompositions of these azo compounds. Decomposition of *Z*-alkylaryldiazo sulphides (**51**) proceeds as shown in the annexed scheme:[184]

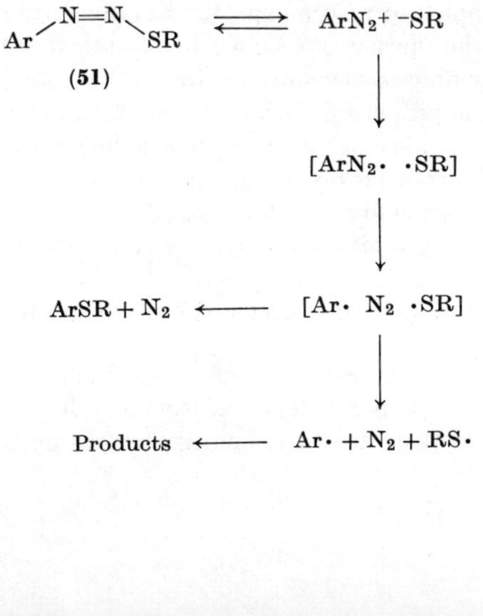

$$Ar{-}N{=}N{-}SR \ \textbf{(51)} \rightleftharpoons ArN_2^+\ {}^-SR$$
$$\downarrow$$
$$[ArN_2\cdot\ \ \cdot SR]$$
$$\downarrow$$
$$ArSR + N_2 \longleftarrow [Ar\cdot\ \ N_2\ \ \cdot SR]$$
$$\downarrow$$
$$\text{Products} \longleftarrow Ar\cdot + N_2 + RS\cdot$$

Spin-trapping studies have shown that both aryl and arylsulphonyl radicals are generated in the thermolyses of arylazo-*p*-tolyl sulphones.[185]

$$\text{ArN=NSO}_2\text{Ar}' \rightarrow \text{Ar}\cdot + \text{N}_2 + \text{Ar}'\text{SO}_2\cdot$$

Thermolysis (or photolysis) of (phenylazo)triphenylsilane appears to afford a convenient route to triphenylsilyl radicals;[186,187] it decomposes at somewhat higher temperatures than does (phenylazo)triphenylmethane.

The unsymmetrical azoalkane, $Me_2C(CN)N{=}NC(OMe)Me_2$, decomposes faster than either of its symmetrical analogues contrary to expectation. The enhanced rate of its decomposition argues against one-bond scission and also against symmetric two-bond scission, but rather for an asymmetric two-bond scission proceeding via the dipolar transition state (**52**).[188] Photolysis of (trifluoromethylazo)methane gives, in addition to

$$\text{Me}-\overset{\overset{\text{CN}}{|}}{\underset{\underset{\text{Me}}{|}}{\text{C}}}{}^{\delta-}\text{---N}\equiv\text{N---}{}^{\delta+}\overset{\overset{\text{OMe}}{|}}{\underset{\underset{\text{Me}}{|}}{\text{C}}}-\text{Me}$$

(**52**)

C_2H_6, $CH_3{-}CF_3$ and C_2F_6 derived from combination of $CH_3\cdot$ and $CF_3\cdot$ radicals; hydrazines result from addition of $CF_3\cdot$, and to a lesser extent of $CH_3\cdot$, radicals to the azo-compound.[189] The hydrazines constitute more than 50% of the products at room

$$\text{CF}_3\dot{\text{N}}\text{N(CH}_3\text{)CF}_3 \longleftarrow \text{CF}_3\cdot + \text{CF}_3\text{N=NCH}_3 \longrightarrow (\text{CF}_3)_2\text{N}\dot{\text{N}}\text{CH}_3$$

$$\text{CF}_3\dot{\text{N}}\text{N(CH}_3\text{)CF}_3 \xrightarrow{\text{CH}_3\cdot} \text{CF}_3(\text{CH}_3)\text{N—N(CH}_3)\text{CF}_3$$

$$\text{CF}_3\dot{\text{N}}\text{N(CH}_3\text{)CF}_3 \xrightarrow{\text{CF}_3\cdot} (\text{CF}_3)_2\text{NN(CH}_3)\text{CF}_3 \xleftarrow{\text{CF}_3\cdot} (\text{CF}_3)_2\text{N}\dot{\text{N}}\text{CH}_3$$

$$(\text{CF}_3)_2\text{N}\dot{\text{N}}\text{CH}_3 \xrightarrow{\text{CH}_3\cdot} (\text{CF}_3)_2\text{NN(CH}_3)_2$$

temperature. A similar process occurs in the thermolysis of 1,1′-azobisformamide, $H_2NCON{=}NCONH_2$, when $\cdot CONH_2$ radicals add to the parent azo-compound.[190] Photolysis of this azo-compound does not lead to radicals. Thermolysis of azobistrimethylsilane (**53**) does not lead to the expected hexamethyldisilane but rather to the products (**54**)–(**60**) by the mechanism shown in the scheme.[191–193] Labelling studies show that a bimolecular process accounts for the formation of (**54**) and most of (**55**).

The values of the enthalpies of formation of a series of azo-compounds suggest that thermolysis of azoalkanes is a less exothermic process than was previously considered.[194] The first instance of chain transfer involving AIBN has been shown to occur during the AIBN-initiated polymerization of methyl methacrylate.[195]

The rates of thermal decomposition of bicyclic compounds (**61**) increase considerably with increasing dihedral angle between the plane of the cyclopropane ring and the rest of the molecule.[196] Photoelectron spectroscopy offers an explanation for these observations.

The rates of decomposition of 1,4-diaryl-1,4-di-*tert*-butyltetrazenes, $ArN(Bu^t)N{=}NN(Bu^t)Ar$, are much less sensitive to electronic effects in the aryl nucleus than are those of their dimethyl analogues $ArN(Me)N{=}NN(Me)Ar$.[82] This is

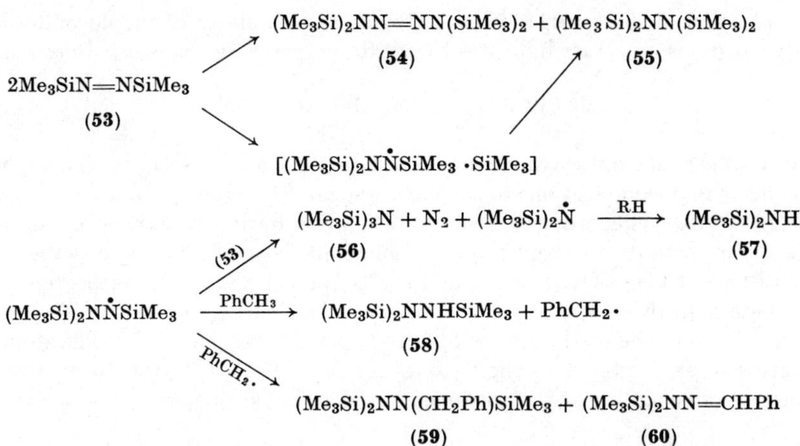

attributed to steric hindrance in the *tert*-butyl compounds, forcing them into the conformation (**62**) which is favourable for decomposition since in this conformation there is maximum stabilization of the unpaired electron forming at the anilino-nitrogen and the aryl group. In the methyl compounds, the ground-state conformation is such

$(CH_2)_n$ N N

(61)

$N{=}\!)_2$ N Bu^t

(62)

that the lone-pair on nitrogen is stabilized by the aromatic ring, and before decomposition can occur this has to undergo conformational change. *N*-Ethylanilino-radicals have been trapped during the photolysis of 1,4-diethyl-1,4-diphenyltetrazene which affords *N*,*N'*-diethylhydrazobenzene as well as products of the photobenzidine rearrangements of this.[196a]

Peroxides

Activation volumes for thermolyses of peresters have been used to assess whether they are undergoing one-bond or two-bond scission.[197] A new method for determining activation volumes of peresters RCO_3Bu^t based on empirical internal solvent pressures supports previous conclusions by suggesting that when R = Ph one-bond scission occurs, whereas when R = Me, Pr^i or $PhCH_2$ two-bond cleavage is favoured.[198] Another study also favours two-bond cleavage for peresters when R = Pr^i or Bu^t;[199] the particularly low activation volumes obtained were taken to indicate a significant polar contribution to the transition state (**63**).

$$\overset{\delta+}{R}\text{---}\overset{\overset{\displaystyle O}{\|}}{C}\text{=\!=\!=}O\text{---}\overset{\delta-}{O}Bu^t$$

(63)

An evaluation of the isotope effects obtained from a study of the decomposition of *tert*-butyl peroxyisobutyrate indicates that it decomposes by a concerted process:[200]

$$Me_2CHCO_3Bu^t \rightarrow [Me_2CH\cdot\ CO_2\ \cdot OBu^t]$$

Isolation of product and scavenger methods have been used to examine the importance of cage effects in decompositions of several peresters RCO_3Bu^t in chlorobenzene and in cumene.[201–203] Cage effects are at least 30% when R is a primary alkyl group, decreasing to 0–12% for tertiary alkyl groups.[201] Significant cage effects are observed in the decompositions of $Ph_3CCO_3Bu^t$ and $Ph_2Bu^tCCO_3Bu^t$ (11 and 4–5%, respectively) even though these radicals abstract hydrogen only slowly.[203] Disproportionation also occurs within the solvent cage to the extent of 5–10% per β-hydrogen of R. "Phantom" cage effects are also encountered in the decomposition of $Ph_2CHCO_3Bu^t$ to give products (**64**) arising from coupling at the *para*-position of a phenyl ring.[202]

(**64**)

A study of the thermolysis of (*S*)-(+)-*tert*-butyl 2-methylperoxybutyrate indicates that the ratio of the rate of tumbling of 2-butyl radicals (k_p) to the rate of their combination with *tert*-butoxy-radicals is much lower than that reported for the fluorenyl–diazenyl radical pair.[204, 205] This is attributed to the intervention of a molecule of carbon dioxide between the pair of radicals.

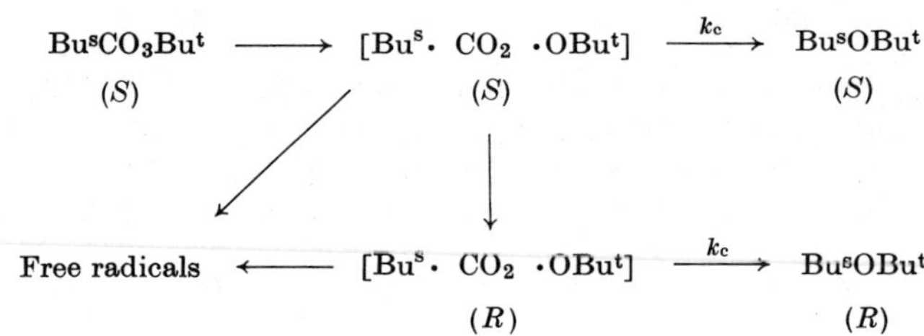

There is some retention of configuration in Bu^sOBu^t obtained from the thermal decomposition of $Bu^tCON(NO)OBu^t$, suggesting that a 2-butyl–*tert*-butoxy-radical pair is formed initially.[206]

Polar effects are important in the decomposition of unbranched aliphatic peroxides and peresters as indicated from a correlation with Taft's σ^* values.[207] With α- and β-substituents, steric effects are also shown to be important. A good Hammett correlation is obtained for the decompositions of substituted β-phenylisovaleryl *p*-nitrobenzoyl peroxides ($\rho = -0.76$), reflecting the importance of a polar transition state.[208]

From kinetic data based on the rates of thermolysis of $CH_2{=}CHCMe_2CO_3Bu^t$ and $HC{\equiv}CCMe_2CO_3Bu^t$ it can be concluded that propargylic radicals are slightly less stable than the allylic radicals.[209] This is attributed to the electron-withdrawing nature of the triple bond, which inhibits development of positive charge in the polar transition state.

Neighbouring-group effects result in an enhanced rate of decomposition of (**65**; X = SPh) relative to (**68**).[210] Decomposition of (**65**) leads to (**67**) via the transition state (**66**). The negative entropy of activation reflects a very constrained transition state and is not compatible with steric acceleration as the correct explanation for the faster decomposition. Similar effects are observed for (**65**; X = I).

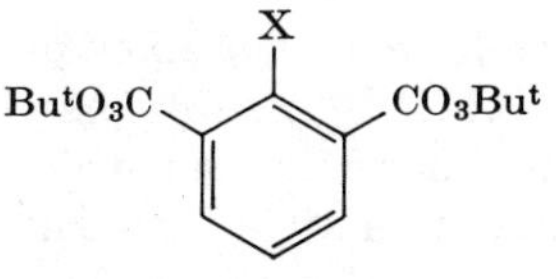

(**65**) 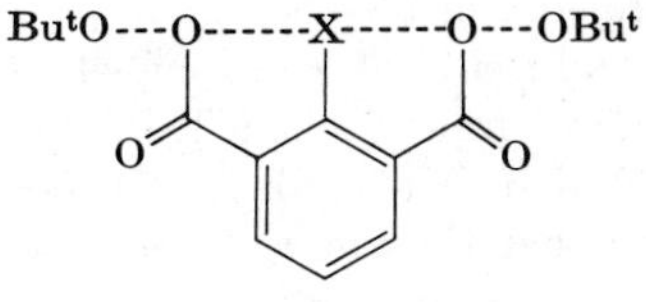 (**66**)

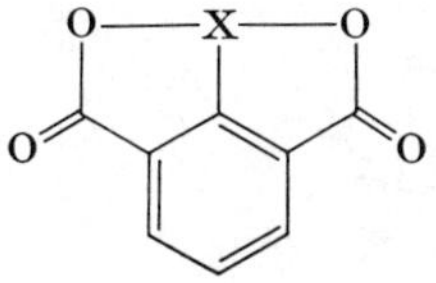

(**67**) 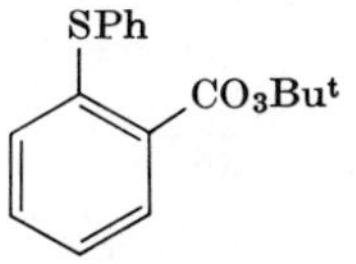 (**68**)

The result of thermolysis of the β-azodiacyl peroxide (**69**) can be interpreted in terms of a two-stage process involving initially O—O bond cleavage followed by C—N bond fission.[211]

$$[Me_2C(CN)N{=}NCMe_2CH_2CO_2]_2 \rightarrow 2Me_2C(CN)N{=}NCMe_2CH_2\cdot + 2CO_2$$
(**69**)
$$Me_2C(CN)N{=}NCMe_2CH_2\cdot \rightarrow Me_2\dot{C}CN + N_2 + Me_2C{=}CH_2$$

Thermolysis of (**70**) proceeds via homolysis of the perester O—O bond to give acetoxyacyloxy-radicals which lose carbon dioxide to form acetoxyalkyl radical;[212] carboxylic acids were also formed, and it was suggested that the acetoxyacyloxy-radicals might abstract hydrogen before decarboxylation occurs.

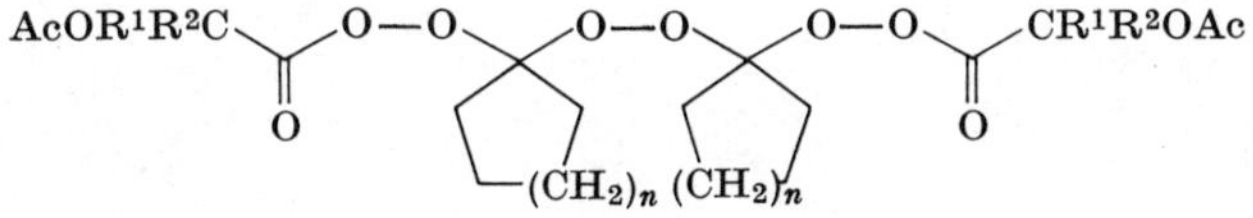

(**70**)

Studies of CIDNP effects on the cage products derived from the decomposition of diacyl peroxides and of peroxycarbonates,[213] and on out-of-cage products from di-*tert*-butyl peroxyoxalate have been carried out.[214]

Photolysis of dibenzoyl peroxide either directly or with singlet sensitizers gives phenyl benzoate in greater than 10% yield, whereas triplet sensitizers give very low yields of phenyl benzoate (~3%), suggesting that phenyl benzoate is derived from cage recombination of phenyl and benzoyloxy-radicals.[215] The benzoyloxy- and phenyl radicals generated in the thermolysis of dibenzoyl peroxide can be trapped by nitrogen dioxide to afford benzoyl nitrate and nitrobenzene, respectively. Benzoyl nitrate cannot be isolated but its presence is indicated by isolation of 2,5-dimethylnitrobenzene from reactions in the presence of *p*-xylene.[216] Aroyl nitrates also undergo thermolysis to afford aroyloxy radicals:[217] $ArCO{-}ONO_2 \rightarrow ArCO_2\cdot + NO_2$.

Radical decompositions of dibenzoyl peroxide in the presence of triethylamine[218] and of triphenylverdazyl[219] have been examined. The rates of decomposition of bis(pentafluorobenzoyl) peroxide are faster than those of dibenzoyl peroxide in both benzene and hexafluorobenzene.[220]

The importance of polar effects in the decomposition of *tert*-alkyl peroxides is demonstrated from a good Taft correlation ($\rho^* = -0.131$).[221] The decompositions of $Bu^tOOCMe_2CH_2OH$ and Bu^tOOCMe_2COOH were, however, anomalously fast, suggesting that they may decompose via a cyclic transition state (**71**). The quantum

O=CMe₂ / Bu^t—O X / H---O

(71)

yield for photolysis of di-*tert*-butyl peroxide was 1.9 ± 0.1; no evidence for quenching of the excited peroxide could be detected.[222] Photolysis of the peroxide (**72**) afforded mainly (**74**) and (**75**), which were formed as shown in the accompanying scheme. The same peroxide in the presence of iron(II) sulphate gave (**76**) as the main product.[223] Although all these products are postulated as derived from the same biradical (**73**), no explanation can be given for why reaction with iron(II) sulphate should give different products. The decomposition of dicyclohexylidene peroxide has been further examined.[224]

Thermolysis of tetrahydro-2-methylfuryl hydroperoxide (**77**) and peroxide gave the keto-alcohol (**79**) as the main product, together with propyl acetate;[225] both products arise from fragmentation of the intermediate alkoxy-radical (**78**).

(**77**) → (**78**)

(**78**) → $MeCOCH_2CH_2CH_2O\cdot \longrightarrow MeCOCH_2CH_2CH_2OH$ (**79**)

(**78**) → $MeCOOCH_2CH_2CH_2\cdot \longrightarrow MeCOOCH_2CH_2CH_3$

Studies on the decompositions of *tert*-butyl hydroperoxide,[226] cyclohexyl hydroperoxide,[227] 1-methylcyclohexyl hydroperoxide[228,229] and dicyclohexyl peroxydicarbonates[230] have also been reported.

Isotope effects require that homolysis of the $C_\alpha{-}CO$ bond accompanies that of the O—O bond in the thermolysis of the γ-peroxylactone (**80**).[231] Cleavage of the $C_\beta{-}C_\gamma$ bond occurs subsequently, as formation of the ethylene is non-stereospecific. The oxetane (**81**) is also obtained as a minor product (6%).

(72) → (73) → → $HO_2C(CH_2)_4$–cyclohexyl–COOH (74)

→ → (75)

→ $HOOC(CH_2)_4CH{=}CH(CH_2)_4COOH$ (76)

(80) $\xrightarrow[-CO_2]{\Delta}$ (81) $-Me_2CO$ $-Me_2CO$

Combination and Disproportionation Reactions

A number of independent studies of the combination of simple carbon-centred radicals in solution have shown that the reaction is normally diffusion-controlled, the rate constant being of the order of 10^9 l mol^{-1} s^{-1} irrespective of the radical structure. This conclusion casts considerable doubt on the much lower rate constants that have been deduced for secondary and tertiary radicals, from indirect evidence obtained from gas-phase studies. Thus, a competitive method employing ESR detection showed that the rate constants for Et•, Pr^i•, Bu^t•, cyclohexyl and CCl_3• radical combination in solution were all essentially equal;[232] the rate constant for *tert*-butyl radical combination was found to be 11×10^9 l mol^{-1} s^{-1} in another solution-phase study; and azoisobutane was shown to be unsuitable as a photoinitiator because of interference from the thermally unstable *cis*-form.[233] Diffusion control of the reaction was established for *tert*-butyl combination and for *tert*-butyl plus pivaloyl combination; a plot of the reciprocal of the combination rate constant against solvent viscosity divided by temperature was found to be linear, in excellent agreement with theory.[234] The *n*-pentyl radical absorption spectrum has been observed, the combination rate constant being 5×10^9 l mol^{-1} s^{-1} in *n*-pentane and slightly less in water.[235] Cyclopentyl radicals produced in the pulse radiolysis of aqueous cyclopentane combine equally rapidly ($k_c = 1 \times 10^9$ l mol^{-1} s^{-1});[236] in the solid phase at 120 K the disproportionation–combination ratio for these radicals was 3.05.[237] The absorption spectra of cyclohexyl and cyclohexenyl radicals have also been recorded; their second-order termination rates were 1.2×10^9 and 1.6×10^9 l mol^{-1} s^{-1}, respectively.[238] Similar measurements on the absorption spectrum of trichloromethyl radicals produced by pulse radiolysis of aqueous carbon tetrachloride gave a termination rate constant[239] of 7.4×10^8 l mol^{-1} s^{-1}. A simple method, based on the Smoluchowski–Waite equation, has been proposed for distinguishing between reactions that are diffusion-controlled and those that are not; the method was applied to the decay of allyl radicals in irradiated polyethylene.[240] Cyclohexyl and other alkyl radicals were found, by the space intermittency method, to diffuse much more slowly than the parent hydrocarbons; benzyl is the only radical so far discovered having a diffusion coefficient equal to that of its parent.[241]

tert-Butoxy-radical combination was found to obey the Noyes equation in individual solvents, but differences from one solvent to another were ascribed to specific interactions of the radicals with the solvent;[242] the extent of cage reaction was insensitive to the bulk viscosity of solutions containing polymers, suggesting that the bulk viscosity gives no measure of the resistance to transmission of motion of the radicals or solvent molecules. Some retention of configuration was observed in 2-butyl *tert*-butyl ether formed by combination of 2-butyl–*tert*-butoxy-radical pairs generated from *S*(+)-*tert*-butyl 2-methylperoxybutyrate; scavengers such as DPPH reduced the yield but raised the optical purity of the product.[243] Phenoxy-radical dimerization in aqueous solution,[244] and 4-alkyl-2,6-di-*tert*-butylphenoxy-radical decay,[245] have also been investigated. The relative yields of alcohol and ketone formed from *sec*-alkylperoxy-radical termination vary markedly with temperature;[246] the radicals behave like *tert*-alkylperoxy-radicals at high temperatures, and for low temperatures a transition state (**82**) has been suggested for tetroxide decomposition as an alternative to the more usual Russell formulation (**83**) since at –90° much more ketone is produced than alcohol, and H_2O_2 is formed. Termination rate constants have been determined for peroxy-radicals derived from hexadecane and decalin.[247]

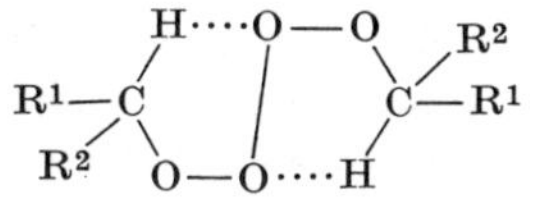

(**82**)

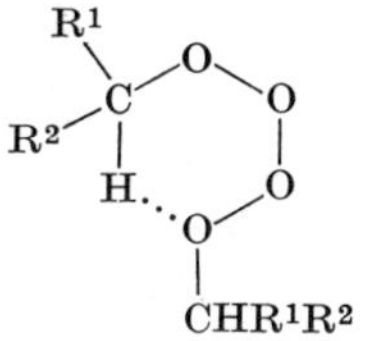

(**83**)

1,1-Dialkylhydrazyl radicals were found to decay rapidly in a second-order reaction producing the tetrazene, which subsequently decomposed to give nitrogen and an amine.[248] The rate of combination of iminyl radicals $R_2C{=}N\cdot$ was drastically reduced by steric effects when the radicals contained bulky substituents such as *tert*-butyl or 1-adamantyl.[249] Unusual activation parameters in the dimerization of *N*,3,5-tri-*tert*-butylanilino-radicals were attributed to severe conformational restraints in the reaction.[250] The rates of bimolecular self-reaction of planar phosphinyl radicals, tetrahedral phosphonyl radicals, trigonal-bipyramidal phosphoranyl radicals and spirophosphoranyl radicals were found to be close to the diffusion limit, except for phosphoranyl radicals with bulky substituents, which were prone to steric retardation.[251] The effect of substituents on the combination and disproportionation of diarylketyl radicals and radical ions has been investigated.[252]

A number of experimental, gas-phase investigations of combination rates of atoms (including the unexpectedly sluggish F atom[253]) and small radicals (such as hydroxy[254] and amino[255]) have been reported, together with a theoretical study employing MINDO/2.[256] A discordant note has been struck in the hitherto harmonious methyl radical story with the report of a combination rate constant, determined by kinetic spectroscopy on three different radical sources, about twice as high as the accepted value.[257] "Low" rate constants for ethyl combination ($10^{8.6}$ l mol^{-1} s^{-1}) and *tert*-butyl combination ($10^{6.6}$ l mol^{-1} s^{-1}) have received support from investigations of the pyrolysis of *n*-butane[258] and isobutane,[259] respectively. A "low" combination rate constant ($\leqslant 10^{7.0}$ l mol^{-1} s^{-1}) has also been reported for cyclohexyl radical combination.[260] The very-low-pressure pyrolysis technique (VLPP),[261,262] on the other hand, gives considerably higher rate constants, Et· $10^{10.4}$, Pr^i· $10^{9.5}$ and Bu^t· $10^{8.5}$ l mol^{-1} s^{-1}. An upward revision to $10^{9.9}$ for the $CCl_3\cdot$ combination rate constant has been suggested on the basis of competitive data, combined with thermochemical estimates for the reaction of trichloromethyl radicals with hydrogen chloride.[263] A shock-tube study of the pyrolysis of 2,4-dimethylhex-1-ene indicates a high rate constant ($10^{10.3}$ l mol^{-1} s^{-1}) for the combination of methyl and isobutenyl radicals.[264]

Cross-disproportionation of alkyl radicals in the condensed phase has been investigated by addition of hydrogen atoms to a mixture of alkenes, one alkene being in great excess so that one of the cross-disproportionation processes could be studied in virtual isolation: steric effects were identified as the most important governing factor.[265] A new determination of the disproportionation–combination ratio for cyclohexyl radicals puts the value at 0.99.[266]

Atom-abstraction Reactions

Hydrogen-atom Abstraction[267–270]

Abstraction and insertion reactions can be characterized as allowed or forbidden by spin and symmetry considerations;[271] free-radical reactions should occur readily when there is no spin degeneracy in the course of the reaction. Interest has again centred on determination of the rate constants of simple abstraction reactions over wide temperature ranges in order to check the linearity of the Arrhenius equation. The abstraction of hydrogen from ethane by hydrogen atoms was found to have a linear Arrhenius plot over 1000 K;[272] non-linear temperature-dependence was, however, found in the Arrhenius plot for the reaction of methyl radicals with molecular hydrogen.[273]

Trifluoromethyl radicals abstract hydrogen from the methoxy-group of methyl esters of perfluoroalkanoic acids two or three times less readily than from non-fluorinated methyl esters; this retardation was attributed to a repulsive polar effect induced by the fluorine substituents.[274] A study of the ratio of the rates of hydrogen abstraction from Group IV tetramethyls by trifluoromethyl and trideuteriomethyl radicals showed a small trend in the order Si > Sn > Ge > C; which follows the Allred–Rochow electronegativities of the atoms.[275] Arrhenius parameters were determined for hydrogen abstraction from C_4 to C_7 cycloalkanes by pentafluoroethyl radicals and used to determine D(C—H) in the cycloalkanes by means of the Evans–Polanyi equation; all the bond strengths were found to be close to the value for the secondary hydrogen atoms in propane.[276]

Deuterium primary kinetic isotope effects were determined for hydrogen abstraction by trichloromethyl radicals from the series of brominated methanes; the decrease in k_H/k_D with increasing bromine substitution was interpreted in terms of the increase in the polar effect.[277] Extensive exchange was observed in a mixture of $PhCD_3$ and $PhCH_3$ undergoing bromination with bromotrichloromethane; this is attributable to the reaction with hydrogen bromide.[278] The exchange was suppressed by adding ethylene oxide, and the Hammett correlation of the relative rate of hydrogen abstraction from substituted toluenes by trichloromethyl radicals then gave $\rho = -0.69$ for σ^+ constants. The relative rates of hydrogen abstraction from ring-substituted benzyl chlorides by trichloromethyl radicals also correlate with σ^+; the ρ-value of -1.02 is greater than with other α-substituted toluenes, probably because the α-chlorine has a destabilizing effect on the transition state.[279] Hydrogen is abstracted only from the benzylic position in 4-methyl-1-phenyl-4-(p-substituted phenyl)pentanes (**84**) by trichloromethyl radicals;[280] the ρ-values calculated for this system and for abstraction from 4-substituted 1-phenylbutanes were quite large considering that the substituents were well removed from the

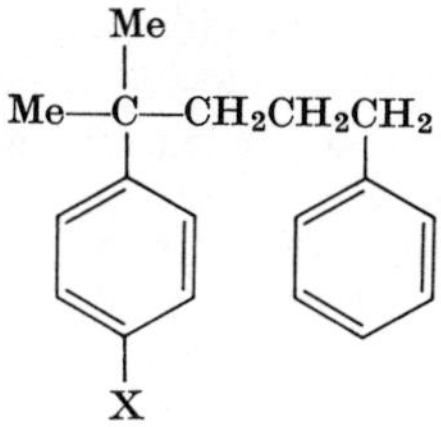

Me
Me—C—$CH_2CH_2CH_2$
X

(84)

reaction centre, and it was proposed that trichloromethyl radicals were complexed by electron-rich centres in the substrate molecules and then transferred to the site of hydrogen abstraction when the proper conformation was attained.

The activation energies for hydrogen abstraction from medium-ring cycloalkanes have been measured for a variety of radicals; the results were rationalized in terms of release or aggravation of conformational strain on formation of the sp^2-hybridized radical centre and, for C_8- to C_{12}-cycloalkanes, in terms of the inaccessibility of carbon—hydrogen bonds.[281] The ρ-value (+0.20) determined for abstraction from substituted toluenes by 2-phenylcyclopropyl radicals indicated little ionic character in the transition state.[282] From a comparison of the rates of hydrogen abstraction by aryl radicals from a series of compounds $CH_3{-}Z{-}R$ ($ZR = O^-$, OH, SMe, NMe_2 or OMe) it was concluded that structures such as $^{-}CH_2{-}\overset{+}{\ddot{Z}}{-}R$ make a significant contribution to resonance stabilization of the product radical and have a concomitant effect on the energy of the transition state.[283] The relative rates of hydrogen abstraction from the hydroxyl group of substituted phenols were shown to correlate with σ^+ constants and with the charge density $q_{c\text{-}1}$ at C-1. An even better correlation was obtained with $Aq_{c\text{-}1} + BE_n$ where E_n (the total energy divided by the number of valence electrons) and $q_{c\text{-}1}$ were calculated by extended Huckel methods.[284]

Interest in atomic fluorination has died down to a very low level, and published work includes only studies of thermal ^{18}F atoms with simple di- and tri-atomic species,[285] fluorination of difluoromethane[286] and some classical three-body trajectory computations.[287]

The main preoccupation in the field of chlorination has been the development of alternatives to molecular chlorine as chlorinating agents. Photo-oximation with nitrosyl chloride was shown to proceed by two mechanisms:[288]

$$NOCl + h\nu \rightarrow NOCl^*$$
$$NOCl^* + RH \rightarrow R\cdot + NO\cdot + HCl$$

and

$$NOCl^* \rightarrow NO\cdot + Cl\cdot$$
$$Cl\cdot + RH \rightarrow R\cdot + HCl$$

The relative importance of the two processes depended on the nature of the substrate and the wavelength of illumination. The main chain-carrying species in the halogenation of alkanes with thionyl chloride was identified as the chlorine atom.[289] Relative rate constants were measured for the addition and abstraction reactions of *N*-bromo- and *N*-chlorosuccinimide with $p\text{-}RC_6H_4(CH_2)_nCMe{=}CH_2$ having $n = 0$ or 1 and a range of substituents.[290] The conclusions were that the addition and abstraction reactions had a common intermediate, and that the transition state with *N*-chlorosuccinimide developed less positive charge than that with *N*-bromosuccinimide. The relative reactivities and selectivities observed in chlorination of alkanes and alkyl halides by trichloromelamine (**85**) suggest that chlorine atoms are the abstracting species in non-polar solvents; in trifluoroacetic acid the protonated amino-radical was the chain carrier.[291] *N*-Chlorosaccharin (**86**) was shown to be useful for allylic and benzylic chlorination, with properties similar to those of *N*-bromosuccinimide.[292]

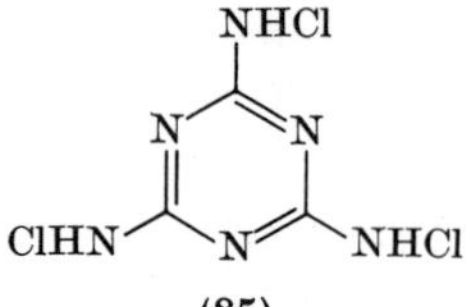

NHCl, ClHN, NHCl, N, N, N

(**85**)

OC—NCl, SO_2

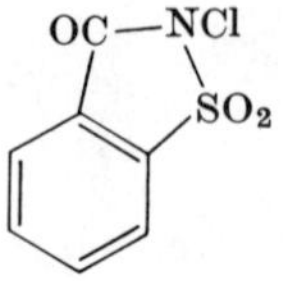

(**86**)

A study of the equilibrium set up on bromination of difluoromethane:

$$CH_2F_2 + Br_2 \rightleftharpoons HBr + CHF_2Br$$

approaching equilibrium from both sides, led to the value $D(CHF_2—Br) = 69\ kcal\ mol^{-1}$. In the analogous system with methyl fluoride it was discovered that the major product was methyl bromide formed by replacement of fluorine.[293] Bromination of methyl iodide produced methyl bromide, probably by a heterogeneous process.[294] The controversy over anchimeric assistance in radical reactions is by no means settled yet; rate enhancements in the radical bromination of alkyl bromides continue to be interpreted in these terms, but no anchimeric assistance was suggested for alkyl chlorides or fluorides.[295] The main product in the photobromination of (+)-2-methyl-1-butanenitrile was 2-bromo-3-methyl-1-butanenitrile, formed in an acid-catalysed ionic reaction; radical bromination giving mainly (±)-3-bromo-3-methyl-1-butanenitrile was, however, observed when a mixture of molecular bromine and *N*-bromosuccinimide was used.[296]

Evidence is accumulating that succinimidyl radicals may be chain carriers under certain conditions in brominations with *N*-bromosuccinimide. Competitive bromination of mixtures of alkanes show considerable rate differences when molecular bromine or *N*-bromosuccinimide is used, although there may be no change in the isomer distribution from a single alkane.[297] Bromination of cyclopropane with Br_2 gave 1,3-dibromopropane,

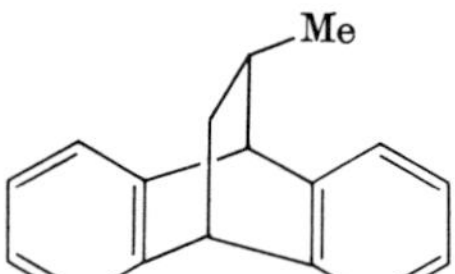

Me

(**87**)

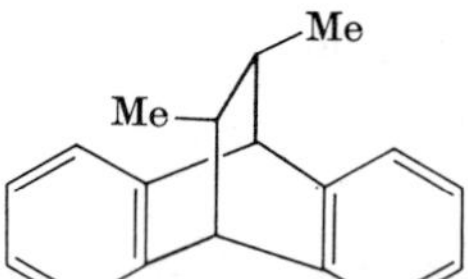

Me, Me

(**88**)

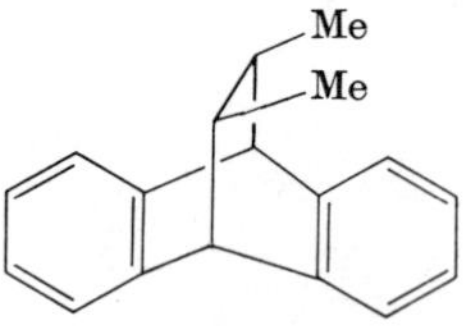

Me, Me

(**89**)

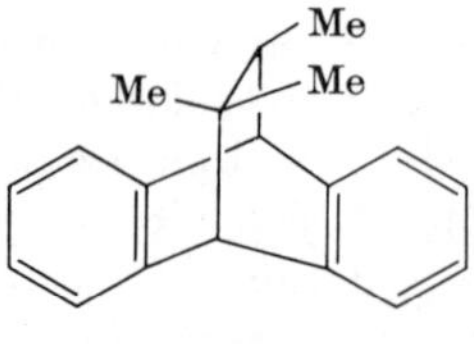

Me, Me, Me

(**90**)

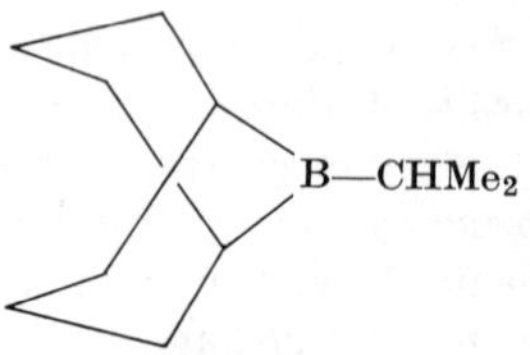

B—$CHMe_2$

(**91**)

whereas cyclopropyl bromide was obtained on use of *N*-bromosuccinimide; different mixtures of *exo*- and *endo*-2-bromides were also obtained from bromination of norbornane with the two reagents.[297] Succinimidyl radicals were also found to be the chain carriers in bromination of alkyl halides by *N*-bromosuccinimide in acetonitrile solution.[298]

The bridged polycyclic compounds (**87**) to (**90**) underwent bromination exclusively at the tertiary bridge positions; after correction for the number of reactive sites available, (**89**) and (**90**) were more than an order of magnitude more reactive than (**87**) and (**88**), indicating back-strain relief of the eclipsed methyl groups as an important factor.[299] The tertiary isopropyl-hydrogen in *B*-isopropyl-9-borabicyclo[3.3.1]nonane (**91**) was 5.5 times as reactive as the tertiary hydrogen in cumene, suggesting that boron confers exceptional stability on an adjacent radical centre by interaction of the unpaired electron with a vacant *p*-orbital of boron.[300] Two derivatives of 2-(bromomethyl)-thiophen, (**92**) and (**93**), are formed in the ratio 1.21:1 on bromination of 3-methoxy-carbonyl-2,5-dimethylthiophen with *N*-bromosuccinimide and this ratio is almost identical with the value found for the corresponding furan derivatives.[301] 3-Phenylindole

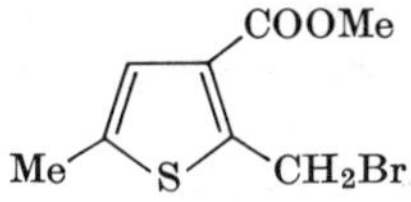

COOMe
Me S CH_2Br
(**92**)

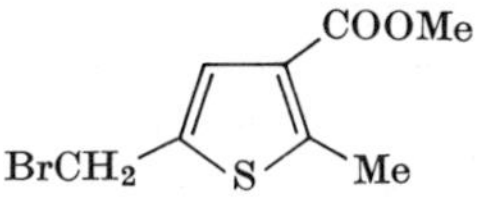

COOMe
$BrCH_2$ S Me
(**93**)

and 9-methyl-3-phenylindole are brominated at the 2-position by *N*-bromosuccinimide, but the *N*-acyl compound does not react.[302] From an investigation of the kinetics of the iodination of CF_3CH_2I, $D(CF_3CH_2{-}I)$ was determined to be 56.3 kcal mol^{-1}, indicating that β-fluorine atoms slightly strengthen the C—I bond.[303]

The relative selectivity of *tert*-butoxy-radicals, generated from *tert*-butyl hypochlorite, with primary and tertiary hydrogens has been reinvestigated, the work being extended to determining relative selectivities for attack at all the sites in 1-fluoro-, 1-chloro- and 1-cyano-butane; the results show that $Bu^tO\cdot$ radicals behave rather like chlorine atoms.[304] An ESR study showed that hydrogen is readily abstracted from the hydroxyl group in alcohols by alkoxy-radicals, *tert*-butoxy being more effective than pentyloxy- or trifluoromethoxy-radicals.[305] Abstraction of hydrogen from several alcohols by *tert*-butoxy-radicals indicated that both hydroxylic hydrogen and hydrogen from side-chain positions could be abstracted.[306,307] For example, with Bu^tCMe_2OH three radicals were detected: $Bu^t\cdot$ [from decomposition of $Bu^tCMe_2O\cdot$], $Bu^t(Me)(\dot{C}H_2)COH$ and $\cdot CH_2CMe_2CMe_2OH$. Kinetic studies indicated that abstraction of hydroxylic-hydrogen may not be important in solution because of hydrogen bonding. An absolute rate constant for the abstraction of hydrogen from toluene by *tert*-butoxy-radicals has been reported.[308] Arrhenius parameters have been determined for abstraction of hydrogen from phenols, aromatic amines and thiophenols by *tert*-butylperoxy- and 2-ethyl-2-propylperoxy-radicals; low activation parameters found for phenols were partly attributed to a mechanism involving equilibrium between the reactants and a hydrogen-bonded free-radical–reactant complex.[309] Low Arrhenius parameters were also determined for hydrogen abstraction from 4-substituted 2,6-di-*tert*-butylphenols by *tert*-butylperoxy-radicals. The deuteriated phenols reacted about an order of magnitude more slowly, again suggesting that the reaction proceeds via a hydrogen-bonded peroxy-radical–phenol complex.[310]

In the photolysis of hydrogen sulphide with Me_2SiD_2, molecular deuterium is formed in a chain reaction.[311]

The activation energies for hydrogen abstraction from (**94**) and (**95**) by 4-hydroxy-2,2,6,6-tetramethylpiperidin-1-oxyl depended on *n* and were correlated with *I*-strain.[312]

(**94**)

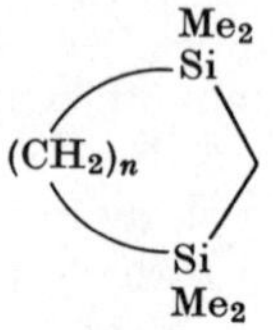

(**95**)

N-Methylacetamido-radicals were formed on flash photolysis of *N*-nitroso-*N*-methylacetamide; absolute rate constants were determined for some hydrogen abstraction reactions of this radical.[313] Acetaldehyde reacted with N_2O_5 to give peroxyacetyl nitrate, and it was demonstrated that propene is converted into acetaldehyde which then gives peroxyacetyl nitrate.[314] Gas-[315] and liquid-phase[316] nitration of methylcyclohexane by nitric acid proceeds by free-radical reactions involving $\cdot NO_2$ radicals to give a complex mixture of products including 1-methyl-1-nitrocyclohexane, adipic, succinic and glutaric acids and gaseous nitric oxide. The photo-excited states of pyrazine and quinoxaline abstract hydrogen from cyclohexane, and the cyclohexyl radicals produced can react with oxygen or dimerize and disproportionate.[317] Free-radical hydrogen atom transfer mechanisms have been suggested for the conversion of glycol into acetaldehyde and of propane-1,2-diol into propanol by the enzyme dioldehydrase.[318a]

The work of Breslow's group on remote functionalization of steroids has progressed towards real synthetic utility with, for example, the conversion shown being effected in ca. 65% yield.[318b] Further improvements in the form of a relay procedure using the *m*-iodobenzoate and iodobenzene dichloride[318c] have led to a partial synthesis of cortisone.[318d]

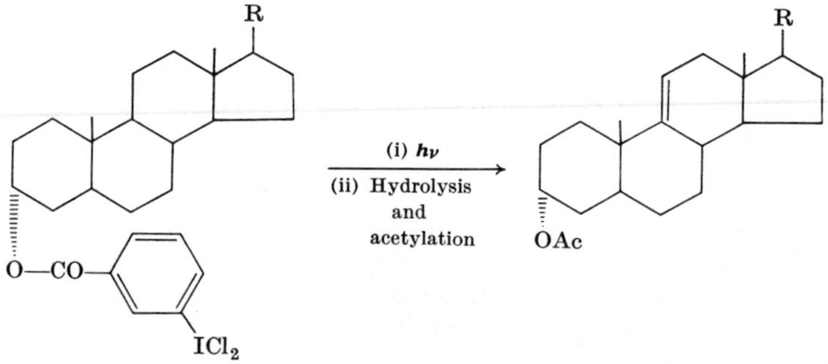

(i) $h\nu$
(ii) Hydrolysis and acetylation

Halogen-atom Abstraction

In the photolysis of liquid cyclohexane with carbon tetrachloride, cyclohexyl chloride is formed in a chain process, together with the products of combination reactions of cyclohexyl and trichloromethyl radicals.[319] Methyl acetate was reduced to ethyl methyl ether by chlorinated hydrosilanes on γ-irradiation in tetrachlorosilane;[320] the rate-determining step in the mechanism was hydrogen-abstraction from the hydrosilane:

$$\cdot SiCl_3 + RC(=O)OR' \longrightarrow R\dot{C}(OSiCl_3)OR' \xrightarrow{HSiCl_3} RCH(OSiCl_3)OR' + \cdot SiCl_3$$

$$RCH(OSiCl_3)OR' \xrightarrow{SiCl_4} RCHClOR' + Cl_3SiOSiCl_3$$

$$RCHClOR' + Cl_3SiOSiCl_3 \xrightarrow{\cdot SiCl_3} R\dot{C}HOR' + SiCl_4 \xrightarrow{HSiCl_3} RCH_2OR' + \cdot SiCl_3$$

Lactones were smoothly reduced to the corresponding ethers by photolysis in trichlorosilane; the lactone carbonyl was selectively reduced in lactone esters such as (**96**) probably because of steric factors:[321]

O, O, OCOMe, OCOMe $\xrightarrow[h\nu]{Cl_3SiH}$ O, OCOMe, OCOMe

(**96**)

The E-isomer of 1,2-dimethyl-1-silacyclobutane (**97**) gave E-1-chloro-1,2-dimethyl-1-silacyclobutane with 95% retention of configuration on treatment with benzoyl peroxide in carbon tetrachloride; similar retention was observed for the Z-isomer and attributed to the pyramidal structure of the silyl radical intermediate.[322] Halosilyl

H, Me, Si, H, Me $\xrightarrow[(BzO)_2]{CCl_4}$ H, Me, Si, Cl, Me

(**97**)

derivatives of transition metals, $L_nMSiRR'Cl$, can be synthesized on treatment of complexes $L_nMSiMe_xCl_{2-x}H$ [where $L_n \equiv Fe(CO)_2(\pi\text{-}C_5H_5)$ or $M(CO)_3(\pi\text{-}C_5H_5)$ where M = Cr, Mo or W] with CCl_4 or CBr_4 in the following free-radical process:[323]

$$CCl_3\cdot + L_nMMe_xCl_{x-2}H \rightarrow L_nM\dot{S}iMe_xCl_{x-2} + CCl_3H$$
$$L_nM\dot{S}iMe_xCl_{x-2} + CCl_4 \rightarrow L_nMSiMe_xCl_{x-1} + CCl_3\cdot$$

Triphenyl- and tri-n-butyl-tin radicals, generated from the corresponding hexaorganoditin, abstract halogen atoms from alkyl halides; the fate of the alkyl radical thus formed can be controlled by the type and concentration of hydrogen-donors present.[324] Tri-n-butyltin radicals abstract chlorine from benzyl and cyclohexyl chloroformate;[325] the resulting radical loses CO_2 in the benzyl case (route a), giving finally toluene, but in the cyclohexyl case the loss of CO_2 would give the much less stable cyclohexyl radical and so fragmentation does not occur (route b).

Complete loss of stereochemistry took place at the 2-position on reduction of 2-substituted 2-halonorbornanes by tri-n-butyltin hydride; the results showed that the presence of an electronegative substituent is not sufficient to permit the radical to

$$\mathrm{ROC(=O)Cl} \xrightarrow{\mathrm{Bu^n{}_3Sn\cdot}} \mathrm{ROC(=O)\cdot}$$

$$\mathrm{ROC(=O)\cdot} \xrightarrow{a} \mathrm{CO_2 + R\cdot} \xrightarrow{\mathrm{Bu^n{}_3SnH}} \mathrm{RH}$$

$$\mathrm{ROC(=O)\cdot} \xrightarrow[b]{\mathrm{Bu^n{}_3SnH}} \mathrm{ROC(=O)H}$$

retain configurational integrity.[326] Investigation of the stereochemistry of the reduction of 1,1-dialkyl-2,2-dibromo-3-methylenecyclopropanes (**98**; $R^1 \neq R^2$) with triorganotin hydrides showed that the major product was that methylenecyclopropyl bromide in which bromine was *cis* to the smaller alkyl group.[327]

R^2 R^1 Br Br

(**98**)

A free-radical mechanism (see Scheme) involving the borane radical anion has been proposed for the conversion of dibromocyclopropanes into monobromides by sodium borohydride:[328]

$$\mathrm{BH_4^- + X\cdot \longrightarrow \cdot BH_3^- + XH}$$

$$\mathrm{\cdot BH_3^-} + \text{7,7-dibromonorcarane} \xrightarrow{\mathrm{S}} \text{7-bromonorcaran-7-yl radical} + \mathrm{H_3B{-}S + Br^-}$$

$$\text{7-bromonorcaran-7-yl radical} + \mathrm{BH_4^-} \longrightarrow \mathrm{\cdot BH_3^-} + \text{7-bromonorcarane (Br, H)} + \text{7-bromonorcarane (H, Br)}$$

Additional evidence in support of a radical mechanism for the demercuration of alkylmercuric halides by sodium borohydride and lithium aluminium hydride comes from the reaction of hex-5-en-1-ylmercuric bromide which gives methylcyclopentane, as expected from a radical intermediate. Almost identical isotope effects were also observed in the reactions with $NaBD_4$ and $LiAlD_4$.[329]

The relative rates of iodine-abstraction from iodonaphthalenes, iodopyridines and iodothiophenes by phenyl radicals correlate with the π-electron density at the carbon atom bearing the iodine; high electron density retards the rate and vice versa.[330] The reactions of $\cdot OH$, $\cdot CH_2OH$, $\cdot CH_2CMe_2H$ and $\cdot CH_3$ radicals with iodine-containing compounds have been investigated by ESR and pulse radiolysis techniques;[331] the trends in the rates were interpreted in terms of the thermochemistry of iodine-abstraction modified by polar effects. The transition state for hydroxymethyl radicals abstracting iodine from iodoacetic acid was depicted thus:

$$HOCH_2^{+} \cdots\cdot \dot{I} \cdots\cdot {}^{-}CH_2COOH$$

α-Toluenesulphonyl iodide loses iodine on photolysis and the resulting radical splits off sulphur dioxide to give the more stable benzyl radical; this then abstracts iodine from the reactant, giving benzyl iodide in good yield.[332]

Addition Reactions

Carbon-centred Radicals

Addition of a radical to ethylene is about 11 kcal mol^{-1} more exothermic than addition to benzene; use of this difference enables a common pattern of reactivity to be established between addition to alkenes and homolytic aromatic substitution.[334] Absolute rate constants have been determined for addition of methyl and ethyl radicals to nitric oxide.[335] Trifluoromethyl radicals generated from trifluoroazomethane readily add to the azo-compound.[336] The critical energy for the addition of methyl radicals at the 2-position in buta-1,3-diene is higher than usual for a radical-addition reaction, probably because of the loss of the conjugation stabilization of butadiene.[337] The relative rates of addition of cyclohexyl radicals to specific sites in chloroethylenes have been determined; the trends in these additions were correlated with the ease of elimination of chlorine from the adduct radicals.[338]

In the photo-addition of bromodichloromethane to ethylene and vinyl fluoride in the gas phase, products from both $CHCl_2\cdot$ and $CCl_2Br\cdot$ radicals were identified; evidently hydrogen is abstracted from the halomethane at about the same rate as bromine.[339] The pattern of the relative rates of addition of bromodifluoromethyl radicals to specific sites in fluoroethylenes is similar to that of other haloalkyl radicals; the results can be correlated by steric and polar effects, but not by considering resonance stabilization in the adduct radicals.[340] Dichlorotris(triphenylphosphine)ruthenium(II) and dichlorotetrakis(triphenylphosphine)ruthenium(II) catalyse the addition of polychloromethanes to alk-1-enes; excellent yields of 1-to-1 adducts are obtained with little telomerization:[341]

$$\cdot CCl_3 + RCH{=}CH_2 \longrightarrow R\dot{C}HCH_2CCl_3$$

$$R\dot{C}HCH_2CCl_3 + {-}Ru{-}Cl \longrightarrow RCHClCH_2CCl_3 + {-}Ru{-}$$

Free-radical addition of carbon tetrachloride to methallyl chloride in the presence of $Fe(CO)_5$ gave nearly equal amounts of $CCl_3CH_2CMeClCH_2Cl$ and $CCl_3CH_2CMe{=}CH_2$.[341a]

Perfluoroalkyl radicals $R_F\cdot$ containing up to seven carbon atoms, generated by electrolysis of polyfluoroalkanoic acids in the presence of alkenes $CH_2{=}CHX$, reacted to give predominantly the dimeric product $R_FCH_2CHXCHXCH_2R_F$, together with other combination and addition products.[342]

A 75:25 mixture of *threo*- and *erythro*-dibromides was obtained on bromination of $CF_3CF{=}CHF$; trifluoromethyl radicals and bromine atoms (from HBr) add predominantly at the CHF site.[343] Methyl trifluorovinyl sulphide was obtained from the photochemical reaction of dimethyl disulphide with trifluoroiodoethylene.[344] The reaction of perfluorovinylsulphur pentafluoride (**99**) with a number of free-radical sources has been investigated;[345] butadiene gave a 1:1 mixture of *cis*- and *trans*-1,1,2-trifluoro-2-(pentafluorothio)-3-vinylcyclobutane (**100**), and trifluoromethyl iodide gave a low yield of the adduct $SF_5CFICF_2CF_3$ together with other products.

$$SF_5CF{=}CF_2 + CH_2{=}CHCH{=}CH_2 \xrightarrow{\Delta}$$

(**99**)

F, F_5S, F_2

(**100**)

trans-Adducts (**102**) were obtained from the addition of CCl_3Br, CCl_4 and PhSH to tricyclo[4.2.2.0^{2,5}]deca-3,7-diene-9,10-dicarboxylic anhydride (**101**).[346]

RX →

(**101**) (**102**)

α-Haloketones were induced to add to conjugated enynes and dienes to give coupled products by zinc in DMSO; addition occurred predominantly at the methylene terminus, and coupling took place preferentially at the less substituted radical centre,[347] as illustrated.

Free-radical telomerization reactions have been reviewed.[348] In the telomerizations of ethylene, propene and vinyl chloride by chloroform, carbon tetrachloride, or bromo-

$$PhCOCH_2Br + CH_2{=}CMeC{\equiv}CR \xrightarrow[DMSO]{Zn\text{–}Cu} PhCOCH_2CH_2MeC{=}C{=}CR\text{–}CR{=}C{=}CMeCH_2CH_2COPh$$

$$+\ PhCOCH_2CH_2MeC(C{\equiv}CR)\text{–}CR{=}C{=}CCH_2CH_2COPh + PhCOCH_2CH_2MeC(C{\equiv}CR)\text{–}CMe(C{\equiv}CR)CH_2CH_2COPh$$

$$PhCOCH_2Br + \text{(diene)} \xrightarrow[DMSO]{Zn\text{–}Cu}$$

$PhCOCH_2CH_2$ CH_2CH_2COPh

H H

(±) and *meso*

trichloromethane at high pressure, the chain carrier was the trichloromethyl radical, and nearly identical Arrhenius parameters were found for all three telogens.[349] The transfer efficiency of various alcohols in the telomerization of ethylene varied remarkably and could be correlated with the reactivity of the hydrogen at the carbon atom α to the hydroxyl group; the order of efficiency was found to be: propan-2-ol > butan-2-ol > propan-1-ol > ethanol > butan-1-ol > 2-methylpropan-1-ol > methanol.[350] Telomerization of ethylene with methyl acetate gave methyl esters of linear and α-branched alkanoic acids and linear and branched alkyl acetates together with hydrocarbons. Linear and α-branched methyl esters of α-cyanoalkanoic acids, but no alkyl cyanoacetates, were obtained from methyl cyanoacetate and ethylene. It was suggested that the branched telomers were formed by intramolecular 1,5-hydrogen transfer in telomer radicals with two ethylene units, followed by chain transfer.[351] Reports of the telomerizations of propene[352] and vinyl chloride[353] by methyl propanoate have appeared. The 1:1 adduct accounts for over 90% of the product in the telomerization of isobutene with carbon tetrachloride, and the adduct is readily converted into 3,3-dimethylacrylic acid by hydrolysis with sulphuric acid.[354] Series of dibromoperfluoroalkanes and iodoperfluoroalkanes were obtained in the gas-phase telomerization of tetrafluoroethylene with dibromodifluoromethane and trifluoroiodomethane, respectively; the transfer constants of the telomer radicals increased with chain length.[355]

On photolysis with propan-2-ol, the (5*H*)-furanone (**103**) gave a 90% yield of both stereoisomers of the adduct (**104**) but reaction was much slower with the α,β-unsaturated lactone (**105**) containing a six-membered ring, only 16% of the adduct being then obtained.[356]

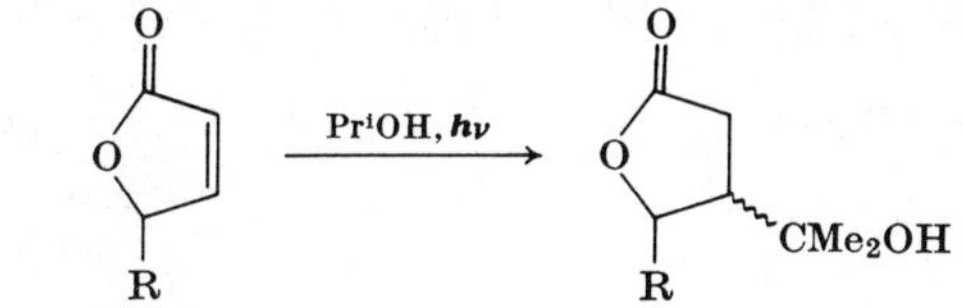

$Pr^iOH, h\nu$

CMe$_2$OH

(**103**) (**104**)

R=H or Me

$Pr^iOH, h\nu$

Me Me CMe$_2$OH

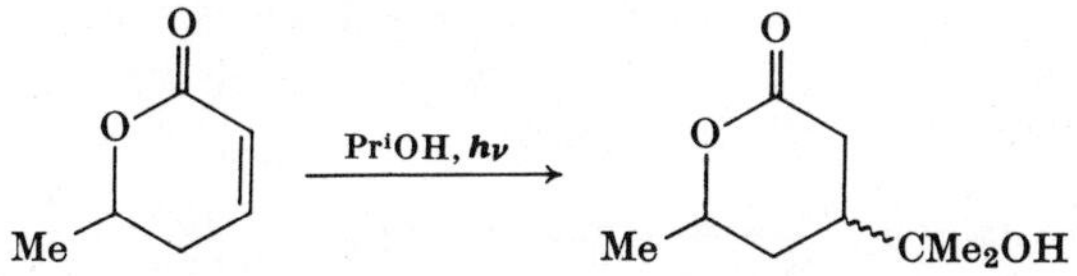

(**105**)

The addition of secondary alcohols to chlorotrifluoroethylene, induced by UV light or radiolysis gave 1:1 and 2:1 adducts $H(CFClCF_2)_nC(OH)R^1R^2$. Treatment of these adducts with aqueous sodium hydroxide gave the fluorinated oxetane (**106**), and the tetrahydropyran derivative (**107**), respectively.[357] Methanol adds to the hexafluoropropene dimer [a mixture of $(CF_3)_2CFCF{=}CFCF_3$ and $(CF_3)_2C{=}CFCF_2CF_3$] to give *erythro*- and *threo*-$(CF_3)_2CFCHFCF(CH_2OH)CF_3$ and $(CF_3)_2CHCF(CH_2OH)CF_2CF_3$ in the ratio 2:2:1; with acetaldehyde the analogous polyfluoroketones were formed.[358] Acetaldehyde adds to pent-1-yne, with peroxide or ultraviolet initiation, to give *trans*-

(106) (107)

hept-3-en-2-one, but this is largely converted into the 2:1 addition product, 3-*n*-propylhexane-2,5-dione (**108**).[359]

$$\mathrm{Me\dot{C}O + Pr^nC{\equiv}CH \longrightarrow Pr^n\dot{C}{=}CHCOMe \xrightarrow{RH} Pr^nCH{=}CHCOMe}$$

$$\xrightarrow{\mathrm{Me\dot{C}O}} \mathrm{MeOC{-}CH(Pr^n){-}\dot{C}H{-}COMe} \xrightarrow{RH} \mathrm{Pr^nCH(CH_2COMe)(COMe)}\ \textbf{(108)}$$

The photolytic addition of acetaldehyde to phenylacetylene gave *cis*- and *trans*-4-phenylbut-3-en-2-one, but with peroxide initiation the 2:1 adducts (**109**) and (**110**) were formed.[360] In the peroxide-induced addition of isobutanal to hex-1-yne the

$$\mathrm{PhCH(CH_2COMe)(COMe)}\ \textbf{(109)} \qquad \mathrm{PhCH_2CH(COMe)(COMe)}\ \textbf{(110)}$$

carbon—hydrogen bond adjacent to the carbonyl group was broken and a mixture of *cis*- and *trans*-β,γ-unsaturated aldehydes was obtained.[361]

Straight-chain aldehydes and β-substituted aldehydes reacted with di-*tert*-butyl peroxide to give low (<10%) yields of secondary alcohols by a mechanism involving addition to the carbonyl groups:

$$\mathrm{Bu^tO\cdot + RCHO \longrightarrow Bu^tOH + R\dot{C}O \longrightarrow R\cdot + CO}$$

$$\mathrm{R\cdot + RCHO \longrightarrow R_2CHO\cdot \xrightarrow{RCHO} R_2CHOH + R\dot{C}O}$$

$$\mathrm{R\cdot + RCHO \longrightarrow R\dot{C}O + RH}$$

α-Substituted aldehydes gave instead small amounts of dialdehydes and enol-ether aldehydes:[362]

$$\mathrm{R_2CHCHO \xrightarrow{R\cdot} [R_2\dot{C}CHO \longleftrightarrow R_2C{=}CHO\cdot] \longrightarrow (R_2C{-}CHO)_2 + R_2C(CHO){-}O{-}CH{=}CR_2}$$

A mechanism involving radical addition to the oxygen of the carbonyl group also accounts for the products observed in the oxidation of the steroid (**111**) with mercury(II) oxide and iodine:[363]

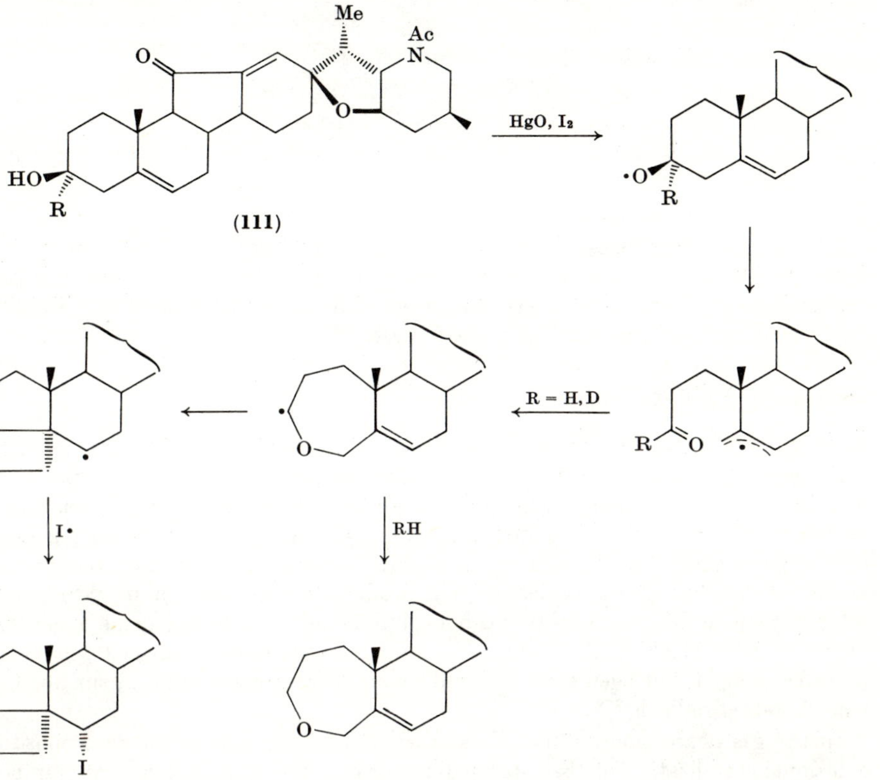

HgO, I_2

(**111**)

R = H, D

I•

RH

Anti-Markovnikov adducts in good yields, accompanied by telomers, were obtained in the argentic oxide-induced addition of acetone to alk-1-enes; yields were lower for internal and cyclic alkenes.[364] The product $HOCMe_2CF_2CHFCl$ obtained from the electrochemical addition of acetone to chlorotrifluoroethylene, could have arisen by a free-radical or a carbanion process.[365] 4-Methyl-4-alkanolides (**112**) can be prepared in good yield by the electrochemical hydrocoupling of alkan-2-ones with ethyl acrylate and acrylonitrile.[366]

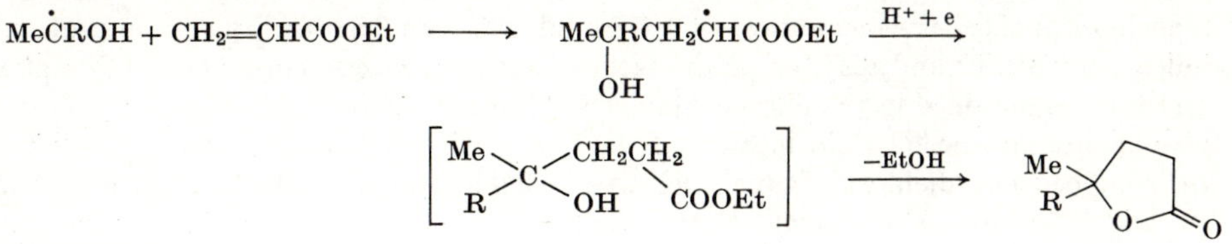

$$\mathrm{Me\dot{C}ROH + CH_2{=}CHCOOEt \longrightarrow MeC(OH)RCH_2\dot{C}HCOOEt \xrightarrow{H^+ + e}}$$

$$\left[\mathrm{Me(R)C(OH)CH_2CH_2COOEt}\right] \xrightarrow{-\mathrm{EtOH}}$$

(**112**)

The reactions of a series of dialkyl and cyclic ethers with methyl trifluoroacrylate have been investigated; cleavage of the C—H bond α to the ether-oxygen atom occurred, and when there were several non-equivalent bonds mixtures of isomeric adducts were obtained.[367] Low yields of the adducts (**113**) and (**114**) were isolated from the peroxide-induced addition of 2,2-dimethyl-1,3-dioxolane and 2,2-dimethyl-1,3-dioxane to hex-1-yne.[368]

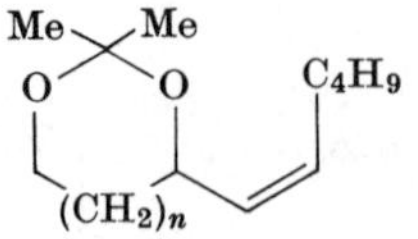

Me Me O O C_4H_9 $(CH_2)_n$

(**113**)

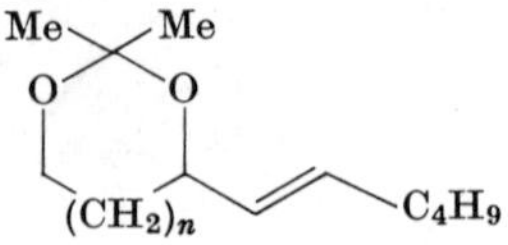

Me Me O O $(CH_2)_n$ C_4H_9

(**114**)

Carbon-centred radicals, formed on hydrogen abstraction by *tert*-butoxy-radicals from the position α to the oxygen in tetrahydrofuran, add to hex-1-yne and phenylacetylene to give a mixture of *cis*- and *trans*-adducts; with tetrahydropyran, adducts are formed from radicals α and β to the oxygen.[369]

Hydrogen and Halogen Atoms[370]

The rate of addition of a hydrogen atom to a fluoroalkene is slower than that to the corresponding hydrocarbon; the trends in the relative rates can be rationalized in terms of inductive and resonance effects.[371] Hydrogen atoms react predominantly with cyclohexa-1,3-diene by addition to the unsaturated system, but with cyclohexa-1,4-diene the major reaction is abstraction, probably because this produces the resonance-stabilized cyclohexadienyl radical.[372] The products of the addition of hydrogen atoms to pent-2-yne in the gas phase are propyne, but-1-yne and buta-1,2-diene, together with buta-1,3-diene which results from isomerization of the initially formed pent-2-en-2-yl radicals by a 1,4-hydrogen shift followed by *cis–trans*-isomerization of the homoallylic pent-2-en-5-yl radicals.[373]

In the gas-phase chlorination of chloroethylenes, the balance of decomposition to collisional stabilization of the intermediate activated chloroalkyl radical is important for *cis*- and *trans*-dichloroethylene and trichloroethylene; but, for tetrachloroethylene, pyrolysis of the intermediate radical is the important factor.[374] The rates of addition of chlorine atoms to chloroethylenes relative to hydrogen abstraction from 1,2-dichloroethane and cyclohexane have been measured.[375] An account of the chlorination of tetrafluoroethylene in the presence of molecular oxygen has appeared.[376] In the gas-phase chlorination of C_4–C_6 alkenes, elevation of temperature favours hydrogen abstraction until addition is practically non-existent at around 350°C; from hex-1-ene chloromethylcyclopentane was obtained by an intramolecular addition.[377]

At –20°C in the dark, Cl_2O reacts with cyclohexene to give *trans*-2-chlorocyclohexyl hypochlorite; there is concomitant molecule-induced homolysis of Cl_2O, giving several radical substitution and addition products plus some ionic reaction products.[378] Complex mixtures are obtained in the photochemical reactions of nitrosyl chloride with tri- and tetra-chloroethylene.[379] Chlorination of various straight-chain and cyclic 1,3-dienes with iodobenzene dichloride gave exclusively dichlorinated products resulting from 1,2- and 1,4-additions of chlorine to the conjugated system; the suggested mechanism involved formation of an allylic radical which retained its configuration during subsequent reaction.[380]

Bromine atoms attack 5-(difluoromethylene)-6,6-difluoronorborn-2-ene (**115**) from the *exo*-side on the internal double bond, giving radicals (**116**) and (**117**) which can pick up bromine to give *cis*- and *trans*-dibromides; intramolecular addition in radical (**117**) occurs, leading to the rearranged product (**118**).[381]

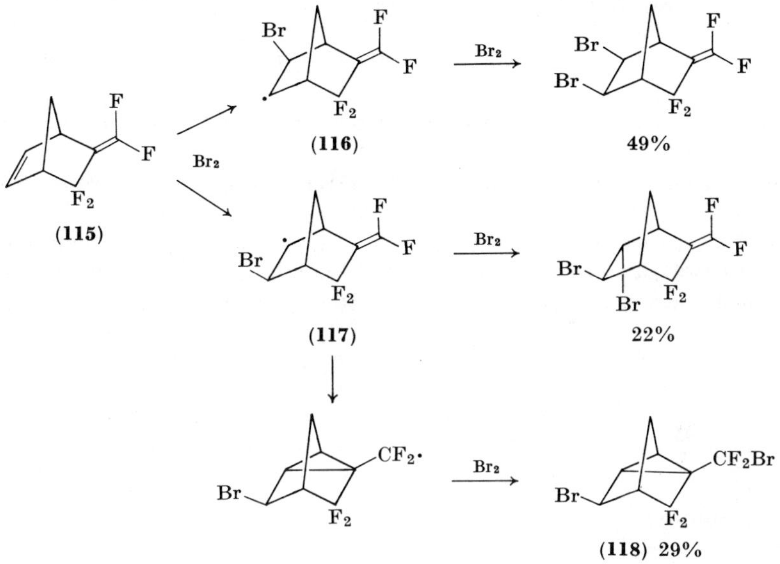

(**115**) (**116**) 49% (**117**) 22% (**118**) 29%

The rates of isomerization of $CH_2{=}CBrCCl_2R$ (for R = Cl, F or Me) into $BrCH_2CCl{=}CClR$ correlate with the rates of bromine-atom addition to the initial alkenes.[382] The stability of bromo-substituted ethyl radicals was determined from the radiation-induced debromination of 1,2-dibromotetrafluoroethylene; the enthalpies of polymerization of $CF_2{=}CF_2$ and $CH_2{=}CH_2$ do not provide a measure of the ease of bromine elimination from $\cdot CF_2CF_2Br$ and $\cdot CH_2CH_2Br$ radicals.[383]

Heteroradicals[384]

The absolute rate constant for the addition of trimethylsilyl radicals to ethylene ($\log k = 7.0 - 2500/2.3RT$ l mol^{-1} s^{-1}) has been measured by the flash-photolysis-ESR method.[385] The ESR spectra of the adduct radicals formed on addition of chlorine atoms and of oxygen-, sulphur- and silicon-centred radicals to fluoroethylene have been reported; interestingly the adduct from triethylsilyl radicals and trifluoroethylene had the structure $Et_3SiCF_2\dot{C}HF$.[386] The stereochemistry of the addition of dichlorosilane to acetylenes has been investigated.[387] Trichlorosilyl radicals gave predominantly central addition with tetrafluoro- and 3,3,3-trifluoro-propyne, whereas trimethylsilyl radicals preferentially attacked the terminus; this was accounted for in terms of a polar effect.[388] A number of long-lived α-aminoalkyl radicals were prepared by multiple addition of silyl or phosphinyl radicals to di-*tert*-butylmethanimine or pivalonitrile,[389] e.g.:

$$Bu^t_2C{=}NH \xrightarrow{R\cdot} Bu^t_2\dot{C}NHR \xrightarrow{-H\cdot} Bu^t_2C{=}NR \xrightarrow{R\cdot} Bu^t_2\dot{C}NR_2$$

Long-lived 1,1,2,2-tetrasubstituted ethyl radicals were prepared in a similar fashion by addition of trimethylsilyl radicals to methyl cyanide, *tert*-butylacetylene, 1,2-di-*tert*-butylethylene, etc., e.g.:

$$\mathrm{Bu^t{}_2C{=}CH_2} \xrightarrow{\mathrm{Me_3Si\cdot}} \mathrm{Bu^t{}_2\dot{C}CH_2SiMe_3} \xrightarrow{\mathrm{Bu^tO\cdot}} \mathrm{Bu^t{}_2C{=}CHSiMe_3} \xrightarrow{\mathrm{Me_3Si\cdot}} \mathrm{Bu^t{}_2\dot{C}CH(SiMe_3)_2}$$

Alternative routes to the same radicals were available through *tert*-butoxy-radical abstraction from silanes.[390]

In the addition of piperidinium radicals $C_5H_{10}\overset{\bullet+}{N}H$ to *meta*- and *para*-substituted styrenes polar effects were caused by ground-state electrostatic attraction between the radical and the π-electrons of the styrenes.[391] Addition of the piperidinium radical to α-pinene at –40°C gave the oxime (**119**), but at 0°C or higher a rearranged product (**120**) was obtained that was oxidized by air to the stable nitroxide (**121**).[392]

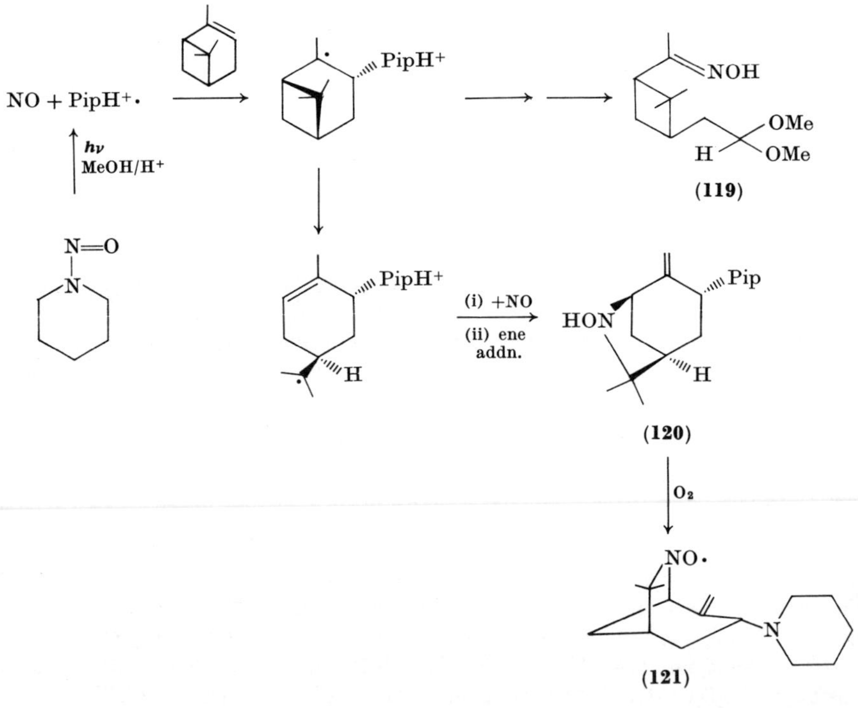

NO + PipH⁺· ; hν, MeOH/H⁺ ; PipH⁺ ; NOH ; OMe ; OMe ; H

(**119**)

N=O ; N ; PipH⁺ ; H ; (i) +NO ; (ii) ene addn. ; HON ; Pip ; H

(**120**)

O_2

NO· ; N

(**121**)

Arrhenius parameters were determined for the isomerization of but-2-enes and pent-2-enes by $\cdot NO_2$ radicals; the rate constants for addition of these radicals to a series of simple olefins were estimated by combining the results with thermodynamic calculations, thus showing that this reaction is unimportant in the bulk consumption of atmospheric pollution.[393] The kinetics of the reaction of N_2O_4 with cyclohexene, giving dinitrocyclohexane, have been investigated.[394] Adducts obtained from the homolytic addition of *N*,*N*,*p*-trichlorobenzenesulphonamide and *N*,*N*-dichlorobenzenesulphonamide to 3,3,3-trichloropropene had the structure $ArSO_2NClCH_2CHClCCl_3$.[395]

The main products from the reaction of PCl_3/O_2/alkene mixtures were the corresponding 2-chloroethylphosphonodichloridates (**122**) and 2-chloroethyl phosphorodichloridates (**123**); the mechanism proposed is shown below.[396, 397]

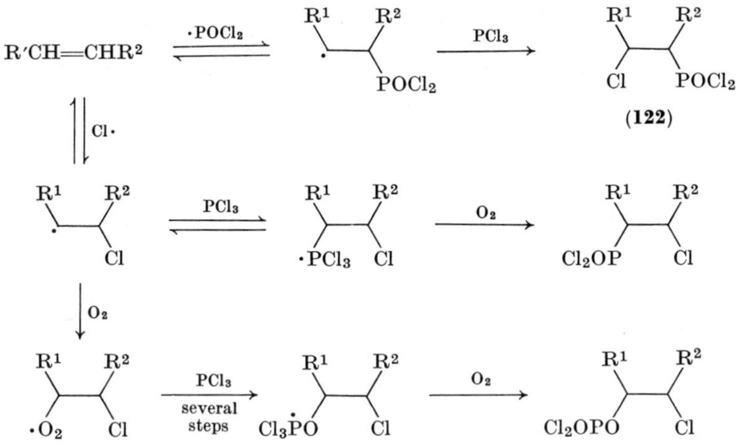

(**122**)

(**123**)

Photochemical or thermal reaction of P_2F_4 with ethylene gives $F_2PCH_2CH_2PF_2$ in >50% yield.[398] Ring-opened products (**124**) and (**125**) were obtained in the radical addition of diethyl phosphite to α- and β-pinene, respectively.[399, 400]

(**124**) (**125**)

Hydroxyl radicals, produced by radiolysis of water, add predominantly to the double bond in unsaturated alcohols but when a conjugated system is present the extent of hydrogen abstraction increases.[401] The absolute rate constant for addition of hydroxyl radicals to acetylene was determined.[402]

In the γ-irradiation of a methanolic solution of methyl oleate, hydrogen atoms and methoxy-radicals were produced, which then added to the ester to give methyl stearate, methyl 9-methoxy- and methyl 10-methoxy-stearate and a trace of dimethoxy-stearate.[403] Rate constants for addition of *tert*-butoxy-radicals to a number of nitrones and nitroso-compounds were measured; these results can be used in spin counting to estimate radical concentrations.[404, 405] The addition of bis(trifluoromethyl) trioxide to fluoroalkenes is a useful method for direct synthesis of fluorocarbon peroxides; products of the type $CF_3OOCR^1R^2CR^3R^4OCF_3$ and $CF_3OCR^1R^2CR^3R^4OOCF_3$ are generally formed in good yield.[406] Perfluoro-*tert*-butoxy-radicals react with alkenes predominantly by addition, in contrast to *tert*-butoxy-radicals which preferentially abstract allylic hydrogen;[407] the ESR spectra of a number of such adduct radicals were described. Valerylperoxy-radicals add faster to cyclohexene than they abstract hydrogen, but cyclohexylperoxy-radicals predominantly abstract.[408]

Methyl and *tert*-butyl mercaptan add to 1,2,3,4,7,7-hexachloronorbornadiene and its 5-chloro-analogue to give (**126**), (**127**) and (**128**): when the 5-substituent X is COOMe

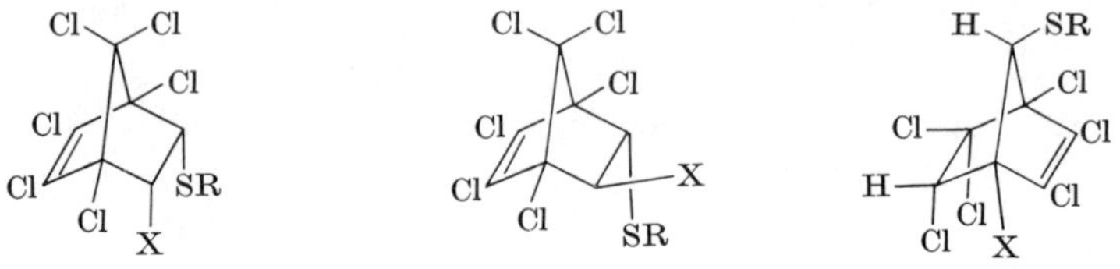

(**126**) (**127**) (**128**)

or Ph, the major or exclusive product is (**126**), whereas when X is methyl only the rearranged product (**128**) is obtained.[409] Sulphonyl iodides add to allene to give 1:1 adducts from both central and terminal attack, but with phenylallene and 3-methylbuta-1,2-diene only the product of central attack was formed.[410] 2,4-Dinitrophenylselenium trichloride reacts in an *anti*-stereospecific fashion with *cis*- and *trans*-1-phenylprop-1-ene to give *threo*- and *erythro-dl-β*-chloroalkyl-2,4-dinitrophenyl selenium dichlorides, respectively.[411]

Intramolecular Additions

Cyclopentylmethyl radicals, generated in the cavity of an ESR spectrometer, did not rearrange to hex-5-en-1-yl radicals, confirming the irreversibility of the cyclization reaction.[412] The rate constant for cyclization of hex-5-en-1-yl radicals, generated from tris(hex-5-enyl)phosphine with *tert*-butoxy-radicals, was measured over a range of temperatures by an ESR technique and found to be in good agreement with earlier estimates.[412] Hex-5-en-1-yl, hept-6-en-1-yl and oct-7-en-1-yl radicals, generated from the appropriate bromoalkene by treatment with tri-*n*-butyltin radicals, all cyclized preferentially to give the corresponding cycloalkylcarbinyl radical (**130**); the enthalpy term was found to be the controlling factor, reactions leading to the thermodynamically

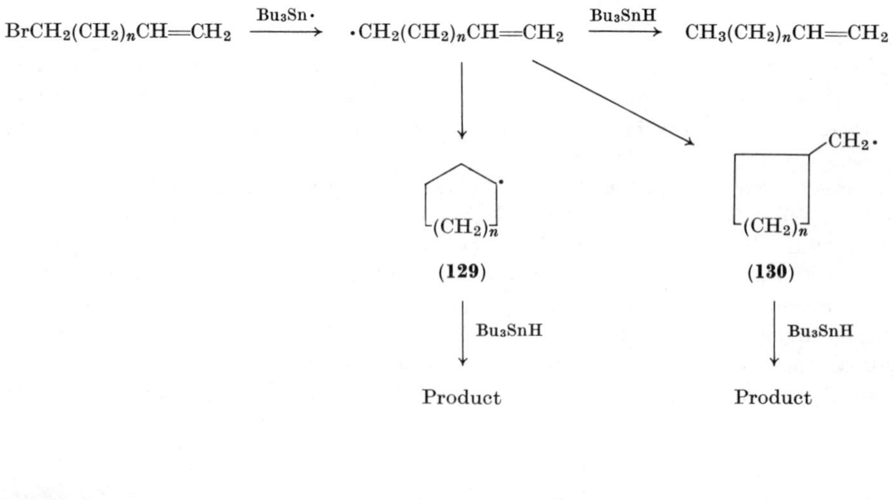

$$BrCH_2(CH_2)_nCH{=}CH_2 \xrightarrow{Bu_3Sn\cdot} \cdot CH_2(CH_2)_nCH{=}CH_2 \xrightarrow{Bu_3SnH} CH_3(CH_2)_nCH{=}CH_2$$

(**129**) (**130**)

Bu_3SnH → Product Bu_3SnH → Product

less stable cycloalkylcarbinyl products (**130**) having smaller $\Delta H^{\ddagger}$ than reactions giving cycloalkyl-type products (**129**).[413] Hept-6-en-2-yl and related radicals (**131**) gave a preponderance of the thermodynamically less stable *cis*-cyclization product (**132**), probably because the transition state for *cis*-cyclization contains an attractive interaction between the alkyl substituent R and the double bond, which is not present in *trans*-cyclization.[414] A substituent at the 5-position in hex-5-en-1-yl radicals greatly

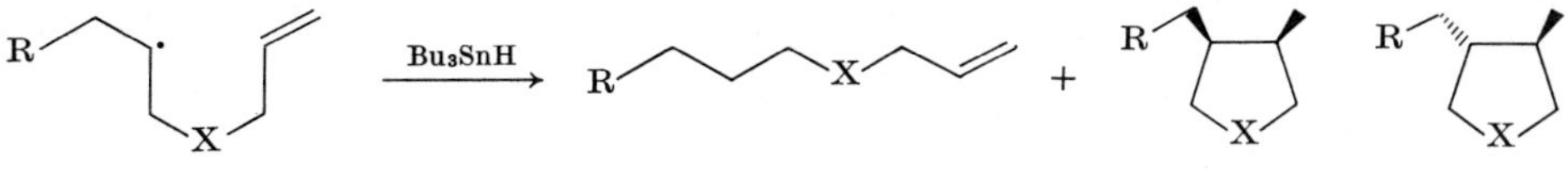

(**131**) $\xrightarrow{Bu_3SnH}$ R–CH₂CH₂CH₂–X–CH₂CH=CH₂ + (**132**)

decreased the rate of cyclization, whereas substituents at the radical centre or at the 6-position had relatively little effect on 1,5-cyclization, suggesting that steric factors are of greatest importance for intramolecular additions;[415] the transition state of the cyclization reaction comprises an unsymmetrical triangular array of centres lying in the same plane as that of the original π-bond which has been substantially broken before formation of the new σ-bond.

Ar_2-6 Cyclization of 4-phenylbut-1-yl radicals was found to be irreversible up to 200°C; the interplay of enthalpy and stereoelectronic factors in the fission of the cyclohexadienyl radicals has been discussed.[416]

The main product from the treatment of the tetrachloroheptene (**133**) with $CuCl_2$ in methyl cyanide containing triethylamine hydrochloride was the six-membered cyclic derivative (**134**). Tetrahydrofuran derivatives were obtained from 3-(trichloromethyl)-

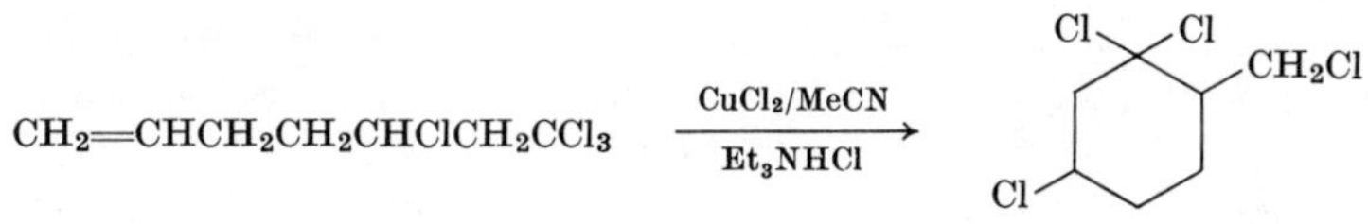

$$CH_2{=}CHCH_2CH_2CHClCH_2CCl_3 \xrightarrow[Et_3NHCl]{CuCl_2/MeCN}$$

(**133**) (**134**) *cis* + *trans*

4-oxa-hept-6-enenitrile on similar treatment and γ- and δ-lactones from allyl trichloroacetate.[417] Hex-5-enyl-type radicals, generated by treatment of 3,3,4,4-tetrafluoro-4-iodobut-1-ene (**135**) with peroxide or metal salts in the presence of an alk-1-ene or alk-1-yne, gave (**136**) and (**137**) together with (**138**), which resulted from addition to the starting material (**135**); polarization of the double bond was important, determining the size of the rings produced.[418] Bicyclization was observed on treatment of *N*-allyl- or

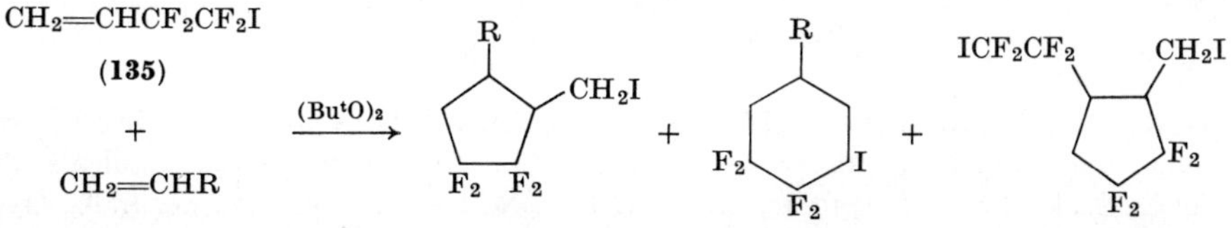

$CH_2{=}CHCF_2CF_2I$ (**135**) + $CH_2{=}CHR$ $\xrightarrow{(Bu^tO)_2}$ (**136**) + (**137**) + (**138**)

N-benzyl-*N*-chloropent-4-enylamine (**139**) with Ti(III) in acetic acid; the pyrrolizidine derivative (**140**) was isolated together with monocyclization products.[419]

(**139**) $\xrightarrow{Ti^{3+}}$ ·N→Ti^{4+} radical → pyrrolidinylmethyl radical → 2-methyl-N-benzylpyrrolidine (10%); 2-(chloromethyl)-N-benzylpyrrolidine (40%); (**140**) (35%)

Fragmentations

The radical-induced isomerization of *N*-vinylsulphonamides has been shown to be a chain reaction involving addition of a sulphonyl radical to the double bond followed by fragmentation of the adduct radical (**141**):[420]

$$PhSO_2\cdot + PhSO_2NRCH{=}CH_2 \rightarrow PhSO_2NR\dot{C}HCH_2SO_2Ph \rightarrow RN{=}CHCH_2SO_2Ph + PhSO_2\cdot$$

(**141**)

The adduct radicals (**142**) and (**143**) formed by addition of *tert*-butoxy- and *tert*-butyl radicals to *tert*-butyl isocyanate[421] undergo fragmentation as does the radical (**144**) obtained by addition of trimethylsilyl radicals to benzyl cyanide.[16]

$$Bu^tO\cdot + Bu^tNC \rightarrow Bu^tN{=}\dot{C}OBu^t \rightarrow Bu^tNCO + Bu^t\cdot$$

(**142**)

$$Bu^t\cdot + Bu^tNC \rightarrow Bu^tN{=}\dot{C}Bu^t \rightarrow Bu^tCN + Bu^t\cdot$$

(**143**)

$$Me_3Si\cdot + PhCH_2CN \rightarrow PhCH_2\dot{C}{=}NSiMe_3 \rightarrow PhCH_2\cdot + Me_3SiCN$$

(**144**)

tert-Butoxy-*tert*-butylnitroxide (**145**) which is formed by addition of *tert*-butoxy-radicals to nitrosobutane undergoes fragmentation to give *tert*-butoxy-radicals at a rate of ca. 0.14 s^{-1} at 40°. Fragmentation with loss of *tert*-butyl radicals occurs by C—N rather than C—O fission with a rate constant of ca. 10^{-3} s^{-1} at 40°.[422]

$$\mathrm{Bu^tO\cdot + Bu^tNO \rightleftharpoons Bu^t{-}\overset{\displaystyle O\cdot}{\overset{|}{N}}{-}OBu^t}\ \textbf{(145)} \quad \begin{cases} \not\rightarrow \mathrm{Bu^t\cdot + Bu^tNO_2} \\ \rightarrow \mathrm{Bu^t\cdot + Bu^tONO} \end{cases}$$

Experiments on the competition between decarbonylation of acyl radicals ($\cdot COR$) and scavenging by nitrosobutane have shown that the rate constants for decarbonylation increase with increasing stability of R· along the series R = Me < Pr^i < 1-adamantyl < $Bu^t\cdot$ < $PhCH_2$.[422] The absolute rate constant for the decarbonylation of $Pr^i\dot{C}O$ was estimated as $3.9 \times 10^3\ s^{-1}$ at 40° by assuming that the rate of addition of this radical to nitrosobutane was the same as that determined for the *tert*-butoxycarbonyl radical ($1.1 \times 10^6\ l\ mol^{-1}\ s^{-1}$). The ease of fragmentation of perfluoropropionyl radicals is considerably greater than that of perfluoroacetyl radicals.[65] CIDNP studies on the decomposition of benzoyl propionyl peroxide in *o*-dichlorobenzene have indicated that the rate of decarboxylation of benzoyloxy-radicals is $1 \times 10^8\ s^{-1}$.[423] Benzoylmethyl radicals $\cdot CH_2COPh$ undergo decarbonylation to benzyl radicals at high temperatures via a cyclic transition state.[54a]

The tendency of an alkoxy-radical to undergo β-scission in competition with hydrogen abstraction from cyclohexane is about 100 times greater for a tertiary alkoxy-radical than for a primary alkoxy-radical.[424] The order correlates with π-bond energy of the carbonyl compound formed and with the stability of the radical eliminated. The transition state for the fragmentation also has significant polar character as indicated by a Hammett ρ value of -1.04 (using σ^+-constants) in the fragmentation of $ArCH_2CMe_2O\cdot$ radicals, and can be represented by (**146**). The relative ease of loss of R· and Me· from

$$\mathrm{ArCH_2{-}\overset{\displaystyle R^1}{\overset{|}{\underset{\underset{\displaystyle R^2}{|}}{C}}}{-}O\cdot \longleftrightarrow ArCH_2^+\ \cdot\overset{\displaystyle R^1}{\overset{|}{\underset{\underset{\displaystyle R_2}{|}}{C}}}{-}O^- \longleftrightarrow ArCH_2\cdot\ \overset{\displaystyle R^1}{\overset{|}{\underset{\underset{\displaystyle R_2}{|}}{C}}}{=}O}$$

(146)

$RCMe_2O\cdot$ is in the order R = COOMe > CH_2OCOMe > CH_2Cl > Me > CH_2Br.[221] Tertiary alcohols containing at least one *tert*-butyl group readily undergo cleavage on reaction with *tert*-butoxy-radicals.[425] It is not clear whether this process proceeds in two stages, involving hydrogen abstraction from the O—H group followed by fragmentation of the resultant *tert*-alkoxy-radical, or whether it is a synchronous process.

The radicals (**147**) obtained by hydrogen abstraction from β-amino-alcohols undergo fragmentation at pH 6.5–7.5 but not at pH 1.[426, 427] The proposed mechanism involves deprotonation of the radical (**147**) to (**148**), followed by fragmentation of the latter.

$$\mathrm{R_3\overset{+}{N}CH_2CH(OH)CH_3 \xrightarrow{HO\cdot} R_3\overset{+}{N}CH_2\dot{C}(OH)CH_3 \xrightarrow{OH^-}}$$

(147)

$$\mathrm{R_3\overset{+}{N}CH_2\underset{\bullet}{\overset{\overset{\displaystyle O^-}{|}}{C}}CH_3 \longrightarrow R_3N + \cdot CH_2COCH_3}$$

(148)

The $PhCOCH_2Br^{\bar{\cdot}}$ and $PhCOBr^{\bar{\cdot}}$ radical anions readily lose bromide to give $PhCOCH_2\cdot$ and $PhCH_2\cdot$, respectively.[158] The rate-determining step in the acid-catalysed dehydration of the hydroxy-methylcyclohexadienyl radical, derived from toluene and hydroxyl radicals, involves protonation of the hydroxyl group.[428] The analogous adduct radical from *p*-toluenesulphonic acid does not undergo dehydration as the electron-withdrawing sulphonic acid group inhibits protonation of the hydroxyl group.[428]

Homolytic Aromatic Substitution

The generation of aryl radicals in the pseudo-Gomberg reaction involving aromatic amines and pentyl nitrite takes place as follows:[429]

$$ArNH_2 \rightarrow ArN{=}NOH \rightarrow (ArN{=}N)_2O \rightarrow Ar\cdot$$

This process is accompanied by a second reaction involving intermediacy of a triazene; the extent of side reactions is increased if more than one equivalent of pentyl nitrite is used.[430] An investigation of the reaction of *m*-nitrobenzenediazonium fluoroborate with nitrite indicated that this was not an electron-transfer process but rather involved the intermediacy of the covalent diazo-nitrite (**149**) which subsequently broke down to generate aryl radicals;[431] the formation of the biaryl and nitroaromatic reaction products is as indicated.

$$ArN_2^+ + NO_2^- \longrightarrow \underset{(\mathbf{149})}{Ar{-}N{=}N{-}NO_2} \longrightarrow Ar\cdot + N_2 + NO_2$$

$$Ar\cdot + C_6H_6 \longrightarrow [Ar.C_6H_6]\cdot \xrightarrow[-HNO_2]{NO_2} C_6H_5Ar$$

$$Ar\cdot + NO_2^- \longrightarrow [ArNO_2]^{\bar{\cdot}}$$

$$ArNO_2^{\bar{\cdot}} + Ar{-}N{=}N{-}NO_2 \longrightarrow ArNO_2 + [Ar{-}N{=}N{-}NO_2]^{\bar{\cdot}}$$

$$[Ar{-}N{=}N{-}NO_2]^{\bar{\cdot}} \longrightarrow Ar\cdot + N_2 + NO_2^-$$

Electron-transfer processes are, however, responsible for the generation of aryl radicals in reactions of diazonium salts with xanthene dyes,[432] formate, cysteine, hypophosphite and phosphite.[433] The generation of phenyl radicals from *N*-nitrosoacetanilide can be suppressed by addition of radical scavengers, such as 1,1-diphenylethylene, to the reaction system.[434] The effect of such scavengers is to remove phenyl radicals as (**150**) from the system thereby reducing very markedly the decomposition of *N*-nitroso-acetanilide by a chain reaction and instead promoting its decomposition via benzyne. The radical (**150**) is less readily oxidized by the diazonium salt than is the [Ph—ArH]· radical.

On the basis of an analysis of thermodynamic data, it has been concluded that addition of phenyl radicals to a benzenoid substrate with the generation of phenylcyclohexadienyl radicals should become reversible only above 200° though similar reactions with benzyl radicals and $Me_3Sn\cdot$, $Me_3Ge\cdot$ or $Me_3Si\cdot$ should be reversible at lower temperatures.[435] The absence of any significant kinetic isotope effect in the material not consumed in the

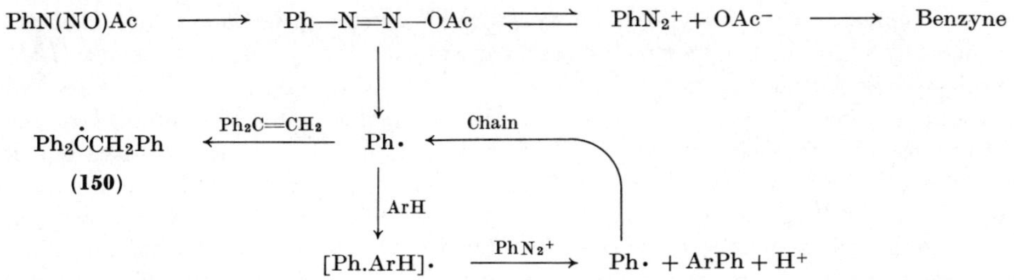

PhN(NO)Ac ⟶ Ph—N=N—OAc ⇌ PhN_2^+ + OAc^- ⟶ Benzyne

Ph—N=N—OAc ⟶ Ph·

Ph· $\xrightarrow{Ph_2C=CH_2}$ $Ph_2\dot{C}CH_2Ph$ (**150**)

Ph· $\xrightarrow{ArH}$ [Ph.ArH]· $\xrightarrow{PhN_2^+}$ Ph· + ArPh + H^+ (Chain)

phenylation of 3-deuterio-4-methylpyridine (**151**) lends support to the irreversible nature of the addition step.[436] There is, however, in this reaction a very substantial isotope effect with respect to the products of reaction (**152**) and (**153**) ($k_H/k_D = 3.7$). This seems to point to a kinetic isotope effect in the subsequent reactions of the intermediate radicals. In contrast to this, formation of the phenyl-2,5-dimethylcyclohexa-

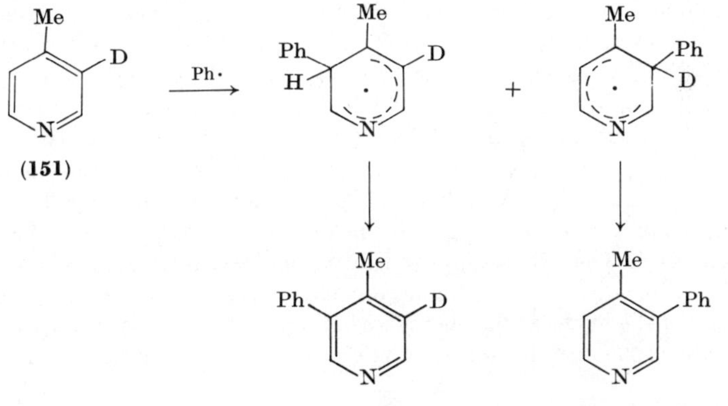

(**151**) $\xrightarrow{Ph\cdot}$ (**152**) + (**153**)

dienyl radical (**154**) in the phenylation of *p*-xylene is claimed to be reversible as the ratio of 4,4′-dimethylbibenzyl (**156**) to 2,5-dimethylbiphenyl (**155**) obtained in this

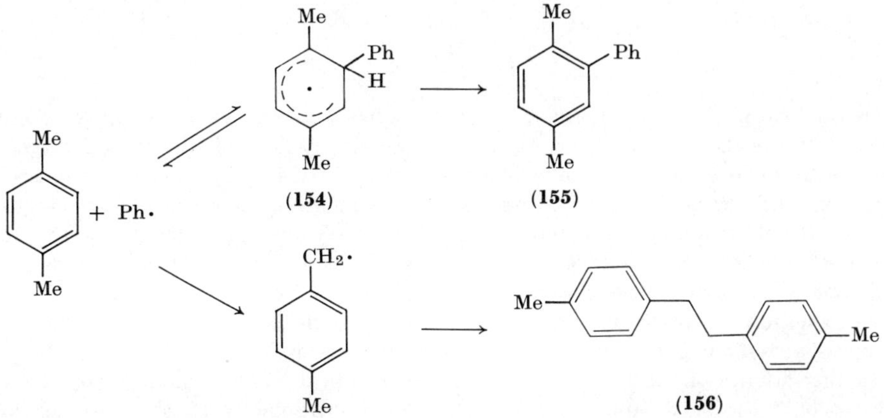

p-xylene + Ph· ⇌ (**154**) ⟶ (**155**)

p-xylene + Ph· ⟶ $CH_2\cdot$ radical ⟶ (**156**)

reaction increases with increasing temperature.[437] Further, the amount of (**155**) is increased by addition of copper(II) salts as a consequence of their ability to oxidize (**154**) before its dissociation.

Doubt has been cast on the validity of isomer ratios of biaryls obtained in the phenylations of aromatic substrates as a fair indication of the relative proportions of phenylcyclohexadienyl radicals obtained initially in such reactions.[438] In the phenylation of 4-methylpyridine it has been shown that while the total yield of biaryl is increased by addition of nitrobenzene the ratio of 3- to 2-phenyl-4-methylpyridine increases substantially. It is suggested that more reliable data concerning the ease of attack at the various positions in an aromatic substrate would be obtained if such reactions were carried out in the presence of an oxidizing additive such as nitrobenzene.

An *ortho*-effect has also been observed in arylations of substituted benzenes with 2-pyridyl radicals[439] and in phenylations of methoxypyridines.[440] Addition of hydrogen atoms to substituted benzenes also occurs preferentially *ortho* to the substituent though this tendency is reduced in the 2,6-d_2 analogues.[441]

The isomer ratios of products obtained in arylations with 2-pyridyl radicals confirm their electrophilic character. Further evidence supporting the weakly nucleophilic character of phenyl radicals was obtained from phenylations of methoxypyridines.[440] Cyclopropyl radicals were concluded to have similar nucleophilicity to phenyl radicals but to be less nucleophilic than cyclohexyl radicals as judged from partial rate factors obtained with substituted benzenes.[442] The relative rates of substitution of 4-substituted pyridines containing electron-withdrawing groups in acid solution increases with increasing nucleophilicity of the radical.[443] It was shown in this way that the nucleophilicity of cycloalkyl radicals increases in the order: cyclopropyl < cyclobutyl < cyclopentyl < cyclohexyl. The cyclopropyl radical is the least nucleophilic as the unpaired electron is in an orbital of much higher *s*-character. α-Alkoxy-groups were similarly shown to increase the nucleophilic character of alkyl radicals; this is consistent with the transition state (**157**) for the reaction having significant charge-transfer character.

(**157**)

However, the nucleophilic character of acyl radicals is reduced by α-alkoxy-groups since acyl radicals are σ- and not π-radicals. The electrophilic character of 2-thiazolyl radicals has been confirmed from measurements of partial rate factors for the substitution of substituted benzenes;[444] electron-withdrawing groups in the 5-position slightly enhance the electrophilic character of these radicals. The rates of trichloromethylation of 9-substituted anthracenes give a Hammett correlation of $\rho = -0.83$ using σ^+-constants, indicating that trichloromethyl radicals have substantial electrophilic character.[445] The rate of reaction of 9-nitroanthracene is higher than would be expected, since the nitro-group is twisted out of the plane of the anthracene system.

Further studies of the reactions of cyclohexyl radicals with *o*-dihalobenzenes have led not only to the 3- and 4-cyclohexyl-1,2-dihalobenzenes with the 3-isomer predominating but also to 1-halo-2-cyclohexylbenzene.[446] There is a marked solvent effect in more polar aprotic media which is interpreted as indicating that reaction proceeds via reversible formation of a π-complex (**158**) having significant charge-transfer character between the cyclohexyl radicals and the dihalobenzene, and that this rearranges to the σ-complexes (**159**)–(**161**). Replacement of halogen also occurs, though

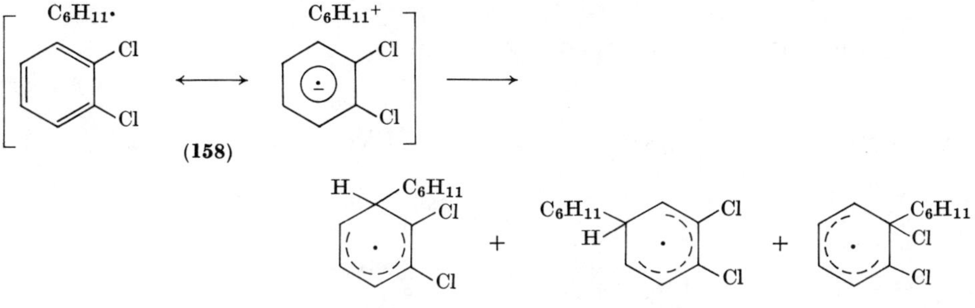

$C_6H_{11}\cdot$ Cl Cl ⟷ $C_6H_{11}^+$ Cl Cl ⟶

(**158**)

H C_6H_{11} Cl Cl + C_6H_{11} H Cl Cl + C_6H_{11} Cl Cl

(**159**) (**160**) (**161**)

to a smaller extent, in reactions of *p*-dihalobenzenes. There is greater relief of steric strain in forming the σ-complex (**161**) in the case of *o*-dihalobenzenes than with *p*-dihalobenzenes.

Homolytic aromatic substitutions by thienyl radicals and of thiophenes and condensed thiophenes have been reviewed.[447] The major products from the reaction of dibenzoyl peroxide and thiophen are 2-thienyl benzoate and 2,2′-bithienyl.[448] The latter results from loss of benzoic acid from the dimer (**163**) obtained from coupling of the radicals (**162**) rather than by coupling of 2-thienyl radicals as had been proposed earlier. An electron-transfer mechanism was also ruled out. Products analogous to (**163**)

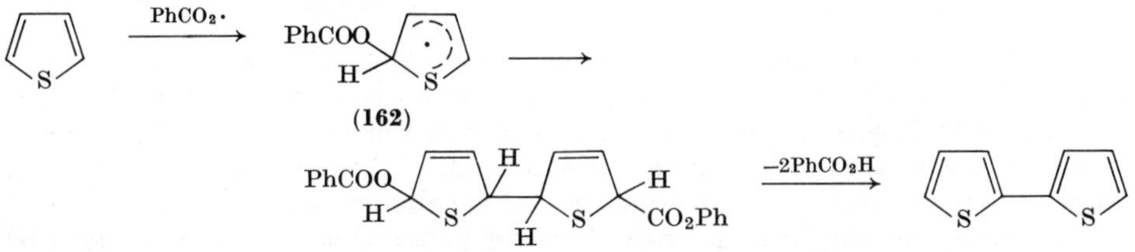

S $\xrightarrow{PhCO_2\cdot}$ PhCOO H S ⟶

(**162**)

PhCOO H S H H S H CO_2Ph $\xrightarrow{-2PhCO_2H}$ S S

(**163**)

can be isolated from reactions of dibenzoyl peroxide and bis(pentafluorobenzoyl) peroxide with perfluoronaphthalene, but not in the reactions with naphthalene as such compounds readily lose benzoic acid to give binaphthyls.[449]

In the reaction of thiophen with di-isopropyl peroxydicarbonate in the presence of copper(II) chloride, the radical (**164**) is oxidized by copper(II) to 2-thienyl isopropyl carbonate (**165**).[450]

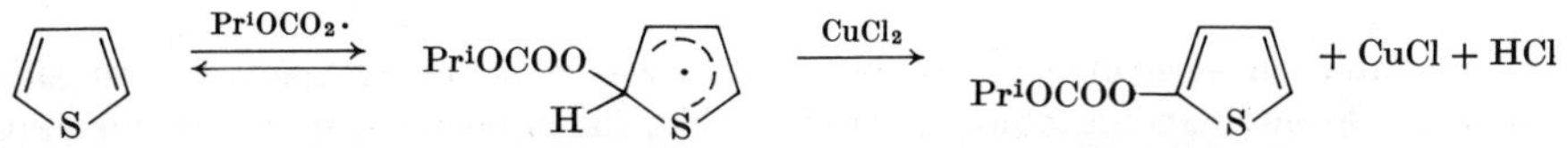

S $\xrightleftharpoons{Pr^iOCO_2\cdot}$ Pr^iOCOO H S $\xrightarrow{CuCl_2}$ Pr^iOCOO–S + CuCl + HCl

(**164**) (**165**)

The scope and synthetic utility of homolytic substitution of protonated heteroaromatics has been reviewed.[451] Reaction of picoline with dibenzylmercury in acetic acid gives exclusively 2-benzyl-4-methylpyridine in 63% yield.[452] Pyridazine undergoes both benzylation and benzoylation in the 4-position.[453]

The homolytic amination of alkylbenzenes and of biphenyl can be effected by using *N*-chloroamines in the presence of iron(II) salts.[454] The partial rate factors for these

$$R_2\overset{+}{N}HCl + Fe^{2+} \longrightarrow R_2\overset{+\bullet}{N}H + Fe^{3+} + Cl^-$$

$$PhH + R_2\overset{+\bullet}{N}H \longrightarrow \left[R_2\overset{+\bullet}{N}H \; C_6H_6 \longleftrightarrow R_2\ddot{N}H \; (C_6H_6)^{+\bullet} \right] \longrightarrow$$

(166)

reactions show the largest polar effects ever observed in homolytic aromatic substitution and suggest that the transition state (**166**) has significant charge-transfer character. Phenylations of imidazo[1,2-*b*]-, *s*-triazolo[4,3-*b*]- and *s*-triazolo[2,3-*b*]-pyridazine occur preferentially at the 8-position.[455] Photosensitized amidations of benzene and alkylbenzenes have been reported:[456]

$$PhH + HCONH_2 \xrightarrow[\text{Me}_2\text{CO or Ph}_2\text{CO}]{h\nu} PhCONH_2$$

Photo-triazinylation of benzene occurs on photolysis of chloro- or bromo-*s*-triazine in benzene.[457] Photochemical hydroxylation of aromatic compounds has also been reviewed.[458] Radiolysis of aqueous benzene involves not only hydroxycyclohexadienyl radicals but also phenyl radicals as intermediates;[459] intervention of the latter is shown by the isolation of small quantities of dihydrobiphenyls. Hydroxylation of phenol and aniline with hydroxyl radicals occurs in the *ortho*- and *para*-positions;[460] reactions occur by addition of the hydroxyl radical followed by loss of water to give phenoxy- and anilino-radicals, respectively, and not by direct hydroxylation, e.g.:

$$PhOH + HO\cdot \rightarrow [HO.PhOH]\cdot \rightarrow PhO\cdot + H_2O$$

$$2PhO\cdot \rightarrow PhO^+ + PhO^-$$

$$PhO^+ + H_2O \rightarrow C_6H_4(OH)_2 + H^+$$

Anodic cyanation of aromatic compounds has been effected by using an emulsified system of the aromatic compound, methylene chloride, aqueous sodium cyanide and a phase-transfer agent.[461] Acyloxylations can similarly be achieved.

Rearrangements

Further studies on the question whether cyclopropyl radicals undergo disrotatory or conrotatory ring-opening have been carried out with bicyclic cyclopropyl radicals, which cannot open in a conrotatory fashion.[462] The isolation of (**170**) from the decomposition of the perester (**167**) indicates that the initial radical (**168**) can undergo disrotatory ring opening to (**169**). The extent of ring opening of (**168**) is, however, much less than

(**167**) (**168**) (**169**) (**170**)

that of the monocyclic *cis*- and *trans*-2,3-diphenylcyclopropyl radicals and it is thus suggested that the last two radicals can ring-open in both a disrotatory and a conrotatory fashion. That the extent of ring opening of (**171**) is greater than that of (**172**) is attributed to their relationship to non-alternant and alternant hydrocarbons, respectively. Theory has now jumped further ahead of experiment with the conclusion

(**171**) (**172**)

that cyclopropyl-radical opening occurs in such a way that rotation at C-2 and C-3 is not synchronized, and that the transition state is reached before significant rotation occurs at one or other of these centres. The activation energy is therefore essentially independent of whether a disrotatory or a conrotatory path is followed.[463]

Cyclobutyl radicals generated in the thermolysis of cyclobutanecarbonyl peroxide in bromoform at 140° give rise to both rearranged and unrearranged products (see below).[464]

The stereochemical requirements of the cyclopropylcarbinyl–homoallyl radical rearrangement have been probed by investigating the ease of rearrangement of the spiro-radicals (**173**) and (**174**).[465] The radical (**173**) which is locked in the bisected conformation (**175**) rearranges to a greater extent than the somewhat more flexible radical (**174**). This is contrary to earlier suggestions that this rearrangement requires

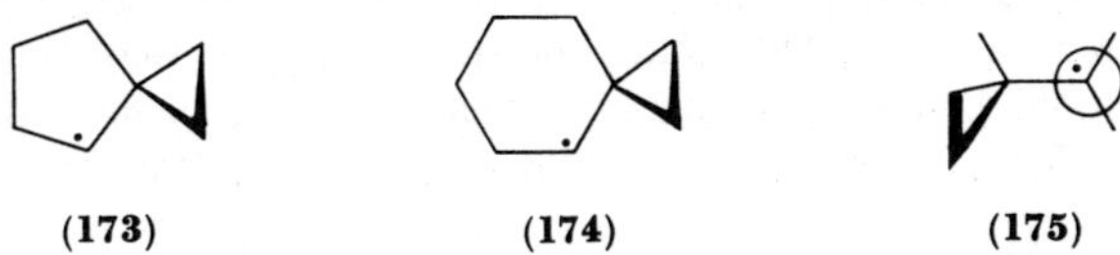

(**173**) (**174**) (**175**)

the cyclopropylcarbinyl radical to be in the perpendicular conformation.[466,467] The aziridinylcarbinyl radical (**176**; R = H) can only be observed below –130°. At higher

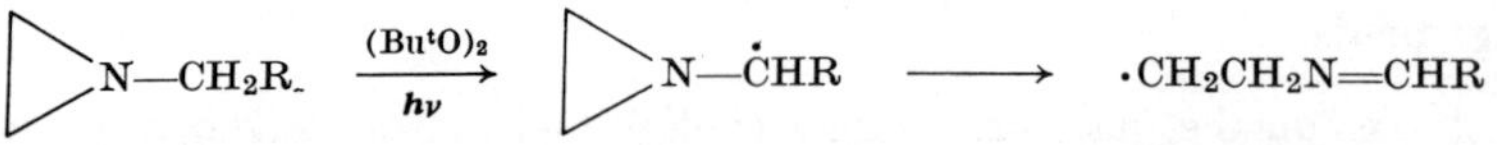

(176) (177)

temperatures this rearranges to (**177**; R = H).[28] The homologous radical (**176**; R = CH_3) rearranges so rapidly that it cannot be observed even at –160°. The same is true of cyclopropylamino-radicals which rearrange to $\cdot CH_2CH_2CH{=}NR$ radicals.

The same bicyclo[3.1.0]hexenyl radical (**181**) is obtained by reaction of either the *endo*- or the *exo*-substituted bicyclo[3.1.0]hexenes (**178**) and (**182**) with *tert*-butoxy radicals.[468] The mechanism of isomerization of (**179**) to (**181**) could possibly involve the cyclohexadienyl radical (**180**) formed in a cyclopropylcarbinyl–homoallyl rearrangement. Products derived from the cyclohexadienyl radical are formed in these reactions at 130° but there is no ESR evidence for (**180**) between –80° and +75° at which temperatures isomerization occurs. The NBS-initiated isomerization of the *endo*-ester (**178**; X = COOMe) is postulated to involve the addition of bromine atoms

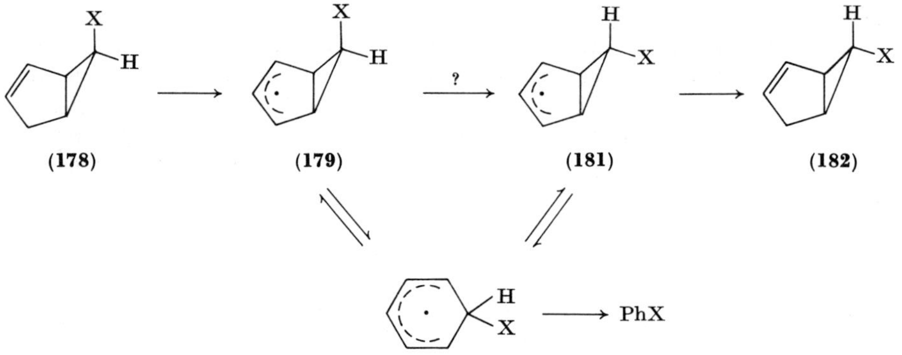

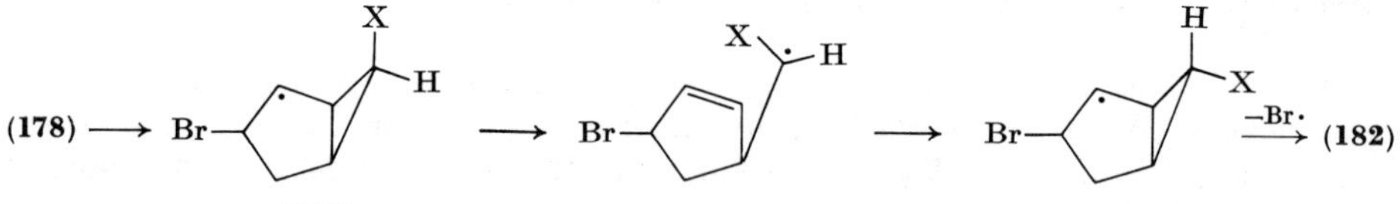

(183) (184)

to give (**183**) followed by ring-opening and cyclization to (**184**) and not by hydrogen abstraction since (**185**) but not (**186**) undergoes the analogous rearrangement.[469] Both these substrates should lead to the same intermediate if hydrogen abstraction were involved. However, the ease of hydrogen abstraction from (**185**) and (**186**) need not be the same.

Cyclopropylmethyl radicals, generated in the thermolysis of *trans*-2-substituted cyclopropyl peroxyacetates, have no existence outside the solvent cage.[470] The energy of isomerization of $\cdot CHMeCH_2CH{=}CH_2$ radicals to $CH_2{=}CHCHMeCH_2\cdot$ radicals

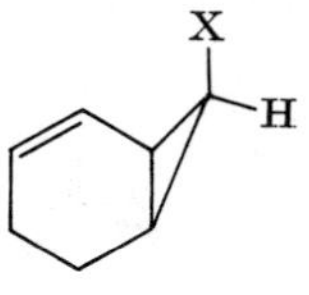

(185) 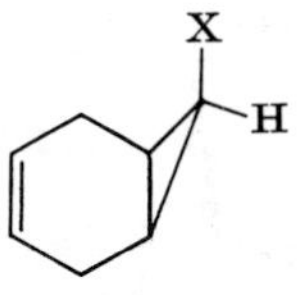 (186)

which proceeds via a cyclopropylcarbinyl radical is less than that for the *cis–trans*-isomerization of homoallyl radicals.[471]

The cyclobutylcarbinyl radicals generated in the addition of $Bu_3Sn\cdot$ to cyclobutyl ketones (**187**) give rise to both rearranged and unrearranged products.[472] The extent of rearrangement is greater with the more stable radical (**188**; R = Ph) than with its methyl analogue (**188**; R = Me).

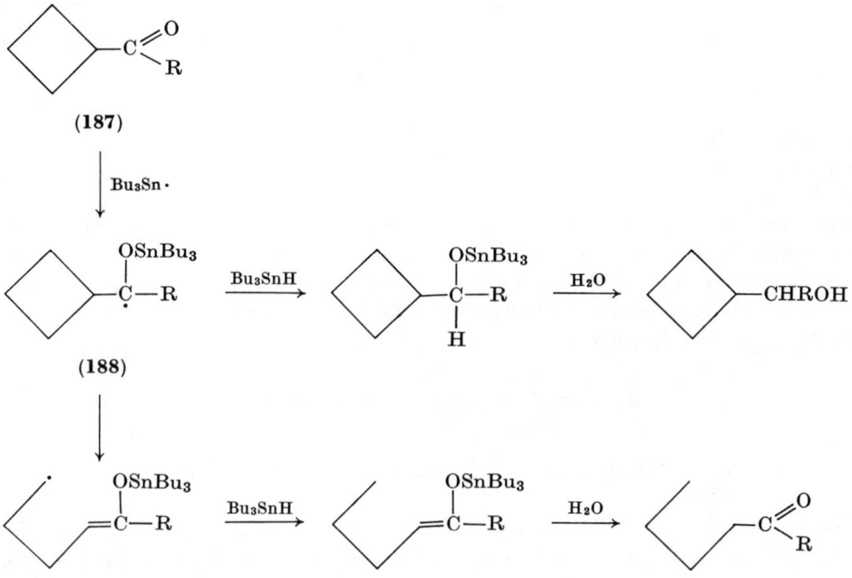

(187)

$Bu_3Sn\cdot$

$OSnBu_3$ Bu_3SnH $OSnBu_3$ H_2O CHROH

(188)

$OSnBu_3$ Bu_3SnH $OSnBu_3$ H_2O

The activation energies for the ring opening of cyclopentyl and cyclobutyl radicals are 34.6 and 31.6 kcal mol^{-1}, respectively, i.e. they are relatively insensitive to ring strain.[473] This is to be expected if the transition state has similar strain to that of the cyclic radical. On the other hand, the results of Shatkina and Reutov[464] are rendered more surprising. Possibly, in their experiments, bromoform acted as an electron-acceptor, and cyclobutyl cations are formed. Alternatively cyclobutyl cations may mediate in a carboxy inversion reaction.

The radical (**189**), generated by treatment of the hydrogen bromide adduct of bullvalene with tributyltin radicals, undergoes rearrangement to (**190**).[474] The rearrangement is reversible since the same products, albeit in different proportions, are obtained when (**190**) is generated from 2-bromobicyclo[4.2.2]deca-3,7,9-triene or its allylic isomer. A similar reversible rearrangement occurs with the radical derived from bullvalene dibromide.

(189) (190)

Rearrangement occurs during the addition of hydrogen bromide to polychloroalkenes if the rearranged radical is more stable than the initial adduct radical.[475] Thus rearrangement occurs during the addition of hydrogen bromide 3,3,3-trichloropropene as (**192**) is more stable than (**191**). Cyclization occurs during the addition of 1,1,1-trichloro-

$$Br\cdot + CH_2{=}CHCCl_3 \longrightarrow BrCH_2\dot{C}HCCl_3 \longrightarrow$$

(191)

$$BrCH_2CHCl\dot{C}Cl_2 \xrightarrow[-Br\cdot]{HBr} BrCH_2CHClCHCl_2$$

(192)

2,2-di-(*p*-chlorophenyl)ethane to styrene.[476] Bromine migrates around the ring in bromocyclohexadienyl radicals generated by the addition of hydrogen atoms to bromobenzene.[477]

The mechanism for the rearrangement of β-keto-alkyl radicals has been shown to occur at least in part by an addition–elimination pathway, as acetyl radicals as well as $MeCOCHMeCH_2\cdot$ and $MeCOCH_2CHMe\cdot$ radicals have been trapped by 4-cyano-pyridine in the persulphate oxidation of (**193**).[478] A concerted process may also be occurring concurrently with this.

$$\underset{(\mathbf{193})}{MeCOCHMeCH_2COOH} \rightarrow MeCOCHMeCH_2\cdot$$

$$MeCOCHMeCH_2\cdot \rightarrow \cdot COMe + MeCH{=}CH_2 \rightarrow MeCOCH_2CHMe\cdot$$

Rearrangement of the *o*-tolyl and *o*-methylbenzoyloxy-radicals to benzyl and *o*-carboxybenzyl radicals, respectively, has been reported to occur in the thermolysis of di-*o*-toluyl peroxide in carbon tetrachloride.[479] The biphenyl-2-carboxamido-radical cyclizes to both γ- and δ-lactam-spirocyclohexadienyl radicals.[480]

In the radical addition of (**194**) to oct-1-ene in the presence of di-*tert*-butyl peroxide, products are derived not only from (**195**) but also from the ring-opened radical (**196**) and from (**197**).[481] (**197**) is formed from (**195**) by either 1,4- or 1,5-hydrogen migration.

$\xrightarrow[h\nu]{(Bu^tO)_2}$ $\longrightarrow$ CHO, $CH_2\cdot$ +

(194) (195) (196) (197)

Vibrationally excited 1-, 2- and 3-pentyl radicals undergo 1,2- or 1,3-hydrogen migrations through either a three- or four-membered cyclic transition state.[482] 1,5-Hydrogen migration is observed during the thermolysis of β-nitro-alkyl nitrates.[483]

Further examples of the photo-Fries rearrangement have been reported.[484] CIDNP studies show that a singlet radical pair is involved in photolysis of *p*-tolyl *p*-chlorobenzoates.[485] An excited singlet state has also been shown to be involved in the rearrangement of phenyl acetate;[486] in this study phenoxy radicals were detected by flash photolysis. The rearrangement of (**198**) to (**199**) constitutes the first example of a *meta*-photo-Fries reaction[487] (although see ref. 485). This reaction only occurs in good

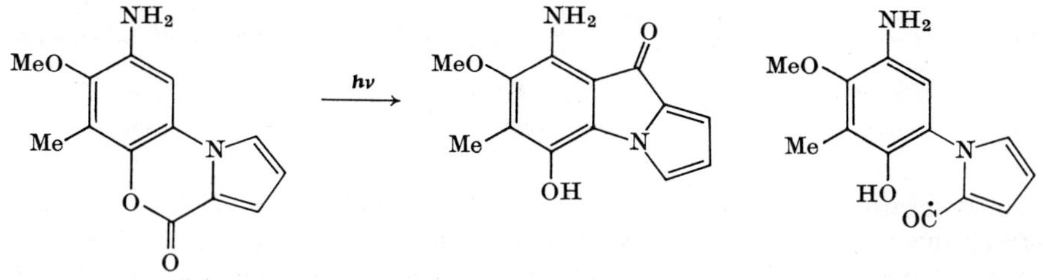

(**198**) (**199**) (**200**)

hydrogen donors in which the radical (**200**) is formed. Phenoxy-1,3,5-triazenes similarly undergo rearrangement to the *ortho*-(1,2,5-triazenyl)phenol.[488] The formation of *m*-alkylphenols arises from the same precursor (**201**) as the *p*-alkylphenol in the photoisomerization of anisole.[489]

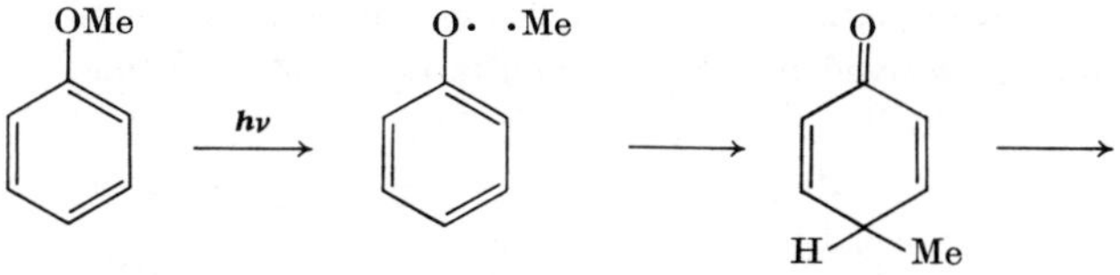

(**201**)

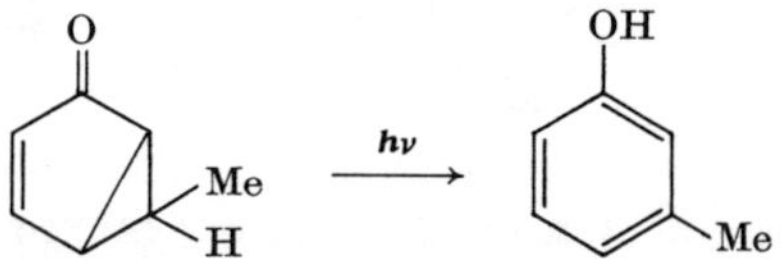

The thermal rearrangement of oxime *O*-allyl ethers occurs at least in part by a radical-pair mechanism involving iminoxy- and allyl radicals.[490] A radical-pair pathway is also involved in the epimerization of *gem*-disulphoxides.[491] The rearrangement of 2-alkoxyquinoline *N*-oxides has been shown to proceed both by a sigmatropic suprafacial mechanism and by a radical-pair mechanism.[492] CIDNP studies indicate that *erythro*- and *threo*-2-(2-methyl-1,2-diphenylbutoxy)quinoline *N*-oxide rearrange by a radical route to the extent of at least 75% and only 20%, respectively. Diphenyliminyl radicals have been detected by ESR spectroscopy in the thermal rearrangement of *O*-methyleneamino thiocarbamates.[493] The reaction shows little sensitivity to changes either in solvent polarity or in the nature of substituents on aromatic nuclei, suggesting a transition state with little charge separation; cross-coupled products can be observed in this rearrangement. The thermal rearrangement of tetra-arylhydrazines to *o*- and *p*-semidines has been shown to proceed by radical coupling of diarylaminyl radicals and not by a radical-substitution process.[494]

Methyl migration occurs in the pyrolysis of 1,5,5-trimethylcyclopentadiene[495] and of enol ethers of unsaturated ketones.[496] In both cases reaction proceeds by an elimination–addition mechanism.

The thermal or photochemical rearrangement of $\alpha\beta$-epoxy-ketones (**202**) to 1,3-diketones (**204**) proceeds via the biradical (**203**) with either consecutive or con-

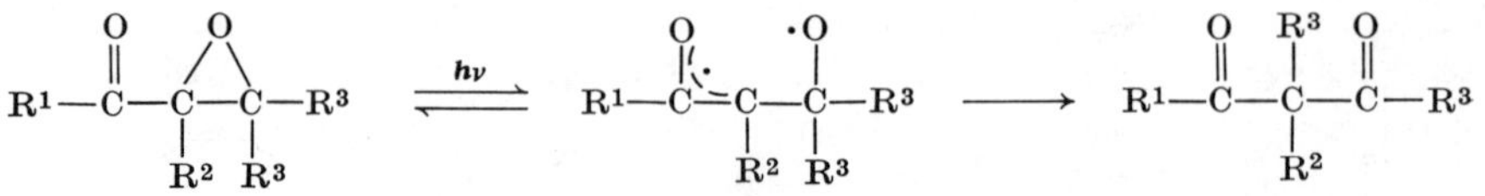

(**202**) (**203**) (**204**)

current migration of an alkyl group;[497] quantum yield data show that the rate of rearrangement is faster than the rate of rotation about the C_2—C_3 bond. The same mechanism operates in the photochemical rearrangements of α-epoxy-spiroketones[498] and of phenalen-1-one oxide.[499] Biradicals are also implicated in the photochemical rearrangements of 4-methylverbenene,[500] 4-methylenebicyclo[3.2.0]hept-2-enes,[501] 2-methylenebicyclo[3.2.1]octa-3,6-diene,[502] 3,3-dimethyl-5,5-diphenylpent-4-en-2-one,[503] and 3-alkylidenecyclohexanones,[504] norbornadienes[505] and in the epimerization of *syn*-9-substituted bicyclo[6.1.0]nonatrienes.[506] Intramolecular hydrogen transfer occurs during the photolysis of 2,4,6-tri-*tert*-butylnitrosobenzene;[507] this compound is used as a spin trap and had been claimed to be photochemically stable. The photo-Beckmann rearrangement of oximes of androsterones leads to mixtures of the epimeric lactams via a biradical.[508] Products derived from the benzyl radicals (**206**), formed in

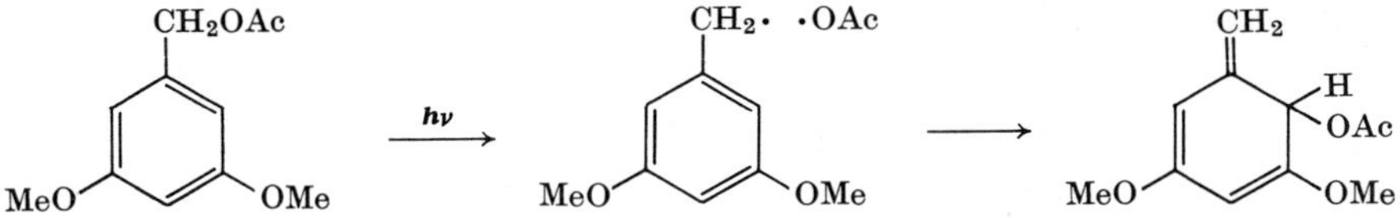

(**205**) (**206**) (**207**)

the photorearrangement of (**205**) to (**207**),[509] can be isolated. The photorearrangement of benzyl benzoate has also been reported.[510] Other examples of rearrangements involving biradicals are dealt with subsequently in the sections on biradicals and photolysis.

Biradicals

Reviews on biradicals[511] and their rearrangements[512] have appeared. Pericyclic reactions proceeding by a biradical mechanism have been shown to be constrained by orbital-symmetry requirements.[513]

Photolysis of the bicyclic azo-compound (**208**) gives the product of retained configuration (**211**) as the major product, the reaction proceeding through a classical biradical intermediate (**209**).[514] Thermolysis, however, gives (**212**) as the major product, this time with inversion. The reaction is believed to proceed by the trimethylene intermediate (**213**) followed by conrotatory ring closure thereof. It is significant that in related systems the degree of inversion increases when a coplanar arrangement of atoms

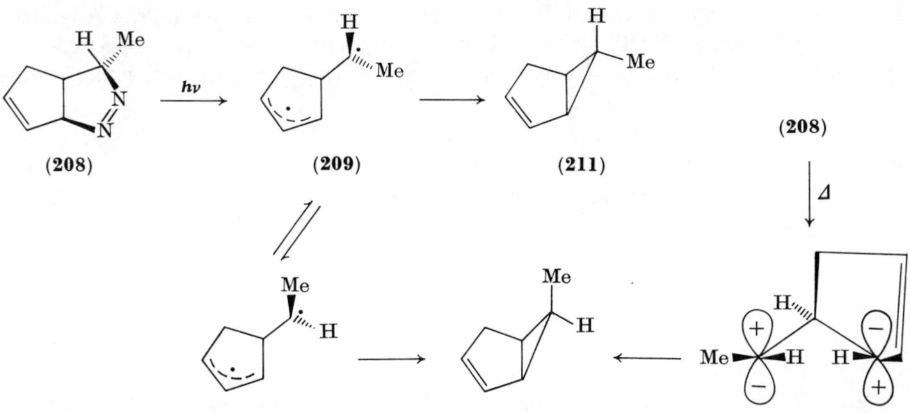

(208) (209) (211) (208) Δ (210) (212) (213)

can occur in this intermediate.[515] Predominant inversion is observed in both the photolysis and thermolysis of (**214**) and (**215**);[516] this reaction is considered to involve C—N cleavage to give the nitrogen-containing biradicals (**216**) and (**217**) and thence predominantly (**218**) and (**219**), respectively. The stereoselectivity of the reaction is

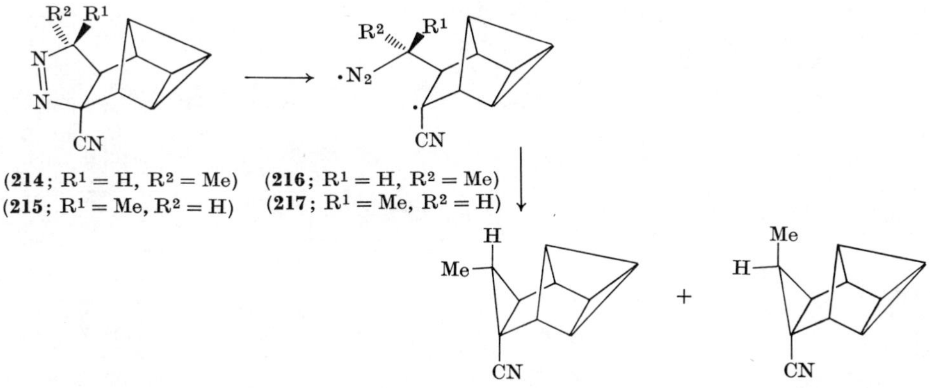

(**214**; $R^1 = H$, $R^2 = Me$) (**216**; $R^1 = H$, $R^2 = Me$)
(**215**; $R^1 = Me$, $R^2 = H$) (**217**; $R^1 = Me$, $R^2 = H$)

(**218**) (**219**)

less with the *endo*-compound (**215**) than with the *exo*-compound (**214**) as the biradical (**217**) is subject to greater steric hindrance to rotation than is (**216**). The products obtained on photolysis and thermolysis of (**220**) are derived from the biradicals (**221**) and (**222**), and from (**223**), respectively.[517]

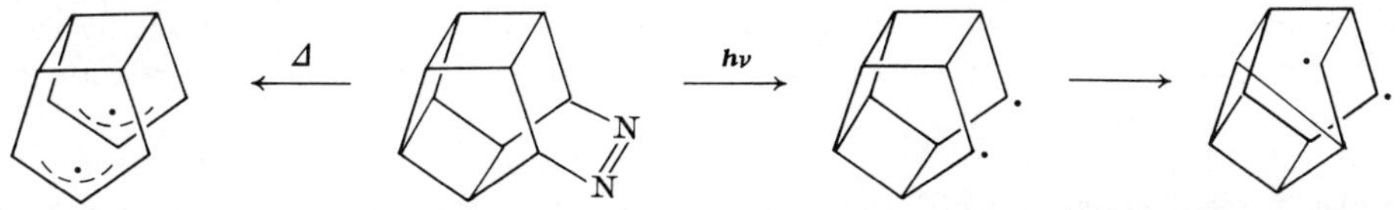

(**223**) (**220**) (**221**) (**222**)

Kinetic studies suggest that a biradical mechanism is involved in the racemization[518] and rearrangement[519] of *N*-arylidene-2,2-diphenylcyclopropylamines (**224**) to 1-pyrrolines (**225**), which is closely related to the vinylcyclopropane–cyclopentene rearrangement.[520] Previous studies have favoured both radical and concerted routes for the latter rearrangement. The postulated biradical intermediate can be generated by thermolysis or photolysis of the appropriate azo-compound. Thus the biradical (**227**) is obtained from the 3-vinylpyrazoline (**226**);[521] the products from this reaction at 120° are (**228**), (**229**) and (**230**). At high temperatures, (**231**) is additionally obtained at the expense of (**230**) by a pericyclic rearrangement. 3,5-Divinylpyrazoline decomposes much more readily than (**226**) as it leads to a highly stabilized diallylic biradical.[522] Kinetic data show that successive introduction of α-vinyl groups in pyrazolines results in a progressive lowering of the activation energy, as would be expected for synchronous C—N bond cleavage.[523]

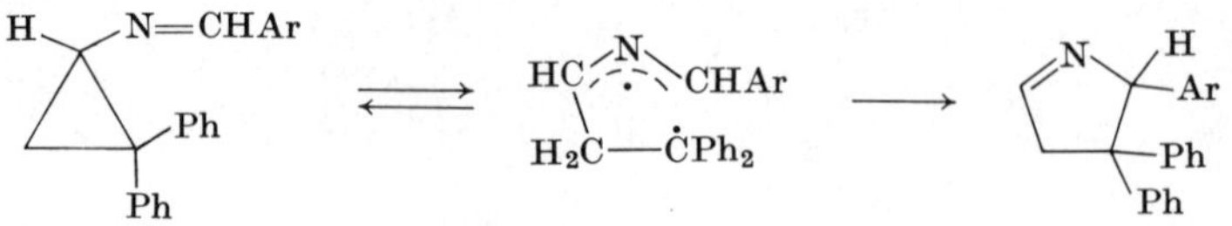

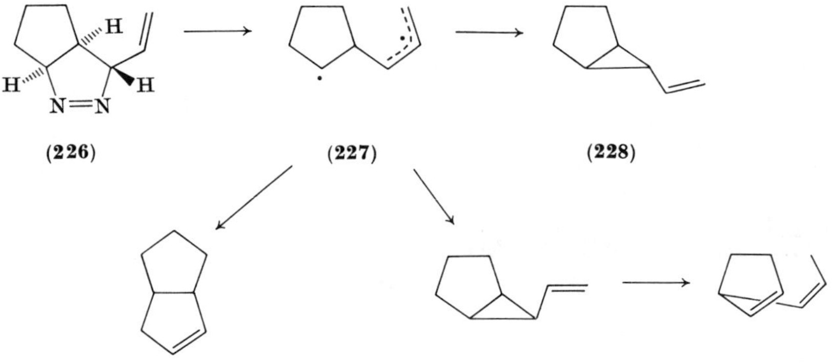

The reactivity of the singlet biradical (**233**), generated from (**232**), towards olefins roughly parallels Diels–Alder dienophilic reactivity, and reactions proceed with high regio- and stereo-selectivity to give fused products (**235**).[524] This is consistent with existence of the biradical in the bisected conformation, which is that predicted for the parent trimethylenemethane.[525] The triplet biradical (**234**) gives the bridged product (**236**), consistently with its being in the planar conformation. The reactivity of the triplet species towards olefins is that to be expected if the transition state resembles that for free-radical addition.

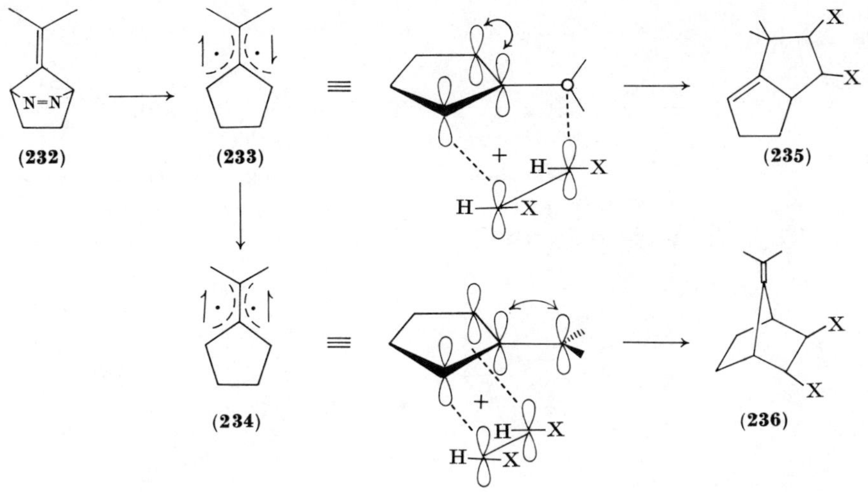

(232) (233) (235)

(234) (236)

Cleavage of the biradical (**239**) is much more stereospecific when it is generated photolytically from (**237**) than thermolytically from (**238**).[526] In the photolysis the biradical has an excess of energy and hence it reacts faster than the thermally equilibrated biradical generated in the thermolysis.

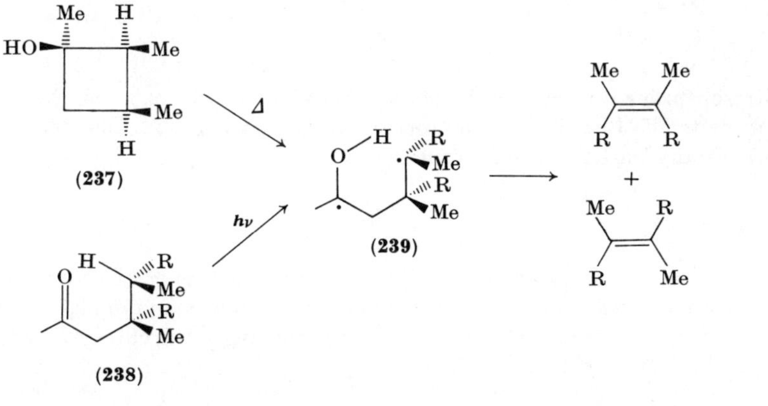

(237) (238) (239)

The biradical (**244**) is generated by addition of the triplet carbene (**242**), derived from (**240**), to 1,1-dicyclopropylethylene (**243**). It subsequently rearranges and undergoes intramolecular 1,5-hydrogen abstraction to give (**245**).[527] This reaction competes with the normal addition of the singlet carbene (**241**) to the olefin.

Biradicals appear to be involved in thermolyses of polyhalopyridazines.[528] CIDNP studies have shown that a biradical is generated from dispiro[2.2.2.2]deca-4,9-diene.[529] The biradical (**247**) has been detected by ESR in the monomerization of 1,2-bis(arylthio)-1,2-dicyanocyclobutane (**246**).[530]

=N$_2$ $\xrightarrow{\Delta \text{ or } h\nu}$ R_2C↑↓ ⟶ R_2C↑↑ $\xrightarrow{(243)}$ $R_2\dot{C}$—CH_2—$\dot{C}$

(240) **(241)** **(242)** **(244)**

(243)

R R

$R_2\dot{C}$—CH_2—C=CH, $H_2\dot{C}$—CH_2

CH=CH_2

(243)

R_2C=CH—C=$CHCH_2CH_3$

(245)

SR, CN, CN, SR $\xrightarrow{\Delta}$ SR, CN, CN, SR ⟶ $2CH_2$=C(SR)(CN)

(246) **(247)**

Nitrosobenzene is formed in the photolysis of 1-hydroxy-1,2,3-triazole via a biradical intermediate.[531] Biradicals also appear to be involved in the decomposition of 4,5,6,7-tetrahydro-1,2,3-benzotriazole.[532]

Nitroxides

Interest in the geometry about the NO group in nitroxides has been maintained. Most nitroxides examined, including (**248**), appear to have a pyramidal arrangement of atoms about the nitrogen.[533–537] Somewhat surprisingly the closely related nitroxides (**249**) and (**250**) are planar.[538, 539]

O COOH

N N N

O· O· O·

(248) **(249)** **(250)**

The a_N value of *p*-methoxyphenyl *tert*-butyl nitroxide is 0.5 G greater than for phenyl *tert*-butyl nitroxide owing to stabilization of the dipolar canonical form (**250a**) by the

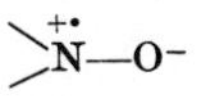

$\gt\overset{+\bullet}{N}—O^-$

(250a)

electron-donating methoxy-group.[540] Introduction of methyl groups at the 3- and 5-positions reduces a_N and also a_H^{OMe} owing to steric inhibition of resonance. The value of a_H^{OMe} is positive, implying transfer of spin by resonance and hyperconjugation interaction. Spin transmission via oxygen (and sulphur) depends on the efficiency of p_π–d_π overlap.[541] The positive coupling constants observed for the β-protons of alkyl groups in *p*-$RC_6H_4N(O\bullet)Bu^t$ are attributed to homohyperconjugative interactions.[540] An *ortho*-fluorine atom in phenyl *tert*-butyl nitroxide is sufficiently large to interact with the *tert*-butyl group, thereby forcing the nitroxide group out of conjugation with the aromatic system.[541]

There is a distinct barrier between the two rotational conformers (**251**) and (**252**) of 2-furyl alkoxy nitroxides.[542] The rotational barrier about the C—N bond is greater for

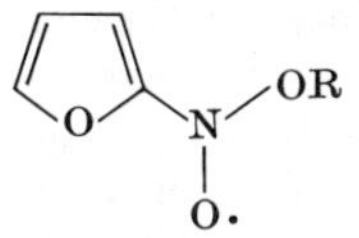

(251) 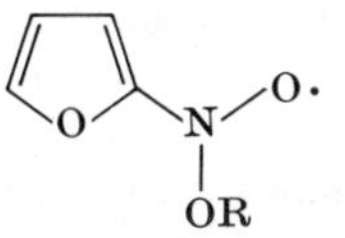 **(252)**

2-thienyl triethylsilyloxy nitroxide, 2-ThN(O•)$OSiEt_3$ than for the 3-thienyl analogue.[543] In 5-substituted 2-thienyl triethylsilyloxy nitroxides electron-donating substituents increase the a_N values.[544]

The a_N values of nitroxides (**253**) derived from 1-hydroxy-2-indolines are low, reflecting destabilization of the dipolar canonical structure (**253b**).[545]

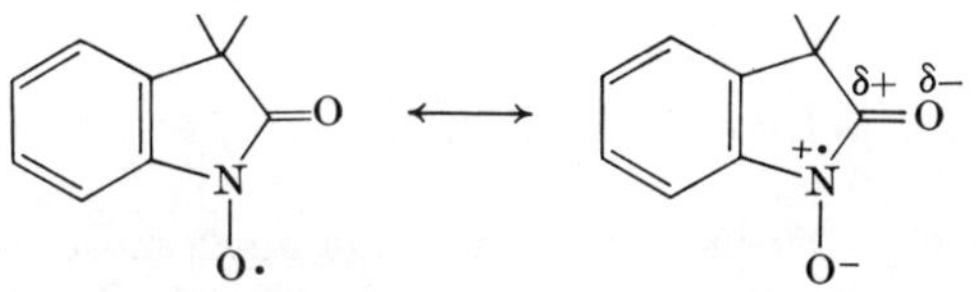

(253a) **(253b)** 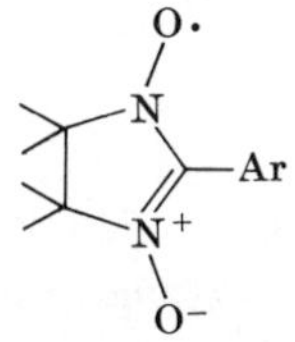 **(254)**

From studies of coupling constants in aryl nitrosyl nitroxides (**254**) it appears that spin polarization factors for C—N and C—C bonds are almost the same.[546] Fluorine-containing aryl nitrosyl nitroxides have been examined.[541] ENDOR studies on the Banfield–Kenyon radical[547] and phenyl *tert*-butyl nitroxide have been reported.[548] The methylene protons are non-equivalent in nitroxides of the type $XYZCCH_2N(O\bullet)Bu^t$.[549, 550] The ESR spectra of cyclic alkoxy nitroxides[551] and of nitroxides with β-chlorine substituents[552] have been reported. The ESR spectra of adducts of $Ph_3Si\bullet$, $\bullet SiHPh_2$ and $\bullet SiH_2Ph$ with phenyl *tert*-butyl nitrone are characterized by a large doublet splitting due to the β-proton.[586] These adducts decay in the dark by a first-order mechanism.

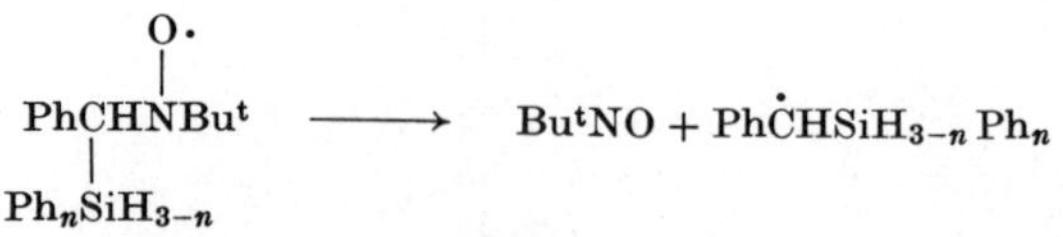

$$\mathrm{PhCH(O^{\bullet}NBu^{t})(Ph_nSiH_{3-n})} \longrightarrow \mathrm{Bu^tNO + Ph\dot{C}HSiH_{3-n}Ph_n}$$

The nitroxide (**255**) exists in the tautomeric forms (**255a**) and (**255b**) with the former predominating in dioxan but the latter in DMSO,[553] whilst the major form of the related porphyrexide in dioxan is (**256**).[554]

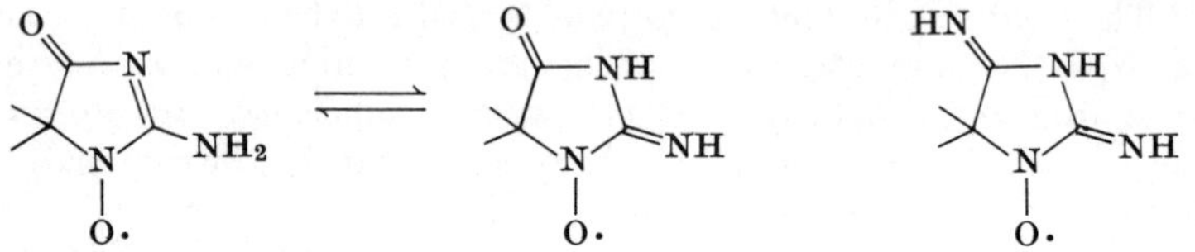

(**255a**) (**255b**) (**256**)

The rate of exchange in binitroxides, though influenced by the nature of the connecting bridge, is not merely related to the number of interconnecting σ-bonds. Thus, although the distance between the radical centres is similar, the rate of exchange of (**257**; X = O; Y = O) is greater than that of (**257**; X = NH; Y = O) or (**257**; X = O; Y = S).[555, 556] A greater exchange rate might be expected in (**258**) owing to the closer proximity of the radical centres and a more specific orientation allowing better overlap of the molecular orbitals containing the unpaired electron; but this is not observed.[555] Related

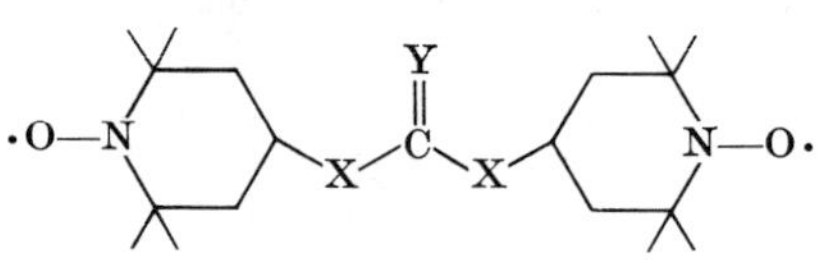

(**257**) 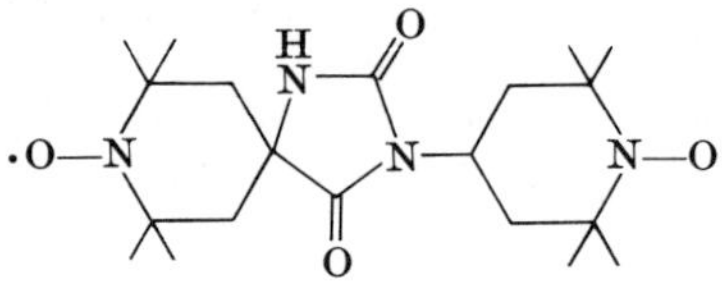 (**258**)

dinitroxides with hydrocarbon bridges have also been examined,[557] as have isoindoline nitroxides.[558]

p-Methoxy- and *p*-phenoxy-phenyl *tert*-butyl nitroxides are much more stable than phenyl *tert*-butyl nitroxide.[559] They dimerize reversibly to give O-*ortho*-C and O-*para*-C coupled products. The latter then breaks down further to give in the case of the methoxy-compound *N*-*tert*-butylanisidine and 2-methoxy-5-*tert*-butylbenzoquinone. The decomposition of both 1- and 2-naphthyl *tert*-butyl nitroxides likewise proceeds via a C—O dimer which breaks down to give the products.[560]

The stability of nitroxides such as (**248**) and di-*tert*-butyl nitroxide is ascribed, not only to steric hindrance towards dimerization and other reactions, but also to the low O—H bond strength in the parent hydroxylamines.[97] Thus hydrogen abstraction by nitroxides will be slow. The factors that influence O—H bond strength in hydroxylamines and hence determine the stability of nitroxides have been measured. Dimerization of several bicyclic nitroxides has been shown to give triplet dimers.[561]

Alkyl *tert*-butyl nitroxides can be conveniently prepared by trapping alkyl radicals, generated from alkyl halides and tributyltin hydride, with nitrosobutane:[562]

$$RX \xrightarrow{Bu_3Sn\cdot} R\cdot \xrightarrow{Bu^tNO} RN(O\cdot)Bu^t$$

Dialkyl nitroxides are generated by use of *tert*-butyl nitrite as the radical initiator.[562]

$$Bu^tONO \rightarrow Bu^tO\cdot + NO$$

$$Bu^tO\cdot + Bu_3SnH \rightarrow Bu^tOH + Bu_3Sn\cdot$$

$$Bu_3Sn\cdot + RX \rightarrow Bu_3SnX + R\cdot$$

$$R\cdot + NO \rightarrow RNO$$

$$R\cdot + RNO \rightarrow R_2NO\cdot$$

Oxidation of *N*-hydroxytriazenes provides a route to alkyl aryl nitroxides:[552, 563]

$$ArN{=}NN(OH)R \xrightarrow{PbO_2} Ar\cdot + N_2 + RNO$$

$$Ar\cdot + RNO \longrightarrow ArN(O\cdot)R$$

Nitrosylsulphinate radical anions, $RN(O\cdot)SO_2^-$,[564] and pentacyanocobalt(III) nitroxide radical anions, $[(NC)_5Co^{III}N(O\cdot)R]^{3-}$ [565] are prepared by trapping SO_2^- and $[Co^{II}(CN)_5]^{3-}$, respectively, with nitroso-compounds. Amidinyl-*N*-oxyls, $R'N{=}CHN(O\cdot)R$, and amidinyl-*N*-oxide-*N'*-oxyls, $R\overset{+}{N}(O^-){=}CHN(O\cdot)R$ are derived through oxidative coupling of amines and hydroxylamines, respectively, with methylene nitrones $CH_2{=}\overset{+}{N}(O^-)R$.[566] Fremy's salt has been used to oxidize hydroxylamines to nitroxides.[567] Oxidation of hydrazyl radicals with oxygen leads to nitroxides.[76] Nitroxides have also been prepared by oxidation of piperidone with sodium tungstate and triton B.[568] The synthesis of $CF_3N(O\cdot)CF_2CF_2N(O\cdot)CF_3$ has been described.[569] Cyclic alkoxy nitroxides[261] are formed as the result of reduction of *o*-nitrobenzenediazonium salts (**259**) in the presence of olefins.[551] The intermediate *β*-*o*-nitroarylalkyl radicals (**260**) cyclize very rapidly, thus accounting for the low yields of Meerwein products in reactions of *o*-nitrobenzenediazonium salts.

NO2 / N2+ —e, −N2→ NO2 (aryl radical) —C=C→

(**259**)

(**260**) → (**261**)

A number of nitroxides that have potential use as spin labels have been prepared. These include an oxazine nitroxide,[570] two steroidal nitroxides[571, 572] and a nitroxide prepared by the photoaddition of *N*-nitrosopiperidine to *α*-pinene.[573] The piperidine in the last-mentioned reaction makes the nitroxide a potentially valuable pH-dependent spin label. Spin-labelled phosphorus compounds have been synthesized,[574] as has a spin-labelled derivative of nicotinamide adenine dinucleotide.[575] The rotational motion of steroid spin labels in lipids has been examined.[576] Limited success in determination of conformations of nitroxides has been achieved by using an octant rule similar to that used for ketones.[577]

Many examples of the use of spin traps as a mechanistic tool in radical chemistry have appeared. 5,5-Dimethylpyrroline 1-oxide is an even more efficient spin trap than nitrosobutane:[578] the rate constants for trapping *tert*-butoxy-radicals are 5×10^8 and 1.5×10^6 mol^{-1} s^{-1}, respectively. Both 5,5-dimethylpyrroline 1-oxide and phenyl *tert*-butyl nitrone have been used to trap $HO\cdot$ and $HO_2\cdot$ produced in the photolysis of hydrogen peroxide.[579] Nitrosobutane has been used as a trap for nucleophilic substitutions involving one-electron transfers.[580] The reaction of alkoxy-radicals with trifluoronitrosomethane has been examined,[581] as has the reaction of $(PhS)_2N\cdot$ radicals with nitroso-compounds.[582] Cycloheptatrienyl radicals produced in the reduction of tropylium salts with zinc have been trapped with nitrosobutane.[583] Radical species are also produced in the reaction of nitrosobutane with potassium *tert*-butoxide.[584]

$$Bu^tNO + Bu^tO^- \longrightarrow Bu^tN(O^-)OBu^t \longrightarrow Bu^tN(O^-)O^- + CH_2{=}CMe_2 + H^+$$

$$Bu^tN(O^-)O^- + 2Bu^tNO \longrightarrow Bu^tNO_2 + 2Bu^t\dot{N}O^-$$

$$Bu^t\dot{N}O^- + O_2 \longrightarrow Bu^tN(O^-)OO\cdot \xrightarrow{ArH} ArN(O\cdot)Bu^t + HO_2^-$$

Spin trapping of radical intermediates has been extended to electrochemical reactions with the use of phenyl *tert*-butyl nitrone as a trap for phenyl radicals generated in the electrochemical reduction of diazonium salts.[585] Aryl *tert*-butyl nitrones, however, decompose at 110° or on photolysis to give $Bu^t_2NO\cdot$ and $ArCHBu^tN(O\cdot)Bu^t$.[587]

Irradiation of benzophenone (methylthiomethyl)imine *N*-oxide (**262**) results in the formation of the unusual nitroxide radical (**263**), which was detected by ESR spectroscopy. This nitroxide decomposed to benzophenone.[588]

$$\underset{(\mathbf{262})}{Ph_2C{=}\overset{+}{N}(O^-)CH_2SMe} \xrightarrow{h\nu} \underset{(\mathbf{263})}{\left[Ph_2C{\cdots}N(O){\cdots}CH_2\right]^{\cdot}} + MeS\cdot$$

$$MeS\cdot \longrightarrow MeSSMe$$

$$(\mathbf{263}) \longrightarrow Ph_2\overset{O}{C{-}N}CH_2\cdot \longrightarrow Ph_2CO + \dot{N}{=}CH_2$$

Oxidation of nitroxide alcohols to the corresponding ketone by *m*-chloroperbenzoic acid,[589] of isoquinuclidinic nitroxide ketones by silver(I),[590] and of "doxyl" derivatives of ketones by nitrogen dioxide[591] has been reported.

Diphenylamine and *p*-ethoxydiphenylamine are obtained by reduction of diphenyl nitroxide with triethyl phosphite in ethanol, and also by decomposition of tetraphenylhydrazine, suggesting that diphenylaminyl radicals are intermediates in both reactions. In hydrogen-donor solvents these abstract hydrogen, whereas in polar solvents they disproportionate (Scheme on p. 129).[592]

Lewis acids have been shown to effect rearrangements of nitroxides.[593] They also

$$Ph_2NO\cdot + P(OEt)_3 \longrightarrow (EtO)_3\dot{P}\text{—}O\text{—}NPh_2 \longrightarrow (EtO)_3PO + Ph_2N\cdot$$

$$Ph_2N\cdot \xrightarrow{RH} Ph_2NH$$

$$Ph_2N\cdot \xrightarrow{EtOH} Ph_2N^- + Ph_2N^+$$

$$Ph_2N^- \longrightarrow Ph_2NH \qquad Ph_2N^+ \longrightarrow p\text{-}EtOC_6H_4NHPh$$

form paramagnetic and diamagnetic adducts with nitroxides. Fremy's salt has been used to bring about the aromatization of tetrahydroisoquinoline.[594] The catalytic hydrogenation of nitroxides has also been reported.[595]

S_H2 Reactions[596]

Tri-*sec*-alkylboranes react with mercury(II) methoxide in tetrahydrofuran to give the corresponding alkylmercuric salt; the mechanism involves homolytic substitution at boron and mercury:[597]

$$MeO\cdot + (R_2CH)_3B \rightarrow MeOB(CHR_2)_2 + R_2CH\cdot$$

$$R_2CH\cdot + Hg(OMe)_2 \rightarrow R_2CHHgOMe + MeO\cdot$$

The kinetics of the reaction of trialkylboranes with some hydroperoxides have also been investigated.[598]

The cleavage of the cobalt—carbon bond in alkylcorrins by chromium(II) may well occur by a homolytic displacement in which the chromium(II) displaces the cobalamin complex from the saturated carbon atom.[599] The products of the atomic fluorination of perfluorocyclobutane consist of a series of perfluoro-*n*-alkanes, and the key step probably involves homolytic displacement by fluorine on the ring-carbon:[600]

$$F\cdot + c\text{-}C_4F_8 \longrightarrow CF_3(CF_2)_3^{*}\cdot \longrightarrow \text{Products}$$

Amino-radicals from *N*-chlorosuccinimide or 1,1,4,4-tetramethyl-2-tetrazene participate in an S_H2 reaction at tin in hexa-alkylditins:[601]

$$R_3Sn\cdot + R'_2NX \rightarrow R_3SnX + R'_2N\cdot$$

$$R'_2N\cdot + R_3SnSnR_3 \rightarrow R'_2NSnR_3 + R_3Sn\cdot$$

An intriguing homolytic substitution reaction has been identified in the reaction of trialkylallyltin compounds with halo-organic substances:[602]

$$R_3Sn\cdot + R'X \rightarrow R_3SnX + R'\cdot$$

$$R'\cdot + R_3SnCH_2CH{=}CH_2 \rightarrow R'CH_2CH{=}CH_2 + R_3Sn\cdot$$

In the reaction with 6-iodohex-1-ene, cyclization gave predominantly (~80%) the five-membered ring product:

$CH_2{=}CH(CH_2)_4I$ + $Bu_3SnCH_2CH{=}CH_2$ $\xrightarrow{-Bu_3SnI}$ [cyclopentyl product] + [open-chain diene] + [cyclohexyl product]

The semi-polar bond in amine oxides and betaines can be cleaved by trialkyltin radicals, giving the corresponding amine:[603]

$$R_3Sn\cdot + C_5H_5N{\rightarrow}Z \longrightarrow R_3SnZ\cdot + C_5H_5N$$

$Z = O$, NCOPh or $N{-}C_6H_4NO_2$-*o*

Sulphoxides and sulphilimines are reduced to sulphides in a rather similar fashion, except that dialkyl sulphoxides react by an alternative route:[604]

$$R{-}S(\rightarrow Z){-}R + Bu_3Sn\cdot \longrightarrow R{-}\dot{S}(ZSnBu_3){-}R \longrightarrow R_2S + Bu_3SnZ\cdot$$

$$R{-}S(\rightarrow Z){-}R + Bu_3Sn\cdot \longrightarrow Bu_3SnR + R{-}\dot{S}{=}Z$$

Polyhaloalkyl radicals react with tri-*n*-butyltin alkoxides by attacking alpha hydrogen borne by the alkoxyl-carbon, followed by rapid expulsion of the $R_3Sn\cdot$ radicals to give the corresponding carbonyl compound; alcohols and polyhalo-derivatives of pentane are formed in side reactions involving attack by $CCl_3\cdot$ either on tin with subsequent elimination of $RO\cdot$ or by abstraction of a hydrogen on the β-carbon atom of the tri-*n*-butyltin group leading to an olefin which then adds a polyhalomethane molecule.[605]

(264) $\xrightarrow[MeCO_2Et]{n\text{-}C_5H_{11}ONO}$ (265) $\xrightarrow{S_Hi}$ (266) + Ph·

(265) $\xrightarrow{RH}$ [diphenyl sulphide with SPh]

(265) $\longrightarrow$ [dibenzothiophene with SPh]

An intramolecular homolytic substitution at sulphur was observed in the reaction of 2-aminophenyl 2-(phenylthio)phenyl sulphide (**264**) with pentyl nitrite. The intermediate aryl radical (**265**) displaces phenyl from the SPh group, giving thianthren (**266**) as the major product; phenyl radicals can be trapped in the system.[606] Similar products were observed in the reaction of 1,2,3-benzothiadiazole with phenyl radicals.

Some further free-radical reactions of platinum complexes[607] have been identified in the oxidative addition of alkyl iodides to $Pt(PPh_3)_3$, a non-chain mechanism was proposed with the formation of radical (**267**) as the rate-limiting step.[608]

$$Pt(PPh_3)_3 \rightleftharpoons Pt(PPh_3)_2 + PPh_3$$

$$Pt(PPh_3)_2 + RI \xrightleftharpoons{\text{Slow}} \underset{(\mathbf{267})}{[PtI(PPh_3)_2]\cdot} + R\cdot$$

$$[PtI(PPh_3)_2]\cdot + R\cdot \xrightarrow{\text{Fast}} trans\text{-}PtI(R)(PPh_3)_2$$

The displaced alkyl radical R• was observed as the spin-adduct with nitrosobutane. In contrast to this a complex cycle of reactions involving a radical-chain process has been suggested[609] for the reaction of alkyl halides with platinum complexes, including examples cited as non-chain in the earlier work.[608] Thiophenol reacts with methylgold(I) and methylplatinum(II), but not with methylgold(III) complexes, by a process involving in part a homolytic substitution at the metal centre:[610]

$$MeM + PhS\cdot \longrightarrow Me\dot{M}SPh$$

$$Me\dot{M}SPh \longrightarrow MSPh + Me\cdot \xrightarrow{PhSH} CH_4 + PhS\cdot$$

$$Me\dot{M}SPh \xrightarrow{PhSH} PhS\cdot + Me(H)MSPh \longrightarrow CH_4 + PhSM$$

Phosphoranyl Radicals

A very small phosphorus hyperfine coupling constant (9.7 G) was observed for the phenylphosphoranyl radical (**268**), indicating a tetrahedral structure at the phosphorus atom with the unpaired electron localized to a large extent in the phenyl ring;[611] it was

Ph$\bar{\cdot}$ — P+ (MeO, OMe, OBu^t)

(**268**)

suggested that the unpaired electron might be captured by either phosphorus or the phenyl ring in a phenylphosphoranyl radical, depending on the electron affinities of the two sites and the amount of distortion needed to form the radical.[612] Phenylphosphoranyl radicals in the series $[Ph_nPX_{3-n}Y]\cdot$ fall into two types; those with X = OEt, Y = OBu^t have low a_p, consistent with tetrahedral structure at phosphorus, whereas those with $X = Cl$, $Y = OBu^t$, $n = 1$ show large a_p and have the normal trigonal-bipyramidal structure.[613] Ligands capable of stabilizing positive charge on phosphorus favour the benzene radical anion, tetrahedral, structure, but electron-withdrawing substituents favour the trigonal-bipyramidal structure.

A distorted trigonal-bipyramidal structure is predicted from INDO calculations for phosphoranyl radicals $\cdot PX_4$ (X = H, F or Cl), the unpaired electron occupying an equatorial position, and the axial P—X bonds being slightly longer than the equatorial bonds.[614] The anisotropic ESR spectrum of $POCl_3^{\overline{\cdot}}$ indicated a trigonal-bipyramidal structure with equivalent chlorine atoms occupying axial positions; MO calculations suggest that the unpaired electron occupies an orbital that can be represented as a Rundle three-centre non-bonding orbital involving the axial ligand orbitals modified by mixing with 3*s* and 3*p* orbitals of phosphorus in an antibonding combination. In this way the change in spin-density distribution in a variety of $\cdot PX_4$ radicals can be explained without recourse to severe distortion of the structure.[615]

The addition of hydrogen atoms to phosphines involves apical attack, giving initially (**269**) which is kinetically favoured over the thermodynamically more stable structure (**270**).[616]

(**269**) (**270**)

An ESR study of the mode of scission of alkoxyalkylphosphoranyl radicals $\cdot PEt_n(OR)_{4-n}$ revealed a maximum stability towards α-scission for $n = 2$; β-scission was thermodynamically favoured because of the strength of the new P=O bond, but was not favoured when an apical C—O bond would be cleaved. Alkyl groups showed a preference for departure from apical positions in the trigonal-bipyramidal phosphoranyl radical.[617]

The reaction of two different mixtures of the stereoisomeric cyclic phosphites (**271**) with *tert*-butoxy-radicals indicated a nearly stereospecific process, showing that the C—O bond of the *tert*-butoxy-group in the intermediate phosphoranyl radical (**272**) or (**273**) is cleaved much more rapidly than pseudorotation can occur.[618]

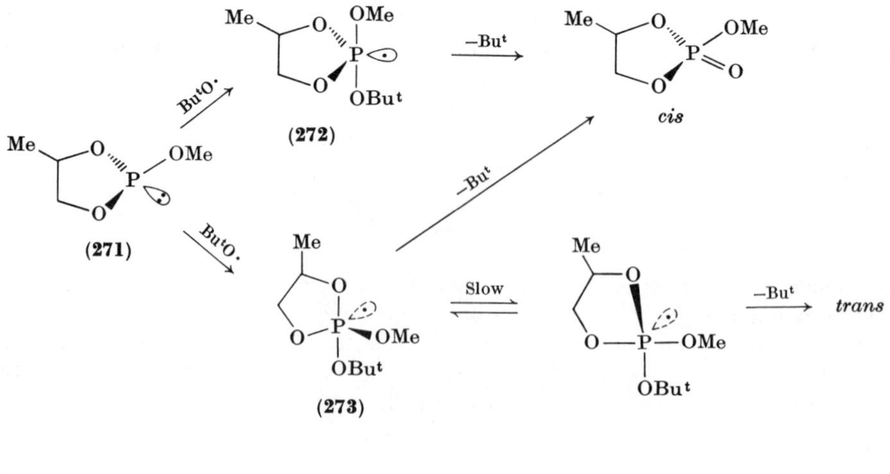

(**271**) (**272**) (**273**) *cis* *trans*

Dimethylamino-radicals were shown to displace alkyl groups from the dioxaphosphorinanes (**274**) with inversion of configuration at phosphorus to give the corresponding phosphoramidite; this indicates that the attacking radical and the substituent Y must be co-axial or co-equatorial in the intermediate phosphoranyl radical.[619]

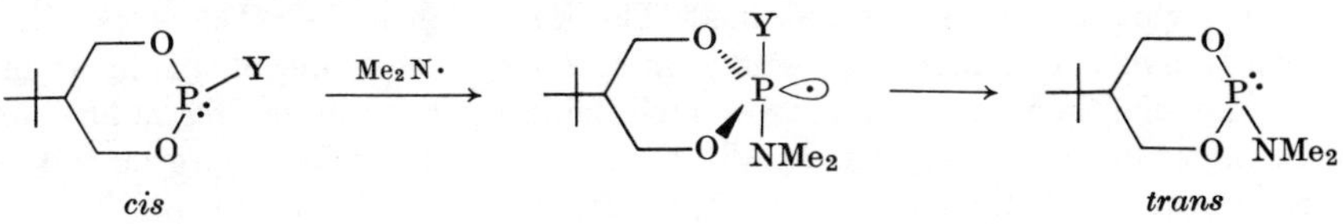

cis (**274**) $\xrightarrow{Me_2N\cdot}$ [phosphoranyl radical with Y and NMe_2] $\longrightarrow$ *trans* (P–NMe_2)

An ESR study of the reaction of alkoxy-radicals with phosphites containing 2- or 3-alkenoxy-substituents showed that the intermediate phosphoranyl radicals may fragment, or may cyclize to give radicals containing an oxaphosphetane (**275**) or an oxaphospholan (**276**) ring.[620]

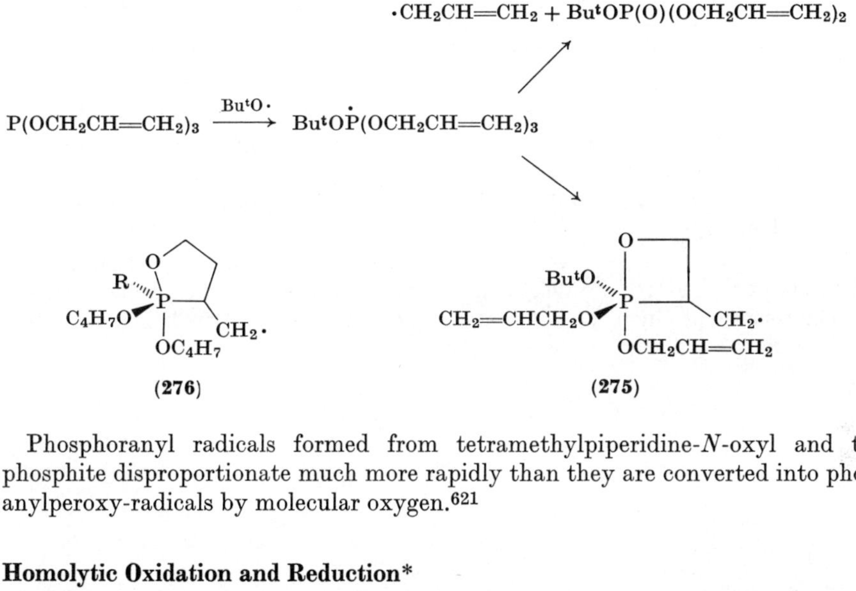

$$P(OCH_2CH{=}CH_2)_3 \xrightarrow{Bu^tO\cdot} Bu^tO\dot{P}(OCH_2CH{=}CH_2)_3 \longrightarrow \cdot CH_2CH{=}CH_2 + Bu^tOP(O)(OCH_2CH{=}CH_2)_2$$

$Bu^tO\dot{P}(OCH_2CH{=}CH_2)_3 \longrightarrow$ (**275**)

(**276**): R, C_4H_7O, OC_4H_7, $CH_2\cdot$

(**275**): Bu^tO, $CH_2{=}CHCH_2O$, $OCH_2CH{=}CH_2$, $CH_2\cdot$

Phosphoranyl radicals formed from tetramethylpiperidine-*N*-oxyl and triethyl phosphite disproportionate much more rapidly than they are converted into phosphoranylperoxy-radicals by molecular oxygen.[621]

Homolytic Oxidation and Reduction*

In general only those publications which involve the study of radicals generated in single electron-transfer reactions will be examined. The metal-ion oxidative cleavage of alcohols has been reviewed.[622] The role of organometallic intermediates, particularly those involving copper, in electron- and ligand-transfer reactions of organic radicals has been reviewed.[623, 624]

The importance of the pH of the solution on the subsequent behaviour of radicals generated in oxidative and reductive processes has been reviewed.[625] The ionized (basic) forms of acids and alcohols have significantly lower redox potentials than the acid forms, making them more powerful reductants. Thus it was found that $E°$ values of

* See also Chapter 4.

$CH_3\dot{C}(OH)COOH$, $CH_3\dot{C}(OH)COO^-$ and $CH_3\dot{C}(O^-)COO^-$ were 0.25, –0.12 and –0.47, respectively.[626] Similar trends were noted for pyrimidine bases.[627] The rate constants for electron-transfer processes involving these species were $\leqslant 5 \times 10^9$ l mol^{-1} s^{-1}.

Oxidation of polymethylbenzenes with iron(III) chloride yields biaryls and diphenylmethanes.[628–630] In mixed oxidations of polymethylbenzenes with naphthalene, good yields of 1-arylnaphthalenes are obtained. The formation of biaryls proceeds by an ECE process involving attack of a radical cation on a second molecule of the aromatic substrate. Diphenylmethane derivatives predominate when the cation radical has a high charge density in a substituted position, as it will readily lose a proton to form a benzylic radical; when the charge density is high in an unsubstituted position, biaryl formation is favoured.

$$ArCH_3 \xrightarrow{-e} ArCH_3^{\dot{+}}$$

$$ArCH_3^{\dot{+}} \xrightarrow{-H^+} ArCH_2\cdot \xrightarrow{-e} ArCH_2^+ \xrightarrow[-H^+]{Ar'H^\cdot} ArCH_2Ar'$$

$$ArCH_3^{\dot{+}} \xrightarrow{Ar'H} [Ar'H.ArCH_3]^+ \xrightarrow[-2H^+]{-e} Ar'C_6H_4CH_3$$

Similar effects are observed in oxidations with cerium(IV) trifluoroacetate.[631] The reaction of toluene with iron(III) chloride has also been examined.[632] Chlorotoluenes are obtained at 70°, whereas under photolytic conditions methyldiphenylmethanes and bibenzyl are formed. Benzyl chloride is obtained when the photolysis is carried out in the presence of water.

Chlorination of toluene, xylenes and mesitylene has been effected by peroxydisulphate in the presence of chloride;[633, 634] the reactions proceed via the aromatic radical cations. Nuclear chlorinated products predominate if the radical cation (**277**) is moderately

Me-C$_6$H$_2$-$R^1R^2R^3$ $\xrightarrow{S_2O_8^{2-}}$ (**277**) radical cation $\xrightarrow{CuCl_2}$ or $\xrightarrow[Cl^-]{S_2O_8^{2-}}$ Me(Cl)C$_6$H-$R^1R^2R^3$

(**277**) $\xrightarrow{-H^+}$ $CH_2\cdot$ (benzyl radical, $R^1R^2R^3$) $\xrightarrow[Cl^-]{S_2O_8^{2-}}$ CH_2Cl (benzyl chloride, $R^1R^2R^3$)

R^1, R^2, R^3 = H or Me

stable, as with mesitylene, whereas side-chain attack occurs exclusively with toluene and *p*-xylene whose radical cations are less stable. Nuclear chlorination occurs exclusively in the presence of copper(II) chloride which is a very effective trap for radical cations.

Electron-transfer has been shown to be involved in the side-chain chlorination of hexamethylbenzene with chlorine in acetic acid.[635]

$$ArMe + Cl_2 \rightarrow ArMe^{\overset{+}{\cdot}} + Cl_2^-$$
$$ArMe^{\overset{+}{\cdot}} \rightarrow ArCH_2\cdot + H^+$$
$$ArCH_2\cdot + Cl_2 \text{ (or } Cl_2^-) \rightarrow ArCH_2Cl + Cl\cdot \text{ (or } Cl^-)$$

Pyridine radical cations are generated by oxidation of pyridine with peroxydisulphate and have been successfully trapped with *tert*-butyl benzylidene nitrone but not with nitrosobutane.[636] The pyridine radical cations thus generated either react with pyridine to give pyridylpyridinium ions (**279**) or, at higher pH, form 2-pyridyl radicals (**280**) and thence bipyridyls; the α-pyridyl radicals have been trapped with nitrosobutane.

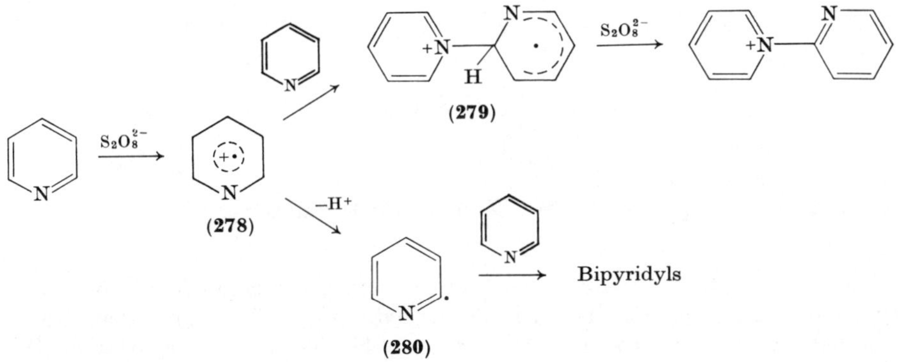

$S_2O_8^{2-}$ $S_2O_8^{2-}$ −H+ Bipyridyls

(**278**) (**279**) (**280**)

The 1,1′-dimer is obtained in the anodic oxidation of phenazine di-*N*-oxide.[689–691] Products derived from the coupling of indolinyl radicals are obtained in the oxidative coupling of indolines and 1,2,3,4-tetrahydroquinolines with permanganate.[719]

The reaction of [2,2]metacyclophanes with copper(II) chloride and dibenzoyl peroxide in acetonitrile leads not only to the nuclear substituted products, 8-cyanomethyl- and 4-benzoyloxy-[2,2]metacyclophanes as a consequence of homolytic aromatic substitution by $PhCO_2\cdot$ and $\cdot CH_2CN$ radicals, but also to pyrene and hydropyrenes.[637] The last two products arise by formation of the radical cation (**281**) which undergoes reversible cyclization to (**282**) followed by oxidation and proton loss. Radical cations

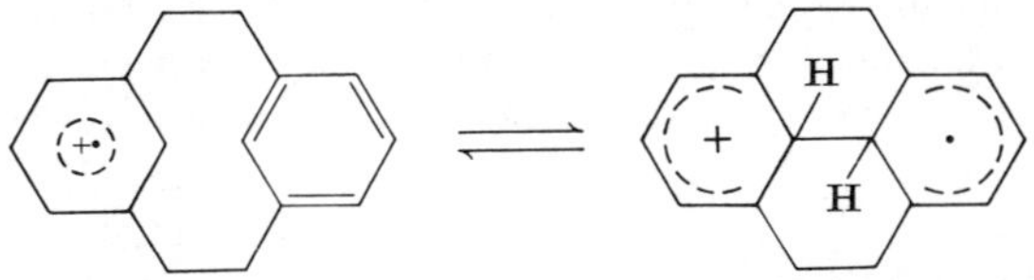

(**281**) (**282**)

are also involved in the oxidation of 7,8,12-trimethylbenz[*a*]anthracene with lead tetra-acetate,[638] in the anodic synthesis of tetrahydro-2,5-dimethyl-2,5-diphenylfuran from α-methylstyrene,[639] in the anodic acetamidation of anthracene,[720] and in the oxidative dimerization of 2-pyrazolines,[640, 641] carbazoles[721] and tertiary aromatic amines.[722]

The chlorination of *cis*- or *trans*-but-2-enes with copper(II) chloride in acetonitrile leads to the same mixture of *meso*- and (±)-dichlorobutanes.[642]

$$\mathrm{MeCH{=}CHMe} \xrightarrow[\mathrm{CH_3CN}]{\mathrm{CuCl_2}} \mathrm{MeCHCl\dot{C}HMe} \xrightarrow{\mathrm{CuCl_2}} \mathrm{MeCHClCHClMe}$$

From a study of the kinetics of the oxidation of alkyl ethers by peroxydisulphate it is apparent that the same type of mechanism is involved as in the oxidation of alcohols.[643] The initially formed α-oxyalkyl radical (**283**) reacts with peroxydisulphate to form a pseudoacetal intermediate (**284**). This either undergoes elimination to give a vinyl ether or hydrolysis to give aldehyde.

$$\mathrm{EtOEt} \xrightarrow{\mathrm{S_2O_8^{2-}}} \underset{(\mathbf{283})}{\mathrm{Me\dot{C}HOEt}} \xrightarrow{\mathrm{S_2O_8^{2-}}} \underset{(\mathbf{284})}{\mathrm{MeCH(OSO_3^-)OEt}}$$

$$\underset{(\mathbf{284})}{\mathrm{MeCH(OSO_3^-)OEt}} \longrightarrow \mathrm{CH_2{=}CHOEt} \quad \text{or} \quad \mathrm{MeCHO + EtOH + HSO_4^-}$$

The rate of oxidation of alkyl phenyl ethers with manganese(III) acetate has been correlated with the electron-density in the aromatic ring.[644] This is less than expected in *tert*-butyl phenyl ether owing to loss of *p*–π overlap. The anodic dimerization of enol ethers proceeds via the enol ether radical cation which either dimerizes or reacts with a further molecule of enol ether.[645] Radicals have been detected in the epoxidation of olefins with *tert*-butyl hydroperoxide in the presence of molybdenum(VI) complexes.[646] Radicals are also involved in the reaction of 7-cumenyl hydroperoxide with copper(II) acetylacetonate.[647]

Nickel peroxide has been used to effect the oxidation of ketones to 1,4-diketones as a result of coupling of α-keto-alkyl radicals formed by abstraction of an α-hydrogen atom.[648] Ease of hydrogen abstraction is in the order: tertiary > secondary > primary.

$$\mathrm{MeCOCHMe_2} \xrightarrow{\mathrm{NiO_2}} \mathrm{MeCO\dot{C}Me_2} \longrightarrow \mathrm{MeCOCMe_2CMeCOMe}$$

On occasions, steric effects are important in determining the reaction product.

Photo-oxidation of alcohols, ethers and carboxylic acids with uranyl cations (UO_2^{2+}) results in abstraction of a hydrogen from the α-carbon atom and is thus similar in effect to that of ·OH and Bu^{t}O· radicals:[649, 650] $\mathrm{RCH_2OH \rightarrow \cdot CHROH + H^+}$. In contrast, photo-oxidation of acids or alcohols with cerium(IV) results in the generation of alkyl radicals from the photolytic breakdown of a cerium(IV) complex.[650] In the presence of oxygen these alkyl radicals are converted into alkylperoxy-radicals.[651]

$$\mathrm{RCH_2OH} \xrightarrow[\mathrm{Ce^{IV}}\downarrow]{h\nu} \mathrm{R\cdot + CH_2O + H^+ + Ce^{III}}$$

$$\mathrm{RCOOH} \xrightarrow[\mathrm{Ce^{IV}}\downarrow]{h\nu} \mathrm{R\cdot + CO_2 + H^+ + Ce^{III}}$$

In contrast to previous reports, $\beta\gamma$-unsaturated acids give products representing the three possible modes of dimerization of the intermediate allylic radical.[652] The non-terminal double bond, as expected, retains the stereochemistry of the starting material.

$$\mathrm{RCH{=}CHCH_2COO^-} \text{ or } \mathrm{RCH(COO^-)CH{=}CH_2} \longrightarrow \mathrm{RCH{=}CHCH_2\cdot} \leftrightarrow \mathrm{R\dot{C}HCH{=}CH_2} \longrightarrow \mathrm{RCH{=}CHCH_2CH_2CH{=}CHR} + \mathrm{RCH{=}CHCH_2CHRCH{=}CH_2} + \mathrm{H_2C{=}CHCHRCHRCH{=}CH_2}$$

The dependence of the product distribution on electrochemical parameters, nuclear substitution and added perchlorate in the anodic oxidation of substituted phenylacetates is consistent with the annexed scheme.[653] The oxidation of *p*-methoxyphenyl-

$$\mathrm{ArCH_2CO_2^-} \rightleftharpoons \mathrm{ArCH_2CO_2^-(ads)} \longrightarrow \mathrm{ArCH_2\cdot} \rightleftharpoons \mathrm{ArCH_2\cdot\ (ads)} \longrightarrow \mathrm{ArCH_2CH_2Ar}$$

$$\mathrm{ArCH_2\cdot\ (ads)} \xrightarrow{-e} \mathrm{ArCH_2^+} \xrightarrow[-\mathrm{H^+}]{\mathrm{MeOH}} \mathrm{ArCH_2OMe}$$

acetate ions proceeds by electron transfer from the aromatic nucleus (pseudo-Kolbe reaction). Alkyl radicals generated during anodic oxidations of carboxylic acids have been trapped by butadiene.[655] γ-Substituted butyrolactones have been synthesized in good yield by the Kolbe oxidation of γ-paraconic acids.[656]

Several detailed studies on the mechanism of the oxidative decarboxylation of organic acids by lead tetra-acetate have been carried out;[657–660] they confirm previous conclusions that the initial step involves generation of an alkyl radical: $\mathrm{Pb^{IV}OCOR \rightarrow R\cdot + Pb^{III} + CO_2}$. In the oxidation of tertiary carboxylic acids, further oxidation to the carbonium ion occurs,[657] whereas with secondary carboxylic acids the less readily oxidizable radical is trapped to form an organolead species and no product derived from cations is produced:[658, 659] $\mathrm{R\cdot + Pb^{IV}OCOR \rightarrow RPb^{IV} + R\cdot + CO_2}$. Thus in the oxidation of 2,3,3-trimethylbutanoic acid the main product is 3,3-dimethyl-2-butyl acetate as a result of an S_Ni decomposition of the organolead intermediate with none of the rearranged products which would have been expected had the 3,3-dimethyl-2-butyl cation been formed.[658] Similar conclusions were reached from studies on bicyclo[3.2.1]octane-2-carboxylic acid and bicyclo[2.2.2]octane-2-carboxylic acid.[659] The product ratios were altered when the oxidations of both tertiary and secondary carboxylic acids were carried out in presence of copper(II) carboxylates, suggesting the intermediacy of organocopper species; where possible, these tend to break down with *cis*-elimination.[657, 658] Thus 2,3,3-trimethylbutanoic acid gives mainly 3,3-dimethylbut-1-ene. The oxidations of $AcO(CH_2)_n\dot{C}HEt$ radicals to $AcO(CH_2)_nCH{=}CHMe$ and $AcO(CH_2)_{n-1}CH{=}CHEt$ by different copper(II) salts has been shown to be little affected by the nature of the counterion.[661] Alkoxy-radicals are generated in oxidations of ω-(2-adamantyl)alkan-2-ols with lead tetra-acetate in benzene but not in pyridine.[662] The radical reactions lead to five-membered cyclic ethers.

The highly strained apotricyclyl radicals (**286**) are generated in the oxidation of the tricyclenic acid with lead tetra-acetate or in the reaction of 1-chloroformyl apotricyclene (**285**) with silver tetrafluoroborate, as shown by the formation of 1-chlorophenylapotricyclene isomers (**287**) in a ratio characteristic of a radical reaction.[663] The corre-

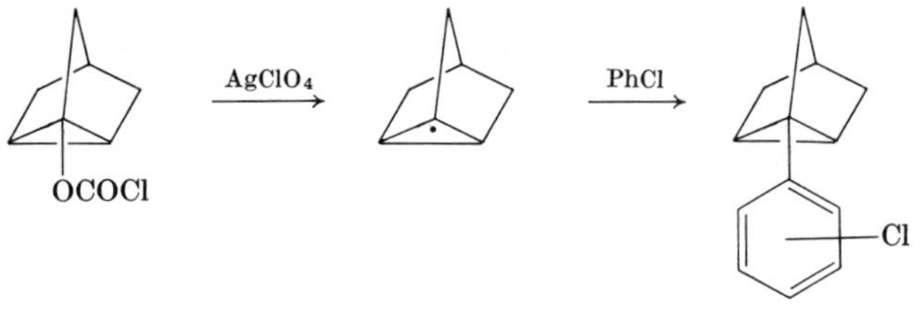

OCOCl —$AgClO_4$→ · —PhCl→ Cl

(**285**) (**286**) (**287**)

sponding reactions of the apocamphor and apobornenyl systems give the corresponding chlorophenyl-substituted systems in an isomer ratio characteristic of electrophilic aromatic substitution. The difference in behaviour is attributed to the fact that the extra cyclopropane ring in the apotricyclyl systems increases the energy of the bridgehead cation to such an extent that it is not formed or, if formed, does not have sufficient time to alkylate the solvent.

Cycloalkyl radicals generated by the peroxydisulphate/silver(I) decarboxylation of cycloalkanecarboxylic acids have been used to alkylate quinones;[664] in the oxidation by this reagent of *cis*-camphoric acid (**288**), the main product (**290**) arises by oxidation of the tertiary carboxylic acid group to give initially the radical (**289**). A new synthesis of γ-phenylbutyrolactone (**292**) based on the peroxydisulphate oxidation of 4-phenyl-

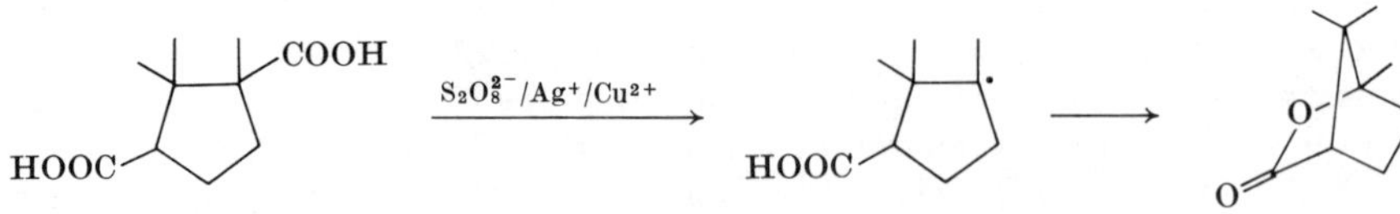

HOOC COOH —$S_2O_8^{2-}/Ag^+/Cu^{2+}$→ HOOC · → O O

(**288**) (**289**) (**290**)

butyric acid has been reported.[665] The reaction proceeds via the radical cation (**291**). In the presence of silver(I) the reaction takes a different course proceeding via the 3-phenylpropyl radical which undergoes fragmentation to a benzyl radical. The radical (**293**) generated in the oxidation of *ortho*-substituted phenoxyacetic acids undergoes

$$Ph(CH_2)_3COOH \xrightarrow{SO_4^{\cdot-}} \overset{+\cdot}{Ph}(CH_2)_3COOH \ (\mathbf{291}) \longrightarrow Ph\dot{C}H(CH_2)_2COOH \xrightarrow{SO_4^{\cdot-}} Ph\overset{+}{C}H(CH_2)_2COOH \xrightarrow{H_2O} (\mathbf{292})$$

$$Ph(CH_2)_3COOH \xrightarrow{Ag^{2+}} Ph(CH_2)_3CO_2\cdot \longrightarrow PhCH_2CH_2CH_2\cdot \longrightarrow CH_2{=}CH_2 + PhCH_2\cdot \xrightarrow{O_2} PhCHO$$

preferential cyclization to (**294**) when R = H, or disproportionates to the anion (**295**) and cation (**296**) when R = Me, in which case steric crowding reduces the stability of (**294**).[666]

(**293**) (**294**) (**295**) (**296**)

The first well-authenticated example of the oxidative coupling of phenols involving a radical substitution mechanism has been reported in the ferricyanide oxidation of 2,3′,4-trihydroxybenzophenone.[667] 4-Alkyl-2,6-di-*tert*-butylphenoxy-radicals decay by first-order kinetics at high radical concentrations and by second-order kinetics at low radical concentrations;[668] this is explained in terms of the reversible formation of the quinol ether dimer (**297**) and disproportionation to phenol and quinone methide. These products could also arise by the molecular decomposition of (**297**).

Oxidation of 3,5-di-*tert*-butyl-2,6-dimethoxyphenol with ferricyanide leads to the unusual product (**302**) formed by coupling of one phenoxy-radical (**298**) with the secondarily formed radical (**299**).[669] The initial coupled product (**300**) is further oxidized in several steps to the biradical (**301**) which cyclizes to (**302**). Reaction of 4,6-disubstituted 2-trialkylsilylphenols with PhHgOH leads to 4,6-disubstituted 2-(phenylmercuri)phenyl trialkylsilyl ethers in a process involving a single-electron transfer followed by migration of the *ortho*-trialkylsilyl group to oxygen.[670]

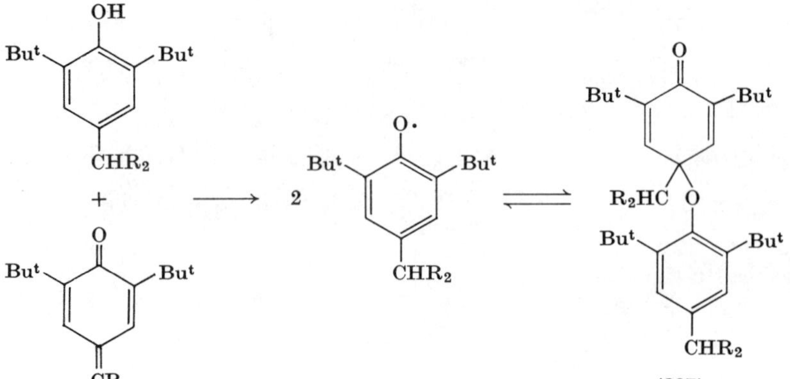

(297)

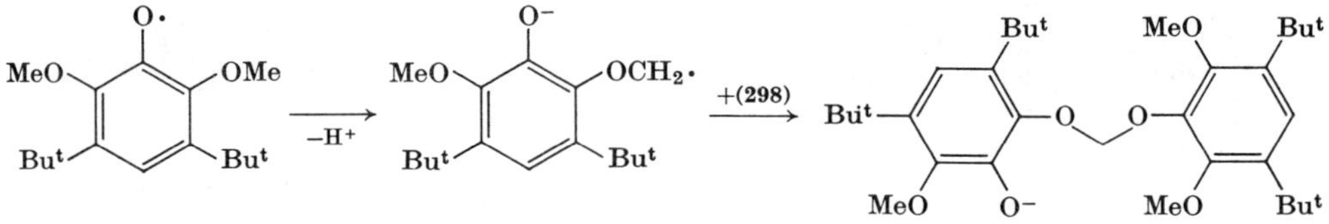

(298) (299) (300) (301) (302)

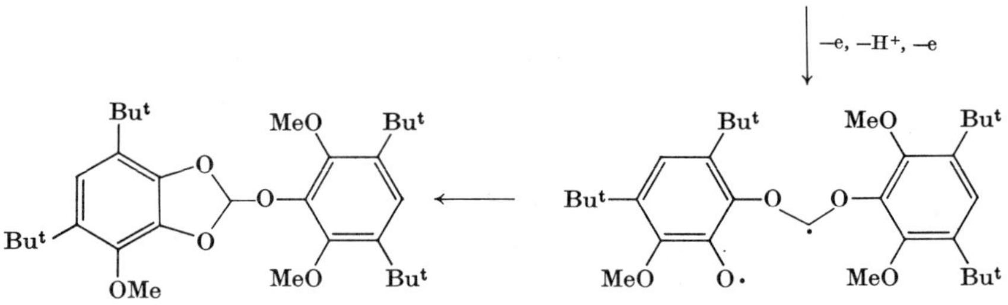

The peroxidase oxidation of 1-(4-hydroxy-3,5-dimethoxyphenyl)ethanol leads to 2,6-dimethoxybenzoquinone; this does not appear to be formed via the acetophenone.[671] The initial step in the oxidation involves formation of the phenoxy-radical, which has been detected by ESR. This radical is also involved in the electrochemical oxidation of 1-(4-hydroxy-3,5-dimethoxyphenyl)ethanol, which in dry acetonitrile leads to the corresponding acetophenone whereas in damp acetonitrile the benzoquinone is obtained.[672] The electrochemical oxidation of the acetophenone has also been examined.[673] Catalytic oxidation of 2,6-di-*tert*-butyl-4-methylphenol with a platinum catalyst leads initially to 3,3′,5,5′-tetra-*tert*-butyl-4,4′-dihydroxybibenzyl and thence to the diphenoquinone.[674] The kinetics of the oxidation of 2,4,6-tri-*tert*-butylphenol by *tert*-butylperoxy-radicals is consistent with reaction involving the formation of a hydrogen-bonded radical–reactant complex followed by hydrogen transfer within the complex.[675] The oxygenation of 4-alkyl-2,6-di-*tert*-butylphenols to the corresponding bis(cyclohexa-2,5-dienone) peroxide by cobalt(II)–Schiff base complexes has been examined.[676] Electron-transfer oxidations of sterically hindered phenols have been reviewed.[677] The rates of coupling of phenoxy-radicals are decreased in the presence of alkali-metal salts.[678] The phenoxy-radical forms complexes with the metal ions Cu^{2+}, Fe^{2+}, Mn^{2+} and Cu^{+}.[678] Phenoxy-radicals are involved in catalytic oxidation of phenol over copper(II) oxide in the aqueous phase.[679] Electrochemical oxidation of phenols gives phenoxy-radicals.[680] Phenol cation radicals are generated in oxidations in acid solution by lead dioxide, hydrogen peroxide or oxygen.[681] Pentafluorophenol is oxidized by lead dioxide to the corresponding perfluoroquinol ether dimer[682] though it is inert to DDQ.[683] It is also oxidized by 4-(aryloxy)-2,4,6-tri-*tert*-butylcyclohexa-2,5-dienones which presumably act as incipient phenoxy-radicals.[683] Autoxidation studies of naphthalenediols,[684] hydroquinones and quinones[685] have been reported; in several instances dimeric radical species can be detected. Substituted phenoxy-radicals behave as electron-donors with vanadium oxychloride, silver perchlorate and bromine.[686] Stable phenoxy-radicals are generated in the reaction of nickel(II) salicylketimine chelates with *tert*-butylperoxy-radicals;[687] this behaviour accounts for the antioxidant properties of these chelates. The reduction of galvinoxyl radicals by metals has been reported.[688]

The reactivities of different C—H bonds in alcohols, ethers and amides towards hydroxyl radicals generated in Fenton's reaction are controlled more by electron densities than by C—H bond strengths, in line with the electrophilic nature of the hydroxyl radical. This behaviour is also indicative of the importance of charge transfer in the transition state.[692] Oxidation of disulphides by Fenton's reagent is an electron-transfer process:[693] $RSSR + HO\cdot \rightarrow [RSSR]^{+\cdot} + HO^{-}$. Electron-transfer is also involved in the reaction of 1-phenylalkanols with both Fenton's reagent and peroxydisulphate.[694] The radical cation (**303**), thus generated, either loses a proton to give the α-hydroxyalkyl radical (**304**) or it undergoes fragmentation to benzaldehyde and alkyl radicals.

$$PhCH(OH)R \xrightarrow{-e} \overset{+\cdot}{Ph}CH(OH)R\ (\mathbf{303}) \xrightarrow{-H^+} Ph\dot{C}(OH)R\ (\mathbf{304}) \xrightarrow[-H^+]{-e} PhCOR$$

$$(\mathbf{303}) \rightarrow PhCHO + R\cdot + H^+$$

The subsequent behaviour of a radical generated in Fenton's reaction depends on its redox potential.[695] Those radicals containing α-OH, OR or NR_2 groups are readily oxidized by iron(III) in an electron-transfer process.[692] The oxidation of $\cdot CH_2OH$ radicals by cobalt(III) hexa-ammine has been reported.[709] α-Alkoxyalkyl radicals are also oxidized readily by hydrogen peroxide;[696] the rates of oxidation of different α-alkoxyalkyl radicals are consistent with a process involving partial charge transfer to hydrogen peroxide in the transition states.

Radicals containing electron-withdrawing groups, e.g. α-carbonyl radicals, may be reduced to the carbanion;[692, 697] the rate of reduction is retarded by electron-releasing groups in the radical. Reduction by iron(II) is much faster than by titanium(III) though the latter is a stronger reductant. Tertiary alkyl radicals are oxidized more rapidly by copper(II) than by iron(III), implicating a different mechanism that probably involves an alkylcopper species (see also p. 138). An alkylcopper species involving the radical (**307**) can explain the preferential formation of the terminal alkene (**308**) in the reduction of the tertiary alkyl hydroperoxide (**305**) by iron(II).[698]

$$\underset{\textbf{(305)}}{\mathrm{Me(CH_2)_3CH(OOH)Me}} \xrightarrow{\mathrm{Fe^{II}}} \underset{\textbf{(306)}}{\mathrm{Me(CH_2)_3CH(O\cdot)Me}} \longrightarrow \underset{\textbf{(307)}}{\mathrm{Me\dot{C}H(CH_2)_2CH(OH)Me}} \xrightarrow{\mathrm{Cu^{II}}} \underset{\textbf{(308)}}{\mathrm{CH_2{=}CH(CH_2)_2CH(OH)Me}}$$

The metal-catalysed decompositions of α-phenylethyl hydroperoxide[699] and perbenzoic acid[700] have also been reported, as have the oxidations of ethanol[701] and propan-2-ol[702] with the titanium(III)-hydrogen peroxide system. In the last case, CIDNP studies indicated that the acetone obtained was derived from reaction of $\cdot CMe_2OH$ radicals with $HO\cdot$ or $HO_2\cdot$ complexed with titanium(IV). The decomposition of hydrogen peroxide by iron(III) had also been examined.[703,704]

In the presence of a proton source the dissolving metal reductions of biphenyls proceed by the transfer of one electron followed by transfer of a proton, whereas in the absence of a proton source two electrons are transferred followed by two protons.[705] The reduction of hex-3-yne with sodium in a protic solvent proceeds via the isomeric vinyl radicals (**309**) and (**310**) which rapidly interconvert (see also p. 146).[706] The reductive acetylation

$$\mathrm{EtC{\equiv}CEt} \xrightarrow[\mathrm{Na^+}]{\mathrm{e}} \mathrm{Et(Na)C{=}\dot{C}Et} \xrightarrow{\mathrm{H^+}} \underset{\textbf{(309)}}{\mathrm{Et(H)C{=}\dot{C}Et}} \rightleftharpoons \underset{\textbf{(310)}}{\mathrm{Et(H)C{=}\dot{C}Et}}$$

$$\textbf{(309)} \xrightarrow{\text{1. e, Na}^+ \text{ 2. H}^+} \textit{trans-}\mathrm{EtCH{=}CHEt} \qquad \textbf{(310)} \xrightarrow{\text{1. e, Na}^+ \text{ 2. H}^+} \textit{cis-}\mathrm{EtCH{=}CHEt}$$

of nitro- and nitroso-compounds leads to *N*-substituted *N,O*-diacetylhydroxylamines (**311**).[707]

$$\mathrm{RNO_2} \xrightarrow{\mathrm{e}} \mathrm{RNO_2^{\overline{\cdot}}} \xrightarrow[-\mathrm{AcO^-}]{\mathrm{Ac_2O}} \mathrm{RN(O\cdot)OAc} \xrightarrow[-\mathrm{AcO^-}]{\mathrm{e}} \mathrm{RNO}$$

$$\mathrm{RNO} \xrightarrow{\mathrm{e}} \mathrm{RNO^{\overline{\cdot}}} \xrightarrow[-\mathrm{AcO^-}]{\mathrm{Ac_2O}} \mathrm{R\dot{N}OAc} \xrightarrow[\mathrm{Ac_2O,\ -OAc^-}]{\mathrm{e}} \underset{\textbf{(311)}}{\mathrm{RN(OAc)COMe}}$$

One-electron reduction of nicotinamide has been reported.[708]

The decomposition of benzenediazonium fluoroborate changes from being exclusively a heterolytic process in pure 2,2,2-trifluoroethanol to being predominantly a homolytic one as pyridine is added to the reaction system.[710] The homolytic process is favoured by the addition of a good nucleophile to the system such that a covalent diazo-compound is formed which is susceptible to facile homolysis. Pyridine is a good nucleophile but only a mediocre homolytic leaving group.

$$PhN_2^+ + Py \rightleftharpoons PhN_2Py^+ \rightarrow Ph\cdot + N_2 + Py^+$$

The requirements are much better met with nitrite, which catalyses the decomposition of diazonium salts by a homolytic process (cf. p. 110). A similar process is involved in the reaction of 1,3-diphenyltriazene with triphenylphosphine.[723] The kinetics and mechanism of the decomposition of diazonium salts by homolytic and heterolytic pathways have been reviewed.[711] The kinetics of the decomposition of diazonium salts by hydroquinone[712] and by *N*,*N*,*N*′,*N*′-tetramethyl-*p*-phenylene-diamine[713] have been studied. The ESR signal of the azobenzene radical cation is observed during the one-electron reduction of diazonium salts:[714] $PhN_2^+ + Ph\cdot \rightleftharpoons [Ph{-}N{=}N{-}Ph]^{+\cdot}$.

Further studies on the fate of 2-benzoylphenyl radicals generated by reduction of diazotised 2-aminobenzophenone by copper(I) have been reported.[715–717] The phenolic product arises in a ligand-transfer reaction with $Cu(H_2O)_n^{2+}$ followed by proton loss.[715] Addition of dioxan decreases the phenol yield owing to micelle formation. The relative amounts of the dimer and fluorenone are pH-dependent with dimer formation favoured at pH > 2.5. At lower pH, arylcopper intermediates, which lead at least in part to dimer,

$ArN_2^+ \xrightarrow{Cu^I} Ar\cdot$ (Ar = 2-benzoylphenyl)

$Ar\cdot \xrightarrow{Cu(H_2O)_n^{2+}} ArOH$

$Ar\cdot \xrightarrow{SH} ArH$

$Ar\cdot \xrightarrow{Cu^{II}} ArCu^{II}$

$Ar\cdot \rightarrow Ar{-}Ar$

$Ar\cdot \rightarrow$ **(311)** $\rightarrow$ **(312)**

(311) **(312)**

decompose. The proportions of benzophenone obtained in reductions of the diazonium salt with copper(I) perchlorate–amine complexes are increased when the amine contains good hydrogen-donor ligands.[716] At high acidity benzophenone is also obtained as a result of hydrogen transfer from the aquated co-ordination sphere of copper(I). It is also derived from reaction with the cyclohexadienyl radical: Ar· + (**311**) → ArH + (**312**).

2,4-Dinitrobenzenediazonium ions react with thiophen and methylthiophens in acidic media to give the corresponding dinitrophenylthiophens.[718] The reaction can be envisaged as involving one-electron transfer to the diazonium salt from the thiophen by reaction of the resultant cation radical with aryl radicals:

$$ArN_2^+ + ThH \rightarrow Ar\cdot + N_2 + ThH^{+\cdot}$$

$$ThH^{+\cdot} + Ar\cdot \rightarrow [Ar.ThH]^+ \rightarrow ArTh + H^+$$

Evidence has been forthcoming in support of the oxidation of phenylcyclohexadienyl radicals by diazonium salts with the concomitant generation of aryl radicals:[437]

$$[Ar.C_6H_6]\cdot + ArN_2^+ \rightarrow [Ar.C_6H_6]^+ + Ar\cdot + N_2$$

$$[Ar.C_6H_6]^+ \rightarrow ArPh + H^+$$

Radical Ions and Electron-transfer Processes

The roles of radical ions as intermediates in oxidation, substitution and electron-transfer reactions have been reviewed.[724]

Interest in S_{RN} reactions continues at a high level.[725] Phenyl and mesityl radicals, generated from halobenzenes either by reaction with solvated electrons or photochemically, react with anions of aliphatic nitriles,[726] α- and γ-picolines[727] and thiophenols.[728]

$$ArI + e \rightarrow ArI^{\cdot -} \rightarrow Ar\cdot + I\cdot$$

$$Ar\cdot + Ar'S^- \rightarrow [ArSAr']^{\cdot -}$$

$$[ArSAr']^{\cdot -} + ArI \rightarrow ArSAr' + ArI^{\cdot -}$$

This type of reaction has been extended to any monosubstituted benzene derivative ArX in which X is more electronegative than carbon.[729] An S_{RN} process seems to be involved in the reductive dehalogenation of haloaromatic compounds by bulky amines.[730] *n*- and *tert*-Butyl radicals can be scavenged by nucleophiles possessing an unsaturated site, e.g. N_3^- and AcO^-, but not by EtO^- or pyrrolidine.[731] The last two, however, have been reported to scavenge cyclohexenyl radicals.[725] The results imply that reaction occurs only if a reasonably low-lying antibonding orbital is available at the reaction site consistent with the intermediacy of a radical anion.

The sulphone group in α-nitro-sulphones is readily displaced by a variety of nucleophiles in a reaction involving radical anion intermediates,[732] as illustrated.

$$R_2C(NO_2)SO_2Ar + R'_2\bar{C}NO_2 \longrightarrow R_2C(NO_2^{\cdot -})SO_2Ar + R'_2\dot{C}NO_2$$

$$R_2C(NO_2^{\cdot -})SO_2Ar \longrightarrow R_2\dot{C}NO_2 + ArSO_2^-$$

$$R_2\dot{C}NO + R'_2CNO_2^- \longrightarrow [R_2C(NO_2)C(NO_2)R'_2]^{\cdot -}$$

$$[R_2C(NO_2)C(NO_2)R'_2]^{\cdot -} + R_2C(NO_2)SO_2Ar \longrightarrow R_2C(NO_2)C(NO_2)R'_2 + R_2C(NO_2^{\cdot -})SO_2Ar$$

The reaction is inhibited by *m*-dinitrobenzene and di-*tert*-butyl nitroxide, consistently with a radical process. Products that can be accounted for by the coupling and cross-coupling of $p\text{-}NO_2C_6H_4\dot{C}Cl_2$ and $p\text{-}NO_2C_6H_4\dot{C}HCl$ radicals are obtained in the reaction of *p*-nitrobenzylidene chloride with sodium hydroxide.[733] There is evidence for an electron-transfer mechanism in the reaction of phenylacetonitrile with aromatic nitro-compounds.[734] 2-Butyl radicals have been trapped by nitrosobutane in the reaction of thiophenoxide with 2-butyl nosylate; this was considered to be consistent with the scheme shown.[735] The reaction of (*S*)-2-octyl nosylate proceeds with complete inversion of configuration.

$$PhS^- + RONs \rightarrow PhS\cdot + RONs^{\cdot-}$$

$$PhS\cdot + RONs^{\cdot-} \rightarrow PhSR + NsO^-$$

$$RONs^{\cdot-} \rightarrow R\cdot + NsO^-$$

CIDNP studies have made possible the detection of radical intermediates in the reactions of butyl-lithium with benzyl halides[735a] and of dibenzyl mercury with triphenylmethyl halides.[735b]

In DMSO 1,2-diaryl-1,2-dichloroethylenes (**313**) undergo base-induced *cis*-elimination of chlorine by a mechanism involving the radical anion (**314**).[736]

$$\underset{(\mathbf{313})}{Ar(Cl)C{=}C(Ar)Cl} \xrightarrow[Et_2O/DMSO]{NaOH} \underset{(\mathbf{314})}{[Ar(Cl)C{=}C(Ar)Cl]^{\cdot-}}$$

$$\xrightarrow{-Cl^-} Ar(Cl)C{=}\dot{C}Ar \xrightarrow{e} Ar(Cl)C{=}C^-Ar \xrightarrow{-Cl^-} ArC{\equiv}CAr$$

Reactions of hydrocarbon radical anions in metallation reactions and in reactions with alkyl halides, ethers and carbonyl compounds have been reviewed.[737,738] The metallation of thiophen by lithium hydrocarbon radical anions occurs in the 2-position by a one-electron transfer process in the presence of diphenylethylene.[739] Metallation of polycyclic aromatic hydrocarbons and nitrobenzene by organo-lithium or -magnesium compounds is greatly facilitated by use of HMPA as solvent.[740] The disproportionation of anthracene and perylene radical anions has been studied.[741]

The rate of electron transfer from aromatic radical anions to alkyl halides is markedly dependent on whether the reaction involves the free ion, a loose ion pair or a tight ion pair.[742] The resulting radicals are reduced rapidly to carbanions. The rate of reduction of hex-5-enyl radicals by sodium naphthalenide is about 1.6×10^9 l mol^{-1} s^{-1}.[743] The nature of the alkali-metal counterion has a very marked effect on the rate of reduction in the order: Li > Na ≫ K ≫ Cs. The carbanions thus derived give either the reduced product RH or, as a result of reaction with alkyl halide, the dimeric product RR. This

$$R\cdot \xrightarrow{NpH^{\cdot-}} R^- \begin{cases} \xrightarrow{SH} RH + S^- \\ \xrightarrow{RX} [R\cdot\ R\cdot\ X^-] \longrightarrow RR + X^- \end{cases}$$

dimer is generally not derived from coupling of alkyl radicals, as shown by the absence of C_8 and C_{10} products in reactions of 1,4- and 1,5-dihaloalkanes.[744] Radical coupling is, however, involved in reactions of benzyl iodide (though not of the bromide or chloride),[745] benzhydryl chloride or triphenylmethyl chloride,[746] since dimeric products are obtained even under conditions that would minimize the concentration of the carbanion. Hexane is one of the major products formed when 1,6-dichlorohexane is reduced by lithium naphthalenide, but not by the sodium compound.[747] It arises via the $^{-}CH_2(CH_2)_4CH_2\cdot$ radical anion. Methylene radical anions are generated in reactions of sodium naphthalenide with methylene halides.[748] Vinyl radicals generated in reductions of vinyl halides undergo equilibration.[706] Radicals are also produced in the reaction of lithium naphthalenide with quaternary ammonium salts.[749] The tetraphenylethane is also derived from coupling of benzhydryl radicals.

$$Ph_2CH\overset{+}{N}Me_3 \xrightarrow{NpH^{\overline{\cdot}}} Ph_2\dot{C}H \xrightarrow{NpH^{\overline{\cdot}}} Ph_2CH^- \xrightarrow{Ph_2CH\overset{+}{N}Me_3} Ph_2CHCHPh_2$$

The electrochemistry of the carbon—halogen bond has been reviewed.[750] Evidence for radical intermediates in the cathodic reduction of 2-halobenzanilides comes from the formation of phenanthridones obtained as a result of radical cyclization.[751] Radicals are also involved in reductions of styrylpyrazoline and styrylpyrazole derivatives,[752] and of 1-chloronaphthalene.[753] 4-Benzoylphenyl radicals (**315**) generated in the cathodic reduction of *p*-bromobenzophenone, couple with thiophenoxide anions to give the radical anion (**316**) and thence *p*-(phenylthio)benzophenone (**317**).[754] The cathodic

Br–C₆H₄–C(=O)Ph ⇌(e) Br–C₆H₄–·C(–O⁻)Ph ⇌(–Br⁻) ·C₆H₄–C(=O)Ph (**315**) →(PhS⁻) PhS–C₆H₄–·C(–O⁻)Ph (**316**) ⇌(–e) PhS–C₆H₄–C(=O)Ph (**317**)

reduction of 1,3-dibromopropane generates the 3-bromopropyl radical which can either abstract hydrogen from the solvent, giving eventually propane, or alternatively may be further reduced to the anion and thence give cyclopropane.[755]

The borohydride reduction of 7,7-dibromonorcarane is essentially an electron-transfer process (cf. p. 96).[757] Electron-transfer processes are also involved in the sodium hydride reduction of azides to amines,[758] and in the reactions of nucleophiles such as sodium naphthalenide with bisduroquinone,[759] of DPPH[760] and isothiuronium salts[761] with hydroxide, or nitro-[762] and nitroso-compounds[763] with methoxide, of pyrylium salts with phenoxide anions and with *N,N,N′,N′*-tetramethyl-*p*-phenylenediamine,[764] of copper(I) phenoxide with carbon tetrachloride,[765] and in certain reactions of azoxy- and azo-benzenes.[766] An electron-transfer mechanism involving the generation of aryl radicals within the solvent shell has been proposed for the Ullmann reaction involving a bromoanthraquinone and aniline.[767]

Reaction of 1-aza-adamantane (**318**) with bromotrichloromethane yields the dimeric product (**319**).[768] It is postulated that this product is formed by coupling of a radical and a radical cation derived from 1-aza-adamantane:

$$\underset{(\mathbf{318})}{\text{N-AdH}} + \cdot CCl_3 \text{ (or Br}\cdot) \rightarrow \text{N-Ad}\cdot + HCCl_3 \text{ (or HBr)}$$

$$\text{N-AdH} + \cdot CCl_3 \text{ (or Br}\cdot) \rightarrow [\text{N-AdH}]^{+\cdot} + CCl_3^- \text{ (or Br}^-)$$

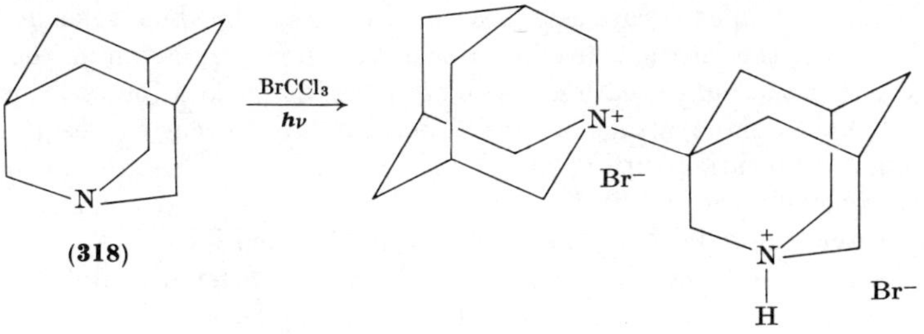

Radical cations are postulated as intermediates in fluorinations of aromatic substrates with cobalt(III) fluoride.[769] The formation of cyclobutanes by the radical cation-induced dimerization of alkenes has been reviewed.[770] A recent example of this reaction is the photochemical dimerization of 9-vinylcarbazole catalysed by copper(II) perchlorate.[771] Oxidation of substrates containing the *p*-dimethylamino-group by triarylaminium radical cations leads initially to the analogous radical cations which subsequently dimerize or undergo C—C or C—H bond scission.[772] *N*-Methylphenothiazine is oxidized to its radical cation by silver(I) salts.[773] The rate-determining step in the reaction between trialkylphosphines and phenanthraquinone involves single-electron transfer with the generation of the $R_3P^{+\bullet}$ radical cation and phenanthraquinone radical anion;[774] in subsequent steps these combine to give an adduct. Electron transfer occurs in the reaction of diethyl malonate with copper isocyanide complexes.[775] The tetracyanoethylene radical anion is generated as a result of charge transfer between *N*,*N*-diethylaniline and tetracyanoethylene.[776]

The disproportionation of radical anions of 9,9-diaza-phenanthrene and anthracene,[777] substituted fluorenones,[778] ketyls,[779] and xanthene dye radicals[780] has been studied. The reversible radical anion–dianion equilibria of species derived from anthracene and nitro-compounds have also been investigated.[781]

p-Nitrophenyl radicals abstract hydrogen 45 times as readily from methoxide ions as from methanol, owing to stabilization of the resultant radical anion $[\cdot CH_2{-}O^- \leftrightarrow {}^-CH_2{-}O\cdot]$:[756]

$$Ar\cdot + MeO^- \rightarrow ArH + CH_2O^{\overline{\cdot}}$$

The trapping of triphenylmethyl anions by nitrobenzene is a very efficient process:[782]

$$Ph_3C^- + PhNO_2 \rightarrow Ph_3C\cdot + PhNO_2^{\overline{\cdot}}$$

The efficiency of the process is greatly increased by conditions which favour tight ion pairing, for then nitrobenzene probably enters the co-ordination sphere of the cation.[782, 783] These anions are also trapped by oxygen. Electron transfer occurs between aromatic hydrocarbon radical anions and dibenzoyl peroxide.[784]

The kinetics and mechanism of unimolecular gas-phase reactions of radical cations have been reviewed.[785] The biochemical significance of porphyrin radical cations has been discussed.[786] The position of equilibrium in the disproportionation of tetrakis-(*p*-methoxyphenyl)ethylene radical cations is very sensitive to solvent effects.[787] The tetraphenylethylene radical cation disproportionates to tetraphenylethylene and its dication,[788] and the latter cyclizes to 9,10-diphenylphenanthrene. The reactions of thianthrene radical cations with ketones,[789] amines[790, 791] and aromatic compounds[792] have been examined; in each case attack occurs at the sulphur but it has not yet been established whether the reactions involve attack by the radical cation on the ketone or amine, or whether the radical cation first disproportionates and the reactions involve the dication. Nucleophilic attack at the 2-position does occur in reactions of the dibenzodioxin radical cation with nitrite and with pyridine.[793] The rates of the reaction of 9,10-diphenylanthracene radical cations with styrene,[794] and of electron-transfer reactions of azaviolene radical cations have been determined.[795]

CIDNP studies provide evidence for direct electron transfer from 1,4-diazabicyclo-[2.2.2]octane to triplet photo-excited 4,4′-dichlorobenzophenone.[796] Electron transfer occurs in the photochemical reaction of *N*,*N*-dimethylaniline with bromobenzene:[797]

$$PhNMe_2 + PhBr \rightleftharpoons [PhNMe_2 \ldots\ldots PhBr] \rightarrow PhNMe_2^{+\cdot} + Ph\cdot + Br^-$$

Products are derived from attack of phenyl radical on the $PhNMe_2^{+\cdot}$ radical cation. Irradiation of carbonyl or heterocyclic compounds, or of aromatic hydrocarbons, in the presence of β-amino-alcohols (**320**) leads to fragmentation products.[798] Exciplex

$$\underset{(\mathbf{320})}{PhNRCH_2CH(OH)Ph} + ArH \xrightarrow{h\nu} Ph\overset{+\cdot}{N}R{-}CH_2{-}CHPh{-}O{-}H \cdots ArH^{-\cdot} \longrightarrow PhNRCH_2\cdot + PhCHO$$

formation with electron transfer also occurs during photoaddition of acrylonitrile to indoles[799] and of benzophenone to aromatic amines.[800] Photoreduction of nitrobenzene in triethylamine occurs by a charge-transfer process,[801] as does the photoreduction of 4-nitropyridine in propan-2-ol.[802] Photochlorination of nitrobenzene with hydrochloric acid has been further studied;[803] the initial step involves electron transfer from chloride ion to the photo-excited nitro-compound:

$$^3PhNO_2^* + Cl^- \rightarrow [PhNO_2^{-\cdot}, Cl\cdot]$$

Trichloromethyl radicals are generated in the photochemical reaction between amines and carbon tetrachloride:[804, 805]

$$RNH_2 + CCl_4 \rightarrow [RNH_2 \ldots\ldots CCl_4] \rightarrow \cdot{}^+RNH_2Cl, Cl^- + \cdot CCl_3$$

Charge transfer occurs within the excited complex. The phenazine radical cation is generated during photolysis of phenazine in acidified methanol;[806] under similar conditions, phthalazine and quinoxaline undergo methylation in the 1- and the 2-position, respectively.[807] Charge transfer occurs during the photolysis of an alkaline solution of sodium 1,2-naphthoquinone-4-sulphonate.[808] 4-Nitrobenzyl radicals are generated in the pulse radiolysis of 4-nitrobenzyl chloride in aqueous acetonitrile.[809]

The role of homolytic processes in the formation and reactions of Grignard reagents have been reviewed.[810–812] Kinetic studies have confirmed that the rate-determining step for the formation of Grignard reagents involves electron transfer from the metal to the alkyl halide and that the radical anion $RX^{-\cdot}$ lies on the principal reaction path to the product.[813]

$$RX + Mg \rightarrow RX^{\overline{\cdot}} + Mg^{+}$$
$$RMgX \leftarrow RX^{\overline{\cdot}} + Mg^{+} \rightarrow [R\cdot \;\; \cdot MgX] \rightarrow RMgX$$

Reaction of the vinyl bromide (**321**) with magnesium leads not only to the expected Grignard reagent (**323**) but also to the rearranged Grignard reagent (**324**) and to (**325**), suggesting that reaction occurs via the radical pair (**322**).[814]

$$\text{(321)} \xrightarrow{Mg} \text{(322)} \longrightarrow \text{(323)}$$
$$\text{(322)} \xrightarrow[-S\cdot]{SH} \text{(325)}$$
$$\text{(322)} \rightarrow PhC{\equiv}CCH_2CH_2\cdot \;\; \cdot MgBr \longrightarrow PhC{\equiv}CCH_2CH_2MgBr \;\; \text{(324)}$$

(**321**) (**322**) (**323**) (**325**) (**324**)

The addition of very pure methylmagnesium bromide to benzophenone gives exclusively the normal 1,2-addition product; however, with less pure Grignard reagent, or in the presence of iron(III) chloride, benzopinacol and other products are formed, which suggests the incursion of a competing electron-transfer process.[815] This competing pathway is even more important in reactions with *tert*-butylmagnesium chloride. The $\cdot C(Ph_2){-}S{-}CH_2CH_2CH{=}CH_2$ radical can be detected in the reaction of but-3-enylmagnesium bromide with thiobenzophenone.[816] In reactions of vinyl Grignard reagents with thioketones the product after a short reaction time is derived from addition to carbon, whereas after a longer period the main product is that derived from addition to sulphur after rearrangement of the radical pair (**326**) to (**327**).[817]

$$R_2C{=}S + CH_2{=}CHMgX \longrightarrow R_2C(CH{=}CH_2){-}SMgX \rightleftharpoons R_2C(CH{=}CH_2){-}S\cdot \;\; \cdot MgX \quad \textbf{(326)}$$
$$\downarrow$$
$$R_2C(MgX){-}S{-}CH{=}CH_2 \rightleftharpoons R_2\dot{C}{-}S{-}CH{=}CH_2 \;\; \dot{M}gX \quad \textbf{(327)}$$

Radicals are also involved in reactions of sterically hindered acyl chlorides with ethylmagnesium bromide in presence of copper(I) chloride.[818]

Electron-transfer processes are involved in reactions of ethyl-lithium with peroxides (see p. 150):[819]

$$EtLi + (Bu^tO)_2 \rightarrow [Et\cdot, Li^+, Bu^tOOBu^{t}\dot{^-}] \rightarrow [Et\cdot \ \cdot OBu^t] + LiOBu^t$$

$$\swarrow \qquad \downarrow \qquad \searrow$$

$$EtOBu^t \qquad C_2H_4 + Bu^tOH \qquad Et\cdot + Bu^tO\cdot$$

the yield of *tert*-butyl ethyl ether is increased by an increase in solvent viscosity; CIDNP effects establish that the ether is derived from radical precursors. The reaction of lead tetra-alkyls with hexachloroiridate in acetonitrile or acetic acid is considered to involve one-electron transfer:[820]

$$R_4Pb \xrightarrow{IrCl_6^{2-}} R_4Pb^{\dot{+}} \longrightarrow R\cdot + R_3Pb^+$$

$$\downarrow IrCl_6^{2-} \qquad \downarrow$$

$$RCl \quad R_3PbOAc$$

Radicals are intermediates in the reaction of alkylmercury(II) halides with sodium borohydride in the presence of oxygen:[821]

$$RHgBr \xrightarrow{NaBH_4} RHgH \xrightarrow{+R\cdot} RH + R\cdot + Hg(0)$$

Copper(I) alkoxides decompose to give alkoxy-radicals.[822] Alkyl radicals are generated in the photolysis but not in thermolysis of $RAgPBu_3$.[823]

Photolysis

Carbonyl Compounds

The primary reaction in the photochemistry of α-branched carbonyl compounds involves type I cleavage of the excited cleavage to give the more stable alkyl radical:[824]

$$Bu^tCOMe \rightarrow Bu^t\cdot + \cdot COMe$$

$$Bu^tCHO \rightarrow Bu^t\cdot + \cdot CHO$$

CIDNP studies show that in the photolysis of propionaldehyde hydrogen abstraction from ground-state propionaldehyde competes with type I cleavage ($^3[RCHO]^* + RCHO \rightarrow \cdot CHR(OH) + \cdot COR$) and with acetaldehyde it is the only process reflecting the order of stability of alkyl radicals: $Bu^t\cdot > Et\cdot > Me\cdot$;[825] both processes also occur with acetone.[826] CIDNP studies likewise show that type I cleavage occurs in the photolysis of cyclic ketones to give biradicals.[827] The photochemistry of butanone and 3-methylbutan-2-one have been studied.[828] Type I cleavage occurs in the photolysis of perfluoroketones in inert solvents.[65] These photoexcited ketones are very efficient hydrogen-abstracting agents, as shown by photoreduction even in poor hydrogen-donors such as cyclopropane. Photoreduction is observed in the photolysis of benzophenone in pentafluorophenol[829] and of biacetyl and pyruvic acid in alcohols (see Scheme on p. 151).[830,831]

From the influence of substituents on triplet lifetimes of deoxybenzoins it is concluded that the transition state for α-cleavage resembles the excited ketone rather than the

$$^3[\mathrm{MeCOCOOH}]^* + \mathrm{Me_2CHOH} \rightleftharpoons [\mathrm{Me\dot{C}(OH)COOH}\ \ \mathrm{\dot{C}Me_2OH}]$$

$$[\mathrm{Me\dot{C}(OH)COOH}\ \ \mathrm{\dot{C}Me_2OH}] \rightarrow \mathrm{CH_2{=}C(OH)COOH + Me_2CHOH}$$

$$[\mathrm{Me\dot{C}(OH)COOH}\ \ \mathrm{\dot{C}Me_2OH}] \rightarrow \mathrm{MeCH(OH)COOH + CH_2{=}C(OH)Me}$$

radical pair.[832] The products derived from photolysis of $PhCOCPh(OMe)_2$ are obtained from the $PhC(O)\cdot\ \cdot CPh(OMe)_2$ radical pair resulting from the excited triplet.[833]

The photochemical decarbonylation of 1,2,3,4-tetraphenylcyclopent-2-ene-1-carbaldehyde proceeds via the n–π^* singlet through the allyl formyl radical pair in a stereospecific process.[834] CIDNP studies indicate that photolysis of acylsilanes, at least in carbon tetrachloride, does not involve typical type I cleavage of the excited ketone but rather proceeds via an exciplex followed by formation of the $[\cdot COCH_3\ \cdot CCl_3]$ singlet radical pair.[835]

$$\mathrm{PhMe_2SiCOMe} \xrightarrow[\mathrm{CCl_4}]{h\nu} [\mathrm{PhMe_2SiCOMe\ \ CCl_4}]^* \longrightarrow [\mathrm{Me\dot{C}O\ \ \dot{C}Cl_3}] + \mathrm{PhMe_2SiCl}$$

$$[\mathrm{Me\dot{C}O\ \ \dot{C}Cl_3}] \rightarrow \mathrm{MeCOCCl_3} \quad\text{and}\quad \text{Free radicals}$$

Photolysis of 2,2-dialkoxy-1,3-diketones proceeds with cleavage of one of the alkoxy-groups.[836] 1,1,4,4-Tetraphenylbutadiene is produced as a result of fragmentation of the biradical (**328**) in the photolysis of the 2,2-diphenylvinyl ketone:[837]

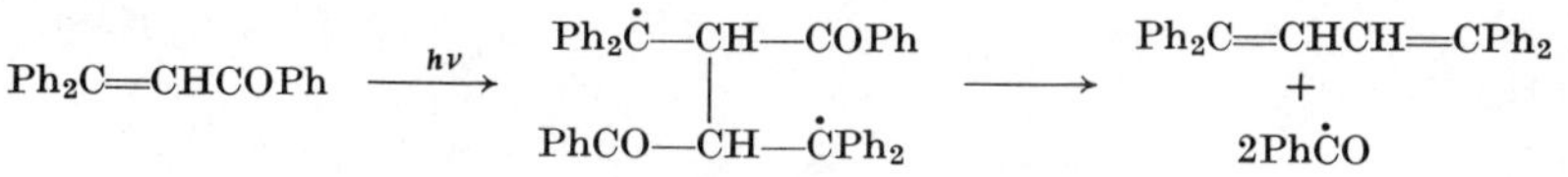

$$\mathrm{Ph_2C{=}CHCOPh} \xrightarrow{h\nu} \begin{array}{c}\mathrm{Ph_2\dot{C}{-}CH{-}COPh}\\ |\\ \mathrm{PhCO{-}CH{-}\dot{C}Ph_2}\end{array} \longrightarrow \mathrm{Ph_2C{=}CHCH{=}CPh_2} + 2\mathrm{Ph\dot{C}O}$$

(**328**)

A diphenoxyl radical is produced on photolysis of 3,3′,5,5′-tetramethyldiphenoquinone.[838]

Both Norrish type I and type II processes are encountered in the photolysis of pentan-2-one;[839] both processes are quenched by hydrogen bromide. The triplet biradical from pentan-2-one behaves similarly in both the gaseous and the liquid phase but the excited singlet undergoes more type II cleavage in the gas phase than in solution.[840] Pentan-2-one triplets as well as singlets are quenched by carbon tetrachloride.[841] Further studies on the photochemical rearrangement of α-methylene ketones have shown that the resultant cyclobutyl ketones are formed as a result of γ-hydrogen abstraction, whilst the accompanying cyclopropyl compounds are derived

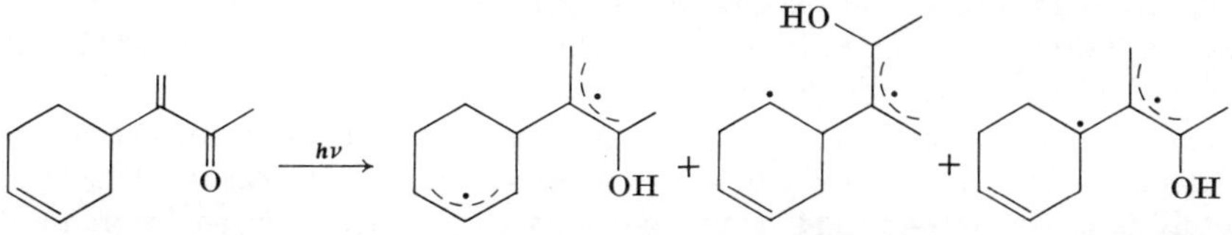

(**329**) $\xrightarrow{h\nu}$ (**330**) + (**331**) + (**332**)

as a result of β-hydrogen abstraction.[842,843] Singlet biradical intermediates appear to be involved. In the photolysis of (**329**), γ-hydrogen abstraction occurs from both allylic and non-allylic positions, as well as β-hydrogen abstraction, to give products derived from the biradicals (**330**), (**331**) and (**332**), respectively.[844]

γ-Hydrogen abstraction also occurs in the photolysis of the hydroperoxy-ketone (**333**), which reacts as the excited triplet to give the biradical (**334**) and thence the dioxetane (**335**) which is photochemically unstable.[845] A similar process is involved in

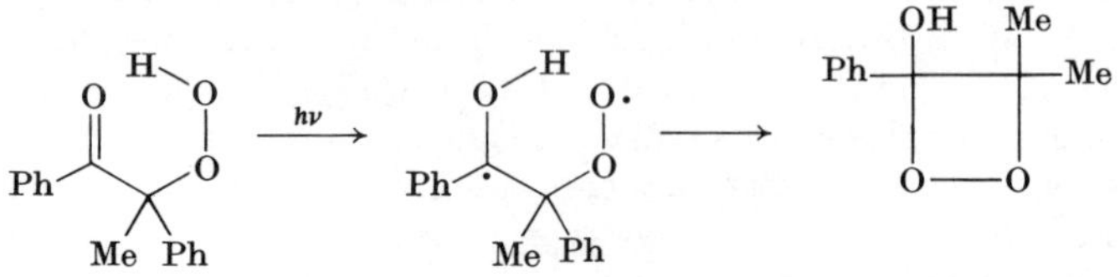

(**333**) (**334**) (**335**)

the generation of the biradical (**336**) from α-allylbutyrophenone.[846] This biradical undergoes elimination, cyclization and rearrangement followed by cyclization to give (**337**), (**338**) and *cis*- and *trans*-(**340**), respectively. From the product ratios it is

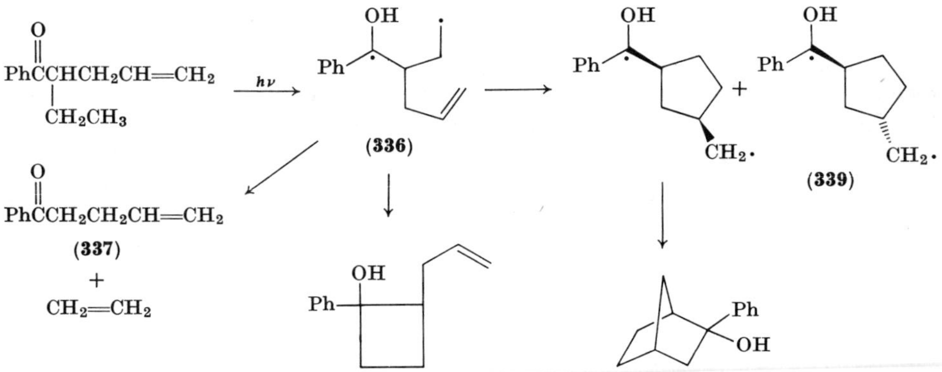

$PhCOCH(CH_2CH_3)CH_2CH{=}CH_2$ $\xrightarrow{h\nu}$ (**336**)

$PhCOCH_2CH_2CH{=}CH_2$ (**337**) + $CH_2{=}CH_2$

(**338**) (**339**) (**340**)

concluded that (**336**) cleaves or cyclizes twelve times as fast as it rearranges to (**339**) and that the rates of cleavage and cyclization are ca. $6 \times 10^5\ s^{-1}$. Medium- and small-ring α-cycloalkoxyacetophenones other than α-cyclopropyloxyacetophenone undergo photochemical Norrish type II reactions to give cyclized and/or elimination products.[847] γ-Hydrogen abstraction from the methoxy-group occurs in the photolysis of 3-methoxychromone.[848] 1,5-Hydrogen abstraction occurs in the photolysis of 1-*o*-tolylpropane-1,2-dione to give (**341**) and thence, as a result of a hydrogen shift, the radical (**342**), which cyclizes to (**343**).[849] In the presence of dimethyl acetylenedicarboxylate (**341**) can be trapped as its Diels–Alder adduct.[850]

A computer simulation study of the intramolecular hydrogen abstraction in the photochemical decomposition of *p*-benzoylbenzoic esters has been carried out.[886]

Photolysis of the ketone (**344**) leads, not only to products derived by non-radical routes, but also to 3,4-dimethylphenol as a result of fragmentation of the radical (**345**)

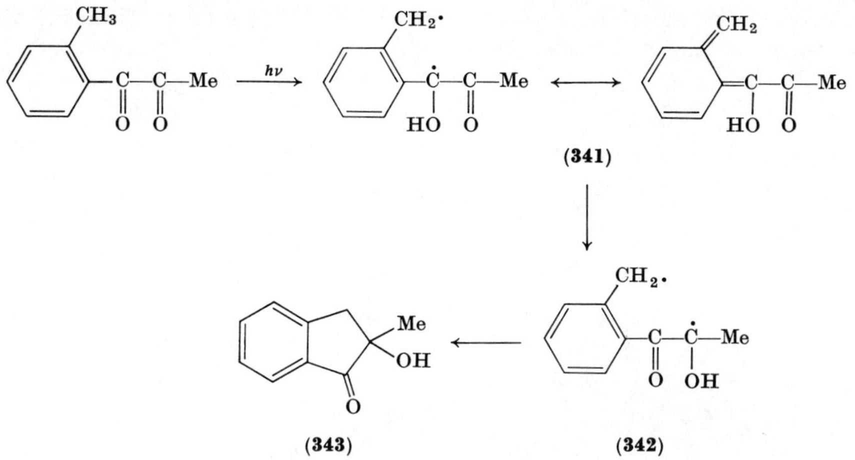

(341)

(343) (342)

which is generated in good hydrogen donors.[851] In poor hydrogen donors products derived from (**346**) are also produced.

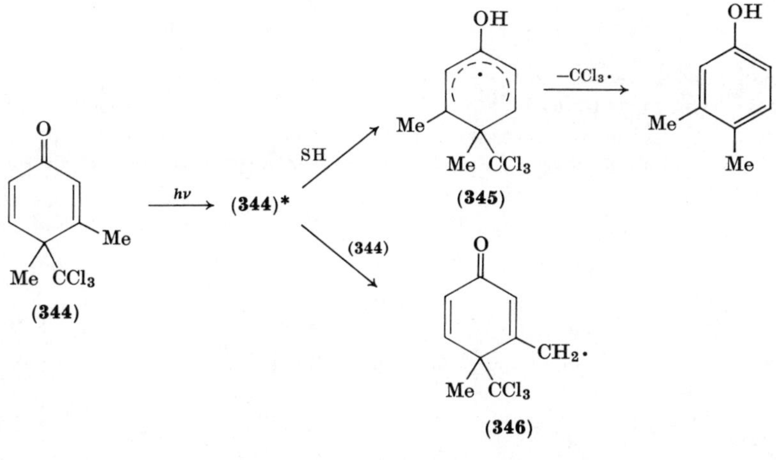

(344) (344)* (345) (346)

Radical intermediates are involved in the photo-addition of acetone to 2,3-dimethylbut-2-ene.[852] Quenching studies indicate that both singlet and triplet acetone are involved. In the photoaddition of methyl pyruvate to 2-methyl- and 2,3-dimethylbut-2-ene, quenching studies indicate that the ether and oxetane are formed via the triplet state of methyl pyruvate whilst alcohols are formed from the excited singlet state.[853] Photo-addition of aromatic ketones to vinylcyclopropanes leads to (**349**) and (**350**) as the principal primary products;[854] they are both derived from the biradical (**347**) which either cyclizes to (**349**) or rearranges and then cyclizes to (**348**). Substituents (R^1) which stabilize the cyclopropylcarbinyl radical centre reduce the rate of cyclization (k_c) and thus result in an increase in the ratio of (**349**)/(**350**), whilst substituents (R^2) which stabilize the benzyl radical centre have the reverse effect.

Diphenylaminyl radicals have been detected by ESR during the irradiation of benzophenone and diphenylamine.[855]

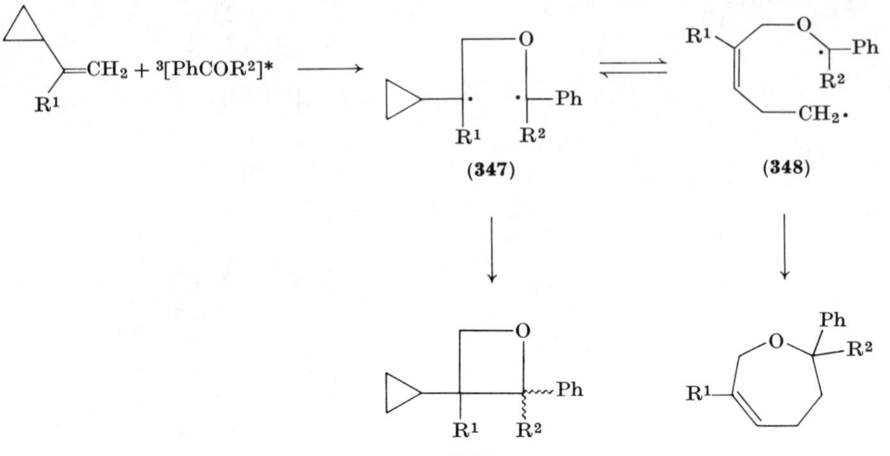

Miscellaneous Compounds

The role of radical intermediates in the photochemistry of olefinic compounds has been discussed.[856] Hot vinyl radicals are produced in the photolysis of ethylene.[857] A hydrocarbon analogue of the Norrish type II process has been observed in the photolysis of (**351**);[858] the same products are derived from (**352**) which affords the same intermediate biradical.

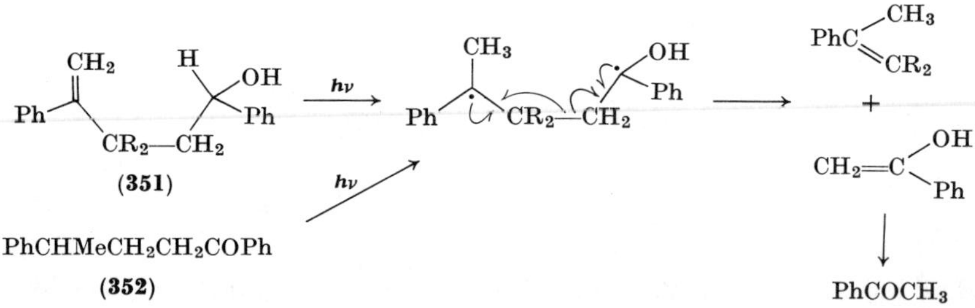

Vacuum-UV photolysis of diethyl ether gives ethyl and ethoxy-radicals.[859] Photolysis of ferrocenyl ethers $FcCR^1R^2OEt$ gives rise to products derived from $Fc\dot{C}R^1R^2$ radicals.[860]

The initial process in the photolysis of 1-iodonorbornane is homolysis of the C—I bond to give a radical pair; this may undergo internal electron transfer to give the corresponding ion pair.[861] Carbon—halogen bond fission with the generation of radicals occurs in the photolyses of 1-chloro-1-nitrosocyclohexane,[862] alkyl haloacetate,[863] polychlorobiphenyls,[864] methyl chloride,[864a] *tert*-butyl chloride[865] and acetyl chloride.[866]

In diethyl ether the last of these reactions affords (**353**) as a result of substitution of diethyl ether within the solvent cage:

$$\text{MeCOCl} \xrightarrow[\text{Et}_2\text{O}]{h\nu} \left[\begin{array}{c}\text{Me}\dot{\text{C}}\text{O}\ \cdot\text{Cl} \\ \text{H} \\ | \\ \text{EtO}\dot{\text{C}}\text{HMe}\end{array}\right] \longrightarrow \underset{(\mathbf{353})}{\text{MeCOCHMeOEt}} + \text{HCl}$$

$$\downarrow$$

$$\text{Me}\dot{\text{C}}\text{O} + \dot{\text{C}}\text{l} \xrightarrow{\text{Et}_2\text{O}} \text{Me}\dot{\text{C}}\text{HOEt} + \text{HCl} + \text{MeCHO}$$

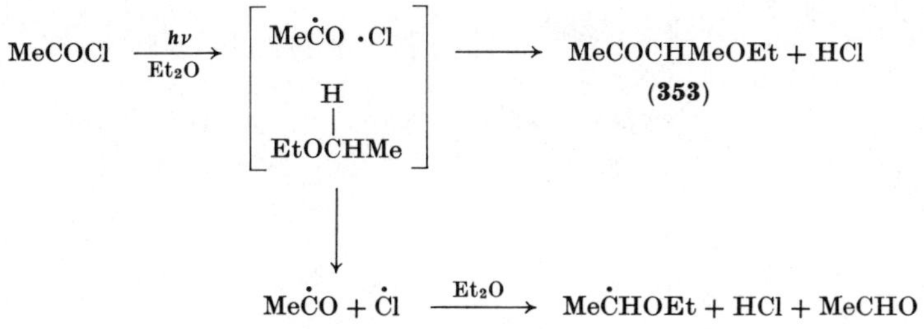

The products of photolysis of *N*-(*m*-methoxyphenethyl)-α-chloroacetamide (**354**) vary markedly with the polarity of the solvent.[867] In polar solvents reaction proceeds via the biradical cation (**355**), whilst in non-polar solvents the radical pair (**358**) is formed. This subsequently forms the biradical (**359**) as a result of hydrogen abstraction by chlorine atom within the solvent cage or it yields (**361**) by hydrogen abstraction from the solvent.

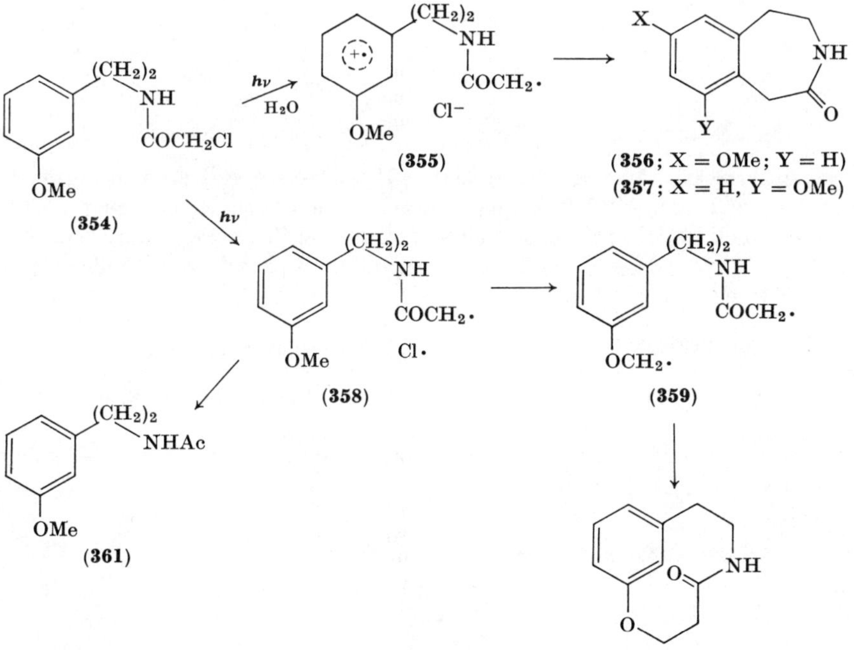

(**354**) $\xrightarrow[\text{H}_2\text{O}]{h\nu}$ (**355**) ⟶ (**356**; X = OMe; Y = H) (**357**; X = H, Y = OMe)

(**354**) $\xrightarrow{h\nu}$ (**358**) ⟶ (**359**) ⟶ (**360**)

(**358**) ⟶ (**361**)

The remote functionalization of steroids has been effected by photolysis of the *m*-iodobenzoate of 3β-cholestanol in the presence of phenyliodine dichloride which generates the radical species (**362**). This affects intramolecular hydrogen abstraction to give (**363**) and thence (**364**)[868] (see p. 156 and also p. 94).

$\xrightarrow[h\nu]{PhICl_2}$ (362) ⟶ (363) $\xrightarrow{PhICl_2}$ (364)

In contrast to the behaviour of ketones which on photolysis readily undergo γ-hydrogen abstraction leading to Norrish type II products, δ-cyclization is the favoured process in aralkyl thiones.[869] This, however, only occurs at short wavelengths. It thus appears that δ-cyclization is a consequence of excitation to a higher excited state.

Naphthylmethyl and benzyl radicals are generated in photolyses of naphthylmethyl phenylacetates:[870]

$$PhCH_2COOCH_2Np \rightarrow [PhCH_2COO\cdot \ \cdot CH_2Np] \rightarrow PhCH_2\cdot + NpCH_2\cdot$$

$$PhCH_2\cdot + NpCH_2\cdot \rightarrow NpCH_2CH_2Np + NpCH_2CH_2Ph + PhCH_2CH_2Ph$$

Photolyses of S-*p*-tolyl thiobenzoates occur with cleavage of the acyl C—S bond to afford acyl and thio-*p*-tolyl radicals.[871, 872]

$$ArCOSC_6H_4Me\text{-}p \longrightarrow Ar\dot{C}O + p\text{-}MeC_6H_4S\cdot$$

$$Ar\dot{C}O \rightarrow ArCHO + ArCOCOAr \qquad p\text{-}MeC_6H_4S\cdot \rightarrow p\text{-}MeC_6H_4SSC_6H_4Me\text{-}p$$

Photolyses of aromatic sulphonyl compounds have been reviewed.[873] Photolysis of nitroalkanes proceeds by two competing pathways involving photoreduction and C—N bond cleavage.[874] The initial step in the photochemical reaction of aromatic nitro-compounds with aromatic amines involves hydrogen abstraction by the photoexcited nitro-compound.[875] Photolysis of *o*-nitrobenzaldehyde affords *o*-nitrosobenzoic acid, the reaction proceeding via a nitroxide.[876] The photolyses of oxaziridines[877] and of azacyclohexadienyl radicals have been examined.[878] The flash photolysis of 2-azidobiphenyl affords carbazole as a result of substitution by triplet nitrene to give a biradical which undergoes 1,2-hydrogen transfer with the formation of carbazole.[879] Biradicals are intermediates in the photofragmentation of phosphorus ylides.[880] Benzophenone has found use as a photoinitiator in the degradation of polyisopropene on account of its ability to abstract hydrogen from the polymer chain.[881] The photolytic decomposition of glucopyranosyl phenyl sulphone acetates proceeds by a free-radical route.[882]

The radical nature of the photo-induced substitution of acridine by alcohols has been established by CIDNP.[883] Reaction involves hydrogen transfer to the photo-excited acridine followed by radical coupling; the same process is involved in the photosubstitution of 2-cyanoquinoline by ethanol.[884] The photoreduction of flavines has been reported.[885]

Pyrolysis

The gas-phase pyrolysis of ethane has been studied by direct measurement of the methyl radical concentration.[888] 1- and 2-Butyl radicals have been invoked in the pyrolysis of *n*-butane.[889] A computer model of the mechanism of isobutane pyrolysis indicated that non-Arrhenius rate constants were important for methyl-radical abstraction and hydrogen-atom addition reactions.[890] The mechanism of thermal ring opening of three-membered cyclic compounds can be characterized with respect to conservation of spin and space symmetry.[891] The rate of conversion of cyclopropane to propene was 1.9 times

Δ

trans

cis

(365)

(367)

(366)

faster with a surface present, probably because of diradical formation at the surface.[892] Thermal cracking of cyclohexene gave ethylene and butadiene via the diradical $\cdot CH_2CH{=}CHCH_2CH_2CH_2\cdot$;[893] aromatic hydrocarbons were also formed, but not from the diene.[894] Cyclohexadiene decomposed thermally by a chain process to give benzene.[894]

A diradical mechanism (p. 157) was suggested for the pyrolysis of *trans*- and *cis*-1,2-diethynylcyclobutane, which gave vinylacetylene (**365**), bicyclo[4.2.0]octa-1,5,7-triene (**366**) and 1,2-dihydropentalene (**367**).[895] Thermal ring cleavage of *cis*- and *trans*-2,3-dimethyloxetane (**368**) gave propene and acetaldehyde or but-2-ene and formaldehyde.[896] The same mixture of but-2-enes was obtained from both oxetane isomers, indicating diradical intermediates. The major products from the pyrolyses of di-, tri-

(**368**)

$C_3H_6 + CH_3CHO$

$C_4H_8 + CH_2O$

and tetra-methyloxetane were also carbonyl compounds and alkenes;[897] 1,4-diradical intermediates were implicated since the same mixture of *cis*- and *trans*-but-2-enes was always observed. Thermal cleavage of *trans*-2-phenyl-3-propyloxetane occurred stereospecifically to give *trans*-1-phenylpent-1-ene and formaldehyde, thus indicating a concerted $2s + 2a$ pathway. The *cis*-isomer of this oxetane on the other hand decomposed more slowly, giving a mixture of *cis*- and *trans*-1-phenylpent-1-ene, suggestive of the more familiar diradical intermediate.[898]

Thermal ring cleavage of 1,3-dioxans was favoured by substituents which facilitate hydrogen-atom transfer from the 2-position.[899] Two diradical intermediates were implicated in the pyrolysis of 2,3-dimethylene-7-oxabicyclo[2.2.1]heptane (**369**), leading to benzocyclobutene (**370**) and 3-oxabicyclo[3.2.0]hepta-1,4-diene (**371**).[900]

The main primary process in the thermolysis of tri-*tert*-butylcarbinol involved loss of a *tert*-butyl radical: $Bu_3^tCOH \rightarrow Bu_2^tCOH\cdot + Bu^t\cdot$. The products di-*tert*-butyl ketone and di-*tert*-butylcarbinol were then formed by disproportionation reactions of the two radicals.[901] Thermal decomposition of 2-methylbutan-2-ol was found to be a first-order process involving elimination of water.[902]

Pyrolysis of phenyl 2,2,2-trideuterioacetate yields ketene and phenol with no *ortho*-

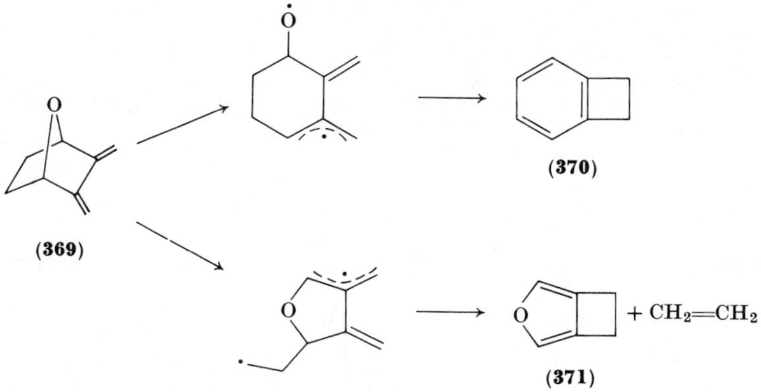

(369) **(370)**

(371) $+ CH_2{=}CH_2$

deuterium, indicating that the reaction is not a concerted sigmatropic rearrangement but a free-radical chain process:[903]

$$PhO\cdot + MeCOOPh \rightarrow PhOH + \cdot CH_2COOPh$$

$$\cdot CH_2COOPh \rightarrow CH_2CO + PhO\cdot$$

Thermolysis of long-chain-alkylphenols or their acetyl derivatives yields a variety of products formed by replacement of OH, OAc or alkyl by hydrogen.[904] Chloronitrobenzenes decompose thermally at 350° by homolytic rupture of a carbon—nitrogen bond.[905] The carbon—iodine bond strength (66.2 kcal mol^{-1}) in C_6F_5I has been determined by the toluene carrier technique.[906]

The key intermediate in the thermal decomposition of dibenz[*b,d*]iodolium iodide (**372**; X = I) was the 2′-iodobiphenyl-2-yl radical which on iodine abstraction gave 2,2′-diiodobiphenyl; a complex mixture of biphenyls was obtained from the bromide or chloride (**372**; X = Br or Cl). An analogous intermediate was formed in the decomposition 10*H*-dibenz[*b,e*]iodininium halides (**373**).[907] A kinetic study of the thermal

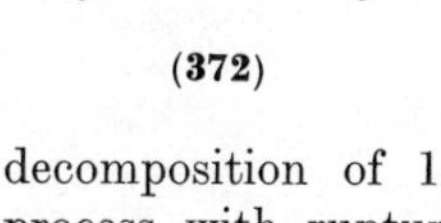

(372) 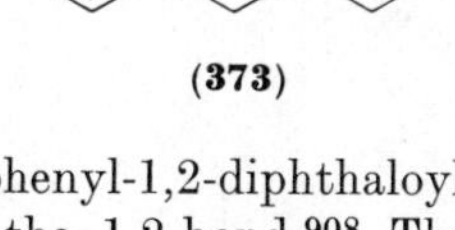 **(373)** 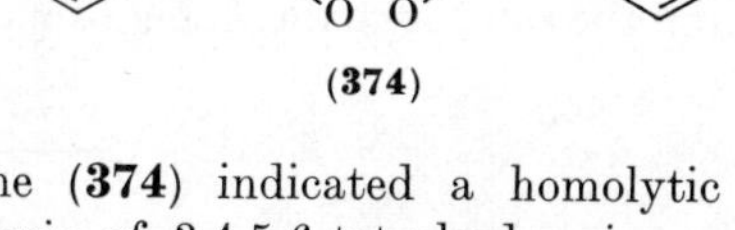 **(374)**

decomposition of 1,2-diphenyl-1,2-diphthaloylethane (**374**) indicated a homolytic process with rupture of the 1,2-bond.[908] Thermolysis of 3,4,5,6-tetrahydro-*cis*- or *trans*-3,6-diphenylpyridazine (**375**) gave styrene and *cis*- and *trans*-1,2-diphenylcyclobutane; it is likely that 1,4-diphenylbuta-1,4-diyl diradicals are the important reactive intermediates.[909] The main reaction route of 1-(2-benzothiazolyl)benzotriazole (**376**) in the thermal, photolytic and electron-impact processes is loss of nitrogen to give (**377**).[910] 1-Phenyl-2-benzimidazolinethione (**378**; X = S) decomposes at 650° to 1-phenylbenzimidazole by a sulphur-loss process and to benzimidazo[2,1-*b*]benzothiazole by loss of hydrogen. 1-Phenylbenzimidazolin-2-one (**378**; X = O) loses formaldehyde or HNCO; good correlation was observed between the pyrolytic and electron-impact processes.[911]

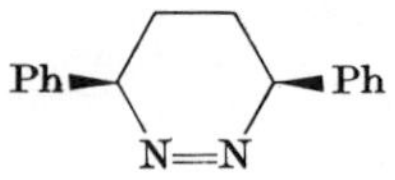

cis (**375**) 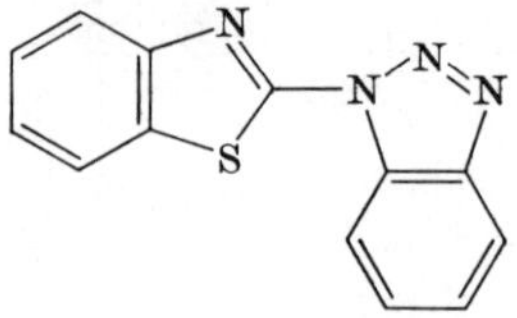 (**376**) 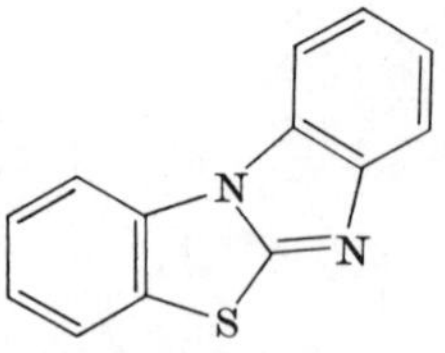 (**377**)

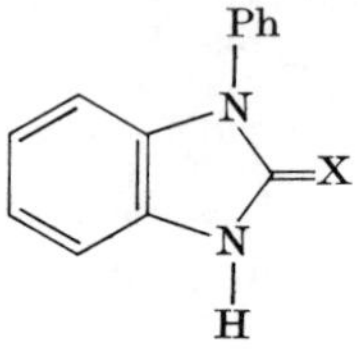

(**378**)

2,4,4-Triaryl-5-(alkylthio)- and -5-(arylthio)-4*H*-imidazoles (**379**) eliminate the sulphur-containing group on thermolysis, and rearrange to 2,4,5-triarylimidazoles by the mechanism shown.[912]

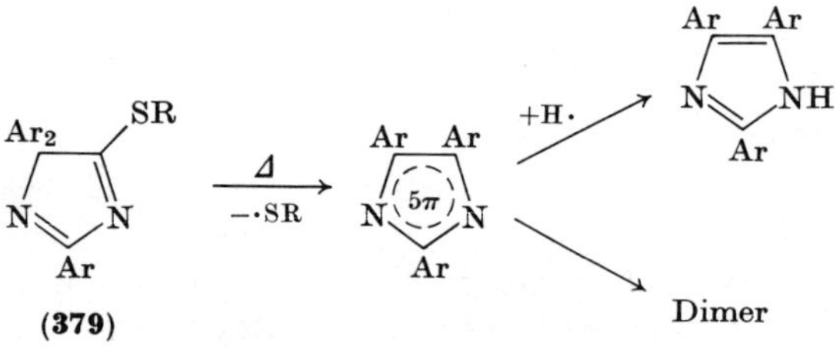

(**379**) $\xrightarrow[-\cdot SR]{\Delta}$ 5π radical $\xrightarrow{+H\cdot}$ 2,4,5-triarylimidazole; → Dimer

In the flow pyrolysis of trimethylgermane both $Me_3Ge\cdot$ and $Me_2HGe\cdot$ radicals were found to be important intermediates; pyrolysis of trimethylsilane was much more complex owing to the formation of silicon–carbon double-bonded intermediates of $Me_2Si(H)CH_2\cdot$ radicals, and the incursion of hydrogen-atom chain reactions.[913] The kinetics of the thermal isomerization of hexamethyldisilane have been investigated, and the radical-chain mechanism confirmed:[914]

$$Me_3SiSiMe_3 \longrightarrow 2Me_3Si\cdot$$

$$Me_3Si\cdot + Me_3SiSiMe_3 \longrightarrow Me_3SiH + Me_5Si_2CH_2\cdot$$

$$Me_5Si_2CH_2\cdot \longrightarrow Me_3SiCH_2SiMe_2\cdot \longrightarrow \text{Dimer}$$

$$Me_3SiCH_2SiMe_2\cdot \xrightarrow{Me_3SiSiMe_3} Me_3SiCH_2SiMe_2H$$

Free radicals were indicated in the thermal decomposition of *n*-alkyl(tri-*n*-butylphosphine)silver(I) reagents;[915] the main products were the *n*-alkyl dimer, tri-*n*-butylphosphine, and Ag(0). Tetrafluoroterephthalic acid salts of cadmium and mercury, but not zinc, decarboxylate to give polymeric organometallic compounds (**380**).[916] The

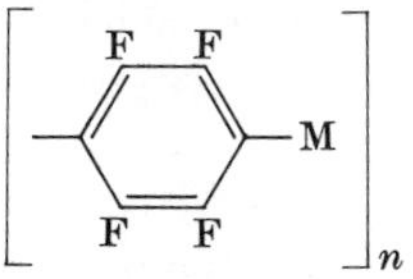

(**380**)

radical $\cdot CH_2CHO$ was proposed as a key intermediate in the thermal decomposition of $ClHgCH_2CHO$.[917]

Radiolysis

Sulphur hexafluoride is frequently used as a moderator in radiolytic experiments, but recent studies with ethane and ethylene containing added sulphur hexafluoride have shown that $\cdot SF_5$ radicals are formed and that they react by combination with alkyl radicals in the system.[918] Product yields have been followed for various experimental conditions in the γ-radiolysis of propane,[919] 3-methylpentane[920] and cyclohexane containing di-*n*-propyl disulphide.[921] On γ-radiolysis of liquid perfluorocyclohexane, the main products are formed by combination and disproportionation of perfluorocyclohexyl radicals.[922] ESR spectra and spin-trapping and bromine-scavenging experiments indicate that the 1-adamantyl radical is formed on γ-radiolysis of adamantane and related compounds.[923,924]

Both $CH_3O\cdot$ and $\cdot CH_2OH$ radicals were shown to be formed in approximately equal proportions in the γ-radiolysis of liquid methanol, by spin trapping with 2-methyl-2-nitrosopropane.[925] The nitroxide formed from $\cdot CH_2OH$ decays very rapidly, so that this radical can easily escape detection. Subsequent studies, using the same spin-trap technique,[926] and direct ESR detection,[927] showed that alkoxy-radicals were formed from a range of alcohols together with α-hydroxyalkyl radicals, except for *tert*-butyl alcohol which gave methyl radicals instead. The ratio of hydroxymethyl to methoxy-radicals formed on γ-radiolysis of methanol was estimated as ca. 1.5 by spin trapping with phenyl-*tert*-butyl nitrone.[928] A tritium-labelling experiment showed that hydrogen was abstracted from the hydroxy-group of the $Me_2C(OH)\cdot$ radical formed in the γ-irradiation of propan-2-ol in hydrogen peroxide.[929] An ESR study of the γ-radiolysis of a range of polyhydric alcohols also indicated two types of primary radical.[930] Hydroxymethyl radicals add to oxygen giving peroxy-radicals $\cdot O_2CH_2OH$ which can ionize to $\cdot O_2CH_2O^-$ in aqueous solution, but which also take part in bimolecular reactions.[931] The most probable structure of the radical formed on γ-radiolysis of glycylglycine was found to be $^+NH_3CH_2CONH\dot{C}HCOO^-$, but glycyl-$\alpha$-alanine gave $^+NH_3\dot{C}HCONHCHMeCOO^-$ on irradiation.[932] The primary step in the γ-radiolysis of aqueous acetaldehyde was shown to be abstraction of the aldehydic hydrogen by hydroxyl radicals.[933]

Three types of radical: $[RCF_2COOH]^{-\cdot}$, $RCF_2\cdot$ and $\cdot CRFCOOH$, were identified by ESR spectroscopy of γ-irradiated fluorinated carboxylic acids.[934] Electron-spin relaxation has been examined in methyl radicals generated by radiolysis of alkali-metal acetates.[935] New evidence shows that the radical formed on irradiation of malonamide is $H_2NCOCH_2CH{=}N\cdot$ and not $H_2NCOCH_2CONH\cdot$ as was previously supposed.[936] The cyano-group in alkyl cyanides is the main target for attack by radicals formed on γ-irradiation of aqueous cyanide solutions:[937]

$$RCN + H\cdot \rightarrow RCH{=}N\cdot + \cdot CR{=}NH$$

$$RCN + \cdot OH \rightarrow RC(OH){=}N\cdot + \cdot CR{=}NOH$$

The ESR spectrum of X-irradiated 5-ethylbarbituric acid indicated that the primary radical was formed by hydrogen abstraction from the 5-position; with the potassium salt, however, radicals were formed by hydrogen abstraction from the ethyl group.[938] In the γ-radiolysis of aqueous solutions of uracil[939] and thymine[940] product distributions could be accounted for satisfactorily in terms of initial addition of H· and ·OH to the 5,6-double bond of the pyrimidine ring; micellar surfactants substantially decreased the yield of products, probably by interacting with the product-forming intermediates. Reduction of *p*-nitroperoxybenzoic acid to *p*-nitrobenzoate anion was brought about by e^-_{aq} on radiolysis in aqueous solution and by 2-hydroxy-2-propyl radicals in propan-2-ol solution.[941] The importance of carbon–nitrogen bond fission on irradiation of nitro-compounds was shown by the isolation of *p*-nitrodiphenylpicrylhydrazine from irradiation of nitromethane or nitrobenzene containing diphenylpicrylhydrazyl (DPPH);[942] diphenylpicrylhydrazine was formed in a second process:

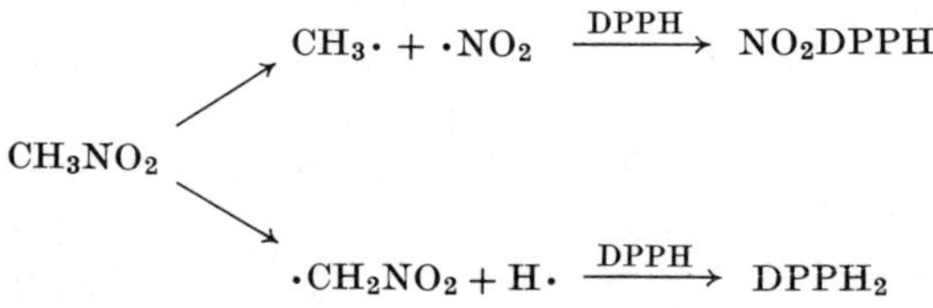

The relative rates of addition of hydrogen atoms to unsaturated substituents in carboranes have been determined from γ-irradiation studies in cyclohexane solution.[943] γ-Irradiation of silver diethyl phosphate at 77K gives $\cdot CH_2CH_2OP(O)_2OC_2H_5^-$ radicals which decay to $\cdot CH(CH_3)OP(O)_2OC_2H_5^-$ on warming.[944] The products of radiolysis of chlorphoxin [o-$ClC_6H_4C(CN){=}NOP(S)(OEt)_2$ and phoxin have been determined.[945]

Miscellaneous

Carbon- and metal-centred radicals are generated in the photolysis of alkyl-manganese pentacarbonyl and have been successfully trapped with nitrosodurene:[946] $RMn(CO)_5 \rightarrow R\cdot + \cdot Mn(CO)_5$. Radical intermediates are generated in the reactions of organocalcium iodides with ketones:[947]

$$RCaI + R'_2CO \longrightarrow R'_2\dot{C}OCaI + R\cdot \begin{cases} \nearrow R'_2C(R)OCaI \\ \searrow R'_2CHOCaI + R(-H) \end{cases}$$

CIDNP studies reveal the intermediacy of radicals in the reactions of 4-picoline *N*-oxide (**381**) with acetic anhydride.[948]

Me Me $\overline{CH_2\cdot \ \cdot OAc}$

Ac_2O

N^+ N^+ N N

O^- OAc OAc

(381)

Et $\overline{CH_2\cdot \ \cdot CH_3}$ Me CH_2OAc

OAc

\+

N N N N

Structures have been assigned to the radical products obtained from the reaction between dehydroascorbic acid and amino-acids.[949, 950] The ESR spectra of the radicals derived during the oxidation of ascorbic acid and related compounds have also been studied.[951, 952] The results indicate that there is significant spin density on the heterocyclic oxygen in the radical derived from α-hydroxytetronic acid.

The oxidative dinitration of 5-nitrouracil occurs on reaction with hydroxyl radicals.[953] Uracyl radicals react with oxygen, cysteamine and cystamine to give isodialuric acid, uracil and 5-(2-aminothioethyl)uracil, respectively.[954] The decay of cysteinyl radicals in both aerated and deaerated solutions has been studied and found to be sensitive to the pH of the solution.[955, 956] ESR studies on the dioldehydrase-oxidation of propane-1,2-diol to propanol are consistent with a hydrogen atom-transfer reaction.[957] The nature and conformations of radicals generated in the irradiation of L-histidine,[958, 959] cytidine, thymidine[961] and caffeine[962] have been reported. Hydrogen-adduct radicals are stabilized by cycloamyloses.[963]

ESR Spectroscopy

The use of ESR spectroscopy to determine conformational and structural changes in radicals during the course of reactions has been reviewed.[964] ESR studies of orientated radicals in applied electric fields have also been reviewed.[965] A computer program for determining all possible ESR spectra with equidistant lines has been published.[966] The influence of fast and slow exchange on line shapes of biradicals has been discussed.[967] Line shapes have also been examined by ELDOR.[968] An improved method for measurement of *g*-factors and splitting constants has been described.[969] The signal-to-noise ratio in ESR spectra can be enhanced by integration of ESR derivative spectra.[970] An improved method for calculation of spin densities involves use of the INDO method with the parameterizations of Kaufman using the projection of the UHF function.[971] Calculations have been carried out on the effect of rotation on the spin polarization contribution to hydroxyl proton coupling constants.[972] The application of NMR spectra to determination of spin densities of stable radicals has been reported.[973] All the hyperfine splitting constants in DPPH and related radicals have been determined by ESR and ENDOR studies.[974] Intermolecular hyperfine coupling between the half-filled orbitals and ^{19}F nuclei has been observed for radicals in substituted pentafluorobenzenes.[975]

CIDNP and CIDEP Methods

The Fourier transform technique has been applied to the determination of spin-lattice relaxation times in chemically polarized species from the decomposition of benzoyl peroxide; ^{13}C enhancement factors were calculated from the relaxation times.[976] When CIDNP effects can be observed for two or more nuclei in a product from a given radical, then the relative signs of the hyperfine coupling constants can be found, and if one sign is known the absolute value of the other can be determined. This method was used to determine the signs of the ^{13}C coupling constants of the carbonyl-carbon in the acylamino-radical (**382**) and thiocarbamoyl radical (**383**) using the known sign of the ^{1}H coupling constant in the NMe group.[977]

$$p\text{-}MeC_6H_4C(=O)\text{—}N(Me)OC(=S)NMe_2 \xrightarrow{\Delta} \overline{p\text{-}MeC_6H_4C(=O)\text{—}\dot{N}(Me)\quad \cdot SC(=O)NMe_2}$$

(**382**) (**383**)

CIDNP has been observed for the first time from a short-chain singlet biradical (**385**) generated by cleavage of the cyclopropyl ring in dispiro[2.2.2.2]deca-4,9-diene (**384**).[978]

$$\xrightarrow{\Delta} \qquad \rightleftarrows \qquad \xrightarrow{RSH} Et\text{—}C_6H_4\text{—}CH_2CH_2SR$$

(**384**) (**385**)

Coefficients of nuclear polarization and lifetimes of radical pairs have been computed for a series of acyl and aroyl peroxides.[979]

CIDNP evidence has again proved valuable in the elucidation of the mechanism of photodecomposition of carbonyl compounds in solution. Acetone was shown to react via an excited triplet state.[980] Propionaldehyde reacted via the triplet to give radical pair (**386**), the ·CHEtOH radical also being identified by ESR. In CCl_4 solution, however, the main reaction was α-cleavage, giving radical pair (**387**), the ethyl radical

$$EtCHO \xrightarrow[EtCHO]{h\nu} \overline{Et\dot{C}HOH\ \ O\dot{C}Et}^{\,T} \longrightarrow \begin{cases} EtCHOHCOEt \\ 2EtCHO \\ Et\dot{C}HOH + Et\dot{C}O \end{cases}$$

(**386**)

$$EtCHO \xrightarrow[CCl_4]{h\nu} \overline{Et\cdot\ \ \cdot CHO}^{\,T} \longrightarrow \begin{cases} EtCHO \\ EtH + CO \\ Et\cdot + \cdot CHO \end{cases}$$

(**387**)

being observed by ESR spectroscopy.[981] The primary decomposition mode of butan-2-one on photolysis in water or *tert*-butyl alcohol was α-cleavage, the CIDNP spectra being explained in terms of radical pair $[CH_3CO\cdot\ \cdot Et]^{T,F}$; the ethyl radical was observed by ESR. Pentan-3-one decomposition followed a similar route.[982] CIDNP evidence and product distributions established type I α-cleavage, predominantly of triplet molecules, as the major primary photolytic process for *tert*-butyl methyl ketone, di-*tert*-butyl ketone, pivaldehyde and isobutyraldehyde in benzene solution; in chlorinated solvents singlet reactions interfere with this process.[983] ^{13}C-CIDNP has been observed from triplet biradicals produced from photolysis of a variety of cyclic ketones; the mechanism is shown for cycloheptanone.[984]

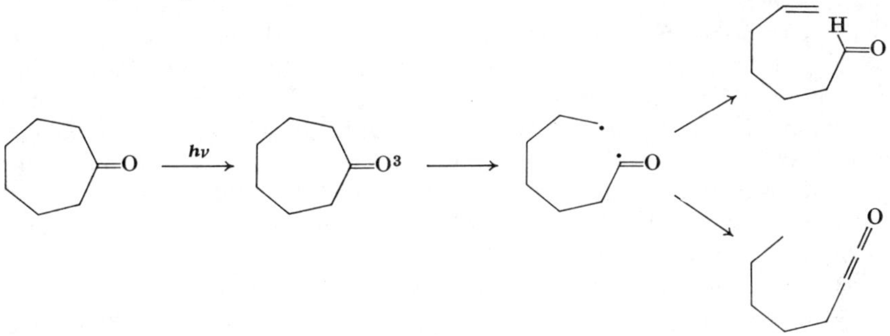

The coupling constants of fluorine atoms in the *ortho*- and *para*-position of monofluorobenzyl radicals were shown to be positive, and that of *meta*-fluorine negative, from a study of the CIDNP spectra of fluorobenzyl phenyl ketones.[985] The ^{19}F-CIDNP spectra observed during UV irradiation of benzophenone containing fluorine-substituted phenols or anilines were explained by a mechanism involving radical pairs such as (**388**); the coupling constants of *ortho*- and *para*-fluorine in phenoxy- and anilino-radicals was positive while that of *meta*-fluorine was negative.[986]

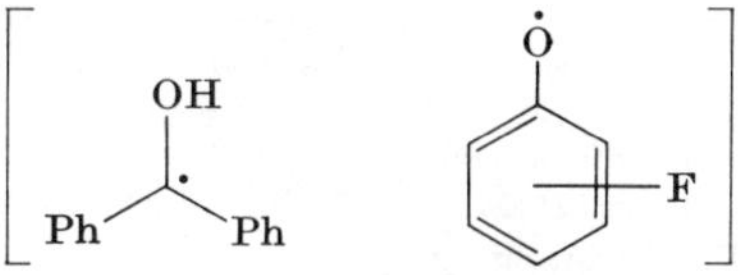

(**388**)

CIDNP evidence from the photolysis of acylsilanes such as $MeCOSiMe_2Ph$ in CCl_4 solution indicated that an exiplex was formed with the solvent which subsequently decomposed to give $PhSiMe_2Cl$ and the key singlet radical pair $[MeCO\cdot\ \cdot CCl_3]$. The product trichloroacetone showed enhanced absorption, and acetyl chloride, formed after diffusion of the acyl radical, showed emission.[987] The polarized spectrum observed during photolysis of trimethylstannyldiethylamine was consistent with direct cleavage

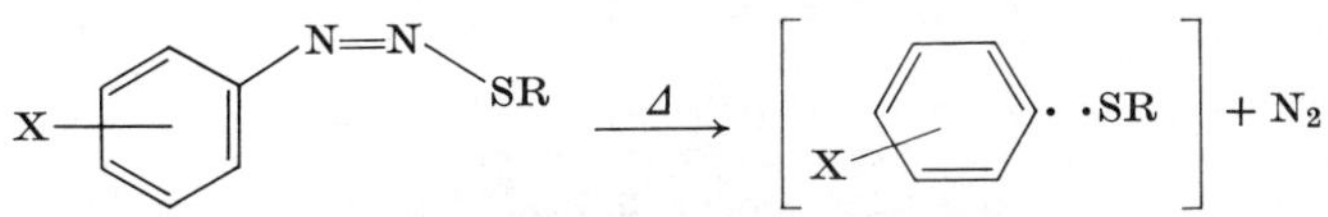

(**389**) (**390**)

of the tin—nitrogen bond, the products being formed from singlet radical pairs [$Me_3Sn\cdot\ \cdot NEt_2$].[988] The key radical pair in the thermal decomposition of *Z*-alkylaryldiazo sulphides (**389**) was shown to be (**390**).[989]

Pentafluorobenzoyloxy-radicals add reversibly, predominantly to the 2- and the 4-position in chlorobenzene, as shown by ^{13}C-CIDNP evidence.[990] The ^{13}C-CIDNP spectra[991,992] and product analysis show that photolysis of methyl diazoacetate in chloromethanes proceeds by cage combination of the radical pair [$Cl_2XC\cdot\ \cdot CHClCOOMe$] which is formed by cleavage of the corresponding chloronium ylide $ClXYC{-}\overset{+}{Cl}{-}\overset{-}{C}HCOOMe$, itself formed by attack of methoxycarbonylcarbene on the solvent. CIDNP has been applied to a study of the kinetics of the reaction of phenyl radicals with carbon tetrachloride.[993]

CIDEP of transient free radicals has been reviewed.[994] The origin of the CIDEP phenomenon is still not clear and two theories are currently in vogue. Pulse radiolysis, and time-resolved ESR of the resulting $\cdot CH_2OH$ radical, were used to show that the observation of CIDEP depends on the isotropic *g*-factor difference of the radical-pair components and hence that $S{-}T_0$ mixing of the electron-nuclear states in radical pairs is the dominant polarization pathway.[995] Further evidence supporting the radical-pair theory was obtained from the distribution of polarized intensities in the CIDEP spectrum of an irradiated alcoholic solution.[996] For alkyl radicals produced by radiolysis two factors were identified as controlling sign of the polarization: (i) which total spin multiplet is the more probable, (ii) the sign of the isotropic hyperfine splitting constant.[997] The rotating triplet model received support from various theoretical and experimental considerations.[998–1000]

References

[1] D. C. Nonhebel and J. C. Walton, *Free-radical Chemistry*, Cambridge, London, 1974.

[2] R. L. Huang, S. H. Goh and S. H. Ong, *The Chemistry of Free Radicals*, Arnold, London, 1974.

[3] J. M. Hay, *Reactive Free Radicals*, Academic Press, London, 1974.

[4] E. S. Huyser in *Organic Reactive Intermediates* (S. P. McManus, Ed.), Academic Press, New York, 1973, p. 1.

[5] G. A. Russell and R. K. Norris in *Organic Reactive Intermediates* (S. P. McManus, Ed.), Academic Press, New York, p. 423.

[6] J. M. McBride, *Tetrahedron*, **30**, 2009 (1974).

[7] R. Zahradnik and P. Carsky, *Progr. Phys. Org. Chem.*, **10**, 327 (1973).

[7a] J. W. Henderson, *Chem. Soc. Rev.*, **2**, 397 (1973).

[8] B. C. Gilbert and R. C. Sealy in *Electron Spin Resonance*, Vol. 2 (R. O. C. Norman, Ed.), Specialist Periodical Reports, The Chemical Society, London, 1974.

[9] G. D. Mendenhall, D. Griller and K. U. Ingold, *Chem. in Britain*, **10**, 248 (1974).

[10] G. D. Mendenhall, D. Griller, D. Lindsay, T. T. Tidwell and K. U. Ingold, *J. Am. Chem. Soc.*, **96**, 2441 (1974).

[11] See *Org. Reaction Mech.*, **1973**, 62.

[12] D. Griller, E. C. Horswill and K. U. Ingold, *Mol. Phys.*, **4**, 1117 (1974).

[13] D. Griller and K. U. Ingold, *J. Am. Chem. Soc.*, **96**, 6715 (1974).

[14] D. Griller, S. Içli, C. Thankachan and T. T. Tidwell, *Chem. Comm.*, **1974**, 913.

[15] D. Griller and K. U. Ingold, *J. Am. Chem. Soc.*, **96**, 6203 (1974).

[16] R. A. Kaba, D. Griller and K. U. Ingold, *J. Am. Chem. Soc.*, **96**, 6202 (1974).

[17] K. S. Chen and J. K. Kochi, *Can. J. Chem.*, **52**, 3529 (1974).

[18] K. S. Chen, P. J. Krusic, P. Meakin and J. K. Kochi, *J. Phys. Chem.*, **78**, 2014 (1974).

[19] P. Meakin and P. J. Krusic, *J. Am. Chem. Soc.*, **95**, 8185 (1973).

[20] K. S. Chen and J. K. Kochi, *Chem. Phys. Letters*, **23**, 233 (1973).

[21] K. S. Chen and J. K. Kochi, *J. Am. Chem. Soc.*, **96**, 794 (1974).

[22] K. S. Chen, P. J. Krusic and J. K. Kochi, *J. Phys. Chem.*, **78**, 2030 (1974).

[23] R. F. Picone and M. T. Rogers, *J. Mag. Res.*, **14**, 279 (1974).
[24] L. J. Aarons, I. H. Hillier and M. F. Guest, *J.C.S. Faraday II*, **70**, 167 (1974).
[25] S. P. Mishra, G. W. Neilson and M. C. R. Symons, *J.C.S. Faraday II*, **70**, 1165 (1974).
[26] K. S. Chen and J. K. Kochi, *J. Am. Chem. Soc.*, **96**, 1383 (1974).
[27] Cf. *Org. Reaction Mech.*, **1972**, 281; **1973**, 63.
[27a] R. V. Lloyd, D. E. Wood and M. T. Rogers, *J. Am. Chem. Soc.*, **96**, 7130 (1974).
[28] W. C. Danen and C. T. West, *J. Am. Chem. Soc.*, **96**, 2447 (1974).
[29] J. K. Kochi, P. J. Krusic and D. R. Eaton, *J. Am. Chem. Soc.*, **91**, 1877, 1879 (1969).
[30] L. Radom, J. Paviot, J. A. Pople and P. von R. Schleyer, *Chem. Comm.*, **1974**, 58.
[31] F. J. Adrian, E. L. Cochran and V. A. Bowers, *J. Chem. Phys.*, **59**, 3946 (1974).
[32] T. Shiga, H. Yamaoka and A. Lund, *Z. Naturforsch., A.*, **29**, 653 (1974).
[33] A. P. Kuleshov, V. I. Trofimov and I. I. Chkheidze, *Zh. Strukt. Khim.*, **14**, 625 (1973); *Chem. Abs.*, **79**, 125402 (1973).
[34] H. R. Falle and F. P. Sargent, *Can. J. Chem.*, **52**, 3401 (1974).
[35] C. Corvaja, G. Giacometti and G. Sartori, *J.C.S. Faraday II*, **70**, 709 (1974).
[36] B. C. Gilbert, R. O. C. Norman and M. Trenwith, *J.C.S. Perkin II*, **1974**, 1033.
[37] R. H. Schuler and L. K. Patterson, *Chem. Phys. Letters*, **27**, 369 (1974).
[38] P. Smith, R. A. Kaba and P. B. Wood, *J. Phys. Chem.*, **78**, 117 (1974).
[39] D. J. Bogan and D. W. Setser, *J. Am. Chem. Soc.*, **96**, 1950 (1974).
[40] A. Hudson, H. J. Kent, R. A. Jackson and R. F. Treweek, *J.C.S. Faraday II*, **70**, 892 (1974).
[41] S. Olivella, M. Ballester and J. Castañer, *Tetrahedron Letters*, **1974**, 587.
[42] K. Schreiner and A. Berndt, *Angew. Chem. Internat. Ed.*, **13**, 144 (1974).
[43] R. V. Lloyd and D. E. Wood, *J. Am. Chem. Soc.*, **96**, 659 (1974).
[44] A. Hinchcliffe, *Chem. Phys. Letters*, **27**, 454 (1974).
[45] D. N. Nanda and P. T. Narasimhan, *Int. J. Quantum Chem.*, **8**, 451 (1974).
[46] T. Okamura, K. Obi and I. Tanaka, *Chem. Phys. Letters*, **26**, 218 (1974).
[47] R. V. Lloyd and D. E. Wood, *J. Chem. Phys.*, **60**, 2684 (1974).
[48] W. R. Roth, G. Rüf and P. W. Ford, *Chem. Ber.*, **107**, 48 (1974).
[49] P. J. Krusic, P. Meakin and B. E. Smart, *J. Am. Chem. Soc.*, **96**, 6211 (1974).
[50] H. Regenstein and A. Berndt, *Angew. Chem. Internat. Ed.*, **13**, 145 (1974).
[51] A. Hinchcliffe and J. C. Cobb, *J. Mol. Struct.*, **23**, 273 (1974).
[52] D. M. Camaioni, H. F. Walter, J. E. Jordan and D. W. Pratt, *J. Am. Chem. Soc.*, **95**, 7978 (1973).
[53] Cf. *Org. Reaction Mech.*, **1973**, 65.
[54] T. Foster, D. Klapstein and P. R. West, *Can. J. Chem.*, **52**, 524 (1974).
[54a] P. H. Kasai, D. McLeod and H. C. McBay, *J. Am. Chem. Soc.*, **96**, 6864 (1974).
[55] G. Cirelli, F. Graf and H. H. Günthard, *Chem. Phys. Letters*, **28**, 494 (1974).
[56] F. Bernardi, C. M. Camaggi and M. Tiecco, *J.C.S. Perkin II*, **1974**, 518.
[57] G. Filby and K. Günther, *J. Phys. Chem.*, **28**, 1521 (1974).
[58] L. R. C. Barclay, D. Griller and K. U. Ingold, *J. Am. Chem. Soc.*, **96**, 3011 (1974).
[59] F. Barigletti, G. Poggi and A. Breccia, *J.C.S. Faraday II*, **70**, 1198 (1974).
[60] P. H. Kasai and D. McLeod, *J. Am. Chem. Soc.*, **96**, 2338 (1974).
[61] E. N. Cain and R. K. Solly, *J. Am. Chem. Soc.*, **95**, 7884 (1973); see also *Org. Reaction Mech.*, **1973**, 466.
[62] E. N. Cain and R. K. Solly, *Chem. Comm.*, **1974**, 148.
[62a] D. Bellus and G. Rist, *Helv. Chim. Acta*, **57**, 194 (1974).
[63] K. Hinrichs, H. Kurreck and W. Niemeier, *Tetrahedron*, **30**, 315 (1974).
[64] H. Kurreck and W. Niemeier, *Tetrahedron Letters*, **1974**, 3523.
[65] P. J. Krusic, K. S. Chen, P. Meakin and J. K. Kochi, *J. Phys. Chem.*, **78**, 2036 (1974).
[66] J. A. Austin, D. H. Levy, C. A. Gottlieb and H. E. Radford, *J. Chem. Phys.*, **60**, 207 (1974).
[67] T. Koenig, J. A. Hoobler, C. E. Klopfenstein, G. Hedden, F. Sunderman and B. R. Russell, *J. Am. Chem. Soc.*, **96**, 4573 (1974).
[68] T. Koenig, R. A. Wielesek and J. G. Huntington, *Tetrahedron Letters*, **1974**, 2283.
[69] A. T. Balaban, N. Negoiță and R. Baican, *Chem. Phys. Letters*, **24**, 30 (1974).
[70] N. Negoiță, R. Baican and A. T. Balaban, *Tetrahedron*, **30**, 73 (1974).
[71] R. Isfratoiu, I. Pascura and A. T. Balaban, *Z. Naturforsch., B*, **1974**, 543.
[72] R. W. Baldock, P. Hudson, A. R. Katritzky and F. Soti, *Heterocycles*, **1973**, 67.
[73] R. W. Baldock, P. Hudson, A. R. Katritzky and F. Soti, *J.C.S. Perkin I*, **1974**, 1422.
[74] A. R. Katritzky and F. Soti, *J.C.S. Perkin I*, **1974**, 1427.
[75] O. B. Donskikh, I. B. Donskikh, B. P. Manannikov, A. A. D'avydov and R. O. Matevosyan, *Dokl. Akad. Nauk SSSR*, **211**, 1341 (1973).

[76] L. Lunazzi and K. U. Ingold, *J. Am. Chem. Soc.*, **96**, 5558 (1974).
[77] V. Malatesta, D. Lindsay, E. C. Horswill and K. U. Ingold, *Can. J. Chem.*, **52**, 864 (1974).
[78] S. F. Nelsen and R. T. Landis, *J. Am. Chem. Soc.*, **96**, 1788 (1974).
[79] R. West and B. Bichlmeier, *J. Am. Chem. Soc.*, **95**, 7897 (1973).
[80] W. Ahrens and A. Berndt, *Tetrahedron Letters*, **1974**, 3741.
[81] M. Ballester, J. Castañer and S. Olivella, *Tetrahedron Letters*, **1974**, 615.
[82] G. R. Underwood and H. S. Friedman, *J. Am. Chem. Soc.*, **96**, 4989 (1974).
[83] K. Hatano, N. Shimamoto, T. Katsu and Y. Fujita, *Bull. Chem. Soc. Japan*, **47**, 4 (1974).
[84] F. A. Neugebauer and S. Bamberger, *Chem. Ber.*, **107**, 2362 (1974).
[85] *Org. Reaction Mech.*, **1972**, 283.
[86] S. F. Nelsen and R. T. Landis, *J. Am. Chem. Soc.*, **95**, 8707 (1974).
[87] E. M. Evleth, P. M. Horowitz and T. S. Lee, *J. Am. Chem. Soc.*, **95**, 7948 (1973).
[88] W. Broser, H. Kurreck, D. Rennoch and J. Reusch, *Tetrahedron*, **29**, 3959 (1973).
[88a] S. L. Vlasova and B. S. Tanaseichuk, *Sb. Nauch. Tr., Ivanov. Energ. Inst.*, **14**, 116 (1972); *Chem. Abs.*, **80**, 59296 (1974).
[89] A. A. Bardina and B. S. Tanaseichuk, *Sb. Nauch. Tr., Ivanov. Energ. Inst.*, **14**, 111 (1972); *Chem. Abs.*, **80**, 59284 (1974).
[89a] P. L. Gravel and W. H. Pirkle, *J. Am. Chem. Soc.*, **96**, 3335 (1974).
[90] F. A. Neugebauer and H. Fischer, *Chem. Ber.*, **107**, 717 (1974).
[91] F. A. Neugebauer and R. Bernhardt, *Chem. Ber.*, **107**, 529 (1974).
[92] F. A. Neugebauer and H. Brunner, *Tetrahedron*, **30**, 2841 (1974).
[93] D. Banks and W. Gordy, *Mol. Phys.*, **26**, 1555 (1973).
[94] J. H. Wieringa, H. Wynberg and J. Strating, *Tetrahedron*, **30**, 3053 (1974).
[95] See also p. 124 for nitroxides and p. 131 for phosphoranyl radicals.
[96] L. J. Aarons, I. H. Hillier and M. F. Guest, *J.C.S. Faraday II*, **70**, 167 (1974).
[97] H. Sakurai, K. Ogi, A. Hosomi and M. Kira, *Chem. Letters (Tokyo)*, **1974**, 891.
[98] D. R. Yarkony, H. F. Schaefer and S. Rothenberg, *J. Am. Chem. Soc.*, **96**, 656 (1974).
[99] K. Ohkubo and F. Kitagawa, *Bull. Chem. Soc. Japan*, **47**, 739 (1974).
[100] E. Mueller, *Rev. Port. Quim.*, **14**, 129 (1972); *Chem. Abs.*, **81**, 119733 (1974).
[101] K. Loth, M. Andrist, F. Graf and H. H. Günthard, *Chem. Phys. Letters*, **29**, 163 (1974).
[102] W. T. Dixon, W. E. J. Foster and D. Murphy, *J.C.S. Perkin II*, **1973**, 2124.
[103] W. T. Dixon, M. Mughimi and D. Murphy, *J.C.S. Faraday II*, **70**, 1713 (1974).
[104] W. T. Dixon, W. E. J. Foster and D. Murphy, *Mol. Phys.*, **27**, 1709 (1974).
[105] W. T. Dixon and D. Murphy, *J.C.S. Perkin II*, **1974**, 1430.
[106] N. N. Bubnov, S. G. Kukes, A. I. Prokof'ev, S. P. Solodovnikov and M. I. Kabachnik, *Dokl. Akad. Nauk SSSR*, **214**, 1319 (1974); *Chem. Abs.*, **80**, 132553 (1974).
[107] E. Melamud, S. Schlick and B. L. Silver, *J. Mag. Res.*, **14**, 104 (1974).
[108] J. A. Howard and E. Furimsky, *Can. J. Chem.*, **52**, 555 (1974).
[109] M. C. R. Symons, *J.C.S. Perkin II*, **1974**, 1618.
[110] T. Gillbro, C. M. L. Kerr and F. Williams, *Mol. Phys.*, **28**, 1225 (1974).
[111] M. Geoffroy, L. Ginet and E. A. C. Lucken, *Mol. Phys.*, **28**, 1289 (1974).
[112] M. Geoffroy, E. A. C. Lucken and C. Mazeline, *Mol. Phys.*, **28**, 839 (1974).
[113] S. A. Fieldhouse, H. C. Starkie and M. C. R. Symons, *Chem. Phys. Letters*, **23**, 508 (1973).
[114] A. V. Ilyasov, Ya. A. Levin and I. D. Morosova, *J. Mol. Struct.*, **19**, 671 (1973).
[115] F. S. Ezra and W. A. Bernhard, *J. Chem. Phys.*, **59**, 3543 (1973).
[116] H. Căldăraru and M. Morani, *J. Am. Chem. Soc.*, **96**, 149 (1974).
[117] G. A. Russell and A. Mackor, *J. Am. Chem. Soc.*, **96**, 145 (1974).
[118] J. J. Zeilstra and J. B. F. N. Engberts, *Tetrahedron*, **29**, 4299 (1973).
[119] S. F. Nelsen, G. R. Weisman, P. J. Hintz, D. Olp and M. R. Fahey, *J. Am. Chem. Soc.*, **96**, 2916 (1974).
[120] S. F. Nelsen and J. M. Bushek, *J. Am. Chem. Soc.*, **96**, 6424 (1974).
[121] A. Ronlán and V. D. Parker, *Chem. Comm.*, **1974**, 33.
[122] K. Scheffler, K. Hieke and H. B. Stegmann, *Tetrahedron Letters*, **1974**, 2949.
[123] I. Baciu, M. Hillebrand and V. Em Sahim, *J.C.S. Perkin II*, **1974**, 986.
[124] P. D. Sullivan and W. W. Paudler, *Can. J. Chem.*, **51**, 4095 (1973).
[125] J. Fajer, D. C. Borg, A. Forman, A. D. Adler and V. Varadi, *J. Am. Chem. Soc.*, **96**, 1238 (1974).
[126] M. Weber-Schaefer and H. Baumgartel, *Ber. Bunsenges. Phys. Chem.*, **78**, 317 (1974).
[127] C. Martino, P. Bruni and L. Greci, *Atti Accad. Sci. Ist. Bologna, Ci. Sci. Fis., Rend.*, **9**, 33 (1973); *Chem. Abs.*, **80**, 107480 (1974).
[128] Y. Ikegami, M. Iwaizumi and I. Murata, *Chem. Letters (Tokyo)*, **1974**, 1141.

[129] G. Bieri, E. Heilbronner, E. Kloster-Jensen, A. Schmelzer and J. Wirz, *Helv. Chim. Acta*, **57**, 1265 (1974).
[130] R. C. Rieke, S. E. Bates, C. F. Meares, L. I. Rieke and C. M. Milliren, *J. Org. Chem.*, **39**, 2276 (1974).
[131] H. Joela, *Chem. Phys. Letters*, **28**, 236 (1974).
[132] R. D. Rieke, D. G. Westmoreland and L. I. Rieke, *Org. Mag. Res.*, **6**, 279 (1974).
[133] N. L. Bauld, F. R. Farr and C. E. Hudson, *J. Am. Chem. Soc.*, **96**, 5634 (1974).
[134] N. L. Bauld and F. R. Farr, *J. Am. Chem. Soc.*, **96**, 5633 (1974).
[135] F. Gerson, J. Jachimowicz and C. Jutz, *Helv. Chim. Acta*, **57**, 1408 (1974).
[136] K. Kimura and M. Imamura, *Bull. Chem. Soc. Japan*, **47**, 1358 (1974).
[137] H. Tylli, *Suom. Kemistiseuran, Tiedon.*, **82**, 70 (1973); *Chem. Abs.*, **80**, 132506 (1974).
[138] F. Gerson, B. Kowert and B. M. Peake, *J. Am. Chem. Soc.*, **96**, 119 (1974).
[139] R. N. Nasirov, S. P. Solodnikov and M. I. Kabachnik, *Bull. Acad. Sci. USSR*, **22**, 2316 (1973).
[140] R. N. Nasirov, A. I. Prokof'ev, S. P. Solodnikov and M. I. Kabachnik, *Bull. Acad. Sci. USSR*, **22**, 1930 (1973).
[141] R. N. Nasirov, S. P. Solodnikov, T. V. Ershova and M. I. Kabachnik, *Bull. Acad. Sci. USSR*, **23**, 439 (1974).
[142] F. Nemoto, F. Shimoda and K. Ishizu, *Chem. Letters* (*Tokyo*), **1974**, 693.
[143] F. Gerson, G. Plattner, A. J. Ashe and G. Maerkl, *Mol. Phys.*, **28**, 601 (1974).
[144] C. Jongsmaa, H. G. de Graaf and F. Bickelhaupt, *Tetrahedron Letters*, **1974**, 1267.
[145] H. J. Sipe and R. West, *J. Organometal. Chem.*, **70**, 353 (1974).
[146] H. J. Sipe and R. West, *J. Organometal. Chem.*, **70**, 367 (1974).
[147] I. G. Makarov and V. M. Kazakova, *Khim. Khim. Tekhnol. Tr. Yubileinoi Konf., Posvyashch. 70-Letiya Inst.*, **1970**, 300; *Chem. Abs.*, **81**, 24610 (1974).
[148] O. P. Anderson, S. A. Fieldhouse, H. C. Starkie and M. C. R. Symons, *Mol. Phys.*, **26**, 1561 (1973).
[149] Z. V. Todres, M. P. Starodubtseva and D. I. Kursanov, *Bull. Acad. Sci. USSR*, **23**, 230 (1974).
[150] C. Miller and W. M. Gulick, *Mol. Phys.*, **27**, 1185 (1974).
[151] L. Echegoyen, H. Hidalgo and G. R. Stevenson, *J. Phys. Chem.*, **77**, 2649 (1973).
[152] G. R. Stevenson, M. Colón, J. G. Concepción and A. McB. Block, *J. Am. Chem. Soc.*, **96**, 2283 (1974).
[153] T. Shiga and A. Lund, *Ber. Bunsenges, Phys. Chem.*, **78**, 259 (1974).
[154] A. G. Evans, J. C. Evans and T. J. Phelan, *J.C.S. Perkin II*, **1974**, 1216.
[155] A. G. Evans, J. C. Evans and C. Bevan, *J.C.S. Perkin II*, **1974**, 1220.
[156] R. L. Blankespoor, *J. Am. Chem. Soc.*, **96**, 6196 (1974).
[157] G. A. Russell, P. R. Whittle, C. S. C. Chung, Y. Kosugi, K. Schritt and E. Goettert, *J. Am. Chem. Soc.*, **96**, 7053 (1974).
[158] T. Shida, S. Iwata and M. Imamura, *J. Phys. Chem.*, **78**, 741 (1974).
[159] M. Guerra, G. F. Pedulli, M. Tiecco and G. Martelli, *J.C.S. Perkin II*, **1974**, 562.
[160] V. M. Kazakova, I. G. Makarov, A. I. Samokhvalova and D. V. Ioffe, *J. Struct. Chem.*, **15**, 205 (1974).
[161] P. B. Koster, M. J. Janssen and E. A. C. Lucken, *J.C.S. Perkin II*, **1974**, 803.
[162] E. Volanschi, C. Vlădescu and C. Volanschi, *Rev. Roum. Chim.*, **19**, 755 (1974).
[163] H. Fujita, S. Kako, J. Yamauchi, H. Ohya-Nishiguchi and Y. Deguchi, *Bull. Chem. Soc. Japan*, **47**, 1541 (1974).
[164] H. Kurreck and S. Oestreich, *Tetrahedron*, **30**, 3199 (1974).
[165] R. Gavars, Yu. A. Bender, L. Baumane and J. Stradins, *Zh. Obsch. Khim.*, **43**, 1584 (1973); *Chem. Abs.*, **80**, 36405 (1974).
[166] G. F. Pedulli, A. Spisui, P. Vivarelli, P. Dembech and G. Seconi, *J. Mag. Res.*, **12**, 331 (1973).
[167] J. Voss, *Annalen*, **1974**, 1231.
[168] J. Voss, *Annalen*, **1974**, 1220.
[169] P. H. Kasai and D. McLeod, *J. Am. Chem. Soc.*, **96**, 2342 (1974).
[170] S. G. Kukes, N. N. Bubnov, A. I. Prokof'ev, S. P. Solodnikov, G. A. Nikiforov and V. V. Ershov, *Bull. Acad. Sci. USSR*, **22**, 2322 (1973).
[171] C. Fabre and B. Tchoubar, *Tetrahedron*, **30**, 1019 (1974).
[172] S. Niizuma, Y. Sato, S. Konishi and H. Kokubun, *Bull. Chem. Soc. Japan*, **47**, 2121 (1974).
[173] E. V. Patton and R. West, *J. Phys. Chem.*, **77**, 2652 (1973).
[174] Y. Kirino, *Chem. Letters* (*Tokyo*), **1974**, 153.
[175] G. F. Pedulli, A. Alberti, L. Testaferri and M. Tiecco, *J.C.S. Perkin II*, **1974**, 1701.
[176] A. H. Reddoch, *J. Mag. Res.*, **15**, 75 (1974).

[177] M. Ogasawara, S. Inaba, H. Yoshida and K. Hayashi, *Bull. Chem. Soc. Japan*, **47**, 1611 (1974).
[178] G. A. Russell and J. L. Gerlock, *J. Am. Chem. Soc.*, **96**, 5838 (1974).
[179] S. W. Mao, K. Nakamara and N. Hirota, *J. Am. Chem. Soc.*, **96**, 5341 (1974).
[180] K. Kano and T. Matsuo, *Chem. Letters (Tokyo)*, **1974**, 11.
[181] Cf. *Org. Reaction Mech.*, **1973**, 14.
[182] K.-G. Seifert and F. Gerhart, *Tetrahedron Letters*, **1974**, 829.
[183] M. Procházka, P. Rejmanová and O. Ryba, *Coll. Czech. Chem. Comm.*, **39**, 2404 (1974).
[184] J. Brokken-Zip and H. van de Bogaert, *Tetrahedron Letters*, **1974**, 249.
[185] M. Kobayashi, E. Akiyama, H. Minato and N. Kito, *Bull. Chem. Soc. Japan*, **47**, 1504 (1974).
[186] H. Watanabe, Y. Cho, Y. Ide, M. Matsumoto and Y. Nagai, *J. Organometal. Chem.*, **78**, C4 (1974).
[187] H. Watanabe, M. Matsumoto, Y. Cho and Y. Nagai, *Org. Prep. Procedures*, **6**, 25 (1974).
[188] N. L. Craig and D. W. Setser, *Internat. J. Chem. Kinetics*, **6**, 517 (1974).
[189] J. E. Herweh and R. M. Fantazier, *J. Org. Chem.*, **39**, 786 (1974).
[190] R. M. Fantazier and J. E. Herweh, *J. Am. Chem. Soc.*, **96**, 1187 (1974).
[191] N. Wiberg and W. Uhlenbrock, *J. Organometal. Chem.*, **70**, 239 (1974).
[192] N. Wiberg and W. Uhlenbrock, *J. Organometal. Chem.*, **70**, 249 (1974).
[193] N. Wiberg, W. Uhlenbrock and W. Baumeister, *J. Organometal. Chem.*, **70**, 259 (1974).
[194] P. S. Engel, J. L. Wood, J. A. Sweet and J. L. Margrave, *J. Am. Chem. Soc.*, **96**, 2381 (1974).
[195] G. Ayrey and A. C. Haynes, *Makromol. Chem.*, **175**, 1463 (1974).
[196] H. Schmidt, A. Schweig, B. M. Trost, H. B. Neubold and P. H. Scudder, *J. Am. Chem. Soc.*, **96**, 622 (1974).
[196a] J.-D. Cheng and H. J. Shine, *J. Org. Chem.*, **39**, 336 (1974).
[197] Cf. R. C. Neumann, *Accounts Chem. Res.*, **7**, 381 (1972).
[198] J. Owens and T. Koenig, *J. Org. Chem.*, **39**, 3153 (1974).
[199] R. C. Neumann and R. P. Pankratz, *J. Am. Chem. Soc.*, **95**, 8372 (1973).
[200] T. Koenig and J. G. Huntington, *J. Am. Chem. Soc.*, **96**, 592 (1974).
[201] J. P. Lorand, *J. Am. Chem. Soc.*, **96**, 2867 (1974).
[202] J. P. Lorand and R. W. Wallace, *J. Am. Chem. Soc.*, **96**, 1398 (1974).
[203] J. P. Lorand and R. W. Wallace, *J. Am. Chem. Soc.*, **96**, 1402 (1974).
[204] T. Koenig and J. M. Owens, *J. Am. Chem. Soc.*, **95**, 8485 (1973).
[205] See *Org. Reaction Mech.*, **1974**, 77.
[206] T. Koenig and J. M. Owens, *J. Am. Chem. Soc.*, **96**, 4052 (1974).
[207] I. S. Voloshanovski and S. S. Ivanchev, *J. Gen. Chem. USSR*, **44**, 860 (1974).
[208] R. C. Lamb, L. L. Vestal, G. R. Cipau and S. Debnath, *J. Org. Chem.*, **39**, 2096 (1974).
[209] P. S. Engel, A. I. Dalton and L. Shen, *J. Org. Chem.*, **39**, 384 (1974).
[210] J. C. Martin and M. M. Chau, *J. Am. Chem. Soc.*, **96**, 3319 (1974).
[211] M. Schulz, G. West and S. Ourk, *Z. Chem.*, **14**, 150 (1974).
[212] A. I. Kirillov and L. F. Ivanova, *Zh. Org. Khim.*, **9**, 2241 (1973); *Chem. Abs.*, **80**, 47244 (1974).
[213] A. V. Kessenikh, A. V. Ignatenko, S. V. Rykov and A. Ya. Shteinshneider, *Org. Mag. Res.*, **5**, 537 (1974).
[214] A. V. Kessenikh, A. V. Ignatenko and S. V. Rykov, *Org. Mag. Res.*, **5**, 533 (1974).
[215] K. Tokumaru, A. Ohshima, T. Nakata, H. Sakuragi and T. Mishima, *Chem. Letters (Tokyo)*, **1974**, 571.
[216] Y. Rees and G. H. Williams, *J.C.S. Perkin I*, **1974**, 2266.
[217] J. Horáček and F. Hrabák, *Coll. Czech. Chem. Comm.*, **39**, 2608 (1974).
[218] O. A. Chaltykyan, S. A. Akopyan, N. M. Betenyan and E. R. Sarukhanyan, *Uch. Zap., Erevan. Univ., Estestv. Nauk*, **1972**, 40; *Chem. Abs.*, **79**, 136175 (1973).
[219] B. M. Yarmolynk, O. M. Polumbrik and G. F. Dvorico, *Nettekhimiya*, **13**, 719 (1973); *Chem. Abs.*, **80**, 59164 (1974).
[220] L. V. Vlasova, L. S. Kobrina and G. G. Yakobson, *Izv. Sib. Otd. Akad. Nauk SSSR, Ser. Khim.*, **1973**, 79; *Chem. Abs.*, **79**, 114889 (1973).
[221] W. H. Richardson, M. B. Yelvington, A. H. Andrist, E. W. Ertley, R. S. Smith and T. D. Johnson, *J. Org. Chem.*, **38**, 4219 (1973).
[222] E. A. Lissi, *Can. J. Chem.*, **52**, 2491 (1974).
[223] E. G. E. Hawkins and R. Large, *J.C.S. Perkin I*, **1974**, 2561.
[224] J. R. Sanderson and P. R. Story, *J. Org. Chem.*, **39**, 3183 (1974).
[225] G. I. Nikishin, V. G. Glukhovtsev and M. A. Nadtochil, *Bull. Acad. Sci. USSR*, **22**, 2224 (1973).
[226] E. P. Tepenitsyna, N. V. Dormidontova, T. N. Sinenko and E. G. Plakhtinskaya, *Org. Khim.*, **1973**, 17; *Chem. Abs.*, **81**, 90852 (1974).

[227] L. K. Kazantseva, V. V. Lipes, V. I. Morozova and M. S. Furman, *Kinet. Katal.*, **14**, 848 (1973); *Chem. Abs.*, **79**, 136270 (1973).
[228] M. N. Puring and V. A. Itskovich, *Zh. Prikl. Khim.* (*Leningrad*), **46**, 1138 (1973); *Chem. Abs.*, **79**, 136263 (1973).
[229] M. N. Puring, V. M. Potekhin, V. A. Itskovich and V. A. Proskuryakov, *Zh. Prikl. Khim.* (*Leningrad*), **46**, 2024 (1973); *Chem. Abs.*, **79**, 145666 (1973).
[230] Z. S. Kartasheva, A. B. Gagarina and N. M. Emanuel, *Dokl. Akad. Nauk SSSR*, **212**, 134 (1973); *Chem. Abs.*, **79**, 136268 (1973).
[231] W. Adam and L. M. Szendrey, *J. Am. Chem. Soc.*, **96**, 7135 (1974).
[232] D. Griller and K. U. Ingold, *Internat. J. Chem. Kinetics*, **6**, 453 (1974).
[233] J. E. Bennett, J. A. Eyre, C. P. Rimmer and R. Summers, *Chem. Phys. Letters*, **26**, 69 (1974).
[234] H. Schuh, E. J. Hamilton, H. Paul and H. Fischer, *Helv. Chim. Acta*, **57**, 2011 (1974).
[235] L. W. Burggraf and R. F. Firestone, *J. Phys. Chem.*, **78**, 508 (1974).
[236] J. Rabani, M. Pick and M. Simic, *J. Phys. Chem.*, **78**, 1049 (1974).
[237] B. Tilquin, J. Allaert and P. Claes, *J. Phys. Chem.*, **78**, 462 (1974).
[238] R. H. Schuler and L. K. Patterson, *Chem. Phys. Letters*, **27**, 369 (1974).
[239] B. Lesigne, L. Gilles and R. J. Woods, *Can. J. Chem.*, **52**, 1135 (1974).
[240] W. Y. Wen, D. R. Johnson and M. Pole, *J. Phys. Chem.*, **78**, 1798 (1974).
[241] R. D. Burkhart and R. J. Wong, *J. Am. Chem. Soc.*, **95**, 7203 (1973).
[242] E. Niki and Y. Kamiya, *J. Am. Chem. Soc.*, **96**, 2129 (1974).
[243] T. Koenig and J. M. Owens, *J. Am. Chem. Soc.*, **96**, 4052 (1974).
[244] I. V. Khudyakov, V. A. Kuz'min, *Khim. Vys. Energ.*, **7**, 331 (1973); *Chem. Abs.*, **79**, 114811 (1973).
[245] R. D. Parnell and K. E. Russell, *J.C.S. Perkin II*, **1974**, 161.
[246] J. E. Bennett and R. Summers, *Can. J. Chem.*, **52**, 1377 (1974).
[247] J. E. Bennett, J. A. Eyre and R. Summers, *J.C.S. Perkin II*, **1974**, 797.
[248] V. Malatesta and K. U. Ingold, *J. Am. Chem. Soc.*, **96**, 3949 (1974).
[249] D. Griller, G. D. Mendenhall, W. VanHoof and K. U. Ingold, *J. Am. Chem. Soc.*, **96**, 6068 (1974).
[250] S. F. Nelsen and R. T. Landis, *J. Am. Chem. Soc.*, **95**, 8707 (1973).
[251] D. Griller, B. P. Roberts, A. G. Davies and K. U. Ingold, *J. Am. Chem. Soc.*, **96**, 555 (1974).
[252] S. G. Cohen, G. C. Ramsay, N. M. Stein and S. Y. Weinstein, *J. Am. Chem. Soc.*, **96**, 5124 (1974).
[253] P. S. Ganguli and M. Kaufman, *Chem. Phys. Letters*, **25**, 221 (1974).
[254] D. W. Trainor and C. W. von Rosenberg, *J. Chem. Phys.*, **61**, 1010 (1974).
[255] R. A. Back and T. Yokota, *Internat. J. Chem. Kinetics*, **5**, 1039 (1973).
[256] R. Zahradnik, Z. Slanina and P. Čársky, *Coll. Czech. Chem. Comm.*, **39**, 63 (1974).
[257] A. M. Bass and A. H. Laufer, *Internat. J. Chem. Kinetics*, **5**, 1053 (1973).
[258] D. G. Hughes, R. M. Marshall and J. H. Purnell, *J.C.S. Faraday I*, **70**, 594 (1974).
[259] R. S. Konar, R. M. Marshall and J. H. Purnell, *Internat. J. Chem. Kinetics*, **5**, 1007 (1973).
[260] J. Currie, H. W. Sidebottom and J. M. Tedder, *Internat. J. Chem. Kinetics*, **6**, 481 (1974).
[261] D. M. Golden, L. W. Piskiewicz, M. J. Perona and P. C. Beadle, *J. Am. Chem. Soc.*, **96**, 1645 (1974).
[262] D. M. Golden, Z. B. Alfassi and P. C. Beadle, *Internat. J. Chem. Kinetics*, **6**, 359 (1974).
[263] I. A. Matheson, H. W. Sidebottom and J. M. Tedder, *Internat. J. Chem. Kinetics*, **6**, 493 (1974).
[264] W. Tsang, *Internat. J. Chem. Kinetics*, **5**, 929 (1973).
[265] R. D. Kelley and R. Klein, *J. Phys. Chem.*, **78**, 1586 (1974).
[266] J. L. Currie, H. W. Sidebottom and J. M. Tedder, *J.C.S. Faraday I*, **70**, 1851 (1974).
[267] M. L. Poutsma, *Free Radicals*, **2**, 113, 159 (1973); *Chem. Abs.*, **80**, 26379, 26386 (1974).
[268] H. Taniguchi and H. Hatano, *Kagaku No Ryoiki*, **27**, 785 (1973); *Chem. Abs.*, **80**, 36381 (1974).
[269] H. Knoll and K. Scherzer, *Kinet. Katal.*, **15**, 565 (1974); *Chem. Abs.*, **81**, 104081 (1974).
[270] T. M. Nagiev, *Azerb. Khim. Zh.*, **1973**, 15; *Chem. Abs.*, **81**, 12610 (1974).
[271] K. Yamaguchi, *Chem. Phys. Letters*, **28**, 93 (1974).
[272] P. Camilleri, R. M. Marshall and J. H. Purnell, *J.C.S. Faraday I*, **70**, 1434 (1974).
[273] P. C. Kobrinsky and P. D. Pacey, *Can. J. Chem.*, **52**, 3665 (1974).
[274] N. L. Arthur and L. F. David, *Austral. J. Chem.*, **27**, 291 (1974).
[275] T. N. Bell, P. Slade and A. G. Sherwood, *Can. J. Chem.*, **52**, 1662 (1974).
[276] D. A. Whytock, J. D. Clark and P. Gray, *J.C.S. Faraday I*, **70**, 411 (1974).
[277] S. Hautecloque, *J. Chim. Phys. Physicochim. Biol.*, **71**, 13 (1974); *Chem. Abs.*, **80**, 107672 (1974).
[278] D. D. Tanner, R. J. Arhart, E. V. Blackburn, N. C. Das and N. Wada, *J. Am. Chem. Soc.*, **96**, 829 (1974).
[279] G. J. Gleicher, *Tetrahedron*, **30**, 935 (1974).
[280] D. D. Newkirk and G. J. Gleicher, *J. Am. Chem. Soc.*, **96**, 3543 (1974).
[281] N. J. Bunce and M. Hadley, *J. Org. Chem.*, **39**, 2271 (1974).

[282] T. Shono and I. Nishiguchi, *Tetrahedron*, **30**, 2183 (1974).
[283] W. J. Boyle and J. F. Bunnett, *J. Am. Chem. Soc.*, **96**, 1418 (1974).
[284] I. Lukovits, J. Kardos and M. Simanyi, *Tetrahedron Letters*, **1974**, 2685.
[285] R. Milstein, R. L. Williams and F. S. Rowland, *J. Phys. Chem.*, **78**, 857 (1973).
[286] A. V. Pariiskaya and V. I. Vedeneev, *Mater, Soveshch. Mekh. Ingibirovaniya Tseprykh Gazov. Reakts., 1st*, **1971**, 55; *Chem. Abs.*, **80**, 70042 (1974).
[287] R. L. Johnson, K. C. Kim and D. W. Setser, *J. Phys. Chem.*, **77**, 2499 (1973).
[288] M. Julliard, *Bull. Soc. Chim. France*, **1974**, 263.
[289] J. M. Krasniewski and M. W. Mosher, *J. Org. Chem.*, **39**, 1303 (1974).
[290] I. V. Bodrikov, N. G. Bronnikova, Z. P. Mamakina and Z. S. Smolyan, *Dokl. Akad. Nauk SSSR*, **217**, 605 (1974); *Chem. Abs.*, **81**, 90971 (1974).
[291] R. L. Neal and M. W. Mosher, *Proc. W. Va. Acad. Sci.*, **45**, 191 (1973); *Chem. Abs.*, **81**, 104342 (1974).
[292] J. M. Bachhawat, A. K. Koul, B. Prashad, N. S. Ramegowda, C. K. Narang and N. K. Mathur, *Indian J. Chem.*, **11**, 609 (1973).
[293] E. N. Okafo and E. Whittle, *J.C.S. Faraday*, **70**, 1366 (1974).
[294] E. N. Okafo and E. Whittle, *Internat. J. Chem. Kinetics*, **5**, 1047 (1973).
[295] K. T. Shea, D. C. Lewis and P. S. Skell, *J. Am. Chem. Soc.*, **95**, 7768 (1973).
[296] D. D. Tanner, T. C. S. Ruo and E. V. Blackburn, *Can. J. Chem.*, **52**, 2242 (1974).
[297] J. G. Traynham and Y.-S. Lee, *J. Am. Chem. Soc.*, **96**, 3590 (1974).
[298] J. C. Day, M. J. Lindstrom and P. S. Skell, *J. Am. Chem. Soc.*, **96**, 5616 (1974).
[299] S. J. Cristol and H. W. Mueller, *J. Am. Chem. Soc.*, **95**, 8489 (1973).
[300] H. C. Brown and N. R. DeLue, *J. Am. Chem. Soc.*, **96**, 311 (1974).
[301] M. Janda, M. Valenta and P. Holý, *Coll. Czech. Chem. Comm.*, **39**, 959 (1974).
[302] T. Hino, M. Tonozuka and M. Nakapawa, *Tetrahedron*, **30**, 2123 (1974).
[303] E.-C. Wu and A. S. Rodgers, *Internat. J. Chem. Kinetics*, **5**, 1001 (1973).
[304] I. K. Stoddart, A. Nechvatal and J. M. Tedder, *J.C.S. Perkin II*, **1974**, 473.
[305] I. H. Elson and J. K. Kochi, *J. Org. Chem.*, **39**, 2091 (1974).
[306] D. Griller and K. U. Ingold, *J. Am. Chem. Soc.*, **96**, 630 (1974).
[307] See also p. 161.
[308] J. P. Lorand and R. W. Wallace, *J. Am. Chem. Soc.*, **96**, 2874 (1974).
[309] J. H. B. Chenier, E. Furimsky and J. A. Howard, *Can. J. Chem.*, **52**, 3682 (1974).
[310] J. A. Howard and E. Furimsky, *Can. J. Chem.*, **51**, 3738 (1973).
[311] A. G. Alexander, R. W. Fair and O. P. Strausz, *J. Phys. Chem.*, **78**, 203 (1974).
[312] K. S. Konoberevskii, E. Sh. Finkel'shtein, N. S. Nametkin and V. M. Vdovin, *Izv. Akad. Nauk SSSR, Ser. Khim.*, **1974**, 61; *Chem. Abs.*, **80**, 119866 (1974).
[313] J. N. S. Tam, R. W. Yip and Y. L. Chow, *J. Am. Chem. Soc.*, **96**, 4543 (1974).
[314] E. D. Morris, Jr. and H. Niki, *J. Phys. Chem.*, **78**, 1337 (1974).
[315] R. El Bacha, J. Cressely and G. Tanielian, *Bull. Soc. Chim. France*, **1974**, 1058.
[316] R. El Bacha, J. Cressely and G. Tanielian, *Bull. Soc. Chim. France*, **1974**, 1053.
[317] A. Lablanche-Combier and B. Planckaert, *Bull. Soc. Chim. France*, **1974**, 225.
[318a] J. E. Valimsky, R. H. Abeles and A. S. Mildvan, *J. Biol. Chem.*, **249**, 2751 (1974).
[318b] R. Breslow, R. J. Corcoran, J. A. Dale, S. Lin and P. Kalicky, *J. Am. Chem. Soc.*, **96**, 1973 (1974).
[318c] R. Breslow, R. J. Corcoran and B. B. Snider, *J. Am. Chem. Soc.*, **96**, 6791 (1974).
[318d] R. Breslow, R. J. Corcoran and B. B. Snider, *J. Am. Chem. Soc.*, **96**, 6792 (1974).
[319] J. Nafisi-Movaghar and Y. Hatano, *J. Phys. Chem.*, **78**, 1899 (1974).
[320] R. Nakao, T. Fukumoto and J. Tsurugi, *Bull. Chem. Soc. Japan*, **47**, 932 (1974).
[321] S. W. Baldwin, R. S. Doll and S. A. Haut, *J. Org. Chem.*, **39**, 2470 (1974).
[322] B. G. McKinnie, N. S. Bhacca, F. K. Cartledge and J. Fayssoux, *J. Am. Chem. Soc.*, **96**, 2637 (1974).
[323] W. Malisch and M. Kuhn, *Chem. Ber.*. **107**, 2835 (1974).
[324] H. C. Kuivila and C. C. H. Pian, *Chem. Comm.*, **1974**, 369.
[325] P. Beak and S. W. Mojé, *J. Org. Chem.*, **39**, 1320 (1974).
[326] J. SanFillipo and G. M. Anderson, *J. Org. Chem.*, **39**, 473 (1974).
[327] G. Leandri, H. Monti and M. Bertrand, *Tetrahedron*, **30**, 283 (1974).
[328] J. T. Groves and K. W. Ma, *J. Am. Chem. Soc.*, **96**, 6527 (1974).
[329] R. P. Quirk and R. E. Lea, *Tetrahedron Letters*, **1974**, 1925.
[330] W. C. Danen, D. G. Saunders and K. A. Rose, *J. Am. Chem. Soc.*, **96**, 4558 (1974).
[331] B. C. Gilbert, R. O. C. Norman and R. C. Sealy, *J.C.S. Perkin II*, **1974**, 1435.
[332] W. E. True and D. L. Heuring, *J. Org. Chem.*, **39**, 245 (1974).

[333] P. I. Abell, *Free Radicals*, **2**, 63 (1973); *Chem. Abs.*, **80**, 26378 (1974).
[334] R. A. Jackson, *Chem. Comm.*, **1974**, 573.
[335] G. Pratt and J. Veltman, *J.C.S. Faraday I*, **70**, 1840 (1974).
[336] N. L. Craig and D. W. Setser, *Internat. J. Chem. Kinetics*, **6**, 517 (1974).
[337] W. P. L. Carter and D. C. Tardy, *J. Phys. Chem.*, **78**, 1245 (1974).
[338] A. Horowitz, A. Mey-Marom and L. A. Rajbenbach, *Internat. J. Chem. Kinetics*, **6**, 265 (1974).
[339] J. C. Gibb, J. M. Tedder and J. C. Walton, *J.C.S. Perkin II*, **1974**, 807.
[340] J. M. Tedder and J. C. Walton, *J.C.S. Faraday I*, **70**, 308 (1974).
[341] H. Matsumoto, T. Nakano and Y. Nagai, *Tetrahedron Letters*, **1973**, 5147.
[341a] B. A. Englin, R. D. Ismailov and R. K. Freidlina, *Izv. Akad. Nauk SSSR, Ser. Khim.*, **1973**, 1498; *Chem. Abs.*, **79**, 114819 (1973).
[342] C. J. Brookes, P. L. Coe, D. M. Owen, A. E. Pedler and J. C. Tatlow, *Chem. Comm.*, **1974**, 323.
[343] R. N. Haszeldine, J. R. McAllister and A. E. Tipping, *J.C.S. Perkin I*, **1974**, 1303.
[344] R. N. Haszeldine, B. Hewison and A. E. Tipping, *J.C.S. Perkin I*, **1974**, 1706.
[345] R. E. Banks, M. G. Barlow, R. N. Haszeldine and W. D. Morton, *J.C.S. Perkin I*, **1974**, 1266.
[346] G. I. Fray, G. R. Geen, D. I. Davies, L. T. Parfitt and M. J. Parrott, *J.C.S. Perkin I*, **1974**, 729.
[347] E. Ghera and S. Shoua, *Tetrahedron Letters*, **1974**, 3843.
[348] C. Starks, *Free Radical Telomerization*, Academic Press, New York and London, 1974.
[349] B. A. Englin, N. A. Grigor'ev and V. M. Zhulin, *Izv. Akad. Nauk SSSR, Ser. Khim.*, **1973**, 1954; *Chem. Abs.*, **80**, 59174 (1974).
[350] L. van Wyk and R. A. Basson, *J.C.S. Faraday I*, **70**, 839 (1974).
[351] T. Fujimoto and I. Hirao, *Bull. Chem. Soc. Japan*, **47**, 1930 (1974).
[352] A. B. Terent'ev, *Izv. Akad. Nauk SSSR, Ser. Khim.*, **22**, 2469 (1973).
[353] A. B. Terent'ev, *Izv. Akad. Nauk SSSR, Ser. Khim.*, **22**, 2472 (1973).
[354] T. Sato, M. Seno and T. Asahara, *J. Synthetic Org. Chem.*, **32**, 184 (1974).
[355] D. S. Ashton, J. M. Tedder and J. C. Walton, *J.C.S. Faraday I*, **70**, 299 (1974).
[356] K. Ohga and T. Matsuo, *J. Org. Chem.*, **39**, 106 (1974).
[357] F. Liška, M. Němec and V. Dědec, *Coll. Czech. Chem. Comm.*, **39**, 580 (1974).
[358] N. Ishikawa, A. Nagashima and S. Hayashi, *Nippon Kagaku Kaishi*, **1974**, 1240; *Chem. Abs.*, **81**, 119432 (1974).
[359] C. Gardrat and R. Lalande, *Compt. rend., C*, **278**, 137 (1974).
[360] E. Montaudon and R. Lalande, *Compt. rend., C*, **278**, 1349 (1974).
[361] E. Montaudon, C. Gardrat and R. Lalande, *Compt. rend., C*, **278**, 367 (1974).
[362] K. Maruyama, M. Taniuchi and S. Oka, *Bull. Chem. Soc. Japan*, **47**, 712 (1974).
[363] H. Suginome, K. Kato and T. Masamure, *Tetrahedron Letters*, **1974**, 1161, 1165.
[364] M. Hájek, P. Šilhavý and J. Málek, *Tetrahedron Letters*, **1974**, 3193.
[365] F. Liška, V. Dědek and M. Němec, *Coll. Czech. Chem. Comm.*, **39**, 689 (1974).
[366] A. Fröling, *Rec. Trav. chim.*, **93**, 47 (1974).
[367] V. P. Šendrik, O. Paleta and V. Dědek, *Coll. Czech. Chem. Comm.*, **39**, 1061 (1974).
[368] R. Lalande, B. Clin and B. Maillard, *Compt. rend., C*, **278**, 963 (1974).
[369] E. Montaudon and R. Lalande, *Bull. Soc. Chim. France*, **1974**, 2635.
[370] C. L. Baumgardner and E. L. Lawton, *Accounts Chem. Res.*, **7**, 14 (1974).
[371] J. P. Kilcoyne and K. R. Jennings, *J.C.S. Faraday I*, **70**, 379 (1974).
[372] K. Furukawa, D. G. L. James and M. M. Papic, *Internat. J. Chem. Kinetics*, **6**, 337 (1974).
[373] W. L. P. Carter and D. C. Tardy, *J. Phys. Chem.*, **78**, 2201 (1974).
[374] P. C. Beadle and J. H. Knox, *J.C.S. Faraday I*, **70**, 1418 (1974).
[375] G. Coppens, G. J. Martens, M. Godfroid, J. Delvaux and J. Verbeyst, *Internat. J. Chem. Kinetics*, **6**, 437 (1974).
[376] V. N. Levashov, V. V. Grigor'eva, V. I. Propoi, O. M. Sarkisov and V. I. Vedeneev, *Izv. Akad. Nauk SSSR, Ser. Khim.*, **1973**, 2202; *Chem. Abs.*, **80**, 194888 (1974).
[377] C. Benjamin, G. Lanchec and B. Blour, *Bull. Soc. Chim. France*, **1974**, 661.
[378] D. D. Tanner, N. Nychka and T. Ochiai, *Can. J. Chem.*, **52**, 2573 (1974).
[379] Y. Calvez, J. Tuaillon and R. Perrot, *Helv. Chim. Acta*, **57**, 1433 (1974).
[380] M.-C. Lasne and A. Thuillier, *Bull. Soc. Chim. France*, **1974**, 249, 1142.
[381] B. E. Smart, *J. Org. Chem.*, **39**, 831 (1974).
[382] R. G. Gasanov, K. A. Kochetkov and R. Kh. Freidlina, *Izv. Akad. Nauk SSSR, Ser. Khim.*, **1974**, 804; *Chem. Abs.*, **81**, 37048 (1974).
[383] M. Ben-Yehuda, M. G. Katz and L. A. Rajbenbach, *J.C.S. Faraday I*, **70**, 908 (1974).
[384] A. J. White, *Chem. Soc. Rev.*, **3**, 17 (1974).
[385] K. Y. Choo and P. P. Gaspar, *J. Am. Chem. Soc.*, **96**, 1284 (1974).

[386] K. S. Chen, P. J. Krusic and J. K. Kochi, *J. Phys. Chem.*, **78**, 2030 (1974).
[387] R. A. Benkeser and D. F. Ehler, *J. Organometal. Chem.*, **69**, 193 (1974).
[388] R. N. Haszeldine, C. R. Pool and A. E. Tipping, *J.C.S. Perkin I*, **1974**, 2293.
[389] R. A. Kaba, D. Griller and K. U. Ingold, *J. Am. Chem. Soc.*, **96**, 6202 (1974).
[390] D. Griller and K. U. Ingold, *J. Am. Chem. Soc.*, **96**, 6203 (1974).
[391] T. Mojelsky and Y. L. Chow, *J. Am. Chem. Soc.*, **96**, 4549 (1974).
[392] H. H. Quon, T. Tezuka and Y. L. Chow, *Chem. Comm.*, **1974**, 428.
[393] J. L. Sprung, H. Akimoto and J. N. Pitts, *J. Am. Chem. Soc.*, **96**, 6549 (1974).
[394] G. B. Sergeev, I. A. Leenson and V. G. Garbusov, *Kinet. Katal.*, **14**, 1394 (1973); *Chem. Abs.*, **80**, 94891 (1974).
[395] N. A. Rybakova, L. N. Kiseleva and N. A. Pochailo, *Bull. Acad. Sci. USSR*, **1974**, 591.
[396] C. B. Boyce, S. B. Webb and L. Phillips, *J.C.S. Perkin I*, **1974**, 1650.
[397] C. B. Boyce, S. B. Webb, L. Phillips and L. R. Ager, *J.C.S. Perkin I*, **1974**, 1644.
[398] K. W. Morse and J. G. Morse, *J. Am. Chem. Soc.*, **95**, 8470 (1973).
[399] H. Francois and R. Lalande, *Compt. rend., C*, **279**, 117 (1974).
[400] R. L. Kenney and G. S. Fisher, *J. Org. Chem.*, **39**, 682 (1974).
[401] M. Simic, P. Neta and E. Hayon, *J. Phys. Chem.*, **77**, 2662 (1973).
[402] A. Pastrama and R. W. Carr, *Internat. J. Chem. Kinetics*, **6**, 587 (1974).
[403] A. Sugii, S. Shimada, K. Nagai and K. Kitahara, *Chem. Letters* (*Tokyo*), **1974**, 211.
[404] E. G. Jansen and C. A. Evans, *J. Am. Chem. Soc.*, **95**, 8205 (1973).
[405] See *Org. Reaction Mech.*, **1973**, p. 103.
[406] F. A. Hohorst, J. V. Paukstelis and D. D. DesMarteau, *J. Org. Chem.*, **39**, 1298 (1974).
[407] A. G. Davies, R. W. Dennis and B. P. Roberts, *Chem. Comm.*, **1974**, 468.
[408] K. E. Simmons and D. F. Von Sickle, *J. Am. Chem. Soc.*, **95**, 7759 (1974).
[409] D. R. Adams and D. I. Davies, *J.C.S. Perkin I*, **1974**, 246.
[410] W. E. Truce, D. L. Heuring and G. C. Wolf, *J. Org. Chem.*, **39**, 238 (1974).
[411] D. G. Garratt and G. H. Schmid, *Can. J. Chem.*, **52**, 3599 (1974).
[412] D. Lal, D. Griller, S. Husband and K. U. Ingold, *J. Am. Chem. Soc.*, **96**, 6355 (1974).
[413] A. L. J. Beckwith and G. Moad, *Chem. Comm.*, **1974**, 472.
[414] A. L. J. Beckwith, I. A. Blair and G. Phillipou, *J. Am. Chem. Soc.*, **96**, 1613 (1974).
[415] A. L. J. Beckwith, I. A. Blair and G. Phillipou, *Tetrahedron Letters*, **1974**, 2251.
[416] M. Julia and B. Malassiné, *Tetrahedron*, **30**, 695 (1974).
[417] P. Piccardi, M. Modena and P. Massardo, *Chim. Ind.* (*Milan*), **55**, 807 (1973); *Chem. Abs.*, **80**, 47176 (1974).
[418] P. Piccardi, P. Massardo, M. Modena and E. Santoro, *J.C.S. Perkin I*, **1974**, 1848.
[419] J.-M. Surzur and L. Stella, *Tetrahedron Letters*, **1974**, 2191.
[420] W. R. Hertler, *J. Org. Chem.*, **39**, 3219 (1974).
[421] L. A. Singer and S. S. Kim, *Tetrahedron Letters*, **1974**, 861.
[422] M. J. Perkins and B. P. Roberts, *J.C.S. Perkin II*, **1974**, 297.
[423] R. E. Schwerzel, R. G. Lawlor and G. T. Evans, *Chem. Phys. Letters*, **29**, 106 (1974).
[424] C. Walling and R. T. Clark, *J. Am. Chem. Soc.*, **96**, 4530 (1974).
[425] I. H. Elson and J. K. Kochi, *J. Org. Chem.*, **39**, 2091 (1974).
[426] T. Foster and P. R. West, *Can. J. Chem.*, **51**, 4009 (1973).
[427] T. Foster and P. R. West, *Can. J. Chem.*, **51**, 3589 (1973).
[428] W. T. Dixon and D. Murphy, *J.C.S. Perkin II*, **1974**, 1630.
[429] P. Hassanaly, G. Vernin, H. J. M. Dou and J. Metzger, *Bull. Soc. Chim. France*, **1974**, 560.
[430] G. Vernin, H. J. M. Dou and J. Metzger, *Bull. Soc. Chim. France*, **1974**, 1079.
[431] H.-J. Opgenorth and C. Rüchardt, *Annalen*, **1974**, 1333.
[432] P. E. Macrae and T. R. Wright, *Chem. Comm.*, **1974**, 898.
[433] J. E. Packer, R. K. Richardson, P. J. Soole and D. R. Webster, *J.C.S. Perkin II*, **1974**, 1472.
[434] J. I. G. Cadogan, C. D. Murray and J. T. Sharp, *Chem. Comm.*, **1974**, 901.
[435] R. A. Jackson, *Chem. Comm.*, **1974**, 573.
[436] S. Vidal, J. Court and J.-M. Bonnier, *Compt. rend., C*, **279**, 797 (1974).
[437] R. Henriquez, A. R. Morgan, P. Mulholland, D. C. Nonhebel and G. G. Smith, *Chem. Comm.*, **1974**, 987.
[438] S. Vidal, J. Court and J.-M. Bonnier, *J.C.S. Perkin II*, **1973**, 2071.
[439] G. Fillipi, G. Vernin, H. J. M. Dou, J. Metzger and M. J. Perkins, *Bull. Soc. Chim. France*, **1974**, 1075.
[440] R. Dufournet, J. Court and J.-M. Bonnier, *Bull. Soc. Chim. France*, **1974**, 1112.
[441] S. Di Gregorio, M. B. Yim and D. E. Wood, *J. Am. Chem. Soc.*, **95**, 8455 (1973).

[442] T. Shono and I. Nishiguchi, *Tetrahedron*, **30**, 2183 (1974).
[443] A. Clerici, F. Minisci and O. Porta, *J.C.S. Perkin II*, **1974**, 1699.
[444] G. Vernin, M. A. Lebreton, H. J. M. Dou and J. Metzger, *Bull. Soc. Chim. France*, **1974**, 1085.
[445] J. C. Arnold, G. J. Gleicher and J. D. Unruh, *J. Am. Chem. Soc.*, **96**, 787 (1974).
[446] J. R. Shelton and A. L. Lipman, *J. Org. Chem.*, **39**, 2386 (1974).
[447] M. Tiecco and A. Tundo, *Internat. J. Sulfur Chem.*, **8**, 295 (1973).
[448] C.-M. Camaggi, R. Leardini, A. Tundo and M. Tiecco, *J.C.S. Perkin I*, **1974**, 271.
[449] L. V. Vlasova, L. S. Kobrian and G. G. Yakobson, *Zh. Org. Khim.*, **10**, 792 (1974); *Chem. Abs.*, **81**, 25440 (1974).
[450] A. P. Manzara and P. Kovacic, *J. Org. Chem.*, **39**, 504 (1974).
[451] F. Minisci and O. Porta, *Adv. Heterocyclic Chem.*, **16**, 123 (1974).
[452] K. C. Bass and P. Nababsing, *Chem. and Ind.* (*London*), **1974**, 574.
[453] G. Heinisch, A. Jentzsch and M. Pailer, *Monatsh. Chem.*, **105**, 648 (1974).
[454] A. Clerici, F. Minisci, M. Perchinunno and O. Porta, *J.C.S. Perkin II*, **1974**, 416.
[455] A. Furlan, M. Furlan, B. Stanovik and M. Tisler, *Monatsh. Chem.*, **105**, 834 (1974).
[456] G. Friedman, *J.C.S. Perkin I*, **1974**, 441.
[457] H. Yamada, H. Shizuka, S. Sekiguchi and K. Matsui, *Bull. Soc. Chem. Japan*, **47**, 238 (1974).
[458] T. Matsuura and K. Omura, *Synthesis*, **1974**, 173.
[459] M. K. Eberhardt, *J. Phys. Chem.*, **78**, 1795 (1974).
[460] P. Neta and R. W. Fessenden, *J. Phys. Chem.*, **78**, 523 (1974).
[461] L. Eberson and B. Helgée, *Chem. Scripta*, **5**, 47 (1974).
[462] A. Barmetler, C. Rüchardt, R. Sustman, S. Sustman and R. Verhülsdonk, *Tetrahedron Letters*, **1974**, 4389; cf. *Org. Reaction Mech.*, **1972**, 324.
[463] M. J. S. Dewar and S. Kirschner, *J. Am. Chem. Soc.*, **96**, 5244 (1974).
[464] T. N. Shatkina and O. A. Reutov, *J. Org. Chem. USSR*, **10**, 876 (1974).
[465] M. Suzuki, S.-I. Murahashi, A. Sonoda and I. Moritani, *Chem. Letters* (*Tokyo*), **1974**, 267.
[466] J. K. Kochi and P. J. Krusic, *J. Am. Chem. Soc.*, **91**, 1877, 1879 (1969).
[467] *Org. Reaction Mech.*, **1973**, 57.
[468] R. Sustmann and F. Lübbe, *Tetrahedron Letters*, **1974**, 2831.
[469] E. Müller, *Tetrahedron Letters*, **1974**, 1835.
[470] T. Shono and I. Nishiguchi, *Tetrahedron*, **30**, 2173 (1974).
[471] W. P. L. Carter and D. C. Tardy, *J. Phys. Chem.*, **78**, 1245 (1974).
[472] J.-Y. Godet and M. Pereyre, *J. Organometal. Chem.*, **77**, C1 (1974).
[473] W. P. L. Carter and D. C. Tardy, *J. Phys. Chem.*, **78**, 1573 (1974).
[474] H.-P. Löffler, *Chem. Ber.*, **107**, 2691 (1974).
[475] D. A. Bochvar, R. G. Gasanov, A. M. Gyul'maliev, I. V. Stankevich and R. Kh. Friedlina, *Dokl. Akad. Nauk SSSR*, **211**, 669 (1973).
[476] N. A. Kuz'mina, E. Ts. Chukovskaya and R. Kh. Friedlina, *Bull. Acad. Sci. USSR*, **23**, 463 (1974).
[477] M. C. R. Symons, *Chem. Comm.*, **1974**, 556.
[478] F. Bertini, T. Caronna, L. Grossi and F. Minisci, *Gazz. Chim. Ital.*, **104**, 571 (1974).
[479] B. C. Childress, A. C. Rice and P. B. Shevlin, *J. Org. Chem.*, **39**, 3056 (1974).
[480] S. A. Gover and A. Gossen, *J.C.S. Perkin I*, **1974**, 2353.
[481] B. Maillard, M. Cazaux and R. Lalande, *Compt. rend.*, *C*, **279**, 701 (1974).
[482] D. C. Tardy, *Internat. J. Chem. Kinetics*, **6**, 291 (1974).
[483] J. M. Larkin, W. M. Cummings and K. L. Krenz, *J. Org. Chem.*, **39**, 714 (1974).
[484] Y. Kanoaka and Y. Hatanaka, *Heterocycles*, **2**, 423 (1974).
[485] W. Adam, *Chem. Comm.*, **1974**, 289.
[486] C. E. Kalmus and D. M. Hercules, *J. Am. Chem. Soc.*, **96**, 449 (1974).
[487] G. J. Siuta, R. W. Franck and A. A. Ortizio, *Chem. Comm.*, **1974**, 910.
[488] Y. Ohto, H. Shizuka, S. Sekiguchi and K. Matsui, *Bull. Chem. Soc. Japan*, **47**, 1209 (1974).
[489] J. J. Houser, M.-C. Chen and S. S. Wang, *J. Org. Chem.*, **39**, 1387 (1974).
[490] A. Eckersley and N. A. J. Rogers, *Tetrahedron Letters*, **1974**, 1661.
[491] C. Y. Myers, L. L. Ho, A. Ohno and M. Kagami, *Tetrahedron Letters*, **1974**, 729.
[492] F. Gerhart and L. Wilde, *Tetrahedron Letters*, **1974**, 475.
[493] R. F. Hudson, A. J. Lawson and K. A. F. Record, *J.C.S. Perkin II*, **1974**, 869.
[494] P. Welzel, L. Günther and G. Eckhardt, *Chem. Ber.*, **107**, 3624 (1974).
[495] U. Schirmer and J. M. Conia, *Tetrahedron Letters*, **1974**, 3057.
[496] M. R. Willcott and I. M. Rathburn, *J. Am. Chem. Soc.*, **96**, 938 (1974).
[497] J. R. Williams, G. M. Sarkisian, J. Quigley, A. Hasiuk and R. Van der Vennen, *J. Org. Chem.*, **39**, 1028 (1974).

[498] J. Muzart and J. P. Pete, *Tetrahedron Letters*, **1974**, 3919.
[499] S. P. Pappas and L. Q. Bao, *J. Am. Chem. Soc.*, **95**, 7906 (1973).
[500] P. S. Mariano and D. Watson, *J. Org. Chem.*, **39**, 2774 (1974).
[501] N. K. Hamer and A. J. Wills, *J.C.S. Perkin II*, **1974**, 88.
[502] H. Hart and M. Kozuya, *Tetrahedron Letters*, **1974**, 1913.
[503] A. C. Pratt, *J.C.S. Perkin I*, **1973**, 2496.
[504] K. G. Hancock and R. O. Grider, *J. Am. Chem. Soc.*, **96**, 1158 (1974).
[505] A. A. Gorman, R. L. Leyland, N. A. J. Rodgers and P. G. Smith, *Tetrahedron Letters*, **1973**, 5085.
[506] J. M. Brown and M. M. Ogilvy, *J. Am. Chem. Soc.*, **96**, 292 (1974).
[507] L. R. C. Barclay and I. J. McMaster, *Tetrahedron Letters*, **1973**, 4901.
[508] H. Suginome and T. Uchida, *Bull. Chem. Soc. Japan*, **47**, 687 (1974).
[509] D. A. Jaeger, *J. Am. Chem. Soc.*, **96**, 6216 (1974).
[510] M. A. Afzal, *Chem. Ind.* (*London*), **1974**, 37.
[511] G. Jones, *J. Chem. Educ.*, **51**, 175 (1974).
[512] R. Kh. Friedlina and A. B. Terent'ev, *Usp. Khim.*, **43**, 294 (1974).
[513] M. J. S. Dewar, S. Kirschner, H. W. Kollmar and L. E. Wade, *J. Am. Chem. Soc.*, **96**, 5242 (1974).
[514] M. Schneider, *Z. Naturforsch.*, *B*, **29**, 290 (1974).
[515] M. Schneider and A. Erben, *Z. Naturforsch.*, *B*, **29**, 288 (1974).
[516] T. Sasaki, S. Eguchi and F. Hibi, *Chem. Comm.*, **1974**, 227.
[517] E. L. Allred and B. R. Beck, *Tetrahedron Letters*, **1974**, 437.
[518] P. Caramella, R. Huisgen and B. Schmolke, *J. Am. Chem. Soc.*, **96**, 2999 (1974).
[519] P. Caramella, R. Huisgen and B. Schmolke, *J. Am. Chem. Soc.*, **96**, 2997 (1974).
[520] Cf. Chapter 13.
[521] M. Schneider and I. Merz, *Tetrahedron Letters*, **1974**, 1995.
[522] M. Schneider and G. Mössinger, *Tetrahedron Letters*, **1974**, 3081.
[523] R. J. Crawford and M. Ohno, *Can. J. Chem.*, **52**, 3134 (1974).
[524] J. A. Berson, C. D. Duncan and L. R. Corwin, *J. Am. Chem. Soc.*, **96**, 6175 (1974).
[525] J. A. Berson, L. R. Corwin and J. H. Davis, *J. Am. Chem. Soc.*, **96**, 6177 (1974).
[526] L. M. Stephenson and T. A. Gibson, *J. Am. Chem. Soc.*, **96**, 5624 (1974).
[527] N. Shimizu and S. Nishida, *J. Am. Chem. Soc.*, **96**, 6451 (1974).
[528] R. D. Chambers, M. Clark, J. A. H. MacBride, W. K. R. Musgrave and K. C. Srivastava, *J.C.S. Perkin I*, **1974**, 125.
[529] T. Tsuji and S. Nishida, *J. Am. Chem. Soc.*, **96**, 3649 (1974).
[530] K.-D. Gundermann and E. Röhrl, *Annalen*, **1974**, 1661.
[531] M. P. Servé, *J. Heterocyclic Chem.*, **11**, 245 (1974).
[532] K. Tsujimoto, M. Ohashi and T. Yonezawa, *Bull. Chem. Soc. Japan*, **46**, 3605 (1973).
[533] P. Murray Rust, *Acta Cryst.*, *B*, **30**, 1368 (1974).
[534] A. H. Cohen and B. M. Hoffmann, *J. Phys. Chem.*, **78**, 1313 (1974).
[535] A. Capiomont and J. Lajzérowicz-Bonneteau, *Acta Cryst.*, *B*, **30**, 2160 (1974).
[536] L. R. Mahoney, G. D. Mendenhall and K. U. Ingold, *J. Am. Chem. Soc.*, **95**, 8610 (1973).
[537] Cf. *Org. Reaction Mech.*, **1974**, 68.
[538] D. Bordeaux and J. Lajzérowicz, *Acta Cryst.*, *B*, **30**, 790 (1974).
[539] J. B. Wetherinton, S. S. Ament and J. W. Moncrief, *Acta Cryst.*, *B*, **30**, 568 (1974).
[540] A. R. Forrester, S. P. Hepburn and G. McConnachie, *J.C.S. Perkin I*, **1974**, 2213.
[541] J. Goldman, T. E. Petersen, K. Torsell and J. Becher, *Tetrahedron*, **29**, 3833 (1973).
[542] C. M. Camaggi, L. Lunazzi and G. Placucci, *J. Org. Chem.*, **39**, 2425 (1974).
[543] C. M. Camaggi, L. Lunazzi, G. F. Pedulli, G. Placucci and M. Tiecco, *J.C.S. Perkin II*, **1974**, 1226.
[544] C. M. Camaggi, R. Leardini and G. Placucci, *J.C.S. Perkin II*, **1974**, 1195.
[545] A. T. Balaban, H. G. Aurich, J. Trösken, E. Brugger, D. Döpp and K. H. Sailer, *Tetrahedron*, **30**, 739 (1974).
[546] J. W. Neely, G. F. Hatch and R. W. Kreilick, *J. Am. Chem. Soc.*, **96**, 652 (1974).
[547] K. Ishizu, A. Nakajima and Y. Deguchi, *Tetrahedron Letters*, **1974**, 1537.
[548] K. Ishizu, H. Nagai, K. Mukai, M. Kohno and T. Yamamoto, *Chem. Letters* (*Tokyo*), **1974**, 1261.
[549] C. Lagercrantz and M. Setaka, *Acta Chem. Scand.*, *B*, **28**, 619 (1974).
[550] *Org. Reaction Mech.*, **1973**, 152.
[551] A. L. J. Beckwith and M. D. Lawton, *J.C.S. Perkin II*, **1973**, 2134.
[552] G. A. Razuvaev, G. A. Abakumov and V. K. Cherkasov, *Dokl. Akad. Nauk SSSR*, **212**, 737 (1974).
[553] H. G. Aurich and J. Trösken, *Tetrahedron*, **30**, 2519 (1974).
[554] H. G. Aurich and J. Trösken, *Tetrahedron*, **30**, 2515 (1974).
[555] E. K. Metzner, L. J. Libertini and M. Calvin, *J. Am. Chem. Soc.*, **96**, 6515 (1974).

[556] V. N. Parmon, A. I. Kokorin, G. M. Zhidomirov and K. I. Zamaraev, *Mol. Phys.*, **26**, 1565 (1973).
[557] E. F. Litvin, L. M. Kozlova, A. B. Shapiro, L. Kh. Friedlin and E. G. Rozantsev, *Bull. Acad. Sci. USSR*, **23**, 40 (1974).
[558] A. M. Giroud, A. Rassat and H. U. Sieveking, *Tetrahedron Letters*, **1974**, 635.
[559] A. R. Forrester and S. P. Hepburn, *J.C.S. Perkin I*, **1974**, 2208.
[560] A. Calder, A. R. Forrester and G. McConnachie, *J.C.S. Perkin I*, **1974**, 2198.
[561] G. Genoud, M.-C. Schouler and M. Decorps, *Chem. Phys. Letters*, **26**, 414 (1974).
[562] E. Flesia and J.-M. Surzur, *Tetrahedron Letters*, **1974**, 123.
[563] G. A. Abakumov, E. P. Sanaeva and G. A. Razuvaev, *Bull. Acad. Sci., USSR*, **23**, 581 (1974).
[564] H. G. Aurich and P. Höhlein, *Tetrahedron Letters*, **1974**, 279.
[565] D. Mulvey and W. A. Waters, *J.C.S. Perkin II*, **1974**, 772.
[566] D. Mulvey and W. A. Waters, *J.C.S. Perkin II*, **1974**, 560.
[567] H. Schlude, *Tetrahedron*, **29**, 4007 (1973).
[568] D. S. Khadzhiev, E. S. Todorova and P. K. Topalova, *Bull. Acad. Sci. USSR*, **22**, 2354 (1973).
[569] R. E. Banks, K. C. Eapen, R. N. Haszeldine, A. V. Holt, T. Myerscough and S. Smith, *J.C.S. Perkin I*, **1974**, 2533.
[570] A. Rassat and P. Rey, *Tetrahedron*, **30**, 3315 (1974).
[571] P. Michon and A. Rassat, *J. Org. Chem.*, **39**, 2121 (1974).
[572] R. Ramasseul and A. Rassat, *Tetrahedron Letters*, **1974**, 2413.
[573] H. H. Quon, T. Tezuka and Y. L. Chow, *Chem. Comm.*, **1974**, 428.
[574] G. Sosnovsky and M. Konieczny, *Z. Naturforsch.*, *B*, **28**, 488 (1973).
[575] W. E. Trommer, H. Wenzel and G. Pfleiderer, *Annalen*, **1974**, 1357.
[576] H. Schindler and J. Sealig, *J. Chem. Phys.*, **61**, 2946 (1974).
[577] R. Ramasseul, A. Rassat and P. Rey, *Tetrahedron*, **30**, 265 (1974).
[578] E. G. Janzen and C. A. Evans, *J. Am. Chem. Soc.*, **95**, 8205 (1974).
[579] J. R. Harbour, Y. Chow and J. R. Bolton, *Can. J. Chem.*, **52**, 3549 (1974).
[580] I. I. Bil'kis and S. M. Shein, *Zh. Org. Khim.*, **10**, 1126 (1974); *Chem. Abs.*, **81**, 49041 (1974).
[581] D. H. R. Barton, R. L. Harris, R. H. Hesse, M. M. Pechet and F. J. Urban, *J.C.S. Perkin I*, **1974**, 2344.
[582] J. Almog, D. H. R. Barton, P. D. Magnus and R. K. Norris, *J.C.S. Perkin I*, **1974**, 853.
[583] K. Okamoto and K. Komatsu, *Bull. Chem. Soc. Japan*, **47**, 1709 (1974).
[584] H. G. Aurich and W. Dersch, *Z. Naturforsch.*, *B*, **28**, 525 (1973).
[585] A. J. Bard, J. C. Gilbert and R. D. Goodin, *J. Am. Chem. Soc.*, **96**, 620 (1974).
[586] B. B. Adeleke, S.-K. Wong and J. K. S. Wan, *Can. J. Chem.*, **52**, 2901 (1974).
[587] K. Sommermeyer, W. Seiffert and W. Wilker, *Tetrahedron Letters*, **1974**, 1821.
[588] J. C. Evans, E. D. Owen and D. A. Wilson, *J.C.S. Perkin II*, **1974**, 557.
[589] J. A. Cella, J. A. Kelley and E. F. Kenehan, *Chem. Comm.*, **1974**, 943.
[590] A. Rassat and P. Rey, *Tetrahedron*, **30**, 3597 (1974).
[591] S. Chou, J. A. Nelson and T. A. Spencer, *J. Org. Chem.*, **39**, 2356 (1974).
[592] J. I. G. Cadogan and A. G. Rowley, *J.C.S. Perkin II*, **1974**, 1030.
[593] C. T. Cazianis and D. R. Eaton, *Can. J. Chem.*, **52**, 2463 (1974).
[594] P. A. Wehrli and B. Schaer, *Synthesis*, **1974**, 288.
[595] E. F. Litvin, L. M. Kozlova, A. B. Shapiro, L. Kh. Friedlina and E. G. Rozaptsev, *Bull. Acad. Sci. USSR*, **23**, 40 (1974).
[596] E. B. Milovskaya, *Russ. Chem. Rev.*, **1973**, 384.
[597] R. C. Larock, *J. Organometal. Chem.*, **67**, 353 (1974).
[598] V. P. Maslennikov, G. I. Makin and Yu. A. Aleksandrov, *Zh. Obshch. Khim.*, **43**, 2289 (1973); *Chem. Abs.*, **80**, 94902 (1974).
[599] J. H. Espenson and T. D. Sellers, *J. Am. Chem. Soc.*, **96**, 94 (1974).
[600] J. B. Levy and R. C. Kennedy, *J. Am. Chem. Soc.*, **96**, 4791 (1974).
[601] P. M. Digiacomo and H. G. Kuivila, *J. Organometal. Chem.*, **63**, 251 (1973).
[602] J. Grignon and M. Pereyre, *J. Organometal. Chem.*, **61**, C33 (1973).
[603] S. Kozuka, T. Akasaka, S. Furumai and S. Oae, *Chem. Ind.* (*London*), **1974**, 452.
[604] S. Kozuka, S. Furumai, T. Akasaka and S. Oae, *Chem. Ind.* (*London*), **1974**, 496.
[605] J.-C. Pommier and D. Chevolleau, *J. Organometal. Chem.*, **74**, 405 (1974).
[606] L. Benati, P. C. Montevecci, A. Tundo and G. Zanardi, *J.C.S. Perkin I*, **1974**, 1271, 1276.
[607] See *Org. Reaction Mech.*, **1973**, 116.
[608] M. F. Lappert and P. W. Lednor, *Chem. Comm.*, **1973**, 948.
[609] A. V. Kramer, J. A. Labinger, J. S. Bradley and J. A. Osborn, *J. Am. Chem. Soc.*, **96**, 7145 (1974).

610 N. G. Hargreaves, A. Johnson, R. J. Puddephatt and L. H. Sutcliffe, *J. Organometal. Chem.*, **69**, C21 (1974).
611 G. Boekestein, E. H. J. M. Jansen and H. M. Buck, *Chem. Comm.*, **1974**, 118.
612 S. P. Mishra and M. C. R. Symons, *Chem. Comm.*, **1974**, 606.
613 A. G. Davies, M. J. Parrott and B. P. Roberts, *Chem. Comm.*, **1974**, 973.
614 A. Hudson and J. T. Whiffen, *Chem. Phys. Letters*, **29**, 113 (1974).
615 T. Gillbro and F. Williams, *J. Am. Chem. Soc.*, **96**, 5032 (1974).
616 M. C. R. Symons, *Mol. Phys.*, **27**, 785 (1974).
617 A. G. Davies, R. W. Dennis and B. P. Roberts, *J.C.S. Perkin II*, **1974**, 1101.
618 H.-W. Tan and W. G. Bentrude, *J. Am. Chem. Soc.*, **96**, 5950 (1974).
619 W. G. Bentrude, W. A. Khan, M. Murakami and H.-W. Tan, *J. Am. Chem. Soc.*, **96**, 5566 (1974).
620 A. G. Davies, M. J. Parrott and B. P. Roberts, *Chem. Comm.*, **1974**, 27.
621 D. G. Pobedimskii, A. L. Buckachenko and V. A. Kurbatov, *Izv. Akad. Nauk SSSR, Ser. Khim.*, **1973**, 2454; *Chem. Abs.*, **80**, 94972 (1974).
622 W. S. Trahanovsky in *Methods in Free-Radical Chemistry*, Vol. IV (E. S. Huyser, Ed.), Dekker, New York, 1974, p. 133.
623 A. E. Jukes, *Adv. Organometal. Chem.*, **12**, 215 (1974).
624 J. K. Kochi, *Accounts Chem. Res.*, **7**, 351 (1974).
625 E. Hayon and M. Simic, *Accounts Chem. Res.*, **7**, 114 (1974).
626 P. S. Rao and E. Hayon, *J. Am. Chem. Soc.*, **96**, 1287 (1974).
627 P. S. Rao and E. Hayon, *J. Am. Chem. Soc.*, **96**, 1295 (1974).
628 K. Nyberg, *Chem. Scripta*, **5**, 115 (1974).
629 K. Nyberg, *Chem. Scripta*, **5**, 120 (1974).
630 *Org. Reaction Mech.*, **1973**, 133.
631 R. O. C. Norman, C. B. Thomas and T. J. Ward, *J.C.S. Perkin I*, **1973**, 2914.
632 H. Inoue, M. Izumi and E. Imoto, *Bull. Chem. Soc. Japan*, **47**, 1712 (1974).
633 A. Ledwith and P. J. Russell, *Chem. Comm.*, **1974**, 291.
634 A. Ledwith and P. J. Russell, *Chem. Comm.*, **1974**, 959.
635 J. K. Kochi, *Tetrahedron Letters*, **1974**, 4305; republished, *ibid.*, **1975**, 41.
636 A. Ledwith and P. J. Russell, *J.C.S. Perkin II*, **1974**, 582.
637 K. Nishiyama, K. Hata and T. Sato, *J.C.S. Perkin II*, **1974**, 577.
638 J. Pataki and R. Balick, *Tetrahedron Letters*, **1974**, 3447.
639 H. Sternerup, *Acta Chem. Scand., B*, **28**, 579 (1974).
640 L. Marchetti, M. Poloni and L. Greci, *Gazz. Chim. Ital.*, **104**, 321 (1974).
641 G. Hilgetag, *Z. Chem.*, **14**, 236 (1974).
642 D. Vofsi, *J.C.S. Perkin II*, **1974**, 857.
643 R. Curci, G. Delano, F. DiFuria, J. O. Edwards and A. R. Gallopo, *J. Org. Chem.*, **39**, 3020 (1974).
644 P. S. Dewar, E. Ernstbrunner, J. R. Gilmore, M. Godfrey and J. M. Mellor, *Tetrahedron*, **30**, 2455 (1974).
645 D. Koch, H. Schäfer and E. Steckhan, *Chem. Ber.*, **107**, 3640 (1974).
646 P. Koch and I. Skibida, *Gazz. Chim. Ital.*, **104**, 225 (1974).
647 S. K. Ivanov, C. Karshalykov and H. Kropf, *Annalen*, **1974**, 1713.
648 E. G. E. Hawkins and R. Large, *J.C.S. Perkin I*, **1974**, 280.
649 H. D. Burrows and T. J. Kemp, *Chem. Soc. Rev.*, **3**, 139 (1974).
650 D. Greatorex, R. J. Hill, T. J. Kemp and T. J. Stone, *J.C.S. Faraday I*, **70**, 216 (1974).
651 T. J. Kemp and M. J. Welbourn, *Tetrahedron Letters*, **1974**, 87.
652 R. F. Garwood, Naser-ud-Din, C. J. Scott and B. C. L. Weedon, *J.C.S. Perkin I*, **1973**, 2714.
653 J. P. Coleman, R. Lines, J. H. P. Uttley and B. C. L. Weedon, *J.C.S. Perkin II*, **1974**, 1064.
654 W. S. Trahanovsky, J. Cramer and D. W. Brixius, *J. Am. Chem. Soc.*, **96**, 1077 (1974).
655 L. A. Mirkind and M. Ya. Fioshin, *Katal. Reakts. Zhidk. Fazc.*, **1972**, 226: *Chem. Abs.*, **79**, 136178 (1973).
656 S. Torii, T. Okamoto and H. Tanaka, *J. Org. Chem.*, **39**, 2486 (1974).
657 A. L. J. Beckwith, R. T. Cross and G. E. Gream, *Austral. J. Chem.*, **27**, 1673 (1974).
658 A. L. J. Beckwith, R. T. Cross and G. E. Gream, *Austral. J. Chem.*, **27**, 1693 (1974).
659 B. C. C. Cantello, J. M. Mellor and G. Scholes, *J.C.S. Perkin II*, **1974**, 348.
660 A. Theine and J. G. Traynham, *J. Org. Chem.*, **39**, 153 (1974).
661 M. I. Katsin, Yu. N. Ogibin and G. I. Nikishin, *Bull. Acad. Sci. USSR*, **23**, 129 (1974).
662 J. Burkhard, J. Janků and S. Lande, *Coll. Czech. Chem. Comm.*, **39**, 1072 (1974).
663 P. Beak and B. R. Harris, *J. Am. Chem. Soc.*, **96**, 6363 (1974).
664 J. Goldman, N. Jacobsen and K. Torssell, *Acta Chem. Scand., B*, **28**, 492 (1974).

[665] A. Clerici, F. Minisci and O. Porta, *Tetrahedron Letters*, **1974**, 4183.
[666] A. R. Forrester, J. Skilling and R. H. Thomson, *J.C.S. Perkin I*, **1974**, 2161.
[667] P. D. McDonald and G. A. Hamilton, *J. Am. Chem. Soc.*, **95**, 7752 (1973).
[668] R. D. Parnell and K. E. Russell, *J.C.S. Perkin II*, **1974**, 161.
[669] R. C. Eckert, H.-M. Chang and W. P. Tucker, *J. Org. Chem.*, **39**, 718 (1974).
[670] G. A. Razuvaer, N. S. Vasileiskaya and N. N. Vavilina, *J. Organometal. Chem.*, **80**, 19 (1974).
[671] M. Young and C. Steelink, *Phytochemistry*, **12**, 2851 (1973).
[672] C. Steelink and W. E. Britton, *Tetrahedron Letters*, **1974**, 2869.
[673] C. Steelink and W. E Britton, *Tetrahedron Letters*, **1974**, 2873.
[674] O. Brauer, P. Köll and J. Voss, *Tetrahedron Letters*, **1974**, 231.
[675] J. A. Howard and E. Furimsky, *Can. J. Chem.*, **51**, 3738 (1973).
[676] A. Nishinga, K. Watanabe and T. Matsuura, *Tetrahedron Letters*, **1974**, 1291.
[677] V. V. Ershov, A. A. Volod'kin, A. I. Prokof'ev and S. P. Solodnikov, *Russ. Chem. Rev.*, **42**, 740 (1973).
[678] I. V. Khudyakov and V. A. Kuz'min, *Khim. Vys. Energ.*, **8**, 171 (1974); *Chem. Abs.*, **81**, 19193 (1974).
[679] A. Sadana and J. R. Katzer, *J. Catalysis*, **35**, 140 (1974).
[680] E. P. Platonova, V. D. Pokhodenko, *Electrokhimiya*, **10**, 278 (1974); *Chem. Abs.*, **80**, 107697 (1974).
[681] M. Ya. Zarubin, A. M. Kutnevich and A. P. Lukashenkov, *Zh. Org. Khim.*, **10**, 402 (1974); *Chem. Abs.*, **80**, 107663 (1974).
[682] L. Denivelle and Hunyh Anh-Hoa, *Bull. Soc. Chim. France*, **1974**, 487.
[683] L. Denivelle and Hunyh Anh-Hoa, *Bull. Soc. Chim. France*, **1974**, 491.
[684] P. Ashworth and W. T. Dixon, *J.C.S. Perkin II*, **1974**, 739.
[685] P. Ashworth and W. T. Dixon, *J.C.S. Perkin II*, **1973**, 2128.
[686] V. D. Pokhodenko, V. A. Khizhniyi and V. G. Koshechko, *Teor. Eksp. Khim.*, **10**, 112 (1974); *Chem. Abs.*, **80**, 132514 (1974).
[687] R. P. R. Ranaweera and G. Scott, *Chem. Ind.* (*London*), **1974**, 774.
[688] V. D. Pokhodenko, V. A. Samarskii and V. A. Khizhnyi, *Zh. Org. Khim.*, **10**, 1335 (1974); *Chem. Abs.*, **81**, 77606 (1974).
[689] A. Stüwe, M. Weber-Schäfer and H. Baumgartel, *Ber. Bunsenges. Phys. Chem.*, **78**, 309 (1974).
[690] M. Weber-Schäfer and H. Baumgartel, *Ber. Bunsenges. Phys. Chem.*, **78**, 317 (1974).
[691] A. Stüwe and H. Baumgartel, *Ber. Bunsenges. Phys. Chem.*, **78**, 320 (1974).
[692] C. Walling, G. M. El-Taliawi and R. A. Johnson, *J. Am. Chem. Soc.*, **96**, 133 (1974).
[693] H. Möckel, M. Bonifačić and K.-D. Asmus, *J. Phys. Chem.*, **78**, 282 (1974).
[694] M. E. Snook and G. A. Hamilton, *J. Am. Chem. Soc.*, **96**, 860 (1974).
[695] Cf. *Org. Reaction Mech.*, **1973**, 142.
[696] B. C. Gilbert, R. O. C. Norman and R. C. Sealy, *J.C.S. Perkin II*, **1974**, 824.
[697] B. C. Gilbert, R. O. C. Norman and R. C. Sealy, *J.C.S. Perkin II*, **1973**, 2174.
[698] Z. Čeković and M. M. Green, *J. Am. Chem. Soc.*, **96**, 3001 (1974).
[699] Sh. K. Bochorishvili, Yu. D. Norikov, E. A. Blyumberg and V. I. Kokochashvili, *Soobshch. Akad. Nauk Gruz. SSR*, **71**, 373 (1973); *Chem. Abs,*, **79**, 136262 (1973).
[700] S. Imamura, T. Banba, H. Teranishi and Y. Takegami, *Nippon Kagaku Kaishi*, **1973**, 2110; *Chem. Abs.*, **80**, 47243 (1974).
[701] V. K. Fedorov and V. M. Berdnikov, *Kinet. Katal.*, **14**, 870 (1973); *Chem. Abs.*, **79**, 145669 (1973).
[702] M. Cocivera, C. A. Fyfe, S. P. Vaish and H. E. Chen, *J. Am. Chem. Soc.*, **96**, 1611 (1974).
[703] C. Walling and T. Weil, *Internat. J. Chem. Kinetics*, **6**, 507 (1974).
[704] Yu. N. Kozlov, A. D. Nadezhdin and A. P. Pourmal, *Internat. J. Chem. Kinetics*, **6**, 383 (1974).
[705] A. J. Birch and G. Nadamuni, *J.C.S. Perkin II*, **1974**, 545.
[706] H. O. House and E. F. Kinloch, *J. Org. Chem.*, **39**, 747 (1974).
[707] L. H. Klemm, P. E. Iversen and H. Lund, *Acta Chem. Scand., B*, **28**, 593 (1974).
[708] U. Brühlmann and E. Hayon, *J. Am. Chem. Soc.*, **96**, 6169 (1974).
[709] K. R. Olson and M. Z. Hoffmann, *Chem. Comm.*, **1974**, 938.
[710] P. Burri, H. Loewenschuss, H. Zollinger and G. K. Zwolinski, *Helv. Chim. Acta*, **57**, 395 (1974).
[711] H. Zollinger, *Accounts Chem. Res.*, **6**, 335 (1973).
[712] A. A. Matnishyan and G. A. Kazaryan, *Arm. Khim. Zh.*, **26**, 991 (1973); *Chem. Abs.*, **80**, 132406 (1974).
[713] B. Ya. Medvedev, L. A. Polyakova, K. A. Bilevich, N. N. Bubnov and O. Yu. Okhlobystin, *Teor. Eksp. Khim.*, **9**, 838 (1973); *Chem. Abs.*, **80**, 70057 (1974).
[714] J. Bargon and K.-G. Seifert, *Tetrahedron Letters*, **1974**, 2265.
[715] A. H. Lewin and R. J. Michl, *J. Org. Chem.*, **39**, 2261 (1974).

[716] A. H. Lewin, N. C. Peterson and R. J. Michl, *J. Org. Chem.*, **39**, 2747 (1974).
[717] Cf. *Org. Reaction Mech.*, **1973**, 141.
[718] M. G. Bartle, R. K. Mackie and J. M. Tedder, *Chem. Comm.*, **1974**, 271.
[719] H. J. Rosenkranz, B. Winkler-Lardelli, H.-J. Hansen and H. Schmid, *Helv. Chim. Acta*, **57**, 887 (1974).
[720] O. Hammerich and V. D. Parker, *Chem. Comm.*, **1974**, 245.
[721] P. Beresford, D. H. Iles, L. J. Kricka and A. Ledwith, *J.C.S. Perkin I*, **1974**, 276.
[722] M. Donbavand and H. Möckel, *Z. Naturforsch., B*, **29**, 403 (1974).
[723] G. De Luca, C. Panattoni, G. Renzi and L. Toniolo, *Tetrahedron Letters*, **1974**, 2463.
[724] G. A. Russell, *Rev. Reactive Species Chem. React.*, **1**, 65 (1973); *Chem. Abs.*, **79**, 114639 (1973).
[725] Cf. *Org. Reaction Mech.*, **1973**, 143.
[726] J. F. Bunnett and B. F. Gloor, *J. Org. Chem.*, **38**, 4156 (1973).
[727] J. F. Bunnett and B. F. Gloor, *J. Org. Chem.*, **39**, 382 (1974).
[728] J. F. Bunnett and X. Creary, *J. Org. Chem.*, **39**, 3173 (1974).
[729] R. A. Rossi and J. F. Bunnett, *J. Am. Chem. Soc.*, **96**, 112 (1974).
[730] E. Farina, L. Nucci, G. Biggi, F. Del Cima and F. Pietra, *Tetrahedron Letters*, **1974**, 3305.
[731] D. Y. Myers, G. G. Stroebel, B. R. Ortiz del Montellano and P. D. Gardner, *J. Am. Chem. Soc.*, **96**, 1981 (1974).
[732] N. Kornblum, S. D. Boyd and N. Ono, *J. Am. Chem. Soc.*, **96**, 2580 (1974).
[733] K. C. Chan, S. H. Golu, S. E. Teoh and W. H. Wong, *Austral. J. Chem.*, **27**, 421 (1974).
[734] M. Makosza, M. Jagusztyn-Grochowska, M. Ludwikow and M. Jawdosiuk, *Tetrahedron*, **30**, 3723 (1974).
[735] S. Bank and D. A. Noyd, *J. Am. Chem. Soc.*, **95**, 8203 (1973).
[735a] R. Z. Sagdeev, Yu. N. Molin, K. M. Salikhov, T. V. Leshina, M. A. Kamha and S. M. Shein, *Org. Mag. Resonance*, **5**, 599 (1973).
[735b] I. P. Beletskaya, S. V. Rykov and A. L. Buchachenko, *Org. Mag. Resonance*, **5**, 595 (1973).
[736] M. Ballester, J. Castañer and A. Ibáñez, *Tetrahedron Letters*, **1974**, 2147.
[737] E. S. Petrov, M. I. Terekhova and A. I. Shatenshtein, *Russ. Chem. Rev.*, **42**, 713 (1973).
[738] N. L. Holy, *Chem. Rev.*, **1974**, 243.
[739] C. G. Screttas, *J.C.S. Perkin II*, **1974**, 745.
[740] E. J. Panek, *J. Am. Chem. Soc.*, **95**, 8460 (1973).
[741] A. Rainis and M. Szwarc, *J. Am. Chem. Soc.*, **96**, 3008 (1974).
[742] B. Bockrath and L. M. Dorfman, *J. Phys. Chem.*, **27**, 2618 (1973).
[743] J. F. Garst and F. E. Barton, *J. Am. Chem. Soc.*. **96**, 523 (1974).
[744] J. F. Garst and J. T. Barbas, *J. Am. Chem. Soc.*, **96**, 3239 (1974).
[745] Y.-J. Lee and W. D. Closson, *Tetrahedron Letters*, **1974**, 1395.
[746] H. E. Zieger, I. Angres and L. Maresca, *J. Am. Chem. Soc.*, **95**, 8201 (1973).
[747] J. F. Garst and J. T. Barbas, *J. Am. Chem. Soc.*, **96**, 3247 (1974).
[748] G. D. Sargent, C. M. Tatum and R. P. Scott, *J. Am. Chem. Soc.*, **96**, 1602 (1974).
[749] I. Angres and H. E. Zieger, *J. Org. Chem.*, **39**, 1013 (1974).
[750] J. Casanova and L. Eberson, in *Chemistry of the Carbon Halogen Bond* (S. Patai, Ed.), Part II, Wiley, London, 1973, p. 979.
[751] J. Grimshaw and J. Trocha-Grimshaw, *Tetrahedron Letters*, **1974**, 993.
[752] J. Grimshaw and J. Trocha-Grimshaw, *J.C.S. Perkin I*, **1974**, 1383.
[753] R. N. Renaud, *Can. J. Chem.*, **52**, 376 (1974).
[754] J. Pinson and J.-M. Savéant, *Chem. Comm.*, **1974**, 833.
[755] K. B. Wiberg and G. A. Epling, *Tetrahedron Letters*, **1974**, 1119.
[756] W. J. Boyle and J. F. Bunnett, *J. Am. Chem. Soc.*, **96**, 1418 (1974).
[757] J. T. Groves and K. W. Ma, *J. Am. Chem. Soc.*, **96**, 6527 (1974).
[758] Y.-J. Lee and W. D. Closson, *Tetrahedron Letters*, **1974**, 381.
[759] A. N. Nesmeyanov, L. S. Isaeva, T. A. Peyanova and A. A. Slinkin, *Bull. Acad. Sci. SSSR*, **23**, 148 (1974).
[760] G. V. Putirskaya and L. Matus, *Magy. Kem. Folyoirat*, **80**, 241 (1974); *Chem. Abs.*, **81**, 119435 (1974).
[761] V. I. Vaganova, V. K. Turchaninov, B. V. Levinskii and T. I. Vaknl'skaya, *Bull. Acad. Sci. USSR*, **23**, 707 (1974).
[762] I. R. Bellobono, P. Govoni and F. Zavattarelli, *J.C.S. Perkin II*, **1974**, 981
[763] F. Zavattarelli, I. R. Bellobono and P. L. Beltrama, *J.C.S. Perkin II*, **1974**, 983.
[764] L. A. Polyakova, K. A. Bilevich, N. N. Bubnov, G. N. Dorofeenko and O. Yu. Okhlobystin, *Dokl. Akad. Nauk SSSR*, **212**, 370 (1973); *Chem. Abs.*, **79**, 145660 (1973).

[765] T. H. Chan, J. F. Harrod and P. van Gheluwe, *Tetrahedron Letters*, **1974**, 4409.
[766] Z. V. Todres and S. P. Avagyan, *Zh. Vses. Khim. Obshchest.*, **18**, 478 (1973); *Chem. Abs.*, **79**, 145479 (1973).
[767] T. D. Tuong and M. Hida, *J.C.S. Perkin II*, **1974**, 676.
[768] W. N. Speckamp and A. W. J. D. Junkins, *Tetrahedron Letters*, **1974**, 1857.
[769] R. D. Chambers, D. T. Clark, T. F. Holmes, W. K. R. Musgrave and I. Ritchie, *J.C.S. Perkin I*, **1974**, 114.
[770] L. J. Kricka and A. Ledwith, *Synthesis*, **1974**, 539.
[771] M. Asai, H. Matsui and S. Tazuka, *Bull. Chem. Soc. Japan*, **47**, 864 (1974).
[772] F. A. Bell, P. Beresford, L. J. Kricka and A. Ledwith, *J.C.S. Perkin I*, **1974**, 1788.
[773] M. M. Litt and J. Radovic, *J. Phys. Chem.*, **78**, 1750 (1974).
[774] G. Boekestein, W. G. Voncken, E. H. J. M. Jansen and H. M. Buck, *Rec. Trav. chim.*, **93**, 69 (1974).
[775] Y. Ito, T. Konoike and T. Saegusa, *Tetrahedron Letters*, **1974**, 1287.
[776] P. G. Farrell and P. N. Ngô, *J.C.S. Perkin II*, **1974**, 552.
[777] A. G. Evans, J. C. Evans and P. J. Pomery, *J.C.S. Perkin II*, **1974**, 1835.
[778] A. G. Evans, J. C. Evans, P. J. Emes and S. I. Haider, *J.C.S. Perkin II*, **1974**, 1121.
[779] S. G. Cohen, G. C. Ramsay, N. M. Stein and S. Y. Weinstein, *J. Am. Chem. Soc.*, **96**, 5124 (1974).
[780] U. Krüger and R. Memming, *Ber. Bunsenges. Phys. Chem.*, **78**, 670, 679, 685 (1974).
[781] B. S. Jensen and V. D. Parker, *Chem. Comm.*, **1974**, 367.
[782] R. D. Guthrie, G. R. Weisman and L. G. Burdon, *J. Am. Chem. Soc.*, **96**, 6955 (1974).
[783] R. D. Guthrie and G. R. Weisman, *J. Am. Chem. Soc.*, **96**, 6962 (1974).
[784] D. L. Akins and R. L. Birke, *Chem. Phys. Letters*, **29**, 428 (1974).
[785] P. J. Derrick and A. L. Burlingame. *Accounts Chem. Res.*, **7**, 328 (1974).
[786] D. Dolphin and R. H. Felton, *Accounts Chem. Res.*, **7**, 26 (1974).
[787] U. Svanholm, B. S. Jensen and V. D. Parker, *J.C.S. Perkin II*, **1974**, 907.
[788] U. Svanholm, A. Ronlán and V. D. Parker, *J. Am. Chem. Soc.*, **96**, 5108 (1974).
[789] K. Kim and H. J. Shine, *Tetrahedron Letters*, **1974**, 4413.
[790] H. J. Shine and K. Kim, *Tetrahedron Letters*, **1974**, 99.
[791] K. Kim, V. J. Hull and H. J. Shine, *J. Org. Chem.*, **39**, 2535 (1974).
[792] K. Kim and H. J. Shine, *J. Org. Chem.*, **39**, 2537 (1974).
[793] H. J. Shine and L. R. Shade, *J. Heterocyclic Chem.*, **11**, 139 (1974).
[794] B. L. Funt and V. Verigin, *Can. J. Chem.*, **52**, 1643 (1974).
[795] C. F. Bernasconi, R. G. Bergstrom and W. J. Boyle, *J. Am. Chem. Soc.*, **96**, 4643 (1974).
[796] H. D. Roth and A. A. Lamola, *J. Am. Chem. Soc.*, **96**, 6270 (1974).
[797] M. Grodowski and T. Latowski, *Tetrahedron*, **30**, 767 (1974).
[798] R. S. Davidson and S. P. Orton, *Chem. Comm.*, **1974**, 209.
[799] K. Yamasaki, T. Matsuura and I. Saito, *Chem. Comm.*, **1974**, 561.
[800] C. Pac, K. Mizuno, T. Tosa and H. Sakurai, *J.C.S. Perkin I*, **1974**, 561.
[801] D. Dopp, D. Müller and K.-H. Sailer, *Tetrahedron Letters*, **1974**, 2137.
[802] A. Cu and A. C. Testa, *J. Am. Chem. Soc.*, **96**, 1963 (1974).
[803] G. G. Wubbels and R. L. Letsinger, *J. Am. Chem. Soc.*, **96**, 6698 (1974).
[804] C. J. Biaselle and J. G. Miller, *J. Am. Chem. Soc.*, **96**, 3813 (1974).
[805] K. G. Hancock and D. A. Dickinson, *J. Org. Chem.*, **39**, 331 (1974).
[806] S. Wake, Y. Otsuji and E. Imoto, *Bull. Chem. Soc. Japan*, **47**, 1251 (1974).
[807] S. Wake, Y. Takayama, Y. Otsuji and E. Imoto, *Bull. Chem. Soc. Japan*, **47**, 1257 (1974).
[808] K. Kano and T. Matsuo, *Tetrahedron Letters*, **1974**, 4323.
[809] H. D. Burrows and E. M. Kosower, *J. Phys. Chem.*, **78**, 112 (1974).
[810] F. Bickelhaupt, *Angew. Chem. Internat. Ed.*, **13**, 419 (1974).
[811] C. Blomberg, *J. Organometal. Chem.*, **68**, 69 (1974).
[812] E. C. Ashby, J. Laemmle and H. M. Neumann, *Accounts Chem. Res.*, **7**, 272 (1974).
[813] R. J. Rogers, H. L. Mitchell, Y. Fujiwara and G. M. Whitesides, *J. Org. Chem.*, **39**, 857 (1974).
[814] J.-L. Derocque and F.-B. Sundermann, *J. Org. Chem.*, **39**, 1411 (1974).
[815] E. C. Ashby and T. L. Wiesemann, *J. Am. Chem. Soc.*, **96**, 7117 (1974).
[816] M. Daggoneau and J. Vialle, *Tetrahedron*, **30**, 3119 (1974).
[817] M. Daggoneau, *J. Organometal. Chem.*, **80**, 1 (1974).
[818] J.-E. Dubois and M. Boussu, *Tetrahedron*, **29**, 3943 (1973).
[819] W. A. Nugent, F. Bertini and J. K. Kochi, *J. Am. Chem. Soc.*, **96**, 4945 (1974).
[820] H. C. Gardner and J. K. Kochi, *J. Am. Chem. Soc.*, **96**, 1982 (1974).
[821] C. L. Hill and G. M. Whitesides, *J. Am. Chem. Soc.*, **96**, 870 (1974).
[822] G. M. Whitesides, J. S. Sadowski and J. Lilburn, *J. Am. Chem. Soc.*, **96**, 2829 (1974).

[823] G. M. Whitesides, D. E. Bergbreiter and P. E. Kendall, *J. Am. Chem. Soc.*, **96**, 2806 (1974).
[824] B. Blank, A. Henne and H. Fischer, *Helv. Chim. Acta*, **57**, 920 (1974).
[825] H. E. Chen, S. P. Vaish and M. Cocivera, *J. Am. Chem. Soc.*, **95**, 7586 (1973).
[826] S. P. Vaish, R. D. McAlpine and M. Cocivera, *J. Am. Chem. Soc.*, **96**, 1683 (1974).
[827] R. Kaptein, R. Freeman and H. D. W. Hill, *Chem. Phys. Letters*, **26**, 104 (1974).
[828] E. A. Lissi, E. Abuin and M. V. Encina, *J. Photochem.*, **2**, 377 (1974).
[829] M. L. Kaplan, M. L. Manion and H. D. Roth, *J. Phys. Chem.*, **78**, 1837 (1974).
[830] P. B. Ayscough and R. C. Sealy, *J.C.S. Perkin II*, **1974**, 1402.
[831] K.-G. Seifert, *Chem. Ber.*, **107**, 2412 (1974).
[832] H.-G. Heine, W. Hartmann, D. R. Kory, J. G. Magyar, C. E. Hoyle, J. K. McVey and F. D. Lewis, *J. Org. Chem.*, **39**, 691 (1974).
[833] M. R. Sandner and C. L. Osborn, *Tetrahedron Letters*, **1974**, 415.
[834] H. Dürr, P. Herbst, P. Heitkämper and H. Leismann, *Chem. Ber.*, **107**, 1835 (1974).
[835] N. A. Porter and P. M. Iloff, *J. Am. Chem. Soc.*, **96**, 6201 (1974).
[836] Y. Otsugi, S. Wake, E. Maeda and E. Imoto, *Bull. Chem. Soc. Japan*, **47**, 189 (1974).
[837] K. Maeda, I. Moritami and A. Sonoda, *Bull. Chem. Soc. Japan*, **47**, 1018 (1974).
[838] S. Tsuruya and T. Yonezawa, *J. Org. Chem.*, **39**, 2438 (1974).
[839] H. E. O'Neal, R. G. Miller and E. Gunderson, *J. Am. Chem. Soc.*, **96**, 3351 (1974).
[840] E. A. Abuin, M. V. Encina and E. A. Lissi, *J. Photochem.*, **3**, 143 (1974).
[841] E. A. Lissi and M. V. Encina, *J. Photochem.*, **3**, 237 (1974).
[842] R. A. Cormier and W. C. Agosta, *J. Am. Chem. Soc.*, **96**, 618 (1974).
[843] *Org. Reaction Mech.*, **1973**, 94.
[844] R. A. Cormier and W. C. Agosta, *J. Am. Chem. Soc.*, **96**, 1867 (1974).
[845] W. H. Richardson, G. Ranney and F. C. Montgomery, *J. Am. Chem. Soc.*, **96**, 4688 (1974).
[846] P. J. Wagner and K.-C. Liu, *J. Am. Chem. Soc.*, **96**, 5952 (1974).
[847] T. R. Darling, N. J. Turro, R. H. Hirsch and F. D. Lewis, *J. Am. Chem. Soc.*, **96**, 434 (1974).
[848] S. C. Gupta and S. K. Mukerjee, *Tetrahedron Letters*, **1973**, 5073.
[849] Y. Ogata and K. Takagi, *Bull. Chem. Soc. Japan*, **47**, 2255 (1974).
[850] Y. Ogata and K. Takagi, *J. Org. Chem.*, **39**, 1385 (1974).
[851] D. I. Schuster and K. V. Prabhu, *J. Am. Chem. Soc.*, **96**, 3511 (1974).
[852] H. A. J. Carless, *J.C.S. Perkin II*, **1974**, 834.
[853] K. Shima, T. Kawamura and K. Tanabe, *Bull. Chem. Soc. Japan*, **47**, 2347 (1974).
[854] N. Shimizu, M. Ishikawa, K. Ishikura and S. Nishida, *J. Am. Chem. Soc.*, **96**, 6456 (1974).
[855] J. H. Marshall, *J. Phys. Chem.*, **78**, 2225 (1974).
[856] J. D. Coyle, *Chem. Soc. Rev.*, **3**, 329 (1974).
[857] H. Hana and I. Tanaka, *Bull. Chem. Soc. Japan*, **47**, 1543 (1974).
[858] J. M. Hornback, *J. Am. Chem. Soc.*, **96**, 6772 (1974).
[859] C. A. F. Johnson and W. M. C. Lawson, *J.C.S. Perkin II*, **1974**, 353.
[860] C. Baker and W. M. Horspool, *Tetrahedron Letters*, **1974**, 3533.
[861] G. S. Poindexter and P. J. Kropp, *J. Am. Chem. Soc.*, **96**, 7142 (1974).
[862] B. G. Gowenlock, J. Pfab and G. Kresze, *J.C.S. Perkin II*, **1974**, 511.
[863] B. Matuszewski and J. Wojtczak, *Rocz. Chim.*, **48**, 821 (1974).
[864] L. O. Ruzo, M. J. Zabik and R. D. Scheutz, *J. Am. Chem. Soc.*, **96**, 3809 (1974).
[864a] A. K. Basak, *J. Indian Chem. Soc.*, **1973**, 767.
[865] A. Harriman, B. W. Rockett and W. R. Poyner, *J.C.S. Perkin II*, **1974**, 485.
[866] A. Nikiforov and U. Schmidt, *Monatsh. Chem.*, **105**, 1044 (1974).
[867] Y. Okuno and O. Yonemitsu, *Tetrahedron Letters*, **1974**, 1169.
[868] R. Breslow, R. J. Corcoran and B. B. Snider, *J. Am. Chem. Soc.*, **96**, 6791 (1974).
[869] P. de Mayo and R. Suau, *J. Am. Chem. Soc.*, **96**, 6807 (1974).
[870] R. S. Givens, B. Matuszewski and C. V. Neywick, *J. Am. Chem. Soc.*, **96**, 5547 (1974).
[871] J. Martens and K. Praefke, *Tetrahedron*, **30**, 2565 (1974).
[872] J. Martens and K. Praefke, *Chem. Ber.*, **107**, 2319 (1974).
[873] O. P. Studzinskii, A. V. El'tsov, N. I. Rtishchev and G. V. Fomin, *Russian Chem. Rev.*, **1974**, 155.
[874] S. T. Reid, J. N. Tucker and E. J. Wilcox, *J.C.S. Perkin I*, **1974**, 1359.
[875] M. Takami, T. Matsuura and I. Saito, *Tetrahedron Letters*, **1974**, 661.
[876] W. G. Filby and K. Günther, *Z. Naturforsch.*, *B*, **29**, 810 (1974).
[877] Y. Kobayashi, *Bull. Chem. Soc. Japan*, **46**, 3467 (1973).
[878] A. Lablanche-Combier, J. P. Quagebeur and H. Ofenberg, *Chem. Comm.*, **1974**, 579.
[879] P. A. Lehman and R. S. Berry, *J. Am. Chem. Soc.*, **95**, 8614 (1973).
[880] M. Schlosser, *Tetrahedron Letters*, **1974**, 3045.

[881] M. U. Amin, B. W. Evans and G. Scott, *Chem. Ind.* (*London*), **1974**, 206.
[882] P. M. Collins and B. R. Whitton, *J.C.S. Perkin I*, **1974**, 1069.
[883] G. Vermeesch, N. Febvay-Garot, S. Caplain and A. Lablanche-Combier, *Tetrahedron Letters*, **1974**, 3127.
[884] N. Hata and T. Saito, *Bull. Chem. Soc. Japan*, **47**, 942 (1974).
[885] D. L. Elliot and T. C. Bruice, *J. Am. Chem. Soc.*, **95**, 7901 (1973).
[886] M. A. Winnick, R. E. Trueman, G. Jackowski, D. S. Saunders and S. G. Whittington, *J. Am. Chem. Soc.*, **96**, 4843 (1974).
[887] D. M. Golden, G. N. Spokes and S. W. Benson, *Angew. Chem. Internat. Ed.*, **13**, 534 (1974).
[888] Yu. P. Yampol'skii, V. M. Rybin and E. J. Molodykh, *Dokl. Akad. Nauk SSSR*, **213**, 394 (1973); *Chem. Abs.*, **80**, 47236 (1974).
[889] D. R. Powers and W. H. Corcoran, *Ind. Eng. Chem. Fundam.*, **13**, 351 (1974).
[890] J. N. Bradley, *Proc. Roy. Soc.*, *A*, **337**, 199 (1974).
[891] K. Yamaguchi and T. Fueno, *Chem. Phys. Letters*, **22**, 471 (1973).
[892] D. M. Kulich, J. E. Taylor and D. A. Hutchings, *Can. J. Chem.*, **52**, 216 (1974).
[893] R. Z. Magaril, N. V. Korzun and N. Jonidis, *Khim. Khim. Tekhnol., Tr. Tyumen. Ind. Inst.*, **1972**, 162; *Chem. Abs.*, **80**, 81784 (1974).
[894] R. Z. Magaril and N. V. Korzun, *Khim. Khim. Tekhnol., Tr. Tyumen. Ind. Inst.*, **1972**, 128; *Chem. Abs.*, **80**, 81783 (1974).
[895] L. Eisenhuth and H. Hopf, *J. Am. Chem. Soc.*, **96**, 5667 (1974).
[896] K. A. Holbrook and R. A. Scott, *J.C.S. Faraday I*, **70**, 43 (1974).
[897] H. A. J. Carless, *Tetrahedron Letters*, **1974**, 3425.
[898] N. Shimizu and S. Nishida, *Chem. Comm.*, **1974**, 734.
[899] D. L. Rakhmankulov, V. S. Martem'yanov, S. S. Zlotskii and V. I. Isagulyants, *Neftekimiya*, **14**, 275 (1974); *Chem. Abs.*, **81**, 24696 (1974).
[900] P. Vogel and M. Hardy, *Helv. Chim. Acta*, **57**, 196 (1974).
[901] J. S. Lomas and J.-E. Dubois, *J. Org. Chem.*, **39**, 1776 (1974).
[902] W. D. Johnson, *Austral. J. Chem.*, **27**, 1047 (1974).
[903] A. C. Barefoot and F. A. Carroll, *Chem. Comm.*, **1974**, 357.
[904] W. Vetler, G. Schill and C. Zürcher, *Chem. Ber.*, **107**, 424 (1974).
[905] V. M. Salakhiev and G. P. Sharnin, *Tr. Kazan. Khim.-Tekhnol. Inst.*, **1973**, 73; *Chem. Abs.*, **80**, 107721 (1974).
[906] M. J. Krech, S. J. W. Price and W. F. Yared, *Internat. J. Chem. Kinetics*, **6**, 257 (1974).
[907] T. Sato, K. Shimizu and H. Moriya, *J.C.S. Perkin I*, **1974**, 1537.
[908] A. B. Gagarina, L. M. Pisarenko, L. I. Murza and N. M. Emanuel, *Dokl. Akad. Nauk SSSR*, **215**, 894 (1974); *Chem. Abs.*, **81**, 3087 (1974).
[909] K. R. Kopecky and J. Soler, *Can. J. Chem.*, **52**, 2111 (1974).
[910] D. C. K. Lin and D. C. DeJongh, *J. Org. Chem.*, **39**, 1780 (1974).
[911] D. C. K. Lin, M. L. Thomson and D. C. DeJongh, *Can. J. Chem.*, **52**, 2359 (1974).
[912] J. Ngitrai and K. Lempert, *Chem. Ber.*, **107**, 1637 (1974).
[913] D. P. Paquin, R. J. O'Connor and M. A. Ring, *J. Organometal. Chem.*, **80**, 341 (1974).
[914] I. M. T. Davidson, C. Eaborn and J. M. Simmie, *J.C.S. Faraday I*, **70**, 249 (1974).
[915] G. M. Whitesides, D. E. Bergbreiter and P. E. Kendall, *J. Am. Chem. Soc.*, **96**, 2806 (1974).
[916] P. Sartori and H. J. Frohn, *Chem. Ber.*, **107**, 1195 (1974).
[917] A. H. Yousufzai and M. M. Zafar, *Proc. Pak. Acad. Sci.*, **10**, 1 (1973); *Chem. Abs.*, **81**, 62834 (1974).
[918] J. Gawlowski and J. A. Herman, *Can. J. Chem.*, **52**, 3631 (1974).
[919] M. Nishikawa, Y.-I. Yamaguchi, K. Fujita, K. Kon and T. Okamoto, *Can. J. Chem.*, **51**, 3966 (1973).
[920] D. D. Mainwaring and J. E. Willard, *J. Phys. Chem.*, **77**, 2864 (1973).
[921] J. A. Stone and J. Esser, *Can. J. Chem.*, **52**, 1258 (1974).
[922] G. A. Kennedy and R. J. Hanrahan, *J. Phys. Chem.*, **78**, 360, 366 (1974).
[923] G. J. Hyfantis and A. C. Ling, *Chem. Phys. Letters*, **24**, 335 (1974).
[924] G. J. Hyfantis and A. C. Ling, *Can. J. Chem.*, **52**, 1206 (1974).
[925] F. P. Sargent, E. M. Gardy and H. R. Falle, *Chem. Phys. Letters*, **24**, 120 (1974).
[926] F. P. Sargent and E. M. Gardy, *Can. J. Chem.*, **52**, 3645 (1974).
[927] F. P. Sargent and E. M. Gardy, *J. Phys. Chem.*, **78**, 1977 (1974).
[928] S. W. Mao and L. Kevan, *Chem. Phys. Letters*, **24**, 505 (1974).
[929] N. S. Kalyazina, K. S. Kalugin and Yu. I. Moskalev, *Khim. Vys. Energ.*, **7**, 471 (1973); *Chem. Abs.*, **79**, 136172 (1973).
[930] A. S. Klimentov, *Sovrem. Prabl. Khim.*, **1973**, 59; *Chem. Abs.*, **81**, 119381 (1974).

931 J. Rabani and D. Klug-Roth, *J. Phys. Chem.*, **78**, 2089 (1974).
932 H. Taniguchi and H. Hatano, *Chem. Letters (Tokyo)*, **1974**, 531.
933 H. Schultze and D. Schulte-Frohlinde, *Z. Naturforsch., B*, **29**, 91 (1974).
934 P. B. Ayscough and K. Mach, *J.C.S. Faraday I*, **70**, 118 (1974).
935 S. Kubota, M. Iwaizumi and T. Isobe, *Chem. Phys. Letters*, **25**, 247 (1974).
936 M. C. R. Symons, *J. Chem. Phys.*, **59**, 5747 (1973); *Chem. Abs.*, **80**, 70034 (1974).
937 I. Draganić, Z. Draganić, Lj. Petković and A. Nikolic, *J. Am. Chem. Soc.*, **95**, 7193 (1973).
938 P. Gulierrez and B. Benson, *J. Chem. Phys.*, **60**, 640 (1974).
939 G. A. Infante, P. Jirathana, E. J. Fendler and J. H. Fendler, *J.C.S. Faraday I*, **70**, 1162 (1974).
940 J. H. Fendler, G. A. Infante, P. Jirathana and E. J. Fendler, *J.C.S. Faraday I*, **70**, 1171 (1974).
941 J. Lilie, E. Heckel and R. C. Lamb, *J. Am. Chem. Soc.*, **96**, 5543 (1974).
942 W. H. Klaus, *Z. Naturforsch., B*, **29**, 286 (1974).
943 R. M. Thibault, D. R. Hepburn and T. J. Klingen, *J. Phys. Chem.*, **78**, 788 (1974).
944 W. A. Bernhard and F. S. Ezra, *J. Phys. Chem.*, **78**, 958 (1974).
945 G. Draeger, *Pflanzenschutz-Nachr. (Amer. Ed.)*, **25**, 32 (1972); *Chem. Abs.*, **80**, 59633 (1974).
946 A. Hudson, M. F. Lappert, P. W. Lednor and B. K. Nicholson, *Chem. Comm.*, **1974**, 966.
947 M. Chastrette and R. Gauthier, *J. Organometal. Chem.*, **66**, 219 (1974).
948 H. Iwamura, M. Iwamura, M. Imanari and M. Takeuchi, *Bull. Chem. Soc. Japan*, **46**, 3486 (1973).
949 M. Namiki, M. Yano and T. Hayashi, *Chem. Letters (Tokyo)*, **1974**, 125.
950 M. Yano, T. Hayashi and M. Namiki, *Chem. Letters (Tokyo)*, **1974**, 1193.
951 Y. Kirino, P. L. Southwick and R. H. Schuler, *J. Am. Chem. Soc.*, **96**, 673 (1974).
952 Y. Kirino, *Chem. Letters (Tokyo)*, **1974**, 153.
953 H. Zemel and P. Neta, *Radiat. Res.*, **55**, 393 (1973); *Chem. Abs.*, **79**, 125530 (1973).
954 E. Gilbert, *Z. Naturforsch., B*, **28**, 805 (1973).
955 J. C. Kertesz, W. Wolf, M. B. Wolf and B. J. Sullivan, *J. Mag. Res.*, **15**, 427 (1974).
956 J. C. Kertesz, M. B. Wolf, W. Wolf and L.-Y. A. Chen, *J. Pharm. Sci.*, **63**, 880 (1974).
957 J. E. Valinsky, R. H. Abeles and A. S. Mildran, *J. Biol. Chem.*, **249**, 2751 (1974).
958 E. Westhof, W. Flossmann, H.-D. Lüdemann and A. Müller, *J. Chem. Phys.*, **61**, 3376 (1974).
959 E. Westhof, W. Flossmann and A. Müller, *Mol. Phys.*, **28**, 151 (1974).
960 D. L. Allison and C. Alexander, *J. Mag. Res.*, **14**, 366 (1974).
961 H. C. Box, E. E. Budzinski and W. R. Potter, *J. Chem. Phys.*, **61**, 1136 (1974).
962 C. Nicolau, *Jerusalem Symp. Quantum Chem. Biochem.*, **4**, 519 (1972); *Chem. Abs.*, **81**, 12695 (1974).
963 J. Bardsley, P. J. Baugh, J. I. Goodall and G. O. Phillips, *Chem. Comm.*, **1974**, 890.
964 K. S. Chen and N. Hirota in *Investigation of Rates and Mechanisms of Reactions, Part II*, 3rd edn. (G. G. Hammes, Ed.), Interscience, New York, N.Y., 1973, p. 595.
965 M. Teodorescu, *Stnd. Cercet. Fiz.*, **26**, 229 (1974).
966 A. T. Balaban and C. Pomponiu, *Rev. Roum. Chim.*, **19**, 581 (1974).
967 V. N. Parmon and G. M. Zhidomirov, *Mol. Phys.*, **27**, 367 (1974).
968 G. V. Bruno and J. H. Freed, *Chem. Phys. Letters*, **25**, 328 (1974).
969 E. Schifflers and R. Debuyst, *J. Mag. Res.*, **14**, 85 (1974).
970 D. W. Posener, *J. Mag. Res.*, **14**, 129 (1974).
971 J. Tiňo and V. Klimo, *Chem. Phys. Letters*, **25**, 427 (1974).
972 A. T. Bullock, *J. Mag. Res.*, **15**, 580 (1973).
973 I. Baxter and J. K. M. Saunders, *Chem. Comm.*, **1974**, 255.
974 N. S. Dalal, D. E. Kennedy and C. A. McDowell, *J. Chem. Phys.*, **61**, 1689 (1974).
975 A. Yalçiner, *Z. Naturforsch., A*, **29**, 1071 (1974).
976 K. A. Christensen, D. M. Grant, E. M. Schulman and C. Walling, *J. Phys. Chem.*, **78**, 1971 (1974).
977 C. Brown, R. F. Hudson and A. J. Lawson, *Chem. Comm.*, **1974**, 831.
978 T. Tsuji and S. Nishida, *J. Am. Chem. Soc.*, **96**, 3649 (1974).
979 Sh. A. Markaryan and A. L. Buchachenko, *Kinet. Katal*, **15**, 497 (1974); *Chem. Abs.*, **81**, 24697 (1974).
980 S. P. Vaish, R. D. McAlpine and M. Cocivera, *J. Am. Chem. Soc.*, **96**, 1683 (1974).
981 H. F. Chen, S. P. Vaish and M. Cocivera, *J. Am. Chem. Soc.*, **95**, 7586 (1973).
982 S. P. Vaish, R. D. McAlpine and M. Cocivera, *Can. J. Chem.*, **52**, 2978 (1974).
983 B. Blank, A. Henne and H. Fischer, *Helv. Chim. Acta*, **57**, 920 (1974).
984 R. Kaptein, R. Freeman and H. D. W. Hill, *Chem. Phys. Letters*, **26**, 104 (1974).
985 J. Bargon and K.-G. Seifert, *J. Phys. Chem.*, **77**, 2877 (1973).
986 M. L. Kaplan, M. L. Manion and H. D. Roth, *J. Phys. Chem.*, **78**, 1837 (1973).
987 N. A. Porter and P. M. Iloff, *J. Am. Chem. Soc.*, **96**, 6200 (1974).

988 M. Lehnig, *Tetrahedron Letters*, **1974**, 3323.
989 J. Brokken-Zijp and H. van de Bogaert, *Tetrahedron Letters*, **1974**, 249.
990 R. Kaptein, R. Freeman and H. D. W. Hill, *Chem. Comm.*, **1973**, 953.
991 H. Iwamura, Y. Imahashi, M. Oki, K. Kushida and S. Satoh, *Chem. Letters* (*Tokyo*), **1974**, 259.
992 H. Iwamura, Y. Imahashi and K. Kushida, *J. Am. Chem. Soc.*, **96**, 921 (1974).
993 L. F. Kasukhin, M. P. Ponomarchuk and A. L. Buchachenko, *Chem. Phys.*, **3**, 136 (1974); *Chem. Abs.*, **80**, 145084 (1974).
994 J. K. S. Wan, S.-K. Wong and D. A. Hutchinson, *Accounts Chem. Res.*, **7**, 58 (1974).
995 A. D. Trifunac and E. C. Avery, *Chem. Phys. Letters*, **27**, 141 (1974).
996 A. D. Trifunac and E. C. Avery, *Chem. Phys. Letters*, **28**, 294 (1974).
997 S.-K. Wong, D. A. Hutchinson and J. K. S. Wan, *J. Chem. Phys.*, **60**, 2987 (1974).
998 P. W. Atkins, J. K. Duggan, K. A. McLauchlan and P. W. Percival, *Chem. Phys. Letters*, **24**, 565 (1974).
999 P. W. Atkins and G. T. Evans, *Chem. Phys. Letters*, **25**, 108 (1974).
1000 P. W. Atkins, A. J. Dobbs, G. T. Evans, K. A. McLauchlan and P. W. Percival, *Mol. Phys.*, **27**, 769 (1974).

CHAPTER 4

Oxidation and Reduction

T. W. Bentley

Department of Chemistry, University College of Swansea

The organization of this Chapter is similar to that of previous years with a short additional section on oxidation and reduction in biological chemistry. One-electron oxidations and reductions are also discussed in Chapter 3. Reactions induced electrochemically or photochemically are not reviewed here as they are included in the Chemical Society's Specialist Periodical Reports.

Oxidation by Metallic Ions

Chromium

During Cr(VI) oxidations of organic substrates direct formation of Cr(III) should be thermodynamically favourable, but evidence for it has only recently been presented. Since oxidation of an alcohol to a ketone requires only two electrons [Cr(VI) $\rightarrow$ Cr(IV)], the three-electron process [Cr(VI) → Cr(III)] involves an additional and simultaneous one-electron oxidation: e.g. oxidation of glycollic acid at high concentrations (6 M), via a 2:1 glycollic acid:Cr(VI) complex [equations (1) and (2)].[1] As might be expected, the normal two-electron oxidation via a 1:1 complex is favoured at low concentrations of

$$2HOCH_2CO_2H + Cr(VI) \rightleftharpoons \text{Complex} \qquad (1)$$

$$\text{Complex} \xrightarrow[\text{limiting}]{\text{Rate-}} OCHCOOH + HO\dot{C}HCOOH + Cr(III) \qquad (2)$$

$$\text{[cyclic Cr complex of two glycollate ligands, HOOC–CH}_2\text{–O–CrO}_2\text{–O–CH}_2\text{–COO}^-\text{]} \longrightarrow HO\dot{C}HCOOH + OCHCOO^- + Cr(III) \qquad (3)$$

glycollic acid (0.1 M), consistently with the observed primary deuterium isotope effect of 6.15. The corresponding isotope effect at high glycollic acid concentrations was 36.5, showing convincingly that the two bond-breaking processes occur simultaneously, possibly as shown in equation (3). Similar results were observed for the co-oxidation of propan-2-ol and glycollic acid.[1]

In contrast to earlier interpretations, there is now good evidence that Cr(IV) and not Cr(V) is responsible for the carbon—carbon bond cleavage occasionally observed during chromic acid oxidations of certain alcohols, e.g. cyclobutanol.[2,3] Several studies indicate that Cr(V) behaves similarly to Cr(VI). Both oxidize alcohols[2,4] and aldehydes[5] in aqueous acetic acid by a two-electron process, whereas in acetic anhydride one-electron oxidation of aldehydes (and possibly alcohols) is observed.[6] The principal difference between aqueous acetic acid and acetic anhydride is that ester formation with the hydrate of the aldehyde [equation (4)] is not possible in acetic anhydride. This led to the proposal that oxidations of aldehydes by both Cr(VI) and Cr(V) may involve initially a one-electron transfer [equation (5)].[5] When an ester is the intermediate this may be followed by intramolecular transfer of a second electron, leading to an apparent two-electron oxidation [equation (6)]. The overall process [equations (4–6)] may present a more convenient way in which Cr(VI) can bring two electrons into two different orbitals in going to Cr(IV); the direct two-electron process would require spin inversion of one of the electrons to give the normal Cr(IV) species.

$$R\text{—}\overset{O}{\overset{\|}{C}}H + AcOCrO_2H \rightleftharpoons R\text{—}\underset{H}{\overset{OH}{C}}\text{—}OCrO_2OAc \qquad (4)$$

$$R\text{—}\underset{H}{\overset{OH}{C}}\text{—}O\text{—}Cr(=O)_2(OAc) \xrightarrow[\text{limiting}]{\text{Rate-}} R\text{—}\overset{OH}{\dot{C}}\text{—}O\text{—}Cr(=O)(OH)(OAc) \qquad (5)$$

$$R\text{—}\overset{OH}{\dot{C}}\text{—}O\text{—}Cr(=O)(HO)(OAc) \longrightarrow R\text{—}\overset{OH}{C}{=}O + Cr(IV) \qquad (6)$$

Further evidence that Cr(IV), produced from an initial rate-limiting two-electron oxidation by chromic acid, causes the carbon—carbon bond cleavage is the observation that Ce(III) or Ce(IV) reduces the rate of chromic acid oxidation of alcohols *and* suppresses the cleavage of phenyl *tert*-butyl carbinol [equations (7) and (8)]. Even at concentrations of 10^{-7} M it appears that cerium catalyses the disproportionation of Cr(IV) to Cr(III) and Cr(VI), and so reduces the cleavage reaction.[7] Details of the study of the relative roles

$$(CH_3)_3CCH(OH)C_6H_5 \xrightarrow[-H\cdot]{Cr(IV)} (CH_3)_3C\cdot + C_6H_5CHO \quad (7)$$

$$(CH_3)_3C\cdot \xrightarrow{Oxidn.} (CH_3)_3COH \quad (8)$$

of Cr(VI), Cr(V) and Cr(IV) show the importance of Cr(IV) in the oxidation of propan-2-ol.[4]

Reactions of Cr(IV) can also be studied by generating it *in situ* from Cr(VI) and V(V). Both aliphatic[8] and aromatic aldehydes[9] appear to react by rate-limiting decomposition of a Cr(IV) complex of the aldehyde hydrate, with hydrogen transfer from the aldehydic carbon atom to the oxidant ($k_H/k_D \sim 2.0$).[8] In chromic acid oxidation of oxalic acid at concentrations where the reaction is first-order in substrate, oxalic acid appears to be very reactive towards Cr(IV) and rather unreactive towards Cr(V) (see Scheme 1).[10]

$$(COOH)_2 + HCrO_2^- + H^+ \rightleftharpoons \text{[cyclic } (O{=}C{-}O)_2CrO_2 \text{ oxalate complex]} \quad (9)$$

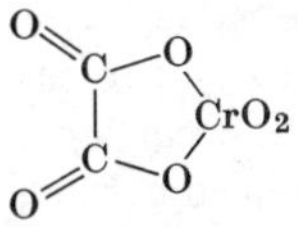

$$\text{[cyclic } (O{=}C{-}O)_2CrO_2 \text{ oxalate complex]} \xrightarrow{\text{Rate-limiting}} 2CO_2 + Cr(IV) \quad (10)$$

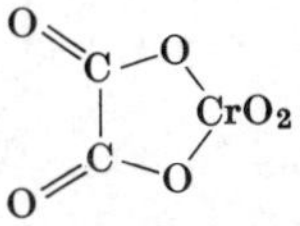

$$(COOH)_2 + Cr(IV) \longrightarrow CO_2 + \cdot CO_2H + Cr(III) \quad (11)$$

$$\cdot COOH + Cr(VI) \longrightarrow CO_2 + Cr(V) \quad (12)$$

$$(COOH)_2 + Cr(V) \longrightarrow 2CO_2 + Cr(III) \quad (13)$$

SCHEME 1. Cr(VI) oxidation of oxalic acid at low concentrations.

$$2(COOH)_2 + Cr(VI) \longrightarrow 3CO_2 + \cdot COOH + Cr(III) \quad (14)$$

$$\cdot COOH + Cr(VI) \longrightarrow CO_2 + Cr(V) \quad (15)$$

$$(COOH)_2 + Cr(V) \longrightarrow 2CO_2 + Cr(III) \quad (16)$$

$$2\cdot COOH \longrightarrow (COOH)_2 \quad (17)$$

SCHEME 2. Cr(VI) oxidation of oxalic acid at high concentrations.

At higher concentrations the three-electron oxidation mechanism [equation (14)] is favoured.[11] The formation and disappearance of Cr(V) (Schemes 1 and 2) has been observed by both ESR and spectrophotometric studies.[12]

Conformational effects in the Cr(VI) oxidation of cyclohex-2-en-1-ols[13] and secondary cycloalkanols[14] have been investigated. Addition of Mn(II) in low concentrations retards the rate of Cr(VI) oxidation of benzoin and acetoin in aqueous acetic acid, whilst at higher Mn(II) concentrations rate enhancement is observed; a mechanism involving reaction via a preformed enol also accounts for the observed first-order dependence on reactant concentrations and $[H^+]$.[15] Other studies of the Cr(VI) oxidations of alcohols,[16] acetophenones ($\rho^+ = -0.75$),[17] and chalcones[18] have been reported. In a reaction analogous to alcohol oxidation, a chromic acid thiolester ($RSCrO_3^-$) was formed rapidly during the Cr(VI) oxidation of $Me_3C(CH_2)_3C(SH)MePr$.[19]

Possibly because of conformational effects, Cr(VI) oxidation of methyl stearate was partially selective, suggesting that certain molecules may be intrinsically susceptible to oxidation at centres generally considered to be unreactive.[20] A detailed study of the chromic acid oxidation of cycloheptatriene to benzaldehyde including isotope effects has been published,[21] and the effect of alkyl substituents in chromic acid oxidation of tetralins,[22] and similar oxidation of toluenes,[23] have been studied. Related work includes a detailed kinetic analysis of the Cr(VI) oxidation of Malachite Green in the presence of oxalic acid,[24] and a comparison of oxidative deoximation by Jones reagent, chromium trioxide–pyridine complex and periodic acid;[25] and a review of chromyl chloride oxidations has appeared.[26]

Lead

Although it is generally agreed that oxidative decarboxylation of acids by Pb(IV) acetate proceeds via the corresponding alkyl radical,[27] the nature of the final oxidative substitution (to alkyl acetate) continues to be investigated so as to assess the relative importance of processes proceeding via carbonium ion and organolead intermediates, e.g. in bicyclic acids[28] and in stereoisomeric 2-methyl-3-phenyl-butyric acids.[29] In the presence of Cu(II) the intermediate radicals appear to be efficiently converted into organocopper species, which may yield olefins by a *cis*-elimination or acetate by an S_Ni displacement.[30, 31]

The primary isotope effect for oxidation of methanol by $Pb(OAc)_4$ ($k^{MeOH}/k^{CD_3OD} = 3.8$ at 25°) confirms that C—H cleavage occurs in the rate-limiting step;[32] formaldehyde, the initial oxidation product thought to be stable in the presence of $Pb(OAc)_4$, is further oxidized via its methyl hemiacetal to methyl formate. Other research on the oxidation of alcohols,[33] and a comparison of the Pb(IV), Tl(III) and Hg(II) oxidations of oct-1-ene have also been reported.[34] Oxidation of the diol (**1**) by $Pb(OAc)_4$ in butadiene yields the adduct (**3**) (57%), possibly via the 2,3-quinone (**2**).[35]

(**1**) (**2**) (**3**)

Manganese[36]

A full report of the oxidative cleavage of cinnamic acid by acidic permanganate has been published,[37] and the kinetics of Mn(VII) oxidations of aliphatic alcohols,[38] 1,3-diols,[39] benzaldehyde,[40] *p*-nitrobenzaldehyde,[41] and ethyl methyl ketone[42] have been investigated.

Oxidation of acetaldehyde by Mn(III) sulphate may proceed by direct attack rather than via the enol.[43] Mn(III) oxidation of deoxybenzoin in 95% acetic acid to give $(PhCOCHPh)_2$ is catalysed by perchloric acid.[44]

Thallium

While oxidation of alkenes by Tl(III) nitrate in methanol is known to yield ketones and glycol dimethyl ethers, a thorough product study has revealed that methoxy-nitrates and dinitrates are also formed during oxidation of dec-1-ene (dinitrate was not detected), 2,3-dimethylbut-2-ene, and *cis*- and *trans*-stilbene.[45] A net *trans*-addition is observed in the formation of the di-ether (**5**), methoxy-nitrate (**6**) and dinitrate (**7**) from *trans*-stilbene, as outlined in Scheme 3. The attacking nitrate ion may be present in the solution

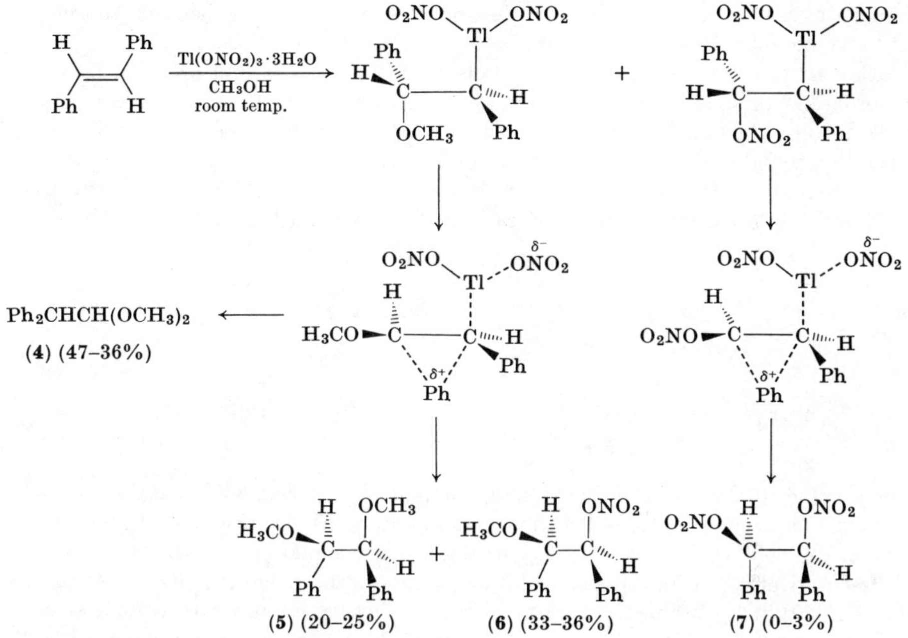

$Tl(ONO_2)_3 \cdot 3H_2O$, CH_3OH, room temp.

$Ph_2CHCH(OCH_3)_2$ (**4**) (47–36%)

(**5**) (20–25%) (**6**) (33–36%) (**7**) (0–3%)

SCHEME 3. Oxidation of *trans*-stilbene by Tl(III) nitrate in methanol.

or alternatively may arise from the co-ordination sphere of the departing thallium. There is also a detailed study of the products of oxidation of oct-1-ene by Tl(III) trifluoroacetate in various solvents.[46]

Kinetic measurements on the Tl(III) acetate oxidation of styrene[47] and acetophenone[48] have been reported, and the stopped-flow technique has been applied to the reaction between Tl(III) and pyrocatechol.[49]

Other Metals

Oxidative decarboxylation of phenylacetic acids by Ce(IV) may proceed via rate-limiting decomposition of a Ce(IV)–carboxylic acid complex to the corresponding benzyl radical and carbon dioxide [equation (18)] through a polar transition state ($\rho^+ = -2.9 \pm 0.3$); products are derived from the benzyl cation [see equation (19)].[50]

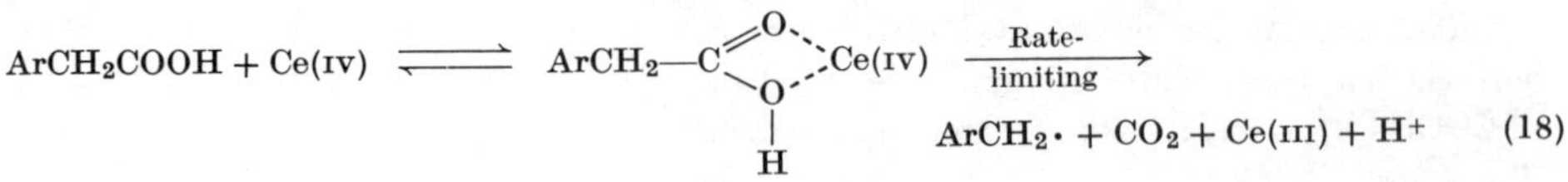

$$ArCH_2COOH + Ce(IV) \rightleftharpoons ArCH_2{-}C\overset{\displaystyle /\!\!/O\cdot.}{\underset{\displaystyle \backslash O\cdot\cdot'}{}}Ce(IV) \xrightarrow[\text{limiting}]{\text{Rate-}} ArCH_2\cdot + CO_2 + Ce(III) + H^+ \quad (18)$$

(the lower O bears H)

$$ArCH_2\cdot \xrightarrow{Ce(IV)} \text{Products of } ArCH_2^+ + Ce(III) \quad (19)$$

In dimethyl sulphoxide, Ce(IV) nitrate oxidation of cyclohexene leads to cyclohex-2-enyl nitrate, while in acetonitrile *N*-(cyclohex-2-enyl)acetamide is formed.[51] A series of aromatic hydrocarbons has been oxidized by Ce(IV) trifluoroacetate,[52] and it has been noted that addition of Mn(II) increases the rate of oxidation of propan-2-ol by Ce(IV) in sulphuric acid, presumably because Mn(III) is formed.[53] The kinetics of Ce(IV) oxidation of formaldehyde,[54] an aldo-sugar,[55] and α-picoline[56] have also been measured. Oxidation of phenylhydrazine by either Ce(IV) or $K_2Cr_2O_7$ yields mixtures of benzene and phenol, but Ce(IV) favours formation of benzene.[57]

Although the rate of oxidation of diols by $Fe(CN)_6^{3-}$ is first-order in substrate, first order in oxidant and second order in alkali, the rate for cycloalkanols is independent of alkali concentration.[58] Similar reactions of cycloalkanones have been studied,[59] and it has been found that low concentrations of dimethyl sulphoxide retard the reaction of acetophenones with alkaline $Fe(CN)_6^{3-}$.[60] Evidence that ferricyanide quantitatively oxidizes the hydroxycyclohexadienyl radical (**8**) to phenol has been presented.[61] The

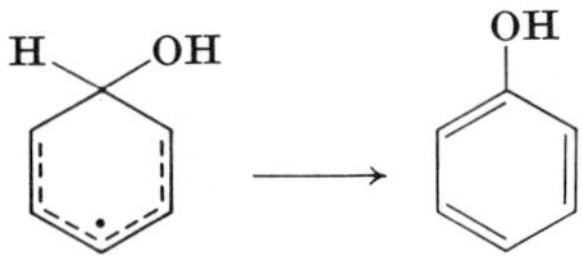

H OH → OH

(**8**)

high ρ value in the oxidation of a series of phenols by $Fe(CN)_6^{3-}$ suggests an ionic mechanism leading to dienone.[62] Oxidation of amines has also been investigated.[63]

From an analysis of the various possible kinetic equations for reaction between $PdCl_4^{2-}$ and ethylene in aqueous solution, it has been concluded that the two-term equation (20) fits experimental results over a wide range of temperature, ionic strength and concentrations of $PdCl_4^{2-}$, Cl^- and H_3O^+.[64] Whilst the first term of equation (20)

$$\text{Rate} = \frac{k[PdCl_4^{2-}][C_nH_{2n}]}{[Cl^-]^2[H_3O^+]} + \frac{k'[PdCl_4^{2-}]^2[C_nH_{2n}]}{[Cl^-]^3[H_3O^+]} \quad (20)$$

has been proposed previously[65] and can be rationalized by equations (21–24), the second term is dominant at higher Pd(II) concentrations, especially if $[Cl^-]$ is low – see equations (25–28). Similar research on isobutene and methylenecyclohexanes[66] and on cyclohexene[67] has been reported. Indirectly related to these studies is the observation that interaction of Pd(II) acetate (normally trimeric) with sodium and lithium acetates in acetic acid yields dimeric species.[68] Detailed product analyses of Pd(II) oxidations of monosubstituted benzenes in trifluoroacetic acid have been carried out.[69]

Further studies of the Co(III) oxidation of toluene,[70] of olefins and aromatic compounds,[71] and of pyridine carboxaldehydes[72] have been reported (see also *Org. Reaction Mech.*, **1973**, 178).

$$PdCl_4^{2-} + C_nH_{2n} \rightleftharpoons \underset{(\mathbf{9})}{PdCl_3C_nH_{2n}^- } + Cl^- \tag{21}$$

$$\mathbf{9} + H_2O \rightleftharpoons \underset{(\mathbf{10})}{PdCl_2C_nH_{2n}(OH_2)} + Cl^- \tag{22}$$

$$\mathbf{10} + H_2O \rightleftharpoons \underset{(\mathbf{11})}{PdCl_2C_nH_{2n}(OH)^-} + H_3O^+ \tag{23}$$

$$\mathbf{11} \xrightarrow{\text{Slow}} \underset{(\mathbf{12})}{\sigma\text{-}PdCl_2(C_nH_{2n}OH)^-} \tag{24}$$

$$\mathbf{12} + H_2O \longrightarrow Cl^- + PdCl^- + C_nH_{2n}O + H_3O^+ \tag{25}$$

$$\mathbf{11} + PdCl_4^{2-} \rightleftharpoons \underset{(\mathbf{13})}{\left[Cl_2Pd(\mu\text{-}Cl)_2Pd(Cl)(OH)(C_nH_{2n})\right]^{2-}} + Cl^- \tag{26}$$

$$\mathbf{13} \xrightarrow{\text{Slow}} \underset{(\mathbf{14})}{Cl_2PdCl_2PdCl(C_nH_{2n}OH)^{2-}} \tag{27}$$

$$\mathbf{14} + H_2O \longrightarrow Cl^- + Pd_2Cl_4^{2-} + C_nH_{2n}O + H_3O^+ \tag{28}$$

There are now two full accounts of the oxidation of alcohols by silver carbonate.[73] On treatment with silver oxide in benzene, *N*-sulphinyl-*N'*-arylhydrazines (RNHNSO) yield RSO_3H, RSSR, RSR and other products.[74]

In a novel reaction, oxidation of the aziridine (**15**) with ruthenium tetroxide gave the amide (**17**), possibly via the α-lactam (**16**).[75] The products of similar oxidation of substituted naphthalenes have been determined.[76]

$$\underset{(\mathbf{15})}{Bu^tCH{-}CH_2(NBu^t)} \longrightarrow \left[\underset{(\mathbf{16})}{Bu^tCH{-}C{=}O\,(NBu^t)}\right] \longrightarrow \underset{(\mathbf{17})}{Bu^tC(=O)NHBu^t} + CO_2$$

Some unusual work on the influence of a magnetic field on ion-catalysed oxidations has been reported.[77] Other studies include the V(V) oxidation of the cyclohexanehexol, inositol,[78] and the Pt(IV)-catalysed oxidation of benzenes.[79] Metal-catalysed autoxidations are discussed in the following Section.

Oxidation by Molecular Oxygen*

Hydrocarbons[80]

Whilst autoxidations of hydrocarbons often require radical initiation, metal-catalysis and/or high temperatures, some hydrocarbons readily give hydroperoxides at room

* Since the term "molecular oxygen" appears to have reached a "transition state", the newer name dioxygen will also be used here.

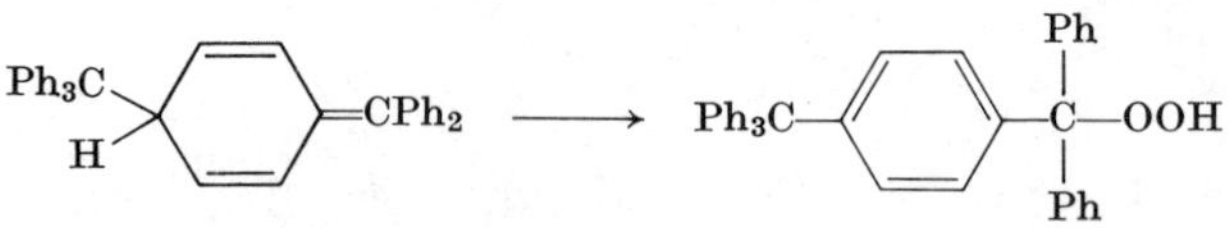

(18)

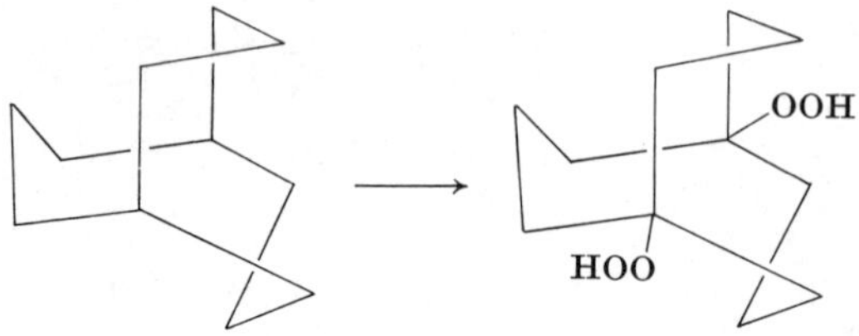

(19)

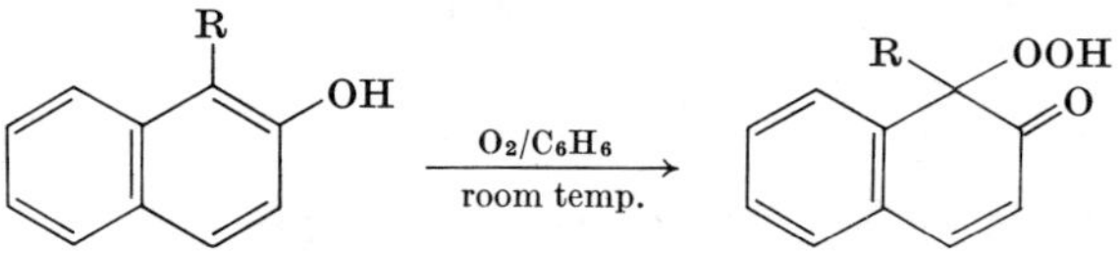

O_2/C_6H_6 room temp.

(20)

temperature or below. Recent examples include the hindered dimer (**18**) of the triphenylmethyl radical,[81] and the bicyclic hydrocarbon manxane (**19**), which has an unusually high bridgehead reactivity.[82] Incidentally, the highly hindered naphthol (**20**, R = Bu^t) absorbs 0.5 mole of dioxygen in 10 min, whereas the less hindered derivative (**20**, R = Me) shows negligible sign of reaction after 3 days.[83]

The products of radical-initiated oxidations of pentane and octane,[84] isobutene and cyclopentene,[85] and α-methylstyrene[86] have been examined. Some comments on the determination of rate constants by the hydroperoxide method have been made, based on a kinetic study of the initiated oxidation of tetralin.[87]

Considerable interest continues in metal-catalysed autoxidations. Rh(I) catalysts, $RhH(CO)(PPh_3)_3$ and $RhCl(PPh_3)_3$, promote oxygenation of terminal olefins to methyl ketones,[88] and the kinetics of similar oxidations suggest that the reaction [equation (29)] has radical character.[89] The process may be diverted along another pathway, favoured by addition of a second catalyst, in which the epoxide (**22**) is formed, indicating that the hydroperoxide (**21**) may then act as an epoxidizing agent [equation (30)].[90]

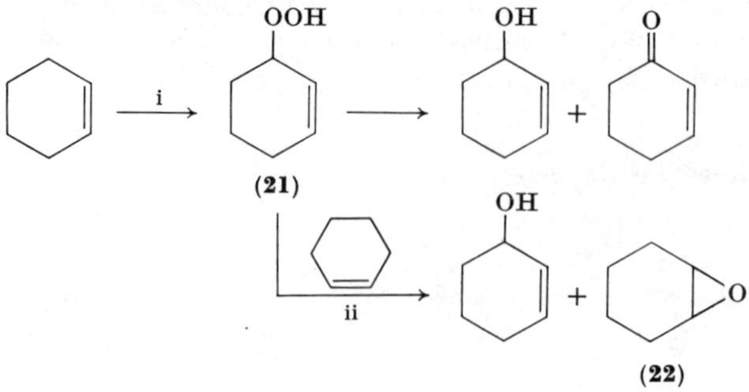

(21) (29)

(22) (30)

i, O_2—$RhCl(PPh_3)_3$. ii, Catalysed by $MoO_5(HMPT)H_2O$ or $V(acac)_3$.

The catalytic effect of bromide ion on Co(II) acetate oxidations may be due to formation of a more soluble complex.[91] In contrast to the autoxidation of cumene initiated by 2,2′-azodi-isobutyronitrile, the rate of which is increased in nitrobenzene, the rate of reaction catalysed by Co(II) acetylacetonate is decreased by addition of nitrobenzene.[92] The rate of autoxidation of pentadecane at 130° was higher when combined Co and Mn or Mn and Cu stearates were used as catalysts than when a single metal was used.[93]

As might be expected, Pd(II)-catalysed oxidative coupling of aromatic hydrocarbons with dioxygen gave different substitution patterns from those reported in Pd(II)/acetic acid or in phenyl radical coupling.[94] Related reports discuss the oxidation of propene catalysed by silver,[95] the oxidative transformations of olefins,[96] metal-catalysed autoxidation of cyclohexane via cyclohexanone to adipic acid,[97] and other autoxidations.[98] There is also a report on the base-catalysed autoxidation of 9,10-dihydroanthracene with electro-generated superoxide (O_2^-).[99]

Other Autoxidations

The problem of the nature of dioxygen-induced chemiluminescence of aryl Grignard reagents has been resurrected with a report that the main emitting species from phenylmagnesium bromide appear to be a mixture of isomeric brominated biphenyls.[100] Autoxidation of the dihydroisoalloxazine [(**23**) → (**24**)] in alkaline solution is also accompanied by chemiluminescence,[101] and research on the kinetics and mechanism of autoxidation of tetrahydropterins continues.[102]

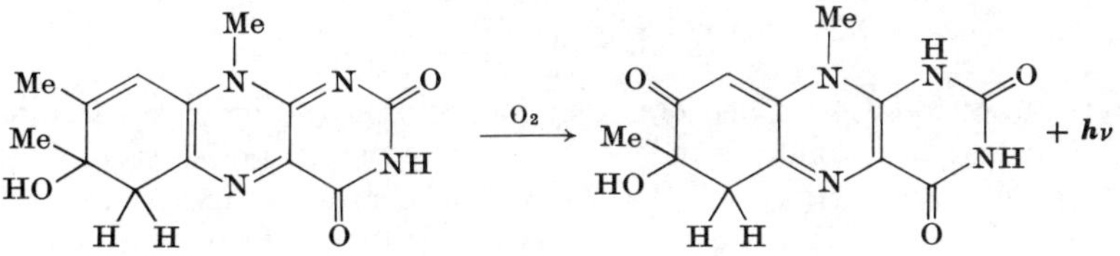

O_2 + $h\nu$

(**23**) (**24**)

After studies of the base-induced oxygenation of the catechol (**25**) and related compounds, all leading to (**26**), a self-consistent mechanism was proposed.[103] In an extension of earlier work it was found that the rates of oxidation of 2-substituted anthranols to the corresponding anthraquinones in alkaline aqueous dioxan give a Hammett ρ value of −1.5.[104] Autoxidation of pyridinium fluorenylides to fluoren-9-ones in benzene and in chloroform may proceed with general-base catalysis via the corresponding hydroperoxide.[105]

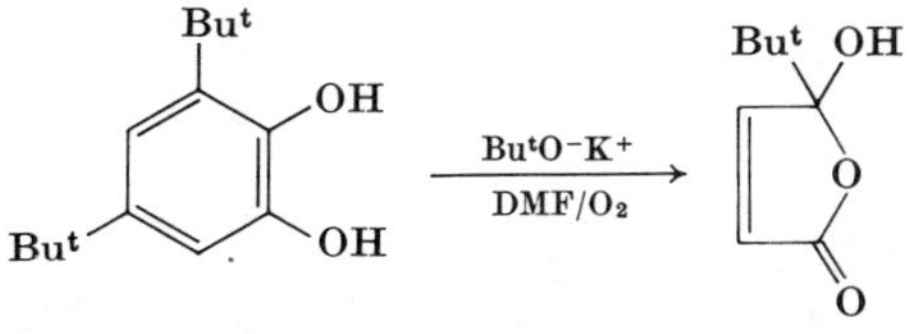

$Bu^tO^-K^+$ / DMF/O_2

(**25**) (**26**)

From a kinetic and isotopic-labelling study, the mechanism (**27**) → (**29**) was proposed for radical-initiated autoxidation of α-methylbenzyl alcohol. The role of main chain-carrier was assigned to the hydroperoxy-radical [equations (31) and (32)].[106] Acetophenone is also produced in reaction (33) and the mechanism has been discussed.[107]

$$\mathrm{Ph(Me)C(H)OH}\ (\mathbf{27}) + \mathrm{rO_2\cdot} \longrightarrow \mathrm{Ph(Me)\dot{C}OH}\ (\mathbf{28}) \xrightarrow{O_2} \mathrm{Ph(Me)C(O{-}O\cdot)OH}\ (\mathbf{29})$$

$$(\mathbf{29}) \longrightarrow \mathrm{Ph(Me)C{=}O} + \mathrm{HO_2\cdot} \qquad (31)$$

$$\mathrm{HO_2\cdot} + (\mathbf{27}) \longrightarrow \mathrm{H_2O_2} + (\mathbf{28}) \qquad (32)$$

$$\mathrm{C_6H_5CH(CH_3)COCOOC_2H_5} \rightarrow \mathrm{C_6H_5COCH_3 + CO_2 + CO + HOOCCOOC_2H_5} \qquad (33)$$

Dibenzoyl peroxide is formed during autoxidation of benzaldehyde in the presence of lithium chloride, which accelerates uptake of dioxygen.[108] Ozone-initiated autoxidation of isobutyraldehyde yields peroxyisobutyric acid,[109] and similar research on simple aldehydes,[110] on steric effects in terpenic aldehydes,[111] and on simple ketones[112] has been discussed. Other autoxidations studied include those of secondary alcohols,[113] amines,[114] *N*-alkylacetamides,[115] dopa,[116] tryptophan,[117] thiols,[118] and a variety of organometallics.[119]

Aspects of the cobalt ion catalysed autoxidations of phenols and flavones,[120] and of benzaldehyde,[121] have been discussed. The rate law [equation (34)] for Co(III)-catalysed oxidation of *p*-toluic acid is similar to the equation obtained for oxidation of ethylbenzene and cumene.[122]

$$\text{Rate} = k[\mathrm{CH_3C_6H_4COOH}][\mathrm{Co(III)}]^2[\mathrm{Co(II)}]^{-1} \qquad (34)$$

The rate of autoxidation of benzoin in methanol, catalysed by nickel acetate to give benzil and hydrogen peroxide, is independent of the partial pressure of oxygen. As a primary isotope effect ($k_H/k_D = 7$) is observed, the rate-limiting step must involve abstraction of a hydrogen atom of the benzoin; substituent effects indicate that at this stage a partial negative charge is introduced into the substrate group. Some intermediate must then react with dioxygen.[123]

Formation of an Fe(III)–enolate complex appears to be the slow step in the iron-catalysed autoxidation of acetoin.[124]

Although various substituted pyrocatechols have been oxidized, e.g. (**25**) → (**26**), the first successful oxidative cleavage of pyrocatechol itself with dioxygen has only recently been reported [equation (35)].[125] While the cuprous chloride/pyridine system

$$\text{pyrocatechol} \xrightarrow{\text{i}} [\text{6-hydroperoxy-6-oxido-cyclohexa-2,4-dienone} \xrightarrow{-\mathrm{H_2O}} \text{muconic anhydride}] \longrightarrow (\mathbf{30})\ \text{(cis,cis-MeOOC–CH=CH–CH=CH–COOMe)} \qquad (35)$$

i, $O_2/CH_3OH/Cu(I)$ chloride in pyridine

(**30**)

usually appears to be a good system for activating dioxygen under mild conditions, addition of methanol is also essential in this delicate case.

Other reports include Cu(II)-catalysed coupling of phenylacetylenes ($\rho = 2.2$),[126] iridium-catalysed oxidation of sulphones,[127] and various aspects of the metal-catalysed autoxidation of aldehydes.[128]

Among the research on antioxidants,[129] there is a study on the substituent effect on the antioxidant properties of 4-X-2,6-di-*tert*-butylphenols.[130]

Ozonation and Ozonolysis

The effect of added aldehyde on ozonation of alkenes continues to receive attention. Both the experimental results themselves and their interpretations are disputed (cf. *Org. Reaction Mech.*, **1973**, 181). Criegee's zwitterion mechanism (Scheme 4, path a) predicts that addition of ^{18}O-labelled aldehyde should lead to the appearance of the labelled oxygen atom at the "epoxy"-oxygen in the ozonide ring, (**33**). An additional pathway, the "aldehyde interchange mechanism" via (**31**), has been proposed which accounts for the possibility that the label appears in the peroxy-ring of the ozonide (**32**).[131,132] In studies involving alkenes of low molecular weight or phenylalkenes, only "epoxy" enrichment (**33**) has been observed,[133] whereas larger alkylalkenes have been reported to yield both (**32**) and (**33**). The earlier report[132] that addition of ^{18}O-labelled acetaldehyde to *trans*-di-isopropylethylene leads to about 70% of (**32**) has recently been disputed, as a value of <10% was obtained under closely similar experimental conditions.[134] Also, addition of aldehyde would be expected to increase artificially the importance of such competing pathways. Clearly some agreement on experimental techniques is necessary before agreement on mechanism can be expected.

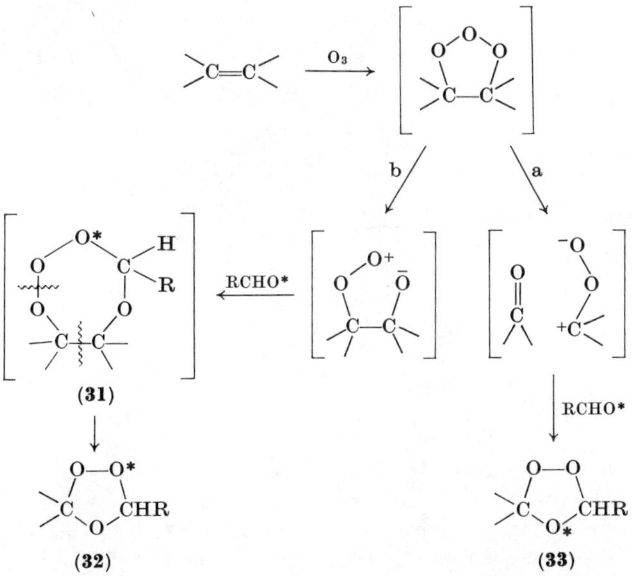

SCHEME 4. Fate of labelled aldehyde in ozonation of olefins.

The major problem with the simplest form of the zwitterion mechanism (Scheme 4, path a) is the observation that stereochemistry of the alkene influences the geometry of the ozonide formed therefrom.[131] To help to explain these results, conformations of ozonides have been studied by microwave spectroscopy[135] and have been calculated theoretically,[136] and a detailed discussion of possible modifications of Scheme 4 to include stereochemical aspects has been presented.[135] Complexes between aromatic hydrocarbons and ozone react with *cis*- and *trans*-alkenes, but give rise to different *cis*/*trans* ratios of ozonides from reaction of ozone alone.[137] However, it should be emphasized that the size of the observed effects (*cis*/*trans* ratios between 2 and 0.5) correspond to very small changes in energy; using the equation $\Delta E_A = \boldsymbol{R}T \ln(k/k')$, one can calculate that a *cis*/*trans* ratio as high as 99/1 corresponds to a difference in activation energies of only 1.1 kcal mol^{-1} at −150°C. Such small effects may arise when solvent affects the reactivity of short-lived intermediates, e.g. the alkene stereochemistry appears to be retained more in isopentane than in CF_2Cl_2.[137] Certainly, in the present state of our knowledge of solution chemistry, detailed discussions of effects of the order of 1 kcal mol^{-1} or less are highly tenuous.

Further evidence for zwitterionic intermediates (e.g. **37**) is provided by the ozonation of the alkene (**34**), which yields a 1:1 mixture of (**35**) and (**36**).[138] Ozonides can also be prepared by oxidation of diazo-compounds with singlet oxygen in the presence of aldehydes.[139] Surprisingly, the rate of ozonolysis of acrylic acid is independent of solvent polarity.[140] Reactions of ozone with alkenes in the gas phase,[141] and with acetylenes,[142] ketenes,[143] and ketenimines[144] have also been studied.

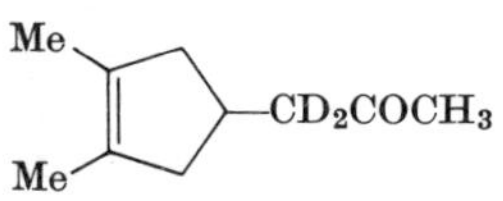

Me, Me, CD_2COCH_3

(**34**)

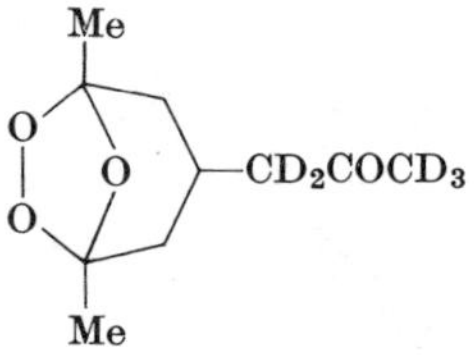

Me, O, O, O, CD_2COCD_3, Me

(**35**)

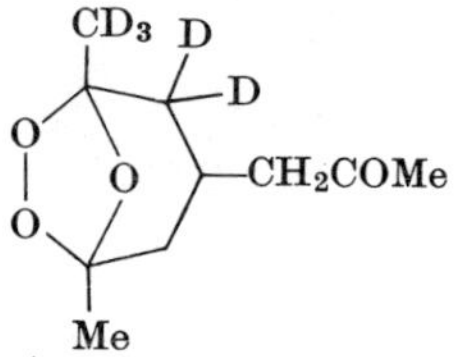

CD_3, D, D, O, O, O, CH_2COMe, Me

(**36**)

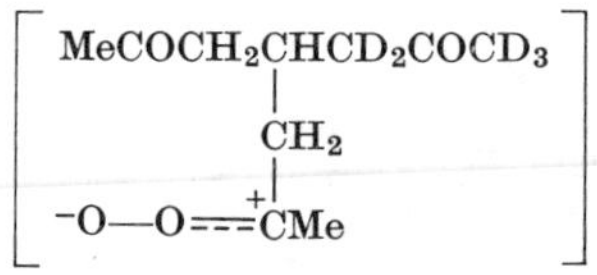

$MeCOCH_2CHCD_2COCD_3$
CH_2
$^-O—O\text{===}\overset{+}{C}Me$

(**37**)

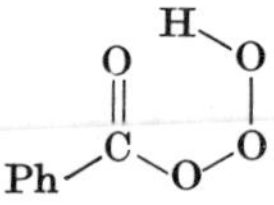

H, O, O, O, O, C, Ph

(**38**)

The first step in the reaction of ozone with saturated hydrocarbons may be formation of a complex.[145] In the presence of ferric chloride, alcohols are formed with almost complete retention of configuration whereas chlorides are formed with almost complete inversion[146] Evidence for the formation of hydrotrioxides (e.g. **38**) during ozonolysis of benzaldehyde or ethers has been presented,[147] and the ozonolysis of acetals has been discussed.[148]

Other Oxidations

Peracids and Peroxides

Possibly because of intramolecular hydrogen bonding, reactivity of substituted peroxycarbamic acids (e.g. **39**) towards alkenes does not appear to depend on solvent, and in tetrahydrofuran these acids react about 200 times faster than peroxybenzoic acid.[149] Rates of epoxidation of simple alkenes by peroxyacetic acid in non-basic solvents correlate with E_T-values, but in basic solvents the rates are independent of solvent polarity.[150] Further studies of the peracid oxidation of substituted cyclopropenes,[151] methylenecyclohexanes,[152] cycloalkenes,[153] and allenes[154] have been reported.

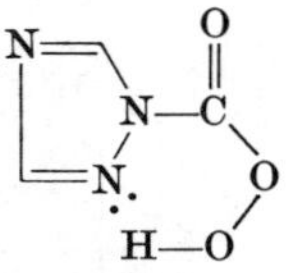

(39)

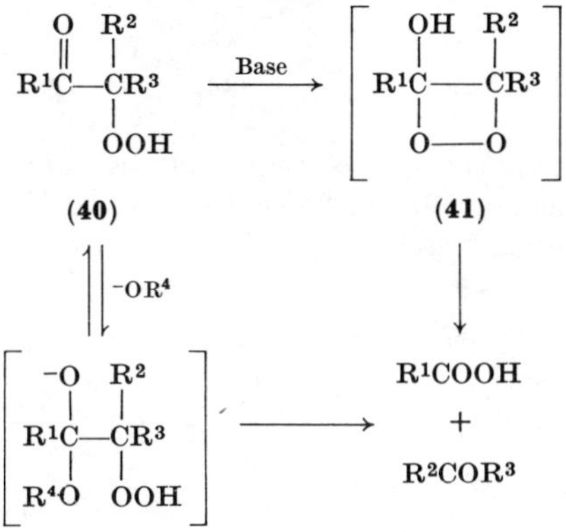

Base

(40) **(41)**

$^{-}OR^4$

R^1COOH + R^2COR^3

(42)

Of the two possible mechanisms for base-catalysed decomposition of α-hydroperoxy-ketones (**40**), the route via the dioxetane (**41**) accounts better for the experimental results, including chemiluminescence;[155] however, the acyclic route via (**42**) may also be important in some cases (e.g. **40**; $R^1 = Pr^i$, $R^2 = R^3 = CH_3$) because methyl iso-butyrate (30%) is formed when the reaction is carried out with sodium methoxide in absolute methanol.[155] Reaction of α-hydroxy-ketones with *m*-chloroperoxybenzoic acid may proceed by epoxidation of the corresponding enediol.[156]

An unusual reaction, conversion of phenyl salicylate (**43**) with perbenzoic acid in ethanol at pH > 9 to *ortho*-ethoxyphenol (**44**), may proceed via an aryl cation or equivalent intermediate;[157] no reaction is observed with phenyl benzoate. The Baeyer–

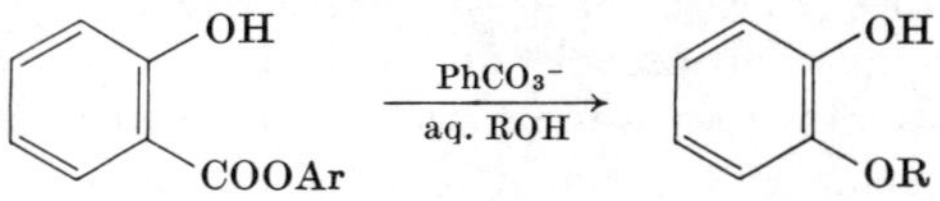

$PhCO_3^-$ / aq. ROH

(43) **(44)**

Villiger reaction continues to receive attention, including examination of cyclic αβ-unsaturated ketones,[158] and of bicyclo[3.3.1]nonan-3-ones,[159] and a theoretical study by CNDO/2.[160] On treatment with SeO_2/H_2O_2 cyclobutanone undergoes either oxidation to γ-butyrolactone or ring contraction to cyclopropanecarboxylic acid, depending on the experimental conditions.[161]

Unexpectedly, oxidation of the amine (**45**) with *m*-chloroperoxybenzoic acid yields the keto-nitroxide (**47**) as well as (**46**). Under such conditions alcohols are not usually affected but, as the reaction is suppressed by dilution of the reaction mixture, intramolecular hydrogen abstraction is probably not responsible. Also the radical (**47**) is produced in good yield from the radical (**46**) upon treatment with one equivalent of *m*-chloroperoxybenzoic acid.[162]

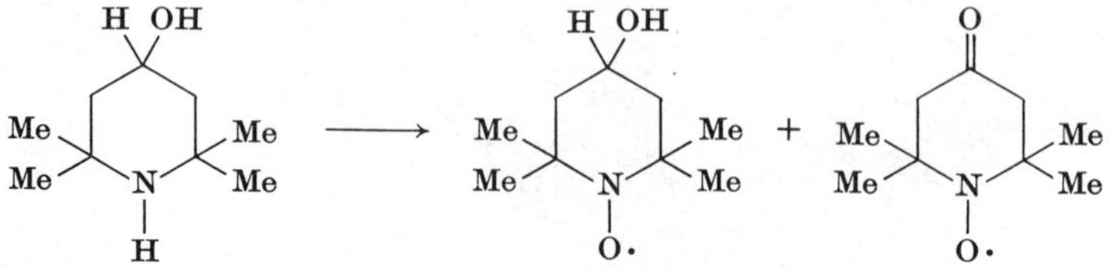

H OH, Me, Me, Me, Me, N, H (**45**) ⟶ H OH, Me, Me, Me, Me, N, O· (**46**) + O, Me, Me, Me, Me, N, O· (**47**)

Intermediates observed in the peracetic acid oxidation of phenol to hexa-2,4-diendioic acid (**48**) and phenoquinone (**49**) can be accommodated by the mechanism shown in Scheme 5.[163] A more detailed study of the initial hydroxylation step (Scheme 5) has

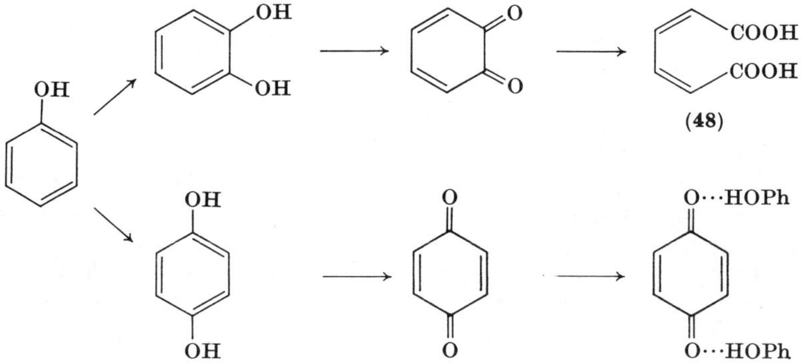

OH ⟶ OH, OH ⟶ O, O ⟶ COOH, COOH (**48**)

OH ⟶ OH, OH ⟶ O, O ⟶ O···HOPh, O···HOPh (**49**)

SCHEME 5. Peracetic acid oxidation of phenol.

been reported for the reaction of permonophosphoric acid with phenol, anisole and toluene;[164] the kinetic evidence suggests that attack by protonated dimeric acid on the aromatic species is the rate-limiting step. There are also studies of the perbenzoic acid oxidation of arylsulphines ($Ar_2C{=}SO$) to the corresponding benzophenone,[165] and of the peracid oxidation of nitrosobenzene[166] and of disilanes.[167]

Synthetically useful improvements to reactions involving oxidations by peracids and peroxides can often be achieved by transition metals, e.g. improved stereoselectivity in epoxidation of acyclic allylic alcohols by *t*-butyl hydroperoxide, catalysed by $Mo(CO)_6$ or $VO(acac)_2$.[168] Mechanistic aspects of similar reactions have also received attention. Since rearrangement was not observed in the molybdenum-catalysed oxidation of acetylenes and in the metal-catalysed decomposition and oxidation of α-diazo-ketones, the formation of oxirenes as intermediates in these reactions was excluded.[169] A kinetic study of the epoxidation of allyl alcohols with peroxy-complexes of molybdenum has been reported.[170] Other aspects include investigations of the kinetics of V(IV)-catalysed oxidation of organosulphur compounds by *t*-butyl hydroperoxide,[171] and the directive effects in azobenzene oxidation by *t*-butyl hydroperoxide, catalysed by Mo(VI).[172]

A highly stereoselective oxidation of cyclohexanol by Fe(II)/H_2O_2 yields mainly the *cis*-1,3-diol, whereas Cu(I)/H_2O_2 yields mainly the 1,4-product. The results appear to be inconsistent with a mechanism involving a free hydroxyl radical, and a cyclic transition state has been proposed.[173]

When a flow system was used to detect the ^{1}H-NMR spectra of short-lived intermediates, enhanced absorption and emission lines were observed during oxidation of propan-2-ol in water by $TiCl_3/H_2O_2$; the role of radical pairs was discussed, supplementing an earlier report on ESR.[174] Research on oxidation of ascorbic acid by Cu(II)/H_2O_2[175] and of *p*-phenetidine by Os(VIII)/H_2O_2[176] has also been carried out.

Halogens

Oscillating reactions provide highly successful lecture demonstrations[177] as well as intriguing mechanistic problems. The reaction of malonic acid in stirred sulphuric acid initially containing potassium bromate and Ce(IV) sulphate has been studied in detail.[178] The oscillations have been monitored by following the concentrations of Ce(IV) and bromide ion potentiometrically, and the following mechanism was proposed: when the solution contains sufficient bromide ion, BrO_3^- is reduced to Br_2 and the malonic acid is then brominated via the enol; the bromide ion concentration decreases to a stage where it becomes too small to remove $HBrO_2$ (from partial reduction of bromate) sufficiently rapidly, so the $HBrO_2$ reacts with BrO_3^- to produce two $BrO_2\cdot$ radicals, which oxidize Ce(III), as in equation (36):

$$BrO_3^- + HBrO_2 + 2Ce(\text{III}) + 3H^+ \rightarrow 2HBrO_2 + 2Ce(\text{IV}) + H_2O \qquad (36)$$

Thus $HBrO_2$ is produced autocatalytically, but indefinite build-up of this species is prevented by its second-order disproportionation; by oxidizing bromomalonic acid, Ce(IV), produced by equation (36), liberates bromide ion which eventually terminates the autocatalytic production of $HBrO_2$ and initiates a repetition of the cycle.[178] The reaction is inhibited by radical scavengers, suggesting the intermediacy of free radicals.[179] Changes of temperature and potential in the reaction of pentane-2,4-dione and bromate/catalyst have been compared with those for malonic acid,[180] and similar studies of acetylacetone have been reported.[181]

The rates of oxidative coupling of benzyl cyanides by halogen or hypochlorite in the presence of strong base, giving α,α'-dicyanostilbenes [equation (37)], may proceed by rate-limiting α-proton abstraction followed by rapid hypohalite attack, giving

$$2PhCH_2CN + 2I_2 + 4RONa \rightarrow PhC(CN){=}C(CN)Ph + 4ROH + 4NaI \qquad (37)$$

α-halobenzyl cyanide.[182] Carbanion stability also appears to influence the products of other hypochlorite oxidations.[183] Halogen or hypohalite oxidation of *N*-benzylanilines to benzylideneanilines in alkaline solution may produce an α-halogeno-derivative, which is then dehydrohalogenated.[184]

Related studies include oxidation of: sodium formate by iodine;[185] mandelic acid by chlorine[186] or by periodate;[187] secondary alcohols by perbromate;[188] and aldehydes,[189] simple ketones,[190] cyclopentanone[191] and allyl alcohol[192] by chloramine T.

Miscellaneous Oxidations

N-Chloroammonium ions have been shown to be highly selective reagents for the introduction of chlorine into closely similar CH_2 groups.[193] In an analogous reaction, trifluoroacetoxy-groups can be introduced by using $Et_2NOH/Fe(II)$ or $Et_3NO/Fe(II)$ in trifluoroacetic acid [equation (38)].[194]

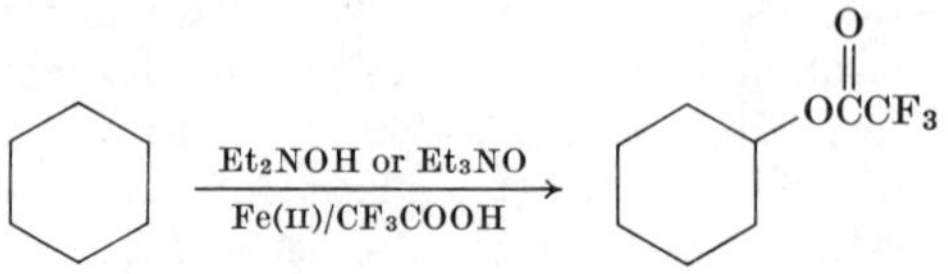

$$\text{cyclohexane} \xrightarrow[\text{Fe(II)/CF}_3\text{COOH}]{\text{Et}_2\text{NOH or Et}_3\text{NO}} \text{C}_6\text{H}_{11}\text{OC(=O)CF}_3 \tag{38}$$

Under similar conditions *n*-octyl trifluoroacetate reacts predominantly (72%) at the 7-position, so the polar selectivity parallels that of the chlorinations.[193] A possible mechanism via nitrogen cation radicals is shown in equations (39) to (42).[194,195]

$$R_3NOH + Fe(II) \rightarrow Fe(III) + R_3'N^{+}\cdot \tag{39}$$

$$R_3'N^{+}\cdot + RH \rightarrow R_3'NH^{+} + R\cdot \tag{40}$$

$$R\cdot + Fe(III) \rightarrow Fe(II) + R^{+} \tag{41}$$

$$R^{+} + CF_3COOH \rightarrow ROCOCF_3 + H^{+} \tag{42}$$

The intramolecular nature of base-induced oxidation of alcohols by dimethyl sulphide and *N*-chlorosuccinimide has been established by deuterium-labelling experiments, and the key intermediate (**51**) may decompose as shown in Scheme 6;[196] these results also

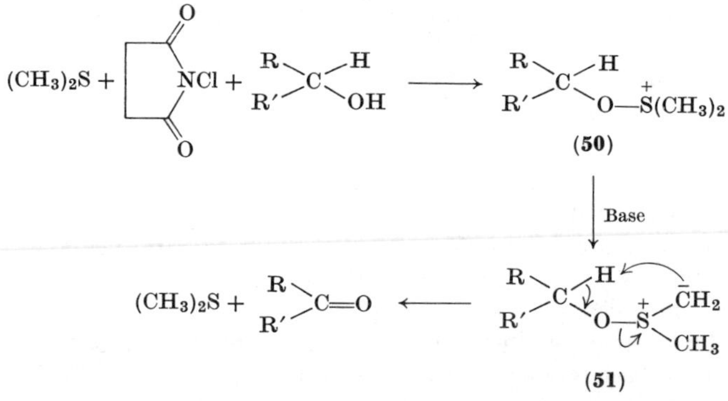

$$(CH_3)_2S + \text{N-chlorosuccinimide} + RR'CH(OH) \longrightarrow RR'CH{-}O{-}\overset{+}{S}(CH_3)_2 \;\textbf{(50)}$$

$$\textbf{(50)} \xrightarrow{\text{Base}} RR'CH{-}O{-}\overset{+}{S}(CH_3)\bar{C}H_2 \;\textbf{(51)} \longrightarrow (CH_3)_2S + RR'C{=}O$$

SCHEME 6. Oxidation of alcohols by dimethyl sulphide.

implicate the oxysulphonium cation intermediate (**50**), which has been proposed to account for *ortho*-quinone formation from catechols.[197] The value of such reactions in selective oxidations has been further increased by using hexamethylphosphoramide at –20° as solvent,[198] and the range of substrates has been extended to vicinal diols and α-ketols,[199] which react without cleavage of the carbon—carbon bond.

Since oxidation of benzyl alcohols to benzaldehydes by dimethyl sulphoxide (DMSO) in the presence of air is acid-catalysed, an alkoxysulphonium ion intermediate has been proposed.[200] Both primary and secondary bromides can be oxidized readily to aldehydes or ketones in the presence of 1.1–1.3 equivalents of $AgBF_4$ at room temperature,[201] although simple alkyl bromides are inert to DMSO even at high temperatures. Similar reactions of tosylates with sodium hydrogen carbonate in DMSO may proceed by initial displacement of the tosylate by hydrogen carbonate, because in some cases (e.g. **52**) cyclic carbonates (**53**) are formed.[202] It may be noted that oxidation of dimethyl sulphide to dimethyl sulphoxide by nitric acid displays CIDNP.[203]

$(CH_3)_2SO$, $NaHCO_3$

(**52**) (**53**)

To exploit the *syn*-elimination of selenoxides to olefins, e.g. in conversion of cyclic ketones and β-dicarbonyl compounds into enones,[204] and in introduction of an α-methylene into lactones,[205] there must be effective methods for introducing selenium, as for example SePh, into substrate molecules. This can be achieved for unactivated olefins (e.g. **54**) by using benzeneselenyl bromide and silver trifluoroacetate in ether at −10°; subsequent mild hydrolysis removes the trifluoroacetoxy-group incorporated in the first reaction to give the β-hydroxyselenide (**55**), which can be converted into the αβ-unsaturated alcohol (**56**).[206] Selenium can also be introduced into nitriles, via the

(**54**) (**55**) (**56**)

i, $PhSeBr/AgOCOCF_3$. ii, $NaHCO_3/H_2O$.

corresponding α-lithiated derivative, but two equivalents of base are necessary so that both the nitrile anion and the anion of the monoselenylated product are formed; this prevents rapid equilibration of the product and starting material leading to the α-bis(selenylated) nitrile.[207] Evidence has been presented for the production of episelenides by reaction of epoxides with tri-*n*-butylphosphine selenide in benzene at room temperature, to which one molar proportion of trifluoroacetic acid was added.[208]

Under nitrogen at room temperature, and in the presence of alkoxides, selenium acts as an unusual oxidant for formates, giving dialkyl carbonates [equation (43)]. In the presence of molecular oxygen, NaSeH is converted into Se [equation (44)] and thus the reaction is catalytic:[209]

$$ROCHO + NaOR' + Se \rightarrow ROCOOR' + NaSeH \quad (43)$$

$$NaSeH + R'OH + \tfrac{1}{2}O_2 \rightarrow Se + H_2O + NaOR' \quad (44)$$

Similar reaction of *N*,*N*-dimethylformamide may proceed via $Me_2NC(O)SeH$, because when methyl iodide was added to the reaction mixture $Me_2NC(O)SeMe$ was produced.[210]

Potassium peroxydisulphate ($S_2O_8^{2-}$) does not react with aromatic compounds in glacial acetic acid unless Pd(II) is present as catalyst; the *meta*-acetoxy-derivative is then the major product.[211] In aqueous media, $S_2O_8^{2-}$ appears to dimerize pyridine oxidatively via the pyridine cation radical.[212] Reactions of hydroxyl radicals (generated from Fenton's reagent) and the sulphate anion radical (generated thermally from $S_2O_8^{2-}$) with various phenyl-substituted alcohols have been examined in detail.[213] A free-radical mechanism involving Ag(II) is consistent with kinetic data for the Ag(I)-catalysed oxidation of phenylacetic acids by $S_2O_8^{2-}$.[214] The kinetics of similarly catalysed reactions of butan-2-ol[215] and of benzaldehyde[216] have been measured. Other $S_2O_8^{2-}$ oxidations include reactions of ketones,[217] ethers,[218] simple amides,[219] amines[220] and nucleic acids.[221]

Oxidative deamination of α-amino acids by the benzoquinone derivative (**57**) leads to the greenish-blue phenoxazine dye (**58**).[222]

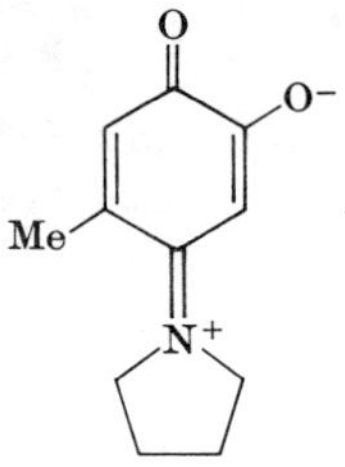

(**57**) 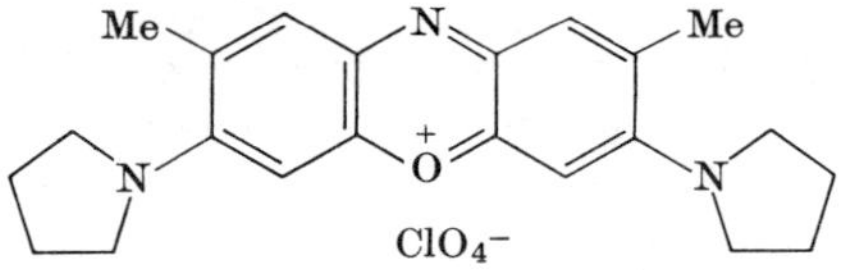 (**58**) 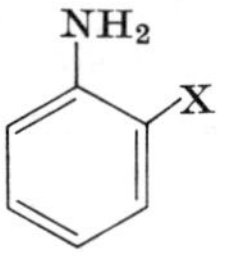 (**59**)

Aspects of the kinetics of reactions of (diacetoxyiodo)benzene, $PhI(OAc)_2$ include oxidation of α-hydroxy-acids,[223] isokinetic relationships for oxidative cyclization of *o*-substituted anilines (**59**),[224] and oxidation of benzamide.[225] Oxidation of alcohols by *N*-iodo-[226] and *N*-chloro-succinimide,[227] of hydroxylamines by Fremy's salt [$(KO_3S)_2NO$],[228] and of 2,4-dinitrophenylhydrazones by NO_2[229] has also been discussed.

The pyrolysis of 1,2-dihydronaphthalene (**60**) at 300° in various solvents (e.g. 1-chloronaphthalene) produced naphthalene and tetralin in 1:1 ratio, possibly in a concerted reaction via a transition state such as (**61**); $\Delta H^{\ddagger} = 35.9 \pm 1.2$ kcal mol^{-1},

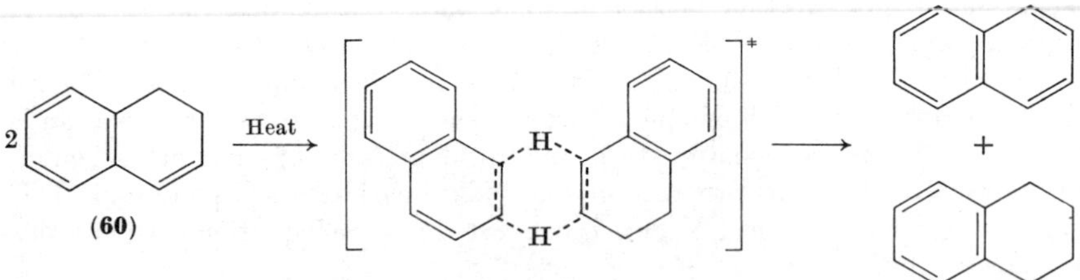

2 (**60**) Heat → (**61**) → naphthalene + tetralin

$\Delta S^{\ddagger} = -17.1 \pm 2.0$ e.u. Although addition of *t*-butyl hydroperoxide did accelerate decomposition, no rate depression was observed in separate runs in the presence of radical inhibitors. However, the reaction was catalysed by the addition of glass powder.[230] Dehydrogenations by double deprotonation to give dianionic intermediates, followed by electron transfer to a suitable acceptor [e.g. Cd(II)],[231] and dehydrogenations by palladium/charcoal on 2-(cyclohex-1-enyl)cyclohexanone have also been discussed.[232]

Reductions

Metal Hydride Reductions

In contrast to the reduction of (**62**) to (**63**), which proceeds readily without significant loss of deuterium, reduction of the labelled dibutyl sulphone (**64**) occurs with difficulty and one or more deuterium atoms are lost, as shown in equation (45);[233] apparently

$$\text{(62)} \xrightarrow[\text{Et}_2\text{O}]{\text{LiAlH}_4} \text{(63)}$$

(62) **(63)**

$$CH_3(CH_2)_2CD_2SO_2CD_2(CH_2)_2CH_3 \xrightarrow[\text{dioxan}]{\text{LiAlH}_4} \text{Dibutyl sulphide} \quad (45)$$

(64) 1-D 8%; 2-D 50%; 3-D 42%.

anion or dianion intermediates are formed before reduction occurs. Anionic intermediates are also involved in a new synthesis of small rings (e.g. **66**) from sulphoxides (e.g. **65**). Similarities with the Ramberg–Bäcklund and Stevens rearrangements are

$$\text{(65)} \xrightarrow{\text{i,ii}} \text{(66)} \quad (12\%\ \text{yield})$$

(65) **(66)**

$$PhCH_2SO_2CH_2Ph \xrightarrow{\text{i,ii}} PhCH{=}CHPh + PhCH_2SCH_2Ph$$

(56%; $c/t = 0.64$) (23%)

(67) **(68)** **(69)**

Reagents: i, Bu^nLi. ii, $LiAlH_4$–dioxan.

apparent from the behaviour of dibenzyl sulphone (**67**), which reacts to give olefins (**68**) as well as the sulphide (**69**).[234]

The reduction of ketones continues to receive attention. Further comments have been made on the necessary role of the alkali-metal cation in metal hydride reductions,[235] and various stereochemical studies of the reduction of "custom-built" hindered ketones have been reported.[236] Regioselectivity in the reduction of cyclohexanes and cyclopentenones has been discussed,[237] and $LiAlH_4$-reduction of β-diketones at low temperatures has been studied.[238]

Interest also continues in reagents capable of highly stereoselective reductions. Dialkylboranes (BR_2H) consistently appear to effect reduction of ketones from the less hindered side to a greater extent than conventional reagents such as $LiAlH_4$ or diborane.[239] The X-ray structure and exceptional stereoselectivity of lithium dimesitylborohydride bis(dimethoxyethane) have been discussed.[240]

It is well known that the reactivity of $LiAlH_4$ as a reducing agent can be greatly modified by addition of metal salts. In a development of studies of $TiCl_3$ as a reducing agent (see also p. 208). it has been shown that the deep black reagent formed from a slurry of one equivalent of $LiAlH_4$ and two equivalents of $TiCl_3$ is able to couple ketones to alkenes reductively (e.g. **70** → **71**; 85% yield).[241] Presumably Ti(II) is responsible

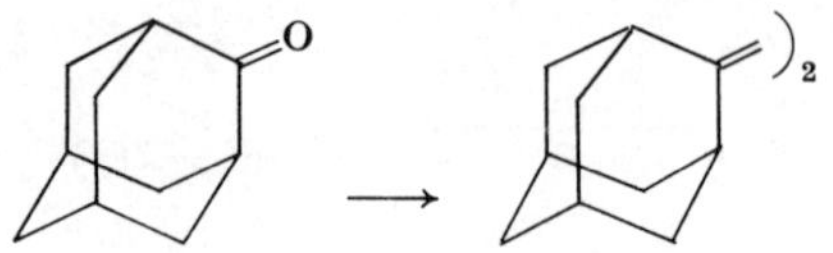

(70) (71)

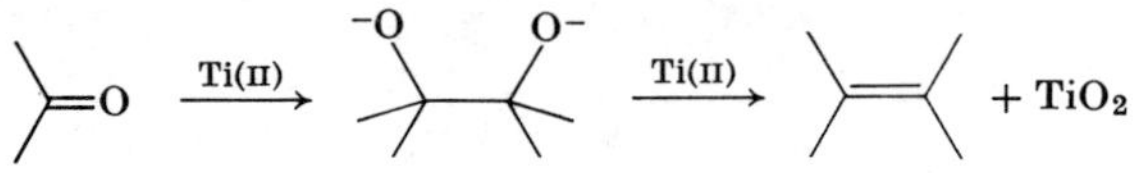

(72)

for the black colour and a possible mechanistic outline via the dianion (**72**) was suggested. An interesting example of the ability of $LiAlH_4/AlCl_3$ to cause hydrogenolysis is the conversion of ethyl *p*-methoxycinnamate (**73**) into the olefins (**74**) and (**75**); the role of the methoxy-group may be to stabilize an incipient carbonium ion before hydrogenolysis, because ethyl cinnamate is *not* hydrogenolysed under the same conditions.[242] Borohydride reduction of α-phenylcinnamates has also been discussed.[243]

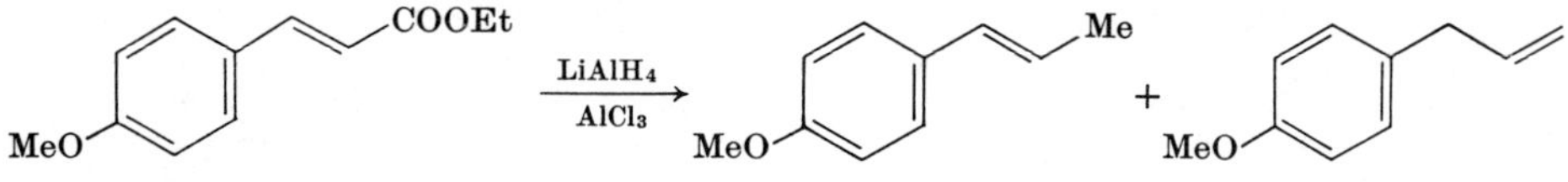

(73) (74) (75)

Labelling studies of the reduction of *N*-(benzyloxy)pyridinium bromide (**76**) with $NaBH_4$ are consistent with the mechanism shown. Deuterium is incorporated into the

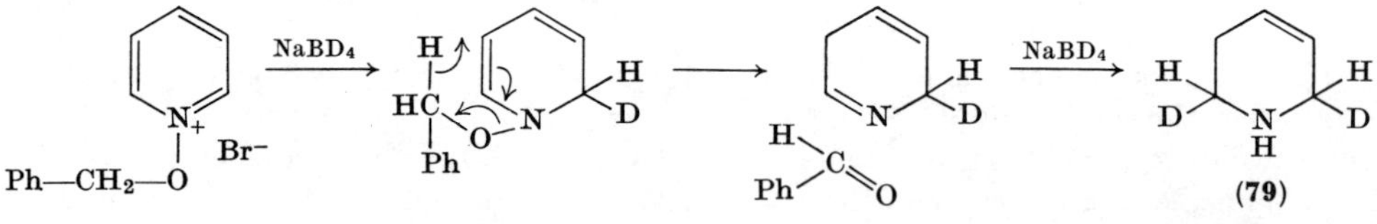

(76) (77) (78) (79) + PhCHDOH

2- and the 6-positions of the tetrahydropyridine derivative (**79**) and into the benzyl alcohol, which can be explained by intramolecular hydrogen transfer in (**77**)—a variation on the usual experiment showed the lack of crossed-products consistent with the intramolecular mechanism.[244]

Another reagent for reducing organic halides, thought to be $(KCuH_2)_n$, is produced by addition of $KBu^s{}_3BH$ (2 equivalents) to CuI.[245] Other aspects of metal hydride reactions include studies of the liberation of N_2 from azides with sodium hydride,[246] $NaBH_4$-reduction of tosylhydrazones,[247] reductive cleavage of acetals and ketals by diborane,[248] and the reductions of azachalcones[249] and various other ketones.[250] The electronic spectra of intermediates in the reduction of anthracene by bis-(2-methoxy-ethoxy)aluminium hydride have been interpreted.[251]

Dissolving Metal Reductions

A variety of dissolving metal reductions can be used to produce *trans*-symmetrical olefins from disubstituted acetylenes. Although the stereochemistry of this reaction has been attributed to the successive formation and rapid protonation of the corresponding radical anions and dianions, it is difficult to reconcile this interpretation with polarographic studies. Even in cases where delocalization of the negative charge is possible, formation of the free dianion is not certain, as highly negative potentials are required to reduce the radical anion. An alternative mechanism, consistent with a recent study of the reduction of hex-3-yne (**80**) with Na in HMPT/THF is shown in Scheme 7.[252] The

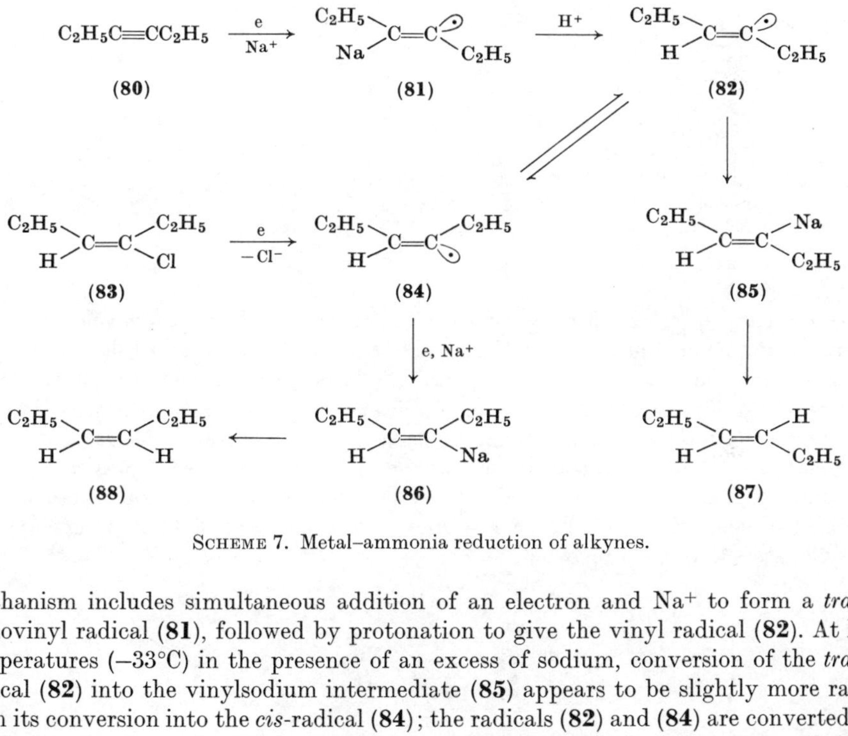

$C_2H_5C{\equiv}CC_2H_5$ (**80**) $\xrightarrow[Na^+]{e}$ $C_2H_5(Na)C{=}\dot{C}C_2H_5$ (**81**) $\xrightarrow{H^+}$ $C_2H_5(H)C{=}\dot{C}C_2H_5$ (**82**)

(**82**) ⇌ (**84**); (**82**) → $C_2H_5(H)C{=}C(Na)C_2H_5$ (**85**) → $C_2H_5(H)C{=}C(H)C_2H_5$ (**87**)

$C_2H_5(H)C{=}C(Cl)C_2H_5$ (**83**) $\xrightarrow[-Cl^-]{e}$ $C_2H_5(H)C{=}\dot{C}C_2H_5$ (**84**) $\xrightarrow{e,\ Na^+}$ $C_2H_5(H)C{=}C(Na)C_2H_5$ (**86**) → $C_2H_5(H)C{=}C(H)C_2H_5$ (**88**)

SCHEME 7. Metal–ammonia reduction of alkynes.

mechanism includes simultaneous addition of an electron and Na^+ to form a *trans*-sodiovinyl radical (**81**), followed by protonation to give the vinyl radical (**82**). At low temperatures (–33°C) in the presence of an excess of sodium, conversion of the *trans*-radical (**82**) into the vinylsodium intermediate (**85**) appears to be slightly more rapid than its conversion into the *cis*-radical (**84**); the radicals (**82**) and (**84**) are converted by similar mechanisms via (**85**) and (**86**) into the *trans*- and *cis*-olefins (**87**) and (**88**), respectively. A key aspect of the mechanism, equilibration of radicals (**82**) and (**84**), is supported by reduction of the *cis*-chloro-olefin (**83**) in deuteriated solvents to olefin, 80% of which was the *trans*-olefin (**87**), and of which over 80% was a monodeuteriated species.

The strong dependence of the product of alkali-metal reduction of 1-phenylbut-1-yne on the nature of the metal has been discussed, but on the basis of the dianion interpretation that is brought in question by the results discussed above.[253] In more conjugated systems, capable of yielding aromatic or homoaromatic species, the evidence for dianions is much more convincing, e.g., the NMR spectra of dianions from alkali-metal reduction of isomeric dimethylcyclo-octatetraenes[254] and *trans*-bicyclo[6.1.0]nona-2,4,6-trienes[255] have been determined. Reduction of quinoline (**89**) by two equivalents of anhydrous liquid ammonia at –33° probably leads to the dianion (**90**), and the monoanion (**91**), which on quenching with NH_4Cl gives 1,4-dihydroquinoline (**93**), *not* 1,2-dihydroquinoline (**92**) as previously accepted.[256] Similar results were noted during reductive alkylation of quinoline.[257]

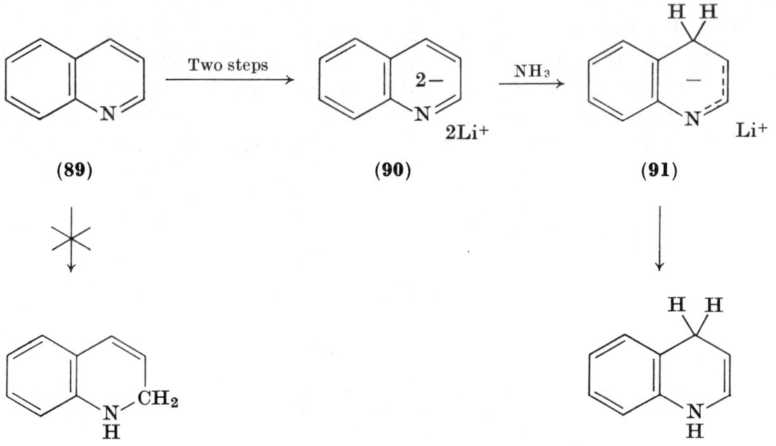

Two steps
2−
2Li+
NH_3
H H
Li+
(**89**) (**90**) (**91**)
CH_2
N H
H H
(**92**) (**93**)

The mechanisms of ring cleavage of acetylcyclopropanes[258] and vinylcyclopropane[259] by metal–ammonia solutions have been investigated. It was found that the ratio of *cis/trans*-stereochemistry of the pent-2-enes, produced from vinylcyclopropane, varied from 80:17 to 35:61 depending on the conditions; the latter ratio was partly due to equilibration of products. It was suggested that the origin of *cis*-stereoselectivity was a solvent-separated ion-pair transition state.[259] The mechanisms of cleavage of 1,4-diketones,[260] and formation of 1,4-diketones (e.g. **95**) from α,α′-dibromo-ketones (e.g. **94**)[261] have been discussed. Further studies of the mechanisms of hydrogenolysis of benzylic esters have been reported.[262]

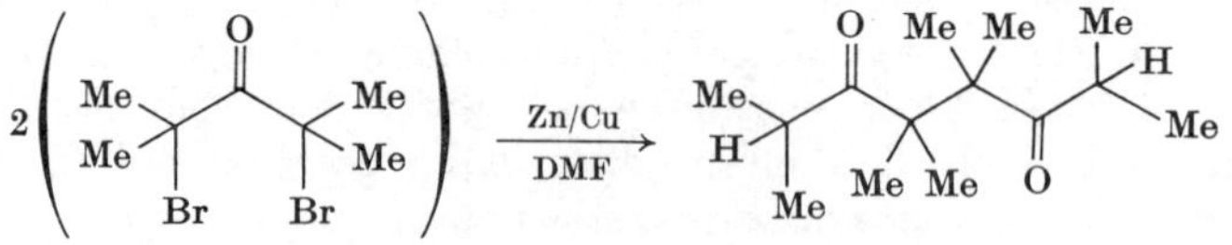

2 Me Me O Me Me Br Br
Zn/Cu DMF
O Me Me Me H Me Me H Me Me Me O
(**94**) (**95**)

Other Reductions

Interest in the organic chemistry of low-valent titanium has increased considerably during the past ten years.[263] Titanocene dichloride effects the reduction of aldehydes, esters and epoxides to alkanes;[264] deuterium-labelling experiments suggest that the reaction involves an intermediate titanium-bound olefin and that the hydrogen arises from the cyclopentadiene ligands attached to the titanium. However, when the reduction of dodecanal was quenched before completion, about 50% of the dodecane contained two atoms of deuterium, while dodec-1-ene and the remaining alkane contained close to natural abundance levels. These results can be accommodated as shown in equation (46).

$$RCH_2CHO \rightarrow RCH{=}CH_2 \rightleftharpoons RCH{=}CH_2{-}[(C_5H_5)_2Ti]_n \rightarrow RCH_2CH_3 \quad (46)$$

In agreement with the suggested route, hydrocarbons are not produced from substrates that would not be expected to form olefins, including aromatic aldehydes and 1-adamantyl derivatives.[264] Aspects of the $TiCl_3$-reductions of enedicarbonyl compounds[265] and nitro-compounds (via the corresponding nitroso-compounds, oximes and imines to ketones)[266] have been discussed. Reduction of benzaldehyde and salicylaldehyde with U(III),[267] and deoxygenation of epoxides with $Co_2(CO)_8$[268] have also been studied.

During column chromatography on alumina with ethanol as solvent, α-ketols were reduced to the corresponding dihydroxy-compounds in a reaction of Meerwein–Ponndorf–Verley type (MPV).[269] Similar MPV reactions for reduction of phenyl trifluoromethyl ketone,[270] and for stereoselective dehydration of β-hydroxy-esters,[271] have been investigated.

N-Alkylnitrilium ions (e.g. **96**) are activated towards nucleophilic attack at carbon and are reduced by $NaBH_4$ to secondary amines, whereas reduction by triethylsilane yields the *N*-alkylaldimine (e.g. **97**).[272] A detailed study of the products and solvent

$$RCN + Et_3\overset{+}{O}\overset{-}{B}F_4 \longrightarrow \underset{\textbf{(96)}}{R{-}\overset{+}{C}{\equiv}NEt \quad BF_4^-} \xrightarrow{Et_3SiH} \underset{\textbf{(97)}}{R{-}CH{=}NEt}$$

effects for reduction of aldehydes and ketones by triethylsilane has also been reported.[273] The "penultimate intermediacy of an ylide-stabilized organozinc carbenoid" has been proposed for the reduction of alicyclic ketones to olefins by chlorotrimethylsilane and zinc in ether.[274]

The results of a kinetic study of the reduction of disulphides by triphenylphosphine can be accommodated by equations (47) and (48).[275] As second-order kinetics are

$$Ph_3P + ArSSAr \rightleftharpoons Ph_3P^+{-}SAr + ArS^- \quad (47)$$

$$Ph_3P^+{-}SAr + H_2O \rightarrow Ph_3PO + ArS^- + 2H^+ \quad (48)$$

observed at both high and low pH the first step may be rate-determining, but the more complex kinetics observed at intermediate pH suggests that reversal of the first step may then be important.[275] Reduction of *N*-(arylsulphonyl)sulphilimines with sodium iodide in aqueous perchloric acid [equation (49)] may proceed by iodide attack on sulphur.[276]

$$R_2S{:}NSO_2Ar + 2I^- + 2H^+ \rightarrow R_2S + I_2 + ArSO_2NH_2 \quad (49)$$

Deoxygenation of *N*-alkyl-α-phenylnitrones by trialkyl phosphites may proceed by nucleophilic attack of the phosphite-P atom on the nitrone-N atom.[277] The effects of substituents on reactivity in the reduction of various nitrobenzenes with benzyl alcohols in the presence of aqueous alkali have been investigated.[278] Attention has also been given to deoxygenation of *N*-oxides induced by alkali[279] or by DMSO (acid-catalysed).[280] There is also a theoretical study of the hydrodimerization of acrolein.[281]

Hydrogenation and Hydrogenolysis[282]

The latest report of the late R. B. Turner's work on heats of hydrogenation presents results for cyclic dienes and trienes.[283] Since thermochemical studies of organic reactions are of fundamental importance and lasting value for the quantitative discussion of both steric and electronic effects, it is unfortunate that relatively few experimental studies are currently carried out in this area.

Treatment of alkali-metal salts of the styrene (**98**) with $RhCl(Ph_3P)_3$ leads to olefin saturation with exclusive production of the *cis*-isomer (**101**).[284] Displacement of both solvent (S) and chloride from the rhodium complex (**99**) is envisaged as leading to the intermediate (**100**), in which hydrogen transfer gives (**101**), whereas heterogeneous hydrogenation leads to appreciable quantities of the *trans*-isomer.[284] Stereochemical aspects of similar reactions have also been discussed.[285]

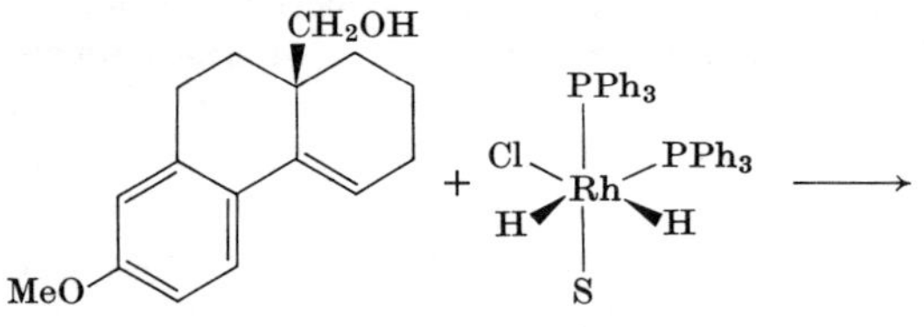

(**98**) (**99**)

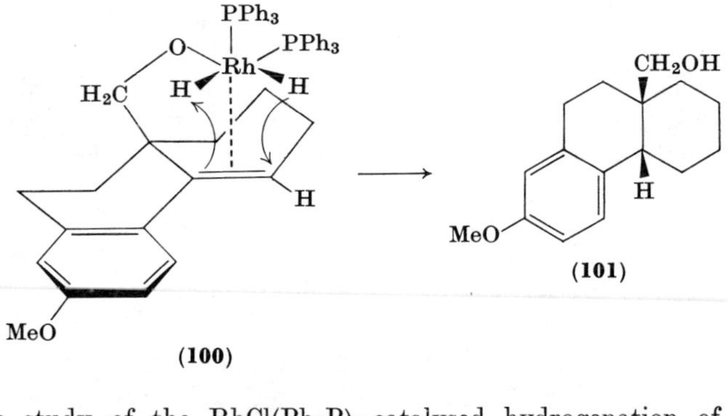

(**100**) (**101**)

After a detailed kinetic study of the $RhCl(Ph_3P)_3$-catalysed hydrogenation of cyclohexene in benzene, it was concluded that a dissociation of the catalyst was slight, that hydrogen added to the undissociated complex but that cyclohexene added to the dissociated complex $RhCl(PPh_3)_2$.[286] A new catalyst, $RhH(DBP)_4$, where DBP is a dibenzophosphole derivative, was found to hydrogenate hex-1-ene seven times more rapidly than $RhCl(PPh_3)_3$.[287]

Interest continues in homogeneous catalysis of hydrogen transfer reactions. A full report of the dioxan/olefin reaction catalysed by $RhCl(PPh_3)_3$ includes kinetic data for octadeuteriodioxan; the isotope effect of about 3 contrasts with reductions by molecular hydrogen in which $k^H/k^D \approx 1.0$.[288] Of the variety of complexes tried, $RhH(PPh_3)_4$ appeared to be the most effective catalyst for hydrogen transfer from alcohols to olefins.[289] Catalysis of hydrogen/deuterium exchange between deuterium and ethanol using $RhCl(PPh_3)_3$,[290] and between primary alcohols and D_2O at 200° using $RuCl_2(PPh_3)_3$ has also been discussed.[291] In other $RuCl_2(PPh_3)_3$ reactions it has been noted that addition of amines affects the rate of reduction of cyclohexanone by propan-2-ol,[292] and that aldehydes inhibit equilibration of alcohols and ketones.[293] Similar reactions of chalcones have also been studied;[294] the complex $Ir(PPh_3)_3H_3$ catalyses hydrogenation of diphenylacetylene to stilbene by hydrogen transfer from ethanol.[295] Hydrogenolysis of carbon tetrachloride can also be achieved, as shown in equation (50).[296]

$$RR'CHOH + CCl_4 \rightarrow RR'CO + CHCl_3 + HCl \qquad (50)$$

The first unequivocal demonstration of catalysis of hydrogenation of aromatic hydrocarbons with a discrete metal complex, η^3-C_3H_5—$Co[P(OCH_3)_3]_3$, has been reported.[297] Hydrogenation by chromium,[298] as well as by other cobalt complexes, have also been investigated.[299] A mixture of molybdenum hexacarbonyl and phenol has been found to be an active and selective homogeneous catalyst for the metathesis of aromatic disubstituted alkynes.[300]

Greater use of catalytic transfer hydrogenation from hydrogen donors such as cyclohexene has been advocated.[301] Whilst little mechanistic work has so far been published, it is clear that the activity of catalyst does not parallel normal heterogeneous hydrogenations and the role of the donor needs to be explored.

A rule of maximum compactness has been put forward as a guide for establishing the stereochemistry of catalytic hydrogenation.[302] Further details, including solvent effects, have been reported of the selective hydrogenations of C—C π-bonds, catalysed by the product from the reaction between $NaBH_4$ and Pd(II).[303] Other aspects of solvent effects[304] and catalyst selectivity[305] have also been studied.

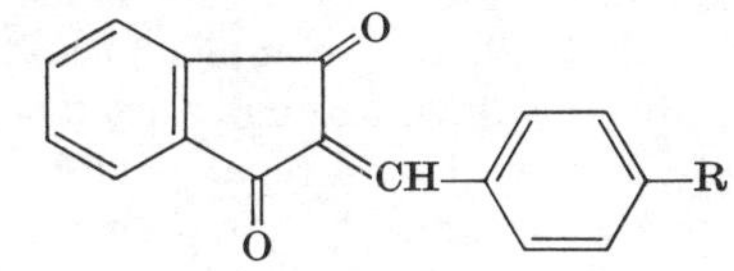

(102)

Surprisingly, the ease of hydrogenation of the indandione (**102**) appears to depend on whether the substituent (R) contains an odd or an even number of carbon atoms.[306] The linear correlation of logarithms of rate constants for catalytic hydrogenation (Pd) of acetophenones with the pH of the medium indicated a rate-limiting electron-transfer process.[307] Electro-catalytic hydrogenation of carbonyl compounds has also received attention,[308] as well as hydrogenations of acetylenes,[309] maleic anhydride,[310] allyl alcohol,[311] benzene[312] and nitrobenzenes.[313]

The mechanisms of deuterium addition to propene over the electron donor–acceptor complex of polynaphthoquinone with potassium have been investigated,[314] and hydrogenations have also been carried out with micro-organism catalysts.[315]

Although it is neither hydrogenation nor hydrogenolysis, the interesting report that oxygen is required for valence isomerization of the bicycloalkene (**103**) to (**104**) and (**105**) catalysed by $RhCl(PPh_3)_3$ is noted here.[316]

As well as the expected products, hydrogenolysis of the nortricyclene derivative (**106**) yields the methyl-shifted product (**107**), which may arise via a metal-bound carbonium

$RhCl(PPh_3)_3$, O_2

MeOOC H COOMe COOMe

(103) **(104)** **(105)**

ion.[317] Research on stereochemical aspects of cyclopropane-ring hydrogenolysis has also been published.[318] Empirical force-field calculations suggest that selective hydrogenolysis of the cyclobutane rings in basketane is thermodynamically controlled.[319]

H_2/Pd on pumice

(106) **(107)**

While hydrogenolysis of strained compounds (e.g. **106**) is common, there has been a recent report of hydrogenolysis of the relatively unstrained carbon—carbon single bonds in cyclohexadienones.[320] Hydrogenolysis of phenols can be achieved via a cyanuric chloride derivative,[321] and related work on three-membered rings,[322] indane,[323] chlorinated *meta*-xylenes,[324] and hydrocarbons[325] has been carried out.

Oxidation and Reduction in Biological Chemistry[326]

There is considerable current interest in the binding of dioxygen to Fe(II) complexes. Only recently has it been shown that reversible complex formation can occur rather than irreversible formation of a complex leading, through autoxidation, to Fe(III). A variety of these models for hemoproteins have been prepared and shown to form a complex reversibly with dioxygen.[327–329] It is thought that their stability is due to reduction of the rate of the bimolecular redox process [equation (51)] by steric hindrance. Less

$$\mathrm{Fe(II)O_2 + Fe(II) \rightarrow Fe(II){-}O{-}O{-}Fe(II) \rightarrow Fe(III)} \qquad (51)$$

hindered systems also show reversible behaviour at low temperatures,[330–332] when reaction (51) is very slow. The suggestion[328] that a neighbouring-group effect of a covalently attached imidazole is a key factor for successful oxygenation has not been substantiated by more recent studies.[330, 333] The complexes appear to be highly dipolar, because their stability is increased in polar, non-co-ordinating solvents.[332, 334] Activation volumes (ΔV^{+})[335] and ESR spectra[336] from studies of the binding of dioxygen to hemoglobin and myglobin have been discussed, and the reactions of Co(II)O_2 complexes with quinol,[337] ferrocytochrome *c* with tris-(1,10-phenanthroline)Co(III),[338] and cytochrome *c*(III) with haxaammineruthenium(II),[339] have also been investigated.

The suggestion[340] that a covalent intermediate was formed in the flavine-catalysed carbonyl-to-carbinol redox reactions has been criticized,[341] and recent results for oxidation of α-ketols are consistent with an electron-transfer mechanism from the 1,2-enediol anion (**108**).[342] The enzymic hydroxylation of flavin mono-oxygenases may involve an oxaziridine as the active oxygenating agent.[343]

(108)

Further evidence against the participation of singlet molecular oxygen in enzymic oxidations has been presented.[344] Studies of isotope effects in hydroxylation of phenethylamine with dopamine β-hydroxylase,[345] of enzymic epoxidation of octadiene,[346] and of hydroxylation of diamantan-1- and -4-ol[347] have been carried out, and the kinetics of oxidation of tryptophan in proteins have been determined.[348] Further research on peroxidase,[349] xanthine oxidase,[350] catalase,[351] lactate oxidase[352] and glutamate dehydrogenase[353] has been reported. There is also a kinetic study of formation of the peroxidic intermediate from deuterioferriheme and hydrogen peroxide.[354]

In continuing work on the electrochemistry of biologically important compounds, polarographic data have been interpreted for 6-substituted purines[355] and for di- and oligo-nucleoside phosphates of adenine and cytosine.[356] One-electron redox reactions of nicotinamide and related compounds were brought about by reaction with hydrated electrons or by electron transfer from acetone ketyl radicals, $(CH_3)_2\dot{C}OH$.[357] Various NADH reductions have also been reported.[358] Since the pH of the solution appears to affect considerably the reduction of acetylene catalysed by molybdenum–thiol complexes,[359] there is now some doubt about the report[360] that ATP is a specific catalyst.

References

[1] F. Hasan and J. Roček, *J. Am. Chem. Soc.*, **96**, 6802 (1974).
[2] F. Hasan and J. Roček, *J. Am. Chem. Soc.*, **96**, 534 (1974).
[3] K. B. Wiberg and S. K. Mukherjee, *J. Am. Chem. Soc.*, **96**, 6647 (1974).
[4] K. B. Wiberg and S. K. Mukherjee, *J. Am. Chem. Soc.*, **96**, 1884 (1974); see also *Org. Reaction Mech.*, **1971**, 530.
[5] K. B. Wiberg and G. Szeimies, *J. Am. Chem. Soc.*, **96**, 1889 (1974).
[6] K. B. Wiberg and P. Lepse, *J. Am. Chem. Soc.*, **86**, 2612 (1964).
[7] M. Doyle, R. J. Swedo and J. Roček, *J. Am. Chem. Soc.*, **95**, 8352 (1973); see also *Org. Reaction Mech.*, **1971**, 530.
[8] J. Roček and C.-S. Ng, *J. Am. Chem. Soc.*, **96**, 1522 (1974).
[9] J. Roček and C.-S. Ng, *J. Am. Chem. Soc.*, **96**, 2840 (1974).
[10] F. Hasan and J. Roček, *Tetrahedron*, **30**, 21 (1974).
[11] F. Hasan and J. Roček, *J. Org. Chem.*, **39**, 2612 (1974).
[12] V. Srinivasan and J. Roček, *J. Am. Chem. Soc.*, **96**, 127 (1974).
[13] K. Kabuto and S. Yamaguchi, *Chem. Letters* (*Tokyo*), **1973**, 29.
[14] J.-C. Richer and J.-M. Hachey, *Can. J. Chem.*, **52**, 2475 (1974).
[15] V. M. Ramanujam, S. S. Sadagopa and N. Venkatasubramanian, *Indian J. Chem.*, **11**, 889 (1973); *Chem. Abs.*, **80**, 36526 (1974).
[16] P. Müller and J.-C. Perlberger, *Helv. Chim. Acta*, **57**, 1943 (1974); K. D. Gupta and D. Guha, *J. Inst. Chem., Calcutta*, **46**, 24 (1974); *Chem. Abs.*, **81**, 104377 (1974).
[17] N. C. Khandual, K. K. Satpathy and P. L. Nayak, *Indian J. Chem.*, **11**, 770 (1973); *Chem. Abs.*, **80**, 36510 (1974); see also P. L. Nayak and N. C. Khandual, *Proc. Indian Acad. Sci., Sect. A*, **78**, 57 (1973); *Chem. Abs.*, **80**, 36528 (1974).
[18] N. C. Khandual, K. K. Satpathy and P. L. Nayak, *J.C.S. Perkin II*, **1974**, 328.
[19] I. Baldea and S. Schoen, *Stud. Univ. Babes-Bolyai, Ser. Chem.*, **18**, 47 (1973); *Chem. Abs.*, **79**, 136207 (1973).
[20] C. R. Eck, D. J. Hunter and T. Money, *Chem. Comm.*, **1974**, 865.

[21] P. Müller and J. Roček, *J. Am. Chem. Soc.*, **96**, 2836 (1974).
[22] J. W. Burnham, W. P. Duncan, E. J. Eisenbraun, G. W. Keen and M. C. Hamming, *J. Org. Chem.*, **39**, 1416 (1974).
[23] S. V. Anantakrishnan and R. Varadarajan, *Indian J. Chem.*, **12**, 373 (1974); *Chem. Abs.*, **81**, 104401 (1974).
[24] A. Granzow, A. Wilson and F. Ramirez, *J. Am. Chem. Soc.*, **96**, 2454 (1974).
[25] H. C. Araújo, G. A. L. Ferreira and J. R. Mahajan, *J.C.S. Perkin I*, **1974**, 2257; see also K. Maeda, I. Moritani, T. Hosokawa and S.-I. Murahashi, *Tetrahedron Letters*, **1974**, 797.
[26] F. Fillmore, *Rev. Reactive Species Chem. React.*, **1**, 37 (1973); *Chem. Abs.*, **79**, 114637 (1973).
[27] J. K. Kochi, *J. Am. Chem. Soc.*, **87**, 3609 (1965); R. A. Sheldon and J. K. Kochi, *Org. Reactions*, **19**, 279 (1970); *Org. Reaction Mech.*, **1965**, 196.
[28] B. C. C. Cantello, J. M. Mellor and G. Scholes, *J.C.S. Perkin II*, **1974**, 348.
[29] A. Thiene and J. G. Traynham, *J. Org. Chem.*, **39**, 153 (1974).
[30] A. L. J. Beckwith, R. T. Cross and G. E. Gream, *Austral. J. Chem.*, **27**, 1673 (1974).
[31] A. L. J. Beckwith, R. T. Cross and G. E. Gream, *Austral. J. Chem.*, **27**, 1693 (1974).
[32] Y. Pocker and B. C. Davis, *Chem. Comm.*, **1974**, 803.
[33] M. Lj. Mihailović, J. Bošnjak and Ž. Čeković, *Helv. Chem. Acta*, **57**, 1015 (1974).
[34] A. Lethbridge, R. O. C. Norman and C. B. Thomas, *J.C.S. Perkin I*, **1974**, 1929.
[35] D. W. Jones and R. L. Wife, *J.C.S. Perkin I*, **1974**, 1.
[36] Y. Ogata, *Kagaku No Ryoiki*, **27**, 941 (1973); *Chem. Abs.*, **80**, 36383 (1974).
[37] D. G. Lee and J. R. Brownridge, *J. Am. Chem. Soc.*, **96**, 5517 (1974); see also *Org. Reaction Mech.*, **1973**, 174.
[38] K. K. Banerji, *Bull. Chem. Soc. Japan*, **46**, 3623 (1973).
[39] P. Nath, K. K. Banerji and G. V. Bakore, *J. Indian Chem. Soc.*, **50**, 30 (1973); *Chem. Abs.*, **79**, 52568 (1973).
[40] S. K. Manjrekar and S. V. Soman, *Indian J. Chem.*, **12**, 307 (1974); *Chem. Abs.*, **81**, 119550 (1974).
[41] S. K. Manjrekar, S. Rajagopalan and S. V. Soman, *Indian J. Chem.*, **11**, 564 (1973); *Chem. Abs.*, **79**, 114847 (1973).
[42] P. V. Subba Rao, *Z. Phys. Chem.* (*Leipzig*), **255**, 382 (1974).
[43] M. S. Murdia, R. Shanker and G. V. Bakore, *Z. Naturforsch.*, **29b**, 691 (1974).
[44] N. C. Khandual and P. L. Nayak, *J. Inst. Chem., Calcutta*, **45**, 95 (1973); *Chem. Abs.*, **80**, 47196 (1974); see also P. L. Nayak and N. C. Khandual, *Proc. Indian Acad. Sci., Sect. A*, **79**, 33 (1974); *Chem. Abs.*, **80**, 145117 (1974).
[45] R. J. Bertsch and R. J. Ouellette, *J. Org. Chem.*, **39**, 2755 (1974); see also J. Grignon and S. Fliszár, *Can. J. Chem.*, **52**, 3209 (1974).
[46] A. Lethbridge, R. O. C. Norman and C. B. Thomas, *J.C.S. Perkin I*, **1973**, 2763; see also ref. 34.
[47] L. Nadon and M. Zador, *Can. J. Chem.*, **52**, 2667 (1974); see also *Org. Reaction Mech.*, **1973**, 176.
[48] P. V. Subramanian, R. Paul and V. Subrahmanyan, *Indian J. Chem.*, **11**, 1074 (1973); *Chem. Abs.*, **80**, 81738 (1974); see also ref. 17.
[49] E. Pelizzetti, E. Mentasti and G. Saini, *J.C.S. Dalton*, **1974**, 721.
[50] W. S. Trahanovsky, J. Cramer and D. W. Brixius, *J. Am. Chem. Soc.*, **96**, 1077 (1974).
[51] C. Briguet, C. Freppel, J.-C. Richer and M. Zador, *Can. J. Chem.*, **52**, 3201 (1974).
[52] R. O. C. Norman, C. B. Thomas and P. J. Ward, *J.C.S. Perkin I*, **1973**, 2914.
[53] P. K. Saiprakash and B. Seethuram, *J. Indian Chem. Soc.*, **50**, 211 (1973); *Chem. Abs.*, **79**, 77768 (1973).
[54] A. K. Wadhawan, P. S. Sankhla and R. N. Mehrotra, *Indian J. Chem.*, **11**, 567 (1973); *Chem. Abs.*, **79**, 114846 (1973).
[55] R. N. Mehrotra and E. S. Amis, *J. Org. Chem.*, **39**, 1788 (1974).
[56] B. C. Singh, P. L. Nayak and M. K. Rout, *Indian J. Chem.*, **12**, 421 (1974); *Chem. Abs.*, **81**, 119587 (1974).
[57] H. M. Singh and B. P. Gyani, *Proc. Nat. Acad. Sci., India, Sect. A*, **52**, 191 (1972); *Chem. Abs.*, **80**, 47202 (1974).
[58] P. S. Radhakrishnamurti and M. K. Mahanti, *Indian J. Chem.*, **11**, 762 (1973); *Chem. Abs.*, **79**, 145700 (1973).
[59] P. S. Radhakrishnamurti and S. Devi, *Indian J. Chem.*, **11**, 768 (1973); *Chem. Abs.*, **79**, 145701 (1973).
[60] P. S. Radhakrishnamurti and S. Devi, *Curr. Sci.*, **42**, 788 (1973); *Chem. Abs.*, **80**, 36515 (1974).
[61] K. Bhatia and R. H. Schuler, *J. Phys. Chem.*, **78**, 2335 (1974).
[62] P. S. Radhakrishnamurti and R. K. Panda, *Indian J. Chem.*, **11**, 1003 (1973); *Chem. Abs.*, **80**, 81729 (1974).

[63] K. Rehse and L. Bergen, *Arch. Pharm.*, **307**, 448 (1974).
[64] I. I. Moiseev, O. G. Levanda and M. N. Vargaftik, *J. Am. Chem. Soc.*, **96**, 1003 (1974).
[65] *Org. Reaction Mech.*, **1966**, 404.
[66] K. Kikukawa, K. Sakai, K. Asada and T. Matsuda, *J. Organometal. Chem.*, **77**, 131 (1974).
[67] E. Bratz, G. Prauser and K. Dialer, *Chem.-Ing.-Tech.*, **46**, 161 (1974); *Chem. Abs.*, **80**, 145120 (1974).
[68] R. N. Pandey and P. M. Henry, *Can. J. Chem.*, **52**, 1241 (1974).
[69] F. R. S. Clark, R. O. C. Norman, C. B. Thomas and J. S. Wilson, *J.C.S. Perkin I*, **1974**, 1289.
[70] R. Kawai and Y. Kamiya, *Nippon Kagaku Kaishi*, **1973**, 1538; **1974**, 933; *Chem. Abs.*, **79**, 125549 (1973); **81**, 24736 (1974); V. N. Sapunov and L. Abdennur, *Kinet. Katal.*, **15**, 30 (1974); *Chem. Abs.*, **80**, 119911 (1974).
[71] R. Kawai and Y. Kamiya, *Nippon Kagaku Kaishi*, **1973**, 1533; *Chem. Abs.*, **79**, 125548 (1973).
[72] R. B. Giuntoli and H. S. Habib, *J. Inorg. Nucl. Chem.*, **36**, 363 (1974); *Chem. Abs.*, **81**, 3123 (1974).
[73] M. Fetizon and P. Mourgues, *Tetrahedron*, **30**, 327 (1974); F. J. Kakis, M. Fetizon, N. Douchkine, M. Golfier, P. Mourgues and T. Prange, *J. Org. Chem.*, **39**, 523 (1974).
[74] C. Carpanelli, G. Leandri and G. Poluzzi Corallo, *Gazz. Chim. Ital.*, **103**, 469 (1973); *Chem. Abs.*, **80**, 47201 (1974).
[75] J. C. Sheehan and R. W. Tulis, *J. Org. Chem.*, **39**, 2264 (1974).
[76] U. A. Spitzer and D. G. Lee, *J. Org. Chem.*, **39**, 2468 (1974).
[77] F. Domka and B. Marciniec, *Ann. Chim. (Paris)*, **7**, 315 (1972); *Chem. Abs.*, **78**, 128892 (1973).
[78] A. Kumar and R. N. Mehrotra, *Int. J. Chem. Kinetics*, **6**, 15 (1974).
[79] J. Garnett and J. C. West, *Austral. J. Chem.*, **27**, 129 (1974).
[80] J. Scheve and E. Scheve, *Z. Chem.*, **14**, 172 (1974); L. Ya. Margolis, *Catal. Rev.*, **8**, 241 (1973); *Chem. Abs.*, **79**, 125343 (1973).
[81] K. J. Skinner, H. S. Hochster and J. M. McBride, *J. Am. Chem. Soc.*, **96**, 4301 (1974).
[82] W. Parker, R. L. Tranter, C. I. F. Watt, L. W. K. Chang and P. von R. Schleyer, *J. Am. Chem. Soc.*, **96**, 7121 (1974).
[83] P. A. Brady and J. Carnduff, *Chem. Comm.*, **1974**, 816.
[84] D. E. Van Sickle, T. Mill, F. R. Mayo, H. Richardson and C. W. Gould, *J. Org. Chem.*, **38**, 4435 (1973).
[85] F. R. Mayo, P. S. Fredericks, T. Mill, J. K. Castleman and T. Delany, *J. Org. Chem.*, **39**, 885 (1974).
[86] F. R. Mayo, J. K. Castleman, T. Mill, R. M. Silverstein and O. Rodin, *J. Org. Chem.*, **39**, 889 (1974).
[87] E. Niki, K. Okayasu and Y. Kamiya, *Int. J. Chem. Kinetics*, **6**, 279 (1974).
[88] C. W. Dudley, G. Read and P. J. C. Walker, *J.C.S. Dalton*, **1974**, 1926.
[89] A. A. Blanc, H. Arzoumanian, E. J. Vincent and J. Metzger, *Bull. Soc. Chim. France*, **1974**, 2175.
[90] H. Arzoumanian, A. A. Blanc, U. Hartig and J. Metzger, *Tetrahedron Letters*, **1974**, 1011.
[91] K. A. Chervinskii, V. F. Vasil'ev and P. A. Sukhopar, *Zh. Prikl. Khim. (Leningrad)*, **47**, 1621 (1974); *Chem. Abs.*, **81**, 90814 (1974); see also V. N. Sapunov, N. G. Digurov, E. F. Selyutina, L. K. Zolotareva and N. N. Lebedev, *Kinet. Katal.*, **15**, 610 (1974); *Chem. Abs.*, **81**, 104387 (1974); V. N. Sapunov, E. F. Selyutina, O. S. Tolchinskaya and N. N. Lebedev, *Kinet. Katal.*, **15**, 605 (1974); *Chem. Abs.*, **81**, 104385 (1974).
[92] S. K. Ivanov and C. Karshalykov, *Annalen*, **1974**, 1713.
[93] M. N. Manakov, V. L. Razin and V. A. Kruchinin, *Neftekhimiya*, **13**, 556 (1973); *Chem. Abs.*, **79**, 136205 (1973).
[94] H. Yoshimoto and H. Itatani, *J. Catal.*, **31**, 8 (1973); *Chem. Abs.*, **79**, 114742 (1973).
[95] L. M. Kaliberdo, F. A. Mil'man and A. S. Vaabel, *Khim. Aromat. Nepredel'n. Soedin*, **1972**, 187; *Chem. Abs.*, **79**, 125560 (1973); L. M. Kaliberdo and V. B. Dorogova, *Khim. Aromat. Nepredel'n. Soedin.*, **1971**, 199; *Chem. Abs.*, **79**, 125559 (1973); L. M. Kaliberdo, V. B. Dorogova and V. V. Chenets, *Khim. Aromat. Nepredel'n. Soedin.*, **1971**, 207; *Chem. Abs.*, **79**, 125558 (1973).
[96] B. Grzybowska and J. Haber, *Wiad. Chem.*, **27**, 825 (1973); *Chem. Abs.*, **80**, 119715 (1974).
[97] G. N. Koshel, M. I. Farberov, T. N. Antonova, L. V. Bedareva, N. G. Vasil'ev and L. V. Ob'edkova, *Neftekhimiya*, **14**, 263 (1974); *Chem. Abs.*, **81**, 24740 (1974).
[98] J. E. Taylor and J. C. Weygandt, *Can. J. Chem.*, **52**, 1925 (1974); L. V. Novaselova, G. S. Popova, V. G. Babel and V. A. Proskuryakov, *Zh. Prikl. Khim. (Leningrad)*, **47**, 641 (1974); *Chem. Abs.*, **80**, 145099 (1974); B. J. Chernyak and A. P. Babii, *Dopv. Akad. Nauk Ukr. RSR, Ser. B*, **35**, 750 (1973); *Chem. Abs.*, **79**, 145674 (1973).
[99] T. Osa, Y. Onkatsu and M. Tezuka, *Chem. Letters (Tokyo)*, **1973**, 99; *Chem. Abs.*, **78**, 110234 (1973).
[100] P. H. Bolton and D. R. Kearns, *J. Am. Chem. Soc.*, **96**, 4651 (1974).
[101] R. Addink and W. Berends, *Tetrahedron*, **30**, 75 (1974).

[102] J. A. Blair and A. J. Pearson, *J.C.S. Perkin II*, **1974**, 80; A. J. Pearson, *Chem. Ind.* (*London*), **1974**, 233; see also *Org. Reaction Mech.*, **1973**, 131.

[103] A. Nishinaga, T. Itahara and T. Matsuura, *Bull. Chem. Soc. Japan*, **47**, 1811 (1974).

[104] Y. Ogata, K. Nate and M. Yamashita, *Bull. Chem. Soc. Japan*, **47**, 174 (1974).

[105] J. P. Saxena and R. Kulshreshtha, *J. Indian Chem. Soc.*, **50**, 654 (1973); *Chem. Abs.*, **80**, 119917 (1974).

[106] T. Vidóczy, E. Danóczy and D. Gál, *J. Phys. Chem.*, **78**, 828 (1974).

[107] D. A. Mayers and J. Kagan, *J. Org. Chem.*, **39**, 3147 (1974).

[108] K. Bühler, W. Diesch and A. K. Ghosh, *Can. J. Chem.*, **52**, 1700 (1974).

[109] S. Miyajima, T. Inukai, H. Harada, R. Yoshizawa, T. Harai, K. Matsunaga and M. Harada, *Bull. Chem. Soc. Japan*, **47**, 2051 (1974).

[110] T. Hara, Y. Ohkatsu and T. Osa, *Bull. Chem. Soc. Japan*, **47**, 156 (1974); D. J. Dixon, G. Skirrow and C. F. H. Tipper, *J.C.S. Faraday I*, **70**, 1078, 1090 (1974); E. A. Oganesyan, I. A. Vardanyan and A. B. Nalbandyan, *Dokl. Akad. Nauk SSSR*, **212**, 153 (1973); *Chem. Abs.*, **79**, 136228 (1973).

[111] R. Caputo, L. Previtera, P. Monaco and L. Mangoni, *Tetrahedron*, **30**, 963 (1974).

[112] I. I. Rif, V. M. Potekhin and V. A. Proskuryakov, *Zh. Prikl. Khim.* (*Leningrad*), **46**, 1860 (1973); *Chem. Abs.*, **79**, 136225 (1973).

[113] A. V. Savitskii, *Zh. Obshch. Khim.*, **44**, 1548 (1974); *Chem. Abs.*, **81**, 119549 (1974); K. Ukegawa and Y. Kamiya, *Nippon Kagaku Kaishi*, **1974**, 753; *Chem. Abs.*, **81**, 104396 (1974).

[114] G. A. Kovtun, V. A. Golubev and A. L. Aleksandrov, *Izv. Akad. Nauk SSSR, Ser. Khim.*, **1974**, 793; *Chem. Abs.*, **81**, 37010 (1974).

[115] T. I. Sapacheva, G. G. Agliullina and A. L. Aleksandrov, *Neftekhimiya*, **14**, 450 (1974); *Chem. Abs.*, **81**, 104405 (1974).

[116] T. E. Young, J. R. Griswold and M. H. Hulbert, *J. Org. Chem.*, **39**, 1980 (1974); C. L. Deasy, L. H. Walling and Sr. A. Alexander, *J. Org. Chem.*, **39**, 1429 (1974).

[117] M. Stewart and C. H. Nicholls, *Austral. J. Chem.*, **27**, 205 (1974).

[118] I. G. Dance, R. C. Conrad and J. E. Cline, *Chem. Comm.*, **1974**, 13; H. P. Misra, *J. Biol. Chem.*, **249**, 2151 (1974).

[119] V. M. Fomin, Yu. A. Alexandrov and V. A. Umilin, *J. Organometal. Chem.*, **61**, 267 (1973); A. Peloso, *ibid.*, **74**, 59 (1974); P. B. Brindley and J. C. Hodgson, *ibid.*, **65**, 57 (1974).

[120] A. Nishinaga, T. Tojo and T. Matsuura, *Chem. Comm.*, **1974**, 896; A. Nishinaga, K. Watanabe and T. Matsuura, *Tetrahedron Letters*, **1974**, 1291; see also *Org. Reaction Mech.*, **1973**, 178.

[121] E. Boga and F. Marta, *Acta Chim.* (*Budapest*), **78**, 105, 193 (1973); **80**, 333 (1973); *Chem. Abs.*, **80**, 26562 (1974); **79**, 145673 (1973); **81**, 90826 (1974); E. Boga, I. Kiricsi, A. Deer and F. Marta, *Acta Chim.* (*Budapest*), **78**, 75, 89 (1973); *Chem. Abs.*, **80**, 2869, 26553 (1974).

[122] M. Kashima and Y. Kamiya, *Bull. Chem. Soc. Japan*, **47**, 481 (1974); Y. Kamiya and M. Kashima, *ibid.*, **46**, 905 (1973).

[123] G. S. Hammond and C.-H. S. Wu, *J. Am. Chem. Soc.*, **95**, 8215 (1973).

[124] A. M. H. P. Van den Besselaar, J. Lubach and W. Drenth, *Rec. Trav. chim.*, **93**, 108 (1974).

[125] J. Tsuji and H. Takayanagi, *J. Am. Chem. Soc.*, **96**, 7349 (1974).

[126] L. G. Fedenok, V. M. Berdnikov and M. S. Shvartsberg, *Zh. Org. Khim.*, **10**, 922 (1974); *Chem. Abs.*, **81**, 63001 (1974).

[127] H. B. Henbest and J. Trocha-Grimshaw, *J.C.S. Perkin I*, **1974**, 607.

[128] T. Hara, Y. Ohkatsu and T. Osa, *Chem. Letters*, **1973**, 953; *Chem. Abs.*, **79**, 125550 (1973); C. V. Gurumurthy, *J. Appl. Chem. Biotechnol.*, **23**, 769 (1973); *Chem. Abs.*, **80**, 107695 (1974); N. N. Lebedev, M. N. Manakov and A. P. Litovka, *Kinet. Katal.*, **15**, 791 (1974); *Chem. Abs.*, **81**, 140382 (1974).

[129] P. Koelewijn and H. Berger, *Rec. Trav. chim.*, **93**, 63 (1974); F. Kerek and L. Both, *Bul. Stiint. Teh. Inst. Politeh. Timisoara, Ser. Chim.*, **18**, 71 (1973); *Chem. Abs.*, **81**, 119570 (1974); K. J. Humphris and G. Scott, *J.C.S. Perkin II*, **1974**, 617; A. B. Gagarina, L. M. Pisarenko and N. M. Emanuel, *Dokl. Akad. Nauk SSSR*, **212**, 653 (1973); *Chem. Abs.*, **79**, 145685 (1973).

[130] T. Miyata, T. Hirashima and O. Manabe, *Kagaku To Kogyo* (*Osaka*), **47**, 150 (1973); *Chem. Abs.*, **79**, 136219 (1973); see also A. J. Peicheva, A. M. Nikolaevskii and T. A. Filippenko, *Dopov. Akad. Nauk Ukr. RSR, Ser. B*, **36**, 345 (1974); *Chem. Abs.*, **81**, 90796 (1974); and O. Brauer, P. Köll and J. Voss, *Tetrahedron Letters*, **1974**, 2311.

[131] P. R. Story, R. W. Murray and R. D. Youssefyeh, *J. Am. Chem. Soc.*, **88**, 3144 (1966); R. W. Murray, R. D. Youssefyeh and P. R. Story, *ibid.*, **89**, 2429 (1967).

[132] P. R. Story, C. E. Bishop, J. R. Burgess, R. W. Murray and R. D. Youssefyeh, *J. Am. Chem. Soc.*, **90**, 1907 (1968).

[133] C. W. Gillies, R. P. Lattimer and R. L. Kuczkowski, *J. Am. Chem. Soc.*, **96**, 1536 (1974).

[134] R. P. Lattimer and R. L. Kuczkowski, *J. Am. Chem. Soc.*, **96**, 6205 (1974).

[135] R. P. Lattimer, R. L. Kuczkowski and C. W. Gillies, *J. Am. Chem. Soc.*, **96**, 348 (1974).

[136] R. A. Rouse, *J. Am. Chem. Soc.*, **96**, 5095 (1974).

[137] P. S. Bailey, J. W. Ward, T. P. Carter, Jr., E. Nieh, C. M. Fischer and A.-I. Y. Khashab, *J. Am. Chem. Soc.*, **96**, 6136 (1974).

[138] R. Criegee and A. Banciu, *Chem.-Ztg.*, **98**, 261 (1974); *Chem. Abs.*, **81**, 37018 (1974).

[139] D. P. Higley and R. W. Murray, *J. Am. Chem. Soc.*, **96**, 3330 (1974).

[140] S. D. Razumovskii, L. V. Berezova and G. E. Zaikov, *Izv. Akad. Nauk SSSR, Ser. Khim.*, **1974**, 223; *Chem. Abs.*, **80**, 107792 (1974).

[141] D. H. Stedman, C. H. Wu and H. Niki, *J. Phys. Chem.*, **77**, 2511 (1973); B. J. Finlayson, J. N. Pitts, Jr. and R. Atkinson, *J. Am. Chem. Soc.*, **96**, 5356 (1974).

[142] N. C. Yang and J. Libman, *J. Org. Chem.*, **39**, 1782 (1974).

[143] J. K. Crandall, S. A. Sojka and J. B. Komin, *J. Org. Chem.*, **39**, 2172 (1974).

[144] J. K. Crandall and L. C. Crawley, *J. Org. Chem.*, **39**, 489 (1974).

[145] H. Varkony, S. Pass and Y. Mazur, *Chem. Comm.*, **1974**, 437.

[146] T. M. Hellman and G. A. Hamilton, *J. Am. Chem. Soc.*, **96**, 1530 (1974); see also S. D. Razumovskii and G. E. Zaikov, *Dokl. Akad. Nauk SSSR*, **212**, 676 (1973); *Chem. Abs.*, **79**, 145675 (1973).

[147] F. E. Stary, D. E. Emge and R. W. Murray, *J. Am. Chem. Soc.*, **96**, 5671 (1974).

[148] P. Deslongchamps, P. Atlani, D. Fréhel, A. Malaval and C. Moreau, *Can. J. Chem.*, **52**, 3651 (1974).

[149] J. Rebek, Jr., S. F. Wolf and A. B. Mossman, *Chem. Comm.*, **1974**, 711.

[150] H. Kropf and M. R. Yazdanbachsch, *Tetrahedron*, **30**, 3455 (1974).

[151] P. J. Kocienski and J. Ciabattoni, *J. Org. Chem.*, **39**, 388 (1974); L. E. Friedrich and R. A. Fiato, *J. Org. Chem.*, **39**, 2267 (1974); *J. Am. Chem. Soc.*, **96**, 5783 (1974).

[152] A. Sevin and J.-M. Cense, *Bull. Soc. Chim. France*, **1974**, 963.

[153] L. E. Friedrich and R. A. Fiato, *J. Org. Chem.*, **39**, 416 (1974).

[154] J. K. Crandall, W. W. Conover, J. B. Komin and W. H. Machleder, *J. Org. Chem.*, **39**, 1723 (1974).

[155] W. H. Richardson, V. F. Hodge, D. L. Stiggall, M. B. Yelvington and F. C. Montgomery, *J. Am. Chem. Soc.*, **96**, 6652 (1974).

[156] T. Greigbrokk, *Acta Chem. Scand.*, **27**, 3365 (1973).

[157] Y. Ogata, Y. Sawaki and M. Furuta, *J. Org. Chem.*, **39**, 216 (1974).

[158] A. DeBoer and R. E. Ellwanger, *J. Org. Chem.*, **39**, 77 (1974).

[159] T. Momose, S. Atarashi and O. Muraoka, *Tetrahedron Letters*, **1974**, 3697.

[160] V. A. Stoute, M. A. Winnik and I. G. Csizmadia, *J. Am. Chem. Soc.*, **96**, 6388 (1974).

[161] J. Salaun, B. Garnier and J. M. Conia, *Tetrahedron*, **30**, 1423 (1974).

[162] J. A. Cella, J. A. Kelley and E. F. Kenehan, *Chem. Comm.*, **1974**, 943.

[163] R. A. G. Marshall and R. Naylor, *J.C.S. Perkin II*, **1974**, 1242.

[164] Y. Ogata, I. Urasaki, K. Nagura and N. Satomi, *Tetrahedron*, **30**, 3021 (1974).

[165] A. Battaglia, A. Dondoni, G. Maccagnani and G. Mazzanti, *J.C.S. Perkin II*, **1974**, 609.

[166] K. M. Ibne-Rasa, J. O. Edwards, M. T. Kost and A. R. Gallopo, *Chem. Ind.* (*London*), **1974**, 964.

[167] H. Sakurai and Y. Kamiyama, *J. Am. Chem. Soc.*, **96**, 6192 (1974).

[168] K. B. Sharpless and R. C. Michaelson, *J. Am. Chem. Soc.*, **95**, 6136 (1973); S. Tanaka, H. Yamamoto, H. Nozaki, K. B. Sharpless, R. C. Michaelson and J. D. Cutting, *J. Am. Chem. Soc.*, **96**, 5254 (1974); see also *Org. Reaction Mech.*, **1973**, 387.

[169] S. A. Matlin and P. G. Sammes, *J.C.S. Perkin I*, **1973**, 2851.

[170] A. A. Achrem, T. A. Timoschtschuk and D. I. Metelitza, *Tetrahedron*, **30**, 3165 (1974).

[171] R. Curci, F. Di Furia, R. Testi and G. Modena, *J.C.S. Perkin II*, **1974**, 752.

[172] N. A. Johnson and E. S. Gould, *J. Org. Chem.*, **39**, 407 (1974).

[173] J. T. Groves and M. Van Der Puy, *J. Am. Chem. Soc.*, **96**, 5274 (1974).

[174] M. Cocivera, C. A. Fyfe, S. P. Vaish and H. E. Chen, *J. Am. Chem. Soc.*, **96**, 1611 (1974); cf. R. James and F. Sicilio, *J. Phys. Chem.*, **74**, 1166 (1970).

[175] E. V. Shtamm and Yu. I. Skurlatov, *Zh. Fiz. Khim.*, **48**, 1857 (1974); *Chem. Abs.*, **81**, 119551 (1974).

[176] V. F. Romanov and G. A. Konishevskaya, *Ukr. Khim. Zh.* (*Russ. Ed.*), **40**, 689 (1974); *Chem. Abs.*, **81**, 119577 (1974).

[177] T. S. Briggs and W. C. Rauscher, *J. Chem. Educ.*, **50**, 496 (1973); see also J. N. Demas and D. Diemente, *ibid.*, p. 357.

[178] R. J. Field, E. Körös and R. M. Noyes, *J. Am. Chem. Soc.*, **94**, 8649 (1972); see also R. M. Noyes and R. J. Field, *Ann. Rev. Phys. Chem.*, **25**, 95 (1974).

[179] Z. Varadi and M. T. Beck, *Chem. Comm.*, **1973**, 30.

[180] E. Körös, M. Orbán and Zs. Nagy, *J. Phys. Chem.*, **77**, 3122 (1973).

[181] D. Janjic, P. Stroot and U. Burger, *Helv. Chim. Acta*, **57**, 266 (1974).
[182] Y. Ogata and K. Nagura, *J. Org. Chem.*, **39**, 394 (1974).
[183] S. K. Chakrabartty and H. O. Kretschmer, *J.C.S. Perkin I*, **1974**, 222.
[184] Y. Ogata and K. Nagura, *J.C.S. Perkin II*, **1974**, 1089.
[185] P. V. Subba Rao, *Z. Phys. Chem.* (*Leipzig*), **255**, 199 (1974).
[186] P. Gupta and K. C. Grover, *J. Indian Chem. Soc.*, **50**, 397 (1973); *Chem. Abs.*, **80**, 36517 (1974).
[187] S. D. Paul and D. G. Pradhan, *Indian J. Chem.*, **11**, 1155 (1973); *Chem. Abs.*, **80**, 132486 (1974).
[188] R. Natarajan and N. Venkatasubramanian, *Tetrahedron*, **30**, 2785 (1974).
[189] R. Sanehi, M. C. Agrawal and S. P. Mushran, *Indian J. Chem.*, **12**, 311 (1974); *Chem. Abs.*, **81**, 119578 (1974).
[190] R. Sanehi, M. C. Agrawal and S. P. Mushran, *Z. Phys. Chem.* (*Leipzig*), **255**, 293 (1974).
[191] A. K. Bose, R. Sanehi and S. P. Mushran, *J. Indian Chem. Soc.*, **50**, 197 (1973); *Chem. Abs.*, **79**, 77767 (1973).
[192] D. S. Mahadevappa and H. M. K. Naidu, *Austral. J. Chem.*, **27**, 1203 (1974).
[193] N. C. Deno, D. G. Pohl and H. J. Spinelli, *Bioorganic Chemistry*, **3**, 66 (1974); see also C. V. Smith and W. E. Billups, *J. Am. Chem. Soc.*, **96**, 4307 (1974), and *Org. Reaction Mech.*, **1971**, 297.
[194] N. C. Deno and D. G. Pohl, *J. Am. Chem. Soc.*, **96**, 6680 (1974).
[195] See also J. R. L. Smith, R. O. C. Norman and A. G. Rowley, *J.C.S. Perkin I*, **1973**, 566; *Org. Reaction Mech.*, **1973**, 138.
[196] J. P. McCormick, *Tetrahedron Letters*, **1974**, 1701.
[197] J. P. Marino and A. Schwartz, *Chem. Comm.*, **1974**, 812.
[198] J. D. Albright, *J. Org. Chem.*, **39**, 1977 (1974).
[199] E. J. Corey and C. U. Kim, *Tetrahedron Letters*, **1974**, 287; J. Defaye and A. Gadelle, *Carbohydrate Res.*, **35**, 264 (1974).
[200] T. M. Santosusso and D. Swern, *Tetrahedron Letters*, **1974**, 4255.
[201] B. Ganen and R. K. Boeckman, Jr., *Tetrahedron Letters*, **1974**, 917.
[202] N. Bosworth, P. Magnus and R. Moore, *J.C.S. Perkin I*, **1973**, 2694; see also *Org. Reaction Mech.*, **1972**, 550.
[203] E. S. Rudakov, V. M. Mastikhin, S. G. Popov and R. I. Rudakova, *Org. Magnet. Res.*, **5**, 343 (1973); *Chem. Abs.*, **79**, 91340 (1973).
[204] H. J. Reich, J. M. Renga and I. L. Reich, *J. Org. Chem.*, **39**, 2133 (1974).
[205] P. A. Grieco and M. Miyashita, *J. Org. Chem.*, **39**, 120 (1974); see also *Org. Reaction Mech.*, **1973**, 186.
[206] D. L. J. Clive, *Chem. Comm.*, **1974**, 100.
[207] D. N. Brattesani and C. H. Heathcock, *Tetrahedron Letters*, **1974**, 2279.
[208] T. H. Chan and J. R. Finkenbine, *Tetrahedron Letters*, **1974**, 2091.
[209] K. Kondo, N. Sonoda and H. Sakurai, *Tetrahedron Letters*, **1974**, 803.
[210] K. Kondo, N. Sonoda and H. Sakurai, *Chem. Comm.*, **1974**, 160.
[211] L. Eberson and L. Jönsson, *Chem. Comm.*, **1974**, 885.
[212] A. Ledwith and P. J. Russell, *J.C.S. Perkin II*, **1974**, 582.
[213] M. E. Snook and G. A. Hamilton, *J. Am. Chem. Soc.*, **96**, 860 (1974).
[214] N. C. Khandual and P. L. Nayak, *J. Indian Chem. Soc.*, **50**, 194 (1973); *Chem. Abs.*, **79**, 77858 (1973); see also *Org. Reaction Mech.*, **1972**, 547.
[215] V. P. Kudesia, *Rev. Roum. Chim.*, **18**, 1373 (1973); *Chem. Abs.*, **79**, 145699 (1973).
[216] S. P. Srivastava, G. L. Maheshwari and S. K. Singhal, *Indian J. Chem.*, **12**, 72 (1974); *Chem. Abs.*, **81**, 24739 (1974).
[217] P. Maruthamuthu and M. Santappa, *Indian J. Chem.*, **12**, 424 (1974); *Chem. Abs.*, **81**, 10402 (1974).
[218] R. Curci, G. Delano, F. DiFuria, J. O. Edwards and A. R. Gallopo, *J. Org. Chem.*, **39**, 3020 (1974).
[219] S. Wasif and W. C. Vasudeva, *Libyan J. Sci.*, **3**, 25 (1973); *Chem. Abs.*, **81**, 24737 (1974).
[220] A. K. Bhattacharya and V. Joshi, *Proc. Nat. Acad. Sci., India, Sect. A*, **43**, 87 (1973); *Chem. Abs.*, **81**, 37026 (1974); L. G. Melik-Ogandzhanyan, N. M. Beileryan and O. A. Chaltykyan, *Kinet. Katal.*, **15**, 569 (1974); *Chem. Abs.*, **81**, 104386 (1974); R. C. Gupta, L. D. Sharma and S. P. Srivastava, *Z. Phys. Chem.* (*Leipzig*), **255**, 317 (1974).
[221] R. C. Moschel and E. J. Behrman, *J. Org. Chem.*, **39**, 1983, 2699 (1974).
[222] G. Prota and G. Curro, *Tetrahedron*, **30**, 3627 (1974).
[223] K. Vaidyanathan and N. Venkatasubramanian, *Indian J. Chem.*, **11**, 1146 (1973); *Chem. Abs.*, **81**, 12839 (1974).
[224] L. K. Dyall, *Austral. J. Chem.*, **26**, 2665 (1973).
[225] K. Swaminathan, K. Vaidynathan and N. Venkatasubramanian, *Int. J. Chem. Kinetics*, **6**, 773 (1974).
[226] T. R. Beebe, A. L. Lin and R. D. Miller, *J. Org. Chem.*, **39**, 722 (1974).

[227] N. S. Srinivasan and N. Venkatasubramanian, *Tetrahedron.*, **30**, 419 (1974).
[228] H. Schlude, *Tetrahedron*, **29**, 4007 (1973).
[229] T. A. B. M. Bolsman and T. J. de Boer, *Tetrahedron*, **29**, 3929 (1973).
[230] C. B. Gill, S. Hawkins and P. H. Gore, *Chem. Comm.*, **1974**, 742.
[231] R. G. Harvey and H. Cho, *J. Am. Chem. Soc.*, **96**, 2434 (1974).
[232] K. Imafuku, J. Oda, K. Itoh and H. Matsumura, *Bull. Chem. Soc. Japan*, **47**, 1201 (1974).
[233] W. P. Weber, P. Stromquist and T. I. Ho, *Tetrahedron Letters*, **1974**, 2595.
[234] J. M. Photis and L. A. Paquette, *J. Am. Chem. Soc.*, **96**, 4715 (1974).
[235] J. L. Pierre and H. Handel, *Tetrahedron Letters*, **1974**, 2317.
[236] B. Caro and G. Jaouen, *Tetrahedron Letters*, **1974**, 3539; M. J. Brienne, D. Varech and J. Jacques, *ibid.*, p. 1233; D. C. Wigfield and D. J. Phelps, *J. Am. Chem. Soc.*, **96**, 543 (1974); O. Štrouf, J. Fusek and O. Červinka, *Coll. Czech. Chem. Comm.*, **39**, 1044 (1974); G. E. Heinsohn and E. C. Ashby, *J. Org. Chem.*, **38**, 4232 (1973); E. C. Ashby and G. E. Heinsohn, *J. Org. Chem.*, **38**, 4343 (1973); M. T. Maurette, C. Benard and A. Lattes, *Tetrahedron*, **30**, 3695 (1974).
[237] J. Durand, N. T. Anh and J. Huet, *Tetrahedron Letters*, **1974**, 2397.
[238] L. Cazaux, G. Chassaing and P. Maroni, *Compt. rend.*, *Ser C.*, **278**, 1305 (1974).
[239] H. C. Brown and V. Varma, *J. Org. Chem.*, **39**, 1631, (1974).
[240] J. Hooz, S. Akiyama, F. J. Cedar, M. J. Bennett and R. M. Tuggle, *J. Am. Chem. Soc.*, **96**, 274 (1974).
[241] J. E. McMurry and M. P. Fleming, *J. Am. Chem. Soc.*, **96**, 4708 (1974); see also S. Tyrlik and I. Wolochowicz, *Bull. Soc. Chim. France*, **1973**, 2147, and T. Mukaiyama, T. Sato and J. Hanna, *Chem. Letters*, **1973**, 1041.
[242] D. C. Wigfield and K. Taymaz, *Tetrahedron Letters*, **1973**, 4841.
[243] J. H. Schauble, G. J. Walter and J. G. Morin, *J. Org. Chem.*, **39**, 755 (1974).
[244] R. E. Manning and F. M. Schaefer, *Tetrahedron Letters*, **1974**, 3343.
[245] T. Yoshida and E. Negishi, *Chem. Comm.*, **1974**, 762; see also *Org. Reaction Mech.*, **1973**, 188.
[246] Y.-J. Lee and W. D. Closson, *Tetrahedron Letters*, **1974**, 381.
[247] S. Cacchi, L. Caglioti and G. Paolucci, *Bull. Chem. Soc. Japan*, **47**, 2323 (1974).
[248] B. Fleming and H. I. Bolker, *Can. J. Chem.*, **52**, 888 (1974).
[249] M. Cussac and A. Boucherle, *Bull. Soc. Chim. France*, **1974**, 1437.
[250] J.-Y. Godet and M. Pereyre, *J. Organometal. Chem.*, **77**, C1 (1974); D. Cabaret and Z. Welvart, *ibid.*, **80**, 185, 199 (1974).
[251] J. Málek, M. Černý and R. Řeřicha, *Coll. Czech. Chem. Comm.*, **39**, 2656 (1974).
[252] H. O. House and E. F. Kinloch, *J. Org. Chem.*, **39**, 747 (1974).
[253] J.-L. Derocque and F.-B. Sundermann, *J. Org. Chem.*, **39**, 1736 (1974).
[254] L. A. Paquette, S. V. Ley, R. H. Meisinger, R. K. Russell and M. Oku, *J. Am. Chem. Soc.*, **96**, 5806 (1974).
[255] S. V. Ley and L. A. Paquette, *J. Am. Chem. Soc.*, **96**, 6670 (1974).
[256] A. J. Birch and P. G. Lehman, *Tetrahedron Letters*, **1974**, 2395.
[257] A. J. Birch and P. G. Lehman, *J.C.S. Perkin I*, **1973**, 2754.
[258] A. J. Bellamy, E. A. Campbell and I. R. Hall, *J.C.S. Perkin II*, **1974**, 1347.
[259] S. W. Staley and A. S. Heyn, *Tetrahedron*, **30**, 3671 (1974).
[260] J. Grimshaw and R. J. Haslett, *Chem. Comm.*, **1974**, 174.
[261] C. Chassin, E. A. Schmidt and H. M. R. Hoffmann, *J. Am. Chem. Soc.*, **96**, 606 (1974).
[262] J. H. Markgraf, W. M. Hensley and L. I. Shoer, *J. Org. Chem.*, **39**, 3168 (1974).
[263] J. E. McMurry, *Accounts Chem. Res.*, **7**, 281 (1974).
[264] E. E. van Tamelen and J. A. Gladysz, *J. Am. Chem. Soc.*, **96**, 5290 (1974).
[265] L. C. Blaszczak and J. E. McMurry, *J. Org. Chem.*, **39**, 258 (1974).
[266] J. E. McMurry and J. Melton, *J. Org. Chem.*, **38**, 4367 (1973).
[267] L. Adamčiková and L. Treindl, *Coll. Czech. Chem. Comm.*, **39**, 1264 (1974).
[268] P. Dowd and K. Kang. *Chem. Comm.*, **1974**, 384.
[269] A. Ichihara and T. Matsumoto, *Bull. Chem. Soc. Japan.* **47**, 1030 (1974).
[270] J. D. Morrison and R. W. Ridgway, *J. Org. Chem.*, **39**, 3107 (1974).
[271] J. A. Katzenellenbogen and T. Utawanit, *J. Am. Chem. Soc.*, **96**, 6153 (1974).
[272] J. L. Fry, *Chem. Comm.*, **1974**, 45.
[273] M. P. Doyle, D. J. DeBruyn, S. J. Donnelly, D. A. Kooistra, A. A. Odubela, C. T. West and S. M. Zonnebelt, *J. Org. Chem.*, **39**, 2740 (1974).
[274] W. B. Motherwell, *Chem. Comm.*, **1973**, 935.
[275] L. E. Overman, D. Matzinger, E. M. O'Connor and J. D. Overman, *J. Am. Chem. Soc.*, **96**, 6081 (1974).

[276] C. Dell'Erba, G. Guanti, G. Leandri and G. Poluzzi Corallo, *Int. J. Sulfur Chem.*, **3**, 261 (1973); *Chem. Abs.* **79**, 125505 (1973).
[277] S. Hashimoto, I. Furukawa and T. Katami, *Nippon Kagaku Kaishi*, **1974** 511; *Chem. Abs.*, **80**, 145023 (1974).
[278] I. Shimao, *Nippon Kagaku Kaishi*, **1974**, 515: *Chem. Abs.*, **80**, 145115 (1974).
[279] M. J. Haddadin, G. E. Zahr, T. N. Rawdah, N. C. Chelhot and C. H. Issidorides, *Tetrahedron*, **30**, 659 (1974); see also *Org. Reaction Mech.*, **1973**, 193.
[280] M. E. C. Biffin, G. Bocksteiner, J. Miller and D. B. Paul, *Austral. J. Chem.*, **27**, 789 (1974).
[281] M. Loutre and Y.-L. Pascal, *Bull. Soc. Chim. France*, **1974**, 1763, 1772.
[282] Advances in Chemistry Series, No. 132, American Chemical Society, 1974.
[283] R. B. Turner, B. J. Mallon, M. Tichy, W. von E. Doering, W. R. Roth and G. Schröder, *J. Am. Chem. Soc.*, **95**, 8605 (1973).
[284] H. W. Thompson and E. McPherson, *J. Am. Chem. Soc.*, **96**, 6232 (1974); see also *Org. Reaction Mech.*, **1971**, 556.
[285] F. J. McQuillin, *Tetrahedron*, **30**, 1661 (1974); see also C. J. Love and F. J. McQuillin, *J.C.S. Perkin I*, **1973**, 2509.
[286] Y. Demortier and I. de Aguirre, *Bull. Soc. Chim. France*, **1974**, 1614, 1619; see also *Org. Reaction Mech.*, **1972**, 563.
[287] D. E. Budd, D. G. Holah, A. N. Hughes and B. C. Hui, *Can. J. Chem.*, **52**, 775 (1974).
[288] T. Nishiguchi and K. Fukuzumi, *J. Am. Chem. Soc.*, **96**, 1893 (1974); see also *Org. Reaction Mech.*, **1973**, 193.
[289] H. Imai, T. Nishiguchi and K. Fukuzumi, *J. Org. Chem.*, **39**, 1622 (1974).
[290] G. Strathdee and R. Given, *Can. J. Chem.*, **52**, 2216, 2226 (1974).
[291] S. L. Regen, *J. Org. Chem.*, **39**, 260 (1974).
[292] L. Kh. Friedlin, V. Z. Sharf, V. M. Krutis and T. V. Lysyak, *Izv. Akad. Nauk SSSR, Ser. Khim.*, **1973**, 1868; *Chem. Abs.*, **79**, 145689 (1973).
[293] Y. Sasson, P. Albin and J. Blum, *Tetrahedron Letters*, **1974**, 833.
[294] M. Dedieu and Y.-L. Pascal, *Compt. rend., Ser. C.*, **278**, 1425 (1974); H. B. Henbest and J. Trocha-Grimshaw, *J.C.S. Perkin I*, **1974**, 601.
[295] R. Zanella, F. Canziani, R. Ros and M. Graziani, *J. Organometal. Chem.*, **67**, 449 (1974).
[296] Y. Sasson and G. L. Rempel, *Tetrahedron Letters*, **1974**, 3221.
[297] E. L. Muetterties and F. J. Hirsekorn, *J. Am. Chem. Soc.*, **96**, 4063 (1974).
[298] M. A. Schroeder and M. S. Wrighton, *J. Organometal. Chem.*, **74**, C29 (1974).
[299] T. Funabiki, S. Kasaoka, M. Matsumoto and K. Tarama, *J.C.S. Dalton*, **1974**, 2043; E. Balogh-Hergovich, G. Speier and L. Markó, *J. Organometal. Chem.*, **66**, 303 (1974); M. N. Ricroch, A. Gaudemer, *J. Organometal. Chem.*, **67**, 119 (1974).
[300] A. Mortreux and M. Blanchard, *Chem. Comm.*, **1974**, 786.
[301] G. Brieger and T. J. Nestrick, *Chem. Rev.*, **74**, 567 (1974).
[302] E. Brown and M. Ragault, *Bull. Chem. Soc. Japan*, **47**, 1727 (1974).
[303] T. W. Russell and D. M. Duncan, *J. Org. Chem.*, **39**, 3050 (1974).
[304] L. Červený, A. Procházka and V. Ružička, *Coll. Czech. Chem. Comm.*, **39**, 2463 (1974); E. F. Litvin, L. Kh. Freidlin, G. K. Oparina, R. N. Gurskii, R. V. Istratova and L. I. Gosteva, *Zh. Org. Khim.*, **10**, 1475 (1974); *Chem. Abs.*, **81**, 90815 (1974).
[305] M. Kajitani, Y. Sasaki, J. Okada, K. Ohmura, A. Sugimori and Y. Urushibara, *Bull. Chem. Soc. Japan*, **47**, 1203 (1974); A. Koppová, V. Ružička, V. Zapletal, J. Soukup, L. Červený and L. Stříz, *Coll. Czech. Chem. Comm.*, **39**, 829 (1974); P. A. Sermon, G. C. Bond and G. Webb, *Chem. Comm.*, **1974**, 417.
[306] L. P. Zalukaev, I. K. Anokhina, O. N. Mittov and N. G. Ermolaeva, *Dokl. Akad. Nauk SSSR*, **215**, 103 (1974); *Chem. Abs.*, **80**, 145101 (1974).
[307] A. V. Finkel'shtein, I. M. Ansimova and L. N. Ansimov, *Zh. Obshch. Khim.*, **43**, 2354 (1973); *Chem. Abs.*, **80**, 59201 (1974).
[308] G. Filardo, M. Galluzzo, B. Giannici and R. Ercoli, *J.C.S. Dalton*, **1974**, 1787.
[309] M. Paez-Pedrosa, I. Schifter and J. E. Germain, *Bull. Soc. Chim. France*, **1974**, 1977; S. P. Mel'nikova, L. N. Vorob'ev and A. A. Petrov, *Zh. Org. Khim.*, **9**, 1586 (1973); *Chem. Abs.*, **79**, 125551 (1973).
[310] J. Kanetaka, *Nippon Kagaku Kaishi*, **1974**, 206; *Chem. Abs.*, **80**, 119942 (1974).
[311] T. Sakai and N. Ohi, *Nippon Kagaku Kaishi*, **1974**, 1152; *Chem. Abs.*, **81**, 77220 (1974).
[312] S. L. Kiperman, B. S. Gudkov and N. E. Zlotkina, *Probl. Kinet. Katal.*, **15**, 110 (1973); *Chem. Abs.*, **79**, 145690 (1973).

[313] P. S. Malpani and S. B. Chandalia, *Indian Chem. J.*, **8**, 15 (1973); *Chem. Abs.*, **80**, 14313 (1974).
[314] Y. Iwasawa, S. Ogasawara, T. Onishi and K. Tamaru, *Bull. Chem. Soc. Japan*, **47**, 1089 (1974).
[315] H. Simon, B. Rambeck, H. Hashimoto, H. Günther, G. Nohynek and H. Neumann, *Angew. Chem. Int. Ed.*, **13**, 608 (1974); B. Rambeck and H. Simon, *ibid.*, p. 609.
[316] K. W. Barnett, D. L. Beach, D. L. Garin and L. A. Kaempfe, *J. Am. Chem. Soc.*, **96**, 7127 (1974).
[317] M. N. Akhtar, W. R. Jackson and J. J. Rooney, *J. Am. Chem. Soc.*, **96**, 276 (1974); M. N. Akhtar, W. R. Jackson, J. J. Rooney and N. G. Samman, *Chem. Comm.*, **1974**, 155.
[318] A. P. G. Kieboom, A. J. Breijer and H. van Bekkum, *Rec. Trav. chim.*, **93**, 186 (1974).
[319] E. Osawa, P. von R. Schleyer, L. W. K. Chang and V. V. Kane, *Tetrahedron Letters*, **1974**, 4189.
[320] B. Miller and L. Lewis, *J. Org. Chem.*, **39**, 2605 (1974).
[321] A. W. von Muijlwijk, A. P. G. Kieboom and H. van Bekkum, *Rec. Trav. chim.*, **93**, 204 (1974).
[322] S. Mitsui, S. Imaizumi, H. Hisashige and Y. Sugi, *Tetrahedron.* **29**, 4093 (1973); S. Mitsui, Y. Sugi, M. Fujimoto and K. Yokoo, *Tetrahedron*, **30**, 31 (1974).
[323] G. Guidot and J. E. Germain, *Ann. Chim. (Paris)*, **1974**, 191.
[324] O. Hinterhofer, *Monatsh. Chem.*, **105**, 279 (1974).
[325] L. Guczi, A. Sárkány and P. Tétényi, *J.C.S. Faraday I*, **70**, 1971 (1974).
[326] G. Henrici-Olivé and S. Olivé, *Angew. Chem. Int. Ed.*, **13**, 29 (1974).
[327] J. E. Baldwin and J. Huff, J. *Am. Chem. Soc.*, **95**, 5757, 8477 (1973).
[328] C. K. Chang and T. G. Traylor, *J. Am. Chem. Soc.*, **95**, 5810 (1973).
[329] J. P. Collman, R. R. Gagné, T. R. Halbert, J. C. Marchon and C. A. Reed, *J. Am. Chem. Soc.*, **95**, 7868 (1973); J. P. Collman, R. R. Gagné and C. A. Reed, *ibid.*, **96**, 2629 (1974).
[330] G. C. Wagner and R. J. Kassner, *J. Am. Chem. Soc.*, **96**, 5593 (1974).
[331] W. S. Brinigar and C. K. Chang, *J. Am. Chem. Soc.*, **96**, 5595 (1974).
[332] D. L. Anderson, C. J. Weschler and F. Basolo, *J. Am. Chem. Soc.*, **96**, 5599 (1974).
[333] J. Almog, J. E. Baldwin, R. L. Dyer, J. Huff and C. J. Wilkerson, *J. Am. Chem. Soc.*, **96**, 5600 (1974).
[334] W. S. Brinigar, C. K. Chang, J. Geibel and T. G. Traylor, *J. Am. Chem. Soc.*, **96**, 5597 (1974).
[335] B. B. Hasinoff, *Biochemistry*, **13**, 3111 (1974).
[336] T. Yonetani, H. Yamamoto and T. Iizuka, *J. Biol. Chem.*, **249**, 2168 (1974).
[337] E. W. Abel, J. M. Pratt, R. Whelan and P. J. Wilkinson, *J. Am. Chem. Soc.*, **96**, 7119 (1974).
[338] J. V. McArdle, H. B. Gray, C. Creutz and N. Sutin, *J. Am. Chem. Soc.*, **96**, 5737 (1974).
[339] R. X. Ewall and L. E. Bennett, *J. Am. Chem. Soc.*, **96**, 940 (1974); D. O. Lambeth, K. L. Campbell, R. Zand and G. Palmer, *J. Biol. Chem.*, **248**, 8130 (1973).
[340] L. E. Brown and G. A. Hamilton, *J. Am. Chem. Soc.*, **92**, 7225 (1970).
[341] S. Shinkai and T. C. Bruice, *J. Am. Chem. Soc.*, **95**, 7526 (1973); S. Ghisla, U. Hartmann, P. Hemmerich and F. Müller, *Annalen*, **1973**, 1388.
[342] S. Shinkai, T. Kunitake and T. C. Bruice, *J. Am. Chem. Soc.*, **96**, 7140 (1974); see also D. Clerin and T. C. Bruice, *ibid.*, p. 5571, and S. B. Smith, N. Brüstlein and T. C. Bruice, *ibid.*, p. 3696.
[343] H. W. Orf and D. Dolphin, *Proc. Nat. Acad. Sci., U.S.*, **71**, 2646 (1974).
[344] L. L. Smith and J. I. Teng, *J. Am. Chem. Soc.*, **96**, 2640 (1974).
[345] L. Bachan, C. B. Storm, J. W. Wheeler and S. Kaufman, *J. Am. Chem. Soc.*, **96**, 6799 (1974).
[346] S. W. May and R. D. Schwartz, *J. Am. Chem. Soc.*, **96**, 4031 (1974).
[347] F. Blaney, D. E. Johnston, M. A. McKervey, E. R. H. Jones and J. Pragnell, *Chem. Comm.*, **1974**, 297.
[348] H. Eckstein, G. Barth, R. E. Linder, E. Bunnenberg and C. Djerassi, *Annalen*, **1974**, 990.
[349] P. B. Baker and B. C. Saunders, *Tetrahedron*, **30**, 337, 3303 (1974); C. F. Phelps, E. Antonini, G. Giacometti and M. Brunori, *Biochem. J.*, **141**, 265 (1974).
[350] P. A. Goodman and J. E. Meany, *Biochemistry*, **13**, 3254 (1974); J. S. Olson, D. P. Ballou, G. Palmer and V. Massey, *J. Biol. Chem.*, **249**, 4363 (1974); R. C. Bray, D. J. Lowe and M. J. Barber, *Biochem. J.*, **141**, 309 (1974).
[351] M. L. Kremer and S. Baer, *J. Phys. Chem.*, **78**, 1919 (1974); E. Zidoni and M. L. Kremer, *Arch. Biochem. Biophys.*, **161**, 658 (1974); R. J. M. Corrall, H. M. Rodman, J. Margolis and B. R. Landau, *J. Biol. Chem.*, **249**, 3181 (1974).
[352] C. Walsh, O. Lockridge, V. Massey and R. Abeles, *J. Biol. Chem.*, **248**, 7049 (1973).
[353] E. Silverstein and G. Sulebele, *Biochemistry*, **13**, 1815 (1974).
[354] P. Jones, K. Prudhoe, T. Robson and H. C. Kelly, *Biochemistry*, **13**, 4279 (1974).
[355] K. S. V. Santhanam and P. J. Elving, *J. Am. Chem. Soc.*, **96**, 1653 (1974).
[356] J. W. Webb, B. Janik and P. J. Elving, *J. Am. Chem. Soc.*, **95**, 8495 (1973).
[357] U. Bruhlmann and E. Hayon, *J. Am. Chem. Soc.*, **96**, 6169 (1974).

[358] J. Fisher and C. Walsh, *J. Am. Chem. Soc.*, **96**, 4345 (1974); M. F. Dunn and J. S. Hutchison, *Biochemistry*, **12**, 4882 (1973); J. W. Jacobs, J. T. McFarland, I. Wainer, D. Jeanmaier, C. Ham, K. Hamm, M. Wnuk and M. Lam, *Biochemistry*, **13**, 60 (1974).

[359] A. P. Khrushch, A. E. Shilov and T. A. Vorontsova, *J. Am. Chem. Soc.*, **96**, 4987 (1974).

[360] D. Werner, S. A. Russell and H. J. Evans, *Proc. Nat. Acad. Sci., U.S.*, **70**, 339 (1973).

CHAPTER 5

Carbenes and Nitrenes

M. S. Baird

Department of Organic Chemistry, University of Newcastle upon Tyne

Reviews on alkoxycarbonylcarbenes,[1] carbenes and their analogues,[2] and the role of nitrenes in chemical reactions,[3] together with annual reviews,[4] have been published. The trapping of carbenoids has also been reviewed,[5] and a brief outline of carbene and nitrene chemistry has been provided.[6]

Structure

An extensive review correlates the reactivity of carbenacycloalkenes with their structure and multiplicity,[7] and another discusses theoretical calculations on carbenes contained in a wide variety of three-membered heterocyclic rings.[8]

A review of the structure of methylene has been published.[9] A MINDO/3 study of the electronic states of methylene gives a S_0–T_1 separation of 8.7 kcal mol^{-1}, in good agreement with experiment;[10] *ab initio* valence bond calculations,[11] comparisons of valence bond and MO studies,[12] and INDO and *ab initio* studies[13] are also reported. The dimerization of methylene has been subjected to a least-motion study,[14] and reactions of the singlet with cyclobutane[15] and of the triplet and singlet with carbon monoxide, but-2-ene and methyl chloride[16] have been examined.

The photolysis of $CH_2{=}CHCHN_2$ in a matrix at 6K is found to give two non-identical ground-state triplets of vinylmethylene; electron distribution suggests that these are best described as allyl radicals together with a σ-unpaired electron largely localized at the bivalent carbon, as in (**1**) and (**2**).[17] An ESR study of triplet vinylcarbenes having phenyl and methoxycarbonyl substituents indicates, however, that (**3**) is a major contributor to the structure.[18]

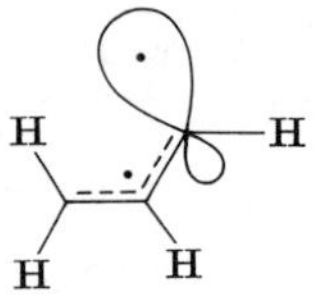

(**1**) 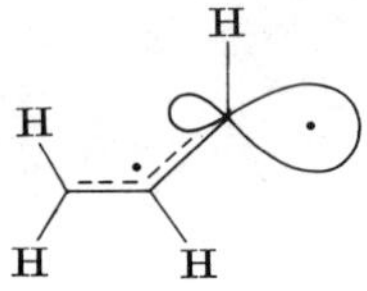 (**2**) 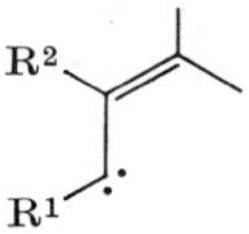 (**3**) 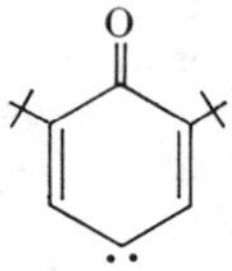 (**4**)

A CIDNP study of the formation of (**4**) shows its reaction with solvent to occur via a singlet,[19] while the related dibenzocarbene is found by EPR to have a triplet ground state.[20]

Theoretical calculations predict the ^1E and $^1A_1$ states of methylnitrene to be 14,200 and 27,700 cm^{-1}, respectively, above the $^3A_2$ ground state.[21]

Generation

The rapid increase in the use of phase-transfer catalysis for carbene generation has continued, and earlier results have been reviewed.[22] Dibenzo[18]crown-6-ether[23] and tertiary amines[24] have been used as catalysts for dichlorocarbene generation, and the method has also been applied to the production of di-iodocarbene,[25] and of (**5**), either from 1-bromo-3-methylbuta-1,2-diene[26] or from 3-chloro-3-methylbut-1-yne.[27]

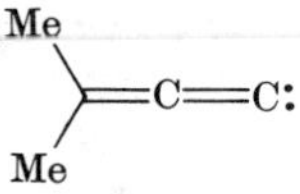

$Me_2C{=}C{=}C{:}$ (**5**)

$R^1R^2C{=}CH(OTf)$ (**6**)

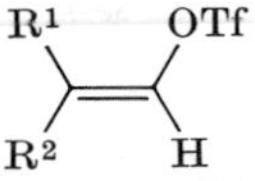

The reaction between chloroform and $LiOC(C_2H_5)_3$ is reported to be an efficient source of dichlorocarbene,[28] which is also obtained by reaction of carbon tetrachloride with pentabutylantimony.[29] Difluorocarbene may be conveniently obtained by reaction of phosphonium salts such as $[Ph_3PCF_2Br]^+Br^-$ with sodium methoxide.[30] Treatment of a range of triflates (**6**; R^1, R^2 = alkyl) with KOBu^t in the presence of alkenes gives good yields of methylenecyclopropanes via $R^1R^2C{=}C{:}$ or the related carbenoid, but, when R^1 or R^2 is hydrogen or aryl, acetylenes are produced.[31]

Thermal decomposition has also been used to obtain a number of carbenes. The formation of difluorocarbene from perfluoropropene oxide is shown to be reversible.[32] Thermolysis of a variety of organomercury reagents, PhHgXYZ (X, Y = H or halogen, Z = SO_2Ph, COOR or $CONMe_2$) provides a variety of carbenes,[33a] and the use of this route to obtain halogenocarbenes is discussed in two reviews,[33b] while thermolysis of Cl_3CSiF_3 in the gas phase leads efficiently to dichlorocarbene.[34]

The photochemical generation of (trimethylsilyl)carbene from the corresponding diazo-compound is reported.[35]

Reactions Postulated to Involve Carbenes or Nitrenes

Carbenes

Thermal decomposition of $2H$-azirines normally leads to C—N bond cleavage; however, compounds (**7**) give styrene, apparently by C—C cleavage to (**8**), intramolecular insertion into the methyl, and loss of RCN.[36]

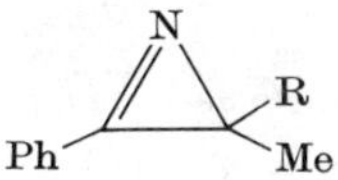

(**7**) 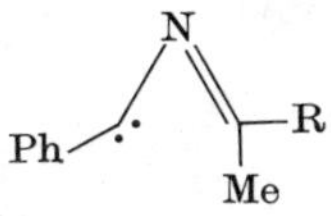 (**8**)

The unusual reactions of electronically excited cyclobutanones have been reviewed.[37] Photolysis of (**9**) in methanol leads to the tetrahydrofurylidene (**10**) which is trapped by solvent; the migrating centre retains its configuration, which allows the formation of diradicals as primary products to be excluded.[38] Photolysis of (**11**) and (**12**) leads to products apparently derived by trapping of the carbenes (**13**) and (**14**).[39]

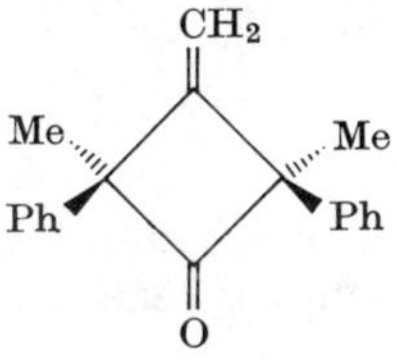

(**9**) 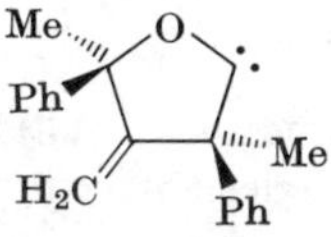 (**10**)

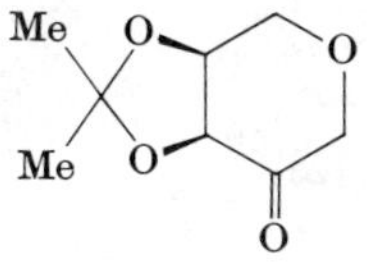

(**11**) 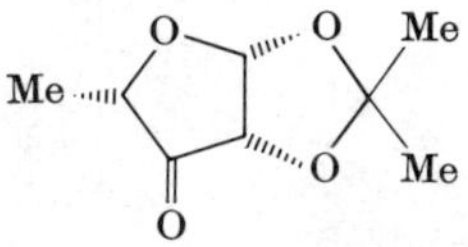 (**12**)

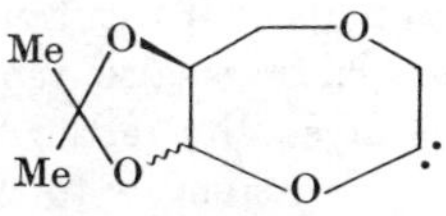

(**13**) 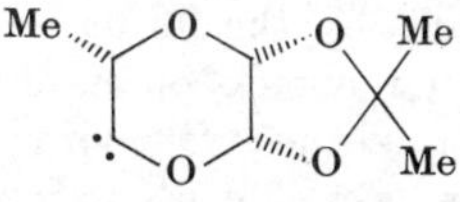 (**14**)

Benzyne and carbon disulphide give products which may be explained by the transient formation of (**15**);[40] this carbene has been shown to be weakly electrophilic.[41]

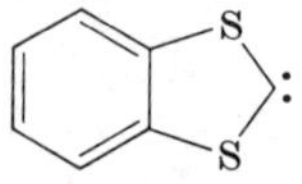

(**15**)

Photolysis of triptycene gives (**16**) as the sole product in inert or photosensitizing solvents; in methanol the product is (**17**), suggesting a reaction via (**18**) rather than a di-π-methane pathway.[42] Photolysis of (**19**) leads to (**20**), possibly via the carbene (**21**).[43]

CH_2OMe CH:

(**16**) 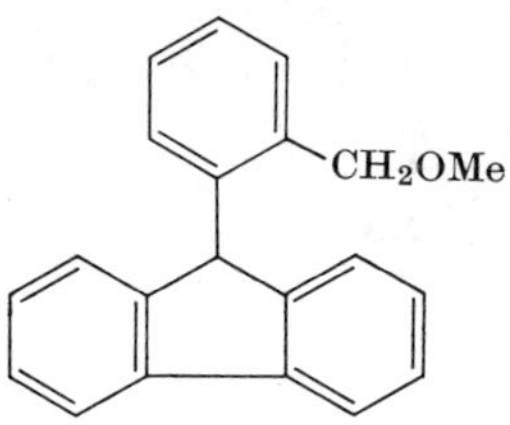 (**17**) 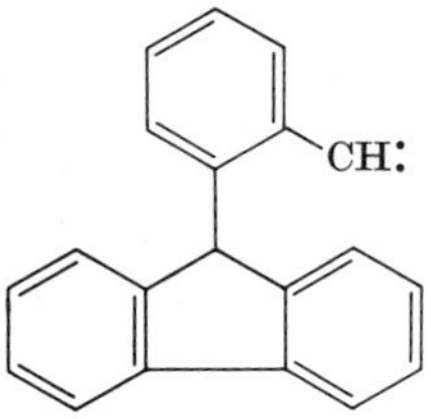 (**18**)

N N CH_2 CH

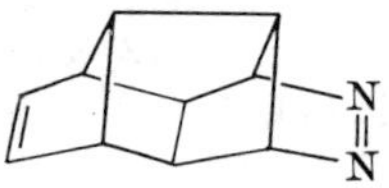

(**19**) 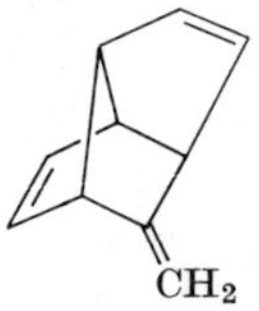 (**20**) 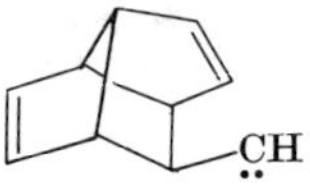 (**21**)

The thermal formation of (**23**) from (**22**) probably involves ions rather than carbenes, as no cyclopropane derivative is produced and the reaction is nearly quantitative.[44]

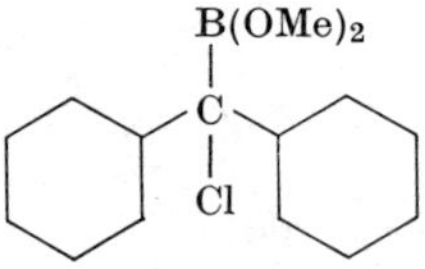

$B(OMe)_2$ C Cl CH

(**22**) 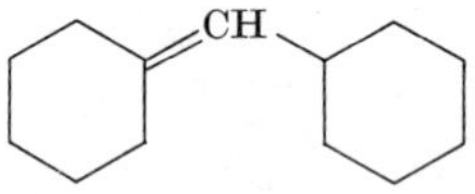 (**23**)

Nitrenes

Strong evidence is presented for the intermediacy of alkylnitrenes in the thermolysis of tertiary azides by the formation of intramolecular aromatic substitution products.[45] Intramolecular attack on naphthalene and thiophene rings is also reported.[46] The formation of azirines from vinyl azides is efficiently catalysed by tertiary amines; the reaction has been shown not to proceed via a triazole, but attempts to trap a nitrene were also unsuccessful;[47] kinetic studies show, however, that a nitrene intermediate is unlikely.[48] Vinylnitrene intermediates in the rearrangement of 2*H*-azirine may be trapped by the formation of heterocycles.[49]

Thermolysis of (**24**) at 100° is shown by trapping to produce phenylnitrene.[50]

$Me_3SiO—NPh—OSiMe_3$
(**24**)

Contrary to earlier reports, toluene-*p*-sulphonamide does not give toluene-*p*-sulphonylnitrene on treatment with $Pb(OAc)_4$, though *N*-aminophthalimide may react partially via a nitrene.[51] Alkaline cleavages of sulphonic acid hydrazides, $PhSO_2NHNRCH_2R^1$ and of $EtOOCNCl_2$ are reported[52,53] to give products derived from aminonitrenes and chloronitrene, respectively, and PhMeNNHSOMe gives dimers of the nitrene PhMeN—N:.[54]

Further studies on the reaction of aromatic nitro-compounds with $P(OEt)_3$ leading to nitrene-derived products have also been reported.[55]

Cycloadditions

Intermolecular

The Simmons–Smith formation of cyclopropanes from alkenes, CH_2I_2, and a zinc–copper couple has been reviewed;[56] iodocyclopropanation may be carried out with a carbenoid derived from iodoform and diethylzinc.[57] A variety of carbenoid addition routes leading to cyclopropanones are discussed.[58] Syntheses of heterocycles involving use of nitrenes have also been reviewed.[59]

Cycloaddition of diacylaminonitrenes to electrophilic and nucleophilic alkenes, which leads stereospecifically to aziridines, has been analysed in frontier orbital terms,[60] and a theoretical study of addition of HN: to ethylene is provided.[61] Application of a simple MO method to the addition of methylene to butadiene shows that both 1,2- and 1,4-processes are allowed, but that the latter is discriminated against by excessive closed-shell interactions.[62]

While addition of dibromocarbene to 1,1-dicyclopropylethylene gives the expected adduct, fluorenylidene and diphenylcarbene give in addition (**25**); the dependence of the relative yields on conditions is interpreted in terms of the spin states of the carbenes.[63]

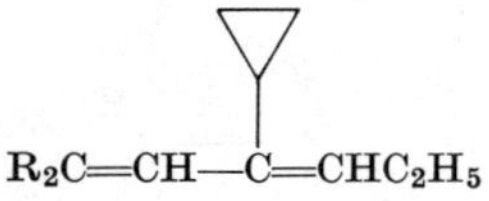

$R_2C{=}CH—C({\text{cyclopropyl}}){=}CHC_2H_5$
(**25**)

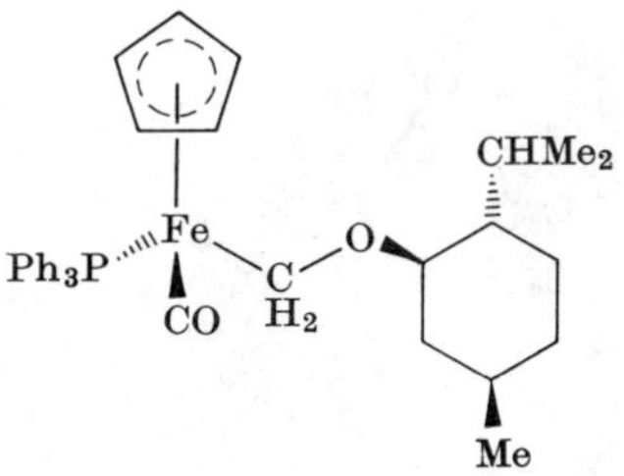

Ph_3P Fe CO $C H_2$ O CHMe$_2$ Me
(**26**)

Optically active cyclopropanecarboxylic acid derivatives may be obtained with an enantiomeric selectivity as high as 70% by the carbenoid reactions of diazoalkanes with olefins catalysed by a chiral cobalt chelate complex.[64] The complex (+)-(**26**) can be used as an optically active carbene-transfer reagent. Thus with *trans*-β-methylstyrene and HBF_4, (−)-*trans*-(1*R*,2*R*)-1-methyl-2-phenylcyclopropane is obtained.[65]

The selectivity of phenylbromocarbene generated from $KOBu^t$ and bromoform towards various alkenes is altered by addition of 18-crown-6-ether to the reaction mixture, then becoming equal to that of the free carbene generated photochemically from phenylbromodiazirine. The selectivity of dichlorocarbene is only slightly changed by addition of the crown ether, supporting the view that it is a free carbene.[66] The selectivity of dichlorocarbene generated from $Ph(CBrCl_2)Hg$ is not solvent-dependent, but when the carbene is generated from $Cl_3CCO_2^-Na^+$ the selectivity increases with the solvent's proton donor ability.[67] A study of the relative rates of addition of $:CF_2$ to alkenes shows cyclohex-2-enone ethylene ketal to be very unreactive (rate 0.02 relative to cyclohexene), whereas the corresponding ketal of cyclohex-3-enone is only marginally less reactive than cyclohexene.[68] The results of comparative rate studies for the addition of dihalogenocarbenes to 1-substituted cycloalkenes are interpreted in terms of steric and electronic effects,[69] and addition of di-iodocarbene to acyclic and cyclic alkenes is reported.[25,70] Dichlorocarbene undergoes predominant addition to the 1,2-bond of 2-phenylbutadiene, in agreement with quantum-mechanical calculations;[71] multiple addition of the carbene to cyclo-octatetraenes,[72] cycloheptatrienes[72] and azepines[72,73] is discussed, and tetraphenylallene is reported to undergo tris-addition of $:CCl_2$ to one aromatic ring.[74] Kinetic studies of the reactions of $:CF_2$ and $:CFCl$ with cyclopentadiene to produce fluorobenzene are reported.[75]

The addition of difluorocarbene or chlorofluorocarbene to norbornadiene is claimed to give the tetracyclic adducts (**27**; X = F or Cl), which may arise by a homo-1,4-addition of the carbene.[76]

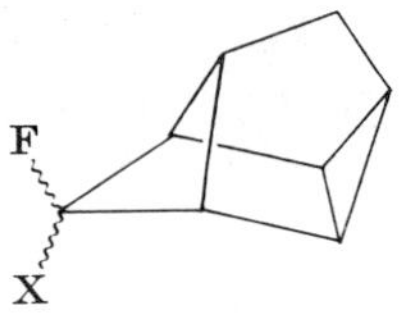

F, X

(27)

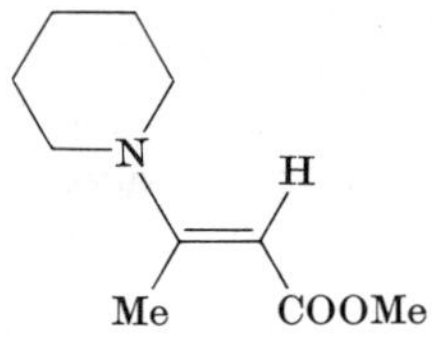

N, H, Me, COOMe

(28)

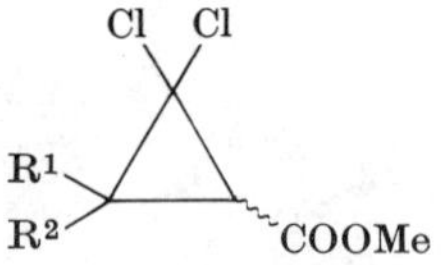

Cl Cl, R^1, R^2, COOMe

(29)

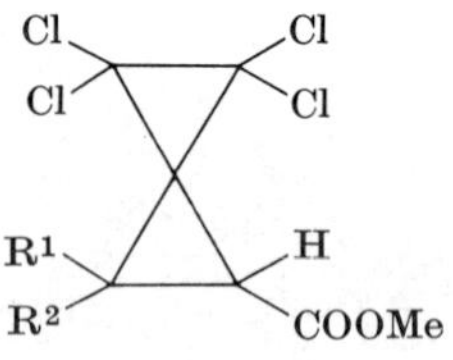

Cl, Cl, Cl, Cl, R^1, R^2, H, COOMe

(30)

Addition of methylene to the enamine ester (**28**) followed by hydrolysis leads to the γ-keto-ester $MeCOCH_2CH_2COOMe$, but dichlorocarbene leads to elimination products such as MeCOCCl=CHCOOEt,[77] the reactions presumably proceed by rearrangement of initially produced adducts. The adduct (**29**) is shown to be an intermediate in the formation of (**30**) in the reaction of phase-transfer generated $:CCl_2$ with α,β-unsaturated esters; the carbene is found to add to the $\gamma\delta$-bond of $\alpha\beta,\gamma\delta$-unsaturated esters.[78]

Addition of dichlorocarbene to $\alpha\beta$-unsaturated ketones,[79] enamines,[80] steroids[81] and uracils[82] is discussed, and a reassignment of the structures of $:CCl_2$ adducts of hexa-methyl-Dewar-benzene is provided.[83]

The nucleophilic carbene, dimethoxycarbene, adds to diethyl fumarate or maleate and to styrene to give cyclopropanes, but with dimethyl acetylenedicarboxylate, phenylacetylene or diphenylketene 1:2 adducts (**31**), (**32**) and (**33**) are observed; the last of these can be reached via (**34**).[84]

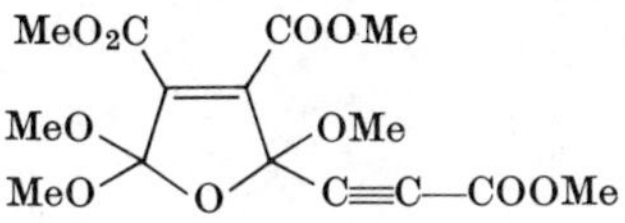

(**31**) 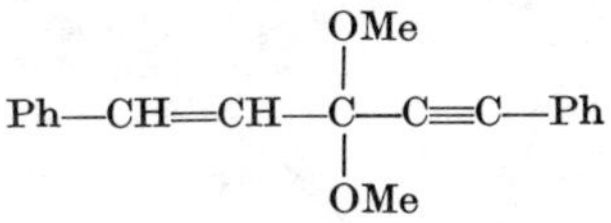 (**32**)

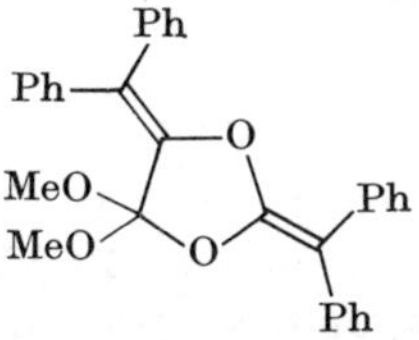

(**33**) 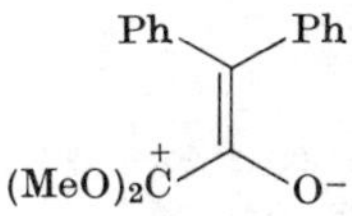 (**34**)

Cycloheptatrienylidene and diphenylcyclopropenylidene are also considered to be nucleophilic carbenes. With tetracyclones the latter gives products (**35**) which are derived by 1,4-addition followed by decarbonylation; however, the former carbene apparently does not undergo 1,4-addition but instead leads initially to (**36**), which is then decarbonylated and gives the observed products (**37**) and (**38**) via (**39**).[85]

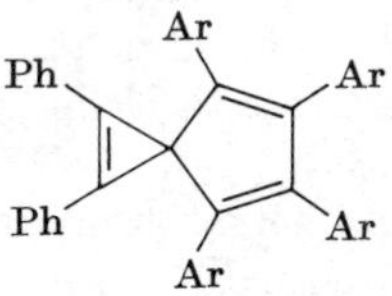

(**35**) 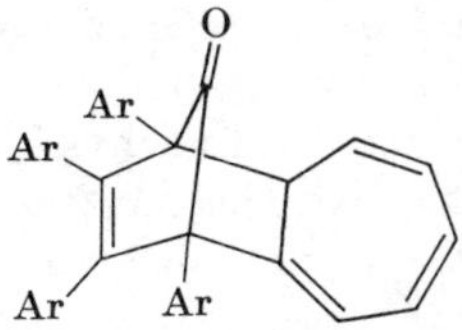 (**36**) 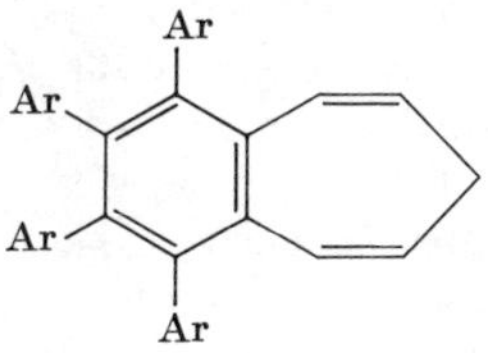 (**37**)

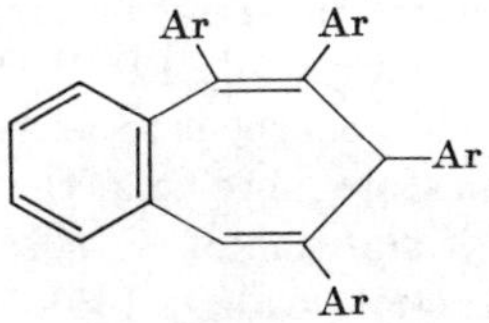

(**38**) 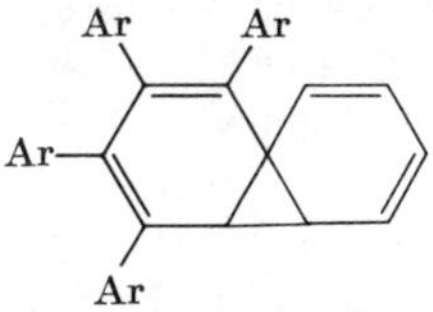 (**39**)

The reaction of cycloheptatrienylidene with 2,3-dicyanobarrelene is also complex; the product, (**40**), is thought to arise either by homo-1,4-addition, or by 1,2-addition to the 2,3-bond, followed by a rearrangement, but a number of other routes may be envisaged.[86] The electrophilic character of singlet cyclopentadienylidene has been proved by comparative studies with substituted styrenes,[87] and the carbene has been identified in a matrix at 20K and shown to dimerize to fulvalene and to react with carbon monoxide to produce (**41**).[88] Additions of the carbene and of fluorenylidenes to acetylenes have been reported;[89] the related species (**42**) undergoes 1,4-addition to dienes.[90]

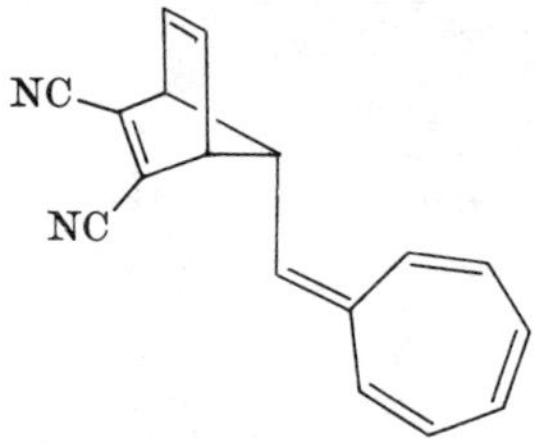

NC NC

(**40**)

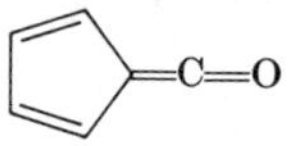

C=O

(**41**)

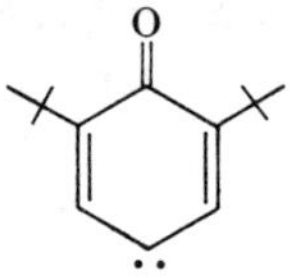

O

(**42**)

The methylenecarbene (**43**) undergoes 1,2-addition to cyclohexa-1,4-diene and norbornadiene but gives (**44**) on reaction with cyclopentadiene; an initial 1,4-addition is possible in the last case, but 1,2-addition followed by a vinylcyclopropane rearrangement seems more likely.[91] Reaction of RCH=CHCHCl$_2$ or ClRC=CHCH$_2$Cl with copper and an isocyanide in the presence of electron-deficient alkenes leads to vinylcyclopropanes.[92]

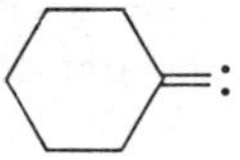

(**43**)

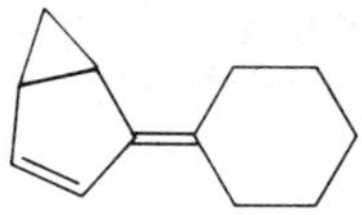

(**44**)

Addition of *N*-phthalimidonitrene to electron-deficient olefins at low temperatures gives the thermodynamically less stable *syn*-products, which equilibrate with the *anti*-isomers on warming.[93] The stereospecificity and substituent effects in the cycloaddition of nitrenes derived from 4-amino-3,5-diphenyl-1,2,4-triazole to olefins,[94] and the addition of nitrenes to steroids,[95] have also been discussed.

A number of carbene additions to X=Y where X and/or Y are heteroatoms are also reported. Phenyl(trihalomethyl)mercury compounds react thermally with highly halogenated ketones, aldehydes and acid chlorides, adding dihalogenocarbene to the carbonyl group to produce oxiranes; oxalyl chloride reacts to produce some oxirane, but the major product is CCl$_3$CO—COCl, while benzil gives PhCOCClPh—COCl, probably via the oxirane.[96] Benzaldehyde adds phase-transfer generated dichlorocarbene, giving PhCH(OH)COOH; again an oxirane is the likely intermediate, while PhCH(OH)CCl$_3$ has been excluded.[97] A Hammett study of the addition of dichlorocarbene to benzylideneanilines indicates an electrophilic carbene with a transition state more polar than in the corresponding addition to styrenes.[98] A full account of the addition of :CCl$_2$ to iminodichlorides and to PhN=NPh is published; the latter reaction leads to (**45**), possibly via loss of PhN: from the initial adduct.[99]

Diethyl azodicarboxylate reacts with carbenes generated from PhHgCX$_3$ to give

$(EtOOC)_2NN{=}CX_2$; an intermediate (**46**) can be isolated when X = Br.[100] A similar reaction occurs between dihalogenocarbenes and azodibenzoyl, PhCON=NCOPh, but further fragmentation leads to (**47**; X = Hal).[101]

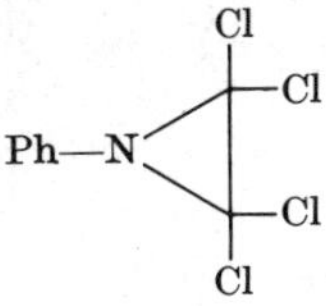

(**45**) 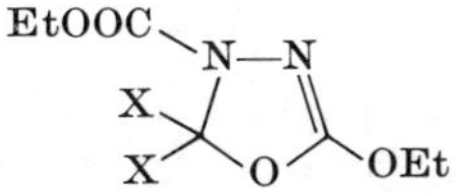 (**46**) 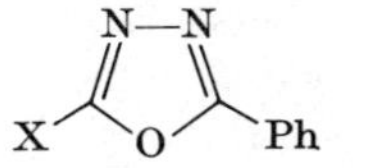 (**47**) 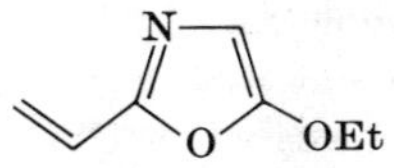 (**48**)

Ethoxycarbonylcarbene undergoes a 1,3-dipolar addition to acrylonitrile, to give (**48**).[102]

Intramolecular

One product of the reaction of (**49**; X = Br) with methyl-lithium is the spiro-compound (**50**), derived by addition of the carbenoid (**49**; X = Li) to the alkene.[103] The carbene obtained from (**51**) and $KOBu^t$ also undergoes intramolecular addition, producing (**52**),[104] though (**53**) gives no addition product but instead those of a C—H insertion and a 1,2-hydrogen shift.[105]

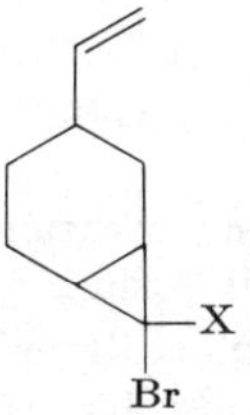

(**49**) 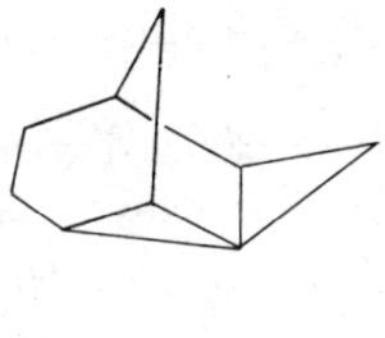 (**50**) 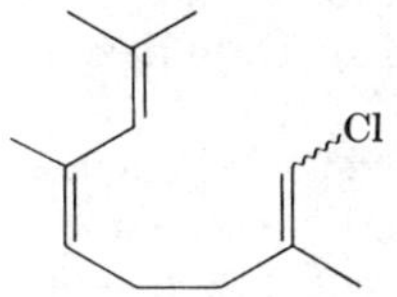 (**51**) 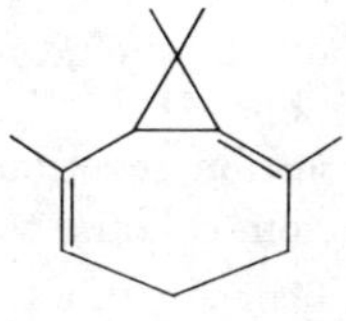 (**52**) 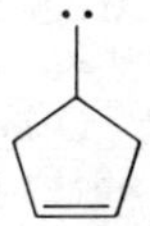 (**53**)

The π-allylpalladium chloride-catalysed decomposition of diazo-ketones (**54**; n and m = 1 or 2) leads to intramolecular carbenoid addition, while (**55**) gives the bis-adduct (**56**).[106] A similar process is apparently involved in the copper-catalysed decomposition of (**57**) to (**58**); an intermediate adduct (**59**) can fragment as shown to the product.[107]

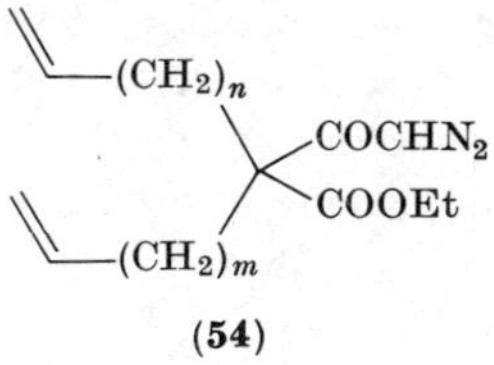

(**54**) 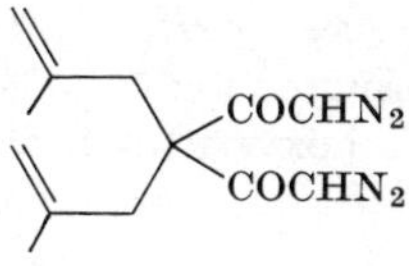 (**55**) 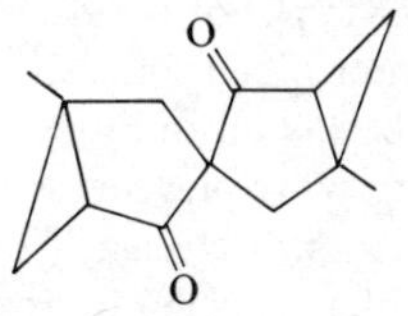 (**56**)

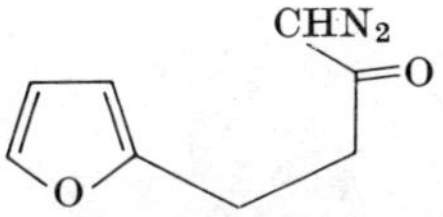

(**57**) 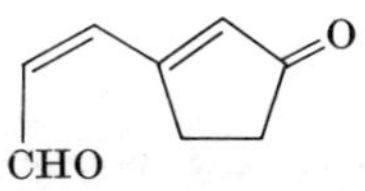 (**58**) 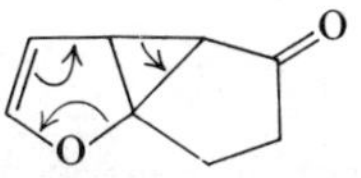 (**59**)

Insertions and Abstractions

Intermolecular

A non-empirical quantum-mechanical study of the reaction of triplet methylene with molecular hydrogen predicts a reaction exothermicity of 5.37 kcal mol^{-1}, in good agreement with the experimental value; the predicted barrier height of 15.5 kcal mol^{-1} is consistent with the lack of observed reaction between $^3B_1$ methylene and hydrogen at 300K.[108] A similar study of the insertion of dichlorocarbene into the C—Cl bond of methyl chloride shows initial electrophilic attack on the bond followed by synchronous transfer of Cl to carbon.[109] A study of $:CX_2$ reaction with CH_2ClF on a solid surface at 550–600° is reported.[110]

Dichloro- and dibromo-carbene generated by the thermal decomposition of $PhHgCX_3$ undergo a synthetically useful insertion into the unactivated C—H bonds of cycloalkanes; in some cases the cycloalkyl halide is a by-product.[111]

Polarization of ^{19}F nuclei during the insertion of triplet $Ph_2C:$ into the benzylic C—H bonds of diasteroisomeric optically active 1-methylheptyl 2-fluoro-2-phenylacetates appears equally in both diastereoisomeric products, indicating that the radical pairs giving rise to CIDNP show no stereochemical preference in their combination.[112] Contrary to an earlier claim, diphenyldiazomethane does insert into the C—H bonds of solvent cyclohexane on photolysis, as does the carbene generated by photolysis of tetraphenylmethane.[113] Similarly, copper-catalysed decomposition of ethyl diazoacetate or diazoacetophenone in cyclohexane gives appreciable C—H insertion, though the copper carbenoid had been thought not to be reactive enough to insert; indeed, other diazo-compounds, including diphenyldiazomethane, give no insertion products under these conditions.[114]

Other insertion reactions reported include the competitive insertion and addition of $:CCl_2$ to homoadamantene,[115] insertion of $:CCl_2$ into germacyclohexanes[116] and into $SiBr_4$ and $GeBr_4$,[117] of $CHF_2CF:$ into Si—H and Si—halogen bonds,[118] and of alkoxycarbonylcarbenes into Si—H bonds of organosilicon hydrides through an only slightly ionic transition state.[119,120]

Some phosphorylnitrenes, $(RO)_2P(=O)N:$, are very resistant to rearrangement to produce a weak $3p$–$2p$ P—N π-bond, and therefore give excellent yields of products of C—H insertion into neighbouring molecules, showing extremely low selectivity towards different types of bonds.[121]*

Perfluorophenylnitrene undergoes insertion into the C—H bonds of benzene and N—H bonds of aniline and fluorinated anilines.[122] The effect of methylene chloride on the singlet–triplet crossing of alkoxycarbonylnitrenes and hence on their insertion reactions is discussed,[123] and insertion of ethoxycarbonylnitrene into various bicyclo-[4.n.0]alkanes is compared.[124]

* Intramolecular insertion reactions in vinylphosphorylcarbenes are also reported, and allylphosphoryl carbenes undergo intramolecular addition to form bicyclo[1.1.0]butanes.[121a]

Intramolecular

Various insertion reactions of the carbenoids derived by treatment of dibromocyclopropanes with an alkyl-lithium have been discussed.* The usual reaction of these carbenoids, ring expansion to an allene, can be suppressed by intramolecular interaction with oxygen as in the formation of (**60**),[125] by tetrasubstitution as in the cases of (**61**)[126] or (**62**),[127] or by fusing the cyclopropane to a second ring containing six carbons, as in the reactions of (**63**) and (**64**) and related systems.[128]

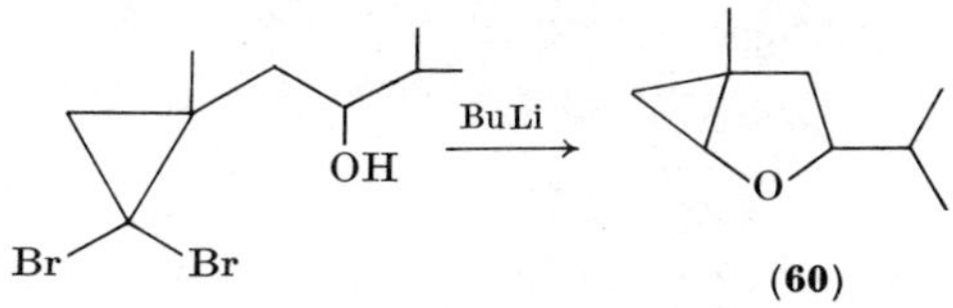

BuLi

OH

Br Br

(**60**)

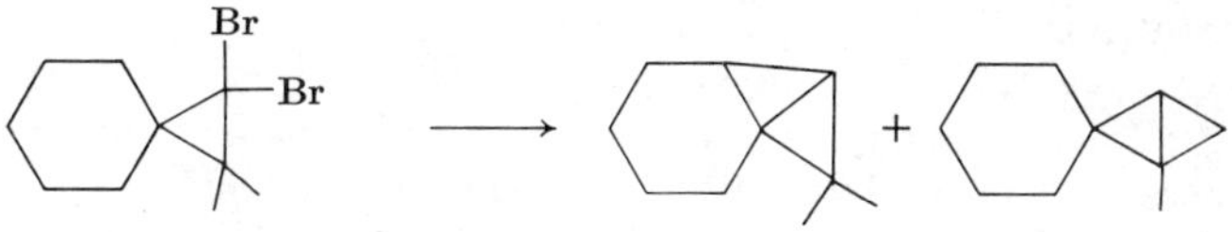

Br

Br

(**61**)

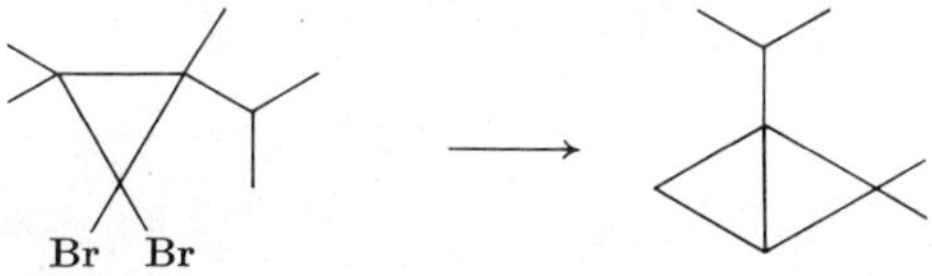

Br Br

(**62**)

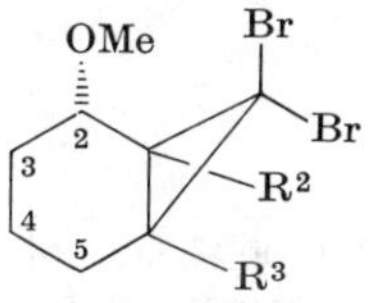

OMe Br Br R^2 R^3

(**63**)

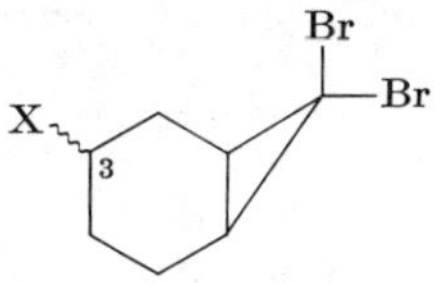

X Br Br

(**64**)

MeO

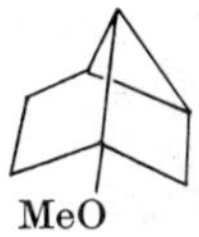

(**65**)

The carbenoid from (**63**; R^2, R^3 = H or alkyl) undergoes predominant insertion into the C^3—H bond, possibly owing to participation by the neighbouring oxygen to offset developing electron deficiency at C-3; insertion into the C^2—H bond is deterred by the electronegativity of the substituent, though some attack on the C^5—H bond does occur. However, the *anti*-isomer of (**64**; X = OMe) does undergo predominant insertion at the ether C^3—H bond to give (**65**); an explanation is provided in terms of the position of transition states. The carbenoid derived by reaction of (**66**) with BuLi inserts into the γ-C—H bond of the substituent to give cyclopentenes (**67**).[129] While photolysis of (**68**) in benzene gives only a dimeric product, thermolysis gives (**69**), possibly by rearrangement of the C—H insertion product (**70**).[130]

* The lithiobromocarbenoids may be trapped at low temperature by reaction with carbonyl compounds.[125a]

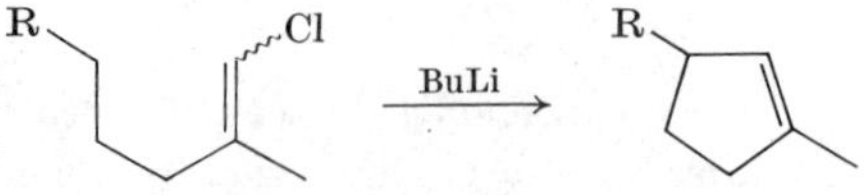

BuLi

(66) **(67)**

R = —CH=CMe$_2$ or —CMe$_2$CH=CMe$_2$

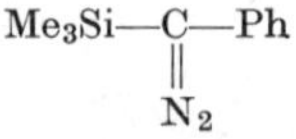

(68) 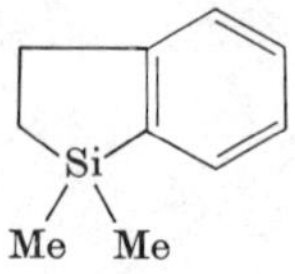 **(69)** 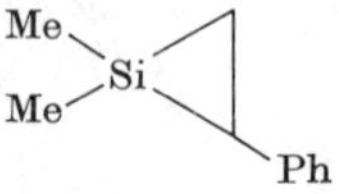 **(70)**

Aromatic Substitution

The Reimer–Tiemann reaction converts 2,2′-bridged phenols into the 2-dichloromethyl derivative, whereas 2,6-di-*tert*-butyl-4-methylphenol gives (**71**).[131]

The carbene derived thermally from (**72**) reacts with monosubstituted benzenes to give cycloheptatrienes, and a Hammett study shows it to be an electrophilic species more closely resembling ethoxycarbonylnitrene than ethoxycarbonylcarbene.[132]

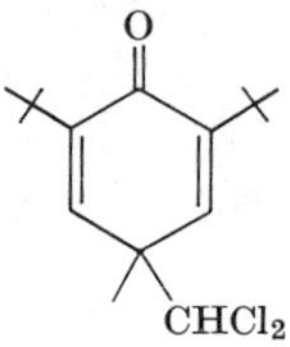

(71) 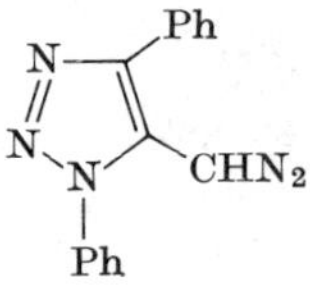 **(72)**

The reaction of sulphonylnitrenes with aromatic compounds having electron-withdrawing substituents leads to *m*-anilides via the singlet species, but crossing to the triplet leads to *o*-anilides.[133] The singlet nitrene is thought to lead to an aziridine which under kinetic control gives an *N*-mesylazepine but under thermodynamic control an *N*-mesylaniline.[134] Intramolecular substitution of a nitrene into aromatics is used to obtain a variety of heterocycles.[135]

Reaction with Nucleophiles

Intermolecular

Diazomethane may be cheaply and conveniently prepared by a modification of a long-known reaction; hydrazine hydrate is treated with dichlorocarbene generated by the phase-transfer process with [18]-crown-6-ether as the catalyst.[136] The kinetics of decomposition of diphenyldiazomethane in the presence of butylamines leading to *N*-(diphenylmethyl)butylamine have been studied, the reaction occurring by attack of singlet carbene on nitrogen followed by a rearrangement, and the triplet carbene leading principally to tetraphenylethylene.[137] A further example of the cleavage of aziridines by dichlorocarbene is reported,[138] and the conversion of amides into nitriles by $:CCl_2$ in a phase-transfer system is discussed.[139]

Ethoxycarbonylnitrene adds to aliphatic imidates to give the previously undescribed aliphatic hydrazonates (**73**) as a mixture of *E*- and *Z*-isomers; the imidate NH reacts less rapidly than the NH of primary or secondary amines, and only 2.3 times as fast as each CH group of cyclohexane.[140]

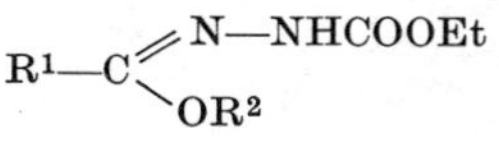

(73) 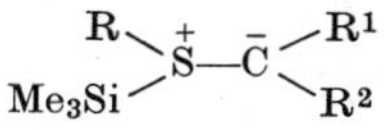 **(74)** 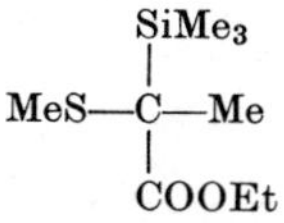 **(75)**

Photochemically generated $:CR_2$ (R = COMe, COOMe or COOEt) undergoes a highly selective reaction with aliphatic and cyclic sulphides to produce sulphonium ylides; thus with 4-*tert*-butylthiacyclohexane only equatorial attack occurs. Nitrenes, however, give equal amounts of axial and equatorial attack, presumably owing to reduced steric requirements.[141] Photolysis of diazo-compounds in alkyltrimethylsilyl sulphides or ethers gives the products of carbene insertion into Si—S and Si—O bonds, respectively, presumably by initial attack on S or O to produce, e.g. (**74**), followed by a 1,2-trialkylsilyl shift.[142] Similarly, photochemically produced trimethylsilyl(ethoxycarbonyl)carbene inserts into the C—S bond of dimethyl sulphide to give (**75**); the presumed intermediate ylide can indeed be isolated at low temperature but does not give (**75**) on warming. Possibly a second photochemical step is involved.[143]

The primary product (**76**) of the reaction between $RC{\equiv}CCH_2SPh$ and $:C(COOMe)_2$ undergoes an efficient 2,3-sigmatropic shift to (**77**). Similarly the carbene converts the allene $PhSCH_2CH{=}C{=}CHC_5H_{11}$ into (**78**).[144]

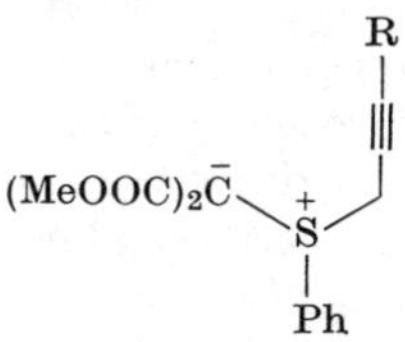

(76) 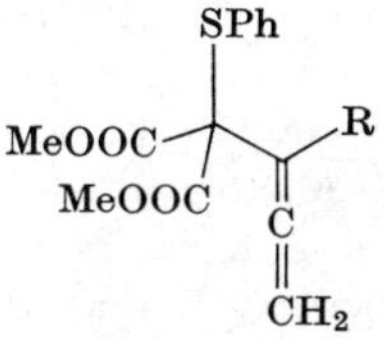 **(77)**

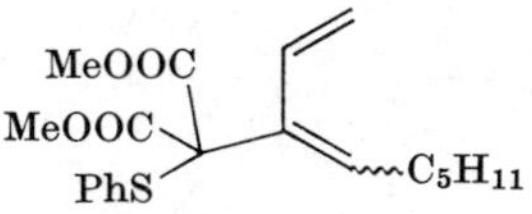

(78) 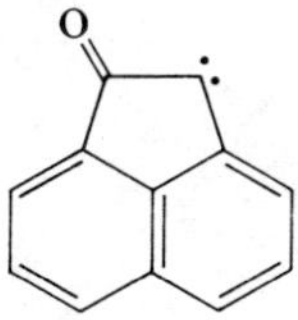 **(79)**

The reaction between methoxycarbonylnitrenes and allylic sulphides or ethers is discussed.[145] There are reports of the formation of $Ar_3As{=}NX$ by reaction of nitrenes with Ar_3As;[146] and a similar reaction occurs with arylcyclopentadienylidenes[147] and between phosphines and (**79**).[148]

Intramolecular

Copper-catalysed decomposition of (**80**) leads largely to (**81**), via the cyclic carbonyl ylide (**82**).[149] Cyclopropenones have been established as intermediates in the photolytic decomposition of 2,6-bis(diazo)cyclohexanone and 1,3-bis(diazo)diphenylpropan-2-one, and in the Ag_2O- and Cu_2Cl_2-catalysed decomposition of the latter; the reactions presumably proceed by interaction of a carbene with the carbon of the second diazo-group, as illustrated.[150]

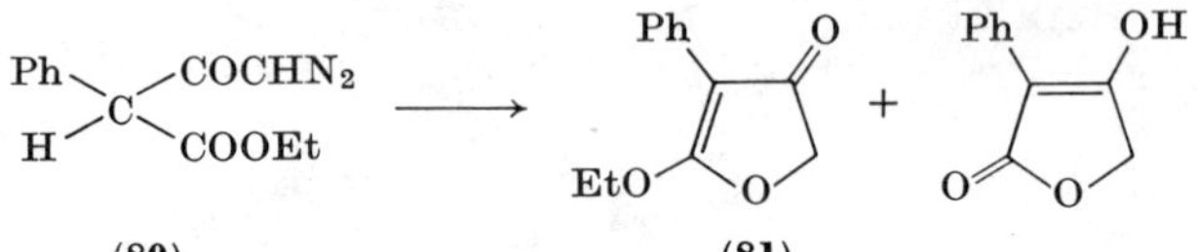

Ph, H, COCHN₂, COOEt → Ph, O, EtO, O + Ph, OH, O, O

(**80**) (**81**)

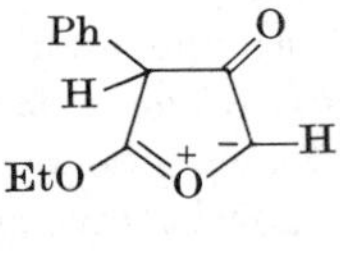

Ph, O, H, H, EtO, O

(**82**)

R, C:, O=C, N₂, C, R′ → R, O=, N₂⁺, R′

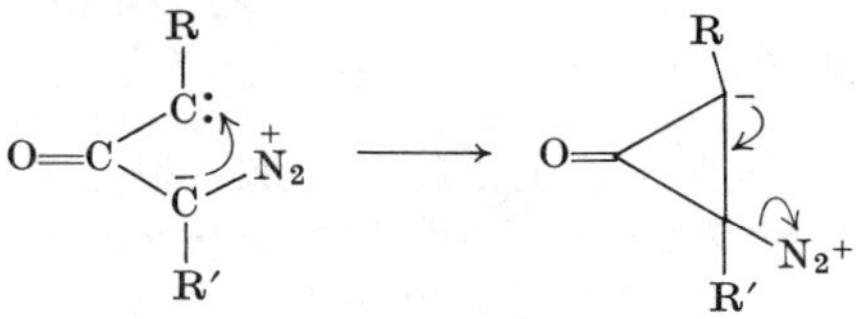

Rearrangements

Alkyl and Hydrogen Shifts

Non-empirical calculations on the singlet ground state of methylcarbene show that the 1,2-hydrogen shift to ethylene involves migration of a methyl-hydrogen *gauche* to the methine-hydrogen.[151] Such shifts are characteristic of singlets rather than triplets, and three geometries may be envisaged, synplanar (**83**), perpendicular (**84**), and antiplanar (**85**), where Z is the migrating group; the perpendicular geometry is found to be considerably favoured in the thermal Bamford–Stevens reaction of the isomeric labelled brexane-5-ones (**86**) (for which the *exo*-hydrogen is effectively perpendicular, and the *endo*-hydrogen approximates to antiplanar). In the corresponding photochemical reaction, the selectivity is far less marked, suggesting a high energy state for the derived carbene or a non-carbenoid pathway.[152]

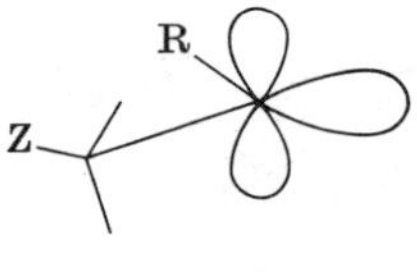

R, Z (**83**) R, Z (**84**) R, Z (**85**)

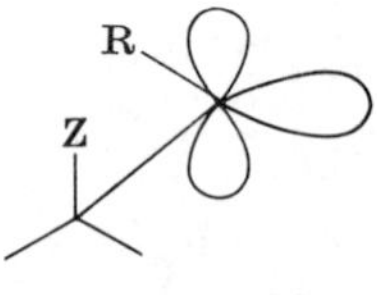

O, H, D

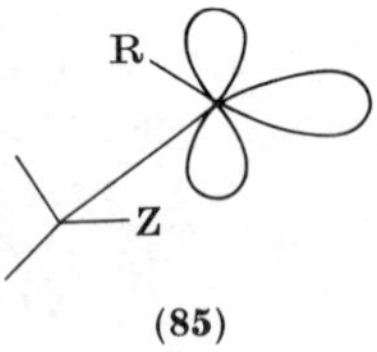

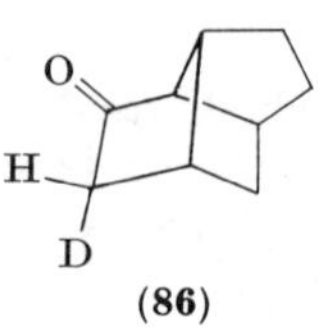

(**86**)

The carbenes Ph_2MeCCH: and PhMeHCCPh: do not rearrange to stilbene via a common intermediate, the singlet (**87**); phenyl migration in the former leads to the *Z*-isomer, whereas H-migration in the latter gives the *E*-isomer.[153]

The rate of thermal decomposition of 3-chloro- or 3-bromo-3-ethyldiazirine is not altered by the solvent. However, in non-polar solvents the derived carbene undergoes a 1,2-shift predominantly to *cis*-1-halogenopropene, whereas in polar solvents the thermodynamically less stable *trans*-alkene is increasingly formed.[154] Photolysis of tetra-substituted alkenes leads to products which may be explained in terms of an alkyl shift to produce a carbene; thus 3,4-dimethylhex-3-ene and 3-ethyl-2-methylpent-2-ene both lead to products apparently derived from (**88**) and (**89**).[155]

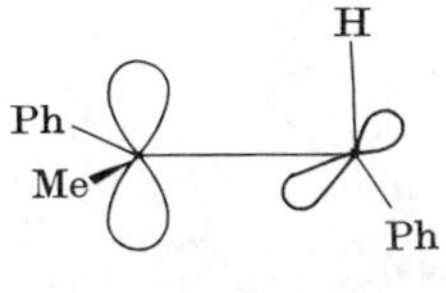

(**87**) 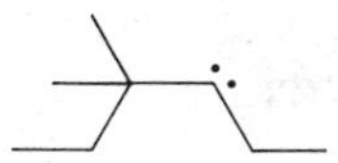 (**88**) 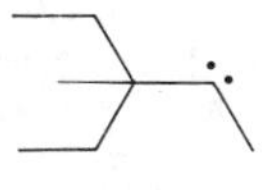 (**89**) 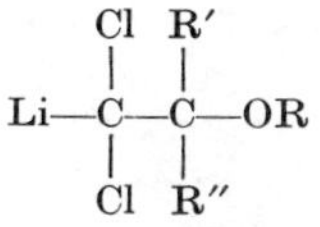 (**90**)

β-Alkoxycarbenoids (**90**) may undergo either α- or β-elimination; the latter is favoured for secondary systems with linear substituents and for tertiary systems, but bulky groups on the γ-position lead to α-elimination. Intramolecular oxygen–lithium interaction may also stop β-elimination.[156] β-Oxidocarbenoids may undergo a synthetically useful rearrangement; thus 1-dibromomethylcyclododecanol is converted into cyclotridecanone (89%) by *n*-butyl-lithium.[157]

Substituent effects in the reactions of (**91**, *exo*) suggest charge transfer from the cyclopropane at the transition state. The tosylhydrazone salt precursor to (**91**, *endo*; X = COOEt) rearranges to the precursor of (**91**, *exo*) under the reaction conditions.[158] The *syn*- and *anti*-isomers of (**92**) rearrange when heated with base to (**93**), via a similar initial fragmentation.[159]

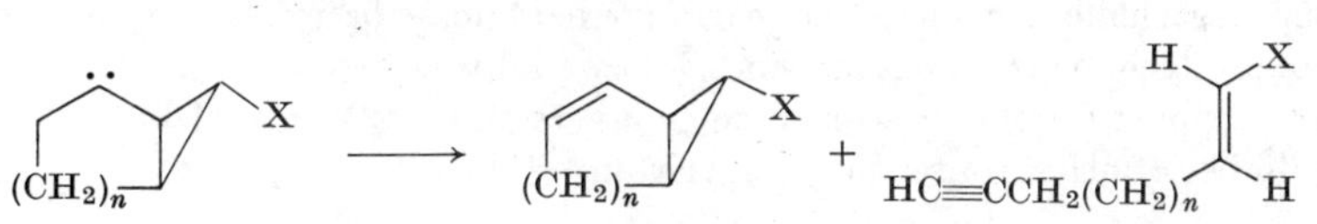

(**91**; *exo*)

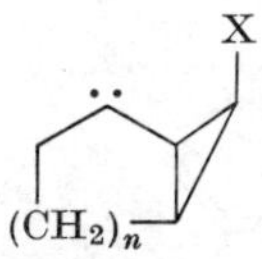

(**91**; *endo*)

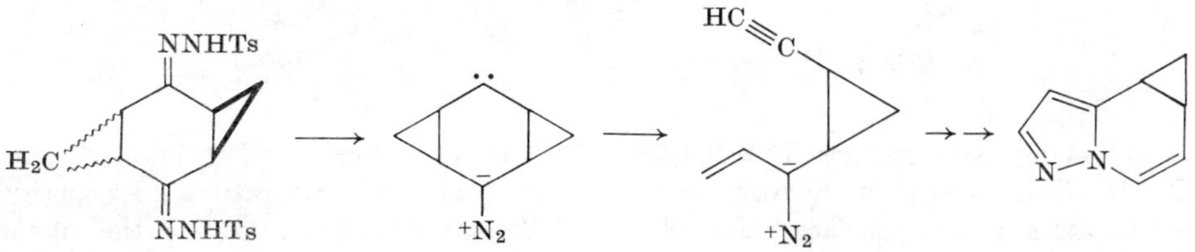

(**92**) (**93**)

Pyrolysis of the sodium salt of (**94**) leads to the pentacyclic hydrocarbon (**95**), presumably by a rearrangment of the initially produced cyclopropylcarbene to a cyclobutene followed by an internal Diels–Alder addition.[160] A 1,2-aryl shift in an intermediate carbenacyclopropane is supposed to occur in the conversion of (**96**) into (**97**) with sodium naphthalene.[161] The synthesis of paracyclophanes by a 1,4-alkyl shift across the cyclic carbenes (**98**) has been extended to [6]paracyclophane, the reactions probably occurring by a cleavage of the carbene to a diradical followed by recombination.[162]

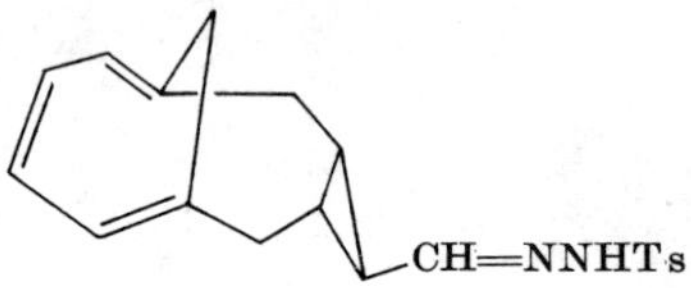

CH═NNHTs

(**94**)

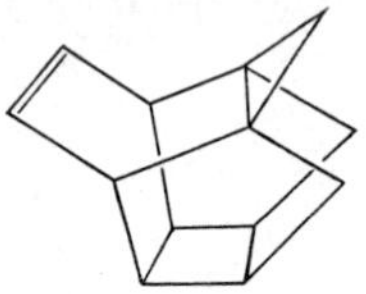

(**95**)

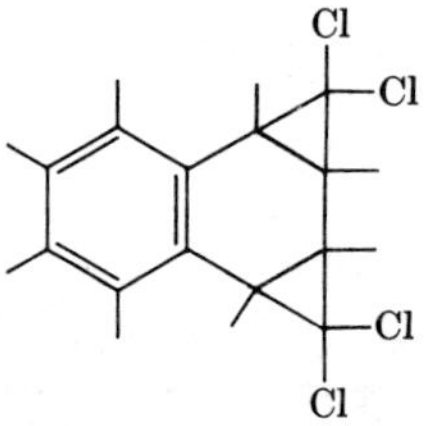

(**96**) 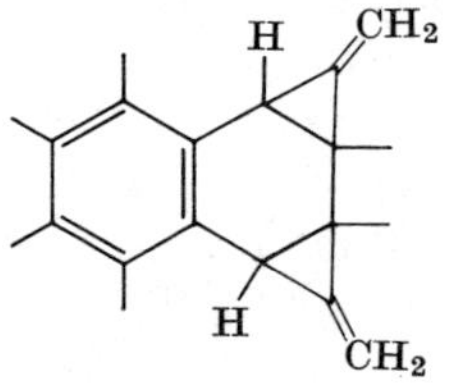 (**97**) 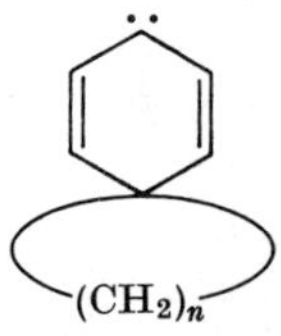 (**98**)

Pyrolysis of Ph^{13}C≡CH at 700° leads to a product with the label equally distributed between the acetylenic carbons; the rearrangement may be explained in terms of a 1,2-shift to produce benzylidenecarbene, followed by a reverse shift of the second substituent. Support for this is seen in the conversion of (**99**) into (**100**) on pyrolysis, though a number of other routes may be written.[163]

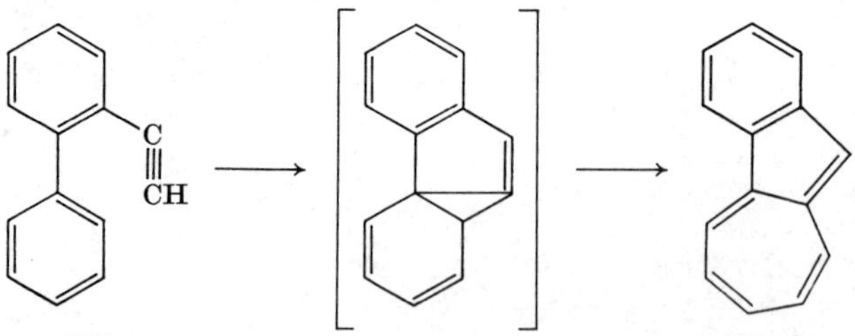

(**99**) (**100**)

At low temperatures PhCH═CHCMe═CHCl is converted into a stable carbenoid PhCH═CHCMe═C(Li)Cl by reaction with *n*-BuLi; at higher temperatures a 1,2-methyl shift occurs to give the acetylene PhCH═CHC≡CMe with retention of the alkene stereochemistry.[164]

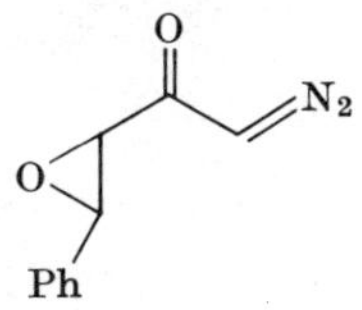

(**101**) 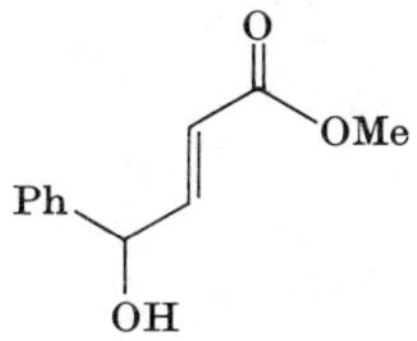 (**102**)

Photolysis of (**101**) in methanol leads to (**102**), presumably by attack of methanol on the ketene derived by Wolff rearrangement of the keto-carbene.[165] The Wolff rearrangement has been used for the contraction of [8]- to [7]-paracyclophanes.[166] Though (**103**) undergoes a normal rearrangement on photolysis in methanol, leading to (**104**), the product of thermal decomposition in the presence of $CuSO_4$ is (**105**); the initially produced carbene might undergo a 2,3-sigmatropic shift to the ketene (**106**), or an intramolecular addition to form (**107**), followed by methanolysis.[167]

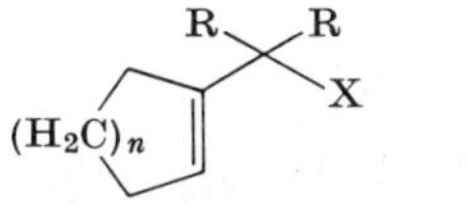

(**103**) X = $COCHN_2$
(**104**) X = CH_2COOMe

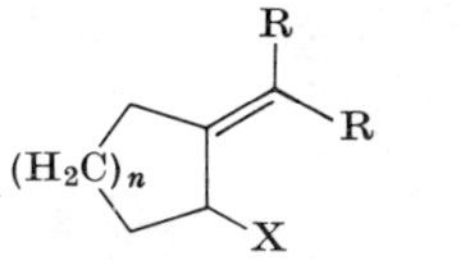

(**105**) X = CH_2COOMe
(**106**) X = CH═C═O

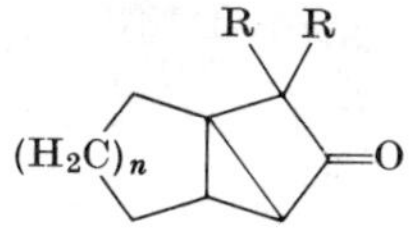

(**107**)

1,2-Migration in Me_2CHON: is also discussed,[168] and an unusual version of the Lossen reaction occurs on treatment of ArCONHOH with triphenylphosphine diethyl azodicarboxylate and ethanol.[169]

The formation of (**109**) on thermolysis of (**108**) may proceed through an intermediate carbene.[170]

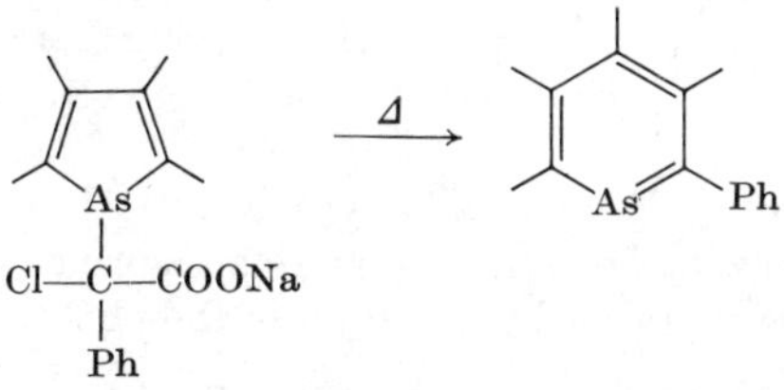

(**108**) (**109**)

Carbene–Carbene Rearrangements

Theoretical calculations on the phenylcarbene to cycloheptatrienylidene interconversion are reported, the latter species apparently being more stable in the cycloheptatetraene form.[171] CNDO/2 and Hückel calculations on the rearrangement of phenylnitrene to (**110**) predict the N to be moving out of the molecular plane on the path of minimum energy; the net charge during the reaction indicates that an interaction between a filled nitrene *p*-orbital and an empty MO of the ring is important in lowering the energy.[172] Similar reasoning is applied to the experimental results for the rearrangement of tolylcarbene.

(110)

Heats of formation were estimated for aromatic carbenes and nitrenes and used to explain experimental data for their rearrangements; chemical activation was found to be very important and its inclusion corrected previous misinterpretations.[173] There have been reports of experiments designed to prove the intermediacy of discrete cyclopropenes or azirenes in such rearrangements. Diels–Alder adducts are reported of the cyclopropene intermediate proposed in the rearrangement of (**111**) to 9-phenanthrylcarbene and for a variety of related systems.[174] The dichloride (**112**; X = Cl, Y = H) is converted by $KOBu^t$–THF into (**113**), presumably derived by insertion of 9-phenanthrylchlorocarbene into the solvent; evidence for a cyclopropene intermediate is obtained by the formation of (**112**; X = H, Y = SMe) on carrying out the reaction with $KOBu^t$–DMSO–MeSH.[175]

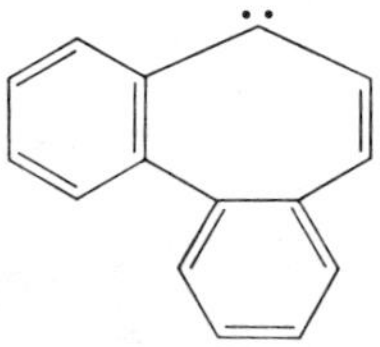

(111) 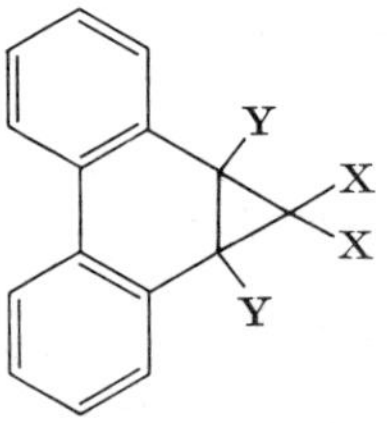 (112) 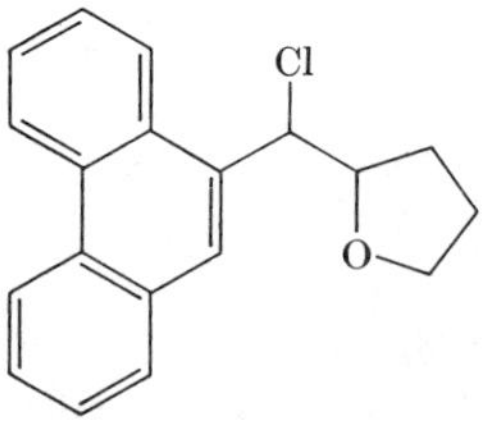 (113)

Substituent effects in the photolytic reactions of 2-biphenylyl azides in dialkylamines are interpreted in terms of a reaction pathway involving rearrangement of a biphenylylnitrene to an azirine which can be diverted to an azepine by the base, or, in its absence, lead to a carbazole.[176]

Practicable yields of 2-alkoxy-3-(alkoxycarbonyl)-3*H*-azepines are obtained by photolysis of *o*-azidobenzoate esters, trapping the azirine with solvent methanol.[177] Azirines are also apparently formed transiently in the photolysis of (**114**).[178]

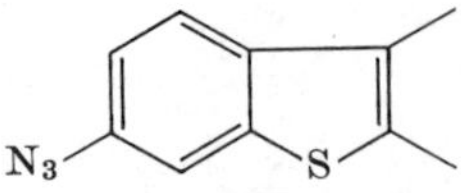

(114)

Thermolysis of aryl azides in benzoyl chloride gives *o*-chlorobenzanilides, possibly derived by trapping of an azirine intermediate itself derived from the singlet nitrene; the triplet nitrene gives the benzanilide itself.[179]

8-Phenyl-2-quinolylnitrene rearranges to 8-phenyl-1-isoquinolylnitrene, presumably by a ring expansion–ring contraction sequence; the second nitrene is trapped by intramolecular attack on the 8-phenyl group.[180]

Thermolysis of the sodium salt of the tosylhydrazone of (**115**) would formally produce (**116**); however, examination of the dimeric products of the reaction suggests that the intermediate might be better described as an allene.[181]

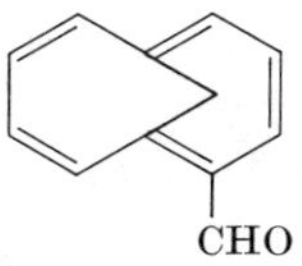

CHO

(**115**) 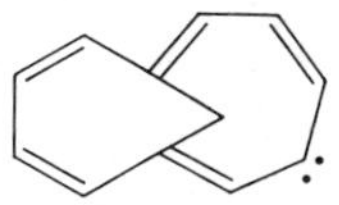 (**116**)

Fragmentation

The photochemical fragmentation of cyclopropyl azides leads to alkenes, nitriles and nitrogen with retention of stereochemistry, probably via nitrenes, in contrast to the thermal reaction which leads to 1-azetines.[182]

Transition-metal Complexes

The proliferation of examples of transition-metal carbene complexes has continued, and the area has been reviewed.[183]

The reaction of isocyanides with amines and $AuCl_4^-$ provides a route to diamino-carbene complexes,[184] and imidoyl chlorides and certain rhodium complexes react in the presence of HCl to give alkyl- and aryl-aminocarbene complexes.[185] Treatment of $Ph_3GeMn(CO)_5$ with MeLi leads to $Ph_3GeMn(CO)_4C(Me)OLi$ which may be alkylated to the ethoxymethylcarbene complex.[186] Carbenoids are also obtained from the reaction of $Ti(NMe_2)_4$ with metal carbonyls[187] and treatment of the complex $FeMe_2$(diphenyl-phosphinoethane)$_2$ with methylene chloride liberates ethylene, suggesting the formation of a carbenoid intermediate.[188]

A variety of alkoxy- and amino-carbene complexes is reported,[189] as are derivatives of methylselenocarbene,[190] 2-oxacyclopentylidene,[191] 2,5-dioxacyclopentylidene,[192] thiazolidinylidenes and pyridinylidenes,[193] and cationic carbene complexes of thiazoles and related systems.[194] Derivatives of vinylmethoxycarbene,[195] carbonylmethylene and thiocarbonylmethylene[196] are also reported.

The first alkylcarbene complex, a derivative of tantalum, is reported.[197] Diphenyl-carbene complexes of rhodium having the ligand in a bridging position are discussed.[198]

The reactions of carbene complexes with acid,[199] alkylation,[200] and ligand rearrangements[201] and displacements have been discussed. A number of diamino- and alkoxyamino-carbene complexes undergo ligand displacement with cyanide ion; thus MeHNC(OMe)AuCl gives MeN=CHOMe.[202,184] Aerial oxidation of a diamino-carbene complex results in the displacement of the ligand as a urea.[203]

The carbene ligand itself may be modified within the complex. Reaction with Br_2PPh_3[204] or BX_3[205] can lead to a carbyne complex, reaction with amines can produce nitrogen ylides,[206] and methoxycarbene derivatives may be transformed to those of phenylthiocarbenes by reaction with NaSPh and HCl.[207] The α-hydrogen atoms of methylcarbene derivatives may be removed by base and the resulting anions alkylated;[208] 1,4-Addition of germylgermylenes and difluoro- and methoxy-phenyl-germylenes to

nucleophilic attack.[209] Tungsten complexes of $EtO\ddot{C}C\equiv CPh$ react with diazomethane to form organometallic derivatives of (**117**).[210]

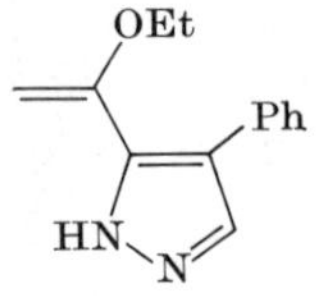

(**117**)

Pentacarbonyl(methoxyphenylcarbene)Cr(0) cleaves 1-vinyl-2-pyrrolidone at the carbon—carbon double bond to produce α-methoxystyrene, via a cyclic transition state.[211]

The reaction of various transition-metal complexes with carbene precursors is discussed by several authors;[212] the carbene derived from $PhC(N_2)COPh$ may be trapped as a Mn complex before the Wolff rearrangement occurs.[213]

Reversible α-H-abstraction from the methyl groups of $CoMePPh_3$ and $RhMePPh_3$ may involve an intermediate carbene–hydride complex.[214]

Silylenes and Germylenes

Reviews of Si—Si bonds and low-valent silicon include sections on silylenes.[215] Selection rules have been applied to the photolysis of polysilanes to generate silylenes,[216] and the loss of $:SiMe_2$ from *cis*- and *trans*-(**118**) has been shown to be highly stereospecific.[217]

Silylene adds to butadiene to give 1-silapent-3-ene;[218] the reacting species is predominantly a triplet, but the ground state is found to be a singlet,[219] in agreement with calculations.[220] The addition to *trans,trans*-hexa-2,4-diene gives both *cis*- and *trans*-isomers of 2,5-dimethyl-1-silacyclopent-3-ene, suggesting a diradical pathway.[221] 1,4-Addition of germylgermylenes and difluoro- and methoxy-phenyl-germylenes to butadiene also occurs.[222] Methoxymethylsilylene does not, however, give the 1,4-adduct with cyclo-octa-1,3-diene, but instead produces (**119**), possibly by diradical cleavage of the 1,2-adduct followed by a 1,5-hydrogen shift.[223]

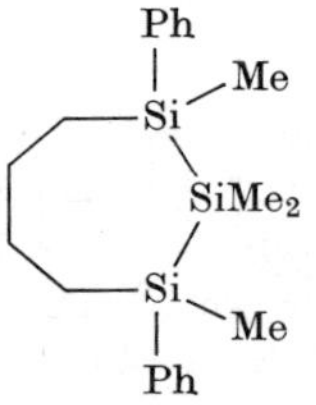

(**118**) 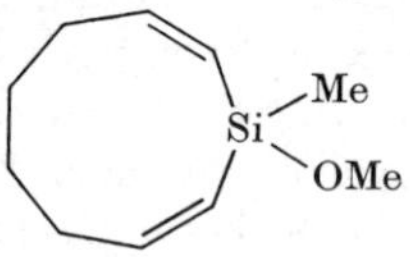 (**119**) 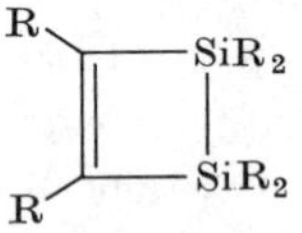 (**120**)

Evidence is provided that the addition of silylenes to acetylenes to give 1,4-disilacyclohexadienes occurs with the initial formation of (**120**) which may open to a 1,4-disilabuta-1,3-diene and then undergo a Diels–Alder reaction with an excess of acetylene.[224]

Relative rates of insertion of GeH_2 and related species into Ge—H and into R—X (R = alkyl or $MeOCH_2$, X = halogen) bonds are discussed.[225]

References

[1] A. P. Marchand and N. MacBrockway, *Chem. Rev.*, **74**, 431 (1974).
[2] O. M. Nefedov and A. I. Ioffe, *Zh. Ves. Khim. Obshchest.*, **19**, 305 (1974).
[3] B. V. Ioffe, V. P. Semenov and K. A. Ogloblin, *Zh. Ves. Khim. Obshchest.*, **19**, 314 (1974).
[4] J. T. Sharp, *Ann. Rep. Progr. Chem., Sect. B*, **69**, 213 (1972); **70**, 177 (1973).
[5] G. Schmid, *Chem. Unserer Zeit*, **8**, 26 (1974).
[6] J. D. Coyle, *Educ. Chem.*, **1974**, 62.
[7] H. Dürr, *Topics Curr. Chem.*, **40**, 103 (1973).
[8] W. A. Lathan, L. Radom, P. C. Hariharan, W. J. Hehre and J. A. Pople, *Topics Curr. Chem.*, **40**, 1 (1973).
[9] J. F. Harrison, *Accounts Chem. Res.*, **1974**, 378.
[10] M. J. S. Dewar, R. C. Haddon and P. K. Weiner, *J. Am. Chem. Soc.*, **96**, 253 (1974).
[11] G. F. Tantardini, M. Raimondi and M. Simonetta, *Int. J. Quantum Chem.*, **1973**, 893.
[12] R. G. A. R. Maclagan and H. D. Todd, *Theor. Chim. Acta*, **34**, 19 (1974).
[13] W. R. Wadt and W. A. Goddard, *J. Am. Chem. Soc.*, **96**, 5996 (1974).
[14] S. Ehrenson, *J. Am. Chem. Soc.*, **96**, 3784 (1974).
[15] H. M. Frey, G. R. Jackson, M. Thompson and R. Walsh, *J.C.S. Faraday I*, **1973**, 2054.
[16] F. S. Rowland, P. S. Lee, D. C. Montague and R. L. Russell, *Faraday Discuss. Chem. Soc.*, **1973**, No. 53, 111.
[17] R. S. Hutton, M. L. Manion, H. D. Roth and E. Wasserman, *J. Am. Chem. Soc.*, **96**, 4680 (1974).
[18] G. E. Palmer, J. R. Bolton and D. R. Arnold, *J. Am. Chem. Soc.*, **96**, 3708 (1974).
[19] G. A. Nikiforov, S. A. Markaryan, L. G. Plekhanova, B. D. Sviridov, T. Pehk, E. Lipmaa and V. V. Ershov, *Izv. Akad. Nauk SSSR*, **1974**, 93.
[20] P. Devolder, P. Goudmand and J. P. Grivet, *J. Chim. Phys. Physicochim. Biol.*, **71**, 899 (1974).
[21] D. R. Yarkony, H. F. Schaefer and S. Rothenberg, *J. Am. Chem. Soc.*, **96**, 5974 (1974).
[22] E. V. Dehmlow, *Angew. Chem. Int. Ed.*, **13**, 170 (1974).
[23] M. Makosza and M. Ludwikow, *Angew. Chem. Int. Ed.*, **13**, 665 (1974).
[24] K. Isagawa, Y. Kimura and S. Kwon, *J. Org. Chem.*, **39**, 3171 (1974).
[25] R. Mathias and P. Weyerstahl, *Angew. Chem. Int. Ed.*, **13**, 132 (1974).
[26] T. B. Patrick, *Tetrahedron Letters*, **1974**, 1407.
[27] T. Sasaki, S. Eguchi and T. Ogawa, *J. Org. Chem.*, **39**, 1927 (1974).
[28] R. H. Prager and H. C. Brown, *Synthesis*, **1974**, 736.
[29] A. N. Nesmeyanov, A. E. Borisov and N. G. Kizim, *Izv. Akad. Nauk SSSR*, **1974**, 1672.
[30] D. J. Burton and D. G. Naae, *J. Am. Chem. Soc.*, **95**, 8467 (1973).
[31] P. J. Stang, M. G. Mangum, D. P. Fox and P. Haak, *J. Am. Chem. Soc.*, **96**, 4562 (1974).
[32] W. Mahler and P. R. Resmck, *J. Fluorine Chem.*, **3**, 451 (1973/74).
[33a] D. Seyferth and R. A. Woodruff, *J. Organometal. Chem.*, **71**, 335 (1974).
[33b] D. Seyferth, *J. Organometal. Chem.*, **62**, 33 (1973); **75**, 13 (1974).
[34] J. M. Birchall, G. N. Gilmore and R. N. Haszeldine, *J.C.S. Perkin I*, **1974**, 2530.
[35] R. N. Haszeldine, D. L. Scott and A. E. Tipping, *J.C.S. Perkin I*, **1974**, 1440.
[36] L. A. Wendling and R. G. Bergman, *J. Am. Chem. Soc.*, **96**, 308 (1974).
[37] W.-D. Stohrer, P. Jacobs, K. H. Kaiser, G. Wiech and G. Quinkert, *Topics Curr. Chem.*, **46**, 181 (1974).
[38] G. Quinkert and P. Jacobs, *Chem. Ber.*, **107**, 2473 (1974).
[39] P. M. Collins, N. N. Oparaeche and B. R. Whitton, *Chem. Comm.*, **1974**, 292.
[40] J. Nakayama, *Chem. Comm.*, **1974**, 166.
[41] G. Scherowsky and J. Weiland, *Annalen*, **1974**, 403.
[42] H. Iwamura and K. Yoshimura, *J. Am. Chem. Soc.*, **96**, 2652 (1974); H. Iwamura, *Chem. Letters*, **1974**, 5.
[43] M. J. Wyvratt and L. A. Paquette, *Tetrahedron Letters*, **1974**, 2433.
[44] J.-J. Katz, B. A. Carlson and H. C. Brown, *J. Org. Chem.*, **39**, 2817 (1974).
[45] R. A. Abramovitch and E. P. Kyba, *J. Am. Chem. Soc.*, **96**, 480 (1974).
[46] R. N. Carde and G. Jones, *J.C.S. Perkin I*, **1974**, 2066, 2072.
[47] M. Komatsu, S. Ichijima, Y. Ohshiro and T. Agawa, *J. Org. Chem.*, **38**, 4341 (1973).
[48] G. L'abbé and G. Mathys, *J. Org. Chem.*, **39**, 1778 (1974).
[49] A. Padwa, J. Smolanoff and A. Tremper, *Tetrahedron Letters*, **1974**, 29.
[50] F. P. Tsui, T. M. Vogel and G. Zon, *J. Am. Chem. Soc.*, **96**, 7144 (1974).
[51] J. I. G. Cadogan and I. Gosney, *J.C.S. Perkin I*, **1974**, 466.
[52] B. V. Ioffe and L. A. Kartsova, *Zh. Org. Khim.*, **10**, 989 (1974).

[53] R. E. White and P. Kovacic, *J. Am. Chem. Soc.*, **96**, 7284 (1974).
[54] S. Mataka and J.-P. Anselme, *Chem. Comm.*, **1974**, 554.
[55] T. Kametani, F. F. Ebetino and K. Fukumoto, *Tetrahedron*, **30**, 2713 (1974); *J.C.S. Perkin I*, **1974** 861.
[56] H. E. Simmons, T. L. Cairns, S. A. Vladuchick and C. M. Hoiness, *Org. Reactions*, **20**, 1 (1973).
[57] S. Miyano and H. Hashimoto, *Bull. Chem. Soc. Japan*, **1974**, 1500.
[58] H. H. Wasserman, G. M. Clark and P. C. Turley, *Topics Curr. Chem.*, **47**, 73 (1974).
[59] T. Kametani, F. F. Ebetino, T. Yamanaka and K. Nyu, *Heterocycles*, **2**, 209 (1974).
[60] H. Person, C. Fayat, F. Tonnard and A. Foucaud, *Bull. Soc. Chim. France*, **1974**, 635.
[61] J. W. Haines and I. G. Csizmadia, *Theor. Chim. Acta*, **31**, 283 (1973).
[62] H. Fujimoto and R. Hoffmann, *J. Phys. Chem.*, **1974**, 1167.
[63] N. Shimizu and S. Nishida, *J. Am. Chem. Soc.*, **96**, 6451 (1974).
[64] Y. Tatsuno, A. Konishi, A. Nakamura and S. Otsuka, *Chem. Comm.*, **1974**, 588.
[65] A. Davison, W. C. Krusell and R. C. Michaelson, *J. Organometal. Chem.*, **72**, C7 (1974).
[66] R. A. Moss and F. G. Pilkiewicz, *J. Am. Chem. Soc.*, **96**, 5632 (1974).
[67] J. Dolanský, V. Bažant and V. Chvalovský, *Coll. Czech. Chem. Comm.*, **38**, 3816 (1973).
[68] R. A. Moss and D. J. Smudin, *Tetrahedron Letters*, **1974**, 1829.
[69] O. M. Nefedov and E. S. Agavelyan, *Izv. Akad. Nauk SSSR*, **1973**, 2045; *Chem. Abs.*, **80**, 59120 (1974).
[70] M. S. Baird, *Chem. Comm.*, **1974**, 196.
[71] R. R. Kostikov, A. P. Molchanov and A. Y. Bespalov, *Zh. Org. Khim.*, **10**, 10 (1974); *Chem. Abs.*, **80**, 107631 197631 (1974).
[72] T. Sasaki, K. Kanematsu and Y. Yukimoto, *J. Org. Chem.*, **39**, 455 (1974).
[73] T. Sasaki, K. Kanematsu and Y. Yukimoto, *Heterocycles*, **1**, 1 (1973).
[74] T. Greibrokk, *Acta Chem. Scand.*, **27**, 3207 (1973).
[75] I. D. Kushina, O. M. Nefedov, A. A. Ivashenko, S. F. Politanskii, V. U. Shevchuk, I. M. Gutor and A. Ya. Shteinshneider, *Izv. Akad. Nauk SSSR*, **1974**, 728; I. D. Kushina, S. F. Politanskii, V. U. Shevchuk, I. M. Gutor, A. A. Ivashenko and O. M. Nefedov, *ibid.*, p. 946; *Chem. Abs.*, **81**, 24623 (1974).
[76] C. W. Jefford, nT. Kabengele, J. Kovacs and U. Burger, *Tetrahedron Letters*, **1974**, 257; *Helv. Chim. Acta*, **57**, 104 (1974).
[77] H. Bieräugel, J. M. Akkerman, J. C. L. Armande and U. K. Pandit, *Tetrahedron Letters*, **1974**, 2817.
[78] E. V. Dehmlow and G. Höfle, *Chem. Ber.*, **107**, 2760 (1974).
[79] R. Bartlet, *Compt. rend.*, *C*, **278**, 621 (1974).
[80] S. A. G. de Graaf and U. K. Pandit, *Tetrahedron*, **30**, 1115 (1974).
[81] R. Ikan, A. Markus and Z. Goldschmidt, *Israel J. Chem.*, **11**, 591 (1973).
[82] H. P. M. Thiellier, A. M. Van den Burg, G. J. Kooman and U. K. Pandit, *Heterocycles*, **2**, 467 (1974).
[83] H. Hart and M. Nitta, *Tetrahedron Letters*, **1974**, 2109.
[84] R. W. Hoffmann, W. Lilienblum and B. Dittrich, *Chem. Ber.*, **107**, 3395 (1974).
[85] T. Mitsuhashi and W. M. Jones, *Chem. Comm.*, **1974**, 103.
[86] K. Saito, Y. Yamashita and T. Mukai, *Chem. Comm.*, **1974**, 58.
[87] H. Dürr and F. Werndorff, *Angew. Chem. Int. Ed.*, **13**, 483 (1974).
[88] M. S. Baird, I. R. Dunkin and M. Poliakoff, *Chem. Comm.*, **1974**, 904.
[89] H. Dürr, B. Ruge and B. Weiss, *Annalen*, **1974**, 1150; H. Dürr, W. Schmidt and R. Sergio, *ibid.*, p. 1132.
[90] G. A. Nikiforov, B. D. Svirdov and V. V. Ershov, *Izv. Akad. Nauk SSSR*, **1974**, 373
[91] M. S. Newman and M. C. V. Zwan, *J. Org. Chem.*, **39**, 761 (1974).
[92] Y. Ito, K. Yonezawa and T. Saegusa, *J. Org. Chem.*, **39**, 1763 (1974).
[93] R. S. Atkinson and R. Martin, *Chem. Comm.*, **1974**, 386.
[94] F. Schröppel and J. Sauer, *Tetrahedron Letters*, **1974**, 2945.
[95] R. P. Gandhi, M. Singh and T. D. Sharma, *Indian J. Chem.*, **12**, 117 (1974).
[96] D. Seyferth, W. Tronich, W. E. Smith and S. P. Hopper, *J. Organometal. Chem.*, **67**, 341 (1974).
[97] A. Merz, *Synthesis*, **1974**, 724.
[98] R. R. Kostikov, A. F. Khlebnikov and K. A. Ogloblin, *Zh. Org. Khim.*, **9**, 2346 (1973); *Chem. Abs.*, **80**, 36488 (1974).
[99] D. Seyferth, W. Tronich and H.-m. Shih, *J. Org. Chem.*, **39**, 158 (1974).
[100] D. Seyferth and H.-m. Shih, *J. Org. Chem.*, **39**, 2329 (1974).
[101] D. Seyferth and H.-m. Shih, *J. Am. Chem. Soc.*, **95**, 8464 (1973); *J. Org. Chem.*, **39**, 2336 (1974).
[102] R. Paulissen, P. Moniotte, A. J. Hubert and P. Teyssie, *Tetrahedron Letters*, **1974**, 3311.
[103] M. S. Baird, *Chem. Comm.*, **1974**, 197.

[104] M. Baumann and G. Köbrich, *Tetrahedron Letters*, **1974**, 1217.
[105] A. Viola, S. Madhavan and R. J. Proberb, *J. Org. Chem.*, **39**, 3154 (1974).
[106] S. Bien and D. Ovadia, *J. Org. Chem.*, **39**, 2259 (1974); *J.C.S. Perkin I*, **1974**, 333.
[107] M. N. Nwaji and O. S. Onyiriuka, *Tetrahedron Letters*, **1974**, 2255.
[108] C. P. Baskin, C. F. Bender, C. W. Bauschlicher and H. F. Schaefer, *J. Am. Chem. Soc.*, **96**, 2709 (1974).
[109] S. P. Kolesnikov, A. I. Ioffe and O. M. Nefedov, *Izv. Akad. Nauk SSSR*, **1973**, 2622.
[110] I. D. Kushina, A. A. Ivashenko and S. F. Politanskii, *Izv. Akad. Nauk SSSR*, **1973**, 2398.
[111] D. Seyferth and Y.-M. Cheng, *Synthesis*, **1974**, 114.
[112] D. Bethell and K. McDonald, *Chem. Comm.*, **1974**, 467.
[113] J. H. Boyer, V. T. Ramakrishnan and K. G. Srinivasan, *Synthesis*, **1974**, 192.
[114] L. T. Scott and G. J. DeCicco, *J. Am. Chem. Soc.*, **96**, 322 (1974).
[115] T. Sasaki, S. Eguchi and M. Mizutani, *Org. Prep. Preced. Int.*, **6**, 57 (1974).
[116] B. C. Pant, *J. Organometal. Chem.*, **66**, 321 (1974).
[117] M. Weidenbruch and C. Pierrard, *J. Organometal. Chem.*, **71**, C29 (1974).
[118] R. N. Haszeldine, A. E. Tipping and R. O'B. Watts, *J.C.S. Perkin I*, **1974**, 2391.
[119] W. Ando, K. Konishi and T. Migita, *J. Organometal. Chem.*, **67**, C7 (1974).
[120] H. Watanabe, T. Nakano, K.-I. Araki, H. Matsumoto and Y. Nagai, *J. Organometal. Chem.*, **69**, 389 (1974).
[121] R. Breslow, A. Feiring and F. Herman, *J. Am. Chem. Soc.*, **96**, 5937 (1974).
[121a] A. Hartmann, W. Welter and M. Regitz, *Tetrahedron Letters*, **1974**, 1825.
[122] R. E. Banks and A. Prakash, *J.C.S. Perkin I*, **1974**, 1365.
[123] R. C. Belloli, M. A. Whitehead, R. H. Wollenberg and V. A. LaBahn, *J. Org. Chem.*, **39**, 2128 (1974).
[124] P. A. Tardella, L. Pellacani, G. DiStazio and M. Virgillito, *Gazz. Chim. Ital.*, **104**, 479 (1974); *Chem. Abs.*, **81**, 104345 (1974).
[125] B. Ragonnet, M. Santelli and M. Bertrand, *Bull. Soc. chim. France*, **1973**, 3119.
[125a] G. Leandri, H. Monti and M. Bertrand, *Tetrahedron*, **30**, 283 (1974); T. Hiyama, S. Takehara, K. Kitatani and H. Nozaki, *Tetrahedron Letters*, **1974**, 3295.
[126] R. B. Reinarz and G. J. Fonken, *Tetrahedron Letters*, **1974**, 441.
[127] D. P. G. Hamon and V. C. Trenerry, *Tetrahedron Letters*, **1974**, 1371.
[128] L. A. Paquette, G. Zon and R. T. Taylor, *J. Org. Chem.*, **39**, 2677 (1974); L. A. Paquette and G. Zon, *J. Am. Chem. Soc.*, **96**, 203, 215 (1974); see also T. Hiyama, T. Mishima, T. Kitatani and H. Nozaki, *Tetrahedron Letters*, **1974**, 3297; K. J. Drachenberg and H. Hopf, *ibid.*, p. 3267.
[129] R. H. Fischer, M. Baumann and G. Köbrich, *Tetrahedron Letters*, **1974**, 1207.
[130] W. Ando, A. Sekiguchi, T. Hagiwara and T. Migita, *Chem. Comm.*, **1974**, 372.
[131] T. Hiyama, Y. Ozaki and H. Nozaki, *Tetrahedron*, **30**, 2661 (1974).
[132] P. A. S. Smith and E. M. Bruckmann, *J. Org. Chem.*, **39**, 1047 (1974).
[133] R. A. Abramovitch, G. N. Knaus and V. Uma, *J. Org. Chem.*, **39**, 1101 (1974).
[134] R. A. Abramovitch, T. D. Bailey, T. Takaya and V. Uma, *J. Org. Chem.*, **39**, 340 (1974).
[135] T. Kametani, F. F. Ebetino and K. Fukimoto, *Tetrahedron*, **30**, 2713 (1974); *J.C.S. Perkin I*, **1974**, 861; R. N. Carde and G. Jones, *ibid.*, p. 2066; G. R. Cliff, G. Jones and J. M. Woollard, *ibid.*, p. 2072.
[136] D. T. Sepp. K. V. Scherer and W. P. Weber, *Tetrahedron Letters*, **1974**, 2983.
[137] D. Bethell, J. Hayes and A. R. Newall, *J.C.S. Perkin II*, **1974**, 1307.
[138] V. I. Markov and A. E. Polyakov, *Zh. Org. Khim.*, **9**, 1759 (1973).
[139] G. Höfle, *Z. Naturforsch.*, **28b**, 831 (1973).
[140] M. F. Betkouski and W. Lwowski, *Chem. Comm.*, **1973**, 932.
[141] D. C. Appleton, D. C. Bull, J. McKenna, J. M. McKenna and A. R. Walley, *Chem. Comm.*, **1974**, 140.
[142] W. Ando, K. Konishi, T. Hagiwara and T. Migita, *J. Am. Chem. Soc.*, **96**, 1601 (1974).
[143] W. Ando, T. Hagiwara and T. Migita, *Tetrahedron Letters*, **1974**, 1425.
[144] P. A. Grieco, M. Meyers and R. S. Finkelhor, *J. Org. Chem.*, **39**, 119 (1974).
[145] W. Ando, H. Fuji, I. Nakamura and T. Migita, *Int. J. Sulphur Chem.*, **8**, 13 (1973).
[146] J. I. G. Cadogan and I. Gosney, *J.C.S. Perkin I*, **1974**, 460.
[147] B. H. Freeman and D. Lloyd, *Tetrahedron*, **30**, 2257 (1974).
[148] J. I. G. Cadogan, R. J. Scott and N. H. Wilson, *Chem. Comm.*, **1974**, 902.
[149] S. Bien and A. Gillon, *Tetrahedron Letters*, **1974**, 3073.
[150] B. M. Trost and P. J. Whitman, *J. Am. Chem. Soc.*, **96**, 7421 (1974).
[151] J. A. Altmann, I. G. Csizmadia and K. Yates, *J. Am. Chem. Soc.*, **96**, 4196 (1974).
[152] A. Nickon, F-c. Huang, R. Weglein, K. Matsuo and H. Yagi, *J. Am. Chem. Soc.*, **96**, 5264 (1974).

153 W. E. Slack, C. G. Moseley, K. A. Gould and H. Shechter, *J. Am. Chem. Soc.*, **96**, 7596 (1974).
154 M. T. H. Liu and D. H. T. Chien, *J.C.S. Perkin II*, **1974**, 937; *Can. J. Chem.*, **52**, 246 (1974).
155 T. R. Fields and P. J. Kropp, *J. Am. Chem. Soc.*, **96**, 7559 (1974).
156 J. Villieras, C. Bacquet and J. F. Normant, *Bull. Soc. Chim. France*, **1974**, 1731.
157 H. Taguchi, H. Yamamoto and H. Nozaki, *J. Am. Chem. Soc.*, **96**, 6510 (1974).
158 S. S. Olin and R. M. Venable, *Chem. Comm.*, **1974**, 104, 273.
159 C. B. Chapleo and A. S. Dreiding, *Helv. Chim. Acta*, **57**, 873, 1259 (1974).
160 E. Vogel, M. Mann, Y. Sakata, K. Müllen and J. F. M. Oth, *Angew. Chem. Int. Ed.*, **13**, 283 (1974).
161 A. Oku and K. Yagi, *J. Am. Chem. Soc.*, **96**, 1966 (1974).
162 V. V. Kane, A. D. Wolf and M. Jones, *J. Am. Chem. Soc.*, **96**, 2643 (1974).
163 R. F. C. Brown, K. J. Harrington and G. L. McMullen, *Chem. Comm.*, **1974**, 123.
164 H. Fienemann and G. Köbrich, *Chem. Ber.*, **107**, 2797 (1974).
165 N. F. Woolsey and M. H. Khalil, *Tetrahedron Letters*, **1974**, 4309.
166 N. L. Allinger, T. J. Walter and M. G. Newton, *J. Am. Chem. Soc.*, **96**, 4588 (1974).
167 A. B. Smith, *Chem. Comm.*, **1974**, 695.
168 E. V. Korolev, *Sovrem. Probl. Khim.*, **1973**, 50; *Chem. Abs.*, **81**, 119841 (1974).
169 S. Bittner, S. Grinberg and I. Kartoon, *Tetrahedron Letters*, **1974**, 1965.
170 G. Märkl, J. Advena and H. Hauptmann, *Tetrahedron Letters*, **1974**, 303.
171 R. L. Tyner, W. M. Jones, Y. Öhrn and J. R. Sabin, *J. Am. Chem. Soc.*, **96**, 3765 (1974).
172 R. Gleiter, W. Rettig and C. Wentrup, *Helv. Chim. Acta*, **57**, 2111 (1974).
173 C. Wentrup, *Tetrahedron*, **30**, 1301 (1974).
174 T. T. Coburn and W. M. Jones, *J. Am. Chem. Soc.*, **96**, 5218 (1974).
175 W. E. Billups, L. P. Lin and W. Y. Chow, *J. Am. Chem. Soc.*, **96**, 4026 (1974).
176 R. J. Sundberg and R. W. Heintzelman, *J. Org. Chem.*, **39**, 2546 (1974); see also B. A. DeGraff, D. W. Gillespie and R. J. Sundberg, *J. Am. Chem. Soc.*, **96**, 7491 (1974).
177 R. K. Smalley, W. A. Strachan and H. Suschitzky, *Synthesis*, **1974**, 503.
178 B. Iddon, M. W. Pickering and H. Suschitzky, *Chem. Comm.*, **1974**, 186, 759.
179 R. K. Smalley, W. A. Strachan and H. Suschitzky, *Tetrahedron Letters*, **1974**, 825.
180 R. F. C. Brown, F. Irvine and R. J. Smith, *Austral. J. Chem.*, **26**, 2213 (1973).
181 R. A. LaBar and W. M. Jones, *J. Am. Chem. Soc.*, **96**, 3645 (1974).
182 A. Hassner, A. B. Levy, E. E. McEntire and J. E. Galle, *J. Org. Chem.*, **39**, 585 (1974).
183 J. A. Connor, *Organometal. Chem.*, **2**, 302 (1973).
184 J. E. Parks and A. L. Balch, *J. Organometal. Chem.*, **71**, 453 (1974).
185 M. F. Lappert and A. J. Oliver, *J.C.S. Dalton*, **1974**, 65; P. B. Hitchcock, M. F. Lappert, G. M. McLaughlin and A. J. Oliver, *ibid.*, p. 68.
186 M. J. Webb, R. P. Stewart and W. A. G. Graham, *J. Organometal. Chem.*, **59**, C21 (1973).
187 W. Petz, *J. Organometal. Chem.*, **72**, 369 (1974).
188 T. Ikariya and A. Yamamoto, *Chem. Comm.*, **1974**, 720.
189 Y. Yamamoto, K. Aoki and H. Yamazaki, *J. Am. Chem. Soc.*, **96**, 2647 (1974); D. F. Christian and W. R. Roper, *J. Organometal. Chem.*, **80**, C35 (1974); H. Fischer, E. O. Fischer and F. R. Kreissel, *ibid.*, **64**, C41 (1973); D. J. Cardin, B. Cetinkaya and M. F. Lappert, *ibid.*, **72**, 139 (1974); R. Zanella, T. Boschi, B. Crociani and U. Belluco, *ibid.*, **71**, 135 (1974); F. Bonati and G. Minghetti, *ibid.*, **59**, 403 (1973); C. P. Casey and C. R. Cyr, *ibid.*, **57**, C69 (1973); C. P. Casey and W. R. Brunsvold, *ibid.*, **77**, 345 (1974); B. Cetinkaya, M. F. Lappert, G. M. McLaughlin and K. Turner, *J.C.S. Dalton*, **1974**, 1591; B. Cetinkaya, P. Dixneuf and M. F. Lappert, *ibid.*, p. 1827; P. R. Branson, R. A. Cable, M. Green and M. K. Lloyd, *Chem. Comm.*, **1974**, 364; E. O. Fischer and H. Fischer, *Chem. Ber.* **107**, 657 (1974).
190 E. O. Fischer, G. Kreis, F. R. Kreissl, C. G. Kreiter and J. Mueller, *Chem. Ber.*, **106**, 3910 (1973).
191 C. H. Davies, C. H. Game, M. Green and F. G. A. Stone, *J.C.S. Dalton*, **1974**, 357; C. H. Game, M. Green, J. R. Moss and F. G. A. Stone, *ibid.*, p. 351.
192 D. W. Bowen, M. Green, D. M. Grove, J. R. Moss and F. G. A. Stone, *J.C.S. Dalton*, **1974**, 1189; J. Daub, U. Erhardt, J. Kappler and V. Trautz, *J. Organometal. Chem.*, **69**, 423 (1974).
193 P. J. Fraser, W. R. Roper and F. G. A. Stone, *J.C.S. Dalton*, **1974**, 760.
194 P. J. Fraser, W. R. Roper and F. G. A. Stone, *J.C.S. Dalton*, **1974**, 102.
195 J. W. Wilson and E. O. Fischer, *J. Organometal. Chem.*, **57**, C63 (1973).
196 E. Lindner and H. Berke, *Chem. Ber.*, **107**, 1360 (1974).
197 R. R. Schrock, *J. Am. Chem. Soc.*, **96**, 6796 (1974).
198 T. Yamamoto, A. R. Garber, J. R. Wilkinson, C. B. Boss, W. E. Streib and L. J. Todd, *Chem. Comm.*, **1974**, 354.

[199] U. Schubert and E. O. Fischer, *Chem. Ber.*, **106**, 3882 (1973); E. O. Fischer, K. R. Schmid, W. Kalbfus and C. G. Kreiter, *ibid.*, p. 3893.
[200] H. L. Conder and M. Y. Darensbourg, *Inorg Chem.*, **1974**, 506.
[201] H. Fischer and E. O. Fischer, *Chem. Ber.*, **107**, 673 (1974); D. J. Cardin, M. J. Doyle and M. F. Lappert, *J. Organometal. Chem.*, **65**, C13 (1974).
[202] J. E. Parks and A. L. Balch, *J. Organometal. Chem.*, **57**, C103 (1973).
[203] G. Minghetti and F. Bonati, *J. Organometal. Chem.*, **73**, C43 (1974).
[204] H. Fischer and E. O. Fischer, *J. Organometal. Chem.*, **69**, C1 (1974).
[205] E. O. Fischer, G. Kreis, F. R. Kreissl, W. Kalbfus and E. Winkler, *J. Organometal. Chem.*, **65**, C53 (1974).
[206] F. R. Kreissl and E. O. Fischer, *Chem. Ber.*, **107**, 183 (1974).
[207] C. T. Lam, C. V. Senoff and J. E. H. Ward, *J. Organometal. Chem.*, **70**, 273 (1974).
[208] C. P. Casey and R. L. Anderson, *J. Am. Chem. Soc.*, **96**, 1230 (1974); *J. Organometal. Chem.*, **73**, C28 (1974).
[209] J. A. Connor and E. M. Jones, *J. Organometal. Chem.*, **60**, 77 (1973).
[210] F. R. Kreissl, E. O. Fischer and C. G. Kreiter, *J. Organometal. Chem.*, **57**, C9 (1973).
[211] E. O. Fischer and B. Dorrer, *Chem. Ber.*, **107**, 1156 (1974).
[212] A. J. Schultz, J. V. McArdle, R. Eisenberg and G. P. Khare, *J. Organometal. Chem.*, **72**, 415 (1974); K. S. Chen, J. Kleinberg and J. A. Landgrebe, *Inorg. Chem.*, **1973**, 2826; A. DeRenzi and E. O. Fischer, *Inorg. Chim. Acta*, **8**, 185 (1974).
[213] W. A. Herrmann, *Angew. Chem. Int. Ed.*, **13**, 599 (1974).
[214] L. S. Pu and A. Yamamoto, *Chem. Comm.*, **1974**, 9.
[215] E. Hengge, *Topics Curr. Chem.*, **51**, 1 (1974); H. Bürger and R. Eujen, *ibid.*, **50**, 1 (1974).
[216] B. G. Ramsay, *J. Organometal. Chem.*, **67**, C67 (1974).
[217] H. Sakurai, Y. Kobayashi and Y. Nakadaira, *J. Am. Chem. Soc.*, **96**, 2656 (1974).
[218] P. P. Gaspar, R.-J. Hwang and W. C. Eckelzman, *Chem. Comm.*, **1974**, 242.
[219] O. F. Zeck, Y. Y. Su, G. P. Gennaro and Y.-N. Tang, *J. Am. Chem. Soc.*, **96**, 5967 (1974).
[220] J. Higuchi, S. Kubota, T. Kumamoto and I. Tokue, *Bull. Chem. Soc. Japan*, **1974**, 2775.
[221] P. P. Gaspar and R.-J. Hwang, *J. Am. Chem. Soc.*, **96**, 6198 (1974).
[222] P. Riviere, J. Satge and D. Soula, *J. Organometal. Chem.*, **63**, 167 (1973); *ibid.*, **72**, 329 (1974); P. Riviere, J. Satge, G. Dousse, M. Riviere-Baudet and C. Couret, *ibid.*, p. 339.
[223] M. E. Childs and W. P. Weber, *Tetrahedron Letters*, **1974**, 4033.
[224] T. J. Barton and J. A. Kilgour, *J. Am. Chem. Soc.*, **96**, 7150 (1974).
[225] M. D. Sefcik and M. A. Ring, *J. Organometal. Chem.*, **59**, 167 (1973); J. Satge, P. Riviere and A. Boy, *Compt. rend.*, *Ser. C*, **278**, 1309 (1974).

CHAPTER 6

Nucleophilic Aromatic Substitution

MICHAEL R. CRAMPTON

Department of Chemistry, Durham University

Nucleophilic substitutions in aromatic systems,[1,2] including photochemical effects,[3] have been reviewed. Further examples of substitutions by the radical chain $S_{RN}1$ mechanism[4] have been reported; reaction is initiated by electron transfer to the substrate, giving a radical anion; this then expels a nucleofugic group to yield an aryl radical which may react with nucleophiles. Thus, in the presence of solvated electrons, halobenzenes give with the anions of aliphatic nitriles the α-phenyl derivative of the nitrile and alkylbenzene in a reaction which may be synthetically useful.[5] The arylation of 2- and 4-picolyl anions is accomplished by treatment with chlorobenzene, phenyltrimethylammonium ions or 2-bromomesitylene under stimulation by potassium metal (a ready source of electrons) or by UV radiation.[6] Related to these reactions is the electrochemically catalysed substitution of *p*-bromobenzophenone by thiophenoxide ions.[7] The most probable mechanism for the cyclization of halogenated α-phenylcinnamic acids and stilbenes with metal amides in liquid ammonia also involves loss of halide ion from an initially formed radical anion.[8]

A preliminary account[9] summarizes evidence showing that, in the absence of a strong base, a reducing agent or light, nucleophilic displacements on benzenediazonium ions proceed by rate-determining cleavage to aryl cations (S_N1) rather than by the bimolecular mechanisms favoured recently. The rates and products of dediazoniation of benzenediazonium tetrafluoroborates in trifluoroethanol/water mixtures are also consistent with a unimolecular mechanism.[10] Further, the exchange of external nitrogen with benzenediazonium ions has been demonstrated.[11]

The S_NAr Mechanism

A detailed kinetic study has been reported[12] of the formation of the spiro-Meisenheimer complex (**3**) from *N*,*N'*-dimethyl-*N*-picrylethylenediamine (**1**); unexpectedly it was found that at low buffer concentrations the formation of (**3**) is subject to general-base catalysis and this indicates that proton-transfer from the zwitterionic (**2**) is rate-limiting and is a consequence of the high rate constant for ring-opening of (**2**) ($k_{-1} = 2 \times 10^5\ \text{sec}^{-1}$). Comparisons of spiro-complexes with their non-cyclic analogues[13] show that values of k_{-1} may be considerably reduced in the non-cyclic compounds normally encountered in

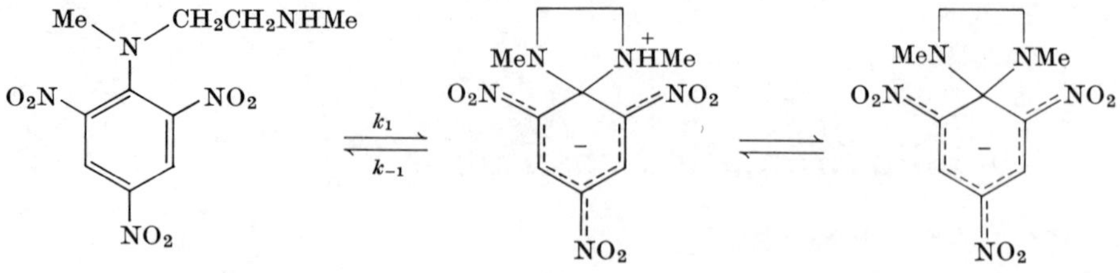

(1) (2) (3)

nucleophilic aromatic substitution. Nevertheless the results of Bernasconi and Gehriger[12] throw light on the mechanism of base catalysis in substitutions by amines; they show that the observation of base catalysis may indicate slow proton-transfer from a zwitterionic intermediate such as (**4**) rather than general-acid catalysis of leaving-group expulsion. The reaction between *p*-anisidine and 1-X-2,4-dinitrobenzenes (X = Cl, Br,

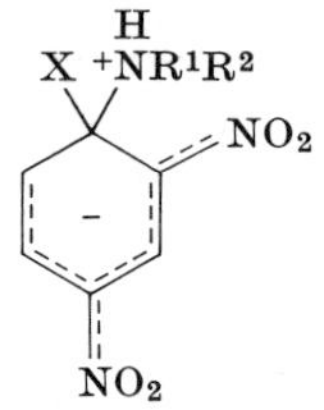

(**4**)

$MeSO_2$ or $PhSO_2$) in benzene is catalysed by DABCO (1,4-diaza[2.2.2]bicyclo-octane) but shows only a small element effect (effect of varying the leaving group).[14] However, rate-determining proton-transfers are even more likely in benzene than in hydroxylic solvents; thus the observation of base catalysis may indicate slow proton-transfer rather than rate-limiting C—X bond fission. The criterion of "absence of element effect" for assessing whether or not bond-breaking occurs in the rate-determining step of nucleophilic substitution reactions thus remains valid.[15] The effects of added salts on the reaction of 4-fluoro- and 4-chloro-nitrobenzenes with trimethylamine in dimethyl sulphoxide (DMSO) have been studied;[16] for the fluoro-compound the results indicate electrophilic catalysis of the decomposition of the reaction intermediate.

Radioisotope-exchange reactions show that in substitution reactions the mobility of the thiocyanate group is similar to that of iodide. The ease of displacement of thiocyanate by azide may result from stabilizing interactions in the transition state.[17] Values for the *o*/*p* ratio for reaction of halonitrobenzenes with alkoxide ions have been interpreted in terms of steric factors operating in the transition states of the *ortho*-isomers.[18] An additional factor may be attack by ion-pairs leading to enhanced reactivity of the *ortho*-isomers.[19]

Nucleophilic substitution reactions have been used to examine substituent effects. Thus the rate constants for reaction of a series of substituted thiophenoxides with 1-chloro-2,4-dinitrobenzene have been measured.[20] The values, which give a measure of the nucleophilic reactivities of the thiophenoxides, correlate well with the corresponding carbon basicities. By use of the reaction with picryl chloride the electronic effects of substituents in aniline have been examined,[21,22] and the importance of conjugation through methylene groups has been studied.[23] Steric effects between adjacent methyl and nitro-groups are important in the reactions of methyl-substituted bromonitrobenzenes with sodium 4-methyl(thiophenoxide).[24]

There have been several kinetic studies of the reactions of bromobenzene and its derivatives with potassium phenoxide; the reaction is catalysed by copper and solvent, and substituent effects have been examined.[25,26] The reactions of halonitrobenzenes with hydroxide have also been studied;[27] in the presence of hydrogen peroxide, substituted phenols and hypochlorite are formed, resulting from an initial attack on the aryl halide by hydrogen peroxide anion.[28] Solvents effects on the reaction of 1-chloro-2,4-dinitrobenzene with pyridine can be correlated with dielectric constant and polarity.[29] Other experimental studies have been made of the reactions of amines with 1-chloro-2,4-dinitrobenzene,[30] 1,3-dihalo-4,6-dinitrobenzenes[31] and dinitrophenyl benzenesulphonates,[32] while theoretical calculations indicate that the formation of π-complexes may facilitate the substitution of picryl chloride by nucleophiles.[33] The reactions of picryl chloride with hydrazino-organophosphorus compounds give picrylhydrazines;[34] rates are faster than with aniline owing to the presence of an unshared electron-pair adjacent to the reaction centre (α-effect). A study[35] of the reaction of *p*-benzoquinones with amino-acids in protic solvents shows that the near-UV absorptions which slowly develop are due to substitution products rather than to charge-transfer complexes as previously postulated.

One of the more interesting recent developments is the use of crown ethers, which have the ability to form complexes with cations, so as to solubilize and activate nucleophilic reagents. Thus, in the presence of crown ether, potassium methoxide reacts with the relatively unactivated *o*-dichlorobenzene to give *o*-chloroanisole,[36] and fluoride displaces chloride from 1-chloro-2,4-dinitrobenzene.[37] The reactivities of nucleophiles may also be enhanced considerably in DMSO relative to protic solvents.[38] In fact, DMSO may itself act as a nucleophile, so that activated chlorine atoms are displaced by the sulphoxide-oxygen when chloro-di- and tri-nitrobenzenes are heated in DMSO.[39] The reaction, in DMSO, between tritium-labelled sodium borohydride and nitrobenzene[40] leads to incorporation of tritium into the nitrobenzene, implying the intermediacy of (**5**), while a stable adduct is formed by hydride addition to 1,3,5-tricyanobenzene.[41]

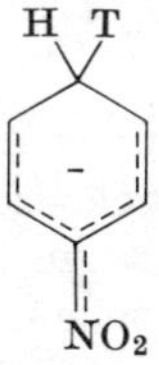

H T

NO_2

(**5**)

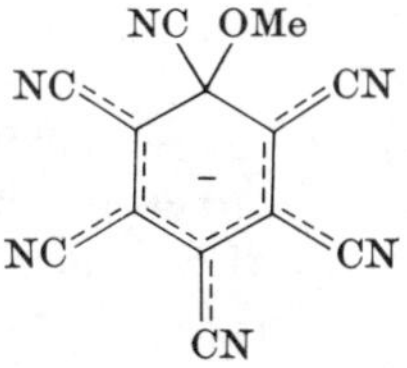

NC OMe

NC CN

NC CN

CN

(**6**)

Hexacyanobenzene reacts with alcohols to give pentacyanophenyl ethers which are dealkylated on heating to give pentacyanophenol. Interestingly the intermediate (**6**) can be isolated by addition either of methoxide ions to hexacyanobenzene or cyanide ions to pentacyanoanisole.[42] Reaction of hexacyanobenzene with amines[43] gives mono- and di-substitution products. Novel *ortho*-substituted benzonitriles are synthesized by nitro-displacement from *o*-nitrobenzonitriles; possible nucleophiles are alkoxides, thiol anions, amines, azides, chloride and hydroxide ions.[44] The incorporation of a dimethylamino-group into an aromatic ring is achieved by heating a suitable precursor with hexamethylphosphoramide;[45] the reaction is thought to involve the intermediate (**7**). Unexpectedly 2-chloro-3,5-dinitrobenzoic acid reacts with phenols and alcohols in pyridine to give aryl or alkyl 3,5-dinitrosalicylates[46] rather than the direct substitution products, suggesting the intermediacy of the β-lactone (**8**). Intramolecular participation, this time by the nitro-group, is also postulated to account for

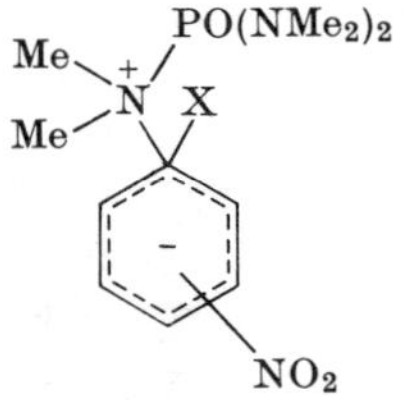

(**7**) 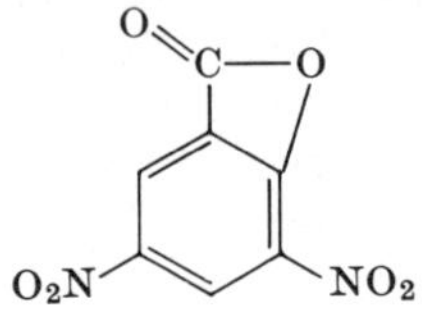 (**8**) 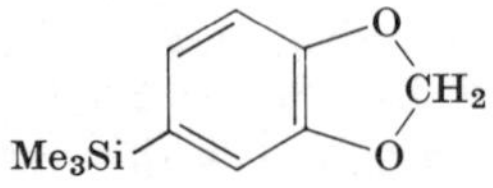 (**9**)

the decomposition products of 2-nitrobenzyl alcohol.[47] Comparison of the rates of alkaline cleavage of 4-(trimethylsilyl)veratrole, 5-(trimethylsilyl)-1,3-benzodioxole and 6-(trimethylsilyl)-1,4-benzodioxan shows the benzodioxole (**9**) to be the most reactive; this is due to hybridization changes at the ring-junction carbon atoms enforced by the strained dioxole ring.[48] Other studies have been reported of the preparation of 1,2,4-triaminoanthraquinones by a Smiles rearrangement,[49] of the formation of 4-nitrosophenol-2-sulphonic acid by reaction of 2-formyl-5-nitrobenzenesulphonic acid with base,[50] and of iodide-exchange in iodobenzoic acid.[51]

A study by visible spectroscopy of the reaction of *p*-dinitrobenzene with sodium hydroxide in aqueous DMSO indicates fast formation of the dinitrobenzene radical anion which may be a precursor in the displacement reaction.[52] Radical mechanisms have also been postulated for the reactions of *m*-dinitrobenzene and *para*-substituted nitrosobenzenes with methanolic methoxide,[53] while the presence of radicals in the reactions of *p*-chloronitrobenzene and *p*-alkoxynitrobenzenes has been demonstrated.[54] The presence of oxygen has an accelerating effect on the reaction of 1-chloro-2,4-dinitrobenzene with triethylamine and results in the formation of 2,4-dinitrophenol, probably by a radical mechanism.[55] The reactions of phenylalkanenitriles with aromatic nitro-compounds in the presence of base may give rise to direct substitution or electron-transfer.[56]

Nucleophilic substitution in polyfluoroaromatic compounds has been excellently summarized.[57] Rate measurements for reactions of polyfluorobenzenes with methoxide and polyfluoropyridines with ammonia show that the activating influence of fluorine decreases in the order $m > o > p$. This order allows a good rationalization of substitution in polyfluoroaromatic compounds.[58] Other kinetic studies have been made of substitutions of ring fluorine by ammonia in a series of perfluoroalkylbenzenes[59] and of substitutions by piperidine;[60] in the former study it was not found necessary to invoke negative hyperconjugation in perfluoroalkyl groups.

Heterocyclic Systems

There have been a number of studies of substitution by the addition-elimination mechanism. Kinetic data for the hydrolysis of 2,4-dichloro-6-heteroaryl-*s*-triazines have been determined over a wide acidity range and are in accord with rate-determining nucleophilic attack on the triazine.[61,62] In the presence of *N*-hydroxysuccinimide,

2-chloro-4-cyclopropylamino 6-isopropylamino-*s*-triazine forms the intermediate (**10**) which then undergoes hydrolysis.[63] 4-Chloro-2-(methylthio)pyrazolo[1,5-*a*]-1,3,5-triazine (**11**) has been prepared; chloride is readily displaced by nucleophiles, but displacement of the methylthio-group requires vigorous conditions.[64] In the case of

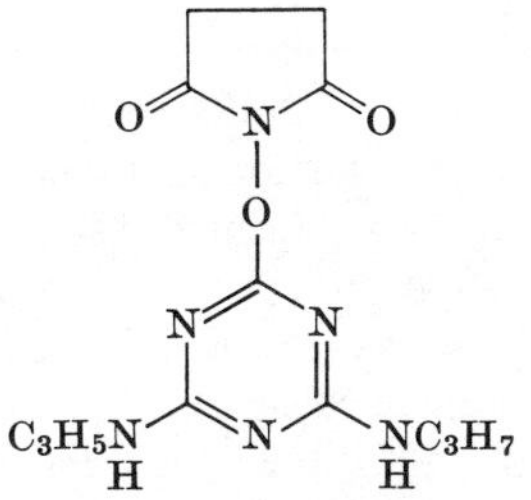

(**10**)

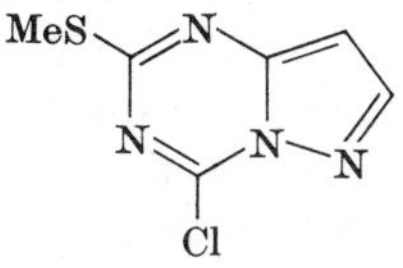

(**11**)

3,5-dichloro-2,6-bis(methylsulphonyl)pyridine the group displaced varies with the nucleophilic reagent, hard nucleophiles such as alkoxides and fluoride displacing methylsulphonyl groups and soft nucleophiles chloride.[65] Kinetic studies have been reported of substitutions of some bromopyridines and bromopicolines with methoxide, phenoxide and thiophenoxide ions.[66] The results confirm a previous conclusion that a 3-methyl substituent activates the 2-position in the case of attack by thiophenoxide (but not with the other three nucleophiles), owing to ion–dipole and dispersion attractive forces between the 3-substituent and the incoming nucleophile. Reductive deiodination competes with nucleophilic substitution for some nitro-activated iodoaromatic compounds and bulky amines, a radical route having been suggested; however, under aza-activation such reductions are not observed; for example 2-iodopyridine when heated with 2-methylpiperidine gives only the normal substitution product and no pyridine.[67] Quaternization of substituted pyridines greatly increases their reactivity towards hydroxide ions; salts with the leaving group in the 2-position are most reactive.[68]

Treatment of pyridine 1-oxides with an imidoyl chloride results in the introduction of a tertiary amide function at the 2-position together with deoxygenation. This reaction involves initial nucleophilic attack by the oxide on the imidoyl chloride followed by intramolecular cyclization.[69] 1,2-Disubstituted pyrazolium salts (**12**) are brominated by *N*-bromoacetamide in the presence of base to give products that undergo substitution to pyrazol-4-in-3-ones.[70] In the presence of neat piperidine, 2- and 4-quinolyl aryl ethers undergo substitution at both the heteroaryl and the aryl carbon atoms;[71] hydrogen isotope effects on the heteroaromatic substitution have been determined. Kinetic studies have also been reported of the reactions of amines with fluoropyrimidines,[72,73] with 5-chloro-1-methyl-3-nitro-1,2,4-triazole[74] and with dihalonaphthyridines.[75] An investigation[76] of the introduction of a piperidyl group into 2-substituted 4-methyl 3,5-dinitro- and 3,5-dinitro-thiophenes shows the presence of a small steric effect only when the leaving group is SO_2Ph.

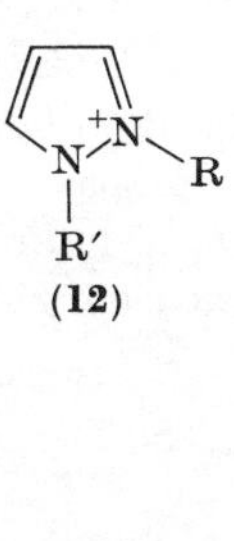

(**12**) 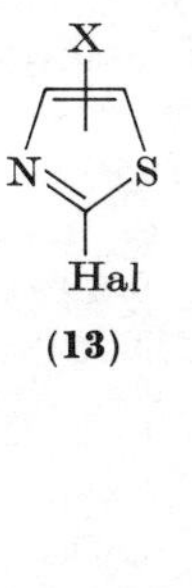 (**13**) 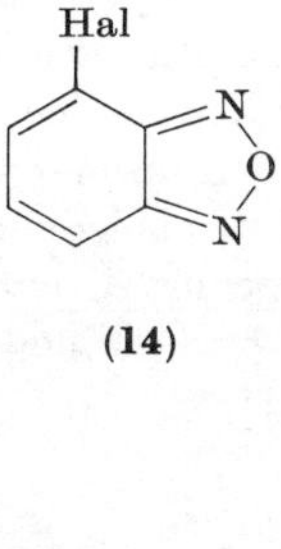 (**14**)

The reactions of un-ionized thiophenols with 4(5)-X-2-halogenothiazoles (**13**) give the normal substitution products. However, the effects of structural changes in the thiazole and in the thiophenol operate in the sense opposite to that normally observed in nucleophilic substitutions;[77] this is due to a pre-equilibrium proton-transfer from the thiophenol to the thiazole which precedes the substitution step. The effects of ring substituents on the nucleophilicities of thiophenoxide ions have been examined in the normal substitution reaction with nitro-activated bromothiophenes.[78] Halobenzofurazans (**14**) give, with sodium thiomethoxide in methanol, in addition to the normal substitution product the product of cine-substitution.[79] These two products are formed by distinct addition–elimination mechanisms. However, cine-substitution in the reaction of iodobenzofurazans with methoxide may result from an elimination–addition mechanism.[80] There has also been a study of the reactions of halogeno-substituted 1,2,3- and 1,3,4-thiadiazoles with methoxide,[81] and a review of benzimidazoles[82] includes a summary of their nucleophilic substitutions.

The interesting and varied possibilities for reaction of heterocyclic compounds in liquid ammonia have been extensively studied by van der Plas and den Hertog and their co-workers. In this solvent, in the presence of potassium amide, 2-bromothiophene and dibromothiophenes undergo bromine migrations from α- to β-positions, while the formation of 3-aminothiophene results from amination of a polybromothiophene to give a product which is subsequently debrominated.[83] Further examples have been cited of substitution by the ANRORC mechanism (Addition of Nucleophile, Ring Opening, Ring Closure); thus it has been shown,[84] by using ^{15}N that the aminations of 2-halo-4-phenyl- and 2-halo-4,6-diphenyl-pyrimidines involve initial amide attack at C-6 with eventual incorporation of the amide ion into the ring of the substitution product. There is also evidence[85] from NMR spectroscopy for amide ion attack at C-6 of 5-bromo-4-R-pyrimidines to give stable σ-adducts (**15**). When the substituent R contains an acidic

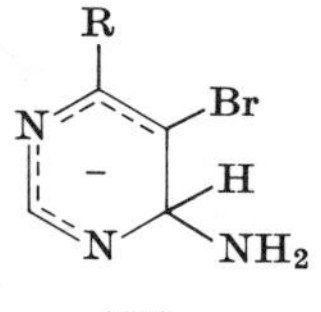

(**15**)

hydrogen atom, deprotonation occurs rather than addition of base. The amination of 3-bromoisoquinoline occurs mainly by an ANRORC mechanism, the remaining substitution, as well as substitution in 1-bromoisoquinoline, occurring by the normal addition–elimination mechanism.[86] In the case of 4-haloisoquinolines, initial amide attack occurs at C-1 and the products include 1-aminoisoquinoline and 1-amino-4-haloisoquinoline.[87] Aminobromoquinolines undergo ring-opening reactions with base in liquid ammonia.[88] Substitution by ANRORC mechanisms has also been postulated in the aminations of quinoazolin-4-one by phenyl phosphorodiamidate[89] and of 2-chloro-4-phenylquinazoline by potassium amide.[84] Halopyridines may react with base by addition–elimination and/or elimination–addition mechanisms.[90, 91] Thus the formation of 3,4-didehydropyridine from 3-bromopyridine has been confirmed by trapping experiments,[92] while 3-bromopyridine 1-oxide reacts via 2,3-didehydropyridine 1-oxide.[93] In the presence of nucleophiles heterocyclics frequently undergo ring-transformations,[94] sometimes in competition with substitution reactions; for example, in liquid ammonia 4-chloro-6-R-pyrimidine 1-oxides (R = Me or Ph) initially give two isomeric σ-complexes

(**16**) and (**17**); however, both these adducts are unstable; (**16**) yields the substitution product, 4-amino-6-R-pyrimidine 1-oxide, together with the ring-contracted 5-amino-3-R-isoxazole (**18**), while (**17**) gives (**18**) exclusively.[95] Pyrimidine and its salts are also converted into isoxazoles by treatment with hydroxylamine hydrochloride.[96] In the

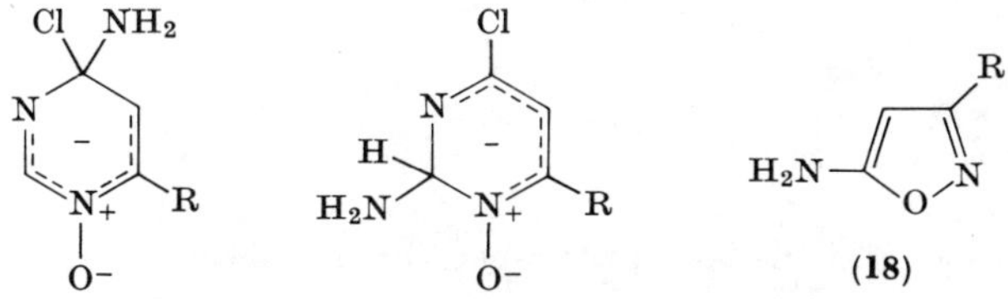

(**16**) (**17**) (**18**)

presence of liquid ammonia, *N*-methylpyridinium salts are demethylated: NMR spectra provide evidence that this reaction involves an ANRORC mechanism, explaining the incorporation of nitrogen from the solvent into the aromatic ring.[97] The conversion of *N*-methylpyridinium salts into pyridines by carbanions follows a similar route.[98] Ring-contractions in liquid ammonia of 4-amino-3-halopyridazine to 4-cyanopyrazole and of 4-amino-3,6-dihalopyridazine to 3-(cyanomethyl)-1,2,4-triazole have also been described.[99] The reaction of 2-nitrothiophene with secondary amines yields open-chain disulphide; the mechanism involves nucleophilic attack at C-5, proton-transfer to yield a sulphonium-type intermediate and ring-opening.[100] Kinetic studies have been reported of the reactions of *N*-(2,4-dinitrophenyl)pyridinium salts with aniline.[101] piperidine[102] and hydroxide ions.[103] Addition of base occurs at the 2-position of the pyridine ring to give (**19**) in the case of aniline and this is followed by ring-opening; in some cases deprotonation of the intermediate such as (**19**) may be rate-limiting.

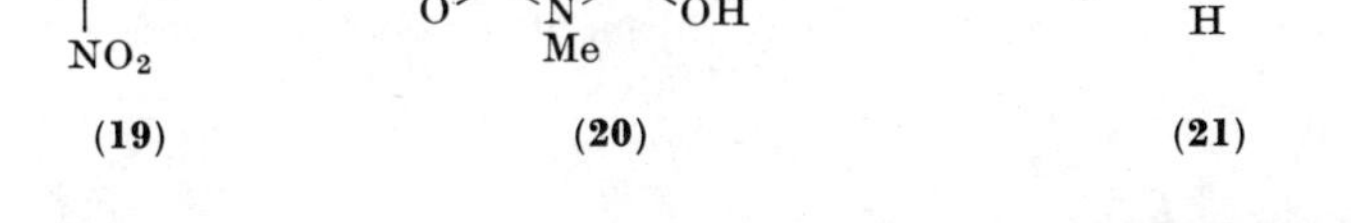

(**19**) (**20**) (**21**)

Several studies in this field have biochemical significance. Thus the displacement of *p*-nitrophenoxide from 5-(*p*-nitrophenoxymethyl)uracil and its *N*-alkylated derivatives, which may involve initial hydroxide attack at C-6 to give (**20**), provides insight into the mechanism of action of thymidylate synthetase.[104] Surprisingly the reaction of 1-substituted or 1,3-disubstituted 5-bromo-6-methyluracils with aromatic amines results in nucleophilic substitution with a concomitant migration to an allylic position, so that 6-arylaminomethyluracils are formed;[105] with aliphatic amines direct displacement of bromine occurs. Mechanistic studies have been reported of the alkaline solvolyses of 5-iodocytosine (**21**)[106] and of 1,3-dimethyl-lumichrome.[107]

A number of kinetic studies involving heteroaryl anion formation have been reported.[108] In the presence of base the pyridinium ylides (**22**) and (**23**) are formed from 3-substituted 1-methylpyridinium ions.[109] The effects of the substituents on the rates

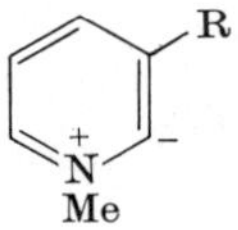

(**22**) 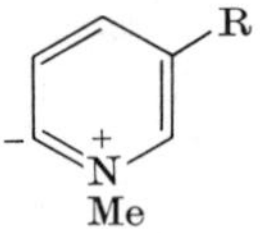 (**23**)

of hydrogen exchange at the 2- and the 6-position have been determined. Studies have also been reported of hydrogen–deuterium exchange in substituted pyridines,[110] pyridine 1-oxides,[111] and diaryl-1,3,4-thiadiazolium salts, e.g. (**24**).[112] Measurements of the rates of detritiation from C-8 of adenine (**25**) and adenosine indicate a mechanism involving rate-determining hydroxide ion attack on the N-7-protonated species.[113] For adenosine at high pH an additional pathway involving hydroxide attack on the neutral compound is operative. Similar mechanisms apply in the case of guanine and related compounds.

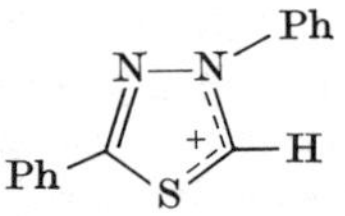

(**24**) 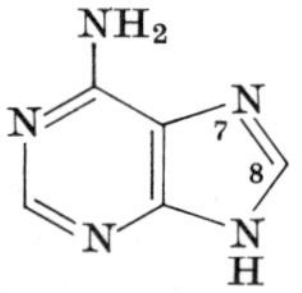 (**25**)

The reactions, including nucleophilic substitutions, of polychlorinated[114] and polyfluorinated[115] heteroaromatic compounds have been reviewed. Pentahalopyridines have been found to react with aliphatic diamines to give *N*-mono- and/or *N,N′*-bis-(tetrahalo-4-pyridyl)diamines,[116] and in some cases bicyclic products are formed by intramolecular attack at the 2-position. The susceptibility of polyhalopyridines to nucleophilic attack is greatly increased by quaternization, which can be achieved by reaction with methyl fluorosulphonate.[117] Perfluoro-3-methylpyridine and perfluoro-3,5-dimethylpyridine are more reactive than perfluoropyridine itself and substitution occurs at the 2- and the 6-position as well as at the 4-position.[118] Polyhalogenated sulphur heterocycles have received less attention though it is reported that reaction of nucleophiles with hexachlorobenzo[*b*]thiophene occurs preferentially at the 2-position.[119]

Meisenheimer and Related Complexes

Particular interest has been shown this year in spiro-complexes, the internal cyclization of *N,N′*-dimethyl-*N*-picrylethylenediamine to give complex (**3**) having been used as a model reaction for nucleophilic substitution by amines.[12] Equilibrium and kinetic data for the formation in methanol[13] and water[120] of complexes, e.g. (**26**), from several ring-activated glycol ethers have been reported. The stabilities of these complexes are considerably greater than those of similarly activated 1,1-dimethoxy-complexes. In moderately concentrated sodium hydroxide solutions the spiro-complexes themselves undergo hydroxide ion additions and there is evidence[120] for the formation of a

1,3-σ-complex (**27**). The ring-opening of two spiro-complexes to give, respectively, 1-(2-hydroxyethoxy)-2,4,6-trinitrobenzene and 1-(2-hydroxyethoxy)-2,4-dinitronaphthalene is subject to general-acid catalysis, suggesting a mechanism involving concerted oxygen-protonation and C—O bond breaking.[121] The enhanced stability of complex

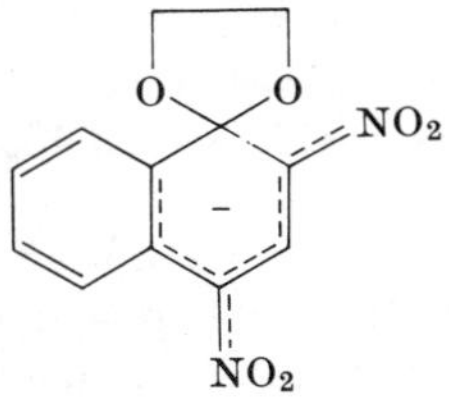

(**26**) 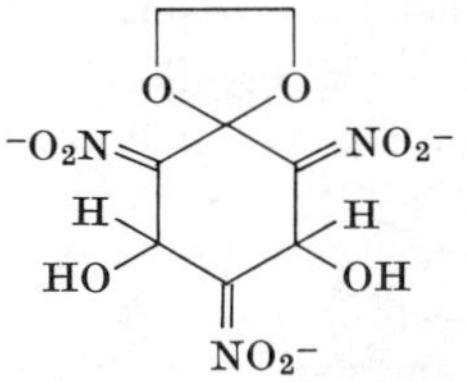 (**27**)

(**28**) relative to (**29**) is attributed[122] mainly to ground-state resonance stabilization in *N*-(2-hydroxyethyl)-*N*-methyl-2,4-dinitroaniline, the precursor of (**29**). Preparations of (**30**)[123] and the sulphur complex (**31**)[124] have been reported.

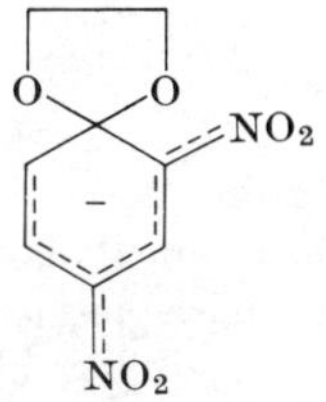

(**28**) 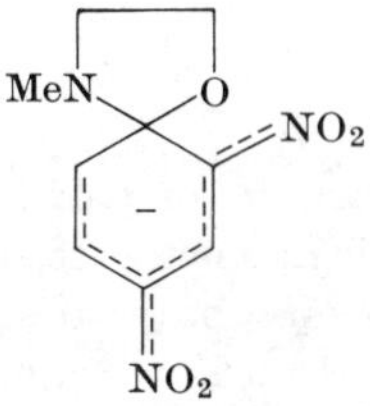 (**29**) 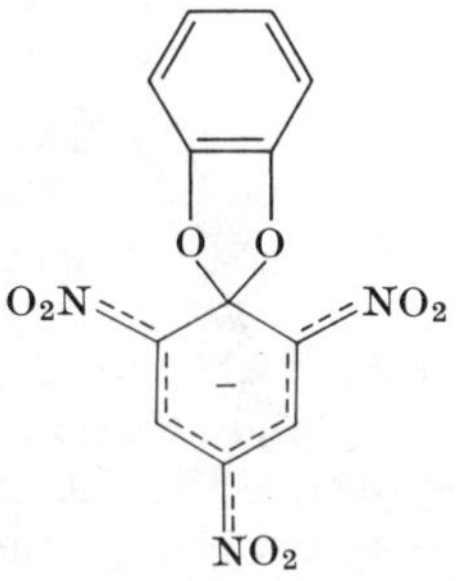 (**30**) 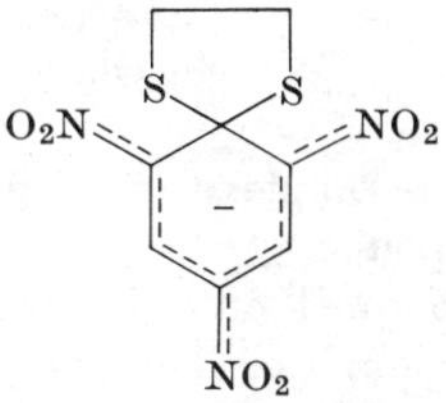 (**31**)

Two possibilities for reaction of 4-methoxy-3,5-dinitrobenzaldehyde with methanolic methoxide are base-addition at C-4 (to give a Meisenheimer complex) or at the aldehyde-carbon atom.[125] There is evidence for stabilization of the former adduct, as with other Meisenheimer complexes,[126] by association with cations. Complexes formed from 2,4,6-trinitroanisole with barium, strontium and calcium methoxides have been reported.[127] Further kinetic studies[128] of the isomeric addition of hydroxide or methoxide ions to 1-X-3,5-dinitrobenzenes (X = CN or CF_3) show that the greater thermodynamic stability of the C-2-adducts, relative to the C-4-adducts, derives mainly from smaller values for the rate constants for their reversal to reactants. Another substrate possessing more than one potentially reactive site is picryl chloride.[129] It has been shown previously[130] that in DMSO initial addition of methoxide ions occurs at a ring-carbon atom carrying hydrogen and that this is followed by the slower replacement of chloride. A kinetic study of these processes in methanol has been reported.[131] The reaction of 2,4,6-trinitrotoluene with base in *sec*- or *tert*-butyl alcohol results in either alkoxide addition at a ring-carbon atom or loss of a methyl-proton;[132,133] in these solvents of low dielectric constant ion-pairing is important. Kinetic studies have been reported of cyanide ion additions to 2,4,6-trinitroanisole and 2,4,6-trinitrotoluene in *sec*-butyl alcohol[134] and to 1,3,5-trinitrobenzene in a series of aliphatic alcohols.[135]

Rate and equilibrium measurements for sulphite additions to picramide and *N*-methyl- and *N*,*N*-dimethyl-picramides indicate that the extent of solvation of the complexes formed plays a major role in determining trends in behaviour.[136] The interaction of sulphite with 2,4,6-trinitrobenzaldehyde[137] is complicated and leads to the formation of isomeric 1:1 and 1:2 complexes. A general formulation of the kinetics of coupled reactions[138] has been applied to 1:2 complex formation between 1,3,5-trinitrobenzene and hydroxide and/or alkoxide ions. The failure to observe, kinetically, *cis–trans*-isomerism in these cases may result from the similar rates of dissociation of the *cis*- and *trans*-complexes.

NMR spectroscopy provides evidence for the formation, in DMSO, of transient 1,3-complexes (**32**) and more stable 1,1-complexes (**33**) from 1-*sec*-amino-2,4-dinitro-naphthalene and sodium ethoxide;[139, 140] and also for the formation at low temperatures of a 1,1-complex from 2,4,6-trinitroanisole and sodium thiophenoxide.[141] A series of σ-complexes, e.g. (**34**), have been characterized from the reactions of anilines[142] or *N*-methylanilines[143] with the methoxide adduct of 1,3,5-trinitrobenzene in DMSO.

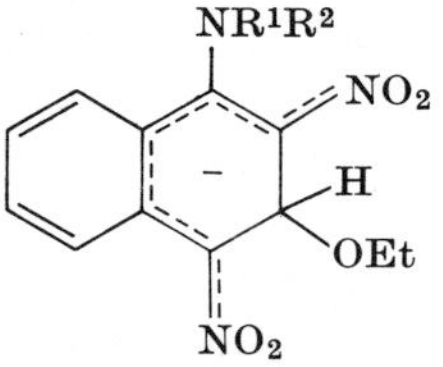

NR^1R^2, NO_2, H, OEt, NO_2

(**32**)

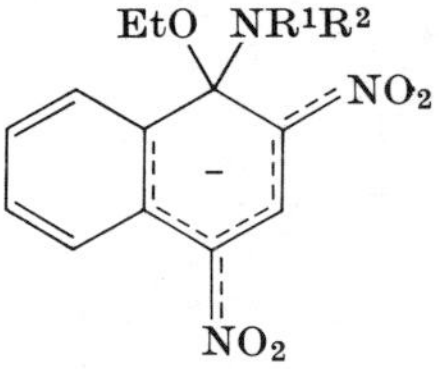

EtO NR^1R^2, NO_2, NO_2

(**33**)

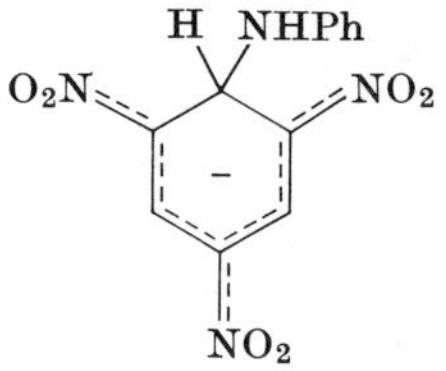

H NHPh, O_2N, NO_2, NO_2

(**34**)

σ-Complexes have also been reported from the reactions of polynitroaromatic compounds with hydroxylamine[144] and alkoxides;[145–147] in some cases anion radicals are formed on their electro-reduction.[148] The fluorescence properties of Meisenheimer complexes have been investigated and preliminary studies indicate the possibility of using such anions as fluorescent biophysical probes.[149] The formation of Meisenheimer complexes in dinitroaniline plant-growth regulators has been investigated.[150]

The formation of (**35**) is reported from 2,4-dinitro-*N*-(*p*-nitrophenyl)pyrrole and methoxide.[151] However, the related *N*-methylated compound undergoes ring-hydrogen exchange rather than addition of base. Methoxide addition to 2-nitrofuran gives (**36**) (X = O); a similar but less stable complex (**36**) (X = S) is formed from 2-nitrothiophene.[152] Considerable increases in complex stability can be brought about by annelation of a benzene ring as in 3-methoxy-2-nitrobenzothiophene.[153] 6-Nitrobenzothiazole gives, with methoxide ions in DMSO, an adduct that with electrophilic reagents yields ring-opened products.[154] The use of Fourier transform NMR shows that potassium

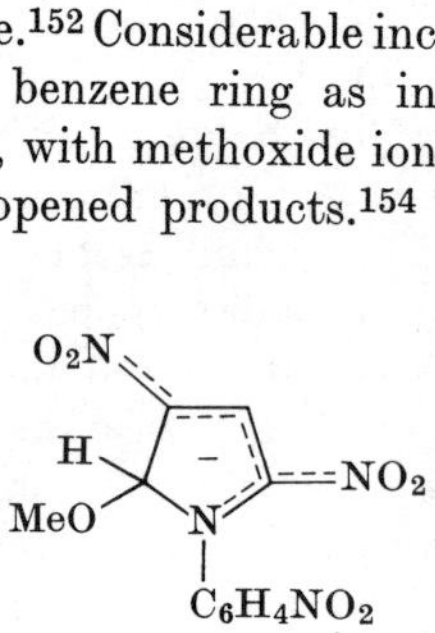

O_2N, H, MeO, N, NO_2, $C_6H_4NO_2$

(**35**)

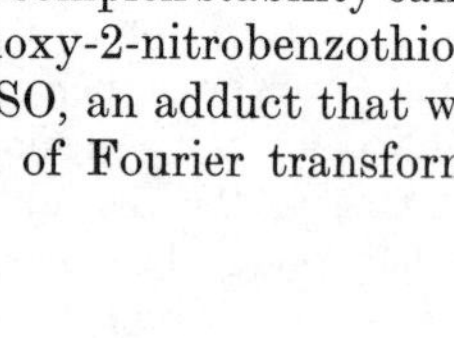

H, MeO, X, NO_2

(**36**)

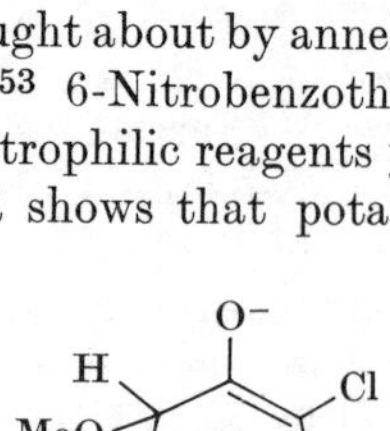

O^-, H, MeO, Cl

(**37**)

methoxide adds at C-7 of 2-chlorotropone to give (**37**); however, this adduct is transitory owing to displacement of chloride by methoxide.[155] Base addition also occurs at C-7 on mixing 2-(ethylthio)tropone with sodium thioethoxide.[156] The products of reaction of azulene with several nitrogen and carbon bases have been identified;[157] addition may occur at C-6 or C-4, while in the case of methylazulenes a further possibility is abstraction of a methyl-proton.

Several interesting papers have appeared reporting kinetic and equilibrium data for the formation of pseudo-bases from heterocyclic cations.[158] For example, in aqueous solution *N*-methyl-3-nitroquinolinium cation yields (**38**) as the kinetically controlled product although (**39**) is thermodynamically more stable, while (**40**) is formed in methanolic methoxide from a dicationic precursor. Neutral adducts are also formed by reaction of *N*-substituted pyridinium salts with cyanide ions,[159] and of selenium heterocyclic cations with several nucleophiles.[160]

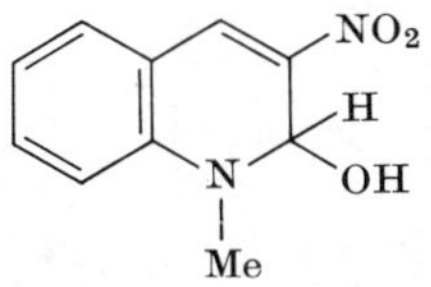

(**38**) 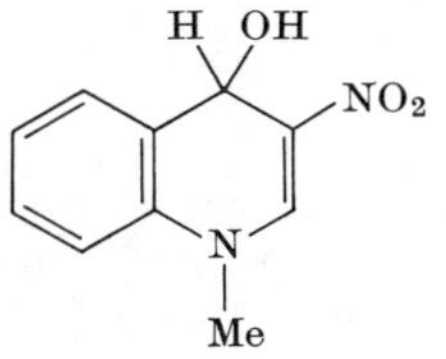 (**39**) 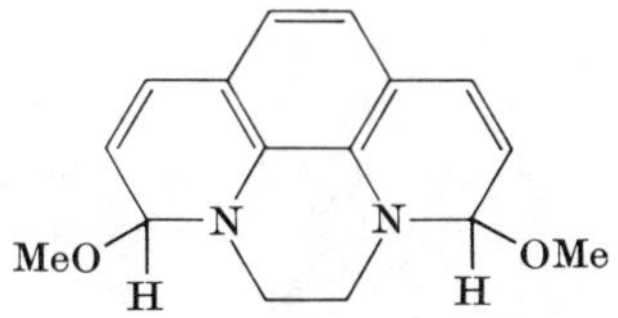 (**40**)

The reaction of 1,3-dinitrobenzene with alkaline acetone to give the intensely coloured adduct (**41**) was first reported by Janovsky and has since been used as a test for active-methylene compounds. Isomeric complexes result from acetonate ion addition at C-2 or C-4 to 1-X-3,5-dinitrobenzenes.[161] Products have also been identified from the reactions of *N*-methyl-*N*,2,4,6-tetranitroaniline[162] and 2,4-dinitroaryl ethers[163] under Janovsky conditions; the former reaction is useful in the analysis of drugs possessing a substituted aniline structure. The reaction of 1,3,5-trinitrobenzene and acetophenone gives complexes formed by carbanion additions at one or two ring positions.[164] The general field of reactions of electron-deficient aromatic compounds with carbanions has been perceptively reviewed by Strauss.[165] In suitable cases, internal cyclization of the

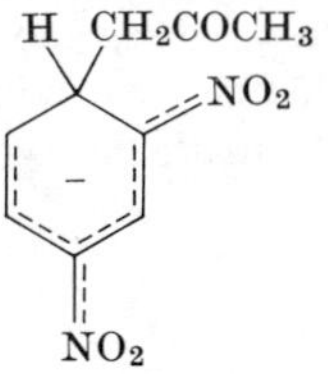

(**41**) 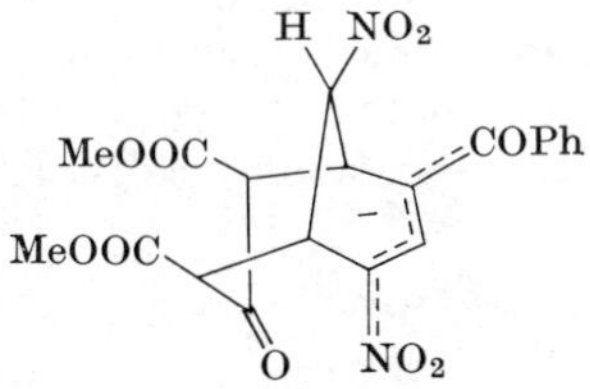 (**42**) 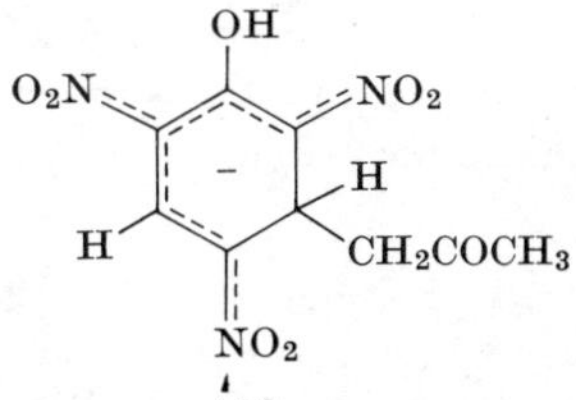 (**43**)

initially formed adducts gives stable bicyclic nitronates. Thus 3,5-dinitrobenzophenone reacts with dimethyl acetone-1,3-dicarboxylate in the presence of diethylamine to give (**42**). However, use of acetone gives 3-diethylamino-5,7-dinitro-1-phenylnaphthalene. The factors favouring this *ortho*-substituent attack over *meta*-bridging have been considered.[166] Picric acid reacts with acetone under alkaline conditions to give an adduct (**43**) which may cyclize to give a *meta*-bridged product.[167] Protonation of (**43**) by acetic acid has been reported.[168] Products similar to (**43**) are formed by acetonate ion addition at unsubstituted ring positions of 3,5-dinitrosalicylic acid[169] and 3-hydroxy-

2,4,6-trinitrobenzoic acid;[170] the latter compound is reported to form isomeric 1:2 hydroxide adducts in concentrated aqueous sodium hydroxide solutions.

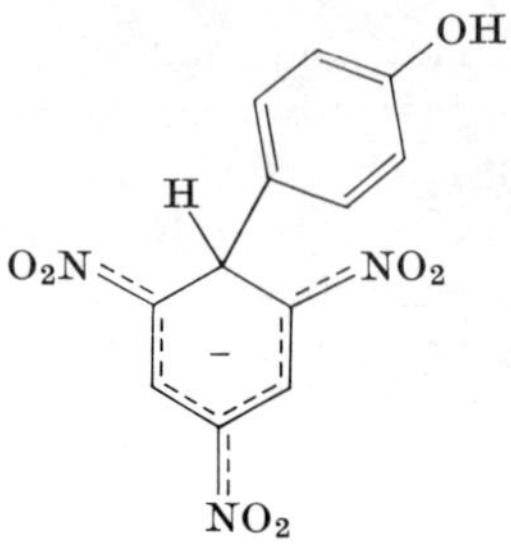

(44)

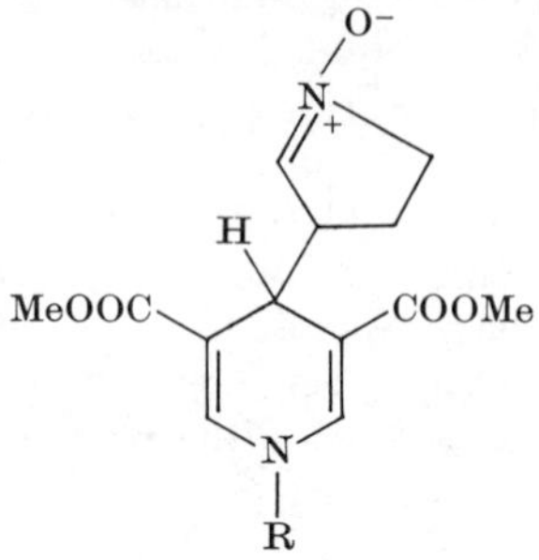

(45)

Thermodynamic data have been determined for some carbanion additions;[171] they indicate unexpected stability for the 1,3,5-trinitrobenzene–cyclopentanone complex. There have been interesting preliminary reports of the formation of (**44**) from 1,3,5-trinitrobenzene and potassium phenoxide[172] and of (**45**) from 3,5-bis(methoxycarbonyl)pyridinium ions and cyclic nitrones.[173] New carbon—carbon bonds are also formed during the reactions of 1,3,5-trinitrobenzene with organotin compounds in DMSO.[174] However, the product of reaction of 9-phenoxyacridine with phenol is probably a π-complex.[175]

Benzyne and Related Intermediates

Arynes have been reviewed.[176] A theoretical study of the three isomeric bisdehydrobenzenes has been made[177] by using MINDO/3 which allows for variations in geometries due to strain energies. The results predict *o*-benzyne to be a singlet with an enthalpy of formation close to that found experimentally. In contrast to previous calculations, *m*-benzyne is predicted to have a stability comparable with that of *o*-benzyne.

It is well known that benzyne is formed by elimination from the benzenediazonium ion induced by acetate ions. Mechanistic studies[178] involving deuterium-labelling show, contrary to a previous report, that a pre-equilibrium type of *E*1c*B* mechanism involving the betaine (**46**) is not operative but they do not allow a distinction to be made between

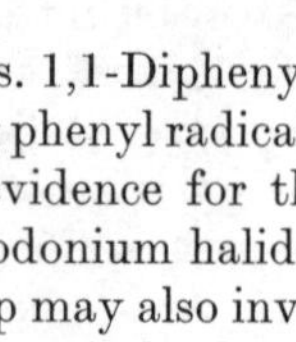

(46)

irreversible *E*1c*B* and *E*2 mechanisms. 1,1-Diphenylethylene and related alkenes act as promoters in this reaction by trapping phenyl radicals and thus suppressing an unwanted radical chain reaction.[179] There is evidence for the intermediacy of benzyne in the arylation of tetrazolides with diaryliodonium halides.[180] The transformations of ylides based on the strained phosphole group may also involve benzyne.[181] The arynes formed from haloanisoles may be reduced to anisoles by hydride-transfer from lithium di-*n*-propylamide;[182] alternative reduction mechanisms involve attack of hydride or amide ions on ring-halogen. The reactions of 1,2,3- and 1,2,4-trichlorobenzenes with primary and secondary amines to give monosubstitution products involve the formation of several isomeric dichlorobenzynes;[183] further reaction to give di- and tri-substituted products may proceed by benzyne or addition–elimination mechanisms.

Recent calculations indicate some stability for 4,5-pyridazyne (**47**). In agreement with this prediction, 4-halo-3-(methoxymethyl)-6-methylpyridazine and potassium

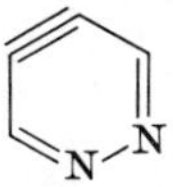

(**47**) 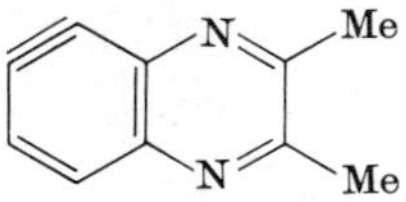 (**48**) 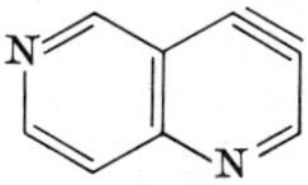 (**49**)

amide in liquid ammonia give a mixture of 4- and 5-aminated products in a ratio that is independent of the halogen present.[184] The hetaryne intermediates (**48**) and (**49**) have been postulated in the aminations of 6-halo-2,3-dimethylquinoxalines[185] and halo-1,6-naphthyridines.[186]

Benzyne usually reacts with olefins to give products via cycloaddition or "ene" mechanisms.[187] Thus reaction with cycloheptatriene[188] gives the [2 + 2]-cycloadduct (**50**) together with the ene product (**51**). A variation on the normal behaviour is afforded

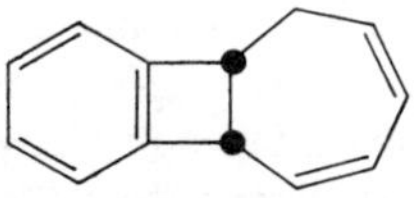

(**50**) 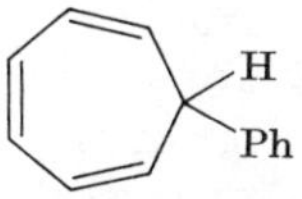 (**51**)

by reaction with camphene[189] which yields a mixture of *E*- and *Z*-ω-phenylcamphenes via a common diradical intermediate (**52**). Steric effects in the ene reaction have been examined in the reactions of benzyne with *cis*- and *trans*-2-methylbut-1-en-1-yl acetates.[190] The [2 + 4]-cycloaddition of benzyne to [2,2](2,5)-furanophane gives a non-internally cyclized adduct which reacts further with benzyne to yield (**53**), the first example of a 2:1 cycloadduct of this diene system.[191]

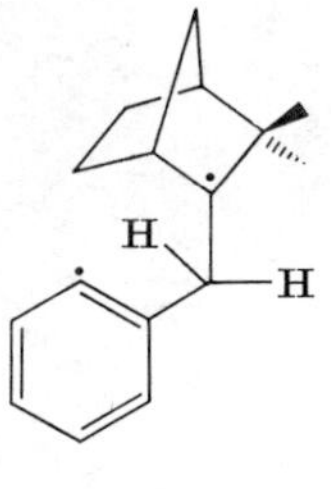

(**52**) 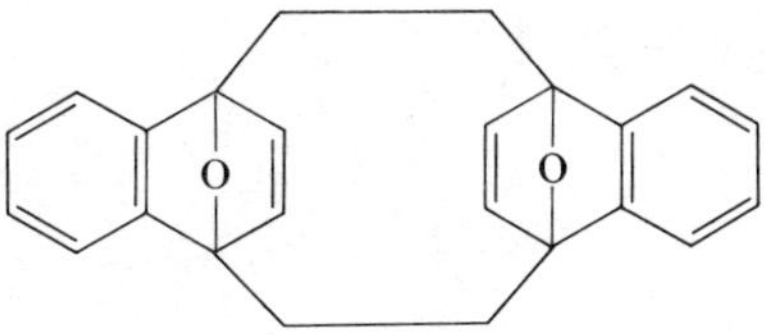 (**53**)

The preparation of benzo-condensed five-membered nitrogen and sulphur heterocycles by cycloaddition of benzyne to mesoionic compounds (sydnones) has been reported;[192] in the reaction with triphenylthiazol-4-one, an intermediate (**54**) was isolated which surprisingly decomposed differently on pyrolysis (giving phenyl isocyanate and 1,3-diphenylbenzo[*c*]thiophene) and photolysis (giving 1,2,4-triphenyl-3-isoquinolone and sulphur). The products of cycloaddition of benzyne with several sulphur heterocycles have also been determined.[193] The reaction of benzyne with pyridine 1-oxides gives β-hydroxyarylated pyridines by a concerted symmetry-allowed $[\sigma 2_s + \pi 2_a + \pi 4_s]$ process.[194] Studies have also been made of the factors affecting the mode of interaction of benzyne with α,β-unsaturated carbonyl compounds,[195] of condensation of 1-naphthalyne with enolates,[196] and of reaction of tetrahalobenzynes with 6,6-dialkylfulvenes.[197]

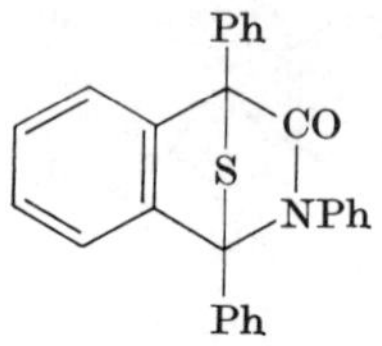

Ph CO S NPh Ph

(54)

Other Reactions

The reaction of cycloheptatrienones carrying a displaceable group at C-2 with amidines results in nucleophilic substitution followed by carbonyl condensation with the other amidine "tooth" to yield 1,3-diazaazulenes.[198] 2-Quinuclidiniotropone reacts with concentrated alkali to give *o*-hydroxybenzaldehyde in a reaction involving a novel carbon-skeletal rearrangement.[199] Fulvenes are subject to nucleophilic attack in the five-membered ring or at the exocyclic carbon atom. Thus 1-bromo-6-dimethylamino-fulvene (**55**) reacts with thiocyanate or pyridine by bromide loss, but with piperidine or phenyl-lithium by dimethylamino-replacement.[200] A kinetic study[201] of the reaction of 6-(*p*-tolylsulphonyloxy)fulvene with piperidine indicates that addition of base at C-6 is rate-determining. Studies of the reactions of 2,*x*-dicyanopyridines with methyl-magnesium iodide[202] show that only products resulting from attack on the nitrile groups are obtained; from 2,3-dicyanopyridine the only product identified (by the unusual NMR combination of shift reagent and nuclear Overhauser effect) was the bicyclic enamine (**56**). In the presence of base, 2-methyl-5-nitropyridine forms an anion which undergoes dimerization by a radical mechanism.[203] Perchloro[4.2.0]octa-1,5,7-

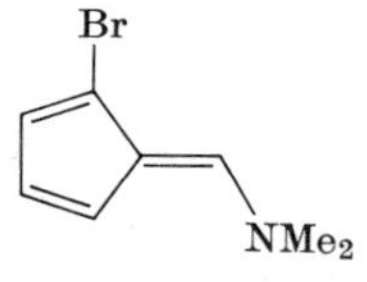

Br NMe_2

(55)

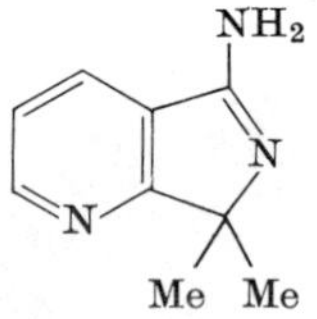

NH_2 N N Me Me

(56)

Cl Cl Cl Cl Cl Cl Cl Cl $\xrightarrow{2X^-}$ Cl Cl Cl X Cl X Cl Cl $+ 2Cl^-$

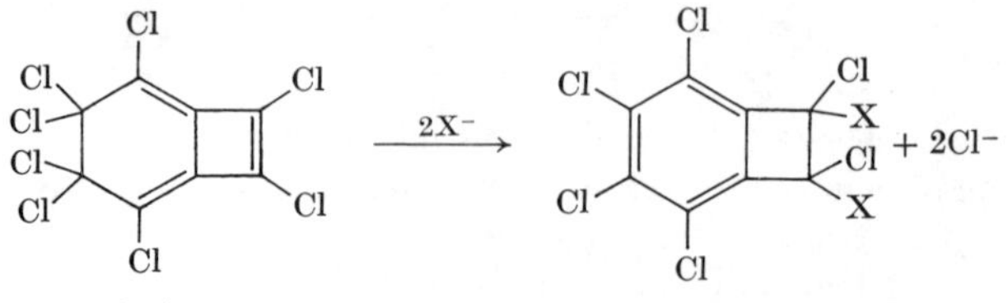

(57) **(58)**

triene (**57**) reacts with nucleophiles under mild conditions to give the aromatized product (**58**).[204, 205]

Other studies have been made of the acid-promoted substitution of nitro- and nitroso-compounds during deoxygenation,[206] nucleophilic substitutions in ferrocenophanes,[207] and *tert*-butylation of 2-naphthol.[208]

References

1 J. Miller, *Cienc. Cult* (*Sao Paulo*), **24**, 121 (1973); *Chem. Abs.*, **81**, 46979 (1974).
2 G. B. Barlin, *Aromat. and Heteroaromat. Chem.*, **2**, (1974).
3 A. Gilbert, *Chem. Soc. Spec. Report, Photochemistry*, **5** (1974).
4 J. F. Bunnett, *Accounts Chem. Res.*, **5**, 139 (1972).
5 J. F. Bunnett and B. F. Gloor, *J. Org. Chem.*, **38**, 4156 (1973).
6 J. F. Bunnett and B. F. Gloor, *J. Org. Chem.*, **39**, 382 (1974).
7 J. Pinson and J.-M. Savéant, *Chem. Comm.*, **1974**, 933.
8 S. V. Kessar, S. Narula, S. S. Gandhi and U. K. Nadir, *Tetrahedron Letters*, **1974**, 2905.
9 C. G. Swain, J. E. Sheats, D. G. Gorenstein, K. G. Harbison and R. J. Rogers, *Tetrahedron Letters*, **1974**, 2973.
10 P. Burri, G. H. Wahl and H. Zollinger, *Helv. Chim. Acta*, **57**, 2099 (1974).
11 R. G. Bergstrom, G. H. Wahl and H. Zollinger, *Tetrahedron Letters*, **1974**, 2975.
12 C. F. Bernasconi and C. L. Gehriger, *J. Am. Chem. Soc.*, **96**, 1092 (1974).
13 M. R. Crampton, *J.C.S. Perkin II*, **1973**, 2157.
14 B. Lamm and J. Lammert, *Acta Chem. Scand.*, **27**, 191 (1973).
15 Z. Rappoport and J. F. Bunnett, *Acta Chem. Scand.*, **B28**, 478 (1974).
16 D. Ayediran, T. O. Bamkole and J. Hirst, *J.C.S. Perkin II*, **1974**, 1013.
17 J. Miller and F. H. Kendall, *J.C.S. Perkin II*, **1974**, 1645.
18 T. O. Bamkole, J. Hirst and E. I. Udoessien, *J.C.S. Perkin II*, **1973**, 2114.
19 F. Del Cima, G. Biggi and F. Pietra, *J.C.S. Perkin II*, **1973**, 110.
20 M. R. Crampton and M. J. Willison, *J.C.S. Perkin II*, **1974**, 238.
21 J. Hirst and K. U. Rahman, *J.C.S. Perkin II*, **1973**, 2119.
22 S. I. Ette and J. Hirst, *J.C.S. Perkin II*, **1974**, 76.
23 V. Mancini, G. Marino and L. Giachetti, *Gazz. Chim. Ital.*, **104**, 549 (1974); *Chem. Abs.*, **81**, 104462 (1974).
24 P. Beltrame and P. Carniti, *Gazz. Chim. Ital.*, **103**, 723 (1973); *Chem. Abs.*, **81**, 62746 (1974).
25 V. V. Litvak and S. M. Shein, *Zh. Org. Khim.*, **10**, 550, 1478 (1974); *Chem. Abs.*, **80**, 132619 (1974); **81**, 90731 (1974).
26 N. Yu. Aronskaya and V. O. Bezuglyi, *Zh. Org. Khim.*, **10**, 268 (1974); *Chem. Abs.*, **80**, 107772 (1974).
27 P. S. Radhakrishnamurti and J. Sahu, *Indian J. Chem.*, **12**, 370 (1974); *Chem. Abs.*, **81**, 119467 (1974).
28 F. M. Vainshtein, N. I. Kamenichnaya and I. I. Kukhtenko, *Ukr. Khim. Zh.* (*Russian Ed.*), **40**, 656 (1974); *Chem. Abs.*, **81**, 104243 (1974).
29 C. R. Das and A. N. Bose, *Z. Phys. Chem.* (*Leipzig*), **255**, 92 (1974).
30 W. Y. Lim, B. I. Lazaro and F. Manligas-Nacino, *Philipp. J. Sci.*, **100**, 261 (1971); *Chem. Abs.*, **80**, 69944 (1974).
31 D. Shamanna and K. S. Siddalingaiah, *Indian J. Chem.*, **12**, 510 (1974); *Chem. Abs.*, **81**, 119450 (1974).
32 E. N. Ozdrovskii, I. M. Ozdrovskaya and R. V. Vizgert, *Zh. Org. Khim.*, **9**, 1948 (1973); *Chem. Abs.*, **79**, 145613 (1973).
33 L. S. Markova, G. L. Ryzhova, B. F. Minaev and A. F. Terpugova, *Tr. Tomsk. Univ.*, **1973**, 184, 194; *Chem. Abs.*, **81**, 12911, 90904 (1974).
34 M. I. Shandruk, N. I. Yanchuk and A. P. Grekov, *Zh. Obsch. Khim.*, **43**, 2198 (1973); *Chem. Abs.*, **80**, 94890 (1974); *Dopov. Akad. Nauk Ukr. RSR, Ser. B*, **3**, 349 (1974); *Chem. Abs.*, **81**, 90759 (1974).
35 R. Foster, N. Kulevsky and D. S. Wanigasekera, *J.C.S. Perkin I*, **1974**, 1318.
36 D. J. Sam and H. E. Simmons, *J. Am. Chem. Soc.*, **96**, 2252 (1974).
37 C. L. Liotta and H. P. Harris, *J. Am. Chem. Soc.*, **96**, 2250 (1974).
38 H. Szmant, *Dimethyl Sulfoxide*, **1**, 1 (1971); *Chem. Abs.*, **80**, 81520 (1974).
39 M. E. C. Biffin and D. B. Paul. *Austral. J. Chem.*, **27**, 777 (1974).
40 V. Gold and V. Nowlan, *Chem. Comm.*, **1974**, 482.

[41] J. Kuthan, V. Skála, M. Ichová and J. Paleček, *Coll. Czech. Chem. Comm.*, **39**, 1872 (1974).
[42] K. Friedrich and S. Oeckl, *Chem. Ber.*, **106**, 3796 (1973).
[43] K. Friedrich and S. Oeckl, *Chem. Ber.*, **106**, 3803 (1973).
[44] J. R. Beck, R. L. Sobezak, R. G. Suhr and J. A. Yahnev, *J. Org. Chem.*, **39**, 1839 (1974).
[45] E. B. Pedersen, J. Perregard and S.-O. Lawesson, *Tetrahedron*, **29**, 4211 (1973).
[46] R. Muthukrishnan, R. Kannan and S. Swaminathan, *J.C.S. Perkin I*, **1973**, 2949.
[47] J. Baake, *Acta Chem. Scand.*, *B*, **28**, 645 (1974).
[48] L. J. Brocklehurst, K. E. Richards and G. J. Wright, *Austral. J. Chem.*, **27**, 895 (1974).
[49] M. S. Simon and J. F. Downey, *Tetrahedron Letters*, **1974**, 3019.
[50] J. Přidal, J. Latinák and V. Komárková, *Coll. Czech. Chem. Comm.*, **39**, 2808 (1974).
[51] V. Spevacek, *Ustav. Jad. Vyzk., Cesk. Akad. Ved.* [*Rep.*], **1973**, 13; *Chem. Abs.*, **80**, 81645 (1974).
[52] T. Abe, *Chem. Letters* (*Tokyo*), **1973**, 1339.
[53] I. Bellobono, P. Goroni and F. Zavaltarelli, *J.C.S. Perkin II*, **1974**, 981, 983.
[54] S. M. Shein, L. V. Bryukhovetskaya, T. M. Ivanova and L. V. Mironova, *Bull. Acad. Sci. USSR*, **22**, 1543 (1973).
[55] S. M. Shein, T. M. Ivanova and N. M. Gavrilova, *Izv. Akad. Nauk SSSR, Ser. Khim.*, **1973**, 2402; *Chem. Abs.*, **80**, 47139 (1974).
[56] M. Makosza, M. Jagusztyn-Grochowska, M. Ludwikow and M. Jawdosivk, *Tetrahedron*, **30**, 3723 (1974).
[57] L. S. Kobrina, *Fluorine Chemistry Reviews*, Dekker, New York, vol. 7, 1974.
[58] R. D. Chambers, W. K. R. Musgrave, J. S. Waterhouse, D. L. H. Williams, J. Burdon, W. B. Hollyhead and J. C. Tatlow, *Chem. Comm.*, **1974**, 239.
[59] R. D. Chambers, J. S. Waterhouse and D. L. H. Williams, *Tetrahedron Letters*, **1974**, 743.
[60] S. M. Shein and P. P. Rodionov, *Kinet. Katal.*, **14**, 1128 (1973); *Chem. Abs.*, **80**, 36561 (1974).
[61] A. F. Cockerill, G. L. O. Davies and D. M. Rackham, *J.C.S. Perkin II*, **1974**, 723.
[62] J. K. Chakrabarti, A. F. Cockerill, G. L. O. Davies, T. M. Hotten, D. M. Rackham and D. E. Tupper, *J.C.S. Perkin II*, **1974**, 861.
[63] N. I. Nakano, E. E. Smissman and R. L. Schowen, *J. Org. Chem.*, **38**, 4396 (1973).
[64] J. Kobe, R. K. Robins and D. E. O'Brien, *J. Heterocyclic Chem.*, **11**, 199 (1974).
[65] T. J. Giacobbe and S. D. McGregor, *J. Org. Chem.*, **39**, 1685 (1974).
[66] R. A. Abramovitch and A. J. Newman, *J. Org. Chem.*, **39**, 2690 (1974).
[67] E. Farina, L. Nucci, G. Biggi, F. Del Cima and F. Pietra, *Tetrahedron Letters*, **1974**, 3305.
[68] G. B. Barlin and J. A. Benbow, *J.C.S. Perkin II*, **1974**, 790.
[69] R. A. Abramovitch and G. M. Singer, *J. Org. Chem.*, **39**, 1795 (1974); R. A. Abramovitch and R. B. Rogers, *ibid.*, p. 1802.
[70] M. Begtrup, N. Conradsen and J. H. Olsen, *Acta Chem. Scand.*, **27**, 2930 (1973).
[71] F. Fiorani, G. Illuminati and G. Sleiter, *J. Org. Chem.*, **39**, 1888 (1974).
[72] D. J. Brown and P. Waring, *J.C.S. Perkin II*, **1974**, 204.
[73] O. P. Shkurko, S. G. Baram and V. P. Mamaev, *Izv. Sib. Otd. Akad. Nauk SSSR, Ser. Khim. Nauk*, **1973**, 104; *Chem. Abs.*, **80**, 14409 (1974).
[74] N. N. Mel'nikova, M. S. Pevzner, N. M. Malysheva and L. I. Bagal, *Zh. Org. Khim.*, **9**, 2535 (1973); *Chem. Abs.*, **80**, 59317 (1974).
[75] N. Czuba and M. Wozniak, *Rec. Trav. chim.*, **93**, 144 (1974).
[76] D. Spinelli, G. Consiglio, R. Noto and A. Corrao, *J.C.S. Perkin II*, **1974**, 1632.
[77] M. Bosco, V. Liturri, L. Troisi, L. Forlani and P. E. Todesco, *J.C.S. Perkin II*, **1974**, 508.
[78] G. Guani, C. Dell'Erba and P. Macera, *J. Heterocyclic Chem.*, **10**, 1007 (1973).
[79] L. di Nunno, S. Florio and P. E. Todesco, *Tetrahedron*, **30**, 863 (1974).
[80] L. di Nunno, S. Florio and P. E. Todesco, *J.C.S. Perkin II*, **1974**, 1171.
[81] B. Modarai, M. H. Ghandehari, H. Massoumi, A. Shafiee, I. Lalezari and A. Badali, *J. Heterocyclic Chem.*, **11**, 343 (1974).
[82] P. N. Preston, *Chem. Rev.*, **74**, 279 (1974).
[83] H. C. van der Plas, D. A. de Bie, G. Geurtsen, M. G. Reinecke and H. W. Adickes, *Rec. Trav. chim.*, **93**, 33 (1974).
[84] A. P. Kroon and H. C. van der Plas, *Rec. Trav. chim.*, **93**, 111, 227 (1974).
[85] J. P. Geerts, C. A. H. Rasmussen, H. C. van der Plas and A. van Veldhuizen, *Rec. Trav. chim.*, **93**, 231 (1974).
[86] G. M. Sanders, M. van Dijk and H. J. den Hertog, *Rec. Trav. chim.*, **93**, 198 (1974).
[87] G. M. Sanders, M. van Dijk and H. J. den Hertog, *Rec. Trav. chim.*, **93**, 273 (1974).
[88] G. M. Sanders, M. van Dijk and H. J. den Hertog, *Rec. Trav. chim.*, **93**, 298 (1974).
[89] A. P. Kroon and H. C. van der Plas, *Tetrahedron Letters*, **1974**, 3201.

[90] H. J. den Hertog, H. Boer, J. W. Street, F. C. A. Vekemans and W. J. van Zoest, *Rec. Trav. chim.*, **93**, 195 (1974).
[91] H. Boer, *Meded. Landbouwhogesch. Wageningen*, **1973**, 88; *Chem. Abs.*, **81**, 62791 (1974).
[92] W. J. van Zoest and H. J. den Hertog, *Rec. Trav. chim.*, **93**, 166 (1974).
[93] R. Peereboom and H. J. den Hertog, *Rec. Trav. chim.*, **93**, 281 (1974).
[94] H. C. van der Plas, *Ring Transformations of Heterocycles*, Academic Press, London, vol. 1, 1973.
[95] R. Peereboom, H. C. van der Plas and A. Koudijs, *Rec. Trav. chim.*, **93**, 58, 277, 284 (1974).
[96] H. C. van der Plas, M. C. Vollering, H. Jongejan and B. Zuurdeg, *Rec. Trav. chim.*, **93**, 225 (1974); H. C. van der Plas and M. C. Vollering, *ibid.*, p. 300.
[97] E. A. Oostveen, H. C. van der Plas and H. Jongejan, *Rec. Trav. chim.*, **93**, 114 (1974).
[98] E. A. Oostveen and H. C. van der Plas, *Rec. Trav. chim.*, **93**, 233 (1974).
[99] D. E. Klinge, H. C. van der Plas, G. Geurtsen and A. Koudijs, *Rec. Trav. chim.*, **93**, 236 (1974).
[100] G. Guanti, C. Dell'Erba, G. Leandri and S. Thea, *J.C.S. Perkin I*, **1974**, 2357.
[101] J. Kaválek and V. Štěrba, *Coll. Czech. Chem. Comm.*, **38**, 3506 (1973); J. Kaválek, A. Barteček and V. Štěrba, *ibid.*, **39**, 1717 (1974).
[102] J. Kaválek, A. Lyčka,V. Macháček and V. Štěrba, *Coll. Czech. Chem. Comm.*, **39**, 2047, 2056 (1974).
[103] J. Kaválek, J. Polanský and V. Štěrba, *Coll. Czech. Chem. Comm.*, **39**, 1049 (1974).
[104] A. L. Pogolotti and D. V. Santi, *Biochemistry*, **13**, 456 (1974); D. V. Santi, C. S. McHenry and H. Sommer, *ibid.*, p. 471.
[105] S. Senda and K. Hirota, *Chem. Comm.*, **1974**, 483.
[106] E. R. Garrett, T. W. Hermann and H.-K. Lee, *J. Pharm. Sci.*, **63**, 899 (1974).
[107] J. Koziol and A. Koziolowa, *Pr. Zabresu. Towarczn. Chem.*, *Wyzsza. Szk. Ekon. Poznaniu, Zesz Nauk, Ser. 1*, **1972**, 167; *Chem. Abs.*, **80**, 70007 (1974).
[108] J. A. Elvidge, J. R. Jones, C. O'Brien, E. A. Evans and H. C. Sheppard, *Adv. Heterocyclic Chem.*, **16**, 1 (1974).
[109] J. A. Zoltewicz and R. E. Cross, *J.C.S. Perkin II*, **1974**, 1363, 1368.
[110] T. Nakamura, M. Soma, T. Onishi and K. Tamura, *Z. Phys. Chem.* (*Frankfurt*), **89**, 122, 130 (1974).
[111] V. P. Lezina, A. U. Stepanyants, V. S. Zhuravlev, L. D. Smirnov and K. M. Dyumaev, *Izv. Akad. Nauk SSSR, Ser. Khim.*, **1974**, 1498; *Chem. Abs.*, **81**, 104240 (1974).
[112] G. Scherowsky, *Chem. Ber.*, **107**, 1092 (1974).
[113] J. A. Elvidge, J. R. Jones, C. O'Brien, E. A. Evans and H. C. Sheppard, *J.C.S. Perkin II*, **1973**, 2138; **1974**, 174.
[114] B. Iddon and H. Suschitzky, *Polychloroaromatic Compounds*, Plenum, London, 1974.
[115] G. G. Yakobson, T. D. Petrova and L. S. Kobrina, *Fluorine Chemistry Reviews*, Dekker, New York, 1974, p. 115.
[116] D. Moran, M. N. Patel, W. A. Tahir and B. J. Wakefield, *J.C.S. Perkin I*, **1974**, 2310.
[117] E. Ager and H. Suschitzky, *J.C.S. Perkin I*, **1973**, 2839.
[118] R. D. Chambers, R. P. Corbally, T. F. Holmes and W. K. R. Musgrave, *J.C.S. Perkin I*, **1974**, 108, 114, 125.
[119] G. M. Brooke and R. King, *Tetrahedron*, **30**, 857 (1974).
[120] M. R. Crampton and M. J. Willison, *J.C.S. Perkin II*, **1974**, 1681.
[121] M. R. Crampton and M. J. Willison, *J.C.S. Perkin II*, **1974**, 1686.
[122] C. F. Bernasconi and H. S. Cross, *J. Org. Chem.*, **39**, 1054 (1974).
[123] V. N. Drozd, V. N. Knyaev and A. A. Klimov, *Zh. Org. Khim.*, **10**, 826 (1974); *Chem. Abs.*, **81**, 25617 (1974).
[124] E. Farina, A. Veracini and F. Pietra, *Chem. Comm.*, **1974**, 672.
[125] M. R. Crampton, M. A. El Ghariani and M. J. Willison, *J.C.S. Perkin II*, **1974**, 441.
[126] S. Ohsawa and H. Nagasue, *Nippon Kagaku Kashi*, **1974**, 79; *Chem. Abs.*, **80**, 94898 (1974).
[127] S. S. Gitis, A. I. Glaz, T. I. Morozova, A. Ya. Kaminskii, Yu. D. Grudtsyn and E. G. Kaminskaya, *Zh. Org. Khim.*, **9**, 978 (1973); *Chem. Abs.*, **79**, 52922 (1973).
[128] F. Millot and F. Terrier, *Bull. Soc. Chim. France*, **1974**, 1823.
[129] G. L. Rhyzhova, S. S. Kravtsova and B. F. Minaev, *Tr. Tomsk. Univ.*, **240**, 11 (1973); *Chem. Abs.*, **80**, 59287 (1974).
[130] M. R. Crampton, M. A. El Ghariani and H. A. Khan, *Tetrahedron*, **28**, 3299 (1972).
[131] L. H. Gan and A. R. Norris, *Can. J. Chem.*, **52**, 18 (1974).
[132] E. Buncel, A. R. Norris, K. E. Russell, P. Sheridan and H. Wilson, *Can. J. Chem.*, **52**, 1750 (1974).
[133] E. Buncel, A. R. Norris, K. E. Russell and H. Wilson, *Can. J. Chem.*, **52**, 2306 (1974).
[134] L. H. Gan and A. R. Norris, *Can. J. Chem.*, **52**, 8 (1974).
[135] L. H. Gan and A. R. Norris, *Can. J. Chem.*, **52**, 1 (1974).
[136] E. Buncel, A. R. Norris, K. E. Russell and P. Sheridan, *Can. J. Chem.*, **52**, 25 (1974).

137 N. Marendic and A. R. Norris, *Can. J. Chem.*, **51**, 3927 (1973).
138 C. F. Bernasconi and R. G. Bergstrom, *J. Am. Chem. Soc.*, **96**, 2397 (1974).
139 S. Sekiguchi, K. Shimozaki, T. Hirose, K. Matsui and T. Itagaki, *Tetrahedron Letters*, **1974**, 1745.
140 S. Sekiguchi, K. Shinozaki, T. Hirose, K. Matsui and K. Sekine, *Bull. Chem. Soc. Japan*, **47**, 2264 (1974).
141 T. V. Leshina, K. V. Solodova and S. M. Shein, *Zh. Org. Khim.*, **10**, 354 (1974); *Chem. Abs.*, **80**, 120431 (1974).
142 E. Buncel and J. G. K. Webb, *Can. J. Chem.*, **52**, 630 (1974).
143 E. Buncel, H. Jarell, H. W. Leung and J. G. K. Webb, *J. Org. Chem.*, **39**, 272 (1974).
144 Yu. D. Grudtsyn and S. S. Gitis, *Zh. Org. Khim.*, **10**, 1462 (1974); *Chem. Abs.*, **81**, 104247 (1974).
145 G. L. Ryzhova, S. S. Kravtsova and B. F. Minaev, *Zh. Fiz. Khim.*, **48**, 1628 (1974); *Chem. Abs.*, **81**, 119706 (1974).
146 G. L. Ryzhova, S. S. Kravtsova and B. F. Minaev, *Tr. Tomsk. Univ.*, **240**, 3 (1973); *Chem. Abs.*, **80**, 59286 (1974).
147 A. Ohsawa, *Nippon Kagaku Kaishi*, **1973**, 1486; *Chem. Abs.*, **79**, 125417 (1973).
148 V. F. Starichenko, V. A. Ryabinin and S. M. Shein, *Zh. Strukt. Khim.*, **15**, 138 (1974); *Chem. Abs.*, **80**, 107492 (1974).
149 S. Farnham and R. Taylor, *J. Org. Chem.*, **39**, 2446 (1974).
150 R. C. Hall and C. S. Giam, *J. Agr. Food Sci.*, **22**, 461 (1974); *Chem. Abs.*, **81**, 24603 (1974).
151 F. De Santis and F. Stegel, *Tetrahedron Letters*, **1974**, 1079.
152 G. Doddi, A. Poretti and F. Stegel, *J. Heterocycl. Chem.*, **11**, 97 (1974).
153 F. De Santis and F. Stegel, *Gazz. Chim. Ital.*, **103**, 649 (1973); *Chem. Abs.*, **81**, 63004 (1974).
154 G. Bartoli, M. Fiorentino, F. Ciminale and P. E. Todesco, *Chem. Comm.*, **1974**, 732.
155 F. Pietra, *Chem. Comm.*, **1974**, 544.
156 C. Veracini and F. Pietra, *Chem. Comm.*, **1974**, 623.
157 R. N. McDonald, H. E. Petty, N. L. Wolfe and J. V. Paukstelis, *J. Org. Chem.*, **39**, 1877 (1974).
158 J. W. Bunting and W. G. Meathrel, *Can. J. Chem.*, **52**, 303, 951, 962, 975, 981 (1974).
159 R. H. Reuss, N. G. Smith and L. J. Winters, *J. Org. Chem.*, **39**, 2027 (1974).
160 A. Tadino, L. Christiaens and M. Renson, *Bull. Soc. Roy. Sci. Liége*, **42**, 146, 505 (1973); *Chem. Abs.*, **79**, 125443 (1973) **81**, 24683 (1974).
161 J. Kaválek, V. Macháček, V. Štěrba and J. Šubert, *Coll. Czech. Chem. Comm.*, **39**, 2063 (1974).
162 K.-A. Kovar and U. Bitter, *Archiv Pharmazie*, **307**, 561, 709 (1974).
163 P. A. Lehmann Feither and G. A. Ciurlizza, *Rev. Latinoamer. Quím.*, **5**, 72 (1974); *Chem. Abs.*, **81**, 36955 (1974).
164 K. Kohashi, T. Kabeya, Y. Ohkura and T. Momose, *Chem. Pharm. Bull.*, **21**, 2187 (1973); *Chem. Abs.*, **80**, 47120 (1974).
165 M. J. Strauss, *Accounts Chem. Res.*, **7**, 181 (1974).
166 M. J. Strauss, *J. Org. Chem.*, **39**, 2653 (1974).
167 T. Kabeya, K. Kohashi, Y. Ohkura and T. Momose, *Chem. Pharm. Bull.*, **21**, 2168 (1973).
168 T. Kabeya, K. Kohashi and Y. Ohkura, *Chem. Pharm. Bull.*, **22**, 711 (1974); *Chem. Abs.*, **81**, 12643 (1974).
169 K.-A. Kovar, *Archiv Pharm.*, **307**, 76 (1974).
170 K.-A. Kovar, *Archiv Pharm.*, **307**, 100 (1974).
171 R. M. Murphy, C. A. Wulff and M. J. Strauss, *J. Am. Chem. Soc.*, **96**, 2678 (1974).
172 E. Buncel and J. G. K. Webb, *J. Am. Chem. Soc.*, **95**, 8470 (1973).
173 R. M. Wilson and A. J. Eberle, *J. Org. Chem.*, **39**, 2804 (1974).
174 I. P. Beletskaya, G. A. Artamkina and O. A. Reutov, *Izv. Akad. Nauk SSSR, Ser. Khim.*, **1973**, 1919; *Chem. Abs.*, **80**, 4699 (1974).
175 A. Ledochowski and S. Skonieczny, *Rocz. Chem.*, **47**, 1577 (1973); *Chem. Abs.*, **80**, 69964 (1974).
176 E. K. Fields, *Org. Reactive Intermed.*, **1973**, 449.
177 M. J. S. Dewar and W. Li, *J. Am. Chem. Soc.*, **96**, 5569 (1974).
178 J. I. G. Cadogan, C. D. Murray and J. T. Sharp, *Chem. Comm.*, **1974**, 133; *J.C.S. Perkin II*, **1974**, 1321.
179 J. I. G. Cadogan, C. D. Murray and J. T. Sharp, *Chem. Comm.*, **1974**, 901.
180 T. Akiyama, Y. Imasaki and M. Kawanisi, *Chem. Letters* (*Tokyo*), **1974**, 229.
181 J. I. G. Cadogan, R. J. Scott and N. H. Wilson, *Chem. Comm.*, **1974**, 902.
182 E. R. Biehl, S. Lapis and P. C. Reeves, *J. Org. Chem.*, **39**, 1900 (1974).
183 P. Caubere and L. Lalloz, *Bull. Soc. Chim. France*, **1974**, 1983, 1989, 1996.
184 D. E. Clinge, H. C. van der Plas and A. Koudijs, *Rec. Trav. chim.*, **93**, 201 (1974).
185 W. Czuba and H. Poradowska, *Rec. Trav. chim.*, **93**, 162 (1974).

[186] W. Czuba and M. Woźniak, *Rec. Trav. chim.*, **93**, 143 (1974).
[187] G. Mehta and B. P. Singh, *Tetrahedron*, **30**, 2409 (1974).
[188] L. Lombardo and D. Wege, *Tetrahedron*, **30**, 3945 (1974).
[189] G. Mehta and B. P. Singh, *Tetrahedron Letters*, **1974**, 4297.
[190] H. H. Wasserman and L. S. Keller, *Tetrahedron Letters*, **1974**, 4355.
[191] L. A. Kapicak and M. A. Battiste, *Chem. Comm.*, **1974**, 930.
[192] S. Nakazawa, T. Kiyosawa, K. Hirokawa and H. Kato, *Chem. Comm.*, **1974**, 621.
[193] J.-M. Decrouen, D. Paquer and R. Pou, *Compt. rend., Ser. C*, **279**, 259 (1974).
[194] R. A. Abramovitch and I. Shinkai, *J. Am. Chem. Soc.*, **96**, 5265 (1974).
[195] A. T. Bowne and R. H. Levin, *Tetrahedron Letters*, **1974**, 2043.
[196] P. Caubere and M. S. Mourade, *Bull. Soc. Chim. France*, **1974**, 1415.
[197] B. Hankinson, H. Heaney, A. P. Price and R. P. Sharma, *J.C.S. Perkin I*, **1973**, 2569.
[198] R. Cabrino, B. Ricciarelli and F. Pietra, *Tetrahedron Letters*, **1974**, 3069.
[199] G. Biggi, F. Del Cima and F. Pietra, *Tetrahedron Letters*, **1974**, 3537.
[200] K. Hafner and F. Schmidt, *Tetrahedron Letters*, **1973**, 5101, 5105.
[201] D. Capocasale, L. di Nunno and F. Naso, *J.C.S. Perkin II*, **1973**, 2078.
[202] V. Skála, J. Kuthan, J. Dědina and J. Schraml, *Coll. Czech. Chem. Comm.*, **39**, 834 (1974).
[203] K. Kurita and R. L. Williams, *J. Heterocycl. Chem.*, **11**, 611 (1974).
[204] A. Roedig, H.-H. Bauer, G. Bonse and R. Ganns, *Chem. Ber.*, **107**, 558 (1974).
[205] A. Roedig, G. Bonse, R. Helm, R. Ganns and U. Kühnel, *Chem. Ber.*, **107**, 920 (1974).
[206] R. J. Sundberg and R. H. Smith, *J. Org. Chem.*, **39**, 93 (1974).
[207] K. C. Y. Sok, G. Tainturier and B. Gautheron, *Compt. rend., Ser. C*, **278**, 1347 (1974).
[208] R. W. Layer, *Tetrahedron Letters*, **1974**, 3459.

Electrophilic Aromatic Substitution

B. V. Smith

Department of Chemistry, Chelsea College, University of London

Introduction

Recent work on electrophilic aromatic substitution has been reviewed by Taylor[1] (C-substitution) and Scriven[2] (heteroatom-substitution). Reactions described as non-conventional, in which addition–elimination or rearrangement sequences occur, have also been surveyed.[3]

The evaluation of aromatic reactivity by solvolysis of model benzylic halides has been extended to the benzo[*b*]thiophene system.[4] Using this approach with the six isomeric 1-(benzo[*b*]thienyl)ethyl chlorides gives the expected order of reactivity, expressed as σ^{+}_{Ar}, as $3 > 2 > 6 > 5 > 4 > 7$ in fair agreement with the observed order (except in the Friedel–Crafts reaction, known to be complicated by rearrangement).

MO calculations have been applied to the hypothetical system benzene–F^+; the open adduct (**1**) should be more stable than bridged structures.[5] MO-LCAO-SCF calculations applied to C_6H_5F predict the observed *p*-substitution in electrophilic substitution.[6] The electron-releasing ability of amino-nitrogen has been related to the UV spectra of substituted anilines.[7]

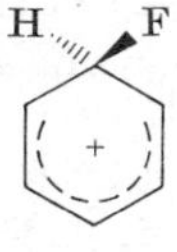

H F

(**1**)

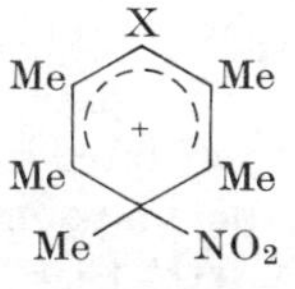

X, Me, Me, Me, Me, Me, NO_2

(**2**)

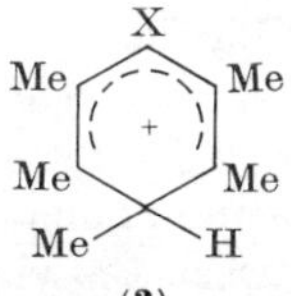

X, Me, Me, Me, Me, Me, H

(**3**)

Considerable attention has been given to the preparation and spectroscopic observation of benzenium ions. The ions (**2**) (generated by interaction of C_6Me_5X and $NO_2^+BF_4^-$ in FSO_3H–SO_2; X = Me, F, Cl or Br) have been compared with the corresponding ions (**3**) (from C_6Me_5X and HF–SbF_5 in SO_2CFCl; X as in **2**) in order to define the charge distribution as measured by ^{13}C- and ^{19}F-NMR spectra.[8] It was concluded that significantly different charge distributions were present in the σ-complexes from the two series of experiments, as reflected in the observation that 1-chloro-2,3,4,5,6-pentamethylbenzene gave one ion with $NO_2^+BF_4^-$ but in protonation three ions were present. It was therefore concluded that the pattern of charge distribution should contribute to kinetic effects in electrophilic substitution. Hexamethylbenzene in FSO_3H gives a stable ion with NO_2^+, and that formed from C_6Me_5OH showed an identical C-1–^{15}N coupling constant, giving no evidence for an ion (**4**).[9] A series of ions (**5**) and (**6**) have been prepared in FSO_3H or FSO_3H–SO_2ClF (R = NO_2, SO_3H, Cl, Br, Me or H) and ^{13}C shifts have been compared.[10]

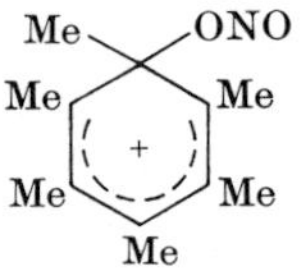

(**4**) 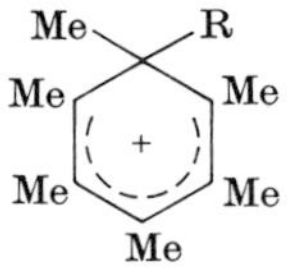 (**5**) 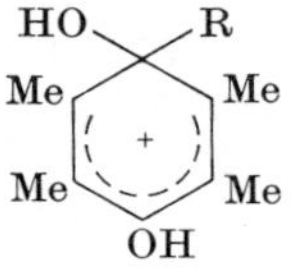 (**6**) 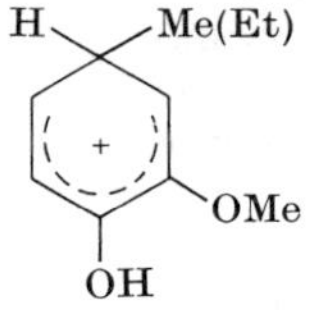 (**7**)

Protonation of alkyl-substituted phenols and alkoxybenzenes (SbF_5–FSO_3H) follows different acidity functions for *C*- and *O*-protonation,[11] and it has been shown that methylphenols unsubstituted at C-4 protonate preferentially at this position. If C-4 is blocked, then C-2-, C-6- and *O*-protonation occur.[12,13] The contribution of inductive, resonance and steric effects on protonation of 2,4,6-trimethoxytoluene (HF–SbF_5 in SO_2ClF) has been analysed, over a range of temperature.[14] The ESR spectrum of a paramagnetic species (not identified) formed from (**7**) has been measured and shown not to be derived from a radical cation, biradical, or impurity.[15]

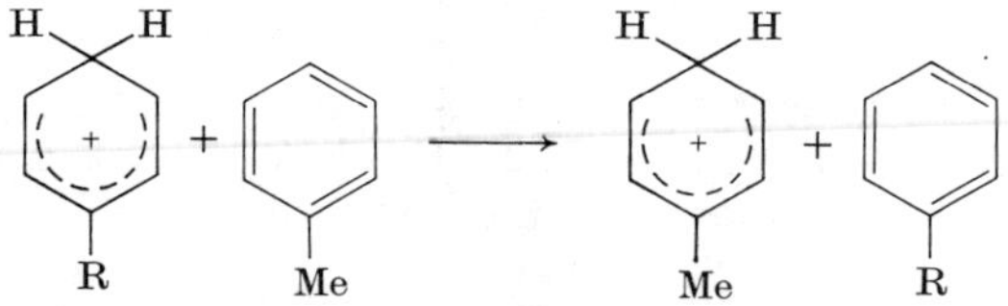

SCHEME 1.

The reaction (Scheme 1) has been followed by pulsed ion cyclotron resonance.[16] The order of *para*-alkyl activation is $Bu^t > Pr^i >$ Et > Me, which is contrary to the solution-phase proton affinities $C_6H_5Me > C_6H_5Bu^t$. This experiment reflects the importance of solvation in substituent effects, and the ground state of $C_6H_7^+$ in the gas phase is hereby deduced to have a classical structure. Ion–molecule reactions, e.g. $C_6H_5Me + C_6H_5CH_2^+$, have been reported[17] and give *p*-$MeC_6H_4CH_2^+$ and C_6H_6, via an intermediate (**8**). Two stage mass spectrometry has been applied to the reactions of CH_3CO^+ and NO_2^+ with benzene in the gas phase.[18] The focused ions, generated from $(CH_3CO)_2O$ and N_2O_4, respectively, enter a vessel containing the aromatic compound at relatively high pressure. There is no correlation between the yield of $XArH^+$ (X = NO_2 or CH_3CO) and relative rates of solution-phase reactions. The ion is not necessarily a σ-complex, and hence no simple relationship would be expected. Methylenation of benzene in the gas phase has been demonstrated from the reaction of $CH_2{=}\overset{+}{O}Me$ (from Me_2O) with aromatic substrates.[19] The reaction is pictured as proceeding via an intermediate (**9**). There is little positional

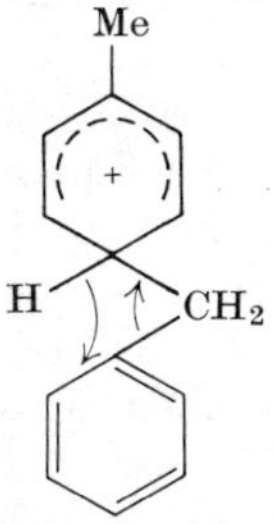

(8) 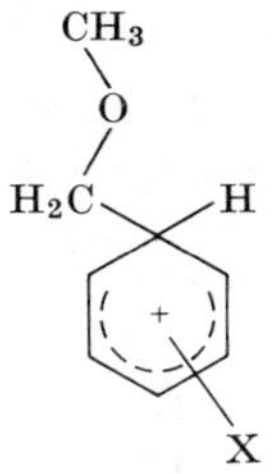 **(9)**

selectivity in substitution of ring-deuteriated toluenes; k_H/k_D, for $C_6H_5Me/C_6D_5Me = 0.9$. It is suggested that the first-formed ion resembles a π-complex, and that this rearranges to a σ-complex. The lack of positional selectivity could arise from statistical formation of σ-complexes or by scrambling of hydrogen (deuterium), but this is by no means certain.

Gas-phase radiolysis (^{60}Co, γ) of C_3H_8–C_6H_6–C_6D_6 mixtures leads to $C_6H_5Pr^i$ composed of d_0 and d_{1-5} species.[19a] The reaction is interpreted as proceeding via a $C_3H_7C_6H_6^+$ species which interacts with another molecule of benzene (π-complex), allowing isotopic interchange. The process is contrasted with Friedel–Crafts isopropylation with regard to description of the intermediate. Added O_2, CO, NO or N_2O inhibit the exchange processes, possibly by proton extraction.

Sulphonation

In the sulphonation of 3-substituted benzenesulphonic acids (fuming H_2SO_4; 25°) the isomer proportions are acidity-dependent;[20] sulphonic acid anhydrides are also formed. In the sulphonation of a series of *ortho*-substituted alkylbenzenes, diphenylalkanes and triphenylmethane, the partial rate factors were measured and variations in the *ortho/para* product ratio were ascribed to retardation of formation of the substituted sulphonic acid σ-complex and/or retardation of conversion of the corresponding sulphonate σ-complex into the product sulphonic acid.[21] A considerable discussion is given of the "conformational" control of these processes. Other reports from the same laboratory describe side-chain sulphonation of 9-methyl- and 9,10-dimethyl-anthracene,[22] the sulphonation of 1,2-dihydrobenzocyclobutene[23] (in which some ipso-attack is noted), the sulphonation of acenaphthene[24] (and its mono- and di-sulphonic acids) and the interaction of H_2SeO_4 (in Ac_2O) with polyalkylbenzenes;[25] this reagent shows high selectivity, and the isomer distribution resembles that obtained by using acetylsulphuric acid.

Sulphonation of phenanthrene (H_2SO_4 in $MeNO_2$, 25°) is interpreted as involving $H_2S_2O_7$ and HSO_3^+; with increasing acid concentration the proportion of 1- and 9-sulphonic acid increases at the expense of the 2- and 3-isomers.[26] Other work with this solvent system has been reported,[27] as well as sulphonation in $C_6H_5NO_2$[28] which gives, with phenanthrene, comparable results to those with $MeNO_2$.

The kinetics of reaction of mesitylenesulphonic acid in dilute fuming sulphuric acid have been measured.[29] Treatment of *p*-BrC_6H_4NO with 95% H_2SO_4 is reported to give *p*-benzoquinone monoxime.[30]

Nitration

There is still a considerable interest in the mechanism of aromatic nitration and much new work has appeared that has as its objective the understanding of fine points of detail.

A theoretical treatment (MO-LCAO-SCF) of the electronic structure of the nitronium ion has appeared.[31]

Nitration of benzene, toluene, *m*-xylene and mesitylene (in Ac_2O), with an increase in the concentration of the aromatic compound, gives a decrease in kinetic order.[32] This is a medium effect and not a change to a rate-determining formation of the electrophile; detailed investigation showed that a first-order dependence of rate on concentration of the aromatic persisted for at least 80% of the reaction. It has been claimed that in aqueous sulphuric acid (74–82% H_2SO_4) the nitration of toluene shows a change from first to zero order with respect to toluene concentration.[33]

The additivity of substituent effects in the nitration of substituted anilinium ions is not borne out in practice.[34] More *o*-substitution is found than is expected. By an analysis of the isomer proportions it was concluded that the inductive and field effects of a nitrogen pole usually deactivate *o*- and *p*-positions more than the *m*-position, but, in the cases studied, electron-releasing groups (OMe, Me) stabilize the transition state for *o*-attack.

The formation of *N*-benzylacetamides during nitration of polyalkyl benzenes (HNO_3–MeCN or CH_2Cl_2) has been interpreted as proceeding through an intermediate benzyl nitrite[35a] and not through a nitromethylenecyclohexadiene.[35b] An analogous process was invoked for pentamethylbenzonitrile, in which complete retention of the CN group was believed to be inconsistent with a diene intermediate. The nitration of 1,2,4-trimethylbenzene in mixed acids has been examined over a range of H_2SO_4 concentrations;[35c] the yields of 3- and 5-nitro-products vary relatively little, showing a small rise with increasing acid concentration; however, the yield of 6-nitro-derivative increases more markedly.

Positional selectivity in encounter-rate nitration was discussed, and was considered to be retained provided that capture of the Wheland intermediate from ipso-attack is

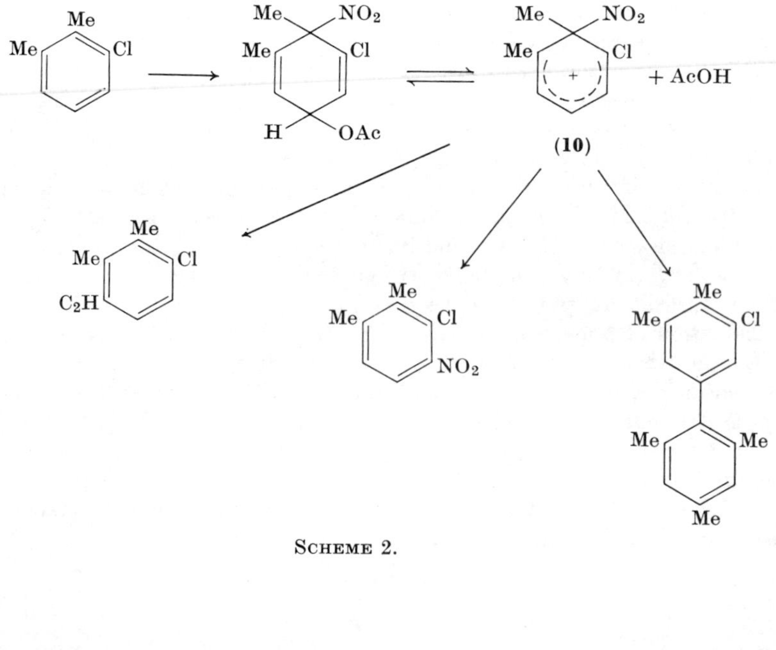

(10)

Scheme 2.

nearly complete. The degree of selectivity thus gives no idea of the attraction between the aromatic compound and the electrophile.

Competitive nitration of nitrobenzene and the mononitrotoluenes has been effected by $NO_2^+PF_6^-$ (in $MeNO_2$ or 96% H_2SO_4).[36] Isomer proportions and rate ratios were measured, and the nature of the transition states was discussed. The product composition and the dependence of rates of reaction on acidity in the nitration of $MeSOC_6H_5$ have been interpreted in terms of reaction via neutral molecules and the conjugate acid.[37] Partial rate factors were calculated for the conjugate acid (98% H_2SO_4).

The partial rate factor for ipso-nitration at a methyl group in toluene or xylenes is estimated to be ca. three times that for *m*-substitution,[38] using a value of $k_T/k_B = 44$. For *o*-, *m*-, *p*- and *i*-positions the partial rate factors are taken as 77, 3.7, 95 and 8.2, respectively, and based on these values the product distribution in nitration of polyalkylbenzenes is calculated and compared with the experimental figures. In an interesting paper on the nitration of 3-chloro-*o*-xylene[39] products consistent with an addition–elimination sequence are reported; in the presence of mesitylene and TFA, a biphenyl derivative, which represents "trapping" of (**10**) is also obtained (Scheme 2). In related work on the nitration of dimethylbenzonitriles,[40] several points of interest emerge. 2,3- and 3,4-Dimethylbenzonitrile react with HNO_3 in Ac_2O to give adducts and substitution products. The adducts (**11**) and (**12**), on thermolysis or decomposition in AcOH, give some starting material together with the 5-nitro-derivative [e.g. (**11**) → (**13**)]

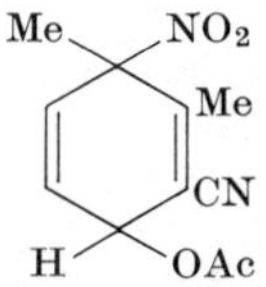

Me NO2 Me CN H OAc

(**11**)

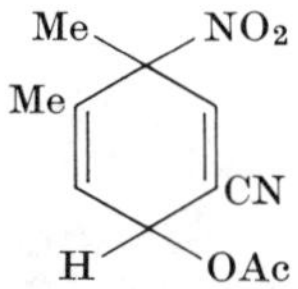

Me NO2 Me CN H OAc

(**12**)

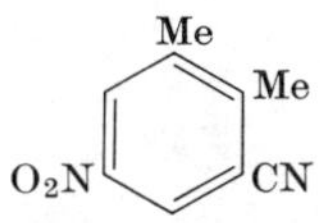

Me Me O2N CN

(**13**)

by a process represented as a 1,3-nitro-shift. The failure to detect the 4-nitro-derivative was held to exclude consecutive 1,2-shifts. It was noted that with the above isomers no ipso-attack occurred at C-2 and C-4, respectively, and that under other conditions adducts (**11**) and (**12**) gave a variety of products which could be fully accounted for on the basis of the structures proposed. Structurally isomeric adducts have been characterized in the nitration of 1,2,4-trimethylbenzene in acetic anhydride, and replacement of acetoxy-groups by other nucleophiles, as well as formation of side-chain substitution products by an elimination–addition–elimination sequence, has been achieved.[41] A reconsideration of the reactivity of *p*-isopropyltoluene has been suggested[42] by the extent of ipso-attack in nitration ($AcONO_2$, 0°); isomeric adducts were recognized by NMR analysis, and a mixture of these in 78% H_2SO_4 gave only 4-isopropyl-2-nitrotoluene; thermolysis of the adducts gave *p*-cymene and thymol acetate. Ipso-attack at the isopropyl position was excluded in adduct formation, but *p*-nitrotoluene formed (10%) is held to result from nitrode isopropylation. The extent of ipso-attack at the methyl position (41%) makes C-1 the most reactive centre.

A steric influence on ipso-attack has been proposed[43] based on the use of (**14**) where $n = 2$, 3 or 4. The β-acetoxy-product (assumed to be at least the amount of ipso-attack) is 3, 19 and 37%, respectively, and relates to the inhibiting effect of the methylene bridge; the corresponding value for indan is 25%.

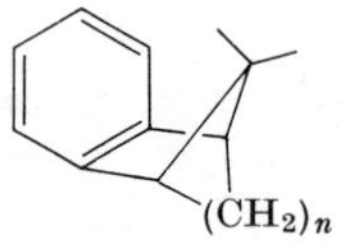

(14) 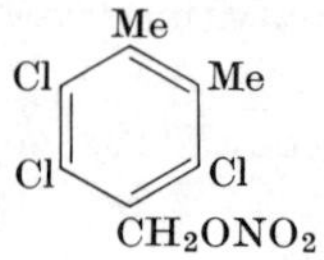 (15) 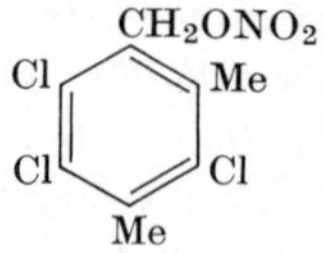 (16)

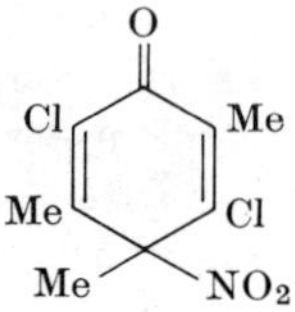

(17) 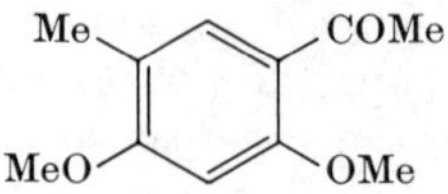 (18)

Side-chain oxynitration of trichloro-1,2,4-trimethylbenzene gives (**15**) and (**16**) and an oxidative process leads to (**17**),[44] by addition–elimination. Nitrodeacylation has been reported with (**18**), together with some nitrodealkylation.[45] Nitrodeiodination of 2-iodo-1,3,5-trineopentylbenzene has been observed to accompany normal substitution.[46] Nitrodephosphonation has been reported.[47] The formation of *N*-methylacetamide and 2,3-dicyano-1,4-naphthoquinone during nitration ($NO_2^+PF_6^-$ in CH_3CN) of a dicyanodimethoxynaphthalene is interpreted as shown in Scheme 3.[48]

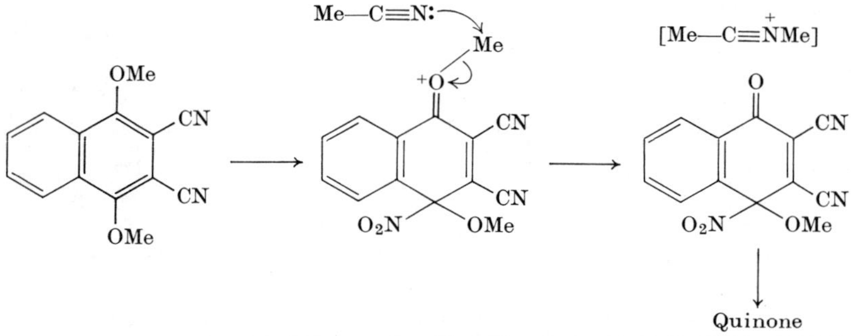

SCHEME 3.

Nitration by $NO_2^+CF_3SO_3^-$ shows high positional selectivity and gives excellent yields.[49] This reagent can be used in CH_2Cl_2 or concentrated H_2SO_4; it is sparingly soluble in CH_2Cl_2 but affords >99% mononitration of toluene (–60°C). The use of $Pd(NO_3)_2$ or $Pd(OAc)_2$–$NaNO_2$ as a nitrating agent gives unusual reactivities; thus, in competition experiments the partial rate factors for *o*-, *m*- and *p*-positions in toluene are 0.3, 1.3 and 3.0 and in chlorobenzene 0.4, 0.3 and 0.7, respectively.[50] A mechanism is proposed which invokes the sequence of Scheme 4.

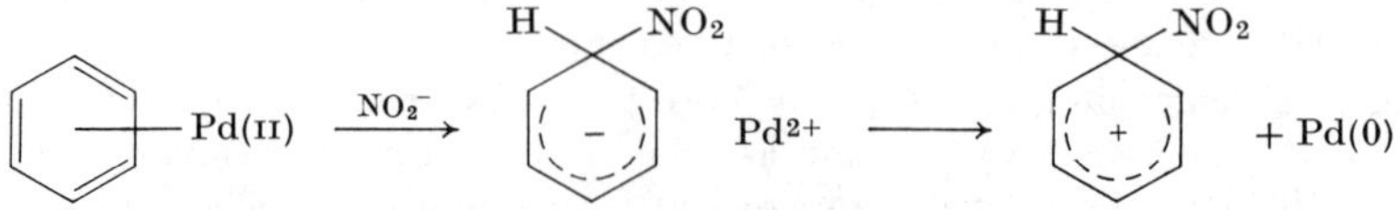

SCHEME 4.

Nitration by N_2O_4 in trifluoroacetic acid gives more typical reactivity patterns, e.g. with toluene, and is thought to involve nitronium ions.[50] Tetranitratotitanium(IV) nitrates aromatic compounds in CCl_4, and competition gives $k_T/k_B = 5$, with a low percentage of *m*-attack.[51] Surprisingly, nitrobenzene gives appreciable yields of nitration product, and aliphatic reactivity has been noted. The mechanism of this intriguing process is not clear, but a π–titanium co-ordination is suggested as the first step. Analogous experiments with $Zr(NO_3)_4$, $Fe(NO_3)_4NO$ and $Cu(NO_3)_2$ have been reported.[51] The last of these compounds is inactive, but the zirconium compound nitrates pyridine (yield = 0.09 mol per mol of reagent) to give principally 4- (64%) and 3-nitropyridine (36%); a trace (<0.01%) of 2-nitropyridine is also formed. Quinoline gives high yields (90%) composed of 7- (96%) and 3-nitroquinoline (4%), when $Zr(NO_3)_4$ is used. Methyl nitrate, with BF_3, nitrates aromatic substances, but the concentration of free NO_2^+ is small and the reaction is believed to proceed by attack on a polarized complex of $MeONO_2$ and BF_3.[52] A lengthy discussion of the mechanism is given in a linear free energy relationship (LFER) analysis; the correlation of transition-state character with π-complex basicity seems unjustified, as the parallel with σ-complex formation accords better with the deductions of the LFER treatment. *N*-Nitration has been related to the basicity of the secondary amine.[53,54] *N*-Denitration of *N*-picrylnitramines shows a linear rate *vs.* acidity function relationship.[55]

Nitration by NO_2–N_2O_4 is believed to involve a mixture of ionic and radical processes.[56] With $C_6H_3(NR_2)_3$ (NR_2 = morpholino- or piperidino) nitrosation was observed.[57]

An infrared method for estimation of nitrating ability of aqueous HNO_3 has been developed.[58] The two-phase nitration of C_6H_5Me[59] and C_6H_5Cl[60] and various nitrations of the three mononitrotoluenes[61] and *o*-dichlorobenzene[61a] have been reported.

Nitration of porphin involved attack at a *meso*-position (5) and further nitration gave the 5,10-dinitro-product.[62] Naphtho[1,2-*b*]thiophen (**19**) is predicted to show a reactivity sequence 3 > 2 > 5; nitration (in common with bromination, formylation, acetylation and metallation) occurred at position 2.[63] Bromination or nitration of the 2-ethoxycarbonyl derivative of (**19**) gave 5-substitution only. The substituted thiatriazacyclazines (**20**) and (**21**) have been prepared; nitration is assumed to occur at position 8.[64]

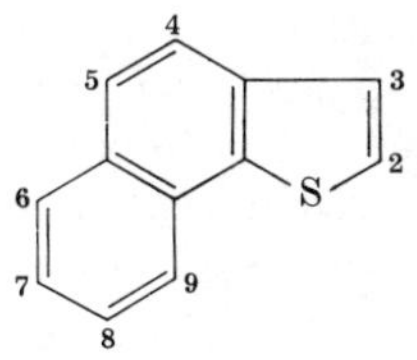

(**19**) 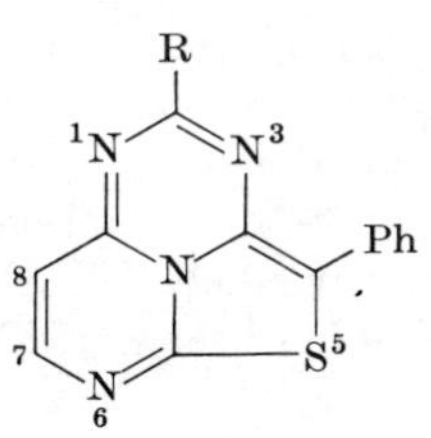 (**20**) 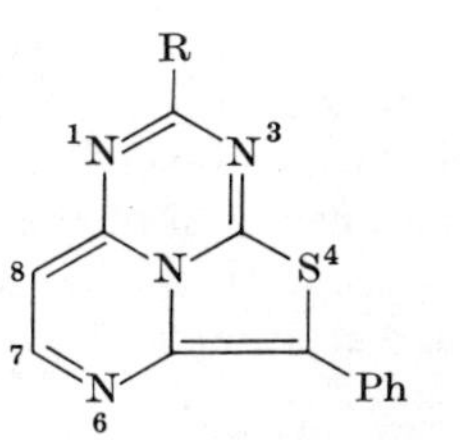 (**21**)

Several papers have appeared from Katritzky and his co-workers on *N*-nitroimides,[65] the nitration of triazoles[66] and pyrazole derivatives.[67,68] Other work reported concerns the kinetics of nitration of carbazole and some of its derivatives,[69,70] the orientation in nitration of the acridizinium ion (and its 6,11-dihydro-derivative),[71] isocytosine 6-acetic acid,[72] phenanthroline derivatives,[73] dibenzodioxans,[74] azulene derivatives,[75] hexafluoropropoxy-benzene and -thiopyran[76] and dimethylnitronaphthalenes.[77]

Nitrosation

Co-ordinated nitrosyl groups show electrophilic behaviour in certain metal complexes.[78] Nitrosation of amines by $[Ru(bipy)_2(NO)X]^{2+}$ leads to incorporation of the amine portion in the complex.[79] A mechanism has been proposed[80] for the transformation of the trithiapentalene (**22**) into an oxadithia-azapentalene (**23**). The nitrosation of proline and

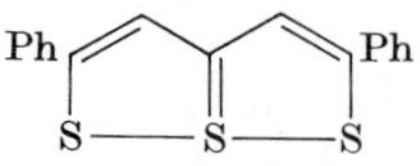

(**22**) 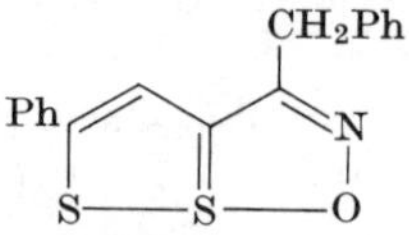 (**23**)

pyrrolidine, in weakly acid buffers, has been found to be independent of pH.[81] Nitrosode-iodination of *p*-iodoanisole (leading to a nitro-compound) gives rise to a coloured intermediate, which is a σ-complex.[82] Nitrosation of aryltrimethylstannanes gives a route to nitro-derivatives.[83]

Azo Coupling

A mathematical model for uncatalysed diazotization[84] and a comparison of experimental results with kinetic equations derived for aniline have been given.[85] The decomposition of $C_6H_5N_2{}^+BF_4{}^-$ in CF_3CH_2OH is described as heterolytic, but in the presence of pyridine homolytic pathways are favoured.[86]

In a comprehensive study of the nature of the transition state in coupling, it has been concluded that the kinetic isotope effect is of greater diagnostic value than the Brønsted β-value, although both are seen to be of limited value in determining the symmetry of the transition state.[87] Rate measurements over a temperature range suggested that tunnelling was not significant. A kinetic study of the reaction of *p*-$MeOC_6H_4N_2{}^+$ and *p*-$MeC_6H_4N_2{}^+$ with 1-naphthol showed general base catalysis was present.[88] The reaction of $ArN_2{}^+$ with PhSH has been compared to that occurring with $PhNH_2$; the initial products (ArN_2SPh) give coupling products with 2-naphthol and it is assumed that the diazo-sulphide dissociates in a manner analogous to the diazoamino system.[89] Diazotized 2,4-dinitroaniline couples with thiophen and methylthiophens, in the latter case giving thiophencarbaldehyde dinitrophenylhydrazones.[90] 2-*tert*-Butylthiophen couples at position 5, however.[91] The case of interaction of diazotized sulphanilic acid with aromatic amines has been related to pK_a values,[92] and coupling has been shown to be faster in dipolar aprotic media than in protic solvents.[93] The displacement of diazonium groups from *o*-arylazophenols by electrophiles has been noted.[94]

Halogenation

A comprehensive review of the pathways in electrophilic halogenation has been presented.[95] Special reference is made to the formation, trapping and reactivity of diene intermediates.

In continuation of earlier work on the naphthalene system, the 1-phenyl derivative forms a tetrachloride, transformed almost exclusively into 2,3-dichloro-1-phenylnaphthalene by base (99.3%). By heating the tetrachloride in CS_2 or treating it with $AlCl_3$–CH_3NO_2, a 2:1 mixture of 2,4- and 2,3-dichloro-1-phenylnaphthalene is formed.[96] From these two observations the structure (**24**) is proposed and NMR evidence supports

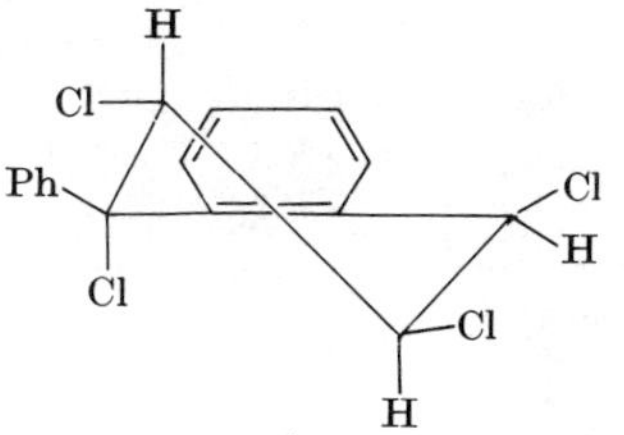

(24)

this. Chlorination of 1,2-dichloronaphthalene gives isomeric tetrachloro-compounds and evidence of adduct formation.[97] Naphthalene, with Cl_2 in CS_2 at low temperatures, gives *trans*-1,2-dichloro-1,2-dihydronaphthalene; heterolytic chlorination gives a new tetrachlorotetralin which loses HCl to give *trans*-1,2,4-trichloro-1,2-dihydronaphthalene.[98] Chlorination of *N*-acetylcarbazole involves electrophilic attack,[99] and not as otherwise claimed.[100]

The chlorination of hexamethylbenzene may proceed via electron transfer,[101] since the ESR spectrum of chlorine–hexamethylbenzene mixtures is consistent with radical cation formation. Anisole, with Cl_2, gives evidence of addition products (NMR and isolation), and other ethers behave similarly.[102] In the solid state, 2,6-dialkylphenols react via a diene-type intermediate at C-4; alkyl groups attached at C-4 migrate.[103] The reaction constant (ρ^+) for chlorination of 2- substituted thiophens is –9.2, as for *p*-chlorination of monosubstituted benzenes,[104] thus meeting a criticism recently levelled at the use of LFERs.[105]

A new type of mechanism for halogenation has been proposed to account for the rapid bromination (Bu^tOBr in DMF–Bu^tO^-) of 1,3,5-tribromobenzene to tetra- and penta-bromo-compounds.[106] It is envisaged that a carbanion is formed which then reacts; the necessity for base was demonstrated.

The bromination ($HOBr–H^+$) of disubstituted benzenes $XC_6H_4(CH_2)_nY$ (X = Me or OMe, $n = 0–3$, Y = N^+Me_3, N^+H_3 or S^+Me_2) has been reported[107] and an analysis made of additivity and substituent effects. Bromination of alkylpyridines (aq. $HClO_4$) is via the conjugate acid.[108] The kinetic effects of *N*-methylation and methoxy-substitution were noted and an estimated partial rate factor for position 3 in pyridinium ions is 6×10^{-13}. Bromination of phenols has been studied by several groups. In water, the rate difference for *p*-chlorophenol and its conjugate base is ca. 6×10^6, and ρ-values are –3.5 and –5.11, respectively; *p*-methoxyphenol forms a benzoquinone derivative.[109] A potentiometric method was used to show that monobromination in the phenol–Br_2 reaction is independent of pH, but the rate of the third bromination increases with pH.[110] The same reaction has been studied in CCl_4[111] with pyridine hydroperbromide,[112] and in the presence of an epoxide that functions as an HBr scavenger.[113] It was shown that the *o*/*p* ratio was dependent on the presence of HBr in, e.g., CCl_4, but not in AcOH. The reaction pathway was considered to involve *ortho*-attack, forming an *o*-quinonoid structure rearranged by HBr to a *p*-quinonoid system. Preparatively, the method is of value, since, with epoxide in CCl_4, 89% of *o*-bromophenol was formed. HCl was less effective than HBr in bringing about the rearrangement. The solvent-dependence of the *o*/*p* ratio when NBS is used was noted.[114] A study of bromination of anilides affords some further evidence for N–H hyperconjugation, as measured by E_a for PhNHAc and PhNMeAc.[115]

Chlorination by a mixture of Bu^tOCl and BF_3 has been reported; with this, toluene gives 70% of *o*-substitution and $k_T/k_B = 21$; the electrophilic species is not Cl_2.[116] The *ortho*-selectivity of Bu^tOCl with phenols has been challenged, and the *o/p* ratio shows dependence on hydrogen bonding.[117] $Bu^tOCl–I_2$ is considered to form Bu^tOI, which iodinates anisole.[118] *N*-Iodo-amides have greater reactivity when an electron-withdrawing group X activates the system, e.g. *p*-XC_6H_4CONHI.[119] A new and potent iodinating agent is I_2–thallium(III) trifluoroacetate in TFA.[120] Stepwise iodination of C_6H_6 (up to tetraiodination!) is probably via organothallium interaction in $C_6H_5Tl(TFA)_2$ which reacts with I_2. High yields (>95%) of *para*-substitution with, e.g., PhHal can be realized by $SbCl_5–Br_2$ (or I_2) mixtures (1:1), possibly via BrCl (ICl).[121] Selective halogenation by using Br_2 (or Cl_2) in molecular sieves gives high yields of *p*-XC_6H_4Alkyl and *p*-XC_6H_4Hal (X = Br or Cl).[122] Electrochemical bromination (NH_4Br in CH_3OH) is interpreted as an electrophilic process; thus 2-methylthiophen afforded >80% of 5-bromination;[123] some ring opening, of an oxidative kind, also occurred.

The "radiochemical" bromination of aromatic compounds effected by nuclear decay of ^{80}Br has been reviewed.[124] For this process, and its counterpart with iodine, low positional selectivity was found, together with evidence for intramolecular rearrangement of the ion formed by attack of Hal^+ on ArH. PhBr with I^+ gave a high proportion of *ortho*-substitution.[125] Radiolysis of CCl_4 gave CCl_3^+, which effected substitution in benzene.[126]

The catalytic effect of pyridine on bromination of mesitylene (in AcOH or $CHCl_3$) has been shown to be due to salt effects; pyridinium salts were less effective than some ammonium salts.[127] Differential scanning calorimetry has been applied to the bromine adducts of pyridinium bromide, and a succession of compounds $PyHBr.nBr_2$ (where $n = \frac{1}{3}, \frac{1}{2}, \frac{2}{3}, 1, 3, 6$ or 18) was identified.[128] *tert*-Butyl-6-methyl pyridine with Br_2 in fuming H_2SO_4 gives the 5-bromo- and 3,5-dibromo-derivatives, and 2,6-dimethyl-3,5-dibromo-pyridine.[129]

Iodination of thiophens proceeds according to the proposed mechanism,[130] and with 2-phenylthiophen the rate becomes independent of concentration, as required. The formation of complexes between $AlCl_3$ and furan or thiophen aldehydes or ketones and their reactivities in bromination have been noted.[131] Benzoselenophen affords 2- and 3-substitution products on halogenation.[132] Chlorination of substituted acenaphthenes,[133] fulvene derivatives,[134] 4-oxo-4*H*-1-benzopyran-3-carbaldehyde,[135] tetrahydrofuran (and its 2,3-dichloro-derivative)[136] and thieno[2,3-*b*]pyridine (with other halogenations)[137] has been reported. In the reaction of the nitrone (**25**) with $COCl_2$, the anil hydrochloride (**26**) is formed, probably via (**27**).[138]

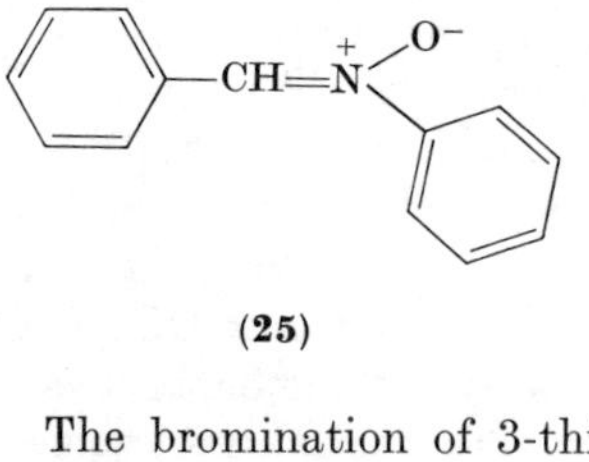

(**25**) 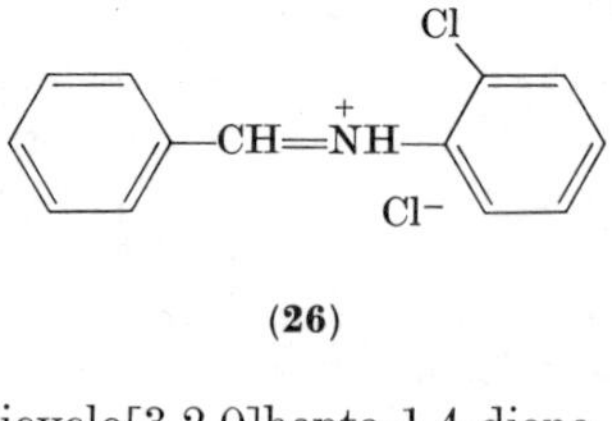 (**26**) 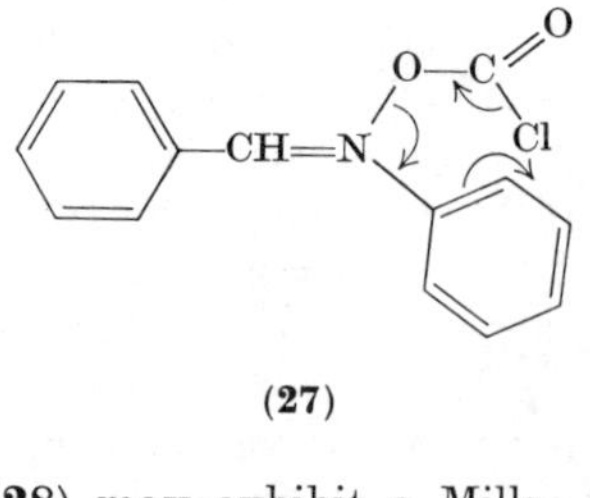 (**27**)

The bromination of 3-thiabicyclo[3,2,0]hepta-1,4-diene (**28**) may exhibit a Mills–Nixon effect in which annelation favours addition to give a tetrabromide;[139] 3,4-dimethylthiophen underwent substitution in the usual way. Benzoylation of (**28**) gave a low yield of ketone.

Other work on bromination includes a study of substitution in indole and its derivatives,[140] toluene- and phenol-sulphonates,[141] various polyalkylbenzenes,[142] imidazole,[143a] indazole,[143b] imidazo[1,2-*a*]pyridine (**29**),[144] benzofurans,[145] the cyclazines

(**30**) and (**31**),[146,147] uracil and its *N*-methyl derivatives,[148] 2(1*H*)-pyrimidinone (**32**) and its *N*-methylated derivatives,[149] dihydroxypyrimidine and methyluracil.[150] Iodination of γ-picoline[151] and protiodehalogenation in the diazepine series[152] are noted.

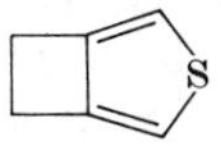

(**28**) 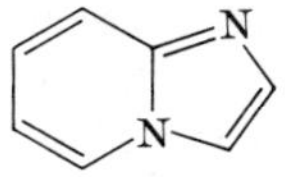 (**29**) 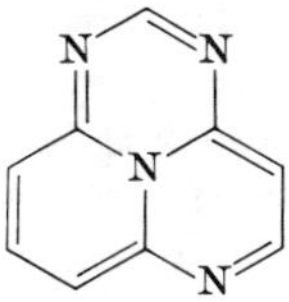 (**30**) 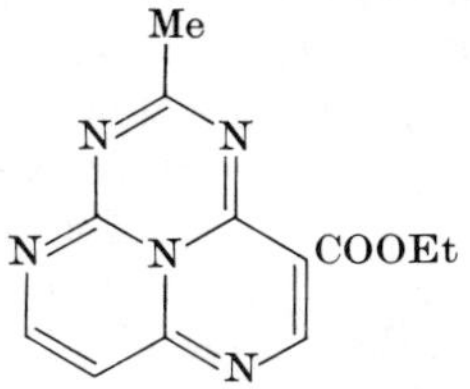 (**31**) 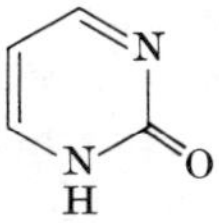 (**32**)

Metal Cleavage

Trifluoroacetic acid cleavage of metal–carbon bonds is exemplified by reactions of $XC_6H_4SnMe_3$ in CF_3COOH–CF_3COOD mixtures. The product isotope effect (PIE) was measured for various substituents. With *m*-CF_3, an anomalous process led to incorporation of more than one D-atom.[153] $XC_6H_4CH_2SnMe_3$ was cleaved by TFA or by bromination (although some anomalies were noted) and substituent effects have been discussed.[154,155] H_2SO_4–AcOH has been used to effect protodesilylation in the system $Me_3SiC_6H_4SOC_6H_5$ (*o*-, *m*- or *p*-positions)[156]. A contrast was made between the requirements of halogenation in PhSOPh and the demetallation. In halogenation it was postulated that the apparent (anomalous) activation of the *p*-position originated from the polarizability of the SOPh group, but the normal electron-withdrawal operated in demetallation. The dual nature of a substituent with a dipolar bond seems surprising. The PIE values for cleavage of $XC_6H_4SnMe_3$ in MeOH–MeOD led to the conclusion that proton-transfer from the solvent to the carbon of the Sn—aryl bond is involved in the rate-determining step.[157]

Metallation

The effect of *o*-alkyl groups has been discussed in relation to anisole/alkyl-lithium interaction.[158] It is proposed that ether-oxygen complex formation with butyl-lithium oligomer is significant, since, in the presence of TMEDA ($Me_2NCH_2CH_2NMe_2$), in which monomers are present, quite high yields of *o*-metallated products are obtained. Base strength is also significant. The reaction of 3,5-diethylanisole with ^{14}C-ethyl-lithium showed that an adduct was not produced; CIDNP did not support radical character in the intermediate.[159] Lithiation occurs *ortho* to a CF_3 group,[160] even when two are present (*meta* to each other). Lithiation of thiophen occurred at C-2, leading to a low yield of 2-carboxylic acid.[161] An investigation of metallation of 3-alkylthiophens (R = Me, Et, Pr^i or Bu^t) and halogen–metal exchange of 3-alkyl-2,5-dibromothiophens by alkyl-lithium (or -magnesium) reagents has shown that steric effects in the reagent and the substituent groups play a significant part.[162] Benzoselenophen reacts with BuLi to form a 3-lithio-derivative.[163] Mercuration of aromatic amines[164] and the reaction of fluorene and R_3SiNa (in THF) have been noted.[165]

Friedel–Crafts and Related Reactions

Some *ab initio* MO calculations on the stability of RCO^+ and XCH_2CO^+ (R = Me, Et, Pr^i or Bu^t; X = Et, vinyl, CH≡C, CN, OH or F) suggest that the stability follows the inductive order of groups R;[166] hyperconjugation from C—C bonds is more effective than that from C—H bonds. The spectroscopic, structural and theoretical investigations of complexes $RCOX.MX_n$ have been reviewed.[167] The rate constant and energy of activation for the degenerate rearrangement of the heptaethylbenzenium ion are ca. 60 s^{-1} and 12.9 kcal mol^{-1}.[168]

Multilayered [2.2]-paracyclophanes, e.g. (**33**), with $AlCl_3$–HCl undergo rearrangement (formally dealkylation followed by substitution), but with $SnCl_4$ two different products are formed, as shown in Scheme 5; as indicated, these are not interconverted under the reaction conditions.[169] The driving force for this process is the relative basicity and the release of strain as the interactions between layered rings are decreased. Alkylation of benzene and anisole with 2,2,4-trimethylpentane and $AlBr_3$ or $AlCl_3$ and an acid catalyst

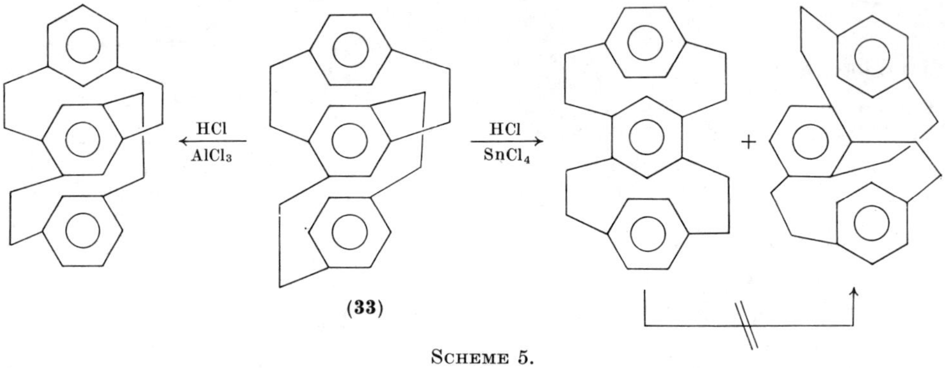

HCl / $AlCl_3$ ← (**33**) → HCl / $SnCl_4$ +

SCHEME 5.

has been shown to yield products of *tert*-butyl substitution (inter alia).[170] Dealkylation with $AlCl_3$–H^+ gives hydrocarbons[171] and with RC_6H_4X (X = Br, R = Pr^i or Bu^t) yields can be 90–100%. Transalkylation has been shown to be easier with *p*-substituted alkylbenzenes.[172]

"Cycloalkylation" with cyclohexanol and a protic acid gives substitution products derived from the methylcyclopentyl skeleton.[173] Amidoalkylation (by $HCONMeCH_2OAc$ in TFA) gives with mesitylene 93% of $C_6H_2Me_3CH_2NMeCHO$, with intermediate formation of $HCONMeCH_2OCOCF_3$.[174] Alkylation of indole and its derivatives with KOH–DMSO gives 1-substitution; 2-methylindole undergoes some 3-alkylation.[175] The system naphthalene–propene has been used as a kinetic model for alkylations.[176] A group of alkyl ester (chlorosulphites, sulphinates, chloro- and fluoro- sulphates, arenesulphonates, triflates, pentafluorobenzenesulphonates and trifluoroacetates) have been shown to have alkylating ability with $AlCl_3$ in $MeNO_2$ or CH_2Cl_2 or with SbF_5 in Freon.[177] Isomerization was extensive with Pr^n and Bu^t groups; k_T/k_B was low (1.8), and the intermediates were discussed in terms of complexing of the reactant with $AlCl_3$ and rearrangement of the complex to form a species $R^+AlCl_3X^-$. It was noted that a high percentage of *meta*-attack characterized some processes, e.g. $C_2H_5OSO_2C_6H_4Me$ was reported to afford 98% of *m*-ethylation of toluene. Some chlorination took place with $ROSO_2Cl$. The character of the attacking species suggests little discrimination in many examples studied. A high selectivity and activity was claimed for benzylation of toluene with a catalyst prepared by calcining the hydrates of Fe(II) and Fe(III) sulphates in air, to yield 42% of *o*- and 51% of *p*-alkylation.[178] Isomerization of polymethylbenzenes was shown[179] to be moderated by the use of cationic zeolite catalysts in which rare-earth metals are present; a similar moderation has been achieved by using $AlCl_3$ in a graphite lattice.[180] In this case, as in the previous one, polysubstitution is less; thus toluene normally affords triethyltoluenes, but with $AlCl_3$/C 52% of monoethylation and 31% of disubstitution were reported. The use as a catalyst of Arene–$Mo(CO)_3$ for homogenous alkylation, acylation and sulphonylation has the advantage that it, too, is milder than $AlCl_3$.[181] An alkylation, catalysed by Pd(II) acetate, is shown in Scheme 6; it is a process subject to steric effects of *ortho*-groups.[182]

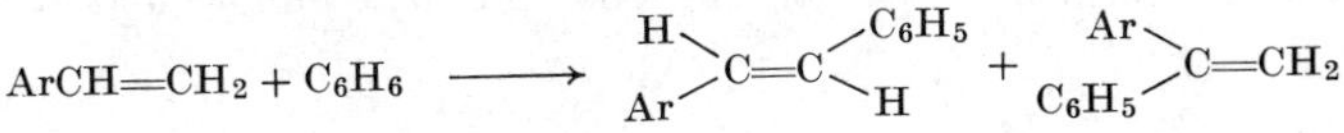

$$ArCH{=}CH_2 + C_6H_6 \longrightarrow (H)(Ar)C{=}C(C_6H_5)(H) + (Ar)(C_6H_5)C{=}CH_2$$

SCHEME 6.

Phenylation, via a cation, is proposed for decomposition of dibenzoyl peroxide in TFA.[183] The comparison of phenylation of anisole for three reactions is shown in the Table. The figures were held to support the polar nature of the Bz_2O_2–CF_3COOH

Products (yields in %) of phenylation of anisole.

	Reagent		
Isomer	Bz_2O_2–CF_3COOH	$PhN_2^+CF_3COO^-$	PhN(NO)Ac
ortho	45	61.6	69
meta	19	9.7	18
para	36	28.7	13

reaction. Alkoxyalkyl halides react with benzene–$AlCl_3$, giving isomeric alkoxyphenylalkanes.[184]

In very strong acids (HF–TaF_5) benzene is transformed into toluene, ethylbenzene and some biphenyl.[185] Better yields are obtained in the presence of hydrogen, and it is proposed that benzene is reduced and that the cyclohexyl framework is isomerized to a methylcyclopentyl structure, or ring-opened to give open-chain ions, which may undergo degradation by successive hydrogen shifts and formation of smaller alkyl fragments; SbF_5, AsF_5 and VF_5 were ineffective, ions of the type $ArH_2^+MF_6^+$ being formed. The complex formed in the HF–TaF_5 reaction is thus thought to differ from the σ-type. Some supporting evidence for the above sequence was the isolation[186] of alkane from reaction of cyclohexane and methylcyclopentane in HF–TaF_5–H_2. Hydrogen transfer from hexane may be indicated by the fairly ready conversion ($t_{0.5} \sim 2$ hr) of benzene (in hexane) to methylcyclopentane and isohexyl products.

Phenol is *C*-methylated by methanol over a ZnO–Fe_2O_3 catalyst,[187] but the pore structure of the catalyst is important.[188]

Acylation of benzene is observed with the mixed acid chlorides (**34**) rather than sulphonylation;[189] the reaction is consistent with the behaviour of electron-withdrawing groups which retard sulphonylation but accelerate acylation. A small amount of C_6H_5Cl is formed as a by-product. A small amount of ipso-attack is noted[190] in the acetylation of (**35**), as well as halogen transfer to give (**36**) (5% of each).

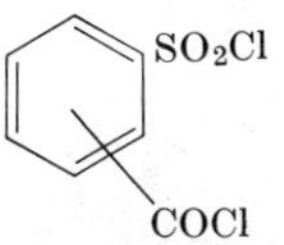

SO_2Cl, COCl

(**34**)

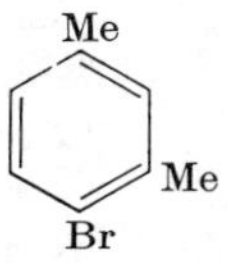

Me, Me, Br

(**35**)

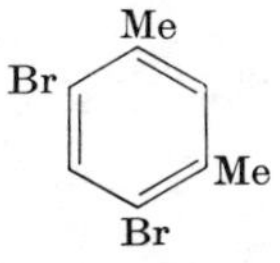

Me, Br, Me, Br

(**36**)

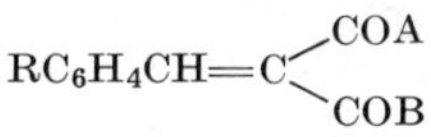

$RC_6H_4CH{=}C(COA)(COB)$

(**37**)

As a tribute to the late C. D. Nenitzescu, a review of condensations of unsaturated acids with aromatic substrates has appeared.[191] Compounds of type (**37**) (R = H, *p*-Me or *p*-OMe; A = B = C_6H_5, Me or Et) have been condensed with anisole; triphenylmethane derivatives were formed.[192] Reaction between 3-cyclopentylacrylic acid and benzene catalysed by $AlCl_3$ was shown to furnish the complex mixture of products indicated in Scheme 7. The formation of these products was discussed in terms of competition between rearrangements of intermediate carbonium ions and their bimolecular reactions with benzene.[193]

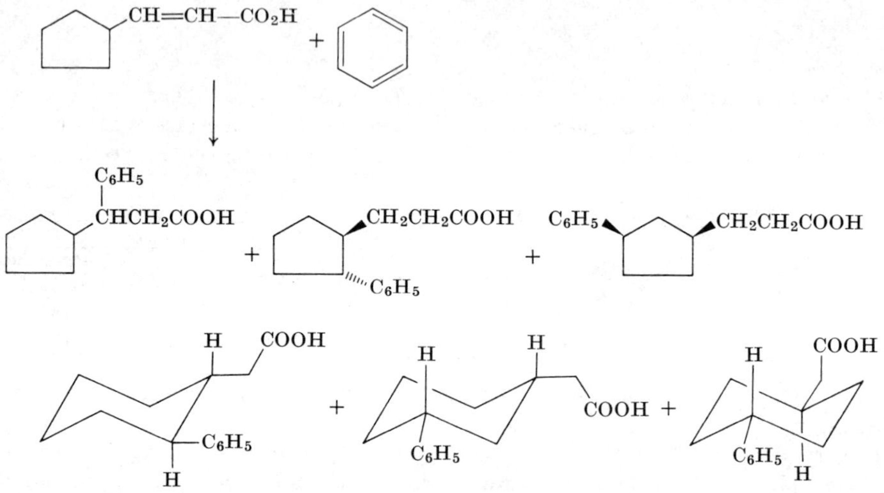

CH=CH—CO_2H + C₆H₆

C_6H_5 CHCH$_2$COOH + CH$_2$CH$_2$COOH / C_6H_5 + C_6H_5 CH$_2$CH$_2$COOH

H COOH / C_6H_5 / H + H H COOH / C_6H_5 + H COOH / C_6H_5 H

SCHEME 7.

The acylation of naphthalene with PhCH=CHCOCl is dependent on temperature and reaction time,[194] and the 2-acylation product is formed in greater yield at elevated temperature and after longer times. Independent experiments failed to effect the rearrangement of the 1-ketone to the 2-isomer; only phenalenone was isolated. In the presence of $AlCl_3$-C_6H_5Cl the 2-ketone added to the styryl double bond, and no 1-isomer was found.

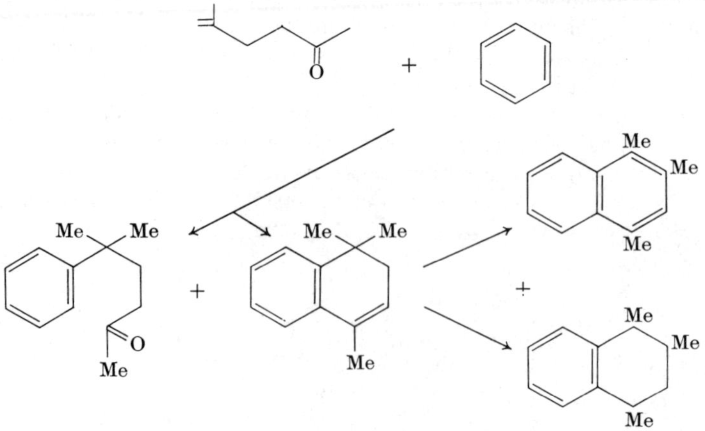

O + benzene

Me Me / O / Me + Me Me / Me → Me / Me / Me + Me / Me / Me

SCHEME 8.

Some cyclization was observed in the reaction of isomeric hexenones and benzene, e.g. 5-methylhex-5-en-2-one gave open-chain ketone (46.5%) and trimethylnaphthalene and its dihydro- and tetrahydro-derivatives.[195] Methyl migration was also observed, as shown in Scheme 8.

Cycloacylation of 3-phenylpropanoic and 4-phenylbutanoic acids in liquid HF is subject to conformational effects when benzylic hydrogens are replaced by methyl groups.[196] The cyclization of mono-, di- and tri-phenylpropionyl chlorides has been discussed.[197]

Acylation of PhX by $ArCH_2COCl/AlCl_3$ is competitive with decarbonylation followed by alkylation by $ArCH_2^+$; the extent of decarbonylation depends on the reactivity of PhX.[198] The inhibiting effect, on benzoylation, of complexing $AlCl_3$ with the product has been reported.[199] 2,7-Dimethylnaphthalene undergoes acetylation to give 1,3- and 1,5-diacetyl derivatives;[200] this is the first recorded example of diacylation in one aromatic ring and probably arises from the steric inhibition of resonance deactivation of the first substituted acetyl group. Steric effects are clearly significant since benzoylation gave 1,5-and 1,8-dibenzoylation. Mesitoylation was slow and occurred at position 1. The reactivity sequence for acetylation or benzoylation of iodotoluenes is $p < o < m$. Side reactions gave di- and tri-iodotoluenes, acetophenone or benzophenone (and methyl derivatives), and with benzene in 1,2-dichloroethane, gave 2-chloroethyl benzoate.[201] Polyphosphoric acid (PPA) brings about benzoylation of naphthalene with benzoic acid;[202] the product ratio was time- and temperature-dependent. 1-Benzoylnaphthalene in PPA isomerized to the 2-isomer, but the reverse reaction was not observed. A mixed benzoic acid–PPA anhydride was considered as an alternative to a benzoylium cation. Concentrated H_2SO_4 effects a general indene synthesis from compounds of the type $Ph_2C(OH)CMe{=}CHPh$, in high yield.[203]

Other work reported includes cyclization of (**38**) by $SnCl_4$–CH_2Cl_2[204] (in the absence of $SnCl_4$ cycloaddition is observed) to give *endo*-2-chlorobicyclo[4.2.2]decan-8-one, acetylation of benzofurans[205] including 2,3-diphenyl derivatives,[206] Fries rearrangement of *p*-$BrC_6H_4SO_3R$ by $AlCl_3$,[207] reaction of 2-alkylpyridines with MeCOCOMe,[208] use of substituted oxetanes,[209] reaction of the furan (**39**) with benzene[210] to afford (**40**), and isomerization and disproportionation of di- and tri-halobenzenes.[211,212]

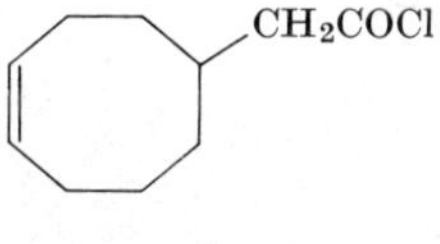

CH_2COCl (**38**) ClOC COCl R O R (**39**) O= =O O (**40**)

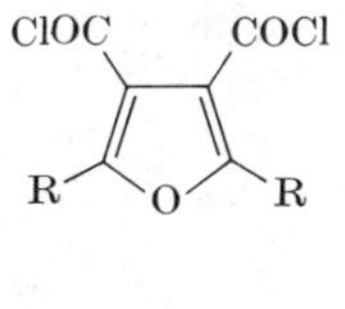

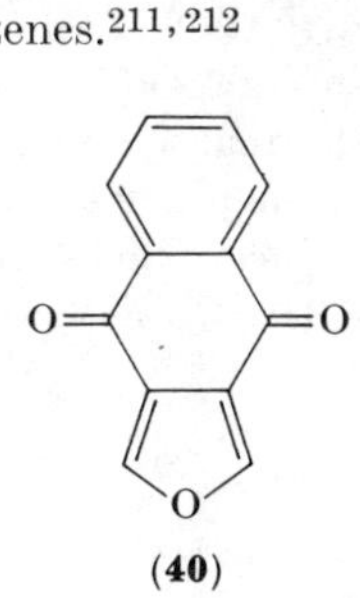

Mesitylene or benzene with $AlCl_3$–CH_3NO_2 gives a mixture of nitrile and aldoxime (*syn*- and *anti*-forms).[213] A species $CH_2{=}N^+(OAlCl_2)_2$ is proposed as the substituting agent. The formation of low yields of Ph_2SbF_3 and Ph_3SbF_2 has been reported, although much polymer is formed, in the reaction of XC_6H_5 with SbF_5.[214] In the presence of BF_3, C_6H_5CHO and C_6H_5Me give benzyltoluenes, yields being respectively 42, 7 and 51% for *o*-, *m*- and *p*-attack.[215]

Experiments with arenesulphinyl halides (ArSOCl) have attracted attention. High k_T/k_B ratios (e.g. 660 for Ar = Ph) and predominant *para*-substitution in toluene (>90%)

were found,[216] indicating a very selective electrophilic species. In other work[217] attention is drawn to the fact that in the presence of $AlCl_3$, $SnCl_4$ or $SbCl_5$ sulphinylation is observed, but with $ZnCl_2$, BF_3–Et_2O or Fe the diaryl sulphide is produced. It was shown that ArSOCl and $SbCl_5$ form a 1:1 complex, and in such complexes the IR spectrum of ArSOCl is modified; this is not true when BF_3 or $ZnCl_2$ is used. Sulphenylation also occurred in the absence of a catalyst and is explained as a disproportionation of ArSOCl to ArSCl and $ArSO_2Cl$. Alkylation was observed on attempted sulphonylation of benzene with $RSO_2Cl(R = Bu^t)$ and is analogous to the behaviour of Bu^tCOCl.[218]

Kinetic data for formylation of thiophen give $\rho = -7.3$; solvent effects are small, but the nature of the amide is critical; thus $COCl_2$:DMA reacts ca. 5×10^3 times slower than does $COCl_2$:DMF.[219] Rate constants and activation parameters were measured for formylation of furan, thiophen, selenophen and tellurophen;[220] within usual errors, $\Delta S^{\neq}$ was identical for all four substrates; an attempt to correlate reactivity with ground-state aromaticities gave an inverse order. Relative reactivities for the series, and for other substitutions, were discussed. Acylation of indoles has been reported, and evidence for two conformations of the *N*-acyl group is presented.[221] Activated styrene derivatives afford indenes under Vilsmeier–Haack conditions.[222] Interestingly, the configuration is crucial, since *Z*-isomers alone give the cyclic product and *E*-isomers give an aldehyde.

The reaction of several benzene derivatives with HCl–CH_2O gives $\rho = -4.97$.[223] Rates for reaction of phenol with CH_2O are also reported.[224] Methoxycarbonylation of benzene with a cobalt carbonyl catalyst probably proceeds via an acylcobalt carbonyl.[225]

Hydrogen Exchange

Hydrogen exchange in the *o*-position of $(C_6H_5)_4C$ is sterically hindered and is the second instance of a steric effect on such a process.[226] Protodetritiation of C_6H_6 in TFA (70–180°C) confirms the earlier value reported for this rate constant,[227] which at 70° is $9.5 \times 10^{-9}\ s^{-1}$, and the recently proposed value is wrong.[228] For the detritiation in TFA at 70°, for the series *o*- or *p*-$Me_3Si(CH_2)_nC_6H_4{}^3H$ (where $n = 1$–4) $\log (f_o/f_p)$ falls in the range 0.865 ± 0.05 applicable to many monosubstituted benzenes, and in agreement with charge distribution in the benzenium ion.[229] The compound (**41**) was examined for exchange in CF_3COOD–D_2SO_4–D_2O under conditions in which sulphonation did not interfere; the reactivity of the two rings was substantially the same and electrostatic effects from the nitrogen pole are not significant for exchange.[230]

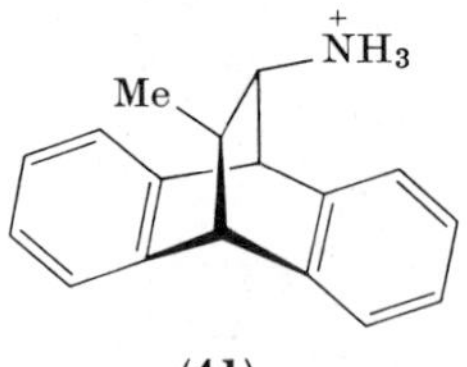

$\overset{+}{N}H_3$ Me

(**41**)

Acid-catalysed detritiation (by aq. H_2SO_4) of benzene and naphthalene follows H_0 with slopes of 1.62, 1.29 ($1\text{-}^3HC_{10}H_7$) and 1.01 ($2\text{-}^3HC_{10}H_7$).[231] The equation

$$\log (\text{partial rate factor}) = -6.3\ \Sigma\ \sigma^+ - 2.1\ \Sigma\, \Pi\ \sigma^+ + 1.6$$

has been proposed (where the term $\Sigma\, \Pi\ \sigma^+$ allows for interaction between substituents)[232] for exchange rate constants for a number of benzene and pyridine derivatives over a reactivity range of 10^{16}.

Exchange in aniline derivatives has been studied by NMR spectroscopy.[233] Conditions of high temperature and low acidity are successfully applied to exchange of aniline hydrochloride in D_2O;[234] in 24 hr only *o*- and *p*-positions are involved in exchange, but with longer reactions, e.g. 35 hr at 250°, substantial *m*-exchange is also noted. Protodetritiation of benzo[*c*]phenanthrene gives an order of reactivities which correlates with Hückel localization energies.[235] Rate constants and partial rate factors are reported for exchange in furan, thiophen and selenophen.[236] Aprotic solvents modify the ratio k_D/k_T in exchange of durene and 2,5-di-*tert*-butylthiophen.[237] Annelation of tropylium with thieno- and furano-rings has a stabilizing influence (in contrast to the effect of benzo-annelation) and exchange occurs at the β-position.[238] The exchange of imidazole via conjugate acid or base has been studied and the relative reactivities of the ring positions have been measured.[239] The rate profile for exchange of 1,2,4-triazole has been analysed in terms of reacting species and fitted to an equation.[240] Protodetritiation of various heteroaromatic species of the purine type has been examined.[241,242]

Deuteriodeprotonation of isoxazole, isothiazole and some methyl derivatives has been used to show a relative degree of bond fixation and aromaticity.[243] Partial rate factors are listed for these and other heteroatom systems. *meso*-Tetraphenylporphyrin with TFA (in $CDCl_3$) shows evidence of exchange of uncharged and diprotonated species;[244] no monocationic species was detected.

A number of papers have dealt with exchange in co-ordinated systems, e.g. tricarbonyl-(π-pyrrolyl)manganese in TFA/C_6H_6,[245] chromium derivatives of the formula *p*-$RC_6H_5(CO)_2CrPPh_3$ (R = Me, MeCO or MeOOC), $C_6H_6Cr(CO)_3$ and $Me_2NC_6H_5$-$Cr(CO)_3$,[246] π-$C_5H_5Mn(CO)_2Ph_2PCH_2PPh_2$,[247] and $C_6H_6Cr(CO)_2PPh_3$ (in which the rate–H_0 dependence was linear up to −2.5 and inverse from −2.5 to −7.0[248] and in which it was shown that k_D/k_T was unity for exchange in the benzene portion),[249] as well as the system $C_6D_6 + C_5H_5Rh(C_2H_4)_2$ (130°) (where exchange occurred in the cyclopentadiene and ethylene units).[250]

Miscellaneous Reactions

$[C_6H_7Fe(CO)_3]^+$ attacks activated aromatic compounds, e.g. 1,3-$(MeO)_2C_6H_4$ and indole;[251] the kinetics of the process have been measured.[252] Hydroxylation of several aromatic systems by micro-organisms[253] and of benzo-α-pyrene in rat liver are reported[254] but no clear evidence of electrophilic attack is given. Cyanuric chloride (trichloro-1,3,5-triazine) with *N*-substituted pyrrole, indole or carbazole systems affords substitution on carbon.[255] Phosphotriazines $N_3P_3Cl_{6-n}(NC_5H_{10})_n$ react with benzene–$AlCl_3$ to give $N_3P_3Ph_mCl_{6-n-m}(NC_5H_{10})$.[256] Acetoxylation in the gas phase shows unusual isomer distribution; Pd(II) species in the catalyst favour *m*-substitution, e.g. C_6H_5Cl gives 70% of *m*-product.[257] Co-oxidants, e.g. $Cr_2O_7^{2-}$ speed the process.[258]

An attempt to identify products from the nitrenium ion was unsuccessful.[259] Reactions of aliphatic acyl cations,[260] oxyphenylation,[261] reactions of 1,3-thiazole anions[262] and thiamine bases[263] with electrophiles, of the pyrazolo[1,5-*a*]-1,3,5-triazine system,[264] of benzofuran with $SnCl_4$–CCl_3COOH[265] and acetoxylation of naphthalene[266] are reported. Reactions of olefins with $Pd(OAc)_2$ are reviewed.[267]

References

[1] R. Taylor in *Aromatic and Heteroaromatic Chemistry*, **2**, 217 (1974).
[2] E. F. V. Scriven in *Aromatic and Heteroaromatic Chemistry*, **2**, 258 (1974).
[3] S. R. Hartshorn, *Chem. Soc. Rev.*, **1974**, 167.
[4] D. S. Noyce and D. A. Forsyth, *J. Org. Chem.*, **39**, 2828 (1974).

[5] W. J. Hehre and P. C. Hiberty, *J. Amer. Chem. Soc.*, **96**, 7163 (1974).
[6] J. Almlof, A. Henriksson-Enflo, J. Kowalewski and M. Sundbom, *Chem. Phys. Letters*, **21**, 560 (1973); *Chem. Abs.*, **80**, 2932 (1974).
[7] P. Fischer, W. Kurtz and F. Effenberger, *Chem. Ber.*, **107**, 1305 (1974).
[8] G. A. Olah, H. C. Lin and D. A. Forsyth, *J. Amer. Chem. Soc.*, **96**, 6908 (1974).
[9] A. N. Detsina, V. I. Mamatyuk, *Izv. Akad. Nauk SSSR, Ser. Khim.*, **1973**, 2337; *Chem. Abs.*, **80**, 59171 (1974).
[10] V. I. Mamatyuk, A. I. Rezvukhin, A. V. Oetsina, V. I. Buraev, I. S. Isaev and V. A. Koptyug, *Zh. Org. Khim.*, **9**, 2429 (1973); *Chem. Abs.*, **80**, 47056 (1974).
[11] J. W. Larsen and M. E-Maksić, *J. Amer. Chem. Soc.*, **96**, 4311 (1974).
[12] S. M. Blackstock, K. E. Richards and G. J. Wright, *Can. J. Chem.*, **52**, 3313 (1974).
[13] R. F. Childs and B. D. Parrington, *Can. J. Chem.*, **52**, 3303 (1974).
[14] G. A. Olah, S. Kobayashi and Y. K. Mo, *J. Org. Chem.*, **38**, 4056 (1973).
[15] M. Ya. Zarubin, A. M. Kutnevich and A. P. Lukashenkov, *Zh. Org. Khim.*, **10**, 400 (1974); *Chem. Abs.*, **80**, 119754 (1974).
[16] W. J. Hehre, R. T. McIver, Jr., J. A. Pople and P. von R. Schleyer, *J. Amer. Chem. Soc.*, **96**, 7162 (1974).
[17] J. Shen, R. C. Dunbar and G. A. Olah, *J. Amer. Chem. Soc.*, **96**, 6227 (1974).
[18] T. P. J. Izod and J. M. Tedder, *Proc. Roy. Soc., A*, **337**, 333 (1974).
[19] R. C. Dunbar, J. Shen, E. Melby and G. A. Olah, *J. Amer. Chem. Soc.*, **95**, 7200 (1973).
[19a] Y. Yamamoto, S. Takamuku and H. Sakurai, *Chem. Letters (Tokyo)*, **1974**, 849.
[20] A. Koeberg-Telder, C. Ris and H. Cerfontain, *J. Chem. Soc. Perkin II*, **1974**, 98.
[21] H. Cerfontain and Z. R. H. Schaasberg-Nieuhuis, *J. Chem. Soc. Perkin II*, **1974**, 536.
[22] A. Koeberg-Telder and H. Cerfontain, *Tetrahedron Letters*, **1974**, 3535.
[23] A. Koeberg-Telder and H. Cerfontain, *J. Chem. Soc. Perkin II*, **1974**, 1206.
[24] H. Cerfontain and Z. R. H. Schaasberg-Nieuhuis, *J. Chem. Soc. Perkin II*, **1974**, 989.
[25] C. Ris and H. Cerfontain, *J. Chem. Soc. Perkin II*, **1973**, 2129.
[26] O. I. Kachurin, E. S. Fedorchuk and V. Ya. Vasilenko, *Zh. Org. Khim.*, **9**, 1940 (1973); *Chem. Abs.*, **79**, 145611 (1973).
[27] O. J. Kachurin, E. S. Fedorchuk, *Ukr. Khim. Zh. (Russian Ed.)* **39**, 527 (1973); *Chem. Abs.*, **79**, 145615 (1973).
[28] O.I. Kachurin, V. Ya. Vasilenko, E. S. Fedorchuk and L. I. Velichko, *Zh. Org. Khim.*, **9**, 1945 (1973); *Chem. Abs.*, **79**, 145612 (1973).
[29] L. D. Abramovich, R. S. Ryabova and M. I. Vinnik, *Zh. Org. Khim.*, **9**, 1967 (1973); *Chem. Abs.*, **79**, 114796 (1973).
[30] E. Yu. Belyaev, L. M. Gornostaev, M. S. Torbis and L. E. Borina, *Zh. Obshch. Khim.*, **44**, 633 (1974); *Chem. Abs.*, **80**, 145226 (1974).
[31] P. Cremaschi, M. Simonetta, *Theor. Chim. Acta*, **34**, 175 (1974); *Chem. Abs.*, **81**, 119730 (1974).
[32] N. C. Marziano, J. H. Rees and J. H. Ridd, *J. Chem. Soc. Perkin II*, **1974**, 600.
[33] J. W. Chapman and A. N. Strachan, *J. Chem. Soc. Chem. Comm.*, **1974**, 293.
[34] R. S. Cook, R. Phillips and J. H. Ridd, *J. Chem. Soc. Perkin II*, **1974**, 1166.
[35a] H. Suzuki and T. Hanafusa, *Bull. Chem. Soc. Japan*, **46**, 3607 (1973).
[35b] E. Hunziker, J. R. Penton and H. Zollinger, *Helv. Chim. Acta*, **54**, 2043 (1971).
[35c] R. B. Moodie, K. Schofield and J. B. Weston, *J. Chem. Soc. Chem. Comm.*, **1974**, 382.
[36] G. A. Olah and H. C. Lin, *J. Amer. Chem. Soc.*, **96**, 549 (1974).
[37] N. C. Marziano, E. Maccarone, G. M. Cimino and R. C. Passerini, *J. Org. Chem.*, **39**, 1098 (1974).
[38] A. Fischer and G. J. Wright, *Austral. J. Chem.*, **27**, 217 (1974).
[39] A. Fischer and C. C. Grieg, *J. Chem. Soc. Chem. Comm.*, **1974**, 50.
[40] A. Fischer and C. C. Greig, *Can. J. Chem.*, **52**, 1231 (1974); *J. Chem. Soc. Chem. Comm.*, **1973**, 396.
[41] A. Fischer and J. N. Ramsay, *J. Amer. Chem. Soc.*, **96**, 1614 (1974).
[42] R. C. Hahn and D. L. Strack, *J. Amer. Chem. Soc.*, **96**, 4335 (1974).
[43] M. W. Galley and R. C. Hahn, *J. Amer. Chem. Soc.*, **96**, 4337 (1974).
[44] H. Suzuki, S. Maruyama and T. Hanafusa, *Bull. Chem. Soc. Japan*, **47**, 876 (1974).
[45] R. Royer, P. Demerseman and S. Risse, *Bull. Soc. Chim. France*, **1974**, 1691.
[46] K. Olsson, *Acta Chem. Scand.*, **28**, 322 (1974).
[47] T. A. Modro and A. Piekos, *Phosphorus*, **3**, 195 (1974); *Chem. Abs.*, **81**, 119726 (1974).
[48] S. Chatterjee and W. D. Kwalwasser, *J. Chem. Soc. Perkin II*, **1974**, 873.
[49] C. L. Coon, W. G. Blucher and M. E. Hill, *J. Org. Chem.*, **38**, 4243 (1973).
[50] R. O. C. Norman, W. J. E. Parr and C. B. Thomas, *J. Chem. Soc. Perkin I*, **1974**, 369.
[51] R. G. Coombes and L. W. Russell, *J. Chem. Soc. Perkin II*, **1974**, 830; *J. Chem. Soc. Perkin I*, **1974**, 1751.

[52] G. A. Olah and H. C. Lin, *J. Amer. Chem. Soc.*, **96**, 2892 (1974).
[53] L. L. Kuznetsov and B. V. Gidaspov, *Zh. Org. Khim.*, **10**, 257 (1974); *Chem. Abs.*, **80**, 107571 (1974).
[54] L. L. Kuznetsov and B. V. Gidaspov, *Zh. Org. Khim.*, **10**, 541 (1974); *Chem. Abs.*, **80**, 132380 (1974).
[55] L. L. Kuznetsov and B. V. Gidaspov, *Zh. Org. Khim.*, **10**, 263 (1974); *Chem. Abs.* **80**, 107573 (1974).
[56] A. F. Smetana and T. C. Castorina, *Explosivstoff*, **21**, 146 (1973); *Chem. Abs.*, **80**, 132456 (1974).
[57] F. Effenberger, W. Kurtz and P. Fischer, *Chem. Ber.*, **107**, 1285 (1974).
[58] D. I. Belkin and M. Ya. Rozkin, *Avtomat. Usoversh. Tekhnol. Khim. Proizvod.*, **1972**, 22; *Chem. Abs.*, **79**, 145793 (1973).
[59] J. W. Chapman, P. R. Cox and A. N. Strachan, *Chem. Eng. Sci.*, **29**, 1247 (1974); *Chem. Abs.*, **81**, 104252 (1974).
[60] C. Hanson, J. G. Marsland and M. A. Naz, *Chem. Eng. Sci.*, **29**, 297 (1974); *Chem. Abs.*, **80**, 119863 (1974).
[61] G. M. Shutov, R. G. Chirkova, M. M. Westerets and E. Yu. Orlova, *Tr. Mosk. Khim.-Tekhnol. Inst.*, **75**, 135 (1973); *Chem. Abs.*, **81**, 119436 (1974).
[61a] E. V. Zakharov, O. P. Murashova and Z. K. Dzhioeva, *Zh. Prikl. Khim.* (*Leningrad*), **47**, 1587 (1974); *Chem. Abs.*, **81**, 90730 (1974).
[62] J. E. Drach and F. R. Longo, *J. Org. Chem.*, **39**, 3282 (1974).
[63] K. Clarke, D. N. Gregory and R. M. Scrowston, *J. Chem. Soc. Perkin I*, **1973**, 2956.
[64] O. Ceder and B. Beijer, *Tetrahedron*, **30**, 3657 (1974).
[65] J. Ersztajn, A. R. Katritzky, E. Lunt, J. W. Mitchell and G. Roch, *J. Chem. Soc. Perkin I*, **1973**, 2622.
[66] A. R. Katritzky and J. W. Mitchell, *J. Chem. Soc. Perkin I*, **1973**, 2624.
[67] A. G. Burton, M. Dereli, A. R. Katritzky and H. O. Tarhan, *J. Chem. Soc. Perkin II*, **1974**, 382.
[68] A. G. Burton, A. R. Katritzky, M. Konya and H. O. Tarhan, *J. Chem. Soc. Perkin II*, **1974**, 389.
[69] G. M. Novikova, V. F. Degtyarev and V. I. Shiskina, *Zh. Fiz. Khim.*, **47**, 2178 (1973); *Chem. Abs.*, **79**, 145600 (1973).
[70] J. Pichlichowski and A. Puszynski, *Monatsh. Chem.*, **105**, 772 (1974).
[71] C. K. Bradsher, L. L. Braun, J. D. Turner and G. L. Walker, *J. Org. Chem.*, **39**, 1157 (1974).
[72] G. N. Mitchell and R. L. McKee, *J. Org. Chem.*, **39**, 176 (1974).
[73] J. Mlochowski and Z. Skrowaczewska, *Rocz. Chem.*, **47**, 2255 (1973); *Chem. Abs.*, **80**, 145013 (1974).
[74] G. Saint-Ruf and B. Lobert, *Bull. Soc. Chim. France*, **1974**, 183.
[75] T. Nozoe, T. Asao and M. Oda, *Bull. Chem. Soc. Japan*, **47**, 681 (1974).
[76] N. Ishikawa, K. Ishikawa and S. Hayashi, *Nippon Kagaku Kaishi*, **1974**, 520; *Chem. Abs.*, **80**, 145042 (1974).
[77] S. R. Robinson, B. C. Webb and C. H. J. Wills, *J. Chem. Soc. Perkin I*, **1974**, 2239.
[78] F. Bottomley, W. V. F. Brooks, S. G. Clarkson and S.-B. Tong, *J. Chem. Soc. Chem. Comm.*, **1973**, 919.
[79] W. L. Bowden, W. F. Little and T. J. Meyer, *J. Amer. Chem. Soc.*, **96**, 5605 (1974).
[80] R. M. Christie, A. S. Ingram, D. H. Reid and R. G. Webster, *J. Chem. Soc. Perkin I*, **1974**, 722.
[81] A. Okany, T. F. Massiah, L. J. Rubin and K. Yates, *Can. J. Chem.*, **52**, 1050 (1974).
[82] A. R. Butler and A. P. Sanderson, *J. Chem. Soc. Perkin II*, **1974**, 1671.
[83] E. Eaborn, I. D. Jenkins and D. R. M. Walton, *J. Chem. Soc. Perkin I*, **1974**, 870.
[84] B. V. Passet, V. A. Sidorov, V. A. Kholodnov and P. A. Kulle, *Zh. Prikl. Khim.* (*Leningrad*), **46**, 1734 (1973); *Chem. Abs.*, **79**, 136229 (1973).
[85] B. V. Passet, V. A. Sidorov, V. A. Kholodnov and P. A. Kulle, *Zh. Prikl. Khim.* (*Leningrad*), **46**, 2057 (1973); *Chem. Abs.*, **79**, 145601 (1973).
[86] P. Burri, H. Loewenschuss, H. Zollinger and G. K. Zwolsinski, *Helv. Chim. Acta*, **57**, 395 (1974).
[87] S. B. Hanna, C. Jermini, H. Loewenschuss and H. Zollinger, *J. Amer. Chem. Soc.*, **96**, 7222 (1974).
[88] S. Kishimoto, S. Kitahara, O. Manabe and H. Hiyama, *Nippon Kagaku Kaishi*, **1973**, 1975; *Chem. Abs.*, **80**, 47096 (1974).
[89] A. B. Sakla, N. K. Masond, Z. Sawiris and W. S. Ebaid, *Helv. Chim. Acta*, **57**, 481 (1974).
[90] M. G. Bartle, R. K. Mackie and J. M. Tedder, *J. Chem. Soc. Chem. Comm.*, **1974**, 271.
[91] S. T. Gore, R. K. Mackie and J. M. Tedder, *J. Chem. Soc. Chem. Comm.*, **1974**, 272.
[92] Y. Hashida, J. Takenaka and K. Matsui, *Bull. Chem. Soc. Japan*, **47**, 507 (1974).
[93] Y. Hashida, H. Ishida, S. Sekiguchi and K. Matsui, *Bull. Chem. Soc. Japan*, **47**, 1224 (1974).
[94] N. J. Bunce, *J. Chem. Soc. Perkin I*, **1974**, 942.
[95] P. B. D. de la Mare, *Accounts Chem. Res.* **7**, 361 (1974).
[96] K. R. Beaford, G. W. Burton, P. B. D. de la Mare and H. Suzuki, *J. Chem. Soc. Perkin II*, **1974**, 459.
[97] H. Suzuki, M. Terakawa and T. Sugiyama, *Bull. Inst. Chem. Res.*, **51**, 373 (1973); *Chem. Abs.*, **81**, 3033 (1974).

[98] G. W. Burton, P. B. de la Mare and M. Wade, *J. Chem. Soc. Perkin II*, **1974**, 591.
[99] P. B. D. de la Mare, M. A. Wilson and O. M. H. El Dusuoqui, *J. Chem. Soc. Perkin II*, **1974**, 634.
[100] L. J. Kricka and A. Ledwith, *J. Chem. Soc. Perkin I*, **1973**, 859.
[101] J. K. Kochi, *Tetrahedron Letters*, **1974**, 4305.
[102] W. D. Watson and J. P. Heeschen, *Tetrahedron Letters*, **1974**, 695.
[103] R. Lamartine and R. Perrin, *J. Org. Chem.*, **39**, 1745 (1974).
[104] R. N. McDonald and J. M. Richmond, *J. Chem. Soc. Chem. Comm.*, **1974**, 333
[105] K. Schofield and C. D. Johnson, *J. Amer. Chem. Soc.*, **95**, 270 (1973).
[106] M. H. Mach and J. F. Bunnett, *J. Amer. Chem. Soc.*, **96**, 936 (1974).
[107] R. Danieli, A. Ricci, H. M. Gilow and J. H. Ridd, *J. Chem. Soc. Perkin II*, **1974**, 1477.
[108] H. M. Gilow and J. H. Ridd, *J. Org. Chem.*, **39**, 3481 (1974).
[109] J. Kulič and M. Večeřa, *Coll. Czech. Chem. Commun.*, **39**, 171 (1974).
[110] G. P. Tsayun and E. E. Yudovich, *Zh. Fiz. Khim.*, **47**, 3036 (1973); *Chem. Abs.*, **80**, 94886 (1974).
[111] N. Sridhar, J. Rajaram and J. C. Kuriacose, *Z. Phys. Chem. (Frankfurt)*, **91**, 235 (1974).
[112] S. Rajan, J. Rajaram and J. C. Kuriacose, *Indian J. Chem.*, **11**, 1152 (1973); *Chem. Abs.*, **80**, 132429 (1974).
[113] V. Calò, L. Lopez, G. Pesce and P. E. Todesco, *J. Chem. Soc. Perkin II*, **1974**, 1192.
[114] V. Calò, L. Lopez, G. Pesce, F. Ciminale and P. E. Todesco, *J. Chem. Soc. Perkin II*, **1974**, 1189.
[115] M. J. Nanjan and R. Ganesan, *Z. Phys. Chem. (Frankfurt)*, **91**, 183 (1974).
[116] C. Walling and R. T. Clark, *J. Org. Chem.*, **39**, 1962 (1974).
[117] W. D. Watson, *J. Org. Chem.*, **39**, 1160 (1974).
[118] S. A. Glover, A. Goosen and H. A. H. Laue, *J. S. Afr. Chem. Inst.*, **26**, 77 (1973); *Chem. Abs.*, **80**, 14224 (1974).
[119] S. A. Glover, A. Goosen and H. A. H. Laue, *J. S. Afr. Chem. Inst.*, **26**, 127 (1973); *Chem. Abs.*, **80**, 14233 (1974).
[120] N. Ishikawa and A. Sekiya, *Bull. Chem. Soc. Japan*, **47**, 1680 (1974).
[121] S. Uemura, A. Onoe and M. Okano, *Bull. Chem. Soc. Japan*, **47**, 147 (1974).
[122] J. van Dijk, J. J. van Daalen and G. B. Paerels, *Rec. Trav. chim.*, **93**, 72 (1974).
[123] M. Němec, M. Janda and J. Šrogl, *Coll. Czech. Chem. Comm.*, **38**, 3857 (1973).
[124] E. J. Knost, *Ber. Kernforschungsanlage Juelich*, **1973**, 83; *Chem. Abs.*, **80**, 47146 (1974).
[125] E. J. Knust, A. Halpern and G. Stöcklin, *J. Amer. Chem. Soc.*, **96**, 3733 (1974).
[126] R. Cipollini, *Radiochem. Radioanal. Letters*, **16**, 193 (1974); *Chem. Abs.*, **80**, 119876 (1974).
[127] G. E. Dunn and B. J. Blackburn, *Can. J. Chem.*, **52**, 2552 (1974).
[128] G. R. Taylor, H. D. Gesser and G. E. Dunn, *Can. J. Chem.*, **52**, 2560 (1974).
[129] L. van der Does, A. van Veldhuizen and H. J. den Hertog, *Rec. Trav. chim.*, **93**, 61 (1974).
[130] A. R. Butler and A. P. Sanderson, *J. Chem. Soc. Perkin II*, **1974**, 1214; cf. A. R. Butler and A. P. Sanderson, *J. Chem. Soc. (B)*, **1971**, 2204.
[131] L. I. Belen'ku, Ya. L. Gol'dfarb and G. P. Gromova, *Izv. Akad. Nauk SSSR, Ser. Khim.*, **1973**, 2733; *Chem. Abs.*, **80**, 94893 (1974).
[132] Tran Quang Minh, L. Christiaens and M. Renson, *Bull. Soc. Chim. France*, **1974**, 2239.
[133] H. Suzuki, K. Kawamura and T. Sugiyama, *Bull. Inst. Chem. Res., Kyoto Univ.*, **51**, 186 (1973); *Chem. Abs.*, **79**, 136151 (1973).
[134] K. Hafner and F. Schmidt, *Tetrahedron Letters*, **1973**, 5101.
[135] A. Nohara, K. Ukawa and Y. Sanno, *Tetrahedron*, **30**, 3563 (2974).
[136] J. Pichler, *Coll. Czech. Chem. Comm.*, **39**, 177 (1974).
[137] L. H. Klemm, R. E. Merrill, F. H. W. Lee and C. E. Klopfenstein, *J. Heterocyclic Chem.*, **11**, 205 (1974).
[138] D. Liotta, A. D. Baker, S. Goldstein, N. L. Goldman, F. W. Lanse, D. F. Reingold and R. Engel, *J. Org. Chem.*, **39**, 2719 (1974).
[139] P. J. Garratt and D. N. Nicolaides, *J. Org. Chem.*, **39**, 2222 (1974).
[140] T. Huio, M. Tonozuka and M. Nakagawa, *Tetrahedron*, **30**, 2123 (1973).
[141] N. P. Kanyaev, *Tr. Ivanov. Khim.-Tekhnol. Inst.*, **1972**, 81; *Chem. Abs.*, **79**, 125498 (1973).
[142] R. A. Capeletti, A. Lenardon, *Rev. Fac. Ing. Quim., Univ. Nac. Litoral*, **40**, 33 (1973); *Chem. Abs.*, **80**, 107616 (1974).
[143] B. Boulton and B. A. W. Coller *Austral. J. Chem.*, **27**, (a) 2331, (b) 2343 (1974).
[144] B. Boulton and B. A. W. Coller, *Austral. J. Chem.*, **27**, 2349 (1974).
[145] T. Okuyama, K. Kunugiza and T. Fueno, *Bull. Chem. Soc. Japan*, **47**, 1267 (1974).
[146] O. Ceder and M. L. Samuelsson, *Acta Chem. Scand.*, **27**, 3264 (1973).
[147] O. Ceder and K. Rosen, *Acta Chem. Scand.*, **27**, 2421 (1973).
[148] S. Banerjee and O. S. Tee, *J. Chem. Soc. Chem. Comm.*, **1974**, 535.
[149] O. S. Tee and S. Banerjee, *Can. J. Chem.*, **52**, 451 (1974).

[150] S. Banerjee and O. S. Tee, *J. Org. Chem.*, **39**, 3120 (1974).
[151] T. Bora, N. N. Dass and I. Hague, *Indian J. Chem.*, **12**, 70 (1974); *Chem. Abs.*, **81**, 24669 (1974).
[152] E. M. Grant, D. Lloyd and D. R. Marshall, *Chem. Ind.* (*London*), **1974**, 525.
[153] C. Eaborn, I. D. Jenkins and D. R. M. Walton, *J. Chem. Soc. Perkin II*, **1974**, 596.
[154] R. Alexander, M. T. Attar-Bashi, C. Eaborn and D. R. M. Walton, *Tetrahedron*, **30**, 899 (1974).
[155] C. J. Moore, M. L. Bullpitt and W. Kitching, *J. Organometal. Chem.*, **64**, 93 (1974).
[156] A. C. Boicelli, R. Danieli, A. Mangini, A. Ricci and G. Pirazzini, *J. Chem. Soc. Perkin II*, **1974**, 1343.
[157] R. Alexander, W. A. Asomaning, C. Eaborn, I. D. Jenkins and D. M. R. Walton, *J. Chem. Soc. Perkin II*, **1974**, 304.
[158] T. E. Harmon and D. A. Shirley, *J. Org. Chem.*, **39**, 3164 (1974); contrast D. W. Slocum and B. P. Koonsvitsky, *Ibid.*, **38**, 1675 (1973).
[159] D. A. Shirley, T. E. Harmon and C. F. Cheng, *J. Organometal. Chem.*, **69**, 327 (1974).
[160] P. Aeberli and W. J. Houlihan, *J. Organometal. Chem.*, **67**, 321 (1974).
[161] C. G. Screttas, *J. Chem. Soc. Perkin II*, **1974**, 745.
[162] S. Gronowitz, B. Cederlund and A.-B. Hörnfeldt, *Chem. Scripta*, **5**, 217 (1974).
[163] Tran Quang Minh, L. Christiaens and M. Renson, *Bull. Soc. Chim. France*, **1974**, 2244.
[164] J. Anient and J. Podstata, *Coll. Czech. Chem. Comm.*, **39**, 955 (1974).
[165] A. G. Evans, G. Salaman and N. H. Rees, *J. Chem. Soc. Perkin II*, **1974**, 1163.
[166] L. Radom, *Austral. J. Chem.*, **27**, 231 (1974).
[167] B. Chevrier and R. Weiss, *Angew. Chem. Internat. Edn.*, **13**, 1 (1974).
[168] D. V. Korchagina, B. G. Derendyaev, B. G. Derendyaev and V. G. Shubin, *Zh. Org. Khim.*, **10**, 1457 (1974); *Chem. Abs.*, **81**, 119601 (1974).
[169] H. Horita, N. Kannen, T. Otsubo and S. Misumi, *Tetrahedron Letters*, **1974**, 501.
[170] R. Miethchen, H. Mann, M. Scheel and I. Stier, *Z. Chem.*, **14**, 354 (1974).
[171] R. Meithchen and K. Ewald, *Z. Chem.*, **14**, 240 (1974).
[172] V. V. Chenets, A. V. Vysotskii, A. D. Snilitskaya, O. K. Sergeeva and V. G. Lipovich, *Neftekhimiya*, **13**, 808 (1973); *Chem. Abs.*, **81**, 12772 (1974).
[173] M. A. Shakhgel'diev, R. M. Shamkhalov and Sh. A. Agdurakhmanova, *Azerb. Khim. Zh.*, **1973**, 109; *Chem. Abs.*, **80**, 107598 (1974).
[174] K. Nyberg, *Acta Chem. Scand.*, **28**, 825 (1974).
[175] R. Garner, P. J. Albisser, M. A. Penswick and M. J. Whitehead, *Chem. Ind.* (*London*), **1974**, 110.
[176] B. N. Bychkov, M. I. Farberov, G. D. Mantynkov and A. I. Katalov, *Kinet. Katal.*, **14**, 1414 (1973); *Chem. Abs.*, **80**, 107580 (1974).
[177] G. A. Olah and J. Nishimura, *J. Amer. Chem. Soc.*, **96**, 2214 (1974).
[178] K. Arata and I. Toyoshima, *Chem. Letters* (*Tokyo*), **1974**, 929.
[179] V. Kruglikov, N. V. Goucharova, O. D. Konovalchiko, and N. I. Ochelkova, *Redkozemel, Metal., Splavy Soedin., Mater. Sovesch.*, **1973**, 328; *Chem. Abs.*, **80**, 145601 (1974).
[180] J.-M. Lalancette, M.-J. Fournier-Breault and R. T. Liffault, *Can. J. Chem.*, **52**, 589 (1974).
[181] J. F. White and M. F. Farona, *J. Organometal. Chem.*, **63**, 329 (1973).
[182] M. Watanabe, M. Yamamura, I. Moritani, Y. Fujiwara and A. Sonoda, *Bull. Chem. Soc. Japan*, **47**, 1035 (1974).
[183] N. Kamiyata, H. Minato and M. Kobayashi, *Bull. Chem. Soc. Japan*, **47**, 894 (1974).
[184] S. Matsuda, *Asahi Garasu Kogyo Gijutsu Shoreikai Kenkyu Hokoku*, **22**, 197 (1973); *Chem. Abs.*, **81**, 77154 (1974).
[185] M. Siskin and J. Porcelli, *J. Amer. Chem. Soc.*, **96**, 3640 (1974).
[186] M. Siskin, *J. Amer. Chem. Soc.*, **96**, 3641 (1974).
[187] T. Kotanigawa, *Bull. Chem. Soc. Japan*, **47**, 950 (1974).
[188] T. Kotanigawa and M. Yamamoto, *Bull. Chem. Soc. Japan*, **47**, 954 (1974).
[189] E. C. Dart and G. Holt, *J. Chem. Soc. Perkin I*, **1974**, 1403.
[190] T. A. Elwood, W. R. Flack, K. J. Inman and P. W. Rabideau, *Tetrahedron*, **30**, 535 (1974).
[191] A. M. Glatz, *Rev. Roumaine Chim.*, **19**, 455 (1974).
[192] J. P. Vecchionacci, J. C. Canévet and Y. Graff, *Bull. Soc. Chim. France*, **1974**, 1683.
[193] A. M. Glatz, V. Wertheimer and F. Badea, *Rev. Roumaine Chim.*, **19**, 87 (1974).
[194] H. J. Williams and R. G. Shotter, *Austral. J. Chem.*, **27**, 685 (1974).
[195] M. F. Ansell and S. A. Mahmud, *J. Chem. Soc., Perkin I*, **1973**, 2789.
[196] D. M. Brouwer, J. A. van Doorn, A. A. Kiffen and P. A. Kramer, *Rec. Trav. chim.*, **93**, 189 (1974).
[197] K. M. Johnston and R. G. Shotter, *Tetrahedron*, **30**, 4059 (1974).
[198] E. Rothstein and F. Vallely, *J. Chem. Soc. Perkin I*, **1974**, 443.
[199] M. I. Farberov, G. S. Mironov, G. N. Timoshenko, V. A. Ustinov and Yu. A. Moskvichev, *Kinet. Katal.*, **15**, 82 (1974); *Chem. Abs.*, **80**, 119862 (1974).
[200] P. H. Gore, A. Y. Miri and A. S. Siddiquei, *J. Chem. Soc. Perkin I*, **1974**, 2936.

[201] P. H. Gore, S. Thorburn and D. J. Weyell, *J. Chem. Soc. Perkin I*, **1974**, 2940.
[202] I. Agranat, Y-S Shih and Y. Bentor, *J. Amer. Chem. Soc.*, **96**, 1259 (1974).
[203] W. G. Miller and C. U. Pittman, Jr., *J. Org. Chem.*, **39**, 1955 (1974).
[204] S. Moon and T. F. Kolesar, *J. Org. Chem.*, **39**, 995 (1974).
[205] P. Demerseman, L. Rene, N. Platzer and R. Royer, *Bull. Soc. Chim. France*, **1973**, 2349.
[206] V. Barboiu, H. Wexler and B. Arventiev, *Org. Mag. Res.*, **6**, 327 (1974).
[207] A. R. Parikh, K. A. Thakar and R. D. Desai, *Indian J. Appl. Chem.*, **35**, 87 (1972); *Chem. Abs.*, **81**, 90847 (1974).
[208] J. D. Henion, M. C. Sammons, C. E. Parker and M. M. Bursey, *Tetrahedron Letters*, **1973**, 4925.
[209] P. O. I. Virtanen, S. Peltonen and J. Hyyppä, *Acta Chem. Scand.*, **27**, 3944 (1973).
[210] L. Mavoungou Gomès, *Compt. rend.*, *Ser. C.*, **278**, 1055 (1974).
[211] J. Suwinski, M. Soltysk and W. Zielinski, *Wiad. Chem.*, **27**, 855 (1973); *Chem. Abs.*, **80**, 119714 (1974).
[212] Yu. G. Erykalov, V. G. Chirtulov and A. A. Spryskov, *Zh. Org. Khim.*, **9**, 2089 (1973); *Chem. Abs.*, **80**, 26579 (1974).
[213] M. Yoshifuji, S. Tanaka and N. Inamoto, *Chem. Letters (Tokyo)*, **1974**, 1245.
[214] G. A. Olah, P. Schilling and I. M. Gross, *J. Amer. Chem. Soc.*, **96**, 876 (1974).
[215] P. R. Stapp, *J. Org. Chem.*, **39**, 2466 (1974).
[216] G. A. Olah and J. Nishimura, *J. Org. Chem.*, **39**, 1203 (1974).
[217] T. Fujisawa, M. Kakutani and N. Kobayashi, *Bull. Chem. Soc. Japan*, **46**, 3615 (1974).
[218] G. A. Olah, J. Nishimura and Y. Yamada, *J. Org. Chem.*, **39**, 2430 (1974).
[219] P. Linda, A. Lucarelli, G. Marino and G. Savelli, *J. Chem. Soc. Perkin II*, **1974**, 1610.
[220] S. Clementi, F. Fringuelli, P. Linda, G. Marino, G. Savelli and A. Taticchi, *J. Chem. Soc. Perkin II*, **1974**, 2097.
[221] A. Chatterjee and K. M. Biswas, *J. Org. Chem.*, **38**, 4002 (1974).
[222] D. T. Witiak, D. R. Williams, S. V. Kakodkar, G. Hill and M-S. Shen, *J. Org. Chem.*, **39**, 1242 (1974).
[223] I. V. Budnii, G. S. Mironov and M. I. Farberov, *Sb. Nauch. Tr., Yaroslav. Tekhnol. Inst.*, **22**, 55 (1972); *Chem. Abs.*, **80**, 59169 (1974).
[224] R. Natesan, L. M. Yeddanapalli, *Indian J. Chem.*, **11**, 1007 (1973); *Chem. Abs.*, **80**, 94879 (1974).
[225] V. A. Dvinin, M. G. Katsnel'son, G. V. Kuz'mina and E. I. Leenson, *Zh. Prikl. Khim. (Leningrad)*, **47**, 48 (1974); *Chem. Abs.*, **80**, 119808 (1974).
[226] H. V. Ansell and R. Taylor, *J. Chem. Soc. Chem. Comm.*, **1973**, 936.
[227] H. V. Ansell and R. Taylor, *J. Chem. Soc. Chem. Comm.*, **1973**, 952.
[228] K. E. Richards, A. L. Wilkinson and G. J. Wright, *Austral. J. Chem.*, **25**, 2369 (1972).
[229] C. Eaborn, T. A. E. Emokpae, V. I. Sidorov and R. Taylor, *J. Chem. Soc. Perkin II*, **1974**, 1454.
[230] A. J. Layton, J. H. Rees and J. H. Ridd, *J. Chem. Soc. Chem. Comm.*, **1974**, 518.
[231] J. Banger, C. D. Johnson, A. R. Katritzky and B. R. O'Neill, *J. Chem. Soc. Perkin II*, **1974**, 394.
[232] S. Clementi, C. D. Johnson and A. R. Katritzky, *J. Chem. Soc. Perkin II*, **1974**, 1294
[233] Y. Sasaki and M. Sugiura, *Chem. Pharm. Bull.*, **22**, 224 (1974); *Chem. Abs.*, **81**, 2926 (1974).
[234] N. H. Werstiuk and T. Kadai, *Can. J. Chem.*, **52**, 2169 (1974).
[235] J. Le Guen and R. Taylor, *J. Chem. Soc. Perkin II*, **1974**, 1274.
[236] A. I. Shatenshtein, A. G. Kamrach, I. O. Shapiro, Yu. I. Ranneva and E. N. Zvyagintseva, *Khim. Seraorg. Soedin., Soderzhashchikhsya Neft. Nefteprod.*, **9**, 121 (1972); *Chem. Abs.*, **80**, 59256 (1974).
[237] A. I. Serebryanskaya, F. S. Yakushin, P. A. Maksimova and A. I. Shatenshtein, *Kinet. Katal.*, **14**, 866 (1973); *Chem. Abs.*, **79**, 145755 (1973).
[238] S. Gronowitz, B. Yom-Tov and U. Michael, *Acta Chem. Scand.*, **27**, 2257 (1973).
[239] J. L. Wong and J. H. Keck, Jr., *J. Org. Chem.*, **39**, 2398 (1974).
[240] J. D. Vaughan, E. C. Wu and C. T. Huang, *J. Org. Chem.*, **39**, 2934 (1974).
[241] J. A. Elvidge, J. R. Jones, C. O'Brien, E. A. Evans and H. C. S. Sheppard, *J. Chem. Soc. Perkin II*, **1973**, 2138.
[242] J. A. Elvidge, J. R. Jones, C. O'Brien, E. A. Evans and H. C. S. Sheppard, *J. Chem. Soc. Perkin II*, **1974**, 174.
[243] S. Clementi, P. P. Forsythe, C. D. Johnson, A. R. Katritzky and B. Terem, *J. Chem. Soc. Perkin II*, **1974**, 399.
[244] R. J. Abraham, G. E. Hawkes and K. M. Smith, *Tetrahedron Letters*, **1974**, 71.
[245] N. V. Kislyakova, N. I. Pyshnograeva, V. F. Sizoi, N. E. Kolobova, V. N. Setkina and D. N. Kursanov, *Dokl. Akad. Nauk SSSR*, **212**, 367 (1973); *Chem. Abs.*, **79**, 145604 (1973).
[246] N. K. Baranetskaya, V. I. Zdanovich, E. I. Mysov, N. E. Kolobova, K. N. Anisimov, V. N. Setkina and D. N. Kursanov, *Izv. Akad. Nauk SSSR, Ser. Khim.*, **1974**, 758; *Chem. Abs.*, **81**, 36934 (1974).
[247] I. B. Nemirovskaya, V. N. Setkina, A. G. Ginzburg and D. N. Kursanov, *Izv. Akad. Nauk SSSR, Ser. Khim.*, **1974**, 762; *Chem. Abs.*, **81**, 36972 (1974).

[248] D. N. Kursanov, N. K. Baranetskaya, V. I. Lousilkina and V. N. Setkina, *Izv. Akad. Nauk SSSR, Ser. Khim.*, **1974**, 1756; *Chem. Abs.*, **81**, 48975 (1974).
[249] V. I. Zdanovich, F. S. Yakushin, V. N. Setkina, N. K. Baranetskaya, D. N. Kursanov and A. I. Shatenshtein, *Izv. Akad. Nauk SSSR, Ser. Khim.*, **1974**, 1655; *Chem. Abs.*, **91**, 90732 (1974).
[250] L. P. Seiwell, *J. Amer. Chem. Soc.*, **96**, 7134 (1974).
[251] C. A. Mansfield, K. M. Al-Kathumi and L. A. P. Kane-Maguire, *J. Organometal. Chem.*, **71**, C11 (1974).
[252] T. G. Bonner, K. A. Holder and P. Powell, *J. Organometal. Chem.*, **77**, C37 (1974).
[253] R. V. Smith and J. P. Rosazza, *Arch. Biochem. Biophys.*, **161**, 551 (1974).
[254] N. H. Sloane and T. K. Davis, *Arch. Biochem. Biophys.*, **163**, 46 (1974).
[255] R. A. Shaw and P. Ward, *J. Chem. Soc. Perkin II*, **1973**, 2075.
[256] S. Das, R. A. Shaw and B. C. Smith, *J. C. S. Dalton*, **1974**, 1610.
[257] L. Eberson and E. Jönsson, *Acta Chem. Scand.*, **28**, 597 (1974).
[258] L. Eberson and E. Jönsson, *Acta Chem. Scand.*, **28**, 771 (1974).
[259] O. E. Edwards, G. Bernath, J. Dixon, J. M. Paton and D. Vocelle, *Can. J. Chem.*, **52**, 2123 (1974).
[260] G. A. Olah, J.-M. Denis and P. W. Westerman, *J. Org. Chem.*, **39**, 1206 (1974).
[261] H. Horino, M. Arai and N. Inoue, *Bull. Chem. Soc. Japan*, **47**, 1683 (1974).
[262] G. Knaus and A. I. Meyers, *J. Org. Chem.*, **39**, 1192 (1974).
[263] A. Takamizawa, I. Makino and S. Yonezawa, *Chem. Pharm. Bull.*, **22**, 286 (1974); *Chem. Abs.*, **80**, 119823 (1974).
[264] J. Kobe, R. K. Robins and D. E. O'Brien, *J. Heterocyclic Chem.*, **11**, 199 (1974).
[265] T. Okuyama, K. Kunugiza and T. Fueno, *Bull. Chem. Soc. Japan*, **47**, 1271 (1974).
[266] G. G. Arzoumanidis and F. C. Ranch, *J. Org. Chem.*, **38**, 4443 (1973).
[267] I. Moritani and Y. Fujiwara, *Shokubai*, **13**, 101 (1971); *Chem. Abs.*, **80**, 132225 (1974).

CHAPTER 8

Carbonium Ions[1]

R. BAKER

Chemistry Department, The University, Southampton

Bicyclic and Polycyclic Systems

Derivatives of Norbornane and Related Compounds

Full details of the ^{13}C and X-ray photoelectron studies of the 2-norbornyl cation, indicating the non-classical character of this ion, have been published.[2] The 2-*exo*-3-dimethyl-2-norbornyl cation has been obtained from a series of precursors and shown to rearrange to the more stable 1,2-dimethyl-2-norbornyl cation above −60°.[3] A similar amount of σ-delocalization was found on generation of the 1,2-diphenylnorbornyl cation in FSO_3H–SO_2ClF at −78° compared to the 1,2-dimethyl-2-norbornyl cation.[4] π-Delocalization into the phenyl groups appears to be reduced owing to unfavourable non-bonded interactions, and σ-delocalization into the C-1—C-6 bond becomes more important. A comparison of the 2-methyl- and 2-phenyl-2-norbornyl cations, however, shows the importance of σ- and p–π delocalization, respectively.

A plot of the NMR chemical shifts of H(1) *vs* those of H(3) for a series of 2-aryl-2-norbornyl cations shows a marked deviation corresponding to non-classical stabilization for norbornyl cations more demanding than the 2-phenylnorbornyl cation.[5] Substituent stabilization energies of α-delocalized ions have been calculated by a perturbational theory.[6] A method, based on the determination of σ^* substituent constants for solvolysis rates of methyl-substituted tertiary cyclic systems, has been developed for the prediction of solvolysis rates of secondary cyclic systems. *exo*-2-Norbornyl toluene-*p*-sulphonate is convincingly shown to involve σ-participation concerted with ionization.[7] An *exo*/*endo* rate ratio of 1120 has been found for the trifluoroacetolysis of 2-norbornyl toluene-*p*-sulphonates at 25° measured with the aid of a new photometric method.[8]

The relative rate of solvolysis of 2-*tert*-butyl-2-*endo*-norbornyl *p*-nitrobenzoate in 80% aqueous acetone at 25° is 39,600 compared to that of 2-methyl-2-*endo*-norbornyl *p*-nitrobenzoate. An even larger difference is found for the comparable 2-adamantyl derivatives confirming that rigid systems are the most appropriate for the investigation of large steric effects.[9] A 95:5 mixture of *exo*- and *endo*-alcohols is obtained from solvolysis of 2-*tert*-butyl-2-norbornyl *p*-nitrobenzoates which was shown to reflect the ground-state energies of the *exo*- and *endo*-products.[10] The selectivity principle and the bridged structure of the 2-norbornyl cation have been discussed. The effect of electron-withdrawing substituents at C-1 of *exo*-norbornyl bromide has been shown substantially to decrease the reactivity of the system.[12]

Concepts from group and graph theory have been used to analyse the multiple rearrangement products from bicyclo[2.2.1]carbonium ion derivatives.[13,14]

Detailed studies on the rates and products of acetolysis have been reported for four 3-phenyl-2-norbornyl, four 3-*p*-methoxyphenyl-2-norbornyl, and the four 7-phenyl-2-norbornyl toluene-*p*-sulphonates.[15,16] A relative rate of 0.00392 for acetolysis of 3-*exo*-phenyl-2-*endo*-norbornyl toluene-*p*-sulphonate compared to that of 2-*endo*-norbornyl toluene-*p*-sulphonate was attributed to steric inhibition to solvation of the developing positive charge by the phenyl group. Transition-state dipolar repulsive interactions were found in solvolysis of 3-*exo*-phenyl-2-*exo*-norbornyl toluene-*p*-sulphonate between the phenyl π-cloud and the developing toluene-*p*-sulphonate anion. The operation of these dipolar repulsive interactions was confirmed by acetolysis studies of the analogous cyclohexylnorbornyl toluene-*p*-sulphonates.[17]

Steric hindrance to ionization and steric acceleration of ionization in (**2**) and (**1**), respectively, have been invoked to explain the *exo*/*endo* rate of 2960:1 on solvolysis in 60% aqueous acetone.[18] This is the reverse of observations on the corresponding secondary toluene-*p*-sulphonates.[19]

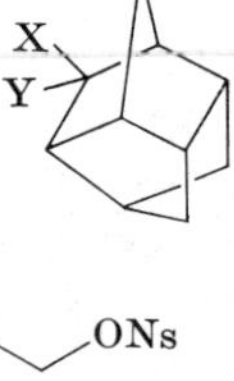

X
Y

(**1**) X = OPNB, Y = Me

(**2**) X = Me, Y = OPNB

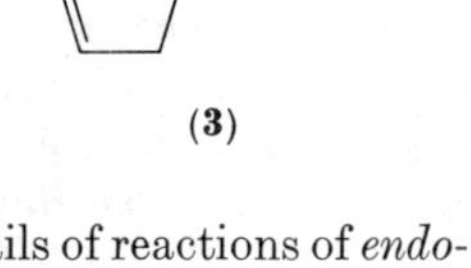

ONs ONs

(**3**) 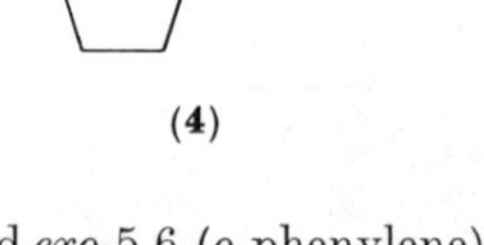 (**4**) 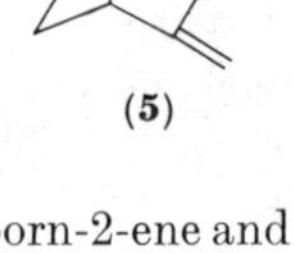 (**5**)

Full details of reactions of *endo*- and *exo*-5,6-(*o*-phenylene)norborn-2-ene and *endo*- and *exo*-5,6(1,8-naphthylene)norborn-2-ene with a palladium chloride–cupric chloride system have appeared.[20]

Clear evidence for π-participation in the acetolysis of α-campholenyl *p*-nitrobenzenesulphonate (**3**) is found in the rate enhancement of 203 compared to its saturated analogue (**4**) at 60° and the almost exclusive formation (99.7%) of cyclized products consisting mainly of (**5**) (94.7%).[21] The composition of the products was similar, but not identical, to those formed by the σ-route from acetolysis of isobornyl and bornyl derivatives; minor differences were attributed either to the differences in position of the counter-ion in ion-pairs or to other small differences of the leaving groups. Two enantiomers of camphene with high optical purities were formed when α-campholenyl *p*-nitrobenzenesulphonate and the isobornyl derivatives were prepared from (+)-camphor.

Anodic oxidation of *exo*- and *endo*-1,7,7-trimethylnorbornane-2-carboxylic acid and *exo*- and *endo*-2,3,3-trimethylnorbornane-2-carboxylic acid,[22] and oxidative decarboxylation of *exo*- and *endo*-bornane-2-carboxylic acid, *exo*- and *endo*-2,3,3-trimethylnorbornane-2-carboxylic acid and α-campholenylcarboxylic acids with lead tetra-acetate in benzene[23] have been investigated.

Products from solvolysis of *endo*-3,3-dimethylnorborn-2-ylmethyl toluene-*p*-sulphonate (**6**) in methanol containing sodium methoxide were camphene (51%) and 2β-methoxy-2α,3,3-trimethylnorbornane; therefore a hydride shift to form ion (**7**) occurs in preference to ring expansion.[24] In contrast, deamination of the corresponding amine and solvolysis of the *endo*-norborn-2-ylmethyl ester both gave ring expanded products. The change in mechanism was discussed in terms of the electronic effects of the 3,3-dimethyl substituent.

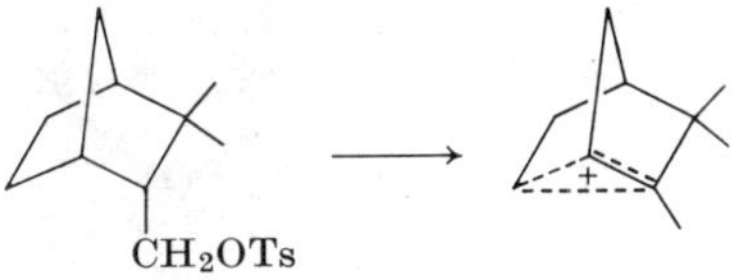

CH_2OTs

(**6**) (**7**)

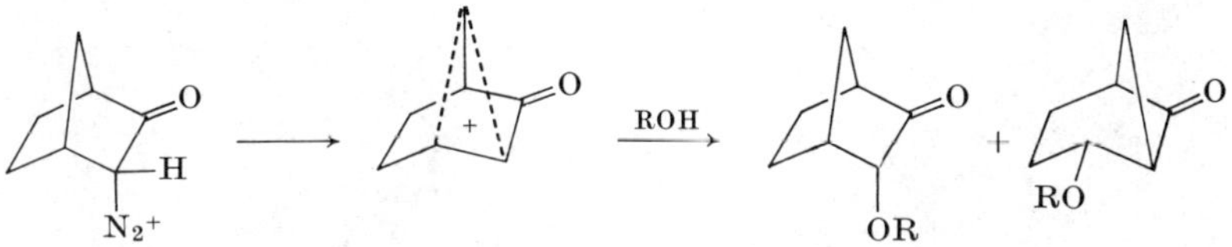

-H

N_2^+

ROH

O O O O

+

RO OR

(**8**) (**9**) (**10**)

The 3-oxo-norbornane-2-*endo*-diazonium ions (**8**) produce derivatives of *endo*-bicyclo[2.2.1]heptane (**9**) and of bicyclo[3.1.1]heptane (**10**), in proportions which depend upon the nucleophilicity of the solvent.[25]

The deamination of 1-norbornylamine, 1-apocamphylamine and 1-adamantylamine yields carbonium ions which are reactive enough to abstract chloride ions from solvents such as methylene chloride.[26] Studies have been reported on the acetolysis of camphene.[27]

Other Bicyclic Systems

Deamination of 2α-amino-6,6-dimethylnorpinane has been shown to involve a high-energy classical ion associated with a counter-ion.[28] A [1,3]-carbon shift has been suggested in the deamination of 2α*H*-10-aminopinane which yields a mixture of rearrangement products including the bicyclo-[2.2.2]-, -[3.2.1]- and -[4.1.1]-octane derivatives.[29] Comparisons of products from methanolysis of pinan-10-yl toluene-*p*-sulphonates with the deamination and deoxidation of the 10-aminopinanes and pinan-10-ols have been rationalized in terms of ion pairs and free and solvated ions.[30]

Introduction of an α-methyl substituent into bicyclo[3.1.1]hept-6-yl and bicyclo-[3.2.0]hept-6-yl 3,5-dinitrobenzoates changes the *endo*/*exo* rate ratios for solvolysis in 80% acetone from 4×10^6 and 500 to 160 and 11, respectively.[31] A comparison was made against the correspondingly substituted cyclobutyl and cyclohexyl derivatives and it was concluded that the *exo*-isomers are good models for unassisted ionization of cyclobutyl derivatives. On the basis of product studies, it was clear that the concerted rearrangement possible in the *endo*-isomers could not be suppressed by α-CH_3 substitution but a *p*-methoxyphenyl substituent did prevent rearrangement of the bicyclo[3.1.1]-heptyl cation. It was concluded that all the evidence would confirm the hypothesis that the cyclobutyl cation is normally less stable than the cyclopropylcarbinyl cation which could be formed by rearrangement.

Solvolysis of *endo*-9-bicyclo[4.2.1]nona-2,4,7-trienyl toluene-*p*-sulphonate (**11**) in acetic acid has been suggested to involve the bicycloaromatic cation (**12**) which rearranges in a specific and symmetrical way to the more stable bishomotropylium ion (**13**) which finally yields a 99:1 mixture of *exo*-dihydroindenyl acetate (**14**) and indene.[32]

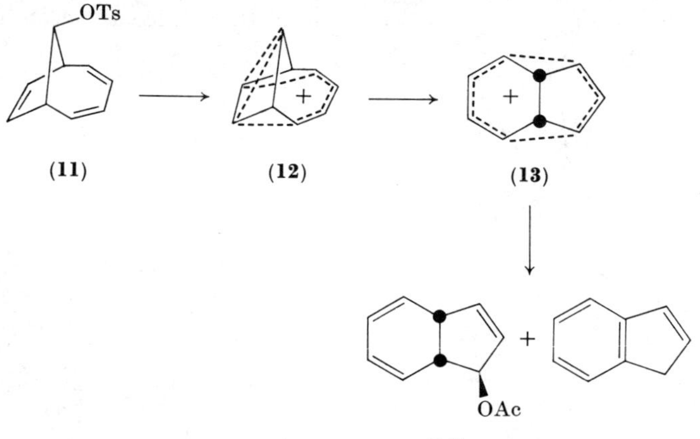

OTs

(**11**) (**12**) (**13**)

+ OAc

(**14**)

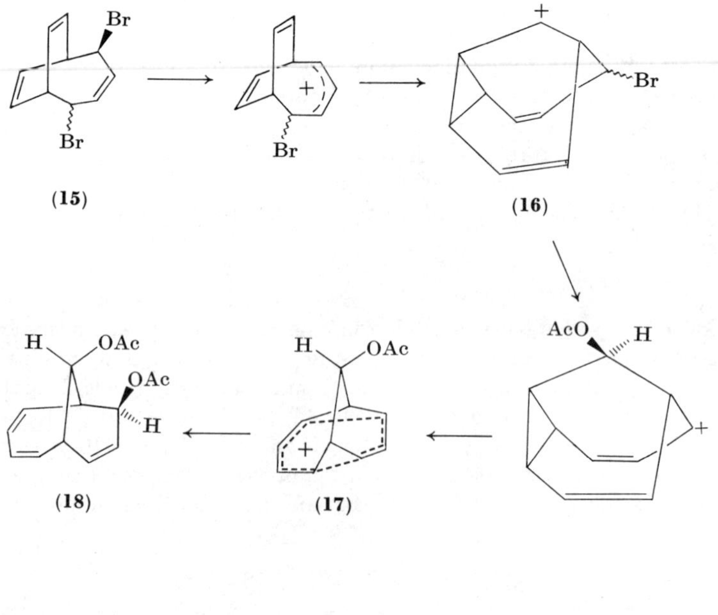

Br Br Br Br

(**15**) (**16**)

H OAc OAc H H OAc AcO H

(**18**) (**17**)

Formation of (**18**) on solvolysis of the dibromide (**15**) has been attributed to intervention of the cyclopropylcarbonium ion (**16**) and the bishomotropylium ion (**17**).[33]

An *exo*/*endo* rate ratio of 515:1 was found for the acetolysis of the epimeric 8-vinylbicyclo[3.2.1]oct-8-yl 3,5-dinitrobenzoates.[34] It appears that carbon–carbon hyperconjugation is still operative in the transition state for formation of a tertiary allylic cation.

Reactions of carbonium ions derived from hexamethyl-Dewar-benzene with triethylamine have been shown to be good synthetic routes to strained compounds with exocyclic methylene groups.[35] It was suggested that abstraction of a proton is a kinetically controlled process occurring at the methyl group adjacent to the carbon atom bearing the highest positive charge; thus (**21**) can be formed by quenching (**20**), derived from (**19**), with triethylamine. The synthesis of 1,2,3,4,5,6-hexamethylbicyclo[2.1.1]hexene[36] and full details of electrophilic additions to hexamethyl-Dewar-benzene[37] have been reported.

HCl/CH_2Cl_2 ; Et_3N

(**19**) (**20**) (**21**)

Acid-catalysed rearrangement of (**22**) to (**23**) has been investigated with deuterium-labelled methyl groups and suggested to proceed as shown.[38] This contrasts with the reaction of similar derivatives with methyl groups substituted on the double bond.[39]

(**22**) (**23**)

The high reactivity of the 1-apocamphyl cation has been demonstrated by its unusually low substrate and positional selectivity in electrophilic aromatic substitution.[40] From a comparison of the products of acid-catalysed rearrangements of the pinan-2-ols in aqueous dioxan and acetic acid, it has been suggested that the water molecule produced affects the course of reaction in poor ionizing solvents.[41]

The influence of C-7 substituents on the cycloheptatriene–norcaradiene equilibrium has been studied.[42] The rearrangements of *endo*-5-halo-1,2,3,4,5,6-hexamethyl[2.2.1]-hexenyl cations,[43] ^{13}C-spectra of hexamethylbicyclo[2.1.1]hexenyl and hexamethylbicyclo[3.1.0]hexenyl cations,[44] and the stereochemistry of bicyclohexenyl cation formation in the photoisomerization of cyclohexadienyl cations,[45] have been reported.

Ab initio molecular-orbital theory has been used to investigate the energetics of circumambulatory degenerate rearrangements in the bicyclo[3.1.0]hexenyl and homotropylium cations.[46] Further rearrangements of the dibenzobicyclo-octadiene system have been studied.[47]

Polycyclic systems

σ-Participation has been confirmed by the additive effect on the rate enhancement by successive methyl substitution in the solvolysis of the series 2-adamantyl-, 1-methyl-2-adamantyl-, 1,3-dimethyl-2-adamantyl toluene-*p*-sulphonates.[48] Only a small rate enhancement was found in the solvolysis of 1,2-dimethyl-2-adamantyl relative to the parent 2-methyl-2-adamantyl bromide, confirming the intervention of a classical intermediate in the former reaction. The NMR spectrum of the 1,3,5,7-tetramethyl-2-adamantyl cation in superacid was consistent with a rapid equilibration between four equivalent bridged structures.

The deamination of 1-adamantylamine has been shown to yield 1-adamantanol and 1-adamantyl acetate;[49] no adamantane,[50] which could have been formed as a result of intermolecular hydride shift, was detected.

The 3-homoadamantyl cation (**24**) is more stable than the 1-cation (**25**).[51] Homoadamantane, 1- and 2-methyladamantane and 1-adamantylmethanol are formed on

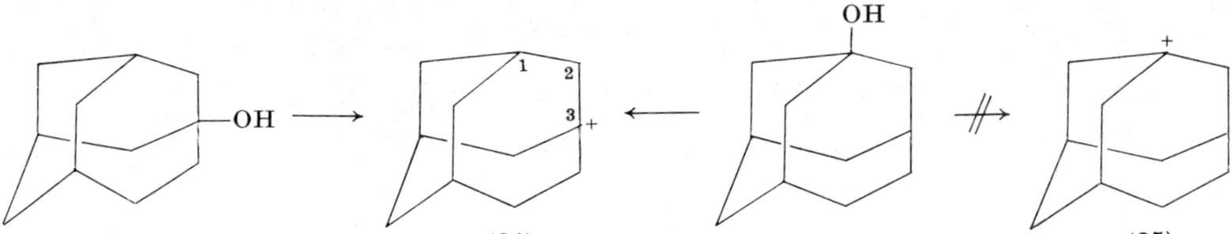

(**24**) (**25**)

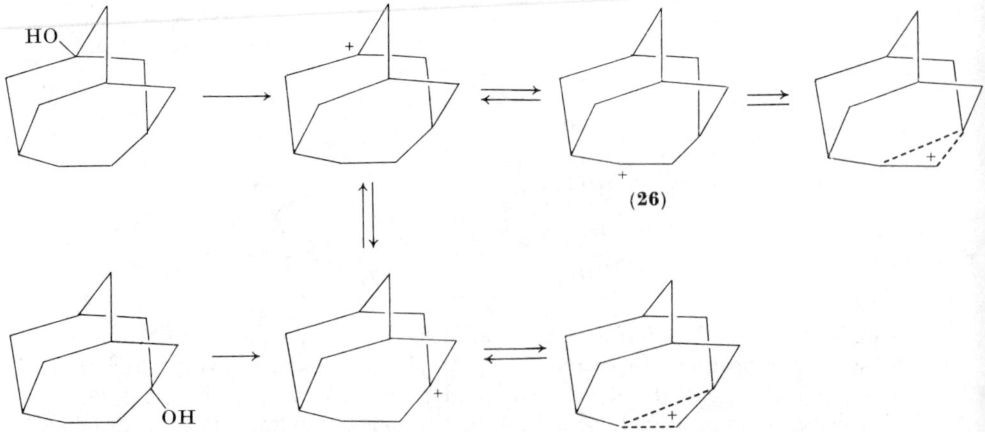

(**26**)

(**27**)

reaction of 1- and 3-homoadamantanol with 70% sulphuric acid.[52] A mechanism was suggested involving hydride transfers and rearrangements of the resulting classical homoadamantyl cations (**26**) and (**27**) into the corresponding bridged ions shown.

A mixture of 4-homoisotwistane (45–50%), homoadamantane (20–25%) and 2-methyladamantane (20–25%) is obtained on reaction of homoadamantene or 4-homoadamantanol with concentrated sulphuric acid in pentene.[53]

Full details of the solvolysis of 4-*exo* and 4-*endo*-protoadamantyl toluene-*p*-sulphonate, demonstrating the considerable anchimeric assistance in the 4-*exo*-derivative, has been published.[54] Solvolysis of 2-methyl-2-adamantyl toluene-*p*-sulphonate and 4-methyl-4-*exo*-protoadamantyl 3,5-dinitrobenzoate leads directly to the common bridged ion (**28**).[55] The same bridged ion is also formed by leakage from a prior formed classical ion from 4-methyl-4-*endo*-protoadamantyl 3,5-dinitrobenzoate. The high α-CH_3/H rate ratio of $10^{7.7}$ found for the 4-*endo*-protoadamantyl system and the figure of $10^{3.8}$ for the 4-*exo*-derivatives confirm the absence of participation in the former.

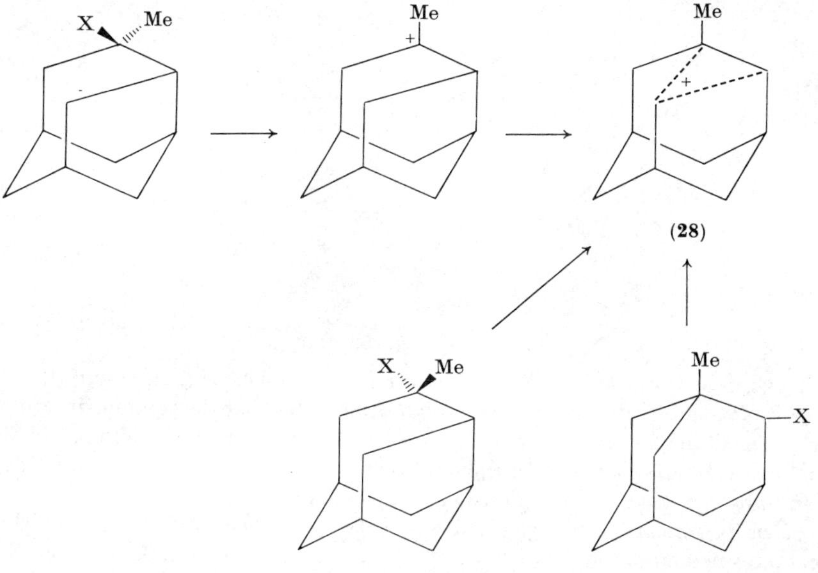

X Me Me Me

(**28**)

X Me Me X

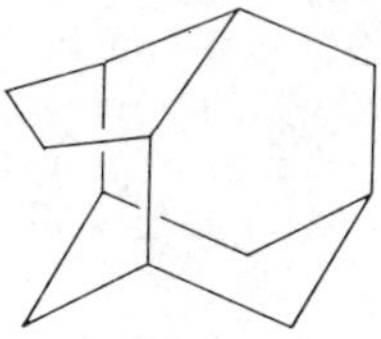

(**29**)

That ethanoadamantane (**29**) is the most stable $C_{12}H_{18}$ hydrocarbon has been confirmed by Lewis-acid-catalysed isomerization studies, and by empirical force-field calculations.[56] The importance of intermolecular hydride shifts has been noted in the Friedel–Crafts reaction of 1-(2-chloroadamant-1-yl)-2-methylaminopropane.[57]

Since the use of σ-values gave a better correlation than σ^+ for the solvolyses of the *exo* toluene-*p*-sulphonates (**30**), the pathway has been suggested to involve intermediates (**31**) and (**32**).[58]

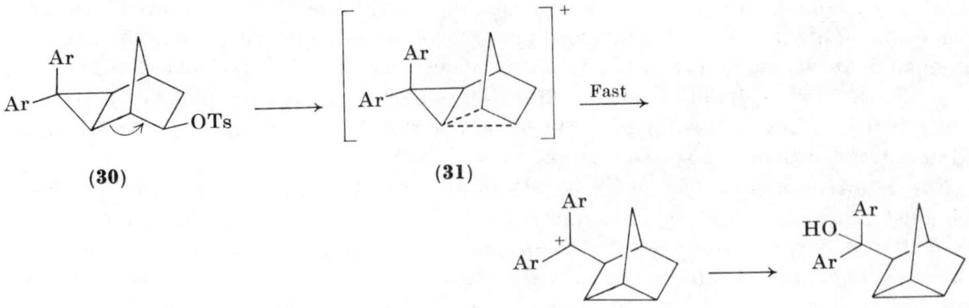

(30) (31) (32)

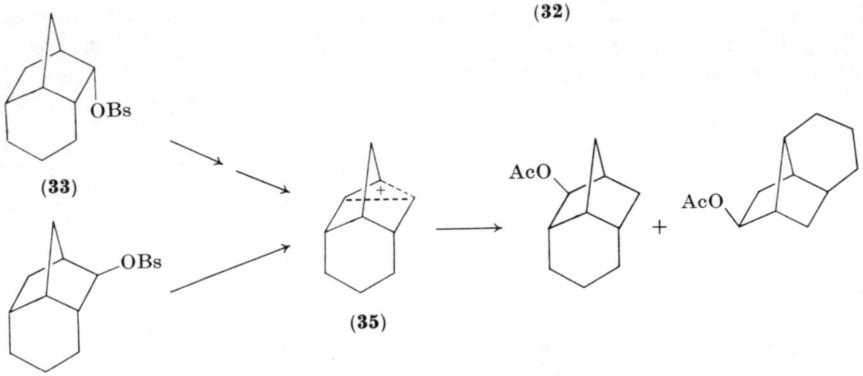

(33) (34) (35)

The solvolysis studies on *exo*-7-isodeltacyclyl *p*-bromobenzenesulphonate have been reported.[59] Formation of the same two products indicates that the common intermediate (**35**) is formed in the acetolysis of 2-*endo*- (**33**) and 2-*exo*-tricyclo[3.3.1.1^{3,9}]decyl (2-homobrendyl) *p*-bromobenzenesulphonates (**34**).[60] An *exo*/*endo* rate ratio of 953:1 was ascribed to both anchimeric assistance and steric hindrance to ionization.

Solvolysis of tertiary 2-tricyclo[3.3.0.0^{3,7}]octyl *p*-nitrobenzoates proceeds through classical intermediates; the amount of σ-participation for the secondary system (determined from rate measurements) has been shown to be in error by 10^4–10^5.[61] Full details of the acetolyses of *syn*- and *anti*-6-methylpentacyclo[5.3.0.0^{2,5}.0^{3,9}.0^{4,8}]dec-6-yl toluene-*p*-sulphonates have appeared,[62] which provide further evidence for the extent of participation in the secondary 1,3-bishomocubyl system. Attempts to prepare the secondary cation in strong acid were not successful and reaction of (**36**) with FSO_3H–SO_2 at 50° gave (**37**).[63]

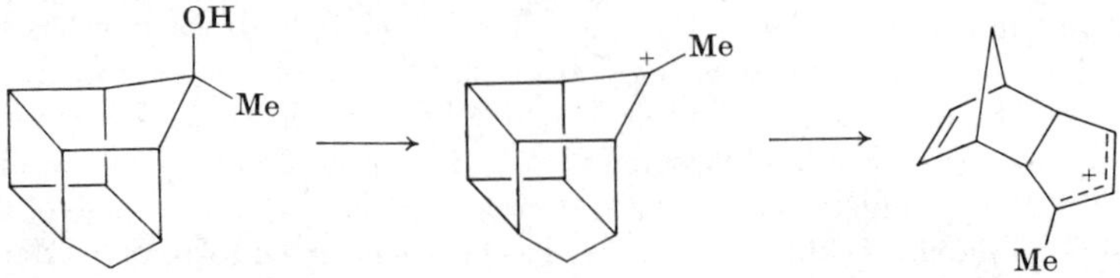

(36) (37)

A convenient synthesis of homobullvalenone has appeared.[64] Further details of the stereospecificity and regioselectivity of the Ag(I)-catalysed rearrangement of tricycloheptanes to bicyclohept-6-enes have been published.[65] The 1,3-dimethyl- and 3-methoxy-1-methyl-tricyclo[4.1.0.0^{2,7}]heptane epimeric pairs lead to substituted bicyclo[3.2.0]-hept-6-ene products. The stereoproximally substituted tricycloheptanes yield only 2-*exo*-bicycloheptenes whilst the stereodistal pair yield the 2-*endo*-isomers. This rearrangement is also regioselective with the 6-methyl isomer arising 1.5–4.0 times more rapidly than the 7-methyl analogue. The data for this γ-type isomerization is consistent with bimolecular reaction between Ag^+ and the strained ring, involving competitive edge attack at bonds b–c or c–d (**38**) followed by 1,2-carbon shift. These two processes lead to the 6-methyl and 7-methyl isomer, respectively.

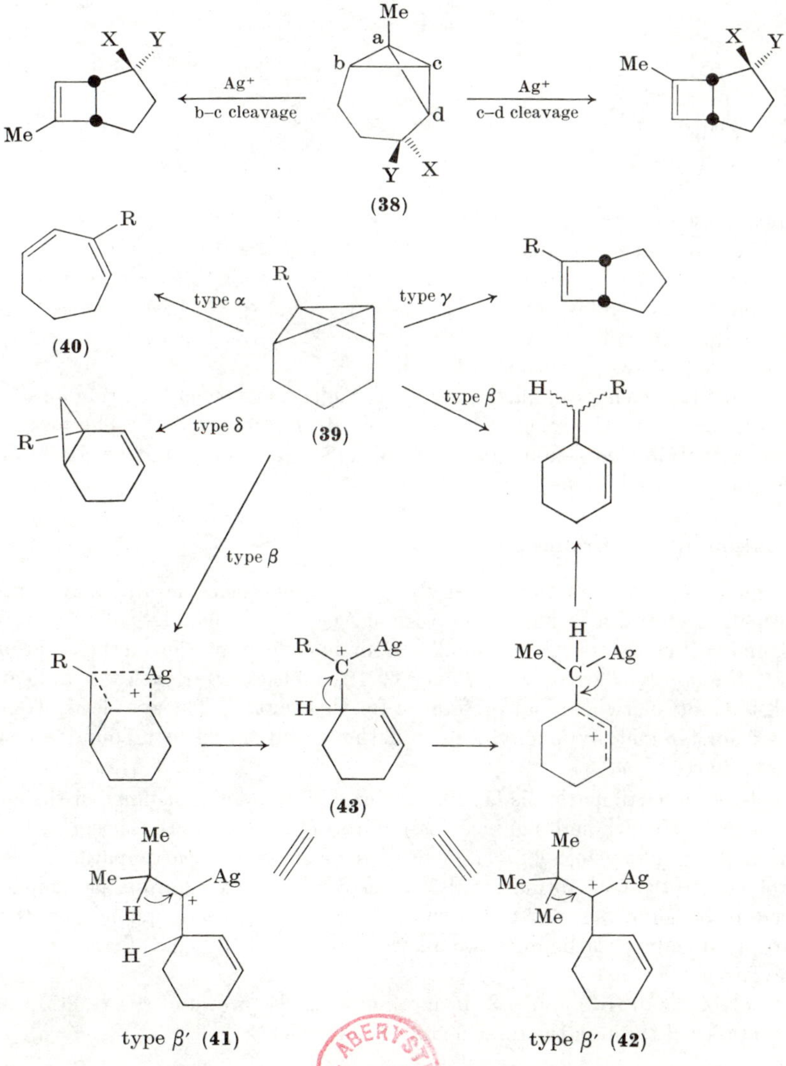

The $AgClO_4$-promoted isomerizations of 1-substituted tricyclo[4.1.0.$0^{2,7}$]heptanes (**39**) have been examined in detail.[66] Increasing steric bulk appears to favour the γ-rearrangement over the α-rearrangement with formation of cyclohepta-1,3-diene (**40**). In some instances 1,2-hydride shifts (**41**) and 1,2-alkyl shifts (**42**) from the R-group in the argentocarbonium ions compete favourably with migration of the allylic cyclohexene hydrogen (**43**).

With methyl tricyclo[4.1.0.$0^{2,7}$]heptane-1-carboxylate the formation of mainly methyl cyclohepta-1,3-diene-2-carboxylate is in contrast to the products with alkyl groups at C-1. This is associated with an electronic effect which decelerates the γ-rearrangement.[67] The type α- and γ-rearrangements are discussed.

Acid-catalysed rearrangement of (**44**) gives (**45**) in high yield at room temperature.[68]

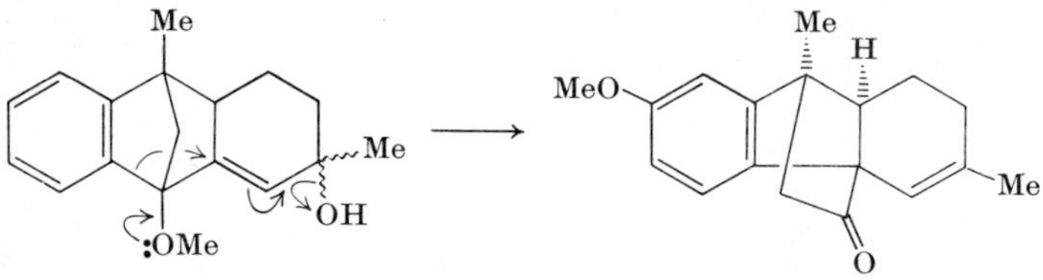

(**44**) (**45**) 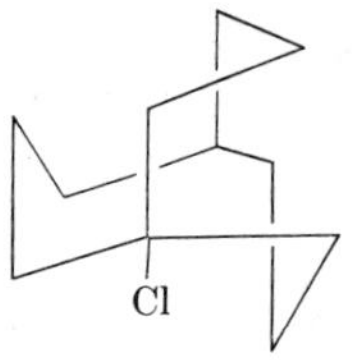 (**46**)

Acid-catalysed rearrangement of dieldrin epoxide[69] and the acetolysis of 4-deuterio-*cis*-4-tetracyclo[7.2.$0^{2,8}0^{3,7}$]dodecanyl *p*-bromobenzenesulphonate[70] have been reported. Semi-empirical computation has shown a system with a C_5 symmetry to be the most stable form of Goldstein and Kline's $C_{11}H_{11}$ cation.[71] Other studies include the pyrolytic elimination of tricyclo[7.3.1.$0^{2,7}$]trideca-2-en-13-yl acetate[72] and the dehydration of tricyclo[7.3.1.$0^{2,7}$]tridecane-2,13-diols.[73]

Strain calculations have confirmed that the high reactivity of the bridgehead chloride (**46**) is due largely to the relief of angle strain during ionization.[74] The mechanism of tetrazole formation from 1-adamantyl arenesulphonates in acetonitrile containing azide has been investigated.[75]

Participation by Aryl Groups

The formolysis of 4-(*p*-methoxyphenyl)butyl *p*-bromobenzenesulphonate appears to yield 6-methoxytetralin through a mixture of Ar_2-5 (69.5%) and Ar_2-6 (30.5%) participation, indicating the relatively small stabilizing effect of the methoxy group.[76] In contrast, the *p*-tolyl derivative involves 69.4% of the Ar_2-6 pathway indicating the general, but not overwhelming, preference for this route.[77] The synthesis of polycyclic ketones from 1-*p*-methoxyphenyl-2-phenylethane and similar compounds in SbF_5/HF has been reported.[78]

Phenyl-group participation is favoured over double-bond assistance in the acetolysis of (**47**) to yield (**48**) although significant amounts of the latter are observed with eventual formation of benzobicyclo[2.2.2]octane derivatives.[79] Phenyl participation is also found in acetolysis of (**49**) with formation (95%) of (**50**).[80] No cyclopropyl participation was observed in the solvolysis of (**51**) although the mixture of products included those with cyclopropyl opening. Carbonium ion rearrangements in dibenzoalkane systems have been reviewed.[81]

Photosolvolysis of (**52**) and (**53**) in methanol yields (**58**), and the reaction has been shown to proceed through the intermediates (**54**)–(**57**).[82]

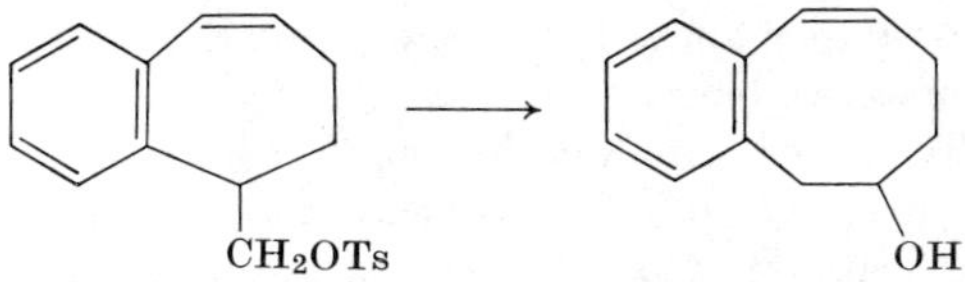

CH2OTs (47) → OH (48)

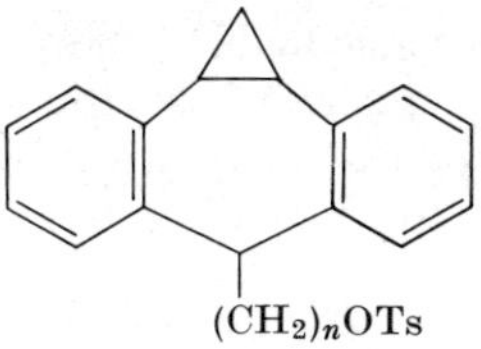

(CH2)nOTs (49) $n = 1$ (51) $n = 2$ — OH (50)

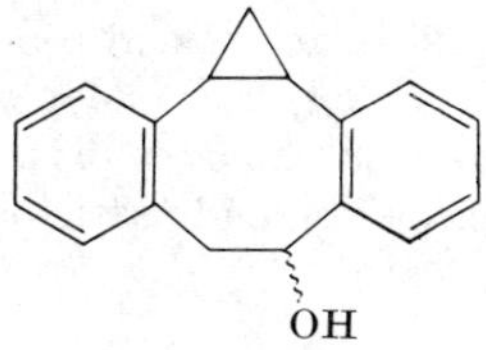

Studies on the ρ-value for the solvolyses of *exo*-3,3-diaryltricyclo[3.2.1.0^{2,4}]oct-*anti*-8-yl toluene-*p*-sulphonate indicate that aryl participation and cyclopropyl ring opening are essentially concerted.[83]

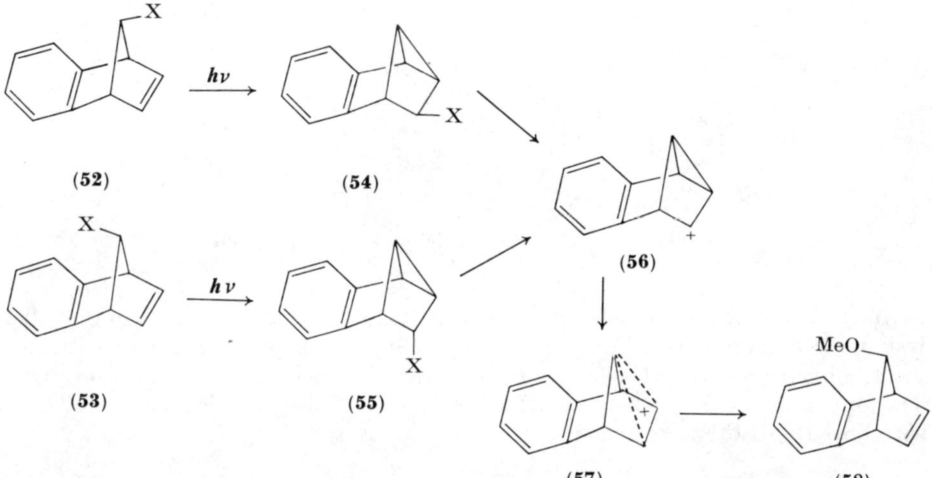

(52) X —$h\nu$→ (54) X → (56) +

(53) X —$h\nu$→ (55) X → (56) → (57) → (58) MeO

A difference of 22.3 kcal/mol has been estimated by a semiempirical MINDO/2 calculation for a tetrahedral and a planar arrangement of the ethylenebenzenium ion.[84] Modified CNDO/2 calculations have been made on different conformations of various phenylpropyl cations.[85] The conformational requirements of cyclodehydration of phenylcyclohexanols to octahydrophenanthrene derivatives have been discussed.[86] Long-lived phenethyl and ethylenebenzenium cations have been studied in the gas phase.[87] Evidence has been presented that β-aryl-group participation leads to a symmetrical transition state,[88] and the effect of pressure on the rate of formolysis of α-(*p*-substituted) phenethyl toluene-*p*-sulphonates has been studied.[89]

A correlation of the rates of reaction between acetyl perchlorate and butyl ethers of 2-arylethanols with σ^+ in acetic acid, including those with electron-withdrawing substituents, indicates the non-importance of a k_s component. This contrasts with the acetolysis of 2-arylethyl toluene-*p*-sulphonates.[90]

Solvolysis of optically active 1,2-dimethylbenzonorbornen-*exo*-2-yl chloride and *p*-nitrobenzoate in methanol gave optically active products which provides evidence for the intervention of a classical intermediate produced by an unassisted pathway.[91]

Participation by Double and Triple Bonds

The solvolysis of (**59**) in acetic acid proceeds with some double-bond participation and leads to formation of (**60**).[92] In solvolysis of the epimeric 7-isopropylidenebenzonorbornen-2-yl toluene-*p*-sulphonates (**61**), the *exo*- and *endo*-derivatives react with assistance from the benzene ring and participation from the double bond, respectively.[93]

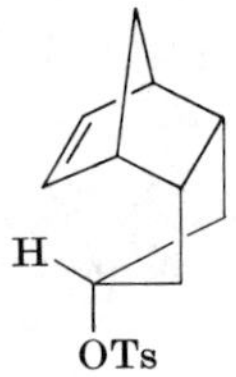

H OTs (**59**) AcO (**60**) H OTs (**61**)

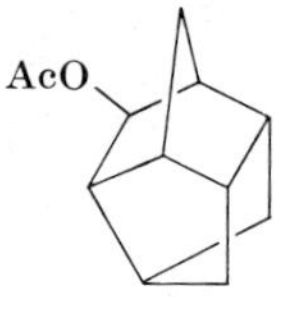

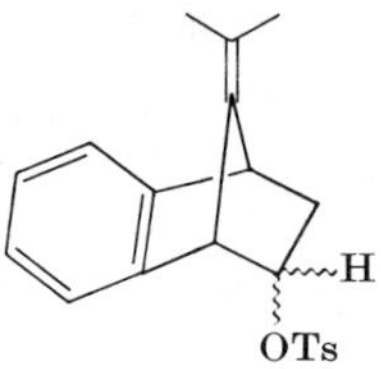

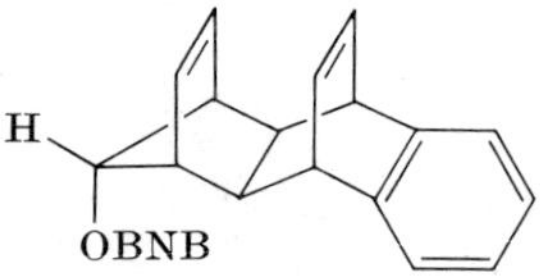

H OBNB (**62**) OTs (**63**) 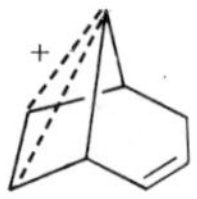 + (**64**)

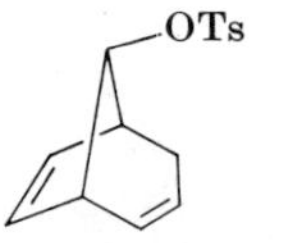

Kinetic evidence has indicated that participation by both double bonds in the solvolysis of (**62**) is unimportant.[94] Methanolysis of (**63**) in the presence of sodium methoxide yields all three methyl ethers expected from the bishomocyclopropenyl cation (**64**).[95] The formation of the 9-*endo*-tricyclo[4.2.1.0^{2,5}]nona-3,7-dienyl cation has been studied with FSO_3H–SO_2ClF as solvent.[96]

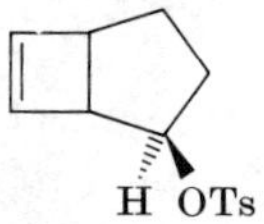

H OTs (**65**) + (**66**)

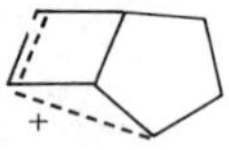

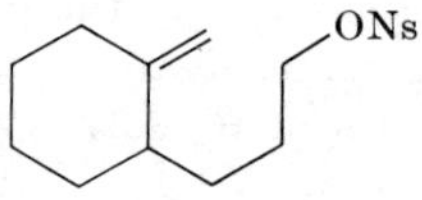

ONs (**67a**) ONB (**67b**) ONs (**68**) NsO (**69**)

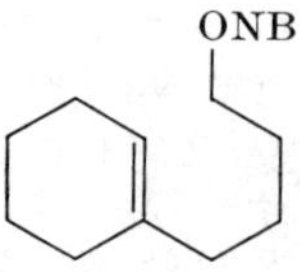

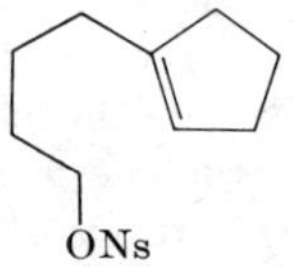

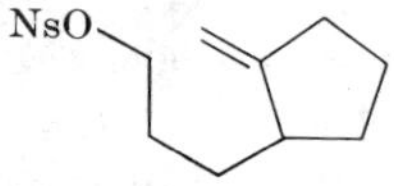

Methyl migration in the 7-*tert*-butylbicyclo[2.2.1]hept-2-en-7-yl cation intermediate has been shown to occur with a preference for the *syn*-face and been attributed to the operation of steric factors.[97]

Solvolysis of *exo*-2-bicyclo[3.2.0]hept-6-enyl toluene-*p*-sulphonate (**65**) involves double-bond participation to form the cation (**66**) which rapidly rearranges to the 7-norbornenyl cation;[98] only products derived from the latter are found so that this system provides an alternative π-route to this cation.

Double-bond participation in the solvolysis of (**67a**)[99] and (**67b**)[100] leads to the 9-decalyl cation, and the 8-hydrindyl cation can be formed from (**68**) and (**69**).[101,102] Acid-catalysed cyclization of 2-alkenyl-1-methylcyclohexanols[103] and the formation of stereoisomeric 1,4-dimethylspiro[4.5]dec-8-en-7-ones[104] have been reported.

Studies into the application of double-bond participation into total synthesis continue. Cyclization of (**70**),[105] (**71**),[106] (**72**)[107] and (**73**)[108] have been examined.

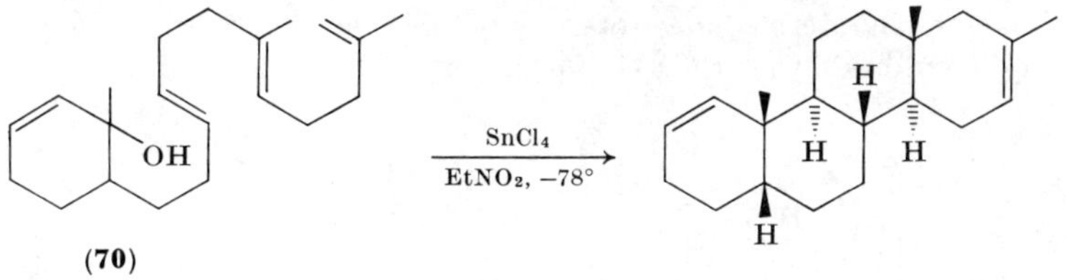

OH SnCl$_4$ / EtNO$_2$, −78° → H H H H

(**70**)

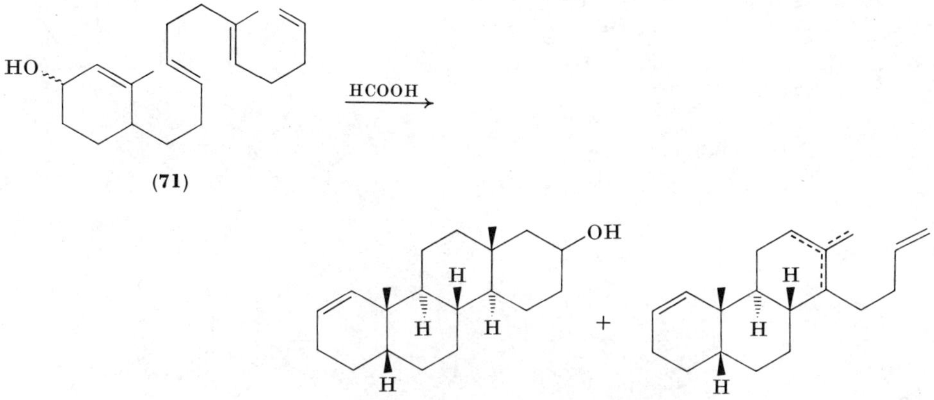

HO HCOOH →

(**71**)

OH H H H H + H H H

OSiMe$_2$Bu^t H H H OH → → OH H H H H HO H

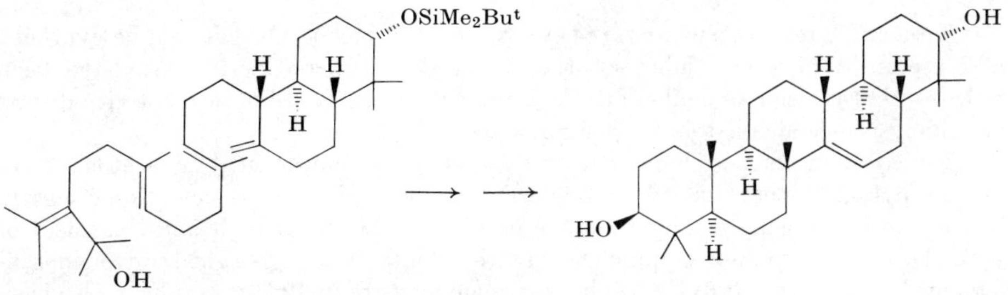

(**72**)

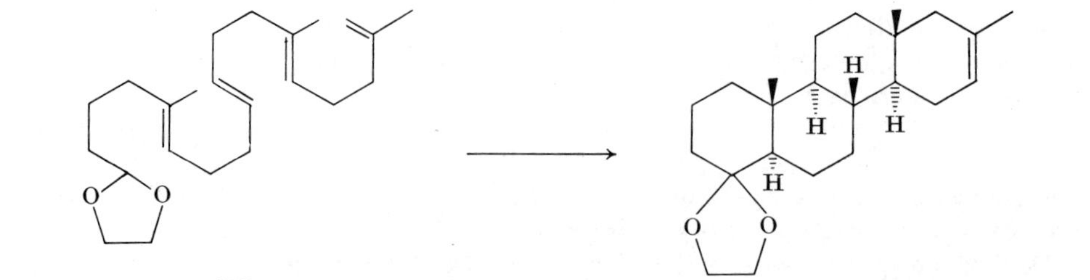

(**73**)

Cyclization of (**74**) and of its A–B antipode leads to (**75**) and (**76**), respectively.[109] Rearrangement through (**77**)–(**79**) must occur after cyclization. In contrast, cyclization of (**80**) cannot involve a similar carbonium ion but proceeds to product through (**81**). It was concluded that in the latter cyclization, the entire A–B–C ring formation must be concerted or involve intermediate frozen mono- or bi-cyclic carbonium ions (**82**)–(**83**), which do not equilibrate with their conformationally isomeric cations.

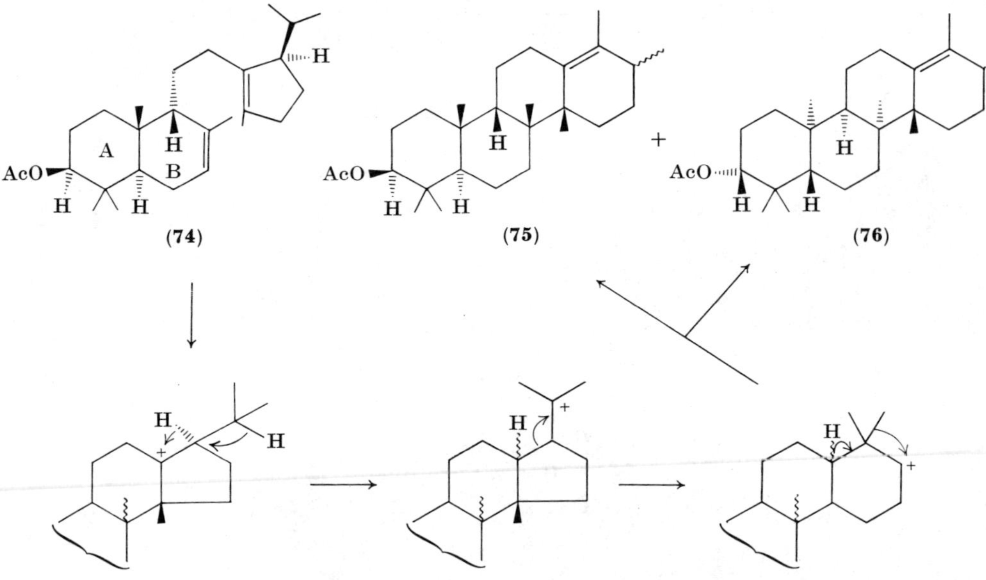

(**74**) (**75**) (**76**)

(**77**) (**78**) (**79**)

An interesting rearrangement of (**84**) to (**85**) may, it has been suggested, be synthetically useful.[110] Further studies on the solvolysis of allenic derivatives have been made.[111–115] Inversion of configuration occurs with retention of optical activity during the solvolysis of γ-allenic toluene-*p*-sulphonates.[116]

Extended treatment of humulene with aqueous sulphuric acid in acetone gives 10-isopropyl-3,7-dimethylbicyclo[5.3.0]deca-2,10-diene.[117] Selective acylation occurs in the ring closure of geranyl acetate in the presence of Lewis acids,[118] and treatment of geraniol or nerol with fluorosulphonic acid at low temperature gave good conversions into a novel iridoid ether (**86**).[119] Allylic rearrangements of linalyl acetate have also been investigated.[120]

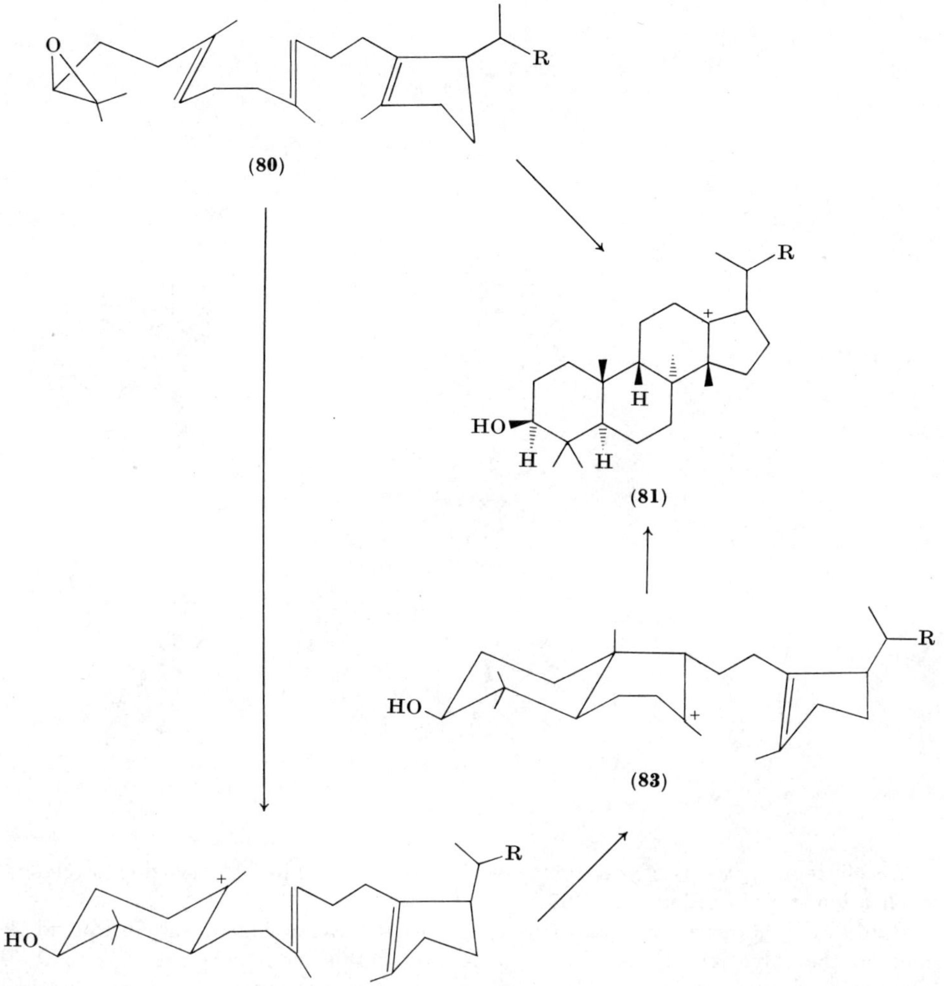

(**80**)

(**81**)

(**83**)

(**82**)

(**84**)

AcOH, Ac_2O

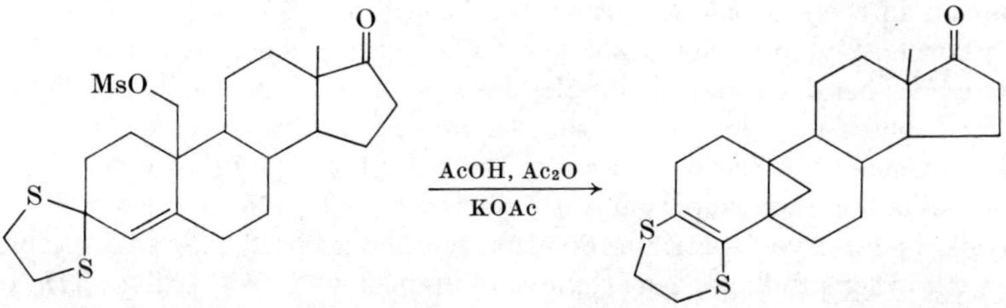

KOAc

(**85**)

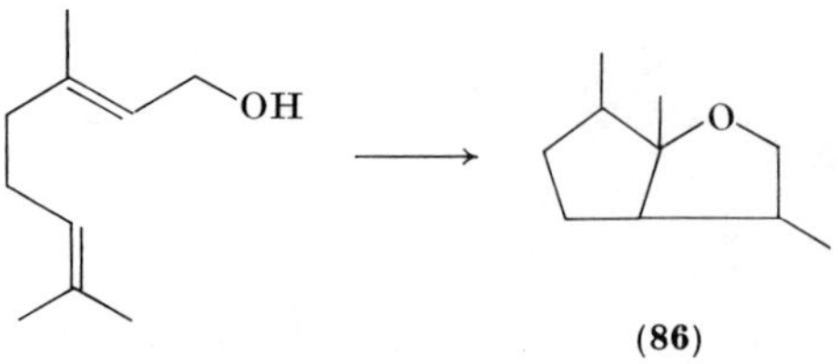

(86)

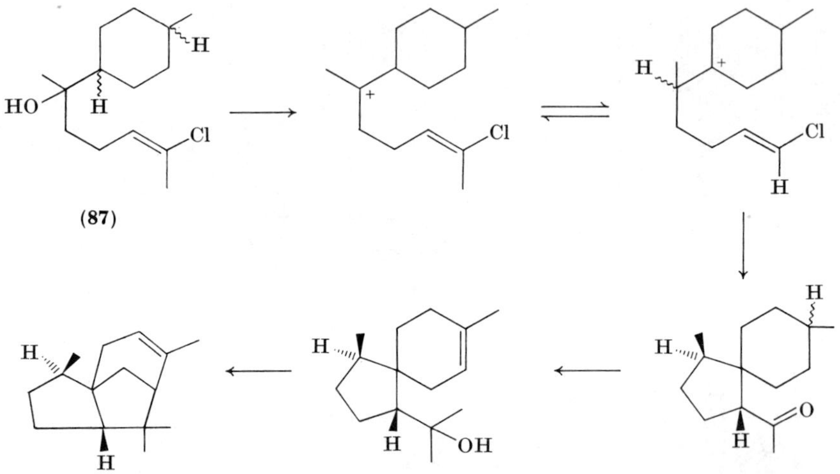

(87)

(88)

An efficient synthesis of α-cedrene (**88**) has been developed by utilizing successive cyclization steps beginning with (**87**).[121]

Studies on the rearrangement of *trans*-tetrahydroabietic acid in sulphuric acid,[122] and on the solvolysis of steroidal cyclodecenyl derivatives, have been reported.[123] Double-bond participation and the formation of bicyclic acetates has been observed in the pyrolysis of cyclohept-4-en-1-ylmethyl acetate under flow conditions.[124] Results from the photolytic deamination of bicyclo[4.2.1]nona-2.4.7-trien-9-one tosylhydrazone indicate that in this reaction, and in the solvolysis of the corresponding toluene-*p*-sulphonate, double-bond participation is unimportant.[125]

The cyclization of 3-methyl-2-penten-4-yn-1-ol to 2,3-dimethylfuran is catalysed by cuprous chloride.[126]

An increase in the *exo*/*endo* rate ratio has been found for the solvolysis of 2-aryl-5-methylnorbornen-2-yl *p*-nitrobenzoates with increasing electron demand of the aryl substituent.[127] This detection of double-bond participation, which is reflected in the formation of increasing amounts of the rearranged product 1-aryl-3-methyl-3-nortricyclanol, contrasts with the behaviour of the parent 2-arylnorbornen-3-yl derivatives where no double bond participation was detected.

Solvolysis of pent-3-yn-1-yl trifluoromethanesulphonate in 2,2,2-trifluoroethanol gave, together with other products, 2-methylcyclobutene-1-yl 2′,2′,2′-trifluoroethyl ether.[128] The extensive cyclization is associated with the low nucleophilicity of the solvent.

Reactions of Small-ring Compounds

Cyclopropylmethyl Derivatives

Both *exo*- and *endo*-tricyclo[3,3,0,$0^{2,8}$]oct-3-yl toluene-*p*-sulphonate (**89**) solvolyse with assistance in aqueous acetone, and similar product ratios indicate a common intermediate (**90**).[129]

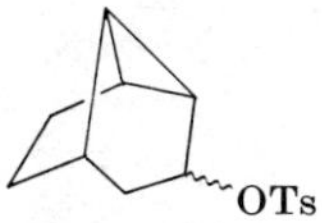

OTs

(**89**) 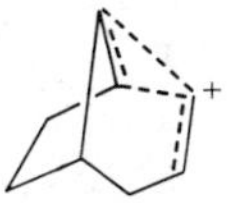 (**90**) 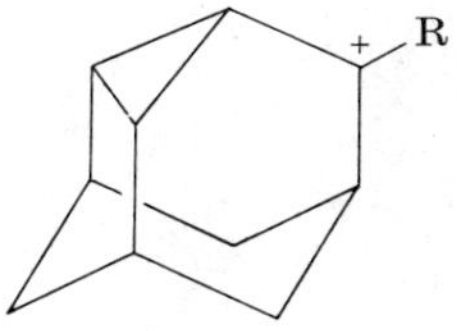 (**91**) 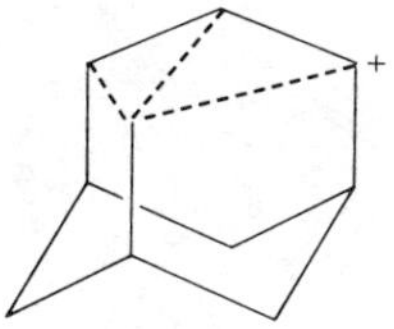 (**92**)

R = H or CH_3

Full details of the reactions of 2-bicyclo[3.1.0]hexyl cations in deamination and solvolysis,[130] and solvolytic studies on the tricyclo[4.2.0.$0^{2,4}$]oct-5-yl derivatives, have appeared.[131]

In contrast to ^{13}C-NMR studies, three experimental comparisons of the rates of solvolysis, of ρ-values and of olefin ionization potentials of cyclopropyl with model systems have indicated the stabilization of a cyclopropylcarbinyl cation is by hyperconjugation (vertical stabilization).[132] Studies on the solvolysis of (*trans*-2-methylcyclopropyl)methyl methanesulphonate has provided evidence for a bisected cyclopropylcarbinyl cation as intermediate with this primary system.[133]

The ^{1}H- and ^{13}C-NMR data for the 8,9-dehydro-2-adamantyl- and 2-methyl-8,9-dehydro-2-adamantyl (**91**) cations indicate that the former is best represented as a partially delocalized equilibrating classical ion and the latter as a static classical ion.[134] On this basis, the non-classical carbonium ion (**92**) was excluded as being the sole representation for the 8,9-dehydro-2-adamantyl cation.

Full details of the reactions of *endo*-2-methoxytricyclo[4.1.0.$0^{3,7}$]heptanes with silver perchlorate[135] and studies on the annulated norcaradienylcarbinyl cations[136,137] have

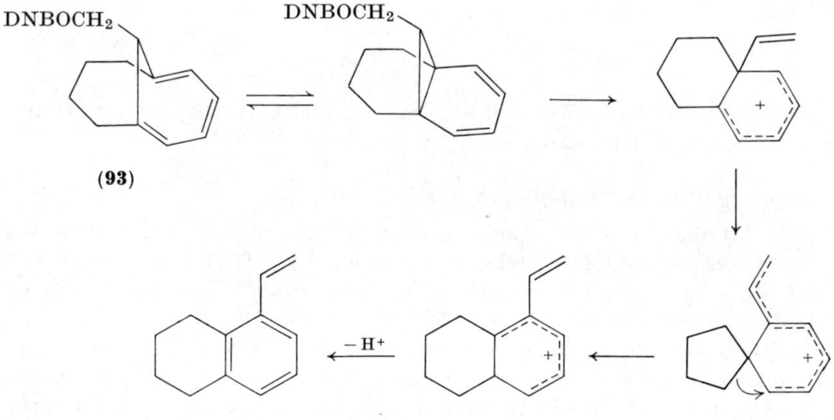

DNBOCH$_2$ DNBOCH$_2$

(**93**)

− H$^+$

(**95**) (**94**)

appeared; it was suggested that the mechanism for almost exclusive conversion of (**93**), or its epimer, into 5-vinyltetralin (**95**) proceeds through (**94**).[136]

No large effect on the rate of solvolysis of (**96**) and (**97**) could be detected compared with that of the saturated analogue (**98**), which is not consistent with the formation of a six-electron cation of unusual stability.[138] Solvolysis studies on allenylcarbinyl chloride and its cyclopropanated homologues have appeared.[139]

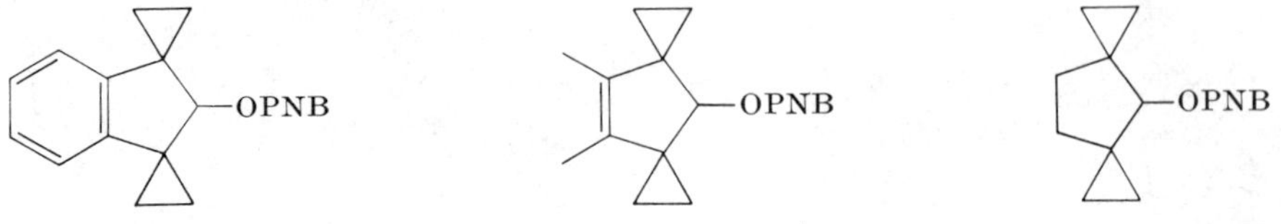

—OPNB —OPNB —OPNB

(**96**) (**97**) (**98**)

Extensive carbon–carbon participation has been observed in the solvolysis of cyclobutylcarbinyl (99%) and cyclopentylcarbinyl (91%) *p*-bromobenzenesulphonates.[140] Cyclobutyl participation was found to compete effectively with aryl participation in the solvolysis of 1-aryl-substituted cyclobutylcarbinyl *p*-bromobenzenesulphonates.[141]

Calculations have been made on degenerate rearrangements in the homocyclopropenyl cation[142] and on the stabilizing and transmitting ability for substituent effects of the cyclopropyl group.[143] *Ab initio* MO calculations for the three interconverting cyclopropylcarbinyl cations indicate that introduction of a methyl group enhances the stability of one form and effectively blocks the possibility of further rearrangement.[144]

The preference for formation of *endo*-product from solvolysis of *endo*- and *exo*-tricyclo-[3.3.0.0^{2,8}]oct-3-yl derivatives has been attributed to the non-planarity of the carbonium ion intermediate.[145] Calculations have indicated that this might be a general phenomenon for this and other geometrically constrained cyclopropylcarbinyl cations.

CH_2 OMs → OAc

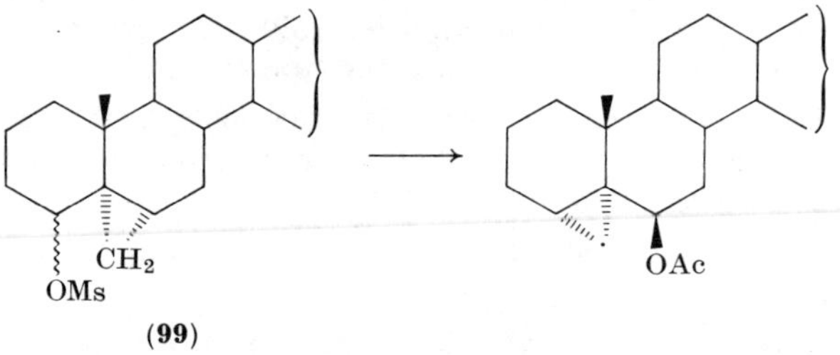

(**99**)

An interesting rearrangement has been observed in the acetolysis of both epimers of (**99**)[146] to the same product, as depicted.

Participation by More Remote Cyclopropane Rings[147]

Solvolysis of (**100**) and (**101**) in aqueous acetone has shown that these systems are the most reactive trishomocyclopropenyl systems [giving, e.g., (**102**)] known, since both are about 10^{14}–10^{15} times more reactive than the 7-norbornyl derivatives.[148] Apart from internal return the major products were (**103**) (75%) and (**104**) (80%) which is an indication that, besides cyclopropyl participation, the relief of ground-state strain is partly responsible for the high reactivity of the system. The amount of anchimeric assistance has been estimated at 10^5 by kinetic studies of the toluene-*p*-sulphonate derivatives of (**103**) and (**104**).[149]

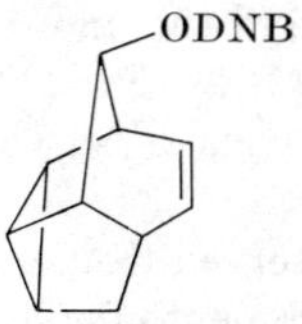

ODNB ODNB + OH OH

(100) 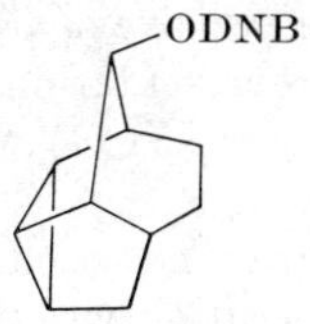 (101) 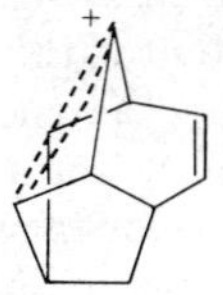 (102) 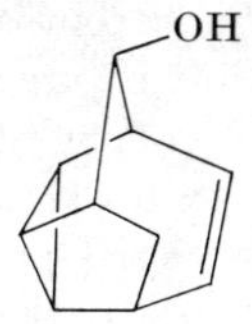 (103) 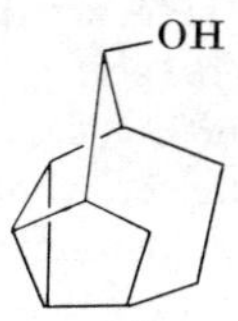 (104)

The cation (**106**) has been generated from (**105**) and the ^{1}H-NMR spectrum has been studied in SbF_5–SO_2ClF at –120°.[150] This spectrum could also be consistent with the rapidly equilibrating trishomocyclopropenyl cations (**107**) and (**108**).

The parent trishomocyclopropenyl cation (**109**)[151] and an ethano-bridged derivative (**110**)[152] have also been prepared in super-acid media. For the former, the large $J(^{13}C–H)$

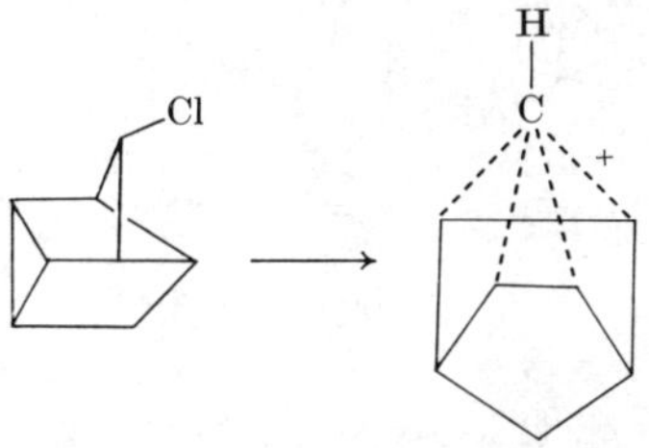

Cl H C +

(105) (106) 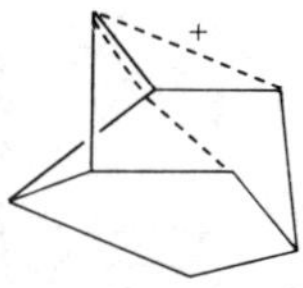 (107) 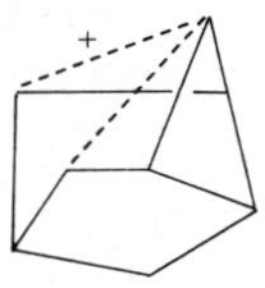 (108)

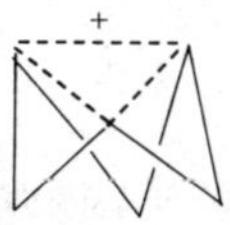

(109) 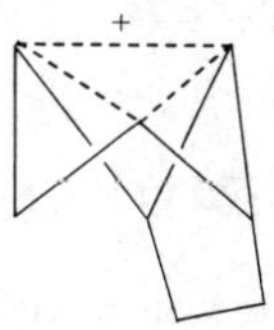 (110) 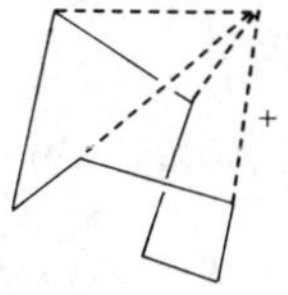 (111)

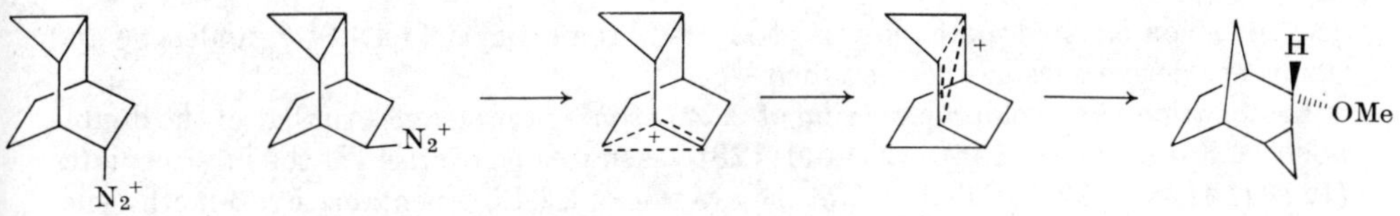

N_2^+ N_2^+ + + H OMe

(112) (113) (114) (115)

MeO OMe MeO H OH H^+ O

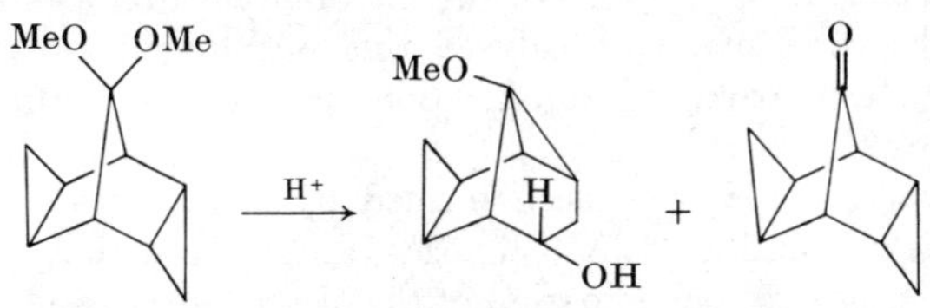

(116) (117) (65%) (35%)

value associated with the carbon atoms carrying the positive charge and the high-field chemical shift of these atoms are consistent with the representation as (**109**). For (**110**), further delocalization was suggested by the spectra and (**111**) was considered to be making some contribution to the overall structure.

Formation of 2,3-dimethyl-1-propylcyclopentyl cation has been observed from 4αH-thujan-3β-ol in SO_2–HSO_3F. In the presence of SbF_5, a cation with the *p*-menthane structure is formed.[153]

No product resulting from participation of the cyclopropyl group was observed in the decomposition of (**112**) whereas Wagner–Meerwein rearrangement took place in the decomposition of (**113**).[154] Formation of (**115**) was attributed to the cyclopropylcarbinyl structure of the rearranged cation (**114**).

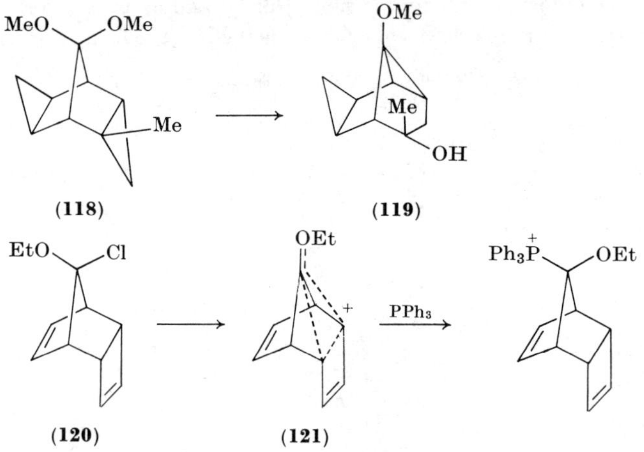

MeO OMe Me → OMe Me OH

(**118**) (**119**)

EtO Cl → OEt + —PPh₃→ $Ph_3\overset{+}{P}$ OEt

(**120**) (**121**)

Formation of (**117**) and (**119**) has been observed in the hydrolysis of the acetals (**116**) and (**118**).[155]

On the basis of its NMR spectrum and its reaction with triphenylphosphine, (**121**) involving edge participation by a β-cyclobutene ring has been suggested to be formed from reaction of (**120**) in liquid SO_2 at –60°.[156]

Reactions of Cyclopropyl and Cyclobutyl Derivatives

The silver-ion-assisted methanolysis of 11,11-dibromotricyclo[4.4.1.$0^{1,6}$]-undecane and -undecene derivatives has been studied.[157]

Protonation of a cyclopropane ring of (**122**) occurs in preference to that of the double bond.[158] Formation of (**126**), (**127**) and (**128**) was suggested to arise via the intermediates (**123**), (**124**) and (**125**). Full details of the reaction of 1,2,4,5,6-pentamethyl-3-methylene-tricyclo[2.2.0.$0^{2,6}$]hexane with acid have been published.[159]

Charge development has been shown to be small in the disrotatory ring opening of cyclopropyl bromides,[160] and the rates of acid-catalysed ring opening of arylcyclopropanols has been measured.[161] Methylenecyclopropyl cation opens to the methylene-allyl cation in a disrotatory mode.[162]

A bicyclo[2.2.1]heptane derivative (**131**) has been formed by acid-catalysed ring opening of (**129**) in nitromethane containing stannic chloride, and the reaction is considered to involve opening of the ring with participation of the olefinic group and generation of a cyclohexyl cation (**130**).[163]

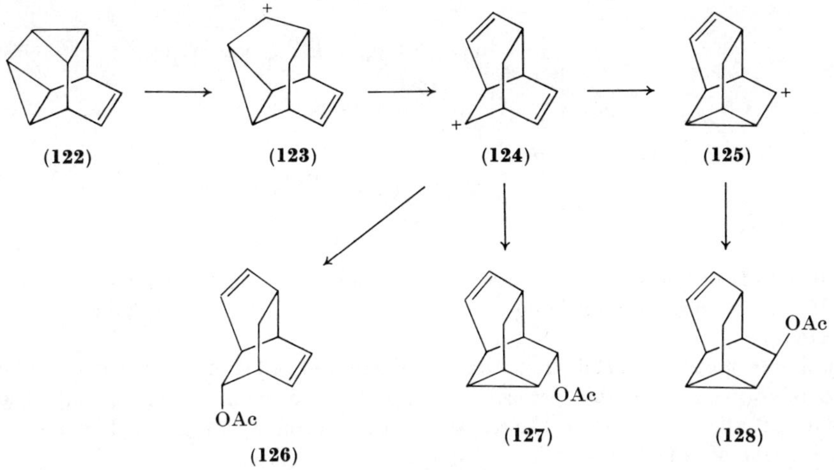

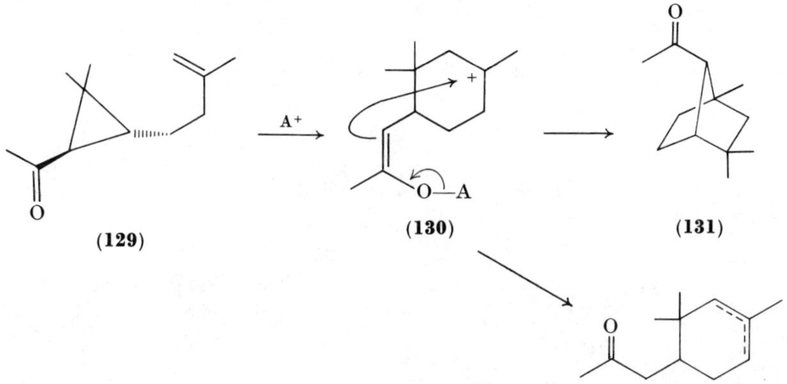

(122) (123) (124) (125)

(126) (127) (128)

(129) (130) (131)

Further evidence that the cyclopropane group does not transmit electronic effects in a similar manner to a double bond has been obtained from studies on the charge-transfer frequencies of 2-methoxy-1-phenylcyclopropanes.[164]

Calculations have been made of the effect of substituents on the stereomutations of allyl cations[161] and of the opening of a cyclopropyl to an allyl cation,[166] and a polarized basis set has been used for calculations on the $C_3H_7^+$ cations.[167]

Solvolyses of C/D *cis*-16α- and 16β-substituted D-norsteroids have been interpreted on the basis that the C-ring exists in a boat conformation.[168]

The rearrangement of tetrasubstituted cyclobutanones in strong acids has been published.[169]

Metallocenylmethyl Cations and Other Derivatives

Reaction of (**132**) with a catalytic amount of 25% sulphuric acid gave ferrocenylbenzyl ketone, indicating that the ferrocenyl group is more effective at stabilization of a vinyl cation (**133**) than is the phenyl group (**134**).[170]

Complete scrambling from C-1 to C-2 in the solvolysis of [1,1-$^2H_2$]-β-*p*-ferrocenylphenethyl toluene-*p*-sulphonate confirm that the ferrocenyl group is electron-donating.[171]

$$\text{FcC}\equiv\text{CPh}\ (\mathbf{132}) \xrightarrow{\text{H}^+} \text{Fc}\overset{+}{\text{C}}\text{=CHPh}\ (\mathbf{133}) \xrightarrow{\text{H}_2\text{O}} \text{FcCCH}_2\text{Ph}\ (\text{C=O})$$

$$\text{FcC}\equiv\text{CPh}\ (\mathbf{132}) \overset{\text{H}^+}{\not\longrightarrow} \text{FcCH=}\overset{+}{\text{C}}\text{Ph}\ (\mathbf{134}) \xrightarrow{\text{H}_2\text{O}} \text{FcCH}_2\text{CPh}\ (\text{C=O})$$

Full details of the torsional barriers in secondary and tertiary ferrocenylalkylium ions[172] and inter-cation exchange rearrangements of ferrocenylalkylium ions have appeared.[173]

Specific formation of (**136**) from the dehydration of (**135**) indicates the considerable charge stabilization by the ferrocenyl group. Even a small change of substituents, however, is sufficient to change the reaction to that of ring-opening and (**138**) is obtained from reaction of (**137**).[174]

Fc, OH (**135**) ⟶ Fc (**136**)

Fc, OH (**137**) ⟶ HO, Fc (**138**)

The structures of α-ferrocenylcarbonium ions have been further examined by a number of physical methods. Studies have been made of the chirality of 1-metallocenyl-2-methylpropyl cations,[175] the ionization of ferrocenyl alcohols in aqueous sulphuric acid,[176] the protonation of ferrocenyl ketones in aqueous sulphuric acid,[177] and the distribution

$$\text{Fc—C(R}^1\text{)(R}^2\text{)—OH} \xrightarrow[\text{Et}_2\text{O}]{\text{HBF}_4} \text{Fc—C}^+\text{(R}^1\text{)(R}^2\text{)}\ (\mathbf{139}) \xrightarrow{\text{R}_2\text{NH}} \text{Fc—C(R}^1\text{)(R}^2\text{)—NR}_2\ (\mathbf{140})$$

$$(\mathbf{139}) \xrightarrow{\text{CN}^-} \text{Fc—C(R}^1\text{)(R}^2\text{)—CN}\ (\mathbf{141})$$

of positive charge from pK_a measurements.[178] Solvolyses of π-(tricarbonylchromium)-cumyl chloride and *p*-[π(tricarbonylchromiumphenyl)]cumyl chloride in aqueous acetone,[179] and the infrared spectra of manganese-stabilized carbonium ions[180] have been studied. *Ab initio* SCFMO calculations on silyl-substituted carbonium ions are reported;[181] these show that the carbonium ion is destabilized by an α-silyl group, but stabilized by a β-silyl group.

Reactions of α-ferrocenylcarbonium tetrafluoroborate (**139**) with amines and cyanide ion yield (**140**) and (**141**), respectively.[182]

Cycloaddition reactions of ferrocenylalkylium ions with cyclopentadiene[183] and the crystal structure of ferrocenyldiphenylcyclopropenium tetrafluoroborate[184] have been studied.

Stable Carbonium Ions and their Reactions[185]

Full details of the preparation, spectral assignments and rearrangements of the cation (**142**) have appeared.[186] Reaction of (**142**) with nucleophiles such as water and methanol takes place exclusively at the apical position to give (**143**). This contrasts with a similar reaction with (**144**) which takes place at the basal position to give (**145**). This was considered to be due to the reaction at the least hindered secondary position in each case.

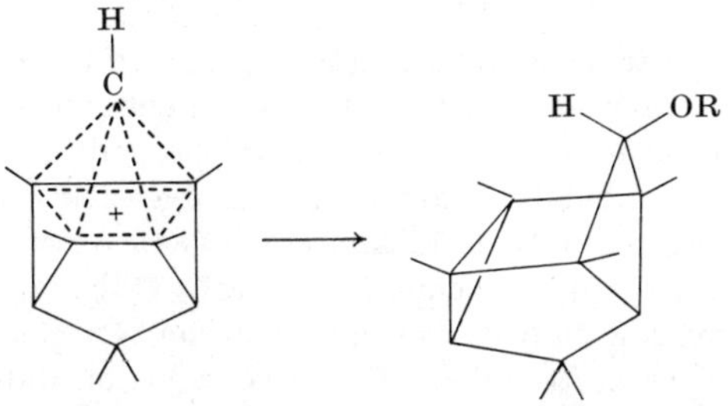

(**142**) (**143**)

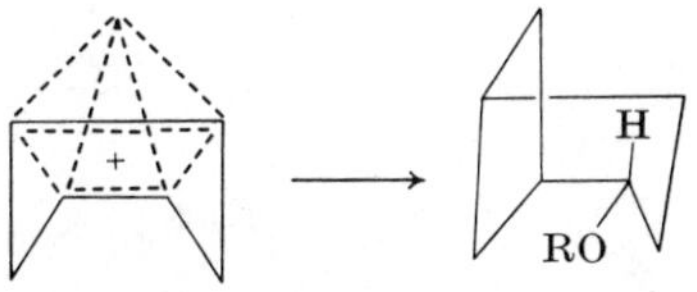

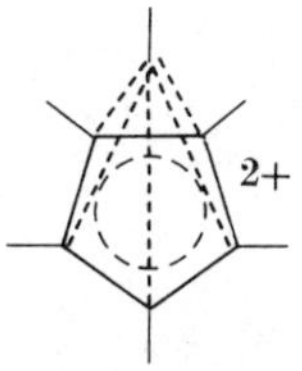

(**144**) (**145**) (**146**)

The dication (**146**) has been prepared from a number of precursors and the assignment as a non-classical ion further discussed and favoured.[187] The absence of any UV absorptions with molar extinction greater than 100 and *ab initio* calculations are also consistent with this structure.[188]

Reaction of (**147**) with FSO_3H at 78° yields the bicyclo[3.1.0]hexenyl cation (**148**).[189] Since the NMR spectrum was consistent with substantial delocalization of the C-1—C-6 and C-5—C-6 cyclopropane bonds, the best representation of the structure was suggested as (**148a**) or (**148b**). Thermal isomerization of (**148**) to (**149**) takes place at –35° with $\Delta F^{\ddagger} = 17.6$ kcal mol^{-1}. Irradiation of (**149**) yields (**148**). The processes were considered to take place involving the making and breaking of the C-1—C-5 bond in electrocyclic processes.

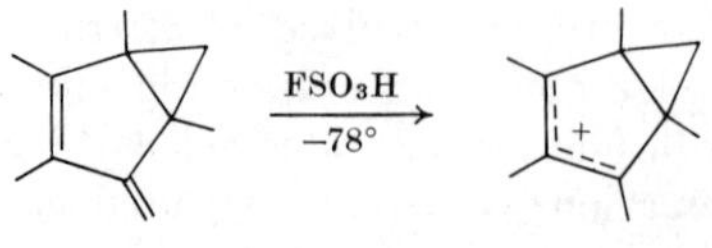

FSO_3H, −78°

(147) **(148)**

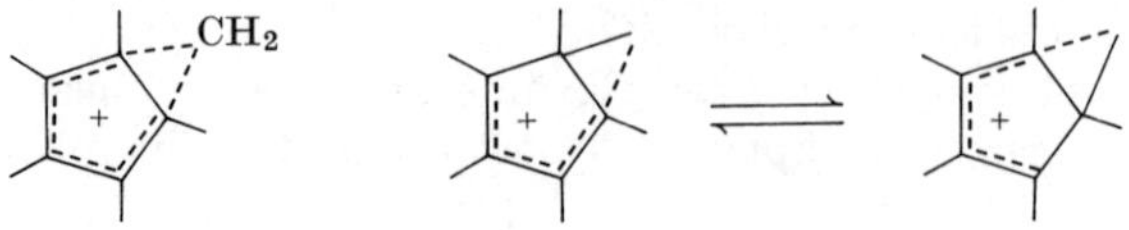

CH_2

(148a) **(148b)**

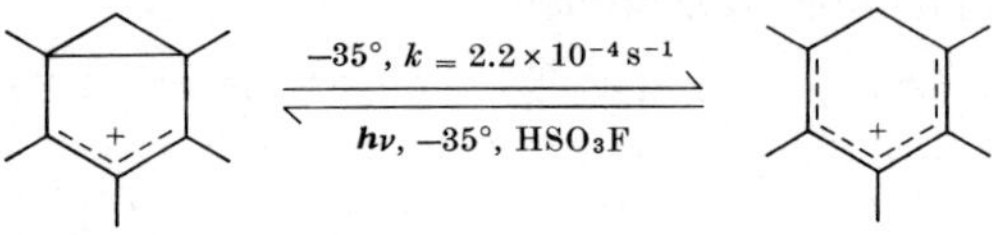

−35°, $k = 2.2 \times 10^{-4}\,s^{-1}$

$h\nu$, −35°, HSO_3F

(148) **(149)**

Protonation of (**150**) yields the cation (**151**) which on irradiation in FSO_3H/SO_2ClF below −90° gave a cation with spectra consistent with the structure (**152**).[190] On warming to −39°, only three signals were observed, with the ratio 15:3:3 and it was evident that a process was occurring that had the effect of moving the cyclopropane group around the five-membered ring. This circumambulatory migration was suggested to take place by successive concerted 1,4-sigmatropic shifts (**152 a–c**, etc.). The ease of migration was shown to depend upon the nature of substituents, particularly at the migrating carbon C-6; the rate of the rearrangement was greater when this carbon bore charge-stabilizing groups, indicating that it becomes electron-deficient in the transition state.

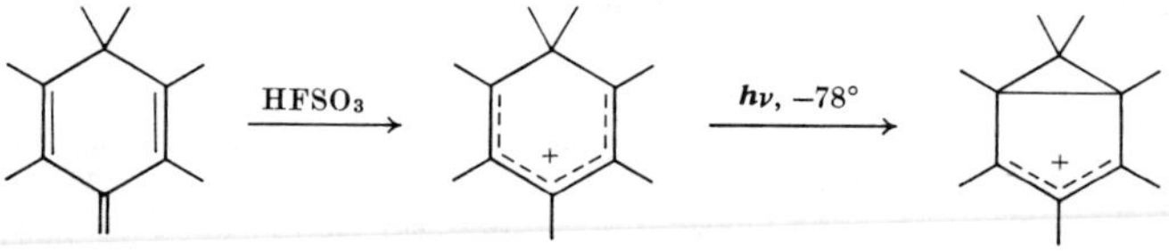

$HFSO_3$ $h\nu$, −78°

(150) **(151)** **(152)**

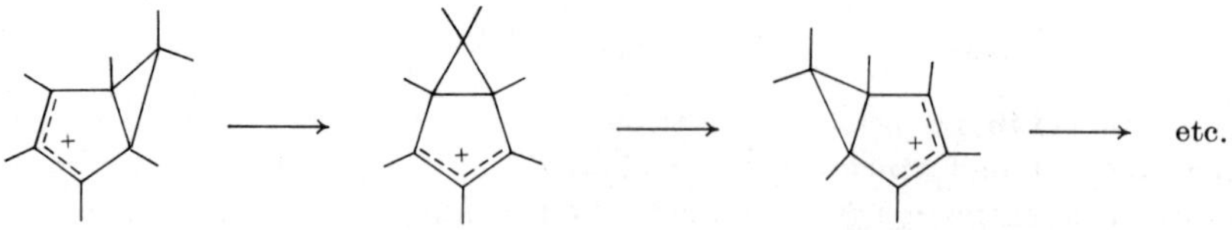

etc.

(152a) **(152b)** **(152c)**

To investigate the intermediates of the above rearrangement further, measurement was made of the rates of the degenerate rearrangement in which the C-6 group migrates around the ring of the zwitterions, formed from reaction of a series of 5-acyl-1,2,3,4,5-pentamethylcyclopentadienes (**153**) with Al_2Cl_6 in CH_2Cl_2.[191] Correlation with σ^+ gave a ρ value of +4.6, indicating the rearrangement rate is inversely dependent on the positive charge-stabilizing ability of the C-6 substituent. A bicyclo[3.1.0]hexenyl

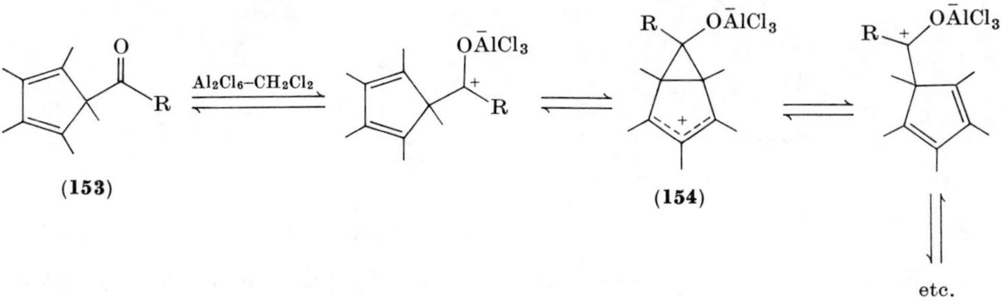

(153)

(154)

etc.

zwitterion (**154**) was suggested as either a high-energy intermediate or a transition state in the reaction.

Accurate rates have been measured for two degenerate rearrangements in a 1,2,4-trialkyl-substituted cyclohexenyl cation.[192]

Generation of the *trans*-pentadienyl cations (**155**) and (**157**) in FSO_3H–SO_2ClF have been studied and rearrangements to (**156**) and (**158**) observed on warming. The last two cations were also obtained by dissolving the alcohols in 100% FSO_3H at –78° or 96% H_2SO_4 at 0°C.[193] This strong driving force to the *cis*-pentahapto-complexes was considered to arise from the enforced planarity of the cyclohexadienyl component which is strain-free compared with a planar acyclic *cis*-dienyl structure. The rate differences were consistent with substantial charge localization at the β-carbon, and (**159**) was considered to make an important contribution to the overall structure.

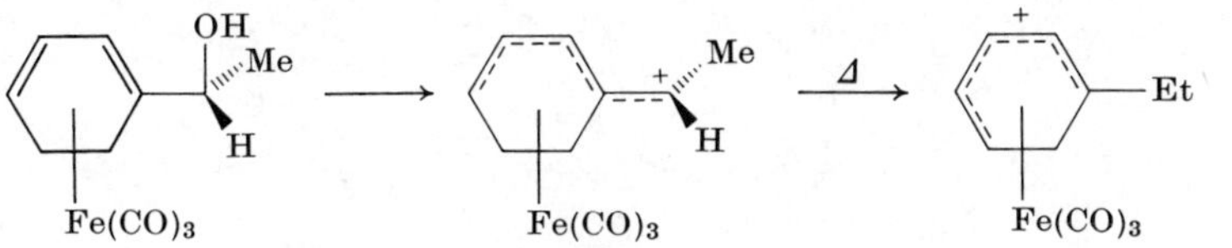

(155)

(156)

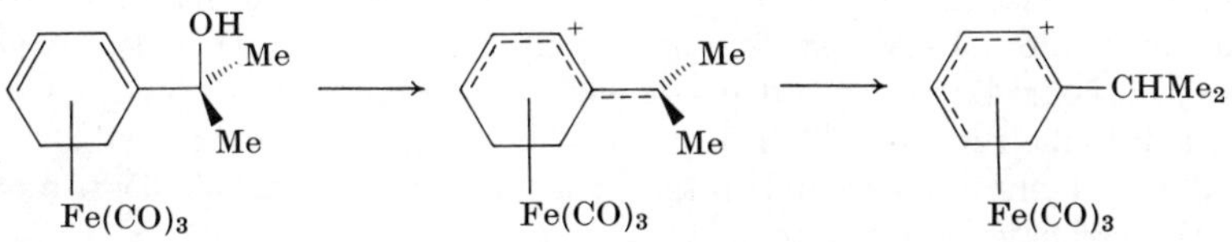

(157)

(158)

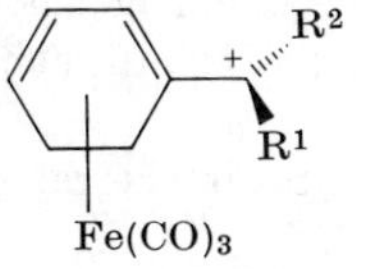

(159)

The cation (**161**) has been obtained by photolysis of the homotropylium ion (**160**), the suggested mechanism being an allowed disrotatory ring closure at C-1 and C-5, followed by hydride migration.[194]

The relative stabilizing ability of a phenyl and a cyclopropyl group to an adjacent carbonium ion centre has been fully discussed.[195] ^{13}C-NMR spectroscopic studies have

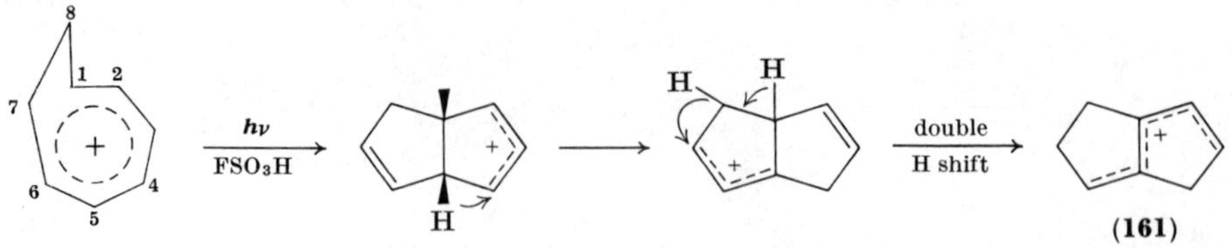

(160) $\xrightarrow[\text{FSO}_3\text{H}]{h\nu}$ ⟶ $\xrightarrow[\text{H shift}]{\text{double}}$ (161)

now shown that the 7-perhydropentalenyl cation is a rapidly equilibrating species in common with the 9-decalyl and 8-hydrindanyl cations.[196]

Three 2-(*p*-trifluoromethylphenyl)-substituted bicyclo-octyl cations (**162**)–(**164**) have been characterized by ^{19}F- and ^{1}H-NMR spectroscopy at –90°. At –60°, the same equilibrium mixture is obtained and the order of stability was established as (**164**) > (**163**) ≫ (**162**).[197]

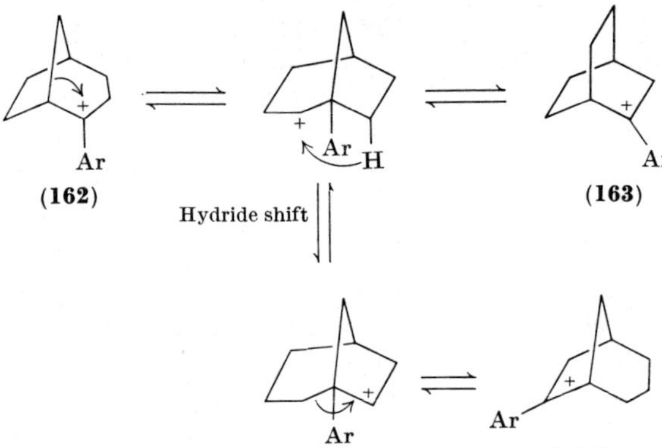

(162) ⇌ ⇌ (163)

Hydride shift

⇌ (164)

A number of methoxy-stabilized phenonium ions have been generated from BF_3 complexes of 1-chloro-di- and -tri-methoxyphenylethanes in SbF_5–SO_2. Oxonium ion formation is favoured over phenyl participation where *ortho*-methoxy groups are not complexed initially with BF_3.[198] Intermolecular exchange reactions with the corresponding alkylcarbonium ions[199] and proton-catalysed transfer from alkanes to methylated benzyl cations[200] have been studied.

1-Methoxycyclo-octatrienylium cation has been prepared from 7,7-dimethoxycyclo-octa-1,3,5-triene by reaction with PCl_5 or $SOCl_2$ in liquid SO_2,[201] and a benzocyclopropenium cation has been generated by ionization of 1,1-dichloro-2,5-diphenylcyclopropabenzene.[202]

Cyclobutadienyl dications have been suggested as intermediates in the rapid equilibration of 4-halo-1,2,3,4-tetrasubstituted cyclobutenyl cations.[203]

A homocyclopropenyl cation (**165**) has been prepared from 3-acetoxycyclobutene in FSO_3H–SbF_5–SO_2ClF at –78°.[204] The temperature-dependence of the NMR spectrum was attributed to a slow ring-inversion process (**165a**)⇌(**165b**) through a planar cyclobutenyl cation. 4-Chlorohomocyclopropenyl cation was also prepared for comparison but did not show the ring-inversion process. It was suggested that the strong 1,3-orbital interaction in the cyclobutenyl cations involves overlap intermediate between σ- and π-bonding.

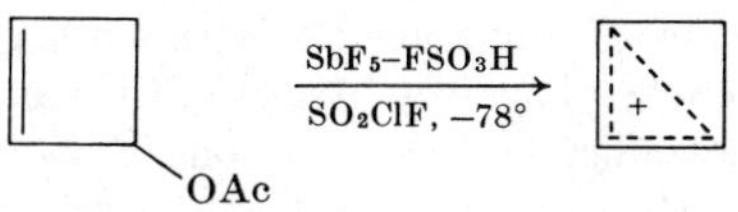

$\xrightarrow[\text{SO}_2\text{ClF, }-78°]{\text{SbF}_5\text{–FSO}_3\text{H}}$

OAc

(165)

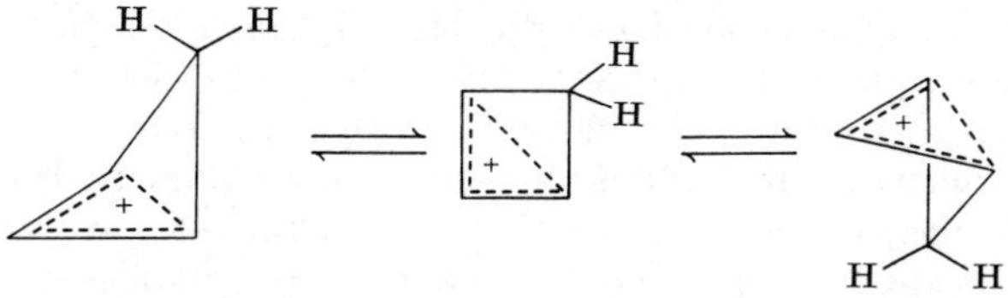

(165a) **(165b)**

The stability of halocarbonium ions,[205] protonation and cleavage of dialkyl pyrocarbonates in $FSO_3H–SbF_5–SO_2$,[206] reactions of alkyl haloformates, thiohaloformates and halosulphites with SbF_5[207] and ^{13}C-spectra of trimethylcarbonium, dihydroxyphenylcarbonium ion and the benzoyl cation[208] have been studied.

A MINDO/3 study of the multiplicity of the pentachlorocyclopentadienyl cation, $(CCl)_5^+$, has shown that the triplet is more stable than the singlet by 1.6 kcal/mol.[209]

Other Reactions[210]

Evidence has been presented that relief of 1,5- and 1,6-non-bonded interactions in congested nitrobenzoates can contribute as much as 10^3–10^4 acceleration in solvolytic reactivity.[211] Thus (**166**) solvolyses 2860 times faster than *trans*-9-decalyl *p*-nitrobenzoates (**167**) owing to the repulsions involving the carbonyl-oxygen, carbonyl-carbon and, in some conformations, the aryl group. These interactions are not present in the corresponding chlorides, and kinetic studies and calculations confirmed the enhanced rate of solvolysis of (**166**). It is, therefore, confirmed that leaving group strain (F strain) can be large in some systems.

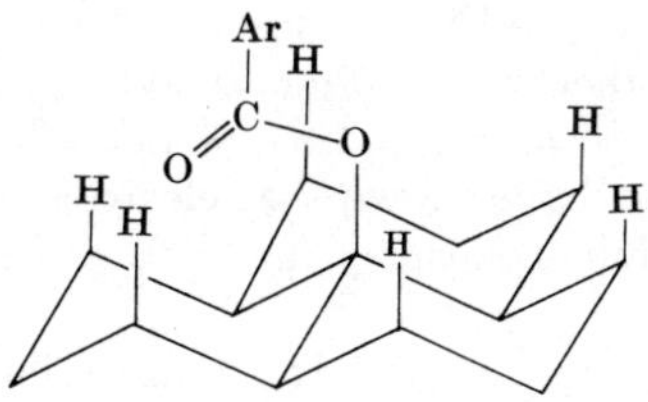

(166) 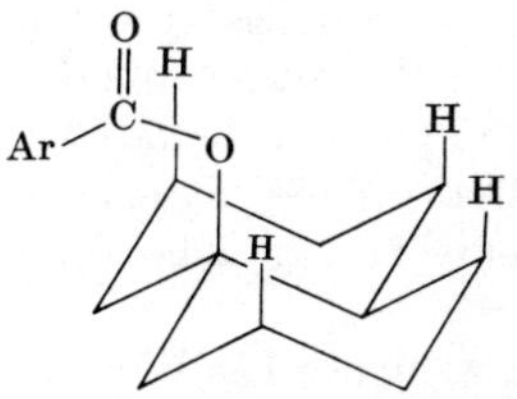 **(167)**

Solvolyses of six isomeric 1-(benzo[*b*]thienyl)ethyl chlorides[212] and of 2-aryl-2-chloropropanes in DMSO, nitromethane, acetonitrile and dimethylformamide[213] have been studied. 6-Methyl-2-naphthalenesulphonate (menasylate) has been suggested as a useful leaving group for trifluoroacetolysis since the rates can conveniently be followed at 326 nm.[214]

Reaction of 5α-cholestane-4α,5-diol 4-acetate with sulphuric acid–acetic anhydride–acetic acid gives cholest-5-en-4α-yl acetate by loss of the 6α-proton.[215]

Similarities have been shown between the reactions of *cis*- and *trans*-4-*tert*-butylcyclohexyl and *erythro*- and *threo*-3-phenyl-2-butyl chloroformate with silver hexafluoroantimonate in acetic acid and corresponding deaminations.[216] Although dehydration of tri-*tert*-butylcarbinol in dimethyl sulphoxide involves a carbonium ion, reaction in hexamethylphosphorotriamide proceeds by a free-radical mechanism.[217] Examination of the oxidation reactions of alcohols with chlorodiphenylmethylium hexachloroantimonates has complemented earlier studies with the triphenylmethyl cation.[218]

Anionic micelles of sodium lauryl sulphate catalyse reaction of the tri-*p*-methoxyphenylmethyl cation with butyl- and hexyl-amine and with 2-methylpyrrolidine.[219] The affinities of weak nucleophiles for incipient methyl cations[220] and reactions of the trithiocyclopropenyl cation with nucleophiles[221] have been investigated.

Spectroscopic studies include the solvent-dependence of absorption and magnetic circular dichroism spectra of the triphenylcarbonium ion,[222] the absorption and MCD spectra of ring-halogenated difluorophenylcarbonium ion,[223] and the electronic spectra 1,1-diarylphenylcarbonium ions.[224] The relative stabilities of several methoxycarbonium ions have been determined,[225] and the α-disulphide carbonium ions have been shown to be far less stable than α-sulphide carbonium ions.[226]

Reactions of the 2-methoxyallyl cation with benzene, toluene, *p*-xylene and mesitylene give varying amounts of bicyclo[3.2.2]nona-6,8-dien-3-ones together with other products.[227] Tricyclic adducts are produced by the debromination of α,α′-dibromocycloalkenones on a column of a modified zinc–copper couple in the presence of furan.[228]

Tetraphenylethylene cation radical undergoes a cyclization reaction giving equimolar quantities of tetraphenylethylene and 9,10-diphenylphenanthrene.[229] Hydride-transfer reactions to hydroxyallyl and tetrahydrofuryl cations,[230] carbonium ion rearrangements in the alkylation of tertiary halides with trimethylaluminium,[231] and the kinetics of carbonium ion formation from anhydroretinol[232] have been studied, and trialkoxycyclopropenyl ions have been prepared.[233]

Electrolytic reactions involving carbonium ions have been reported, including the anodic oxidation of alkyl iodides[234] and phenylacetate ions[235] and the anodic acetamidation of esters.[236] Carbonium ion formation by oxidation of hydrocarbons by fluorosulphonic acid[237] and the kinetics of the reduction of substituted cyclopropenium ions with Cr(II)[238] have been studied.

Geometry-optimized theoretical calculations in the INDO approximation on the cyclopropenyl, azirinyl, and diazirinyl cation[239] and on six cyclobutadienyl dications,[240] and semiempirical calculations of the CNDO/2 type for the 1,1,3,3-tetrakis(dimethylamino)allyl cation[241] have been made. Calculations have shown that electron-releasing substituents favour π-bonding between the terminal carbons in 2-substituted allyl cations.[242]

Strong evidence has been presented that in the gas phase $C_7H_7^+$ ions derived from toluene exist as the benzyl cation; these are stable as such for a period of seconds at least (together with the tropylium ion). From reactions with labelled toluene derivatives, it has been suggested that the reaction:

$$C_7H_7^+ + C_6H_5CH_3 \rightarrow C_8H_9^+ + C_6H_6$$

proceeds through (**168**).[243]

Formation of the benzyldimethyloxonium ion in a mass spectrometer has been shown to take place by benzyl cation transfer in the collision complex formed between benzyl

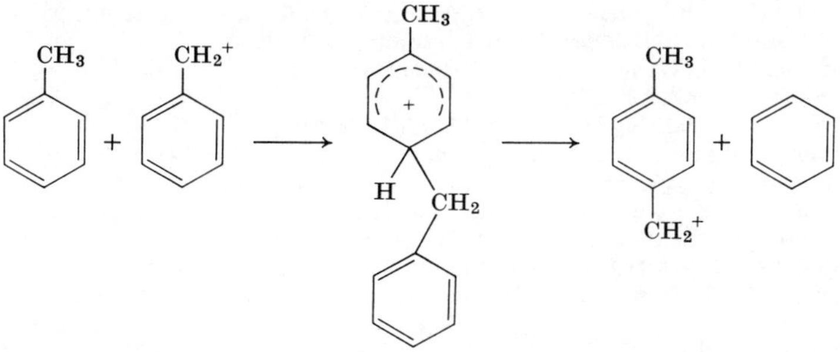

(**168**)

methyl ether and dimethyl ether.[244] Evidence has been presented that the almost total loss of positional identity of the carbon atoms in toluene on formation of the tropylium ion in a mass spectrometer occurs after its formation.[245]

Collisional activation spectra for the gaseous tropylium, benzyl and norbornadienyl cations,[246] and the stabilities of fluorocarbonium ions[247] have been examined. Reactions of methyl cations and methyl halides have been studied.[248]

A cyclotron resonance technique has been used to estimate the relative stabilities of cyclic and acyclic carbonium ions, acyl cations and cyclic halonium ions.[249] The following conclusions emerged from a comparison with heats of ionization of corresponding bromides obtained in HSO_3F at −60°: (*a*) relative to acyclic alkylcarbonium ions and acyl cations, cyclic bromonium ions are significantly more stable in the gas phase than in solution, and (*b*) whilst of comparable instability in solution, the 1-adamantylcarbonium ion is significantly more stable than the *tert*-butylcarbonium ion in the gas phase (ca. 11 kcal/mol).

References

1 "Carbonium Ions", S. P. McManus and C. U. Pittman, in *Reactive Intermediates*, Academic Press, N.Y., 1973, p. 193; J. W. Henderson, "Chirality in Carbonium Ions, Carbanions and Radicals", *Chem. Soc. Rev.*, **2**, 397 (1973); P. W. Cabell-Whiting and H. Hogeveen, "Photochemistry of Carbonium Ions", *Adv. Phys. Org. Chem.*, **10**, 129 (1973); H. Hogeveen, "Reactivity of Carbonium Ions towards Carbon Monoxide", *ibid.*, p. 29; J. M. Harris, "Solvolytic Substitution in Simple Alkyl Systems", *Progr. Phys. Org. Chem.*, **10**, 89 (1974); D. J. Cram and J. M. Cram, "Carbanions and Carbonium Ions, Stereochemical Analogs", **7**, 1 (1973).

2 G. A. Olah, G. Liang, G. D. Mateescu and J. L. Riemenschneider, *J. Amer. Chem. Soc.*, **95**, 8698 (1973).

3 G. A. Olah and G. Liang, *J. Amer. Chem. Soc.*, **96**, 189 (1974).

4 G. A. Olah and G. Liang, *J. Amer. Chem. Soc.*, **96**, 195 (1974).

5 D. G. Farnum and A. D. Wolf, *J. Amer. Chem. Soc.*, **96**, 5166 (1974).

6 C. F. Wilcox, L. M. Loew, R. G. Jesaitis, S. Belin and J. N. C. Hsu, *J. Amer. Chem. Soc.*, **96**, 4061 (1974).

7 J. M. Harris and S. P. McManus, *J. Amer. Chem. Soc.*, **96**, 4693 (1974).

8 J. E. Nordlander, R. R. Gruetzmacher, W. J. Kelly and S. P. Jindal, *J. Amer. Chem. Soc.*, **96**, 181 (1974).

9 E. N. Peters and H. C. Brown, *J. Amer. Chem. Soc.*, **96**, 263 (1974).

10 E. N. Peters and H. C. Brown, *J. Amer. Chem. Soc.*, **96**, 265 (1974).

11 H. C. Brown and E. N. Peters, *Proc. Nat. Acad. Sci., U.S.*, **71**, 132 (1974).

12 D. Lenoir, *Tetrahedron Letters*, **1974**, 1563.

13 C. K. Johnson and C. J. Collins, *J. Amer. Chem. Soc.*, **96**, 2514 (1974).

14 C. J. Collins, C. K. Johnson and V. F. Raaen, *J. Amer. Chem. Soc.*, **96**, 2524 (1974).

15 D. C. Kleinfelter, E. S. Trent, J. E. Mallory, T. E. Dye and J. H. Long, *J. Org. Chem.*, **38**, 4127 (1973).

[16] D. C. Kleinfelter, M. B. Watsky and W. E. Wilde, *J. Org. Chem.*, **38**, 4134 (1973).
[17] D. C. Kleinfelter and J. M. Miller, *J. Org. Chem.*, **38**, 4142 (1974).
[18] I. Rothberg and R. L. Garnick, *J. C. S. Perkin II*, **1974**, 457.
[19] See *Org. Reaction Mech.*, **1970**, 7.
[20] R. Baker and D. E. Halliday, *J. C. S. Perkin II*, **1974**, 208.
[21] G. E. Gream, D. Wege and M. Mular, *Austral. J. Chem.*, **27**, 567 (1974).
[22] G. E. Gream and C. F. Pincombe, *Austral. J. Chem.*, **27**, 589 (1974).
[23] G. E. Gream, C. F. Pincombe and D. Wege, *Austral. J. Chem.*, **27**, 603 (1974).
[24] P. I. Meikle and D. Whittaker, *J. C. S. Perkin II*, **1974**, 322.
[25] R. Siegfried, *Chem. Ber.*, **107**, 1472 (1974).
[26] E. M. White, R. H. McGirk, C. A. Aufdermarsh, H. P. Tiwari and M. J. Todd, *J. Amer. Chem. Soc.*, **95**, 8107 (1973).
[27] Y. Castanet, F. Petit and M. Evrard, *Bull. Soc. Chim. France*, **1974**, 1097.
[28] H. Indyk and D. Whittaker, *J. C. S. Perkin II*, **1974**, 646.
[29] P. I. Meikle and D. Whittaker, *J. C. S. Perkin II*, **1974**, 318.
[30] E. Chong-Sen, R. A. Jones and T. C. Webb, *J. C. S. Perkin II*, **1974**, 38.
[31] K. B. Wiberg and W.-F. Chen, *J. Amer. Chem. Soc.*, **96**, 3900 (1974).
[32] A. F. Diaz, J. Fulcher, M. Sakai and S. Winstein, *J. Amer. Chem. Soc.*, **96**, 1264 (1974).
[33] H.-P. Loffler, *Tetrahedron Letters*, **1974**, 2121.
[34] G. D. Sargent and T. J. Mason, *J. Amer. Chem. Soc.*, **96**, 1063 (1974).
[35] H. Hogeveen and P. W. Kwant, *J. Org. Chem.*, **38**, 2624 (1974).
[36] H. Hogeveen, P. W. Kwant and J. Thio, *Rec. Trav. chim.*, **93**, 152 (1974).
[37] H. Hogeveen, P. W. Kwant, E. P. Schudde and P. A. Wade, *J. Amer. Chem. Soc.*, **96**, 7518 (1974); see *Org. Reaction Mech.*, **1973**, 4.
[38] H. Hart and I. C. Huang, *Tetrahedron Letters*, **1974**, 3245.
[39] See *Org. Reaction Mech.*, **1971**, 24, 27.
[40] P. Beak and B. R. Harris, *J. Amer. Chem. Soc.*, **96**, 6363 (1974).
[41] H. Indyk and D. Whittaker, *J. C. S. Perkin II*, **1974**, 313.
[42] W. Betz and J. Daub, *Chem. Ber.*, **107**, 2095 (1974).
[43] I. S. Isaev, V. I. Buraev, G. P. Novikov, L. M. Sosnovskaya and V. A. Koptyng, *Zh. Org. Khim.*, **9**, 2430 (1973); *Chem. Abs.*, **80**, 47228 (1974).
[44] V. I. Mamatynk, A. I. Rezvukhim, I. S. Isaev, V. I. Buraev and V. I. Koptyng, *Zh. Org. Khim.*, **10**, 662 (1974); *Chem. Abs.*, **80**, 132336 (1974).
[45] J. W. Pavlik and R. J. Pasteris, *J. Amer. Chem. Soc.*, **96**, 6107 (1974).
[46] W. J. Hehre, *J. Amer. Chem. Soc.*, **96**, 5207 (1974).
[47] S. J. Cristol and R. J. Bopp, *J. Org. Chem.*, **39**, 1336 (1974).
[48] D. Lenoir, P. Mison, E. Hyson, P. von R. Schleyer, M. Saunders, P. Vogel and L. A. Telkowski, *J. Amer. Chem. Soc.*, **96**, 2157 (1974).
[49] J. Sabo, J. Hiti and R. C. Fort, *J. Org. Chem.*, **39**, 250 (1974).
[50] B. A. Kazanskii, E. A. Shokova and S. I. Knopova, *Izv. Adad. Nauk SSR, Ser. Khim.*, **1968**, 2645.
[51] S. A. Godleski, W. D. Graham, T. W. Bentley, P. von R. Schleyer and G. Liang, *Chem. Ber.*, **107**, 1257 (1974).
[52] J. Janjatovic, D. Skare and Z. Majerski, *J. Org. Chem.*, **39**, 651 (1974).
[53] K. M. Majerski and Z. Majerski, *Tetrahedron Letters*, **1973**, 4915.
[54] D. Lenoir, R. E. Hall and P. von R. Schleyer, *J. Amer. Chem. Soc.*, **96**, 2138 (1974); see *Org. Reaction Mech.*, **1970**, 20; **1971**, 21.
[55] D. Lenoir, D. J. Raber and P. von R. Schleyer, *J. Amer. Chem. Soc.*, **96**, 2149 (1974).
[56] D. Farcasiu, E. Wiskott, E. Osawa, W. Thielecke, E. M. Engler, J. Slutsky, P. von R. Schleyer and Kent, *J. Amer. Chem. Soc.*, **96**, 4669 (1974).
[57] J. K. Chakrabarti, M. R. J. Jolley and A. Todd, *Tetrahedron Letters*, **1974**, 391.
[58] J. W. Wilt and J. R. Flanyak, *J. Org. Chem.*, **39**, 716 (1974).
[59] P. K. Freeman, B. K. Stevenson, D. M. Balls and D. H. Jones, *J. Org. Chem.*, **39**, 546 (1974).
[60] J. G. Henkel and L. A. Spurlock, *J. Amer. Chem. Soc.*, **95**, 8339 (1973).
[61] R. R. Sauers and E. M. O'Hara, *J. Amer. Chem. Soc.*, **96**, 2510 (1974); see *Org. Reaction Mech.*, **1972**, 12; **1970**, 20.
[62] W. L. Dilling and J. A. Alford, *J. Amer. Chem. Soc.*, **96**, 3615 (1974); see *Org. Reaction Mech.*, **1971**, 16; **1699**, 18.
[63] W. L. Dilling, R. A. Plepys and J. A. Alford, *J. Org. Chem.*, **39**, 2856 (1974).
[64] J. T. Groves and K. W. Ma, *Tetrahedron Letters*, **1973**, 5225.
[65] L. A. Paquette and G. Zon, *J. Amer. Chem. Soc.*, **96**, 215 (1974).

[66] L. A. Paquette and G. Zon, *J. Amer. Chem. Soc.*, **96**, 203 (1974); see *Org. Reaction Mech.* **1973**, 7; **1972**; 9; **1971**, 12.
[67] L. A. Paquette and G. Zon, *J. Amer. Chem. Soc.*, **96**, 224 (1974).
[68] K. Wiesner, P.-T. Ho and H. Ohtani, *Can. J. Chem.*, **52**, 640 (1974).
[69] J. W. ApSimon, J. A. Buccini and A. S. Y. Chau, *Tetrahedron Letters*, **1974**, 539.
[70] T. Svensson, *Chem. Ser.*, 1974, **6**, 22.
[71] C. Trindle, J. T. Hwang and D. Hough, *Tetrahedron*, **30**, 1393 (1974).
[72] N. Barbulescu and M. Leca, *Rev. Roumaine Chim.*, **19**, 443 (1974).
[73] N. Barbulescu and M. Leca, *Rev. Roumaine Chim.*, **19**, 233 (1974).
[74] W. Parker, R. L. Tranter, C. I. F. Watt, L. W. K. Chang and P. von R. Schleyer, *J. Amer. Chem. Soc.*, **96**, 7121 (1974).
[75] D. N. Kevill and C.-B. Kim, *J. Org. Chem.*, **39**, 3085 (1974).
[76] M. Gates, D. L. Frank and W. C. von Felton, *J. Amer. Chem. Soc.*, **96**, 5138 (1974).
[77] L. M. Jackman and V. R. Haddon, *J. Amer. Chem. Soc.*, **96**, 5130 (1974).
[78] J. P. Gesson, J.-C. Jacquesy and R. Jacquesy, *Tetrahedron Letters*, **1974**, 4119.
[79] I. Dragutan, C. Draghici, A. Haines, A. Dumitriu, E. Cioranescu and C. D. Nenitzescu, *Rev. Roumaine Chim.*, **19**, 67 (1974).
[80] A. Banciu, M. Eiian, A. Bucur, D. Georgescu, E. Cioranescu and C. D. Nenitescu, *Rev. Roumaine Chim.*, **19**, 213 (1974).
[81] A. Banciu, *Stud. Cercet. Chim.*, **21**, 773 (1973); *Chem. Abs.*, **80**, 2687 (1974).
[82] S. J. Cristol, T. D. Ziebarth, N. J. Turro, P. Stone and P. Scribe, *J. Amer. Chem. Soc.*, **96**, 3016 (1974).
[83] J. W. Wilt, T. P. Malloy, P. K. Mookerjee and D. R. Sullivan, *J. Org. Chem.*, **39**, 1327 (1974); see *Org. Reaction Mech.*, **1970**, 24.
[84] W. W. Schoeller, *Chem. Comm.*, **1974**, 872.
[85] H. Griengl and P. Schuster, *Tetrahedron*, **30**, 117 (1974).
[86] F. Brisse, A. Lectard and C. Schmidt, *Can. J. Chem.*, **52**, 1123 (1974).
[87] N. M. M. Nibbering, T. Nishishita, C. C. Van der Sande and F. W. McLafferty, *J. Amer. Chem. Soc.*, **96**, 5668 (1974).
[88] D. F. Eaton and T. G. Traylor, *J. Amer. Chem. Soc.* **96**, 7109 (1974).
[89] A. Sera, C. Yamayami and K. Maruyama, *Bull. Chem. Soc. Japan*, **47**, 704 (1974).
[90] A. M. Avedikan, A. Kergomard, J. C. Tardivat and J. P. Vuillerme, *Bull. Chem. Soc. France*, **1974**, 2653.
[91] H. L. Goering, C. S. Chang and J. V. Clevenger, *J. Amer. Chem. Soc.*, **96**, 7602 (1974).
[92] I. Rothberg, J. Fraser, R. Garnick, J. C. King, S. Kirsch and H. Skidanow, *J. Org. Chem.*, **39**, 870 (1974).
[93] J. Ipaktschi, M. N. Iqbal and D. Lenoir, *Chem. Ber.*, **107**, 1126 (1974).
[94] L. A. Paquette and I. R. Dunkin, *J. Amer. Chem. Soc.*, **96**, 1220 (1974).
[95] G. W. Klumpp, G. Ellen and J. J. Vrielink, *Tetrahedron Letters*, **1974**, 2991.
[96] A. F. Diaz, D. Harris and S. Winstein, *Rev. Latinoamer. Quim.*, **4**, 69 (1973); *Chem. Abs.*, **80**, 3006 (1974).
[97] F. R. S. Clark and J. Warkentin, *Can. J. Chem.*, **51**, 4090 (1973).
[98] B. J. A. Cooke and P. R. Story, *Tetrahedron Letters*, **1974**, 1705.
[99] G. E. Gream and A. K. Serelis, *Austral. J. Chem.*, **27**, 629 (1974).
[100] K. B. Becker, A. F. Boshung, M. Geisel and C. A. Grob, *Helv. Chim. Acta*, **56**, 2747 (1973).
[101] G. E. Gream, A. K. Serelis and T. I. Stoneman, *Austral. J. Chem.*, **27**, 1711 (1974).
[102] K. B. Becker, A. F. Boshung and C. A. Grob, *Helv. Chim. Acta*, **56**, 2733 (1973).
[103] K. E. Harding and R. C. Ligon, *J. Org. Chem.*, **38**, 4345 (1973).
[104] H. Wolf, R. Jurb and K. Claussen, *Chem. Ber.*, **107**, 2887 (1974).
[105] R. L. Carney and W. S. Johnson, *J. Amer. Chem. Soc.*, **96**, 2549 (1974).
[106] K. E. Harding, E. J. Leopold, A. M. Hudrlik and W. S. Johnson, *J. Amer. Chem. Soc.*, **96**, 2540 (1974).
[107] G. D. Prestwich and J. N. Labovitz, *J. Amer. Chem. Soc.*, **96**, 7103 (1974).
[108] W. S. Johnson, K. Wiedhaup, S. F. Brady and G. L. Olson, *J. Amer. Chem. Soc.*, **96**, 3979 (1974).
[109] E. E. van Tamelen, G. R. Lees and A. Grieder, *J. Amer. Chem. Soc.*, **96**, 2255 (1974).
[110] J. R. Williams and G. M. Sarkisian, *Tetrahedron Letters*, **1974**, 1109.
[111] M. Santelli and M. Bertrand, *Tetrahedron*, **30**, 227 (1974); see *Org. Reaction Mech.*, **1973**, 16; **1971**, 23; **1969**, 46.
[112] M. Santelli and M. Bertrand, *Tetrahedron*, **30**, 235 (1974).
[113] M. Santelli and M. Bertrand, *Tetrahedron*, **30**, 243 (1974).
[114] M. Santelli and M. Bertrand, *Tetrahedron*, **30**, 251 (1974).
[115] M. Santelli and M. Bertrand, *Tetrahedron*, **30**, 257 (1974).

[116] B. Ragonnet, M. Santelli and M. Bertrand, *Helv. Chim. Acta*, **57**, 557 (1974).
[117] D. Baines, J. Forrester and W. Parker, *J. C. S. Perkin I*, **1974**, 1598.
[118] S. Kumazawa, Y. Nakano, T. Kato and Y. Kitahara, *Tetrahedron Letters*, **1974**, 1757.
[119] D. V. Banthorpe, P. A. Boullier and W. D. Fordham, *J. C. S. Perkin I*, **1974**, 1637.
[120] K. Kogami and J. Kumanotani, *Bull. Chem. Soc. Japan*, **47**, 226 (1974).
[121] P. T. Lansbury, V. R. Haddon and R. C. Stewart, *J. Amer. Chem. Soc.*, **96**, 896 (1974).
[122] B. E. Cross, M. R. Firth and R. E. Markwell, *Chem. Comm.*, **1974**, 930.
[123] L. Lorenc, M. J. Gasic, I. Juranic, M. Dabovic and M. J. Mihailovic, *Tetrahedron Letters*, **1974**, 395.
[124] R. W. Thies and L. E. Shick, *J. Amer. Chem. Soc.*, **96**, 456 (1974).
[125] W. Kirmse and G. Voigt, *J. Amer. Chem. Soc.*, **96**, 7598 (1974).
[126] G. P. Chernysh, A. A. Mel'nikova, E. G. Koreshkova, O. I. Ivanishchenko, T. M. Filippova and G. I. Samokhvalov, *Khim. Farm. Zh.*, **8**, 31 (1974); *Chem. Abs.*, **81**, 104248 (1974).
[127] H. C. Brown, M. Ravindranathan and E. N. Peters, *J. Amer. Chem. Soc.*, **96**, 7351 (1974); see *Org. Reaction Mech.*, **1973**, 1.
[128] H. Stutz and M. Hanack, *Tetrahedron Letters*, **1974**, 2457.
[129] R. G. Buckeridge, K. J. Frayne and B. L. Johnson, *Austral. J. Chem.*, **27**, 1235 (1974).
[130] A. S. Bloss, P. R. Brook and R. M. Ellam, *J. C. S. Perkin II*, **1973**, 2165.
[131] L. A. Paquette, O. Cox, M. Oku and R. P. Henzel, *J. Amer. Chem. Soc.*, **96**, 4892 (1974); see *Org. Reaction Mech.*, **1973**, 12.
[132] D. F. Eaton and T. G. Traylor, *J. Amer. Chem. Soc.*, **96**, 1226 (1974); see *Org. Reaction Mech.*, **1973**, 25.
[133] C. D. Poulter and C. J. Spillner, *J. Amer. Chem. Soc.*, **96**, 7591 (1974).
[134] G. A. Olah, G. Liang, K. A. Babiak and R. K. Murray, *J. Amer. Chem. Soc.*, **96**, 6794 (1974).
[135] G. Zon and L. A. Paquette, *J. Amer. Chem. Soc.*, **96**, 5478 (1974); see *Org. Reaction Mech.*, **1972**, 22.
[136] G. L. Thompson, W. E. Heyd and L. A. Paquette, *J. Amer. Chem. Soc.*, **96**, 3177 (1974); see *Org. Reaction Mech.*, **1973**, 5.
[137] P. Warner and S. Lu, *Tetrahedron Letters*, **1974**, 3455.
[138] H. J. Reich and J. M. Renga, *Tetrahedron Letters*, **1974**, 2747.
[139] A. T. Bottini and J. E. Christensen, *Tetrahedron*, **30**, 393 (1974).
[140] D. D. Roberts and C. H. Wu, *J. Org. Chem.*, **39**, 1570 (1974).
[141] D. D. Roberts, *J. Org. Chem.*, **39**, 1265 (1974).
[142] A. J. P. Devaquet and W. J. Hehre, *J. Amer. Chem. Soc.*, **96**, 3644 (1974).
[143] C. F. Wilcox, L. M. Loew and R. Hoffmann, *J. Amer. Chem. Soc.*, **95**, 8192 (1973).
[144] W. J. Hehre and P. C. Hiberty, *J. Amer. Chem. Soc.*, **96**, 302 (1974).
[145] R. D. Bach, J. H. Siefert, M. T. Tribble, R. A. Greenyard and N. A. LeBel, *J. Amer. Chem. Soc.*, **95**, 8182 (1973).
[146] L. Kohout and J. Fajkos, *Coll. Czech. Chem. Comm.*, **39**, 1601 (1974).
[147] J. Haywood-Farmer, "Long-range Interactions of Cyclopropyl Groups with Carbonium Ion Centres", *Chem. Rev.*, **74**, 315 (1974).
[148] G. Ellen and G. W. Klumpp, *Tetrahedron Letters*, **1974**, 2995.
[149] G. Ellen and G. W. Klumpp, *Tetrahedron Letters*, **1974**, 3637.
[150] A. V. Kemp-Jones, N. Nakamura and S. Masamune, *Chem. Comm.*, **1974**, 109.
[151] S. Masamune, M. Sakai, A. V. Kemp-Jones and T. Nakashima, *Can. J. Chem.*, **52**, 855 (1974).
[152] S. Masamune, M. Sakai and A. V. Kemp-Jones, *Can. J. Chem.*, **52**, 858 (1974).
[153] C. M. Holden and D. Whittaker, *Chem. Comm.*, **1974**, 353.
[154] W. Kirmse and K. H. Wahl, *Chem. Ber.*, **107**, 2768 (1974).
[155] R. Bicker, H. Kessler and A. Steigel, *Chem. Ber.*, **107**, 3053 (1974).
[156] P. Schipper, P. B. J. Driessen, J. W. de Haan and H. M. Buck, *J. Amer. Chem. Soc.*, **96**, 4706 (1974).
[157] D. B. Ledlie, J. Knetzer and A. Gitterman, *J. Org. Chem.*, **39**, 708 (1974); D. B. Ledlie and J. Knetzer, *Tetrahedron Letters*, **1973**, 5021.
[158] H. Bos and G. W. Klumpp, *Tetrahedron Letters*, **1974**, 3641.
[159] H. Hogeveen and P. W. Kwant, *J. Org. Chem.*, **39**, 2626 (1974); see *Org. Reaction Mech.*, **1973**, 15.
[160] J. H. Ong and R. E. Robertson, *Can. J. Chem.*, **52**, 2660 (1974).
[161] C. H. DePuy, R. A. Klein and J. P. Clark, *J. Org. Chem.*, **39**, 483 (1974).
[162] G. Leandri, H. Monti and M. Bertrand, *Tetrahedron*, **30**, 289 (1974).
[163] P. A. Grieco and R. S. Finkelhov, *Tetrahedron Letters*, **1974**, 527.
[164] R. S. Brown and T. G. Traylor, *J. Amer. Chem. Soc.*, **95**, 8025 (1973).
[165] L. Radom, J. A. Pople and P. von R. Schleyer, *J. Amer. Chem. Soc.*, **95**, 8193 (1973).
[166] P. Coffey and K. Jug, *Theor. Chim. Acta*, **34**, 213 (1974); *Chem. Abs.*, **81**, 119731 (1974).
[167] P. C. Hariharan, L. Radom, J. A. Pople and P. von R. Schleyer, *J. Amer. Chem. Soc.*, **96**, 599 (1974).

[168] J. Meinwald and A. J. Taggi, *J. Amer. Chem. Soc.*, **95**, 7663 (1973).
[169] F. Bourelle-Wargnier, *Tetrahedron Letters*, **1974**, 1589.
[170] D. Kaufman and R. Kupper, *J. Org. Chem.*, **39**, 1438 (1974).
[171] C. C. Lee, R. G. Sutherland and B. J. Thomson, *J. Org. Chem.*, **39**, 406 (1974).
[172] T. D. Turbitt and W. E. Watts, *J. C. S. Perkin II*, **1974**, 177; see *Org. Reaction Mech.*, **1973**, 22.
[173] T. D. Turbitt and W. E. Watts, *J. C. S. Perkin II*, **1974**, 189; see *Org. Reaction Mech.*, **1972**, 39.
[174] W. M. Horspool and B. J. Thomson, *Tetrahedron Letters*, **1974**, 3529.
[175] V. I. Sokolov, P. V. Petrovskii, A. A. Koridze and O. A. Reutov, *J. Organometal. Chem.*, **76**, C15 (1974).
[176] G. Cerichelli, B. Floris and G. Ortaggi, *J. Organometal. Chem.*, **78**, 241 (1974).
[177] G. Cerichelli, B. Floris, G. Illuminati and G. Ortaggi, *Gazz. Chim. Ital.*, **103**, 911 (1973); *Chem. Abs.*, **81**, 12972 (1974).
[178] T. D. Turbitt and W. E. Watts, *J. C. S. Perkin II*, **1974**, 185.
[179] S. P. Gubin, V. S. Khandakarova and A. Z. Kreindlin, *J. Organometal. Chem.*, **64**, 229 (1974).
[180] A. G. Ginzberg, V. N. Setkina and D. N. Kursanov, *J. Organometal. Chem.*, **77**, C27 (1974).
[181] C. Eaborn, F. Feichtmayr, M. Horn and J. N. Murrell, *J. Organometal. Chem.*, **77**, 39 (1974).
[182] S. Allenmark, *Tetrahedron Letters*, **1974**, 371.
[183] T. D. Turbitt and W. E. Watt, *J. C. S. Perkin II*, **1974**, 195.
[184] R. L. Sime and R. J. Sime, *J. Amer. Chem. Soc.*, **96**, 892 (1974).
[185] M. Tashiro, "Superacids and a few of their salts", *Kagaku* (*Kyoto*), **29**, 513 (1974); *Chem. Abs.*, **81**, 119297 (1974).
[186] H. Hart and M. Kuzuya, *J. Amer. Chem. Soc.*, **96**, 6436 (1974); see *Org. Reaction Mech.*, **1973**, 23.
[187] H. Hogeveen and P. W. Kwant, *J. Amer. Chem. Soc.*, **96**, 2208 (1974); see *Org. Reaction Mech.*, **1973**, 25.
[188] H. Hogeveen and P. W. Kwant, *Tetrahedron Letters*, **1974**, 4351.
[189] R. F. Childs, M. Sakai, D. B. Partington and S. Winstein, *J. Amer. Chem. Soc.*, **96**, 6403 (1974).
[190] R. F. Childs and S. Winstein, *J. Amer. Chem. Soc.*, **96**, 6409 (1974).
[191] R. F. Childs and M. Zeya, *J. Amer. Chem. Soc.*, **96**, 6418 (1974).
[192] R. Cone, R. P. Haseltine, P. Kazmaier and T. S. Sorensen, *Can. J. Chem.*, **52**, 3320 (1974); see *Org. Reaction Mech.*, **1972**, 43.
[193] C. R. Jablonski and T. S. Sorensen, *Can. J. Chem.*, **52**, 2085 (1974); see *Org. Reaction Mech.*, **1970**, 47; **1971**, 37; **1972**, 38.
[194] P. A. Christensen, Y. Y. Huang, A. Meesters and T. S. Sorensen, *Can. J. Chem.*, **52**, 3424 (1974).
[195] G. A. Olah, P. W. Westerman and J. Nishimura, *J. Amer. Chem. Soc.*, **96**, 3548 (1974); see *Org. Reaction Mech.*, **1973**, 18.
[196] G. A. Olah, G. Liang and P. W. Westerman, *J. Org. Chem.*, **39**, 367 (1974).
[197] A. D. Wolf and D. G. Farnum, *J. Amer. Chem. Soc.*, **96**, 5175 (1974).
[198] J. A. Manner, J. A. Cook and B. G. Ramsey, *J. Org. Chem.*, **39**, 1199 (1974).
[199] G. A. Olah and Y. K. Mo, *J. Amer. Chem. Soc.*, **96**, 3560 (1974).
[200] P. Pelt and H. M. Buck, *Rec. Trav. chim.*, **93**, 206 (1974).
[201] J. R. de Dobbelaere and H. M. Buck, *Rec. Trav. chim.*, **93**, 159 (1974).
[202] B. Halton, A. D. Woolhouse, H. M. Hugel and D. P. Kelly, *Chem. Comm.*, **1974**, 247.
[203] A. E. van der Hout-Lodder, J. W. de Hann and H. M. Buck, *Rec. Trav. chim.*, **93**, 156 (1974).
[204] G. A. Olah, J. S. Staral and G. Liang, *J. Amer. Chem. Soc.*, **96**, 6233 (1974).
[205] G. A. Olah, G. Liang and Y. K. Mo, *J. Org. Chem.*, **39**, 2394 (1974).
[206] G. A. Olah, Y. Halpern, P. W. Westerman and J. L. Grant, *J. Org. Chem.*, **39**, 2390 (1974).
[207] G. A. Olah, P. Schilling, J. M. Bollinger and J. Nishimura, *J. Amer. Chem. Soc.*, **96**, 2221 (1974).
[208] G. A. Olah and P. W. Westerman, *J. Amer. Chem. Soc.*, **96**, 2229 (1974).
[209] M. J. S. Dewar and R. C. Haddon, *J. Amer. Chem. Soc.*, **96**, 255 (1974).
[210] K. Shudo, "Nitrenium Ions", *Yuki Gosei Kagaku. Kyokai Shi.*, **31**, 395 (1973); *Chem. Abs.*, **80**, 46978 (1974).
[211] J. Slutsky, R. C. Bingham, P. von R. Schleyer, W. C. Dickason and H. C. Brown, *J. Amer. Chem. Soc.*, **96**, 1969 (1974).
[212] D. S. Noyce and D. A. Forsyth, *J. Org. Chem.*, **39**, 2828 (1974).
[213] T. Uchida, S. Mauri, Y. Miyagi and K. Maruyama, *Bull. Chem. Soc. Japan*, **47**, 1549 (1974).
[214] T. Ando, Y. Saito, J. Yamawaki, H. Morisaki, M. Sawada and Y. Yukawa, *J. Org. Chem.*, **39**, 2465 (1974).
[215] E. T. J. Bathurst, J. M. Coxon and M. P. Hartshorn, *Austral. J. Chem.*, **27**, 1505 (1974).
[216] P. Beak, J. T. Adams and J. A. Barron, *J. Amer. Chem. Soc.*, **96**, 2494 (1974).
[217] J. S. Lomas and J. E. Dubois, *J. Org. Chem.*, **39**, 1776 (1974).

[218] D. H. R. Barton, P. D. Magnus, J. A. Garbarino and R. N. Young, *J. C. S. Perkin I*, **1974**, 2101.
[219] C. A. Bunton, S. Chan and S. K. Huang, *J. Org. Chem.*, **39**, 1262 (1974); C. A. Bunton and S. K. Huang, *J. Amer. Chem. Soc.*, **96**, 515 (1974).
[220] C.-P. Wong, L. M. Jackman and R. G. Portman, *Tetrahedron Letters*, **1974**, 921.
[221] Z. Yoshida, S. Yoneda, T. Miyamoto and S. Miki, *Tetrahedron Letters*, **1974**, 813.
[222] Y. K. Mo, R. E. Linder, G. Barth, E. Bunnenberg and C. Djerassi, *J. Amer. Chem. Soc.*, **96**, 3309 (1974).
[223] Y. K. Mo, R. E. Linder, G. Barth, E. Bunnenberg and C. Djerassi, *J. Amer. Chem. Soc.*, **96**, 4569 (1974).
[224] G. Heublein and M. Helbig, *Tetrahedron*, **30**, 2533 (1974).
[225] R. P. Quirk and C. R. Gambill, *Chem. Comm.*, **1974**, 503.
[226] E. Block, *J. Org. Chem.*, **39**, 734 (1974).
[227] A. E. Hill and H. M. R. Hoffman, *J. Amer. Chem. Soc.*, **96**, 4597 (1974).
[228] J. G. Vinter and H. M. R. Hoffman, *J. Amer. Chem. Soc.*, **96**, 5466 (1974).
[229] U. Svanholm, A. Ronlan and V. D. Parker, *J. Amer. Chem. Soc.*, **96**, 5108 (1974).
[230] D. M. Brouwer and A. A. Kiffen, *Rec. Trav. chim.*, **92**, 1335 (1973).
[231] E. G. Melby and J. P. Kennedy, *J. Org. Chem.*, **39**, 2433 (1974).
[232] V. C. Bulgin and G. L. Lookhart, *J. Amer. Chem. Soc.*, **96**, 6077 (1974).
[233] E. V. Dehmlow, *Angew. Chem. Internat. Edn.*, **13**, 209 (1974).
[234] A. Laurent, E. Laurent and R. Tardivel, *Tetrahedron Letters*, **1973**, 4861; *Tetrahedron*, **30**, 3423, 3431 (1974).
[235] J. P. Coleman, R. Lines, J. H. P. Uttley and B. C. L. Weedon, *J. C. S. Perkin II*, **1974**, 1064.
[236] L. L. Miller and V. Ramachandran, *J. Org. Chem.*, **39**, 369 (1974).
[237] J. W. Larsen, P. A. Bouis, C. R. Watson and R. M. Pagni, *J. Amer. Chem. Soc.*, **96**, 2284 (1974).
[238] K. Okamoto, K. Komatsu and A. Hitomi, *Bull. Chem. Soc. Japan.*, **46**, 3881 (1973).
[239] C. U. Pittman, A. Kress, T. B. Patterson, P. Walton and L. D. Kispert, *J. Org. Chem.*, **39**, 373 (1974).
[240] C. U. Pittman, A. Kress and L. D. Kispert, *J. Org. Chem.*, **39**, 378 (1974).
[241] H. U. Wagner, *Chem. Ber.*, **107**, 634 (1974).
[242] B. K. Carpenter, *J. C. S. Perkin II*, **1974**, 1.
[243] J. Shen, R. C. Dunbar and G. A. Olah, *J. Amer. Chem. Soc.*, **96**, 6227 (1974).
[244] A. P. Bruins and N. M. M. Nibbering, *Tetrahedron Letters*, **1974**, 2677.
[245] A. Siegal, *J. Amer. Chem. Soc.*, **96**, 1251 (1974).
[246] F. W. McLafferty and J. Winkler, *J. Amer. Chem. Soc.*, **96**, 5182 (1974).
[247] R. J. Blint, T. B. McMahon and J. L. Beauchamp, *J. Amer. Chem. Soc.*, **96**, 1269 (1974).
[248] J. M. S. Henis, M. D. Loberg and M. J. Welch, *J. Amer. Chem. Soc.*, **96**, 1665 (1974).
[249] R. D. Wieting, R. H. Staley and J. L. Beauchamp, *J. Amer. Chem. Soc.*, **96**, 7552 (1974).

CHAPTER 9

Nucleophilic Aliphatic Substitution[1]

I. D. R. STEVENS

Department of Chemistry, University of Southampton

Ion-pair Phenomena and Borderline Mechanisms

The Winstein ion-pair scheme has been reviewed and critically discussed in the light of recent developments.[2]

The stability–selectivity relationship for the solvolysis of secondary alkyl chlorides in 70% ethanol has been examined in the light of the Winstein scheme. It is suggested that such relationships arise from a blend of attack on all four species, i.e.

$$S_{\text{obs}} = \sum_{\text{i=I}}^{\text{IV}} a_{\text{i}} S_{\text{i}}$$

where S = selectivity, and a = fraction of total reaction. Evidence already presented by Koskikallio[3] and by Ritchie[4] suggest that S_{I} and S_{IV} are constant, the former where leaving groups are of the same type; and that this is probably true also for S_{II}. The compounds studied by Ritchie all react *via* (**III**) and show a stability–selectivity relationship, and it is therefore proposed that only S_{III} is variable and that it is proportional to the stability of the solvent-separated ion-pair (**III**). It is further suggested that the majority of the variations in stability–selectivity relations are caused by a change in

$$\underset{\textbf{(I)}}{RX} \rightleftharpoons \underset{\textbf{(II)}}{R^+X^-} \rightleftharpoons \underset{\textbf{(III)}}{R^+/\!/X^-} \rightleftharpoons \underset{\textbf{(IV)}}{R^+ + X^-}$$

$$\downarrow k^{I} \qquad \downarrow k^{II} \qquad \downarrow k^{III} \qquad \downarrow k^{IV}$$

Products

the values of the fractions (a_i) of product formed from each species.[5] The relationship observed for solvolysis of 2-adamantyl arenesulphonate in 70% ethanol is shown to fit the analysis proposed, the observed selectivity for water being due to the sulphonate anion, as previously proposed. For solvolysis in 50% ethanol of benzhydryl benzoates, substituted in the leaving group, the k_{EtOH}/k_{H_2O} value is constant; that this is not due to reaction solely *via* separated ions (**IV**) is shown by the change in the k_{EtOH}/k_{H_2O} ratio on changing to 70% ethanol. Further, the ^{18}O exchange and racemization data of Goering and his collaborators is only consistent with reaction *via* (**III**). The lack of change in k_{EtOH}/k_{H_2O} is therefore ascribed to a compensating effect of the change in S_{III} and those in a_{III} and a_{IV} in the balance of reaction as expressed in the equation $S_{obs} = a_{III}S_{III} + a_{IV}S_{IV}$.[6]

OTs

(**1**)

The evidence for "hidden" ion-pair return[7] has been called seriously into question, as a result of work on the solvolysis of (**1**). The results are compared with those for pinacolyl *p*-bromobenzenesulphonate (corrected for the change in leaving group) in Table 1. For (**1**), rearrangement of the intermediate ion-pair cannot prevent ion-pair return, which has been argued to occur for pinacolyl, and therefore the ratio (**1**)/Pin should vary as it does for isopropyl, which is clearly not the case. Further, a comparison of results published by other authors suggest that the ratio of solvolysis rates for pinacolyl and isopropyl

Table 1. Relative rates of solvolysis of (**1**) and pinacolyl OBs

Solvent	97% TFE	HCOOH	AcOH	50% EtOH	80% EtOH
Rate ratio [(**1**)/Pin]	3.6	2.6	0.73	0.89	0.62

toluene-*p*-sulphonates in trifluoroacetic acid is 160 to 190 (at 25°), and not 2800 (at 12°) as given by Shiner.[7] Finally, a plot of trifluoroacetolysis rates of isopropyl, pinacolyl and nine other secondary toluene-*p*-sulphonates against Taft's σ^* is linear for all eleven compounds. It is therefore suggested that the differences observed by Shiner require no other explanation than that they are due to nucleophilic solvent participation and that "hidden" ion-pair return is a myth. The low α-deuterium isotope effect observed in trifluoroacetolysis of pinacolyl *p*-bromobenzenesulphonate is due, it is proposed, to the distortion of the C—D bending modes by the bulky *tert*-butyl group. Trifluoroethanolysis of (**1**) also shows a very low α-deuterium isotope effect (1.107).[8]

The volumes of activation for formolysis of five secondary alkyl toluene-*p*-sulphonates correlate with Taft's σ^*, and all are negative. That all are more negative than that for the 2-adamantyl compound is taken to show that a k_s process is, to some extent, competitive with the k_c one. A similar study in methanol suggests that unhindered secondary compounds react exclusively *via* a k_s process, whereas hindered ones react by a k_c one.[9] The volume of activation for formolysis of 2-aryl-1-methylethyl toluene-*p*-sulphonate has been partitioned into ΔV_{Δ}^{+} and ΔV_s^{+} components. With the exception of the *p*-nitrophenyl and perhaps the *p*-chlorophenyl compounds, the points do not lie on the line found for the wholly aliphatic compounds.[10]

Two important contributions have appeared arguing against the application of the Sneen ion-pair mechanism[11] in the 2-octyl system in 25% and 30% dioxan. The first shows that for 2-octyl methanesulphonate in 25% and 30% aqueous dioxan the rates are linear in activity of sodium azide, and that the curvature observed by Sneen is due to the difference between a_{NaN_3} and $[NaN_3]$. Product plots are linear against both activity and concentration. However, because the slopes of the product plots differ from those of the rate plots, it is argued that the mechanism is indeed "merged" and of a type described in an inaccessible report of the Greek Atomic Energy Commission.[12] The second contribution describes the use of sodium perchlorate to buffer the ionic strength at the high value of 2.0M. Variation of sodium azide concentration from 0 to 0.31M causes a 4.7-fold increase in rate, and the ratio of azide to alcohol product is linear with azide concentration and allows evaluation of the ratio of $k_{azide}/k_{solvolysis}$. Use of this ratio to calculate $k_{observed}/k_0$ gives excellent agreement with the experimental results, whereas use of the Sneen scheme does not, unless the retio of $k_{-1}/k_s = 40$ (a value equivalent to a pure S_N2 reaction) or unless sodium azide has a different salt effect from sodium perchlorate. This was checked by using benzhydryl chloride in 80% acetone, where the ratio of azide product to alcohol is linear with azide concentration but the rate is independent of it. McLennan[13] therefore suggests that the Sneen scheme has nothing to offer.

Exchange rates for 1-arylethyl bromides with $Li^{82}Br$ in acetone give U or V-shaped plots against σ or σ^+. For *p*-methoxyphenyl and *p*-tolyl the ρ-value is −2.2 (σ^+) or −9 (σ), whereas for phenyl, *p*-bromophenyl and *p*-nitrophenyl ρ is +1 (σ). The reaction is bimolecular for all compounds studied, and it is suggested that the activating substituents react *via* ion-pairs, but that the others react *via* a classical S_N2 mode with formation of the new bond leading breakage of the old one, and hence a net negative charge on the carbon undergoing substitution.[14]

The role of stereoisomeric ion-pairs in the solvolysis of tertiary chlorides has been elegantly demonstrated by Grob and his co-workers in a study of *cis*- and *trans*-1-chloro-1,2-dimethylcyclohexane, *cis*- and *trans*-9-chlorodecalin, and *cis*- and *trans*-8-chloroperhydroindane in 80% ethanol and 50% acetone.[15–17] The results for the decalins are illustrated on p. 350.[16] For the two bicyclic cases, the ions were also generated by the π-participation route. In each case the *trans* ion-pair was the more stable of the two, a notable inversion of stability with respect to the ground state for the perhydroindane case.[17] Similar conclusions regarding isomeric tertiary ion-pairs have been drawn from studies on *p*-nitrobenzoate solvolysis of the *cis*- and *trans*-perhydroindan-8-ols and the related [4,4]spirononan-2-ols.[18]

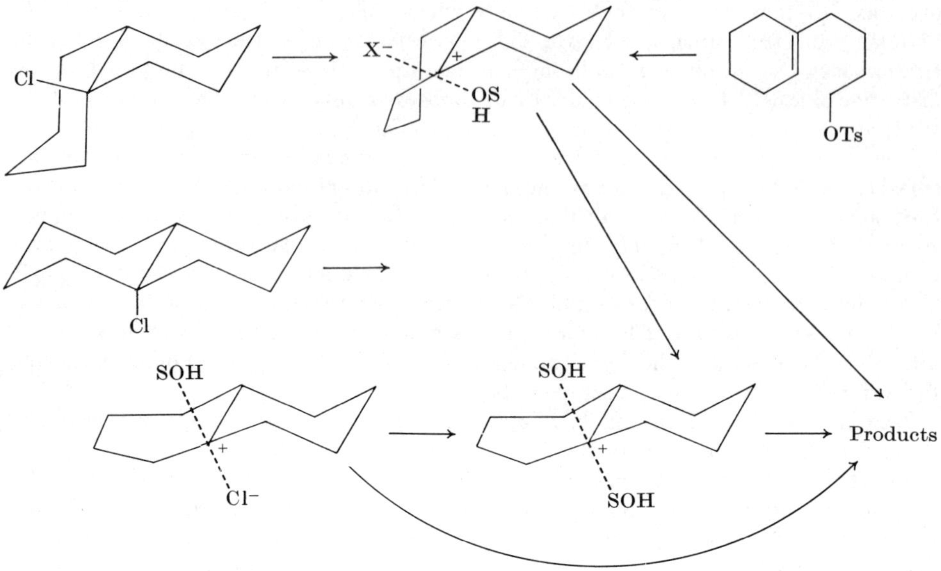

Cl X⁻ OS H OTs Cl SOH Cl⁻ SOH SOH Products

Isomeric, though symmetrically substituted, allylic ion-pairs are involved in the solvolysis of (**2**), (**3**) and (**4**). The *trans*-dinitrobenzoates (**2**) and (**3**) are both solvolysed about twice as fast as (**4**), and in no case is there significant evidence for allylic delocalization of charge. The [1-$^2H_1$]-compounds are converted into the alcohols in aqueous acetone without deuterium scrambling and with only 8% racemization for (**4**). The lack of double-bond

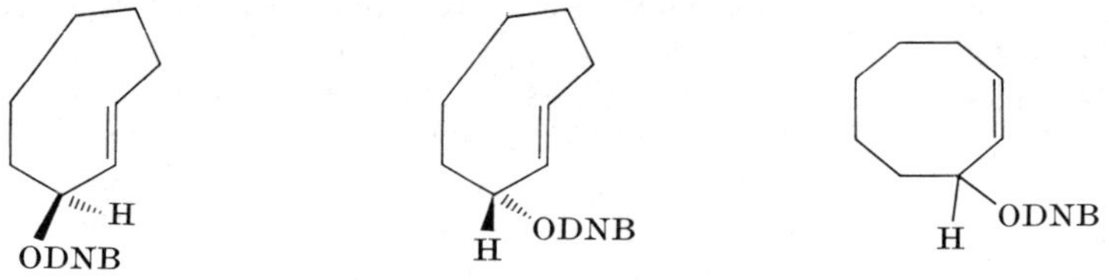

H ODNB (**2**) H ODNB (**3**) H ODNB (**4**)

participation is ascribed to conformational twisting and is shown to be related to ring size, decreasing from total participation for cyclopentenyl ($k_{\text{relative}} = 10^4$) to *cis*-cyclooctenyl ($k_{\text{relative}} = 1$).[19] The silver nitrate-assisted solvolysis of allyl chloride and bromide

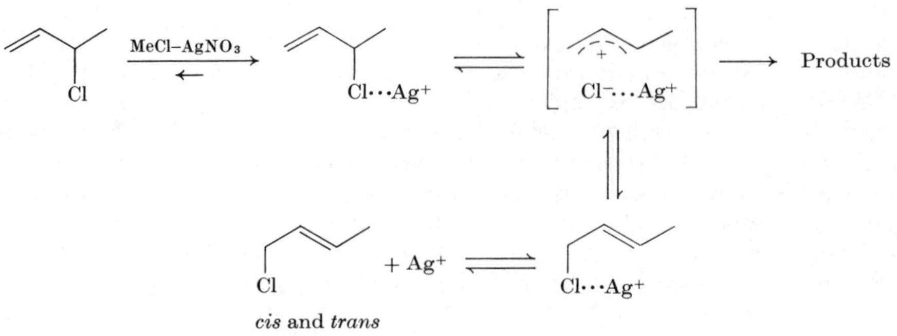

Cl MeCl–AgNO₃ Cl···Ag⁺ Cl⁻···Ag⁺ Products Cl + Ag⁺ Cl···Ag⁺ *cis* and *trans*

in acetonitrile is of order 1.5 with respect to silver nitrate. Reaction with silver perchlorate is 1600 times slower and it is suggested that rate-determining attack by nitrate ion on a complexed ion-pair [R^+ AgX NO_3^-] is involved.[20] The reversibility of such ion-pair formation has been demonstrated for the system shown in the scheme, where starting with 3-chlorobut-1-ene, the concentration of 1-chlorobut-2-ene is observed to build up to a maximum and then decline.[21] The intermediate formed by reaction of butadiene monoepoxide with dimethyl sulphoxide and 2,4,6-trinitrobenzenesulphonic acid exchanges with [D_6]-DMSO by an ion-pair mechanism.[22] α-(2-Indenyl)benzyl halides are now thought to react with amines by an ion-pair process rather than by S_N2–S_N2'.[23]

The presence of a special salt effect induced by cobaltic nitrate in the hydrolyses of benzhydryl and *p*-chlorobenzhydryl chloride in 80% acetone, together with common-ion depression and a recemization rate greater than k^0_{ext}, have been used to show that all three steps of the Winstein scheme are involved.[24] A comparison of the acetolysis, racemization and exchange rates for α-methylbenzyl chloride with the addition of HCl to styrene in acetic acid suggests that >98% of the R^+ Cl^- ion-pairs return to covalent chloride, and that racemization occurs at the intimate ion-pair stage rather than at the solvent separated one as with arenesulphonate solvolysis.[25] Similar studies, in acetone, on the ionization of α-methylbenzyl chloride catalysed by mercuric or cobaltous chloride show that the mercury reaction is both bi- and tri-molecular, while the cobalt reaction is only bimolecular, and the exchange rate is two-thirds of the racemization rate; this is what would be expected for a simple ion-pair mechanism.[26] Phenolyses of optically active α-methylbenzyl chloride in a series of phenols gives rise to alkyl phenyl ethers with retention of configuration, to *o*-alkylphenol also with retention, and to *p*-alkylphenol with inversion. Experiments with ^{18}O-labelled ether showed that under acid catalysis it rearranges with, respectively, 45% and 26% retention of oxygen label for *o*- and *p*-alkyl products, and with optical retention for *o*- and inversion for *p*-products; an ion-pair process is proposed.[27] The effects of pressure on the solvolysis rates of 1-aryl-1-methylethyl chlorides shows a correlation of $\Delta V_0^{\ddagger}$ with σ^+. The negative sign of $\Delta V_0^{\ddagger}$ shows that electrostriction at the transition state occurs, and the negative value of the change in $\Delta V_0^{\ddagger}$ with temperature shows that the volume of the transition-state solvation shell is less responsive to temperature-variation than is that of the ground state, suggesting a very tight, polar transition state.[28]

The hydrolysis of benzyl nitrates gives a non-linear correlation with Hammett's σ but otherwise resembles that of benzyl chlorides.[29] The rates of reaction of benzyl arenesulphonates correlate with σ and give the same value of ρ for aqueous ethanolysis, NaN_3 and LiCl in DMF, and KSCN in acetonitrile; it is therefore proposed that all the reactions proceed by way of an ion-pair S_N1 process.[30] The hydrolyses of benzyl bromides, benzylidene bromides, benzotribromides and benzotrichlorides have been studied. The rates are proportional to the water concentration, but the kinetic order varies with substituent from 2.3 for *p*-$NO_2C_6H_4CH_2Br$ to about 8 for the *p*-tolyl compounds; it is suggested that the first value is consistent with an S_N2 process, the high one with a LIM process, and values around 5 with a borderline one.[31] Concurrent solvolysis and rearrangement of the furylmethyl sulphinates (**5**) is unaffected by added arenesulphinate anion, and the rates vary with solvent Y with an m of 1.5; the formation of two different ion-pairs is proposed to account for the two different sulphones (**6**) and (**7**).[32] 1-(Toluene-*p*-sulphonyloxy)- and 1-bromo-12,13-benzo-16-chloro[10]-2,4-pyridinophanes are very unreactive as compared with benzyl compounds in solvolysis. The toluene-*p*-sulphonates (*syn* and *anti*) react stereospecifically but by attack at sulphur. In the presence of silver acetate, the bromides are acetolysed slowly, the *syn*-isomer with retention of configuration and the *anti* one with 3:1 inversion over retention; recovered bromide is a mixture of isomers, and an ion-pair process seems most probable.[33]

R–(furan)–CH_2OSO_2Ar (5) $\xrightarrow{EtOH}$ R–(furan)–CH_2SO_2Ar (6) + R–(furan, 3-SO_2Ar, 2-Me) (7)

The acetolysis of 2,4,7-trinitrofluoren-9-yl toluene-*p*-sulphonate, in the presence of aromatic π-donors, has been analysed in terms of the reaction of complexed and uncomplexed substrate. The rates of reaction of π-complexed substrate correlates with the energy of the highest occupied molecular orbital of the donor, and the extent of charge-transfer to the fluorenyl system at the transition state is estimated as equivalent to 11–14% of one electron.[34]

The non-linearity originally observed by Hughes in the exchange of α-methylbenzyl bromide with radiobromide ion, and ascribed to borderline behaviour, has been shown to be due to ion association of the lithium bromide in acetone. When lithium nitrate is used as a buffer salt, a plot of [LiBr] against rate is linear with a zero intercept when the ionic strength is kept constant, clearly indicating that only a second-order reaction is involved.[35]

The exact form of the equation used to interpret the conductance data for salt dissociation has been shown to have a large effect on the estimate of the activity of ion-pairs in nucleophilic displacements. In particular the Fuoss–Onsager treatment leads to negative values for the reactivity of the ion-pair in the Acree equation:

$$k_{\text{total}} = k_{\text{ion}}\alpha + k_{\text{pair}}(1 - \alpha)$$

particularly when the degree of association is small. Use of the Fuoss–Hsia equation in the form proposed by Fernandez and Prini leads to positive values for k_{pair} and, although $k_{\text{ion}} > k_{\text{pair}}$, the ratio is only of the order of 3–10 for alkali metal and n-Bu_4N^+ halides.[36]

A reaction field theory developed for the calculation of solvent effects on equilibria has been applied to solvent effects on the $\Delta G^{\neq}$ for the solvolysis of *tert*-butyl chloride. All the results fit well if the dipole moment of the transition state is taken as 7.4 D, corresponding to a charge development of 0.67 unit on the *tert*-butyl cation. A better fit is obtained if the non-polar and the polar solvents are treated separately, leading to estimated charges of 0.64 and 0.8 unit, respectively.[37] *tert*-Butyl chloride, benzyl chloride and other tertiary and allylic chlorides react rapidly and quantitatively with sodium iodide in benzene in the presence of ferric or mercuric chloride; it is suggested that intimate Lewis-acid catalysis and exchange in an ion-quartet are involved.[38]

An analysis of isotope effects and the ratio of *syn/anti* elimination in the solvolysis of *cis*- and *trans*-[2-$^2H_1$]cyclopentyl *p*-bromobenzenesulphonate has been used to suggest that reaction in ethanol takes place at the intimate, in trifluoroethanol at the solvent-separated, and in dioxan at both the intimate and the solvent-separated ion-pair stage.[39]

The correlation of the rates of hydrolysis of substituted benzhydryl bromides against $\Sigma\sigma^+$ gives two parallel lines, one for the 3- and 3,3′-substituents and another for the 3,5-substituents.[40] A stability selectivity relation has been observed for the reaction of alkyl chlorides with *p*-nitrobenzoate anion.[41]

Solvent and Medium Effects[42]

Phase-transfer catalysis has been reviewed,[43] and so has the complexing effects of macrocyclic ligands on first- and second-row cations.[44] The synthesis of macrocyclic bicyclic polyethers, analogues of the cryptates, has been reported.[45] Both *trans-anti-trans-* and *trans-syn-trans*-dicyclohexyl-18-crown-6 polyethers give stable complexes with sodium bromide.[46] The use of "crown" ethers as phase-transfer catalysts has been reported to be very successful for S_N2 reactions, with nucleophilicities falling in the order I > Br > Cl > F.[47] The use of phase-transfer catalysis for the preparation of alkyl fluorides from other halides and methanesulphonates,[48] and for the synthesis of primary and secondary dialkyl and alkyl aryl sulphides has been described.[49] Phenacyl kojate has been shown to complex sodium halides in a manner similar to "crown" ethers, and attention is drawn to a number of other natural compounds that act similarly.[50] The enhanced reactivity of the anions solubilized in benzene by complexing of the cations with "crown" ethers has been startingly exemplified for acetate,[51] substituted acetate[52] and fluoride ions;[53] the effect has been ascribed to the result of having "naked" anions.[51,53] That potassium bromide and iodide are more reactive towards *n*-butyl *p*-bromobenzenesulphonate in acetone than the corresponding $n\text{-Bu}_4N^+$ salts when they are complexed by dicyclohexyl-18-crown-6 has been shown to be due to the higher degree of dissociation of the complexed salts.[54]

The validity of normal salt effects as medium effects has been questioned on the basis of results for 1-methylheptyl methanesulphonate in 30% dioxan. Plots of rate against salt concentration are curved, but they are linear against salt activity. Modified *b* values are derived for the four salts NaBr, $NaNO_3$, $NaClO_4$ and $LiClO_4$, and are 1.75, 0.94, 0.28 and 0.52, respectively. Because these values differ from each other, and because of the linearity of rate *vs* activity, it is suggested that the effects are specific and that the normal practice of dividing the effect of a nucleophilic salt into a "normal" salt effect and a nucleophilic effect is invalid.[55]

Y and *N* values have been determined for hexafluoropropan-2-ol, and show it to be more ionizing and less nucleophilic than formic acid, and not much different from trifluoroacetic acid, except that, as a solvent, it is much less acidic than the latter, facilitating kinetic measurements. By making the assumptions that $k_{\text{titrimetric}} = k_c$ in trifluoroacetic acid (TFA), the ratio

$$\frac{[k_{\text{ROTs}}/k_{\text{2-AdOTs}}\ (\text{solvent})]}{[k_{\text{ROTs}}/k_{\text{2-AdOTs}}\ (\text{TFA})]}$$

(where 2-Ad = 2-adamantyl) has been used to probe for solvent-assisted rate k_s. The values of this ratio for hexafluoropropan-2-ol, with cyclohexyl and isopropyl as R groups, are 0.62 and 0.58, respectively, indicating a limiting k_c process for each; the corresponding values in acetic acid are 28 and 472, showing the extent of nucleophilic participation by solvent.[56] The interpretation given by Nordlander, and discussed last year,[57] for the above ratio when R = pinacolyl has been criticized. In particular, the suggestion that ion-pair return from 2-adamantyl toluene-*p*-sulphonate increases with decreasing solvent *Y* is in direct contrast to Winstein's classic results; it is suggested that the alternative explanation of a k_s component in the pinacolyl case is the valid one, a result in line with variation of the ratio with β-methyl-substitution in R.[58] Plots of log *k* (1-adamantyl bromide) against log *k* (RX) for reactions in aqueous ethanol and trifluoroethanol fall on the same line for substrates where there is no nucleophilic participation by solvent (R =

2-adamantyl or 2-norbornyl); but give two lines where a bimolecular component may be involved ($RX = Bu^tCl$, cyclo-$C_6H_{11}OTs$ or Pr^iOBs); it is suggested that the failure of the mY correlation with Bu^tCl as a model is due to this bimolecular component, the solvent probably functioning as a base, and that mY correlations in aqueous ethanol are satisfactory because water and ethanol have about the same N value, whereas they break down in aqueous trifluoroethanol because of the large differences in N between water and trifluoroethanol; this means that the second term in the $mY + lN$ equation becomes significant.[59]

The use of tetra-alkylammonium tetra-alkylborate salts of m.p. $< 50°$ as polar non-interacting solvents has been proposed.[60] The scale of reactivity of Et_4N^+ halides in molten n-$C_6H_{11}N^+Et_3$, n-$C_6H_{11}B^-Et_3$ towards methyl toluene-p-sulphonate has been shown to be compressed as compared with solvents acetone and DMF, although the order $Cl > Br > I$ is the same; it is suggested that this is due to the greater energy of interaction of the small X^- ions with the R_4N^+ of the solvent, and that their reactivity is more akin to those of associated ions, as in the normal Acree treatment.[61]

The effects of micelles on the nucleophilicity of deuteroxide ion in the presence of alkyldimethylsulphonium salts has been studied.[62] Common-ion depression is observed in the reaction of sodium 4-*tert*-butylphenoxide with allyl bromide in DMF.[63] Solvent effects on the reaction of benzyl bromide with sodium thiosulphate in aqueous alcohols, acetone or dioxan show that rate is inversely proportional to dielectric constant and is slowed down by increased ionic strength.[64] Bi-(1,3-dioxan-2-yl) is as effective a catalyst for the reaction of sodium n-butoxide with benzyl chloride in THF as are dibenzo-18-crown-6 and tetraglyme.[65] Solvent effects on the rate of substitution of chloride and nitrate by sodium azide on the nitrate of ethylene chlorhydrin have been reported; nitrate is replaced from two to three times faster than chloride.[66] The use of enthalpies of transfer for hydroxide ion to evaluate the degree of charge of the transition state for the reactions of OH^- with methyl iodide and ethylene oxide have been reported; the method shows that an important negative pole is present at the transition state for the latter reaction, comparable to that found in ester hydrolysis.[67] The reaction of potassium trifluoroethoxide with n-hexyl iodide in THF is inhibited by added trifluoroethanol; the variation with concentration shows that two molecules of alcohol solvate each alcoholate anion.[68] Hydrogen peroxide facilitates the hydrolysis of *tert*-butyl chloride by lowering both ΔH^{+} and $T\Delta S^{+}$, largely owing to a destabilizing influence on the ground state, like organic and other co-solvents, but with hydrogen peroxide the effect of ΔH^{+} just dominates that on $T\Delta S^{+}$. However, poly(vinylpyrrolidone) slows the reaction slightly, behaving like other organic co-solvents and apparently holding water in a polymer mesh without grossly modifying its structure, rather than as previously postulated by immobilizing it in an ice-like structure.[69]

The equilibrium transfer of a methyl group to solvent from methyl trifluoromethanesulphonate has been measured and used to relate solvent nucleophilicity to that of HMPT. Trimethyl phosphate and dimethyl ether stand out as being 10^4 times less nucleophilic than HMPT and about 3 times less than $CF_3SO_3^-$.[70] The nucleophilicity values (N) for various aqueous trifluoromethanol mixtures have been determined from the equation

$$N = \log(k^{\text{MeOTs}}/k_0^{\text{MeOTs}}) - 0.30Y$$

reported in 1972.[71] The low N value (–2.59) for 97% trifluoroethanol has considerable use when one wishes to enhance a participation over a solvent-assisted solvolysis route.[72] Solvation effects on the $HgCl_2$-catalysed hydrolysis of 4-methoxybenzyl monophthalate in water and in aqueous acetonitrile have been measured.[73] It has been pointed out that the use of solvent variation to alter the free-energy change during a proton transfer reaction does not lead to normal Brønsted β-values; thus the pK_a's of acids may be modified by solvation change and this may not show up at the transition state; clearly such an effect will vary with the charge type of the species involved.[74]

The enthalpies of transfer for the single ions Ba^{2+}, Zn^{2+}, Cd^{2+} and $CF_3SO_3^-$ from water to five other solvents have been determined.[75] The influence of aqueous micelles and of solvent restrictions on ligand exchange for vitamin B_{12a} in benzene and in water have been reported.[76] Heats of transfer of salts from water to micellar solutions of CTAB, DTAB and sodium dodecane-1-sulphonate have been measured.[77] Conductance measurements of solvation and ion-pairing in the alkali-metal *tert*-butoxides and methoxides in DMSO solution show that for the butoxides the ion-pair equilibrium constants are 10^8, 10^6, 270 and 200 for the metals Li, Na, K and Cs, respectively. The methoxide–metal ion-pairs tend to be strongly associated.[78] The effects of added *N*,*N*-dimethylacetamide, *N*-methylacetamide, DMSO, and HMPT on the conductance of n-Bu_4N^+ picrate in *o*-dichlorobenzene have been measured.[79] The solvating ability of DMSO, dioxan, methanol and acetonitrile have been discussed in the light of determinations of the enthalpies of transfer of H^+ and OH^- from water to them.[80] Solvation enthalpies have been measured for tetra-alkylammonium bromides in water–dioxan mixtures; the results are interpreted to suggest that the change from ethyl to *n*-butyl increases the degree of structure of the water.[81] Thermodynamic properties for the transfer of single ions between protic and diploar aprotic solvents have been reported,[82] as have medium effects for single ions.[83] The activity coefficients of formate and hexanoate ions in aqueous sodium chloride have been determined.[84] The temperature-dependence of the critical micelle concentration for 26 ionic and 48 non-ionic surfactants has been measured and the ΔH and ΔS for micelle formation have been derived.[85] Solubility studies on ion-pairs in organic solvents have been carried out.[86] See also refs. 300 and 301 (p. 361).

Isotope Effects

The exact integrated rate equations for the scheme shown have been calculated. The result is not a simple first-order expression even when k_4 is greater than all other rate

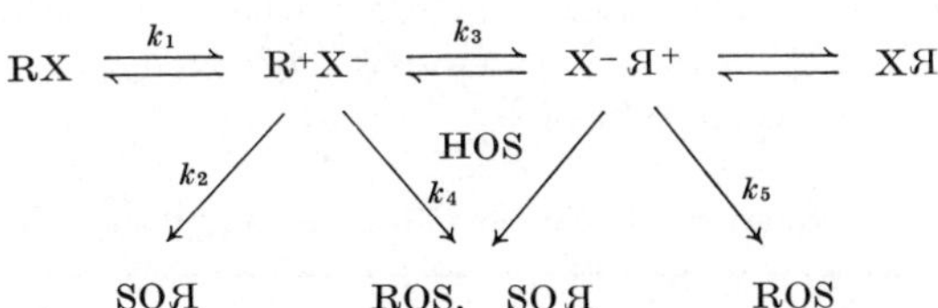

constants. An analysis of the system for isotope effects has also been carried out, and it is shown that no simple relation emerges.[87] Model calculations of the kinetic isotope effects (k.i.e.) in S_N reactions of benzyl bromide have been carried out, and the variations with bond order to entering and leaving groups (n_1 and n_2, respectively) and total bond order (n_t) have been determined. For $n_1 = 0.4n_t$, the $^{12}C/^{13}C$ k.i.e. and the k_H/k_D vary almost linearly with n_t, being 0.98 and 1.18, respectively, when $n_t = 0$ and 1.05 and 1.00 when $n_t = 1$. While the $^{12}C/^{13}C$ k.i.e. varies markedly with the ratio n_1/n_2, and this variation is greatest when $n_t = 0.6$, the $^1H/^2H$ k.i.e. is almost invariant in this respect and depends only on the bending force constants. It is suggested that the temperature-variation of the

$^{12}C/^{13}C$ k.i.e. may be the best probe for a borderline mechanism, for this is predicted to decrease with temperature when $n_t = 1$ but to increase with temperature for $n_t = 0.1$.[88] Partition function ratios for isotopically substituted methyl halides have been calculated.[89]

β-Deuterium kinetic isotope effects on the reaction of cyclohexyl toluene-*p*-sulphonate and bromide with tetra-*n*-butylammonium acetate in acetone are of the same size as those for reaction with the corresponding chloride, thiophenoxide, phenoxide and azide salts. The magnitude is such as to suggest that the transition state is very loose and rather S_N1-like.[90]

α-Deuterium k.i.e.'s for the solvolysis of a series of substituted benzyl nitrates vary from 0.988 for *m*-CF_3, through 1.114 for H, to 1.181 for the 2,6-dimethyl compound. The magnitudes of the k.i.e.'s are larger than those observed for the corresponding chlorides, but the difference decreases for electron-withdrawing groups. It is suggested that the transition state is therefore more ionic for the nitrates than for the chlorides, with the *m*-CF_3 reacting by an S_N2 process and the others by a borderline one.[91] α-Deuterium and $^{12}C/^{14}C$ k.i.e.'s have been determined for the hydrolysis of benzyl chloride, 4-methylbenzyl chloride and 1-methylheptyl *p*-bromobenzenesulphonate in the presence of sodium azide; the effects have been determined separately for the alcohol and azide products, and are significantly different between them in each case (see Table 2). It is considered that these

Table 2. $^{12}C/^{14}C$ and H/D kinetic isotope effects in the hydrolysis of $PhCH_2Cl$, *p*-MeC_6H_4Cl and 2-octyl benzenesulphonate[92]

	$PhCH_2OH$	-N_3	*p*-MeC_6H_4Cl	-N_3	2-OctOH	-N_3
$^{12}C/^{14}C$	1.085	1.130	1.061	1.096	1.063	1.105
H/D	1.003	1.033	0.996	1.004	1.097	1.106

figures argue strongly for competitive S_N1-S_N2 reactions rather than for a Sneen-type scheme.[92] On the other hand, evidence from $^{35}Cl/^{37}Cl$ k.i.e.'s on the borderline solvolysis of 4-methoxylbenzyl chloride in the presence of sodium azide has been argued to indicate the operation of a Sneen scheme. Classical mixed reaction should lead to the k.i.e. tending to 1 as the azide concentration increases, whereas a Sneen type scheme tends to the limit of the k.i.e. of the ionization step as the azide ion concentration increases. The observed values are shown in Table 3.[93]

Table 3. $^{35}Cl/^{37}Cl$ kinetic isotope effects for 4-methoxybenzyl chloride hydrolysis[93]

NaN_3 (M)	0	0.1	0.15	0.25
$10^4\,k$ (sec.$^{-1}$)	2.38	5.87	7.27	9.76
k.i.e.	1.00786	1.00936	1.00985	1.01049

Further calculations on the $^{35}Cl/^{37}Cl$ k.i.e. in the methanolysis of *tert*-butyl chloride have been reported. The model chosen for the transition state had the carbon—methyl and carbon—chlorine bond lengths and the *z*-co-ordinate of the methyl groups as parameters. The C—Cl bond was along the *y*-axis. The geometry and force-field were then varied to fit the k.i.e. at 10° and also to fit the temperature variation from 10° to 60°. The best-fitting model gave excellent agreement with the observations and predicts a transition-state geometry that is markedly non-planar at the central carbon with a C—Cl distance of 1.89 Å and the methyl groups with a *z*-displacement of –0.16 Å compared to their initial position of –0.47 Å.[94]

A revised value for the fractionation factor, *l*, for L_3O^+ in water–deuterium oxide mixtures has been proposed.[95]

Neighbouring Group Participation

Participation by Ether and Hydroxyl Groups

That a limit to the acceleration caused by O-3 participation exists has been demonstrated by using the isomeric cyclopentane derivatives (**8**) and (**9**). Whereas variation of the aryl group in (**9**) and in cyclopentyl sulphonates results in a linear $\rho\sigma$ plot for formolysis, the

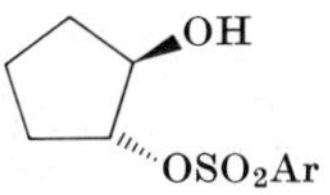

OH OSO$_2$Ar

(**8**)

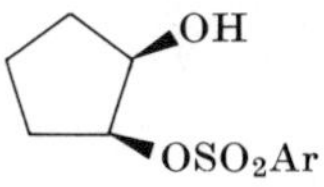

OH OSO$_2$Ar

(**9**)

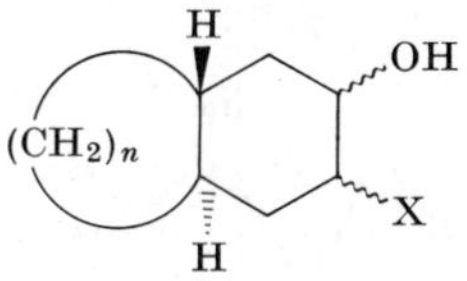

H OH (CH$_2$)$_n$ X H

(**10**)

X = F, Cl or Br

rate for (**8**) is linear to *p*-BrC$_6$H$_4$ and then drops for *m*-CF$_3$C$_6$H$_4$ and *p*-NO$_2$C$_6$H$_4$. The effect is more noticeable when k_{trans}/k_{cis} is plotted against σ, the ratio rising from 208 (*p*-MeO) to 253 (*p*-Br) and then falling to 155 (*p*-NO$_2$).[96] The rates of ring closure to epoxide from the halohydrins (**10**) has been studied as a function of *n* (= 2, 3 or 4) and of orientation of OH and X. The diaxial compounds react from 1.3×10^3 to 7.4×10^3 times faster than the diequatorial ones, and faster for $n = 2$ than for $n = 4$ (17:1).[97] *myo*-Inositol 1,4-di(toluene-*p*-sulphonate) reacts with sodium azide in 2-methoxyethanol to give the *allo*-1,5-diazido-compound, by way of epoxide closure and diaxial opening. On the other hand, the corresponding tetra-acetate affords the *muco*-1,3-diazido tetra-acetate. Other examples of O-3 participation in inositol di(toluene-*p*-sulphonates) have also been reported.[98]

Condensation of benzaldehyde with α-chlorophenylacetonitrile leads to a mixture of the *E* and *Z* glycidic nitriles, whose composition varies with the solvent and base employed. Thus sodium *tert*-butoxide in benzene gives an 85:15 *E*–*Z* mixture, while in HMPT a 2:98 mixture is formed. Independent preparation of the intermediate *erythro*- and *threo*-halohydrins and their submission to the reaction conditions, showed that the condensation step is *always* reversible, but more so in HMPT, and also when small cations are used Li > Na > K. The *E*/*Z* ratio depends on the rate of ring closure of the intermediate β-chloroalkoxides.[99] With potassium isoproxide in propan-2-ol, dichloro- and dibromo-acetonitrile react with ketones to give α-haloglycidic nitriles and imidoyl esters. At –78°, the intermediate bromohydrins are isolable when lithium ethoxide is used as base, or when aldehydes are employed, whereas the chlorohydrins are stable up to –20°. The rate of closure therefore depends on the halogen, the metal ion, and the degree of substitution at the alkoxide.[100] Small amounts of O-3 participation are observed in the alkaline hydrolysis of 2α-hydroxy-3α-tosyloxypinane, but the major products are formed by ring contraction and methyl migration.[101] That nearly the same product mixture is formed from the reactions of 1-halopropan-2-ols and 2-halopropan-1-ols with KOAc at 120° suggests that the major path is *via* propylene oxide.[102]

The olefinic epoxide (**11**) reacts with participation by the epoxy-group to form (**12**) when treated with iodine in aqueous KI solution.[103] On the other hand, solvolysis of (**13**) occurs with participation of carbon rather than of oxygen, as indicated.[104] There is no

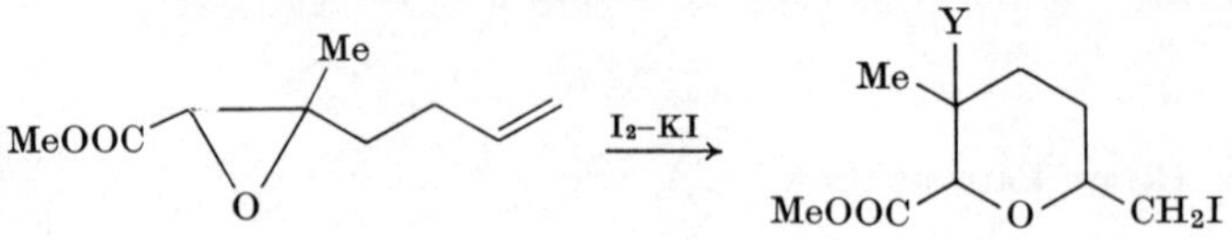

(**11**) (**12**)

Y = OH or I

evidence for participation by the epoxide in either *cis*- or *trans*-(**14**), even though the epoxide ring is opened during solvolysis.[105] The *endo*-oxide bridge in (**15**) does, however,

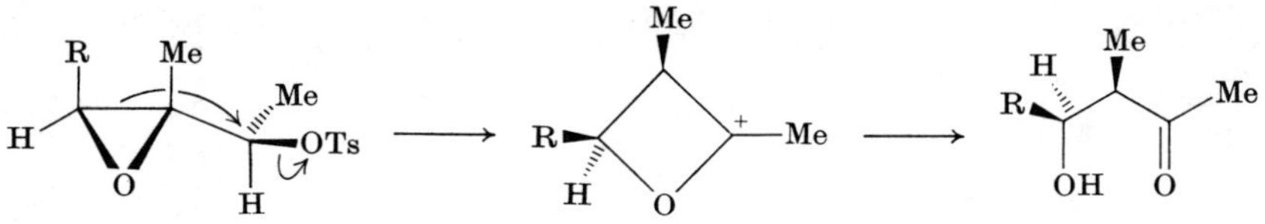

(**13**)

R = H or Me

participate strongly, as shown by the acceleration of 50 over (**16**), and more clearly by the activation parameters: $\Delta H^{\neq} = 19.2$ and 30.8 kcal mol^{-1}, and $\Delta S^{\neq} = -18.0$ and 7.7 e.u. for (**15**) and (**16**), respectively.[106]

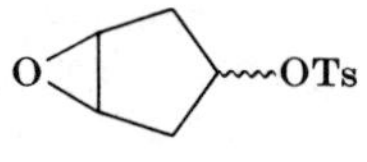

(**14**) 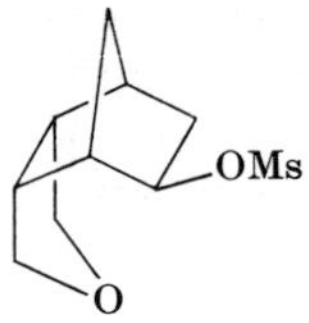 (**15**) 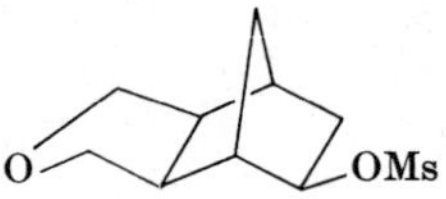 (**16**)

In agreement with the Ruzicka hypothesis, the major effect controlling the relative rates of cyclization in (**17**) as *n* increases from 6 to 10 is in the $\Delta S^{\neq}$ term.[107] Whereas O-5 participation might be expected to be preferred to O-6 in the cyclization of (**18**) as is

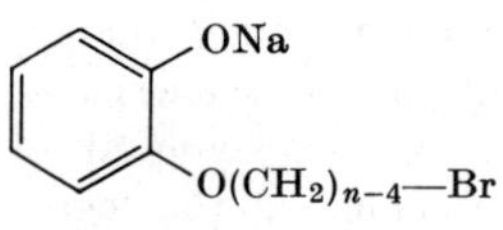

(**17**) 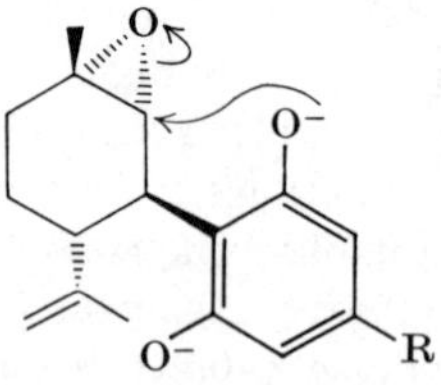 (**18**) 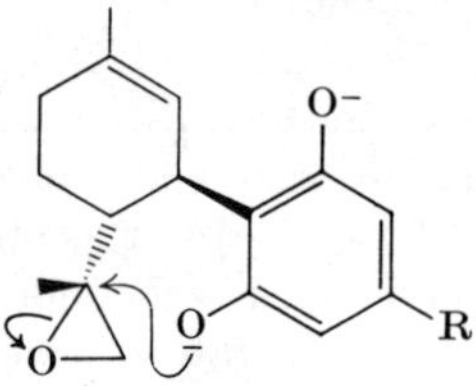 (**19**)

found, that O-6 should occur to the exclusion of O-7 in the cyclization of (**19**) indicates an unexpected preference for six-membered ring formation despite the difficulty of attack at the tertiary centre.[108] *o*-Hydroxyacetophenone reacts with iodine in base to form coumaran-3-one, and the rates of cyclization of the intermediate ω-haloacetophenones for chlorine and bromine have been measured. Comparison with the alkylation rate for reaction of *o*-hydroxyacetophenone with ethyl bromoacetate suggests that the effective phenoxide concentration for O-5 participation is about 400M.[109] Other work involving oxygen-participation has also been reported.[109a]

Participation by Thioether and Thiol Groups

The thiirane group in (**20**) also participates exclusively by carbon migration [cf. (**13**) above], to give 3-acetoxythietane as the sole product of acetolysis. The rate of acetolysis is about 70 times that of cyclopropylmethyl chloride, indicating the additional electron release due to sulphur.[110] Whereas (**21**) and (**22**) react with substantial acceleration

(**20**) (**21**) (**22**) (**23**)

[5.9×10^6 (R = H) and 5.4×10^7, respectively] and exclusively by S-3 participation even when R = Ph, no S-4 participation is detectable in the solvolysis of (**23**).[111] A suggestion that neighbouring-group participation should be divided into S_N2-participation, in which the charge is mainly on the neighbouring group, and S_N1-participation, in which the charge is mainly on the carbon, but with solvation by the neighbouring group, has been investigated by using additions to (**24**) as a model; acids add to give the ions (**25**;

(**24**) (**25**)

Z = H); formic acid adds more slowly than to the corresponding methoxy-compound, but more rapidly than to allylbenzene, and an S_N1 type of S-5 participation is involved. Iodine also adds to (**24**) to give (**25**; Z = I), although not to *o*-allylanisole or to allylbenzene, again indicating participation by the sulphur at the carbonium ion stage rather than by addition to the double bond followed by S_N2 type displacement by sulphur.[112]

Participation by Halogen

Further ^{13}C-NMR work on participation by halogen has shown that the ions (**26**; X = Br, Cl or I) are all bridged and symmetric, as is the 1,2-dimethyl ion (**27**): the same mixture of *cis*- and *trans*-isomers is formed independently of the stereochemistry of the starting 2,3-dibromobutane. The tetramethyl ion (**28**) is an interconverting mixture of

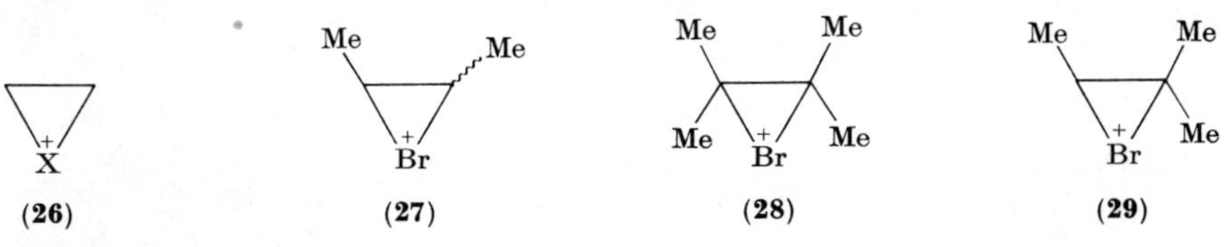

(**26**) (**27**) (**28**) (**29**)

bridged and open tertiary ions, while the trimethyl one (**29**) exists as an equilibrating unsymmetrical bridge and the open tertiary carbocation. Further results on the five-membered ring halonium ions are in accord with the results reported last year.[113] *Ab initio* calculations on the system $C_2H_4X^+$ (X = F or Cl) indicate that the staggered ion (**30**) has no barrier to conversion into the halonium ion (**31**), which is a local minimum; (**30**) is also less stable than the eclipsed ion (**32**), which is converted into (**33**), also without

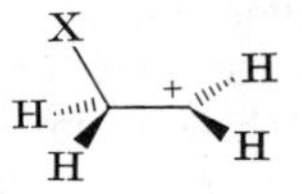

(**30**) 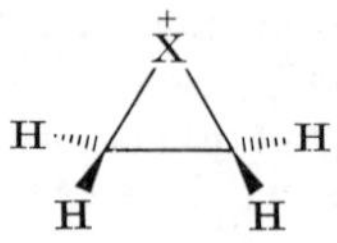 (**31**) 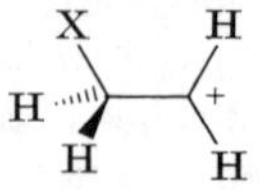 (**32**)

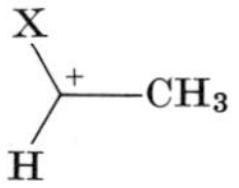

(**33**) 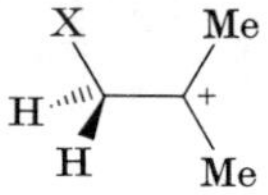 (**34**)

any barrier. For both fluorine and chlorine, (**33**) is a local minimum and is more stable than (**31**) by 19.6 and 4.5 kcal $mole^{-1}$, respectively for X = F or Cl. The geometry and charges on the bridged ions (**31**) are found to be similar to those in oxirane and thiirane. The effects of methyl substitution were also calculated. For X = F, (**33**) remained the most stable form, although the tertiary ion (**34**) was of comparable stability. For X = Cl, substitution of two, three or four methyl groups makes the tertiary ion (**34**) more stable than (**33**), but less stable than (**31**) when three or four methyl groups are present.[114] Participation by chlorine and bromine has been investigated by ^{14}C-labelling experiments in the deamination of 2-chloro- and 2-bromo-ethylamine. In all the five solvents employed, participation by bromine was greater than that by chlorine.[115] Similar experiments with *p*-bromo- and *p*-nitro-benzenesulphonate as leaving group gave similar results, with more participation in acetic than in formic or trifluoroacetic acid.[116]

Deamination of (**35**) and (**36**) has been investigated at 0° and 50°; reaction of (**35**) takes place by competitive X-3 participation to give (**37**) or ring-expansion to give (**38**), and reaction of (**36**) proceeds either *via* the tertiary ion (**37**) or by hydrogen-shift affording (**39**); the proportion reacting by each path varies with temperature and halogen. On the assumption that lower temperature should increase selectivity, it is argued that decrease

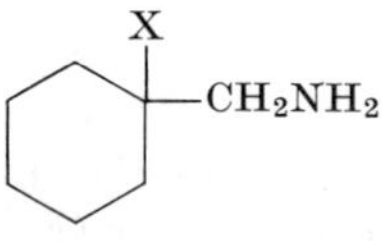

(**35**) 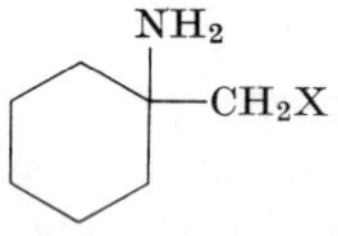 (**36**)

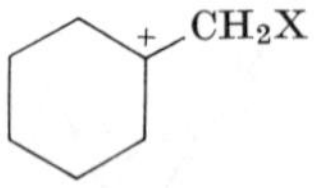

(**37**) 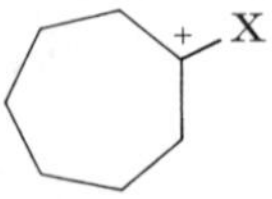 (**38**) 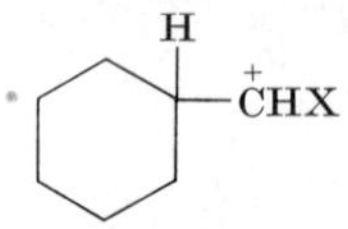 (**39**)

X = Cl or Br

in the fraction of products from (**37**) in the deamination of (**35**; X = Cl) at 50° means that the tertiary cation is more stable than the α-chloro-carbocation (**38**; X = Cl). Similarly the increase in the proportion of products formed *via* (**37**) from (**35**; X = Br) indicates that (**38**; X = Br) is more stable than (**38**; X = Cl).[117] When (**40**) is treated with halogens, the two halolactones (**41**) and (**42**) are formed. Because of the difference in nucleophilicity of COOH and COOMe, it is argued that the ratio of (**41**) to (**42**) shows the selectivity of

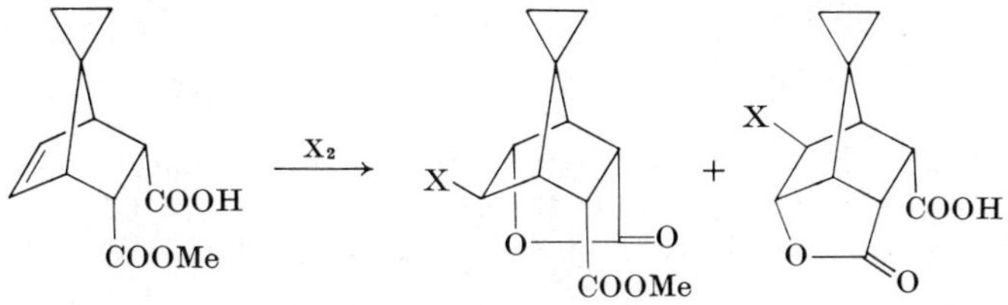

(**40**) (**41**) (**42**)

the intermediate cation, and hence the degree of bridging, if any, by the halogen. In both acetonitrile and methanol, the selectivity is found to be a linear function of the electronegativity of the halogen concerned, and hence a symmetrical bridged halonium ion is proposed.[118] The enthalpies of solvation of the three- and five-membered cyclic bromonium ions in FSO_3H have been determined by a procedure involving ion cyclotron resonance measurements of their gas-phase bromide affinities.[119] Competing substitution mechanisms are evident in the conversion of bromohydrins into dibromides by PBr_3 and $SOBr_2$; for example, mixed S_N1 and Br-3 participation could explain the fact that (+)-(1*S*, 2*S*)-*trans*-2-bromocyclohexanol gives the *trans*-1,2-dibromocyclohexane with only 20% of optical retention from treatment with PBr_3 and about 80% optical retention from reaction with $SOBr_2$.[120] The thermal racemization of *trans*-1,2-bromocycloalkanes and some open-chain dibromides has been studied kinetically at several temperatures; the results parallel the iodide ion-catalysed debromination, and intermediacy of ion-pairs involving Br-3 participation is proposed.[121]

Acetolysis of 3-bromo[1-^{14}C]propyl *p*-nitrobenzenesulphonate occurs with 4.3% of label shift to the 3-position, but none to the 2-position, showing that about 8% of Br-4 participation occurs.[122]

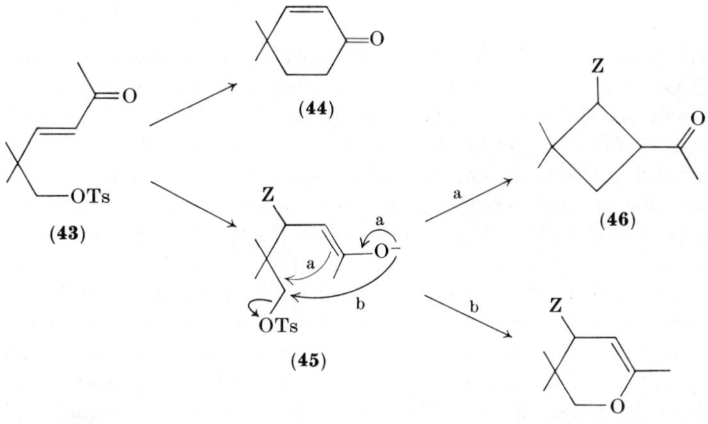

(**43**) (**44**) (**45**) (**46**) (**47**)

Participation by Carbonyl Groups

The keto toluene-*p*-sulphonate (**43**) reacts with methoxide ion by enolization and by addition followed by enolate ring closure at carbon (path a) affording (**44**) and (**46**; Z = OMe). With KCN, however, the intermediate enolate (**45**) ring-closes both at carbon and at oxygen to give a 1:2 mixture of (**46**; Z = CN) and (**47**; Z = CN).[123] On the other hand, (**48**) participates exclusively through oxygen, to give (**49**).[124]

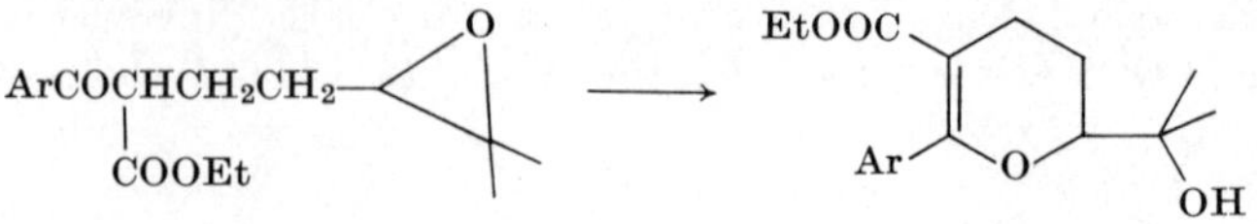

ArCOCHCH$_2$CH$_2$— (epoxide), COOEt → EtOOC, Ar, O, OH

(**48**) (**49**)

Participation by Ester, Carboxyl and Amide Groups

The 2-chloro-amine hydrochlorides (**50**) react with sodium carbonate in DMSO to give the cyclic urethanes (**52**); reaction probably occurs at the solid base and goes by way of the carbamate (**51**).[125] A similar mechanism has been ruled out for the conversion of (**53**) into (**54**) with NaHCO$_3$ in DMSO, for K$_2$CO$_3$ gives no reaction and sodium hydride followed by CO$_2$ gives no (**54**).[126] The neighbouring carboxyl group interacts with the sulphoxide when *o*-carboxyaryl sulphoxides are treated with acetic anhydride.[127] The halolactonization of cyclohexeneacetic acids has been studied.[128]

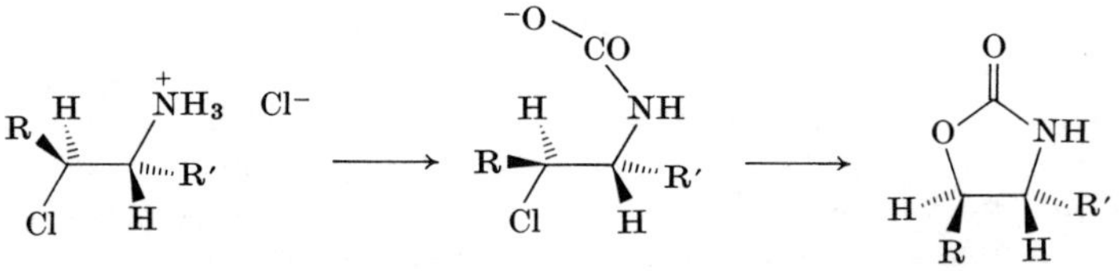

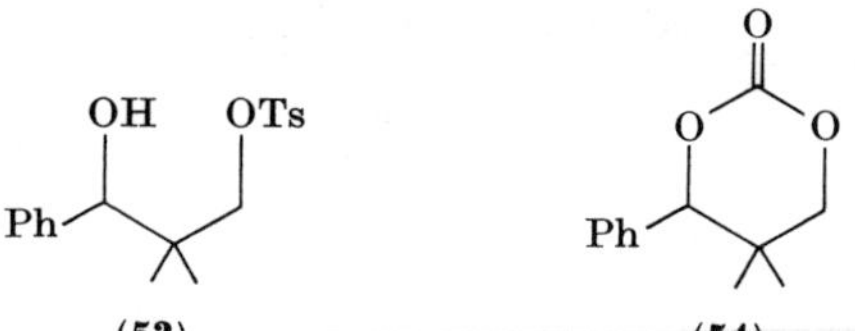

(**50**) (**51**) (**52**)

(**53**) (**54**)

Tri-*O*-acetyl-1,6-anhydro-*β*-D-galactopyranose reacts with trifluoromethanesulphonic acid by competitive *cis*-AcO-5 participation of the 3-acetoxy-group at C-4 and by *trans*-AcO-5 participation of the 2-acetoxy-group at C-3. Further transformations occur which lead to ring contraction to the furanose sugar with retention of configuration at C-4 and C-5.[129] Similar transformations occur in the *β*-D-allopyranose series.[130] ^{18}O-Labelled acetate has been used to show that AcO-5 and -6 participation occur in the conversion of (**55**; $n = 2$ or 3) into (**56**), and subsequently as shown; when $n = 4$, no participation occurs.[131] Participation by the nitrogen in ArSO$_2$—NH—CH$_2$CH$_2$Y (Y = Br or Cl) occurs with a ρ for cyclization of -0.58 when Y = Br.[132] The oxime group in (**57**) participates through nitrogen to give the azetine 1-oxide (**58**).[133] The ratio of oxazoline to

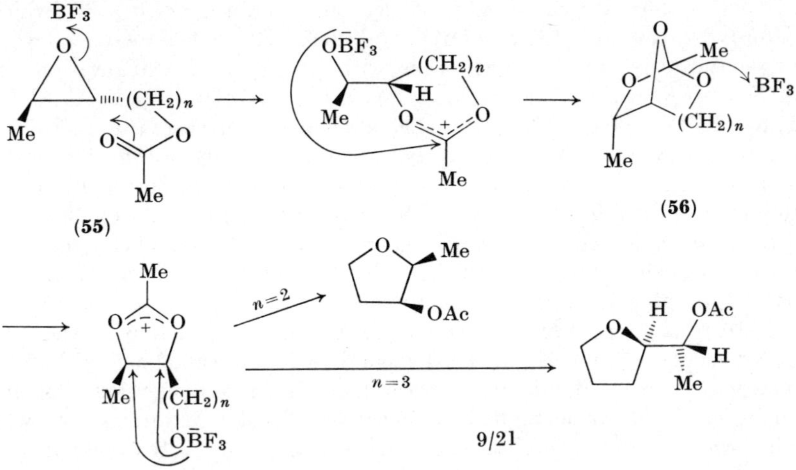

(55)

(56)

9/21

oxazine formed when (**59**) is treated with bromine depends on the stability of the intermediate cation.[134] It is suggested that the semicarbazide group participates in the ring

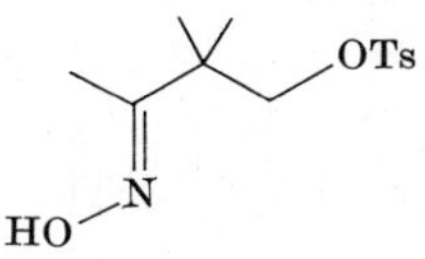

(57) 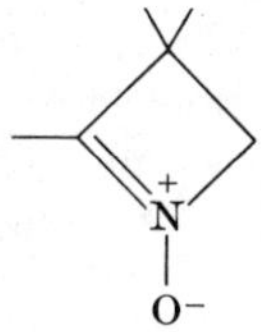 (58) 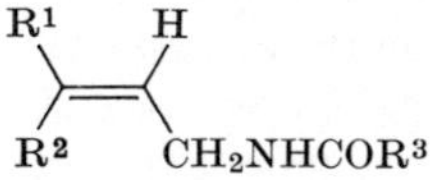 (59)

opening when the semicarbazone of 2,3-epoxycyclohexanone is hydrolysed with nitrous acid to form the *cis*-2,3-diol.[135] An extensive study of amide-6 participation in the 2-(benzamidomethyl)-4-*tert*-butylcyclohexyl series has been reported.[136]

Participation by Neighbouring Amino and Phosphorus Groups

A study of the rates of hydrolysis of 2-chloro- and 2-bromo-ethylamine as a function of acid concentration has permitted determination of the rates for the free bases and their conjugate acids. With chloride as leaving group, the accelerations are of the order of 5×10^4–10^5, depending on the group on nitrogen, while with bromide they are much less, 90–700.[137] A study of the conjugate acids with methyl substituents on C-1 and C-2 showed that an isokinetic relationship held for ten compounds studied.[138] The rates of solvolysis of compound (**60**; R = Et, X = Cl) correlate with the Kosower *Z* value, and the previously postulated intermediate has been isolated as the perchlorate (**62**).

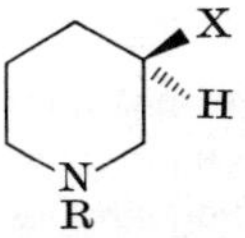

(60) 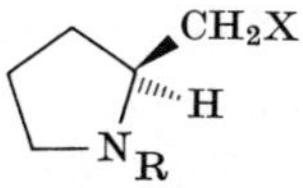 (61) 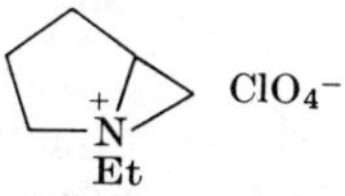 (62)  (63)

Optically active (**60**; R = Et, X = Cl) of (*R*)-configuration gives (*S*)-(**62**) which with water affords (*R*)-(**60**; R = Et, X = OH) and (*S*)-(**61**; R = Et, X = OH), showing that both ring-closure and ring-opening occur with inversion of configuration. (*S*)-(**61**; R = Et, X = Cl) likewise gives the same mixture of (*R*)-(**60**; R = Et, X = OH) and (*S*)-(**61**; R = Et, X = OH).[139] The rates of alkaline methanolysis of (**60**) and (**61**), when R = H or Me, and X = Cl or Br, are strongly accelerated [ca. 10^4 for (**60**) and 10^7 for (**61**)]. Studies of (**63**) also show accelerated rates (490 and 214 times, respectively, for R = H, with Me and X = Cl), but this is caused by fragmentation and not by N-4 participation.[140] Other examples of N-3 participation are found in references 141 and 142. The all-*cis*-triepoxide of benzene reacts with *N,N'*-dimethylhydrazine to give 3,4-epoxy-6,7-dimethyl-6,7-diazabicyclo[3.2.1]-octane-2,8-diol by N-5 participation.[143] N-(1,2,4-Triazol-3-yl)hydrazonyl bromides react in aqueous organic solvents by N-5 participation through N-2 at pH 3–6, but through N-4 when the solution is alkaline.[144] The formation of 1-azatwistanes from decahydro-*N*-methylisoquinolin-6-yl methanesulphonates is faster in methanol than in acetonitrile, unlike intermolecular quaternizations, which are slower in protic solvents. This has been ascribed to the difficulty of hydrogen-bond formation to nitrogen in the preferred conformation of the *cis*-decalin type system.[145]

Ion-cyclotron resonance (i.c.r.) and trapped ion i.c.r. have been used to investigate the reaction of oxirane with phosphine in the presence of protons. The results suggest nucleophilic ring-opening, followed by loss of water to form the phosphiridinium ion.[146] The evidence relating to participation by phosphorus-containing groups has been analysed, and it is suggested that whether participation or fragmentation occurs depends on the electronegativity of the atoms attached to phosphorus; carbon leads to participation, whereas oxygen (e.g. phosphoryl) leads to fragmentation.[147] As a result of rate and product studies of (**64**) and (**65**), it is concluded that the results for participation of

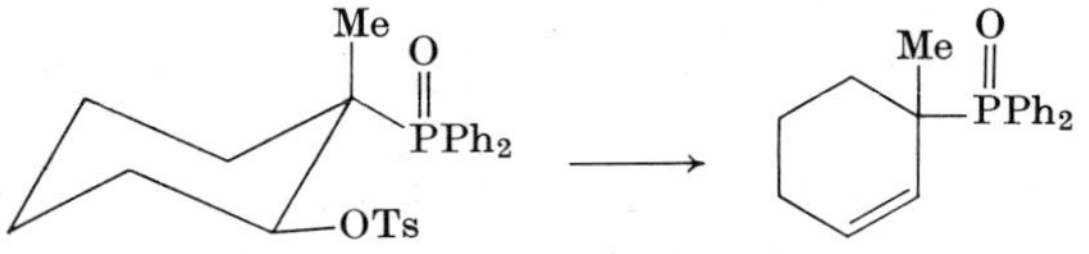

(**64**)

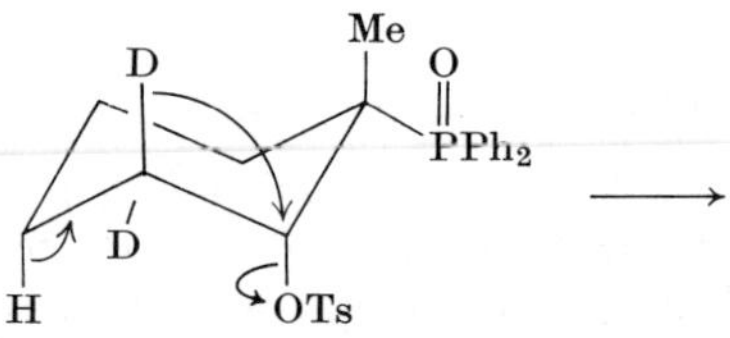

(**65**)

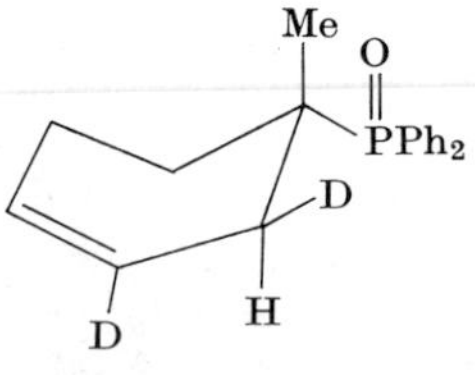

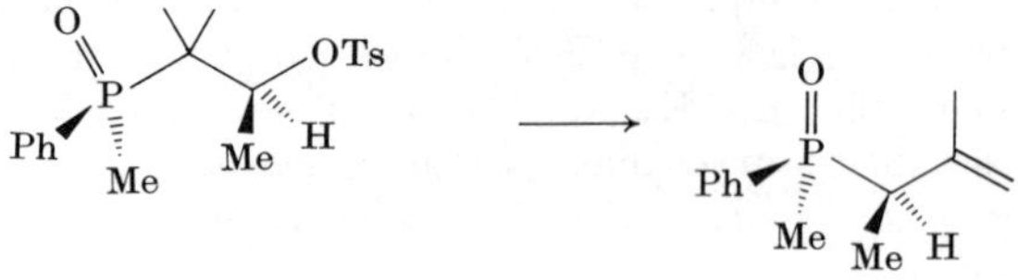

(**66**) (**67**)

diphenylphosphinyl reported last year are the result of electronic and not conformational factors.[148] *X*-Ray crystal-structure determinations have shown that (**66**) is converted into (**67**) with retention of configuration at phosphorus and inversion at carbon, as would be expected for a concerted process.[149]

Solvolysis rate studies on *o*- and *p*-nitrobenzhydryl bromide in trifluoroethanol and and benzene–trifluoroethanol mixtures show that the *ortho*/*para* rate ratio increases with increasing concentration of benzene (from 30 to 900). However, the corresponding *o*- and *p*-bromo-compounds show a falling rate ratio (from 0.18 to 0.055). Correlation with Y values gives m values of 0.39, 1.65, 2.08 and 1.73 for the *o*-NO_2, *p*-NO_2, *o*-Br and *p*-Br compounds, respectively, and it is suggested that ion-pair return accounts for the results, being large for all except the *o*-NO_2 compound where the nitro-group acts as an internal nucleophile.[150]

Participation by Neighbouring Carbanion and Carbon—Metal Bonds[151]

An extensive investigation of the Ramberg–Bäcklund reaction of compounds (**68**) to (**71**) has shown that H-to-D exchange occurs essentially completely before significant amounts of halide are liberated. The high k_{Br}/k_{Cl} ratio (170–200), and the negative ρ values for loss of halide (–2.5 for (**69**) and –1.7 for (**71**)) lead to the conclusion that the

$PhCHX—SO_2—CHMe_2$ (**68**)

$ArCH_2—SO_2—CH_2Br$ (**69**)

$ArCHBr—SO_2—Me$ (**70**)

$ArCHX—SO_2—CH_2Ar$ (**71**)

Me, Ph, Br–C–SO_2–C–Me, Ph, Br (**72**)

O_2 S Br (**73**)

reaction is two-stage, with equilibrium carbanion formation preceding rate-determining ring-closure to episulphone.[152] *erythro*- and *threo*-Sulphone (**72**) react stereoselectively, giving essentially *cis*- and *trans*-olefin, respectively; in both cases H-to-D exchange occurs without epimerization, but *threo*-(**72**) reacts at a rate independent of methoxide concentration; it is therefore suggested that in this case the loss of SO_2 from the episulphone is rate-determining. The cyclic sulphone (**73**) has exchange, kinetic and thermodynamic parameters very similar to those for (**71**; Ar = Ph, X = Br), showing that, even when the geometry is favourable, the reaction is not concerted.[153] An extensive investigation of the conversion of *meso*- and *dl*-*n*-$PrCHBrCH_2CHXPr$-*n* (X = Br, F or I) into 1,2-di-*n*-propylcyclopropane with a variety of metals and metal alkyls gives results that are stereoconvergent, indicating a two-step process of intermediate carbanion formation, followed by displacement.[154] Use of (–) (*RR*)-4,6-dibromononane with lithium amalgam afforded (*S*,*S*)-1,2-*trans*-di-*n*-propylcyclopropane, showing that the reaction occurs with inversion at both carbon atoms.[155] The variation of E/Z ratio as a function of solvent has been investigated for the cyclization products of the reactions of the sodium salt of phenylacetonitrile with 1,2-dibromopropane and 1,3-dibromobutane.[156]

O, C̄ (**74**)

O, C̄ (**75**)

The difficulty of getting the three centres collinear [cf. (**74**) and (**75**)] has been suggested to account for the 20-fold slower cyclization of (**76**; $n = 2$) than of (**76**; $n = 3$);[157] when the degree of substitution at each end of the epoxide is the same, as in (**77**) or (**78**), then ring-closure occurs to give the smaller of the two possible rings for $n = 1, 2, 3$ or 4. As a result of the stereochemistry of cyclization of (**79**), it is suggested that a *trans-anti*-geometry as (**80**) is preferred for the reaction.[158] Similar conformational factors have been invoked to account for the preferential formation of a *cis*-cyclopentane on reaction of the lithium salt of 2,4,4,-trimethyloxazoline with 1,4-di-iodopentane.[159] ^{13}C-NMR studies of the cations derived from $Ar_2C(OH)CH_2HgX$ in sulphuric and trifluoroacetic acid suggest very little participation by the metal.[160]

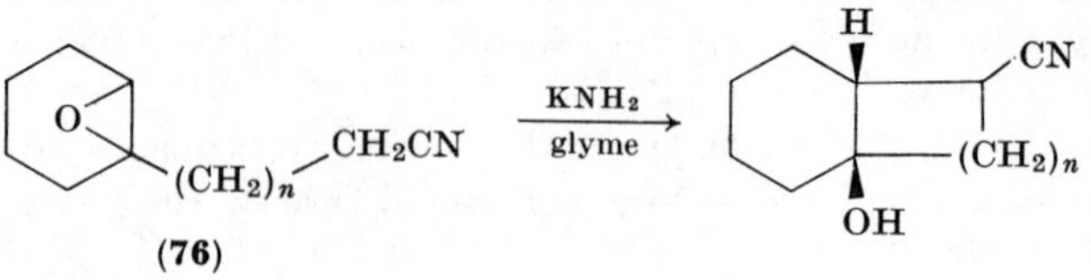

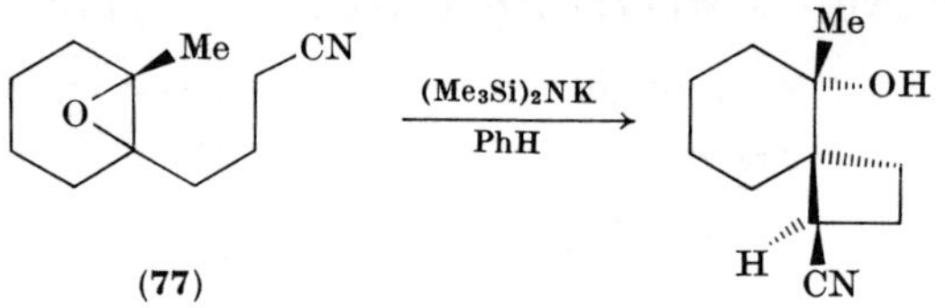

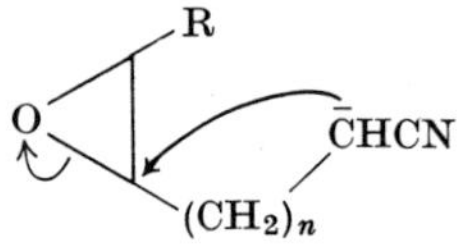

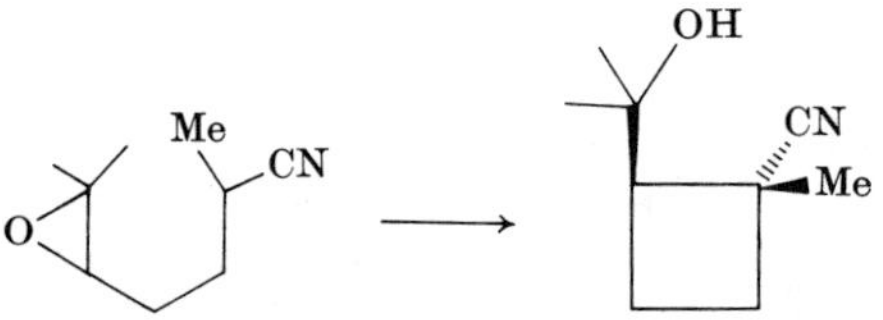

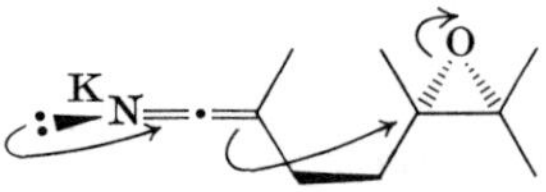

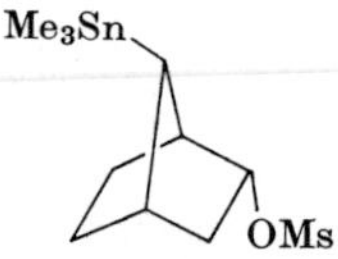

Acetolysis rates and activation parameters show that the carbon—tin bond in (**81**) participates strongly, affording (**82**) as the sole product. The *syn-endo*-isomer (**83**) also gives some (**82**) but is not accelerated, indicating the importance of the W geometry.[161]

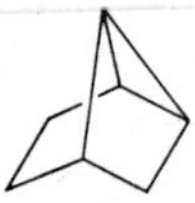

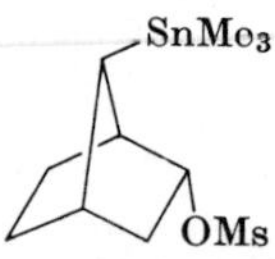

Neighbouring Carbon and Hydrogen

The inclusion of *d*-orbitals in the basis set for *ab initio* calculations on the $C_3H_7^+$ system reduces the overall difference in stability between *iso*-Pr^+ and the other possibilities and makes a few differences in the order of stabilities. Thus edge-protonated cyclopropane is only 6 kcal mol^{-1} above the corner-protonated isomer, which in turn is now calculated to be more stable (by 1 kcal mol^{-1}) than *n*-Pr^+. The inclusion of *p*-functions in the hydrogen basis set has not been carried out, but extrapolations from $C_2H_5^+$ suggest that protonated propene might lie only 11 kcal mol^{-1} above *iso*-Pr$^+$ and that there would be no barrier for conversion of *n*-Pr$^+$ into *iso*-Pr$^+$.[162] FSGO calculations on $C_2H_5^+$ and $C_3H_7^+$ have also been reported. They suggest that *n*-Pr$^+$ lies at a shallow minimum with a 2.5 kcal barrier for conversion into *iso*-Pr$^+$ and a 3.6 kcal one for conversion into protonated cyclopropane, which lies about 6 kcal mol^{-1} above *iso*-Pr$^+$ in energy.[163]

The small isotope effect observed in the trifluoroethanolysis of [2,2,4-$^2H_3$]menthyl toluene-*p*-sulphonate compared to that observed for 4-*tert*-butyl[2,2,6,6-$^2H_4$]cyclohexyl toluene-*p*-sulphonate ($k_H/k_D = 1.27$ and 2.42, respectively), and the fact that the major alcohol product is menthol, have been used to suggest that, whereas the cyclohexyl compound reacts *via* a twist conformation,[164] the menthyl compound does so in a chair conformation. This would result in the two axial deuterium atoms exerting a negligible effect owing to their incorrect orientation. It is suggested that the change is due to the substituents at C-2 and C-5, which also hinder attack from the rear and favour retained products.[165] A cyclic orthoacetate is proposed as an intermediate in the Serini reaction of (**84**), as a result of the experiments shown in the scheme; the reaction is catalysed by acetic acid and 97% of the deuterium present in (**84**) is found in the product. The allylic

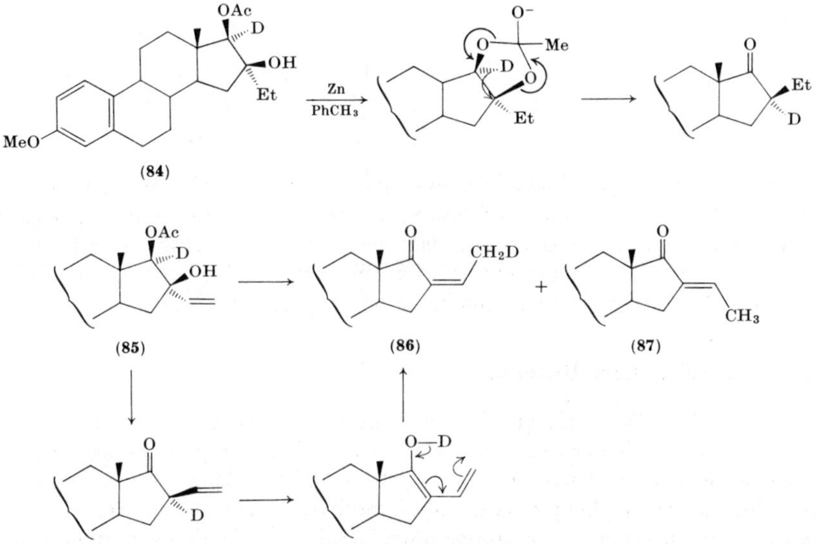

(**84**)

(**85**) (**86**) (**87**)

alcohol (**85**) gives a 3:1 mixture of (**86**) and (**87**). The (**86**) formed contains 33% of deuterium and a specific 1,4-hydrogen shift is suggested,[166] but the route shown seems more probable. The epoxy-acetate (**88**) affords exclusively the diol (**90**) when treated with an excess of MeMgI, while its 20-epimer gives only the 17-epimer of (**90**); a modified Serini reaction with conformational control by co-ordination to the magnesium is proposed.[167] Conformational factors are also invoked to account for the fact that 2-noradamantanone is converted into 5-protoadamantanone by diazomethane, whereas hydration of 2-methylenenoradamantane gives 4-protoadamantanone.[168] Similar effects are also invoked to explain the stereochemistry of the products from the reaction of diazoethane with norcamphor,[169] while the preference for migration of methylene over methine in the reaction of diazomethane with norcamphor has been ascribed to the principle of least motion.[170] Backbone rearrangement of 4β-acetoxycholestan-5α-ol in DOAc–Ac_2O in the presence of D_2SO_4 occurs without incorporation of deuterium, showing that methyl migration is concerted, or, if not, that proton loss and re-addition to a protonated cyclopropane is slow compared with migration.[171] Further

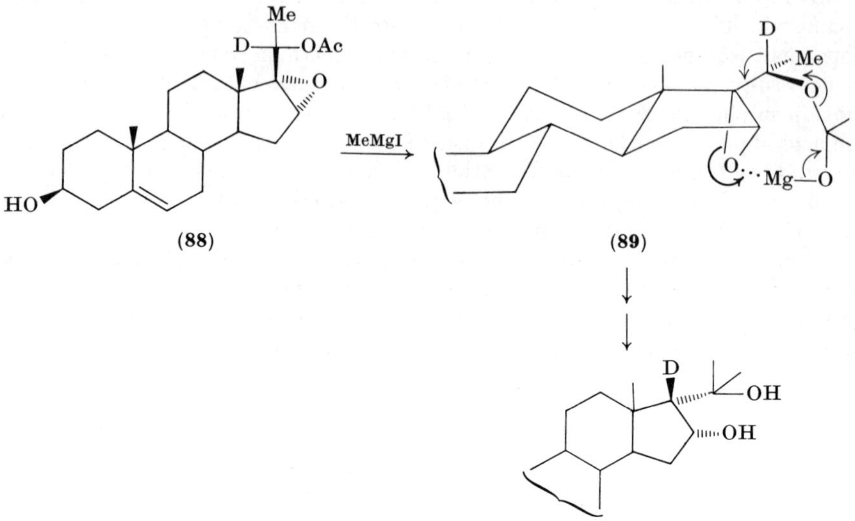

(88) MeMgI → **(89)** → → **(90)**

studies on the steroidal alkaloids holamine and dimethylholaphylline have shown that hydride migration, and proton loss followed by re-addition, are competing paths in the backbone rearrangement.[172] Deuterium isotope effects in the methyl group are 1.144 in the dienone–phenol rearrangement of 4,4-dimethylcyclohexa-2,5-dienone.[173] Other results on the dienone–phenol reaction have been reported.[174]

Deamination and Related Reactions

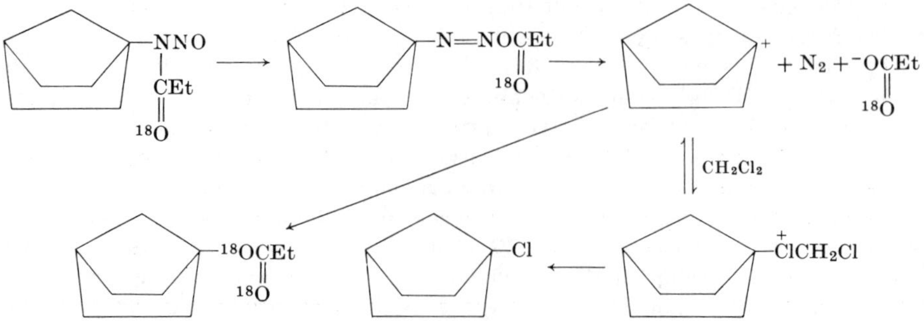

The reactions of octane-2-diazotate with lithium salts, acetic and trifluoroacetic anhydride, and trimethylchlorosilane in HMPT are readily interpreted on the basis of the Moss scheme reported in 1970,[175] with Li^+, Ac or Me_3Si taking the role of the electrophile.[176] Reactions of 1-phenylethane- and 1-methylpropane-1-diazotate with Grignard reagents also fit the scheme, the proportion of product of retained configuration being greater the more readily it can bear a positive charge.[177] The full paper on the deamination of *cis*-myrtanylamine has been published.[178] The deamination of micellar n-$C_{12}H_{25}NH_3^+$ in aqueous nitrous acid is catalysed by thiocyanate; inhibition of this catalysis by other anions is due to competition for sites at the micelle surface.[179] 1-Norbornyl and 1-adamantyl diazopropionate, formed by the nitrosourethane route, decompose in all solvents to form the bridgehead carbonium ions; in chlorocarbon solvents, chlorine abstraction (to form 1-chloronorbornane and 1-chloroadamantane) accompanies collapse of the intermediate ion pair to give the 1-propionates, as shown; the norbornyl cation is more reactive than the adamantyl one, as shown by the higher percentage of chloride formed (56% as against 2% in CH_2Cl_2); in benzene, 1-phenylnorbornane is formed and 1-ethoxynorbornane in ether, and while the adamantyl compound also gives 1-ethoxyadamantane, no 1-phenyladamantane is formed.[180] Deamination of 1-adamantylamine with isopentyl nitrite in benzene or anisole also gives no 1-aryladamantane,[181] contrary to previous reports.[182]

N-Cyclopropylnitrosourea gives 30% of *N*-ethyl-*N*′-cyclopropyltriazene when treated with ethylamine and, when ^{15}N-labelled compound (in the nitroso-group) is treated with LiN_3, the cyclopropyl azide contains 68% of the ^{15}N label at N-2. When similarly labelled *N*-(*n*-butyl)nitrosourea is treated with LiN_3 no ^{15}N remains in the *n*-butyl azide formed. These results are taken to show that the reactivity of the cyclopropanediazonium ion lies between those of aromatic and aliphatic ones.[183]

Protonation of 2-diazohexafluoropropane in FSO_3H at −70° gives the corresponding diazonium ion, whose rate of disappearance (between −5° and +5°) is of the first order in diazonium ion and first order in $FSO_3^-K^+$; the lack of effect of the non-nucleophilic salt $Me_4N^+BF_4^-$ is taken to show that the reaction with FSO_3^- is of the S_N2 type.[184] The formation and reactions of alkanediazonium ions from *N*-alkylnitrosoureas has been carried out in the weakly polar, strongly nucleophilic solvent dimethylamine: in pure dimethylamine, isobutanediazonium ion gives only *N*,*N*-dimethylisobutylamine; but in 1:1 Me_2NH/MeOH, the isobutyl carbocation and its rearrangement products are formed to a small extent. 1-Methylprop-2-ene-1-diazonium ion reacts mainly by an S_N2' process, while the primary but-2-ene-1-diazonium ion reacts mainly by a direct S_N2 route.[185] The reactions of the *syn*- and *anti*-norbornene-7-diazonium ions have been investigated and compared with those of the *exo*- and *endo*-bicyclo[3,2,0]hept-6-ene-2-diazonium ions.[186] Photolysis of the toluene-*p*-sulphonylhydrazone (**91**; R = Ac) in alkaline methanol gives (**92**), while treatment of the hydroxy-tosylate (**93**) with potassium

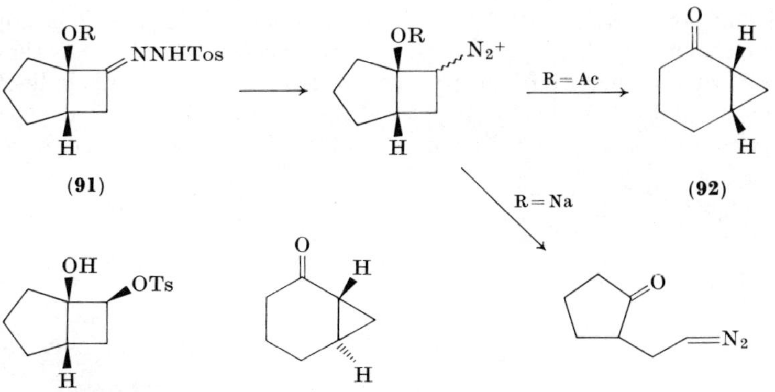

OR NNHTos H (**91**) → OR N_2^+ H —R = Ac→ O H H (**92**)

R = Na ↘

OH OTs H (**93**) O H H (**94**) O $=N_2$ (**95**)

tert-butoxide gives the *trans*-isomer (**94**), and photolysis of (**91**; R = H) leads to fragmentation and products derived from the diazoketone (**95**).[187]

The reactions of *cis*- and *trans*-4-*tert*-butylcyclohexyl chlorocarbonate, and of *erythro*- and *threo*-3-phenyl-2-butyl chlorocarbonate, with $AgSbF_6$ in acetic acid lead to product ratios that strongly resemble those derived by deamination of the corresponding amines. The results are interpreted in terms of a k_c process with some slight leakage into k_s and k_Δ processes.[188]

Reactions of Aliphatic Diazo-compounds

A study of the effects of the mole fractions of water and of added salts on the water- and acid-catalysed reactions of azibenzil in water/*tert*-butyl alcohol have been reported.[189] The rates and yields of the reaction of *m*-chloroperbenzoic acid with α-diazocycloalkanones vary very little with the size of the ring concerned, showing that the transition state differs little from the ground state in geometry.[190] Diazo-ketones are readily converted into the corresponding α-fluoro-ketones by treatment with HF in pyridine.[191]

Fragmentation Reactions

The variation in the proportion of fragmentation in the reaction of (**96**) at an H_0 of −1.35 with variation in group R has been studied. Log $k_{frag.}$ correlates well with Taft's E_s

Me R But Me AcOH H₂SO₄ Me Me + R But Me Fragmentation Me R Me Me + ⁺Buᵗ

(**96**) (**97**)

Rearr.

Buᵗ R Me + Me Me → Products

(**98**)

constants, but very poorly with the combination of σ^* and E_s ($\rho = 0.6 \pm 5.0$), whereas $k_{rearr.}$ correlates well with σ^* ($\rho^* = -8.0$) and rather poorly with E_s; thus there appears to be a clean separation, with fragmentation affected only by the steric bulk of R, while rearrangement is affected only by its polar effect [presumably the steric effects in (**97**) (**98**) compensate one another].[192] Fragmentation of (**99**) depends both on the nucleophile

O Nu⁻ X CONu

(**99**)

O⁻ OH O O

(**100**)

and on the leaving group; when X = OTs, reaction with methoxide, sodium acetylide or methyl-lithium leads to fragmentation (as shown), while that with sodium cyanide occurs by direct substitution; the more active leaving group OSO_2CF_3 favours direct substitution, even with acetate anion.[193] Zinc–ethanol reduction of 15-bromolongibornane-8,9-dione results in a ring-opening fragmentation.[194]

Epoxides react with alkaline hydrogen peroxide to give peroxides, as (**100**), which fragment as shown when the epoxide is a hindered one.[195]

17β-Pregnane-3,20-diones isomerize at C-17 and C-13 by a fragmentive ring-opening when dissolved in HF–SbF_5. Reaction occurs without loss of deuterium from the [17-$^2H_1$]-compounds, and without epimerization at C-5.[196] Work on fragmentation is also reported in ref. 140, see p. 344.

Displacements at Elements Other than Carbon

Silicon, Germanium, Tin and Lead

A second example of the effects of angle strain on the stereochemistry of displacement at silicon confirms that reported last year. 1-Chloro-1,2-dimethyl-1-silacyclobutane is reduced by $LiAlH_4$ with retention of configuration, and a process of equatorial attack to form a 5-co-ordinate silicon atom and apical loss therefrom are proposed.[197] The much faster displacement of halogen from (**101**) than from (**102**) or an open-chain analogue

X
Si

(**101**)

X = Cl or F

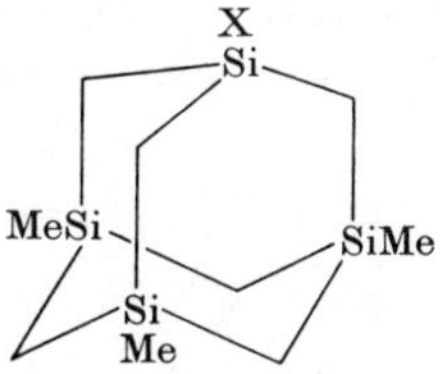

X
Si
MeSi
SiMe
Si
Me

(**102**)

has also been interpreted as due to relief of angle strain in an S_N2-Si transition state for an S_N2–5 reaction (which proceeds with retention).[198] The methyl groups of the acetylacetone residue in (**103**) are identical in the ^{1}H-NMR spectrum even at –142°. At 100° (**103**) is rapidly converted into (**105**), and it is suggested that both reactions are readily explained by formation and pseudorotation or migration in (**104**) as shown.[199] The shift of silicon from one oxygen to the other in (**106**) is also rapid as shown by ^{1}H- and ^{13}C-NMR; however, the two methyl groups are always distinct (diastereotopic), showing that rearrangement takes place with $>10^7$:1 retention of stereochemistry at silicon.[200] Rate-determining formation of a 5-co-ordinate silicon atom, followed by rapid loss of a ligand, has also been proposed to account for the retention of stereochemistry observed in the reaction of (**107**; X = OMe) with alkyl-lithiums; it is suggested that whether methoxy or menthoxy is lost depends on the size of the group, the smaller group co-ordinating to the metal more readily and therefore being lost preferentially.[201] The stereochemistry of reaction of Grignard reagents with (**108**; X = F or OMe) changes from inversion to retention as the basicity of the solvent increases (Et_2O, THF or DME), and the relative rates of reaction increase in the same order; these results are taken to confirm the above hypothesis of rate-determining addition followed by rapid loss of X^- and to rule out the S_Ni-Si route.[202] Further work on the racemization of chloro- and

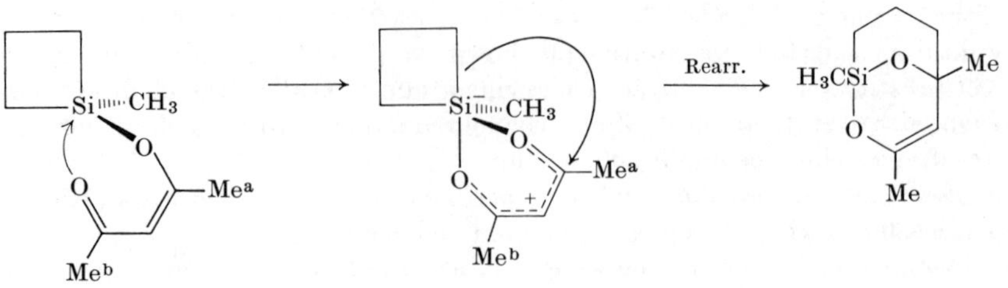

(103) **(104)** **(105)**

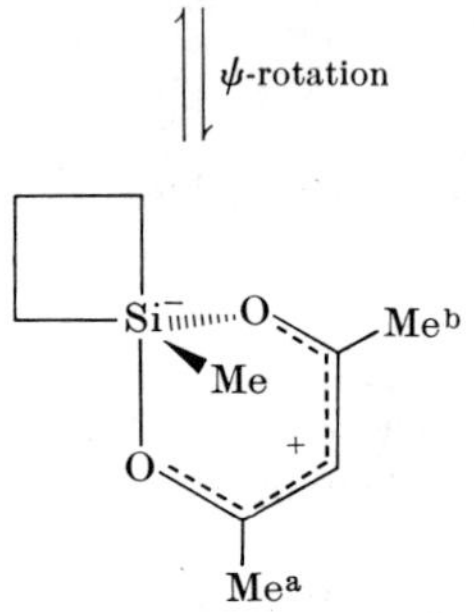

ψ-rotation

bromo-silanes by HMPT tends to confirm the mechanism reported in 1971.[203] The full paper on the reaction of optically active silanes with trialkyltin alkoxides has been published.[204]

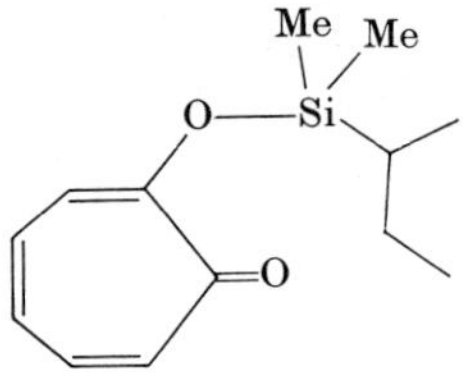

(106) 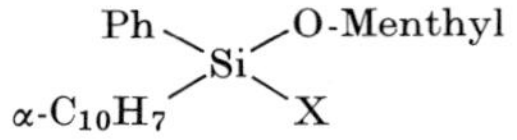 **(107)** 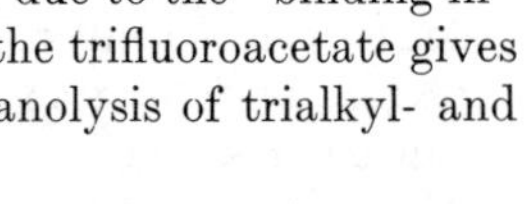 **(108)**

Displacement at silicon in (**109**) by $R_4N^+F^-$ is thought to be due to the "binding in" of fluoride ion, for the corresponding acetate fails to react and the trifluoroacetate gives the allene.[205] The kinetic isotope effect observed in the methanolysis of trialkyl- and

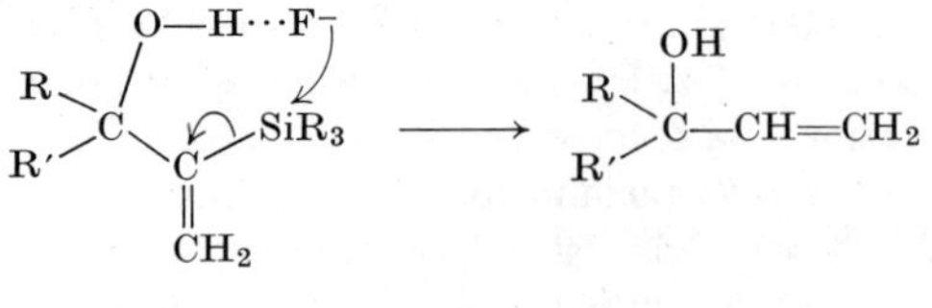

(109)

triaryl-silanes increases with increasing reactivity of the substrate, as does the acid-catalysis in the rate-determining step; pre-equilibrium formation of pentaco-ordinate silicon with rate-determining cleavage is therefore proposed.[206] A similar mechanism is also proposed for the alkaline hydrolysis of substituted benzyltrimethyl silanes and -stannanes.[207] Other reactions where basically similar mechanisms are thought to operate are the alkaline cleavage of aryl(dimethyl)-4-chlorobenzylsilanes and aryldimethyl-2-benzofurylsilanes,[208] the hydrolysis of aryltrimethylsilanes in alkali,[209] the solvolysis of *N*-(trialkylsilyl)anilines in alkaline methanol,[210] and the hydrolysis of *N*-(aryldimethylsilyl)pyrrole in ethanolic sulphuric acid or in amine buffers.[211] The uncatalysed cleavage of *N*-(aryldimethylsilyl)-*N*-methylaniline shows $\rho(\sigma) = -0.96$ and it is suggested that this argues against a mechanism involving 5-co-ordinate silicon.[212] A study of the hydrolysis of (*p*-methoxyphenoxy)triphenylsilane at constant ionic strength has shown that at low hydroxide ion concentration the rate is second-order in $[OH^-]$, leading to the conclusion that the breakdown of the 5-co-ordinate intermediate is base-catalysed.[213] Alkaline methanolysis of the same compound and its *p*-chloro-analogue has been shown to be general-base catalysed. The change in solvent isotope effect with mole fraction of MeOD shows that for the proton being transferred the fractionation factor is between 0.25 and 0.50, i.e. $k_H/k_D = 2$–4. However, the overall solvent isotope effect lies between 1.16 and 1.25, showing that the kinetic isotope effect for proton transfer is nearly cancelled out by an inverse isotope effect from solvation of ground and transition states.[214] Octet expansion by silicon has also been postulated for the acid-catalysed hydrolysis of (aryloxy)tributylsilanes.[215] The rate of cleavage of Me_3SiCHX_2 (X = Cl, Br or I) depends on the stability of the anion CHX_2^- which is lost.[216] Other kinetic studies include the cleavage of Me_3SiOPh,[218] the hydrolysis of octamethylcyclotetrasiloxane,[219] and the hydrolysis of tetramethylsilane and phenyltrimethylsilane catalysed by platinum complexes.[220] Chlorinated silicon metal is converted into chlorotrimethylsilane by methyl chloride, *via*, it is proposed. a five-co-ordinate intermediate.[221]

The acid-catalysed hydrolysis of phenoxy-tributyl- and -triphenyl-germane is some 10^5 times faster than for the corresponding silane. The kinetic order with respect to water concentration is $\geqslant 2$, and a mechanism involving protonation on phenoxy-oxygen followed by nucleophilic attack by water is proposed.[222] The acid and the neutral hydrolysis of the series $ArSMR_3$, where M = Si, Ge, Sn or Pb, have been studied in aqueous dioxan. The reactivity order is Si > Ge ≪ Sn < Pb, with relative rates $2:1:2.6 \times 10^5:6.6 \times 10^5$. It is suggested that the silicon compounds react by pre-equilibrium protonation on sulphur, with rate-determining attack by water, while the tin and germanium ones form a pentaco-ordinate species in equilibrium with the products ($ArSH + HOMR_3$) in neutral media. The lack of observable hydrolysis under neutral conditions is due to the ready reversibility of this reaction, and hydrolysis is observed in the presence of HCl owing to formation of ArSH, R_3MCl, and water from the $ArSMR_3 \cdot OH_2$ intermediate.[223] The kinetics of solvolysis of germanium tetrachloride in a series of six alcohols have been measured.[224]

Phosphorus

Nucleophilic displacement at tetraco-ordinate P(v) compounds has been reviewed, with particular emphasis on the evidence in favour of the turnstile rotation mechanism and its role in permutational isomerization.[225]

A number of NMR studies of permutational mechanisms have been reported. Use of ^{1}H-NMR on stable pentacovalent phosphoranes has shown that the strain factor involved in placing a five-membered ring diequatorial is larger when N or O is bonded to the phosphorus than when carbon is.[226] A critical examination of all the available NMR data has shown it to be consistent with the mechanisms for topological change.[227] The temperature-dependence of the ^{31}P-NMR spectra of Me_2NPF_4 and $MePF_4$ shows that the interchange of groups is consistent with a Berry pseudorotation mechanism. A mathematical analysis shows that even if a square pyramid were involved it could not be detected by line-shape analysis at the present resolution. The rate of pseudorotation is found to vary with substituents on the phosphorus in the order F, Me > Cl > H, SR > NMe_2.[228] ^{1}H-NMR line-shape analysis suggests that (**110**; Ar = 2-$Pr^iC_6H_4$)

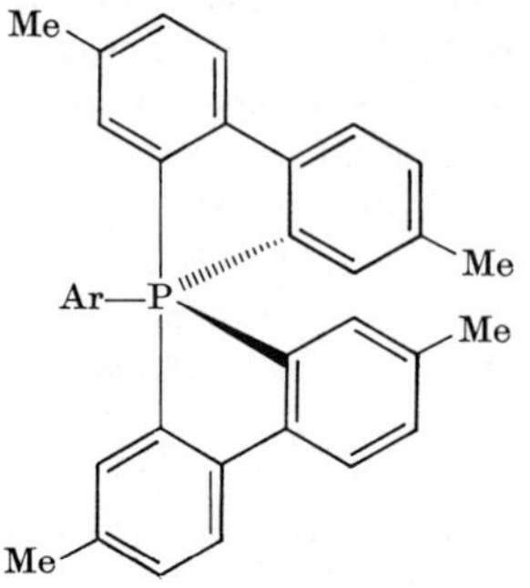

(**110**) 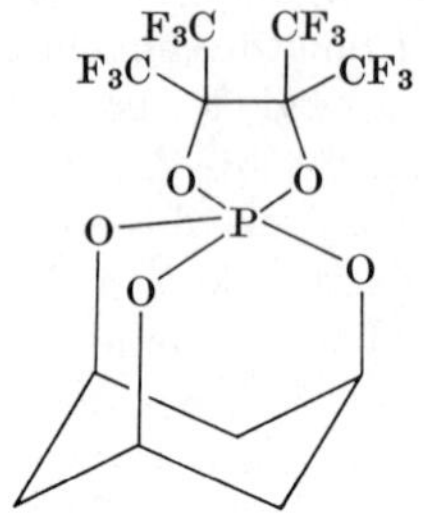 (**111**)

pseudorotates by way of a tetragonal prism, a mechanism that accounts for the free rotation of the isopropyl group that accompanies the permutational isomerization. When Ar = 2,4,6-$Pr_3{}^iC_6H_2$, pseudorotation is by way of a trigonal bipyramid with the aryl group axial, and the ΔG^{+} increases from 17.8 to 26.0 kcal mol^{-1}.[229] The low barrier ($\Delta G^{+} \leqslant 5$ kcal mol^{-1}) observed in NMR for the permutational isomerization of (**111**) is held to be consistent only with a turnstile (as opposed to Berry) rotation process.[230] Permutational isomerization in spirocyclic phosphoranes has also been studied by epimerization and racemization techniques, which reveal that the barrier depends not only on the atoms directly attached to phosphorus, but also on more distant groups.[231] The apicophilicities of groups have been determined by ^{19}F-NMR measurements.[232]

The isolation of a stable oxyphosphorane anhydride has shown that P(v) intermediates may be involved in the interconversion of ADP and ATP.[233] The stable five co-ordinate P(v) compound (**112**) with only two oxygen ligands has been prepared.[234] The spirophosphoranes (**113**) react with pyridine or phenol–triethylamine to give stable six-co-ordinate salts (**114**).[235] Tris-spirocyclic salts analogous to (**114**) have also been

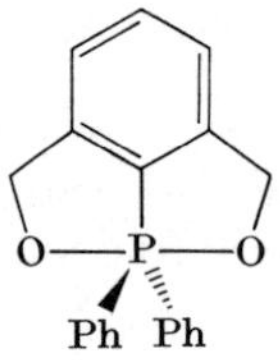

(**112**) 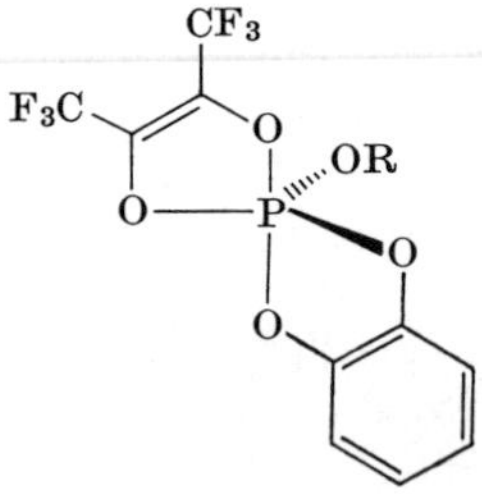 (**113**) 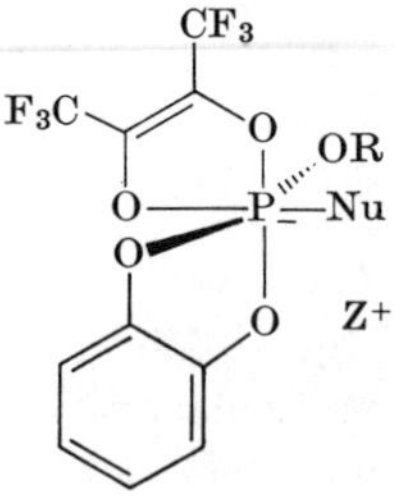 (**114**)

R = Me or Ph

reported,[236, 237] and it is suggested that they are intermediates in the normal substitution reactions of five-co-ordinate phosphorus compounds.[237]

The rate–pH profile for hydrolysis of (**115**) is such as to suggest the mechanisms of equations (1) and (2). In 0.05M-HCl, the solvent isotope effect ($k_{H_2O}/k_{D_2O} = 3.5$) and ΔS^{+}

(–41 eu) both show that proton-transfer is involved at the transition state, and hence that k_3 is rate-determining; at neutral pH, general-base catalysis is observed, with k_1 now rate-determining, while above pH 9, the rate is of first order in OH^- and the process

$$R_3POMe^+ + H_2O \underset{-H^+}{\overset{k_1}{\rightleftharpoons}} R_3P(OH)OMe \xrightarrow{k_3[H^+]} R_3PO + MeOH \qquad (1)$$

$$R_3POMe^+ \xrightarrow{k_2[OH^-]} R_3P(OH)OMe \xrightarrow{k_4} R_3PO + MeOH \qquad (2)$$

of equation (2) takes over.[238] Reaction of (**116**) with toluene-*p*-sulphonyl isocyanate gives the corresponding phosphinimine (**116**; O replaced by NTs) with retention of configuration. Open-chain phosphine oxides react, but with racemization, and a mechanism

PF_6^-

(**115**) (**116**)

identical with that discussed last year for the sulphoxide-to-sulphilimine conversion is postulated.[239] The kinetics of ^{18}O-exchange, and their variation with pH, have been studied for cyclic and bicyclic phosphine oxides.[240] Neighbouring-group participation by the solvated aldehyde has been proposed to account for the ready conversion of (**117**) into (**118**).[241] The ease of cleavage of 2-(1-methylpyrrolyl) from phosphonium salts lies between that of benzyl and phenyl, and 2-(1-methylpyrrolyl)methyl is also cleaved less readily than benzyl.[242] Hydrolysis of tetraphenylphosphonium chloride is of the second order in hydroxide ion and is 5×10^7 times faster in 80% dioxan than in water.[243] Hydrolysis of *O*-menthyl *S*-methyl methylphosphonodithioate results in loss of the *S*-methyl group with inversion of configuration.[244]

CHO, $O(CH_2)_4\overset{+}{P}Ph_3$ $\xrightarrow[MeOH]{MeONa}$ $CH(OMe)_2$, $O(CH_2)_4P(O)Ph_2$

(**117**) (**118**)

The acid-catalysed hydrolysis of diphenylphosphinanilides goes by an A-1 route, while that of methylphenylphosphinanilide goes with inversion of configuration and is therefore more associative;[245] a more extended study indicates a merged A-1 and A-2 process for the methylphenyl compounds with a tendency to shift towards A-1 as the acid concentration increases.[246] Change of leaving group from anilines to dialkylamines causes a change of mechanism from A-1 to A-2, and it is suggested that the greatly increased rate of hydrolysis of phosphinamides over carboxamides is due to the relative ease of protonation on nitrogen. Because the rate of hydrolysis of 1-dimethylamino-2,2,3,4,4-pentamethylphosphetane 1-oxide is 10^3 times slower than that of di-isopropyl-*N*,*N*-dimethylphosphinamide, it is suggested that the former reacts by an S_N2-P process rather than the normal one *via* a pentaco-ordinate intermediate.[247] The kinetics of alkaline hydrolysis, and ^{18}O-exchange for dialkyl phosphinates have been measured in 60% aqueous dimethoxyethane and 33% dioxan.[248]

Methanolysis of 1-chloro-1,3,2-oxazaphospholanes occurs with retention of configuration at phosphorus, while ring-cleavage of the 1-methyl and 1-phenyl compounds leads to breakage of the P—N bond with inversion of configuration.[249] Stereochemistry, solvent, and salt effects have been studied in the reaction of 2-X-2-thio-1,3,2-dioxaphosphorinanes with sodium phenoxides; in addition to the mechanism of addition–pseudorotation–loss (proposed last year for the oxygen analogues), it is suggested that a direct S_N2-P process also operates.[250] The stereochemistry of ring opening in the bicyclic

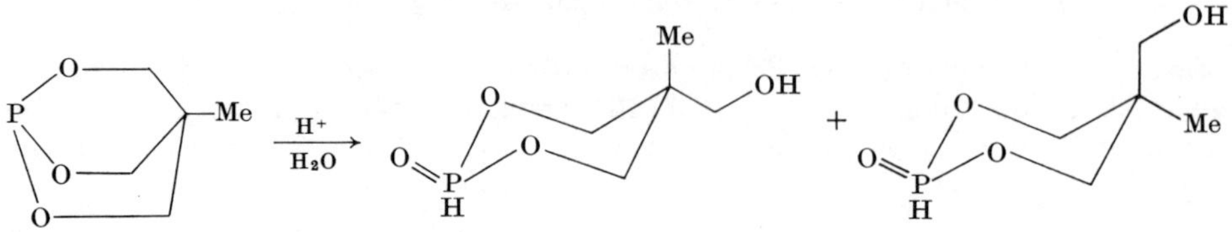

(**119**)

phosphite (**119**) has been determined, and it has been shown that the oxygen on phosphorus comes from the water.[251] Acyclic, optically active phosphine oxides are reduced by phenylsilane with retention of configuration in 85–96% yields; a four-centre cyclic mechanism is proposed.[252]

Sulphur, Selenium and Antimony

The alkaline hydrolysis of 2-NO_2-4-CF_3-C_6H_3SX, where X = Br, Cl, OEt, SAr or S(O)Ar, has been investigated by classical kinetics and by ^{19}F-NMR spectrometry. The overall rate is $k[\text{ArSX}][\text{OH}^-]$, but the rate of loss of ArSO^- is proportional to $[\text{ArSO}^-]^2[\text{OH}^-]^{-1}$. The mechanism proposed is outlined in equations (3)–(9).

$$\text{ArSX} + \text{OH}^- \rightarrow \text{ArSOH} + \text{X}^- \quad (3)$$

$$\text{ArSOH} + \text{OH}^- \rightarrow \text{ArSO}^- + \text{H}_2\text{O} \quad (4)$$

$$\text{ArSO}^- + \text{ArSX} \rightarrow \text{ArS(O)SAr} + \text{X}^- \quad (5)$$

$$\text{ArS(O)SAr} + 2\text{OH}^- \rightarrow \text{ArSO}_2^- + \text{ArS}^- + \text{H}_2\text{O} \quad (6)$$

$$\text{ArS}^- + \text{ArSX} \rightarrow \text{ArSSAr} + \text{X}^- \quad (7)$$

$$\text{ArS}^- + \text{ArS(O)SAr} \leftrightharpoons \text{ArSSAr} + \text{ArSO}^- \quad (8)$$

$$\text{ArSO}^- + \text{ArSOH} \leftrightharpoons \text{ArS(O)SAr} + \text{OH}^- \quad (9)$$

When X = Cl or Br, step (5) is too fast for measurement and the overall rate is determined by the rate of hydrolysis of diaryl disulphide; but, when X = OEt, step (5) does not take place. When X = SCN, attack occurs at each of the sulphur atoms; but when X = CN, attack occurs exclusively at carbon with loss of thiophenoxide.[253] The exchange of ArSSAr with thiophenoxide occurs with a rate of 1.5×10^4 l mol^{-1} sec^{-1} at 70°.[254] The reactivity of L-ergothioneine and other sulphur heterocycles and thioamides as nucleophiles at sulphur in disulphides has been investigated.[254a] The reaction of triphenylphosphine with diaryl disulphides is irreversible only at high or low pH; *meta*-substituents give a ρ of 2.94, while the *para*-substituents lie above this line, as reported last year for cyanide as nucleophile; direct displacement at sulphur is proposed.[255] The reaction of tris(trimethylsilyl) phosphite with diphenyl disulphide has also been studied.[256] 4-Methoxybenzyl trichloromethanesulphenate is cleaved by ethoxide with attack at benzylic carbon, while 4-methoxybenzyl *o*-nitrobenzenesulphenate reacts exclusively at sulphur and very much more slowly.[257] The Cu_2Cl_2-catalysed alcoholysis of *p*-nitro- and *p*-chloro-*N*-sulphinylaniline has been investigated.[258] Alkyl isothiocyanates react with carbanions such as CCl_3^-, generated by phase-transfer catalysis, to give the alkyl sulphides and cyanide ion.[259] The four-membered ring in 1,8-endothionaphthalene is cleaved by $LiAlH_4$ or methyl-lithium with attack at sulphur.[260] Alkylmagnesium halides and alkyl-lithiums react with 2,1,3-thia- and -selena-diazoles by nucleophilic attack at sulphur or selenium.[260a]

The conversion of (–)-(**120**) to (**121**) occurs with retention of configuration at sulphur, and it is suggested that this is due to the electrophilic character of Hg^{2+}, with exchange axial-equatorial on the trigonal bipyramid as shown. *tert*-Butyl mercaptan converts (**121**) into (+)-(**120**) by an S_N2-S process.[261] Sulphoxides RS(O)R′ react with methyl- and *n*-butyl-lithium to give the methyl (or *n*-butyl) sulphoxide, with displacement of the

Me_2N—S(=O)—SBu^t ⟶ [Me_2N–S(=O)–$\overset{+}{S}Bu^t$ bridged by Cl—Hg—Cl] ⟶ Me_2N—S(=O)—Cl

(**120**) (**121**)

more electronegative of R and R′ as its lithium salt, exactly as found in alkaline hydrolysis of phosphonium salts; reaction takes place with inversion of configuration and an S_N2-S process is preferred to one involving addition and pseudorotation in a five-co-ordinate species.[262] The mechanism proposed in 1972 for racemization and ^{18}O-exchange of sulphoxides in concentrated H_2SO_4 has been confirmed by further studies.[263] The racemization and fragmentation of *tert*-butyl phenyl sulphoxide takes place by the mechanism of equation (10).

$$Bu^tS(O)Ph \rightleftharpoons [Bu^tS(OH)Ph]^+ \rightleftharpoons tert\text{-}Bu^+ + HOSPh \qquad (10)$$

The kinetic isotope effect for racemization is 1.37 per CD_3 group [2.6 for $C(CD_3)_3$], which is larger than that found by Shiner for ethanolysis of *tert*-butyl chloride.[263a] *N*-Toluene-*p*-sulphonylsulphilimines are hydrolysed by NaOMe to α-methoxy-sulphides (as well as sulphoxides) if they contain an α-hydrogen atom; rate-determining loss of toluene-*p*-sulphonamide from an intermediate enolate is postulated.[264] DMSO reacts with epoxides in the presence of trinitrobenzenesulphonic acid to give salts (**122**). When

R₂C(–O–)CR′R″ $\xrightarrow[ArSO_3H]{DMSO}$ $R_2C(OH)$–$CR'R''(\overset{+}{O}SMe_2)$ $ArSO_3^-$

(**122**)

Ar = 2,4,6-$(NO_2)_3C_6H_2$

(**123**)

R′ or R″ = H, these salts decompose thermally to give ketone and dimethyl sulphide and react with methanol by attack at sulphur. However, when R′ and R″ are both alkyl groups, the reaction with methanol takes place exclusively at carbon.[22] The spiro-sulphurane (**123**) and its *S*-oxide have been prepared,[265] and their structures proved by

X-ray crystallography.[266] Other studies of reaction at sulphur in S(IV) species are reported in refs. 267 and 268.

The only recorded example of an S_N1-S reaction, the hydrolysis of 2,4,6-trimethylbenzenesulphonyl chloride,[269] has been shown to be of the first and not zero order in $[OH^-]$, both in alkaline and in neutral solution.[270] The alkaline hydrolysis of 4-substituted arenesulphonyl fluorides has a ρ of 2.79, and an addition–elimination mechanism with rate-determining attack is proposed.[271] Rate-determining loss of fluoride is suggested for the reaction of anilines with thiophen-2-sulphonyl fluorides, although the bromides and chlorides have the same rate-determining step as in hydrolysis,[272] as does the reaction of thiophen-3-sulphonyl chloride with anilines.[273] Water acts as both nucleophile and general base in the hydrolysis of arenesulphonyl chlorides in aqueous acetonitrile.[274] Halide exchange of arenesulphonyl bromides and methanesulphonyl chloride in acetone, acetonitrile or methanol is suggested to take place by an S_N2-S process.[275] Extended Hückel calculations on displacement at sulphur in RSX, RS(O)X, and RSO_2X show that the best reactivity index is given by the frontier electron density.[276] The HSAB principle has been applied to account for the change in order of nucleophilicities of halide ions when the substrate is Me_2NSO_2Cl rather than $PhSO_2Cl$.[277] An A-1 mechanism has been proposed for the acid- and Cu^{2+}- catalysed hydrolysis of 8-quinolyl sulphate,[278] while alkyl and aryl sulphamates are hydrolysed by an A-2 process.[279] The toluene-*p*-sulphonimidoyl chloride (**124**) has been prepared, and its absolute configuration determined by the sequence (**124**) → (**125**) → (**126**), whose configuration is known; the

N_2H_4 MeMgBr Cl_2

(**124**) (**125**) (**126**) (**127**)

R = (−)-Menthyl

conversion of (**125**) into the other diastereomer (**127**) of the starting material shows that the reduction step involves inversion, and a displacement (with inversion) to the hydrazide followed by loss of nitrogen is proposed; this is consistent with the inversion observed in the reaction of (**124**) with ammonia and with *p*-methylphenoxide.[280]

The episelenonium salt formed by Se-3 participation in $Se(CH_2CH_2Br)_2$ is attacked by cyanide ion at selenium to give $BrCH_2CH_2SeCN$ and ethylene, and by iodide ion to form $BrCH_2CH_2SeI$ which is aerially oxidized to the symmetrical diselenide. Sulphur or selenium nucleophiles, however, react at carbon to give, for example, $PhSCH_2CH_2Se$-CH_2CH_2Br.[281] *p*-Tolyltetraphenylantimony undergoes alcoholysis in neutral solution by an S_N1-Sb mechanism, while in alkaline solution an S_N2-Sb process, by way of an "ate" complex, operates.[282]

Other Elements

In an extensive investigation, the reactivity of nucleophiles at sp^3-hybridized tervalent nitrogen in $H_2NSO_3^-$ has been shown to cover a range of $10^{6.2}$, and the second-order rate constants are given in Table 4. The rates show some correlation against the E_n of the Edwards oxibase equation, and also against S_N2 rates with methyl iodide, but the best correlation is against displacement rates at platinum in $Pt(py)_2Cl_2$, showing that nitrogen is a soft electrophile intermediate between carbon and platinum.[283] Cyanide ion reacts at nitrogen in *N*,*N*-dibenzylhydroxylamine *O*-*p*-nitrobenzoate, but other nucleophiles promote elimination.[284]

Table 4. Reactivity of nucleophiles at nitrogen in $H_2NSO_3^-$

Nucleophile	EtS^-	Ph_3P	$S_2O_3^{2-}$	I^-	Et_3N	$HONH_2$	OH^-
$5 + \log k$	6.51	5.29	4.74	3.84	3.34	2.15	0.3

Ambident Nucleophiles

Ratios of *C*- to *O*-alkylation in the ethylation of the sodium enolate of ethyl acetoacetate vary with leaving group in EtX in a way consistent with the HSAB principle; however, these ratio are independent of total reactivity, as shown by the leaving groups OTs, FSO_3, and CF_3SO_3 with relative rates $1:4 \times 10^4:10^6$, and by the percentages of *C*-alkylation 21, 17 and 13%, respectively.[285] Solvent and structural effects have been studied for the alkylation of 2-phenyl-4,5;6,7-dibenzocyclohepta-4,6-diene-1,3-dione.[286] Dimedone reacts with diethyl(bromomethyl)maleate by initial *O*-alkylation (in acetone/K_2CO_3), and the product readily rearranges by a Cope reaction to the *C*-alkyl product.[287] Benzyl- and allyl-lithium react with cyclohexyl bromide and iodide in good yield; but-2-enyl- and 3-methylbut-2-enyl-lithium react at both C-1 and C-3.[288] The lithium salts of dienamino-ketones, esters and nitriles are alkylated at the position γ to the ketone (ester or nitrile) in THF.[289] Dienamines are alkylated mainly on the double bond nearest the nitrogen by but-2-enyl and cinnamyl bromide; apparent alkylation at the end of the system is mainly due to subsequent Cope rearrangement.[290] A copper-chelated Cope reaction has been suggested to account for the change from mainly α-allylation to an excess of γ-allylation when the enolate of ethyl β-*n*-propylcrotonate is changed from the lithium to the copper chelate.[291]

2-Nitro-4-(trifluoromethyl)benzenesulphenate anion reacts with methyl iodide exclusively at sulphur, whereas methylation with methyl fluorosulphenate gives a 15:1 mixture of the *O*-methyl and *S*-methyl products, in line with the HSAB principle.[292]

Substitution at Vinylic Carbon

The chemistry of vinyl and allenyl cations has been reviewed.[293] Extended Hückel MO calculations have been carried out on the three types of nucleophilic substitution at ethynyl-carbon, with alkanethiolate as nucleophile and halogen as leaving group. The reactivity pattern fits well with the experimental observations of an Ad_N-E mechanism for fluoride, competitive Ad_N-E, and β-addition–elimination–migration for chloride, Ad_N-E and attack at halogen for bromide, and exclusive attack at halogen for iodide.[294]

The rather rare β-addition–elimination–rearrangement has been observed in the reaction of (**128**) with butyl- and phenyl-lithium.[295]

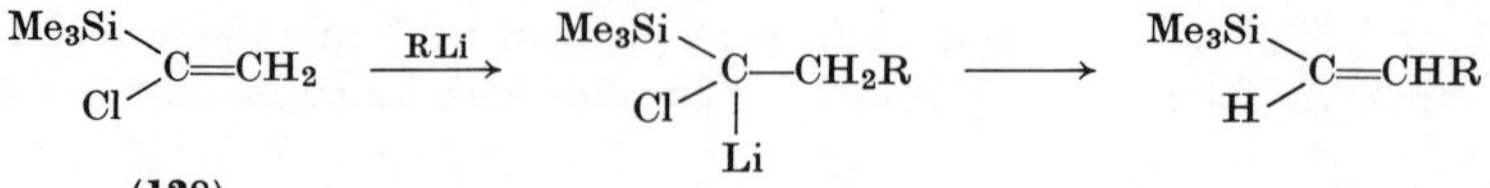

$$\mathrm{(Me_3Si)(Cl)C{=}CH_2 \xrightarrow{RLi} (Me_3Si)(Cl)C(Li){-}CH_2R \longrightarrow (Me_3Si)(H)C{=}CHR}$$

(**128**)

^1H and ^{13}C-NMR spectra of the vinyl cations from p-AnCF=CH_2, and p-AnCF=CMe_2 (An = p-methoxyphenyl) have been obtained in SbF_5/SO_2ClF solution at –78°. The two methyl groups in the latter give identical shifts, showing that the ion is linear.[296] A linear geometry has also been ascribed to the 1-(ferrocenyl)vinyl cation on the basis of the ^1H-NMR shifts and coupling constants; this cation is less stable than the corresponding 1-ferrocenylmethyl carbocation.[297]

Kinetically, alkyl-substituted vinyl cations appear to be bent, and a mechanism involving ion-pairs that are captured by solvent with net inversion of stereochemistry has been proposed on the basis of studies on *Z*- and *E*-(**129**), (**130**) and (**131**). Trifluoroacetolysis of (**129**) gives 7.4% net inversion, while acetolysis gives 53% inversion in the

$MeCH{=}CMeOSO_2CF_3$ (**129**)

$DCH{=}CBu^nOSO_2CF_3$ (**130**)

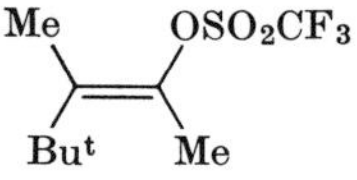

$(Me)(Bu^t)C{=}C(OSO_2CF_3)(Me)$ (**131**)

acetate product. Acetolysis of (**130**) occurs with 80% inversion, and 32% inversion is observed even for (**131**) where solvent capture by the "free" ion, generated by addition, occurs exclusively *trans* to the *tert*-butyl group.[298]

The use of k_{OTs}/k_{Br} rate ratios as a mechanistic criterion for carbonium-ion character of the transition state in a solvolysis reaction has been strongly criticized as a result of studies on the vinyl systems (**132**) and (**133**). Hoffmann had proposed that values of

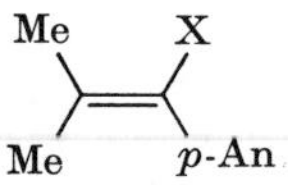

$Me_2C{=}C(X)(p\text{-An})$ (**132**)

X = Br or OTs

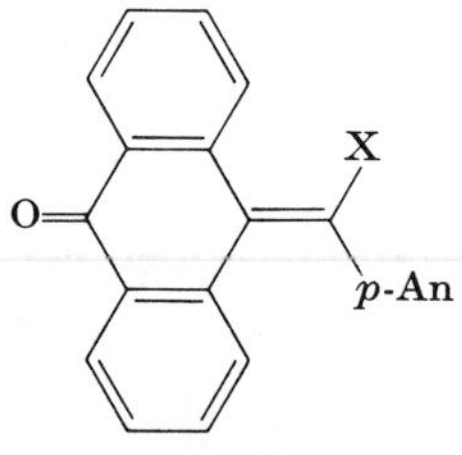

(**133**)

$k_{OTs}/k_{Br} < 1$ indicated an S_N2 type reaction, while values above 10^3 showed a limiting S_N1 process, with a general idea that the faster the reaction the more ionic it is and hence the larger the value of k_{OTs}/k_{Br}. However, (**133**; X = Br) reacts 300 times faster than 2-adamantyl bromide in acetic acid but has $k_{OTs}/k_{Br} = 5.15$ compared to 1844 for the 2-adamantyl system. Similarly (**132**; X = Br) acetolyses at one-fifth of the rate of (**133**; X = Br) and has k_{OTs}/k_{Br} 149 times larger. Further, trifluoroethanolysis of (**133**) gives a k_{OTs}/k_{Br} of 0.157. The vinyl systems both show common-ion depression and hence react by way of free ions. The investigators suggest that k_{OTs}/k_{Br} ratios reflect steric and solvent effects and not the degree of ionization at the transition state; they suggest that (**133**; X = Br) is sterically crowded and therefore accelerated, while when X = OTs it is not crowded and reacts at a retarded rate for some unclear reason.[299] The variation of solvolysis rate for (**132**; X = OTs, OBs, Br or Cl) and its *o*-methoxyphenyl isomer (X = OTs) with the mole fraction of water in aqueous trifluoroethanol shows a minimum at n_{H_2O} = ca. 0.25. It is shown that the *Y* value of aqueous trifluoroethanol mixtures depends strongly on the model compound, and it is suggested that the minimum observed in the rate for (**132**) is due to opposing effects on ionization rate and the partition of ion-pairs between free ions and return.[300] An explanation based on the differing nucleophilicities of water and trifluoroethanol as advanced in ref. 59 would seem more reasonable. A study of solvent effects on the solvolysis of vinyl bromide and toluene-*p*-sulphonate has shown that the stabilization of the transition state (even when LIM) by solvation is much less than that observed in the hydration of acetylenes which should give the same vinyl cation.[301]

Further work on isotope effects in vinyl cations has confirmed the results reported last year on the strong geometric requirement for interaction. Thus (**134**) and (**135**) give a

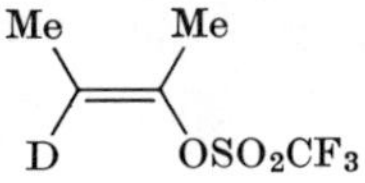

Me Me D OSO_2CF_3

(**134**)

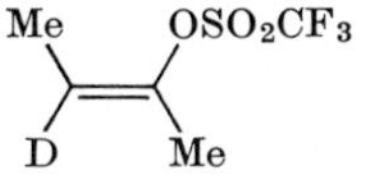

Me OSO_2CF_3 D Me

(**135**)

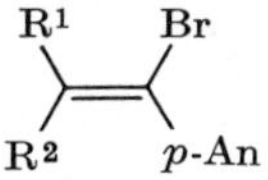

R^1 Br R^2 *p*-An

(**136**)

k_H/k_D of 1.25 and 2.01, respectively, in 60% ethanol at 75°. The product isotope effects for elimination are 1.26 and 1.95, respectively, showing that the transition states for elimination must be extremely close to the vinyl cation.[302] Although (**136**; R^1 = H, R^2 = D, and R^1 = D, R^2 = H) show no detectable isotope effect in 50% ethanol, the *E*-isomer has k_H/k_D = 1.32 in acetic acid and the *Z*-isomer one of 1.05, in line with the results on (**135**) and (**134**).[303] A systematic study of variation of the alkyl group in aqueous ethanolysis and in trifluoroethanolysis of vinyl trifluoromethanesulphonates has shown that a group *trans* to the leaving group exerts a much larger effect (ρ^* = −4.2) than when it is *cis* (ρ^* = −1.7). Similarly, the effects of groups on the α-carbon are enhanced when a methyl group is *cis* to the leaving group (ρ^* = −4.7 *vs* −2.0 for the *trans*-isomers).[304]

The full paper on the solvolysis of substituted bromomethylenecyclopropanes has been published. As reported last year, when R = Ph in (**137**), no cyclobutane is formed, but decreased stabilization of (**138**, R = H or Me) results in major amounts of cyclobutanone (the exclusive product when R = H).[305] The cyclobutenyl cation (**139**) is trapped as the vinyl ether when trifluoroethanol is used as solvent, and this also allows observation of its formation by the π-route from pent-3-ynyl trifluoromethanesulphonate, which gives

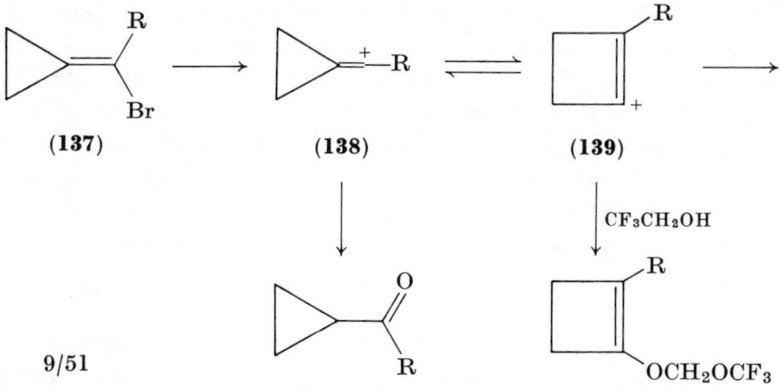

R Br

(**137**)

+ R

(**138**)

R +

(**139**)

R O

CF_3CH_2OH

O R

R OCH_2OCF_3

(**140**)

86% of (**140**) and only 4% of the acetylenic product.[306] Methyl-substitution on the ring of (**141**) results in a rate increase for either the *Z*- or the *E*-isomer; the identical nature of the products from each indicates rate-determining formation of (**142**) with subsequent partitioning three ways as shown, the major path involving the more substituted bond a.[307] The chlorolium ion (**143**) is formed in the silver-ion-assisted acetolysis of *E*,*Z*-1-bromo-2-chloro-1,4-diphenylbutadiene and of its *Z*,*Z*-isomer; chlorine participation occurs subsequently to formation of the vinyl cation.[308] Similarly, acetolysis of (**144**; X = Cl, Br or I) shows that chlorine and bromine do not participate, whereas iodine provides anchimeric acceleration and product stereospecificity.[309] It has been suggested that the rate ratio of ca. 6 between the *trans*- and *cis*-isomers of (**145**) is due to hydrogen participation in the former; however, the products show the same percentage of cyclopropyl ethyl ketone from each isomer.[310] The amount of ^{14}C-scrambling in solvolysis of 1,2,3-triphenyl[2-^{14}C]vinyl trifluoromethanesulphonate varies with the solvent (HOAc, HCOOH or CF_3COOH) in a way consistent with phenyl shift in an ion-pair whose lifetime varies with nucleophilicity of the solvent.[311] A kinetic study of displacement by alkoxide in triarylvinyl chlorides and bromides has been carried out.[312] The allenyl (vinyl) cation is a substantial contributor to the resonance hybrid of ethynylcarbinyl cations as evidenced by ^{13}C- and ^{1}H-NMR studies.[313] Allenyl bromides react with acetoacetate anions by S_N2' displacement. Even when the γ-carbon is disubstituted, only 40% of vinyl substitution is observed.[314] S_N2' is also the exclusive reaction between dialkylamines and α,γ-disubstituted allenyl bromides.[315] Vinyl toluene-*p*-sulphonates of the type RC≡CC(OTs)=CHR′ react with morpholine by an S_N2' process.[316]

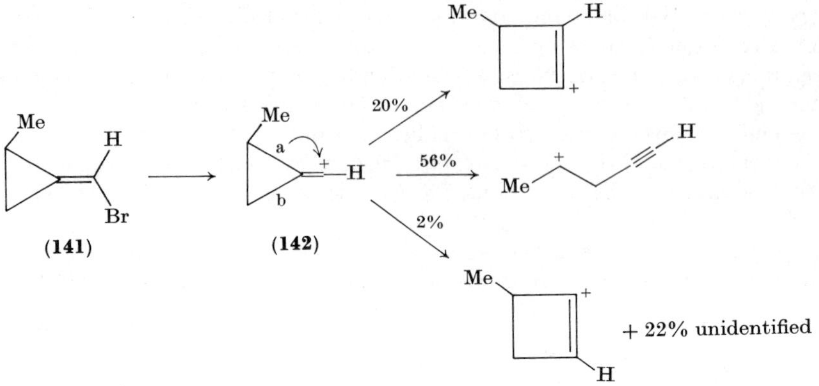

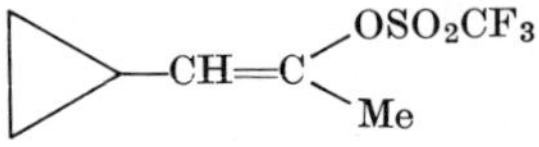

The *E*- and the *Z*-isomers of ethyl 3-halo-2-methylbut-2-enoate (halogen = Cl, Br or I) react with ethoxide and *tert*-butyl mercaptide by an elimination–addition route, as shown by incorporation of deuterium from the solvent. In contrast, reaction with ethane- and benzene-thiolates occurs by an Ad_N-E process without incorporation of deuterium and with 90–95% retention of configuration.[317] 3-Halobut-3-enoates and their 2-methyl homologues react with RSM (R = Et, Bu^t or Ph; M = metal) to give mainly the non-conjugated esters by the elimination–addition process; but amines give mainly the conjugated product.[318] Sulphur nucleophiles react with 2,3-dihalobut-2-enonitrile to displace the 3-halogen (with 80–100% retention) when one equivalent is used, but both halogen atoms are removed when two equivalents are employed.[319] 3-Halobicyclo[3.2.1]-oct-3-en-2-ones react with thiolates by an Ad_N-E process with migration of sulphur from C-4 to C-3.[320] The rate of displacement of halogen by amines in ArCOCH=CHX is only slightly affected by solvent when the amine is primary or secondary, but markedly so when the amine is tertiary; it is suggested that this is due to intramolecular proton-transfer in the Ad_N step for primary and secondary amines with little charge development, whereas a zwitterionic transition state is formed in the tertiary case.[321] Dimethyl-copperlithium displaces thio-groups which are β on $\alpha\beta$-unsaturated ketones and esters.[322] Further work on the elimination–addition route in cycloheptenyl,[323] and bicyclo-octenyl halides has been reported.[324] Other vinyl displacements are recorded in refs. 325 and 326.

Reactions of α-Halocarbonyl Compounds

α-Keto-carbocation and related reactions have been reviewed.[327] A comparison of the energy difference between CH_3CH_3 and $CH_3CH_2^+$ and that between CH_3CHO and $^+CH_2CHO$ has been carried out by perturbational and *ab initio* methods at the STO-3G level. When standard geometries are used, the second reaction is 15 kcal mol^{-1} more difficult than the first. Minimization with respect to bond lengths results in a net stabilization for (**146**) of 1.2 kcal mol^{-1} and puts (**146**) lower than (**147**) by 10 kcal mol^{-1}. However,

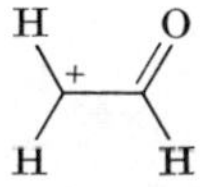

H O H H 85° 130°

(**146**) 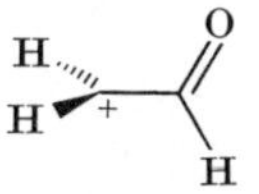 (**147**) 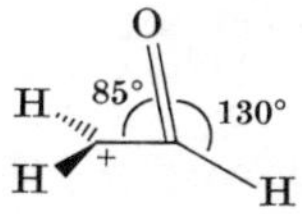 (**148**)

angular distortion in (**147**) leads to (**148**) as an energy minimum with significant overlap of the oxygen lone-pair with the vacant *p*-orbital, and this reduces the difference between $C_2H_6 \rightarrow C_2H_5^+$ and $CH_3CHO \rightarrow$ (**148**) to only 6.5 kcal mol^{-1}.[328] 2-Bromodecanoic acid and its methyl ester are isomerized to (**149**) and (**150**), as well as to their straight-chain homologues, by $AgSbF_6$ in tetrachloroethylene. In order to account for the difference from acid-catalysed isomerization of olefinic acids, a mechanism involving conversion of a zwitterionic α-carboxy-carbocation into the protonated cyclopropane (**151**), and

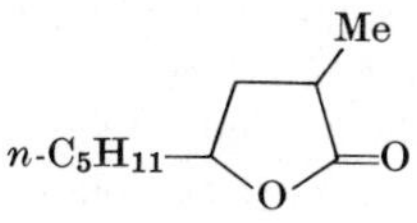

Me n-C_5H_{11} O O Me O O Bu^n n-C_6H_{13}—CH_2---CH—C O CH_2 O

(**149**) 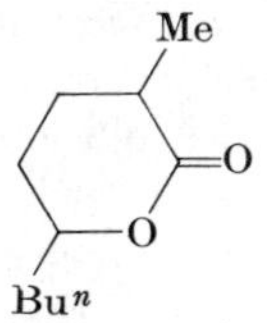 (**150**) 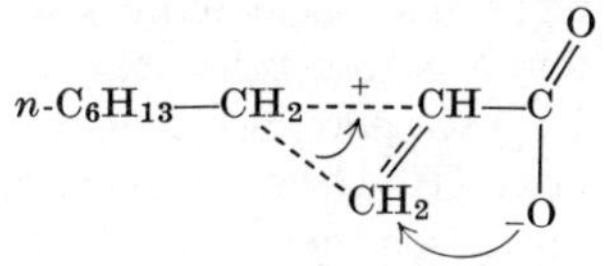 (**151**)

subsequent ring-opening *via* a β-lactone is proposed.[329] The Favorskii reaction of 1-benzoyl-1-bromo-4-*tert*-butylcyclohexane catalysed by $AgSbF_6$ in alcohol probably does not involve an α-keto-carbocation.[330]

Acetolysis of the chloro-ketone (**152**) gives mainly rearranged acetate (**153**) when R = H or Me but gives unrearranged acetate when R = Ph. A mechanism involving reaction by way of initial enolization towards R, and rate-determining attack on the ion-pair produced by subsequent ionization of the allylic chloride, is supported by the lack of reaction of (**152**) when R = Bu^t, which enolizes extremely slowly even in alkaline solution.[331] (**154**) also acetolyses by way of the enol, whose formation is rate-determining.[332] Base-catalysed displacement in 2-chloro-4-isopropylidene-2,3,3-trimethylcyclobutanone and 10-chloro-10-methylbicyclo[7.2.0]undec-1-en-11-one occurs by way of a dienol.[333]

$PhSCHCOCH_2Cl$ (with R on the CH)

(**152**)

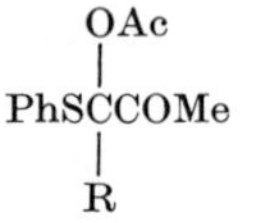

$PhSC(OAc)(R)COMe$

(**153**)

O, O, O, Br

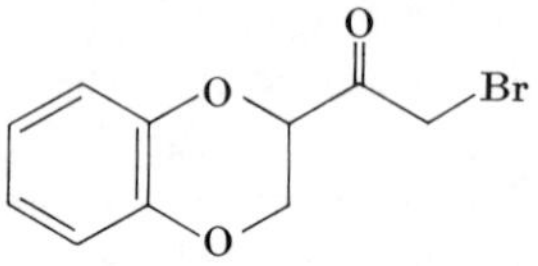

(**154**)

Rate-determining attack at the carbonyl group of chloral and α,α,α-trichloroacetophenone in the Perkow reaction with acyclic, cyclic and bicyclic phosphites has been adduced on the basis of the correlation of reactivity with their reactions with alkyl halides and with biacetyl.[334]

Rates and activation parameters have been determined for the 50% ethanolysis of 2-(bromoacetyl)-thiophens and -selenophens.[335] Semibenzilic Favorskii reactions have been suggested to account for the products of reaction of 2-bromo-2-methyldimedone and 2,2-dibromodimedone with sodium acetate.[336] The reactions of the oxime of phenacyl bromide and of its *O*-alkyl ethers with nucleophiles has been studied.[337]

S_N2 Processes and Other Reactions

Extended Hückel calculations for S_N2 reactions at a cyclopropane-carbon atom predict that they should occur by way of a trigonal-bipyramidal complex and with retention of configuration at carbon.[338] EHMO calculations have also been made for the chloride ion–methyl bromide reaction.[339]

The flowing afterglow technique has been used to follow the reactions of anions generated in the gas phase and follow their reactions with methyl chloride and fluoride; comparison between the observed rate and the capture rate enable activation energies to be measured. For $H^- + CH_3F$ a value of 3.6 kcal mol^{-1} is found, in good agreement with that (3.8) determined by *ab initio* calculations and reported in 1972. The value found for the $CN^- + CH_3F$ reaction does not agree so well (5 kcal mol^{-1} found *vs* 17 calculated).[340] Gas-phase nucleophilic reactions have also been investigated by the trapped-ion, pulsed-ion cyclotron resonance method. Reactions occur with 90% inversion of configuration (±15%); and absolute rates are less than the collision frequency, showing that a barrier exists. However, the results do not correlate with either the ion-dipole or the Gioumousis–Stevenson theory.[341] Chemical ion-mass spectroscopy of alkyl halides (Cl, Br or I) allows observation of the RX_2^- ions and evaluation of their heats and entropies of gas-phase association; comparison with the Finkelstein reactions in solution shows that nucleophilicity and alkyl-group effects are due entirely to solvation.[342]

The work reported in 1972 on reaction of nucleophiles with neopentyl toluene-*p*-sulphonate in HMPT has been extended.[343] Evidence has been presented in favour of an S_N2 reaction at tertiary carbon in the compounds (**155**)–(**158**). In DMF each reacts cleanly with LiN_3 to form the corresponding azide; (**156**) does not form any rearranged product and reacts with NaSCN in methanol to give solely thiocyanate and no solvolysis product; neither (**155**) nor (**156**) gives solvolysis products in 60% methanol, and (**157**) reacts with azide ion nine times faster than (**155**); finally, (**158**) gives azide of inverted configuration, whereas the corresponding acid reacts to afford azide of retained configuration.[344]

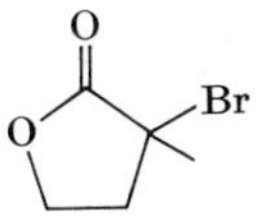

(155)	(156)	(157)	(158)
O, Br	$PhCOCMe_2$ / Br	O, Br	Me / EtCCOOEt / Br

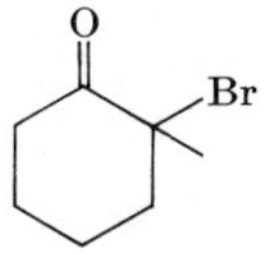

The three-states criterion has been advocated for the detection of nucleophilic reactivity in the conjugate acid, HNu, of an anionic nucleophile Nu^-.[345] The rates of reaction of 14 amines with ethylene oxide in aqueous solution are correlated by the N_+ values of Ritchie and Virtanen,[346] but not by their pK_a values.[347] The rates of attack on methyl iodide by heterocyclic sulphur nucleophiles correlate with the ionization potential of the *n* orbital, but not with that of the π one.[348] The nucleophilicity of the anions from dimethyl phosphate, *N,N,O*-trimethylphosphoramide, methyl methylphosphonate and *O,O′*-dimethyl phosphothioate towards 2-bromopentane have been studied.[349] Other studies of the nucleophilicity of P(V) acids have also been carried out.[350] Rates of reaction of thio-, seleno- and telluro-cyanate anions with benzyl bromide have been measured in acetonitrile.[351] Thiols and glutathione react with benzene oxide with a Brønsted $\beta = 0.22$, indicating an early transition state for the trapping of the zwitterion by thiolate anions.[352] The relative rate of reaction of dibutylcopperlithium with alkyl bromides shows the same profile as that for a typical S_N2 process, and therefore this mechanism has been proposed.[353] Hydrogen peroxide anion is 55 times more nucleophilic than hydroxide ion towards butyl methanesulphonate.[354] Trifluoromethanesulphinate anion is nucleophilic towards primary bromides and gives sulphones.[355] Methyl esters are cleaved in the presence of ethyl esters by S_N2 attack of CN^- in HMPT. The rates correlate with σ and σ^* for aromatic and aliphatic acids, respectively.[356] Kinetic asymmetric induction is observed when the enolate of a chiral oxazoline reacts with two equivalents of a secondary alkyl iodide; both recovered iodide and the acid formed by hydrolysis of the oxazoline are optically active.[357] Bicarbonate anion in DMSO is extremely nucleophilic (see ref. 126). Sodium diphenylphosphide reacts with both menthyl and neomenthyl chloride with complete inversion of configuration.[358] Complexes of iron, cobalt and molybdenum act as nucleophiles at the metal atom and react with 95, 95 and 80% inversion, respectively.[359] The reaction of copper(I) alkoxide and phenoxides with alkyl halides has been investigated.[360] Dimethylcopperlithium reacts with cyclic homoallyl toluene-*p*-sulphonates to give products of methylation of the cyclopropylcarbinyl system, e.g. cholesteryl toluene-*p*-sulphonate gives 6β-methyl-i-cholestane; however, acyclic compounds react normally and simple cyclopropylcarbinyl ones give the homoallyl products.[361] *n*-Butylhydridocopperlithium reacts as an H^- nucleophile and reduces alkyl bromides in the presence of esters, but not in presence of ketones.[362] Mercuric salts enhance the nucleophilicity of anions towards alkyl halides.[363] The reactivity of mono-*tert*-butyl phosphite towards alkyl halides has been measured,[364] Trialkylarsine oxides react at oxygen with butyl iodide, and their reactivity suggests an S_N2 reaction.[365] Alkanethiolate anions may be generated by electrochemical reduction of dithianes in DMF or acetonitrile.[366]

It has been suggested that the transition state for S_N2 reactions is a hybrid of the canonical forms (**159**) as a result of studies on benzenethiolate with 1-BuX and 2-BuX in aqueous ethanol. The ratio of rate constants $k_{1\text{-Bu}}/k_{2\text{-Bu}}$ varies with X (Br, I, OTs, OMs or ONs) in the same way that the $E_{1/2}$ for reduction of MeX does and correlates with it.

$$\mathrm{Nu^- \;\; C{-}L \longleftrightarrow Nu{-}C \;\; L^- \longleftrightarrow Nu^- \;\; C^+ \;\; L^- \longleftrightarrow Nu^{\bullet} \;\; {}^{\bullet}C \;\; L^-}$$

(**159**)

Reaction of PhS^- with 1-methylpropyl *p*-nitrobenzenesulphonate in the presence of *tert*-butyl phenyl nitrone gives spin-trapped 1-methylpropyl radical and in the presence of styrene leads to polymer formation, yet the reaction occurs with 99–100% inversion of configuration.[367] Rates and activation parameters have been measured in a study of the effects of solvent and leaving group on the displacement of sulphonate by azide in monosaccharides.[368] Dibenzo[18]crown-6 has been used to carry out the silver nitrate-catalysed reaction of glycosyl bromides with alcohols under homogeneous conditions.[369] Primary, secondary and tertiary alkyl salicylates are converted into the corresponding chlorides in 80–95% yield by reaction with PCl_5,[370] and triphenylphosphine dibromide reacts with alkyl acetates to give the alkyl bromides.[371] Alcohols react with $Me_2NCH(OCH_2Bu^t)_2$ in the presence of acetic acid to give acetates of inverted configuration.[372] *N*-(Arenesulphonyl)methylsulphilimines react with arenethiolates to form the methyl thioethers; with 1-methylheptylsulphilimine, 94% inversion of configuration was found.[373] Hexahydro-1,1-dimethyl-3*H*-2,4,7-ethanylidyne-1*H*-cyclopenta[*c*]thiapyrilium bromide reacts with nucleophiles exclusively at the methylene group.[374] Dealkylation of trialkyl phosphites by HCl involves protonation on phosphorus followed by nucleophilic attack of HCl_2^- on the alkyl group.[375] Ether cleavage by acetyl chloride with either stannic or zinc chloride follows an S_N1 route, the less substituted alkyl group forming alkyl acetate, while the more substituted one gives alkyl chloride with loss of configuration.[376] The dealkylation of phenolic ethers[377] and of *O*-methylcoumarins by pyridine hydrochloride has been investigated.[378] The validity of linear free-energy relationships at various temperatures and their relation to the isokinetic temperature have been discussed.[379] Activation parameters have been determined for bromide exchange and hydrolysis of five benzoylglycosyl bromides in acetone; the former is an S_N2 and the latter an S_N1 process.[380] The effect of H_0 and counterion on the $E_a^{\neq}$ and $\Delta S^{\neq}$ for hydrolysis of α- and β-anomers of methylglucopyranosides have been studied.[381] 2,6-Dioxabicyclo[2.2.2]octane gives a stereoregular polymer by an S_N2 process when the catalyst is SiF_4, PF_5 or BF_3, but a stereorandom one by an S_N1 reaction when it is $MeSO_3H$ or CF_3COOH.[382] Benzyltrimethylammonium salts are useful phase-transfer catalysts for the alkylation of phenols in methylene chloride–water mixtures.[383] The reaction of diethyl pyrocarbonate with nucleophiles has been investigated.[384] Activation parameters have been determined for the hydrolysis of bis(chloromethyl) ether in neutral, acid and alkaline water.[385] Rates and equivalent conductances of potassium ethoxide have been determined in a study of the effects of pressure on the reaction of KOEt with ethyl bromide.[386] The kinetics of the hydrolysis of 2-chloroethanol and 3-chloropropane-1,2-diol in aqueous sodium hydrogen carbonate have been reported.[387] Activation energies have been determined in a study of solvent and medium effects on the solvolysis of *tert*-butyl chloride.[388] The alkaline hydrolysis of trifluoromethyl groups in a series of trifluoromethylarenes has been studied kinetically.[389] The activation parameters for the hydrolysis of chloroacetate anion in aqueous alkaline methanol go through a minimum value when the mole fraction of water is about 0.7.[390] A study of the alkaline hydrolysis of 2-chloropropionic acid has been reported.[391]

Allylic alcohols and acetals are reduced by an S_N2' process, and with *syn*-faciality by lithium aluminium hydride.[392] The σ-palladium complex (**160**) is converted into the

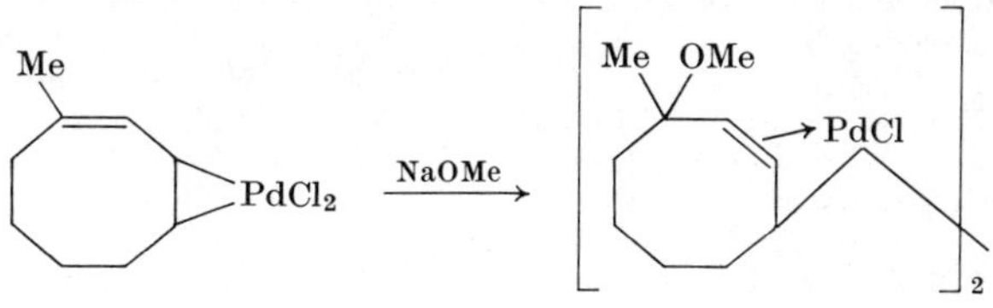

(**160**)

π-allyl complex by an S_N2' process.[393] Tertiary allylic alcohols afford mixtures of the *E*- and *Z*-primary allyl chlorides when treated with thionyl chloride.[394] The bisethynyl bromide $Bu^tC{\equiv}CCHBrC{\equiv}CBu^t$ reacts with nucleophiles by direct and not S_N2' displacement.[395] 1,3-Dialkyl-5-bromouracil reacts with nucleophiles to afford the uracil substituted in the 6-methyl group by isomerization to the 6-methylene-5*H*-compound and S_N2' displacement of bromide.[396]

Quaternization continues to attract considerable interest, and a wide variety of aromatic and aliphatic amines have been studied.[397]

Ring-opening of epoxides continues to extract major efforts; about equal interest has been displayed in nucleophilic[398] and in electrophilic[399] opening reactions. The reactions of arene oxides have been reviewed.[400] There has also been much work on the ring-opening of aziridines and aziridinium ions,[401] and nucleophilic ring-opening of thiirans and thiiranium ions has been studied.[402]

Arylhydroxamate and aryl-*N*-methylhydroxamate anions show α-effects of 66 and 150, respectively, relative to phenoxide anions.[403] Through-bond coupling of the nitrogens in 1,4-diazabicyclo[2.2.2]octane leads to enhanced reactivity, as compared to quinuclidine, in nucleophilic reactions; orbital diagrams suggest splitting of the HOMO level and hence a larger perturbational stabilization of the transition state.[404]

A further example of nucleophilic ring-opening of a cyclopropane has been reported.[405] A σ2s + σ2a pericyclic mechanism has been proposed for the reaction of alcohols with the Ph_3P/CCl_4 reagent; this would explain the lack of competition by external nucleophiles and the differences in the products from the reaction of the corresponding toluene-*p*-sulphonates with lithium chloride; however, it is clear that the sigmatropic reaction must be competitive with ionization of the $ROPPh_3Cl$ intermediate, for cholesterol gives 34% of the i-cholesteryl phosphorane,[406] and *cis*-5-hydroxy-2-phenyl-1,3-dioxan gives 93% of a mixture of *cis*- and *trans*-4-bromomethyl-2-phenyl-1,3-dioxalane and only 7% of the *trans*-5-bromo-1,3-dioxan when treated with $Ph_3P{-}CBr_4$.[407] Sugars may be conveniently benzylated by using benzyl trifluoromethanesulphonate.[408]

References

[1] *Hydrolysis of Alkyl Halides in the Presence of Metal ions*—M^+-S_N2, *and* M^+-S_N1 *Reactions*, E. S. Rudakov, I. V. Kozhevnikov and V. V. Zamaschikov, *Usp. Khim.*, **43**, 707 (1974); *Chem. Abs.*, **81**, 24536 (1974).
[2] J. M. Harris, *Progr. Phys. Org. Chem.*, **11**, 89 (1974).
[3] See *Org. Reaction Mech.*, **1972**, 58, ref. 18.
[4] See *Org. Reaction Mech.*, **1972**, 59.

[5] J. M. Harris, D. C. Clark, A. Becker and J. F. Fagan, *J. Am. Chem. Soc.*, **96**, 4478 (1974).
[6] J. M. Harris, A. Becker, J. F. Fagan and F. A. Walden, *J. Am. Chem. Soc.*, **96**, 4484 (1974).
[7] See *Org. Reaction Mech.*, **1970**, 59.
[8] T. W. Bentley, S. H. Liggero, M. A. Imhoff and P. von R. Schleyer, *J. Am. Chem. Soc.*, **96**, 1970 (1974).
[9] A. Sera, C. Yamagami and K. Maruyama, *Bull. Chem. Soc. Japan*, **46**, 3864 (1973).
[10] A. Sera, C. Yamagami and K. Maruyama, *Bull. Chem. Soc. Japan*, **47**, 704 (1974).
[11] *Org. Reaction Mech.*, **1973**, 295; **1972**, 53; **1971**, 50; **1970**, 63; **1969**, 72; **1966**, 44.
[12] G. A. Gregoriou, *Tetrahedron Letters*, **1974**, 233.
[13] D. J. McLennan, *J.C.S. Perkin II*, **1974**, 481.
[14] A. R. Stein, *Tetrahedron Letters*, **1974**, 4145.
[15] K. B. Becker and C. A. Grob, *Helv. Chim. Acta*, **56**, 2723 (1973).
[16] K. B. Becker, A. F. Boschung, M. Geisel and C. A. Grob, *Helv. Chim. Acta*, **56**, 2747 (1973); cf. *Org. Reaction Mech.*, **1971**, 12.
[17] K. B. Becker, A. F. Boschung and C. A. Grob, *Helv. Chim. Acta*, **56**, 2733 (1973).
[18] G. E. Gream, A. K. Serelis and T. I. Stoneman, *Austral. J. Chem.*, **27**, 1711 (1974).
[19] K. B. Wiberg and T. Nakahira, *Tetrahedron Letters*, **1974**, 1769, 1773.
[20] D. N. Kevill and L. Held, *J. Org. Chem.*, **38**, 4445 (1973).
[21] D. N. Kevill and C. R. Degenhardt, *Chem. Comm.*, **1974**, 662.
[22] M. A. Khuddus and D. Swern, *J. Am. Chem. Soc.*, **95**, 8393 (1973).
[23] G. Maury and N. H. Cromwell, *Bull. Soc. Chim. France*, **1974**, 623.
[24] A. F. Diaz and N. Assamunt, *Tetrahedron*, **39**, 797 (1974).
[25] A. F. Diaz and M. Blanco, *J. Org. Chem.*, **39**, 1313 (1974).
[26] C. Monge, J. L. Palmer and A. F. Diaz, *J. Org. Chem.*, **39**, 1920 (1974); A. F. Diaz, C. Monge, J. L. Palmer and A. Santos, *Rev. Latinoamer. Quim.*, **4**, 161 (1973); *Chem. Abs.*, **80**, 145237 (1974).
[27] K. Okamoto, T. Kinoshita and O. Makino, *Bull. Chem. Soc. Japan*, **47**, 1770 (1974).
[28] A. Sera, T. Miyazawa, T. Matsuda, Y. Togawa and K. Maruyama, *Bull. Chem. Soc. Japan*, **46**, 3490 (1973).
[29] K. M. Koshy and R. E. Robertson, *Can. J. Chem.*, **52**, 2485 (1974).
[30] J. Delhoste, *Bull. Soc. Chim. France*, **1974**, 133.
[31] F. Quemeneur, B. Bariou and M. Kerfanto, *Compt. rend., Ser. C*, **278**, 299 (1974).
[32] S. Braverman and T. Globerman, *Tetrahedron*, **30**, 3873 (1974).
[33] W. E. Parham, P. E. Olson, K. R. Reddy and K. B. Sloan, *J. Org. Chem.*, **39**, 172 (1974).
[34] A. K. Colter, A. L. McKenna and M. A. Kassem, *Can. J. Chem.*, **52**, 3748 (1974).
[35] A. R. Stein, *J. Org. Chem.*, **38**, 4022 (1973).
[36] P. Beronius, *Acta Chem. Scand., A*, **28**, 77 (1974).
[37] M. H. Abraham and R. J. Abraham, *J.C.S. Perkin II*, **1974**, 47.
[38] J. A. Miller and M. J. Nunn, *Tetrahedron Letters*, **1974**, 2691.
[39] K. Humski, V. Sendijarević and V. J. Shiner, *J. Am. Chem. Soc.*, **96**, 6187 (1974).
[40] J. Mindl and N. Večeřa, *Coll. Czech. Chem. Comm.*, **38**, 3496 (1973).
[41] K. Okamoto and T. Kinoshita, *Chem. Letters (Tokyo)*, **1974**, 1037.
[42] M. H. Abraham, *Progr. Phys. Org. Chem.*, **11**, 1 (1974).
[43] E. V. Dehmlow, *Angew. Chem. Internat. Ed.*, **13**, 170 (1974).
[44] C. Kappenstein, *Bull. Soc. Chim. France*, **1974**, 89.
[45] A. C. Coxon and J. F. Stoddart, *Chem. Comm.*, **1974**, 537.
[46] J. F. Stoddart and C. M. Wheatley, *Chem. Comm.*, **1974**, 390.
[47] D. Landini, F. Montanari and F. M. Pirisi, *Chem. Comm.*, **1974**, 879.
[48] D. Landini, F. Montanari and F. Rolla, *Synthesis*, **1974**, 428.
[49] D. Landini and F. Rolla, *Synthesis*, **1974**, 565.
[50] C. D. Hurd, *J. Org. Chem.*, **39**, 3144 (1974).
[51] C. L. Liotta, H. P. Harris, M. McDermott, T. Gonzalez and K. Smith, *Tetrahedron Letters*, **1974**, 2417.
[52] H. D. Durst, *Tetrahedron Letters*, **1974**, 2421.
[53] C. L. Liotta and H. P. Harris, *J. Am. Chem. Soc.*, **96**, 2250 (1974).
[54] D. J. Sam and H. E. Simmons, *J. Am. Chem. Soc.*, **96**, 2252 (1974).
[55] P. J. Daïs and G. A. Gregoriou, *Tetrahedron Letters*, **1974**, 3827.
[56] F. L. Schadt, P. von R. Schleyer and W. Bentley, *Tetrahedron Letters*, **1974**, 2335.
[57] See *Org. Reaction Mech.*, **1973**, 296.
[58] A. Pross and R. Koren, *Tetrahedron Letters*, **1974**, 1949.
[59] J. M. Harris, D. J. Raber, W. C. Neal and M. D. Dukes, *Tetrahedron Letters*, **1974**, 2331.

[60] W. T. Ford, R. J. Hauri and D. J. Hart, *J. Org. Chem.*, **38**, 3916 (1973).
[61] W. T. Ford, R. J. Hauri and S. G. Smith, *J. Am. Chem. Soc.*, **96**, 4316 (1974).
[62] Y. Yano, T. Okonogi and W. Tagaki, *J. Org. Chem.*, **38**, 3912 (1973).
[63] G. Auzou and R. Uzan, *Compt. rend., Ser. C.* **279**, 163 (1974).
[64] G. Punnaiah and E. V. Sundaram, *Z. Phys. Chem.* (*Frankfurt*), **89**, 188 (1974).
[65] F. Chastrette and M. Chastrette, *Compt. rend., Ser. C*, **277**, 1251 (1973).
[66] B. V. Gidaspov, N. I. Alekseev, Y. M. Kalmykov and N. V. Margolis, *Izv. Vyssh. Uchebn. Zaved., Khim. Khim. Tekhnol.*, **17**, 1030 (1974); *Chem. Abs.*, **81**, 119448 (1974).
[67] S. Villermaux and J. J. Delpuech, *Bull. Soc. Chim. France*, **1974**, 2541.
[68] M. Chastrette, H. Gauthier-Countani and R. Gauthier, *Bull. Soc. Chim. France*, **1974**, 229.
[69] M. J. Blandamer and J. R. Membrey, *J.C.S. Perkin II*, **1974**, 1400.
[70] C.-P. Wong, L. M. Jackman and R. G. Portman, *Tetrahedron Letters*, **1974**, 921.
[71] *Org. Reaction. Mech.*, **1972**, 61.
[72] D. J. Raber, M. D. Dukes and J. Gregory, *Tetrahedron Letters*, **1974**, 667.
[73] K. Saramma, *Tetrahedron Letters*, **1974**, 2775.
[74] B. G. Cox and A. Gibson, *Chem. Comm.*, **1974**, 638.
[75] G. R. Hedwig and A. J. Parker, *J. Am. Chem. Soc.*, **96**, 6589 (1974).
[76] J. H. Fendler, F. Nome and H. C. Van Woert, *J. Am. Chem. Soc.*, **96**, 6745 (1974).
[77] J. W. Larsen and L. M. Magid, *J. Am. Chem. Soc.*, **96**, 5774 (1974).
[78] J. H. Exner and E. C. Steiner, *J. Am. Chem. Soc.*, **96**, 1782 (1974).
[79] H. W. Aitken and W. R. Gilkerson, *J. Am. Chem. Soc.*, **95**, 8551 (1973).
[80] S. Villermaux and J. J. Delpuech, *Bull. Soc. Chim. France*, **1974**, 2534.
[81] O. M. Bhatnagar and A. N. Campbell, *Can. J. Chem.*, **52**, 203 (1974).
[82] B. G. Cox, G. R. Hedwig, A. J. Parker and D. W. Watts, *Austral. J. Chem.*, **27**, 477 (1974); R. Alexander, D. A. Owensby, A. J. Parker and W. E. Waghorne, *Austral. J. Chem.*, **27**, 933 (1974).
[83] N. Matsuura and K. Umemoto, *Bull. Chem. Soc. Japan*, **47**, 1334 (1974).
[84] S. Backlund, F. Eriksson and R. Friman, *Acta Chem. Scand.*, **27**, 3234 (1973); S. Backlund and K. Palm, *ibid.*, *A*, **28**, 595 (1974).
[85] C. Jolicoeur and P. R. Philip, *Can. J. Chem.*, **52**, 1834 (1974).
[86] T. Takamatsu, *Bull. Chem. Soc. Japan*, **47**, 1285, 1287 (1974).
[87] P. C. Vogel, *Can. J. Chem.*, **52**, 1937 (1974).
[88] J. Bron, *Can. J. Chem.*, **52**, 903 (1974).
[89] J. Bron, *Can. J. Chem.*, **52**, 3078 (1974).
[90] D. Cook, R. E. J. Hutchinson and A. J. Parker, *J. Org. Chem.*, **39**, 3029 (1974).
[91] K. M. Koshy and R. E. Robertson, *J. Am. Chem. Soc.*, **96**, 914 (1974).
[92] V. F. Raaen, T. Juhlke, F. J. Brown and C. J. Collins, *J. Am. Chem. Soc.*, **96**, 5928 (1974).
[93] D. G. Graczyk and J. W. Taylor, *J. Am. Chem. Soc.*, **96**, 3255 (1974).
[94] R. C. Williams and J. W. Taylor, *J. Am. Chem. Soc.*, **96**, 3721 (1974); cf. *Org. Reaction Mech.*, **1973**, 300.
[95] A. J. Kresge and W. J. Albery, *Chem. Comm.*, **1974**, 507.
[96] F. C. Haupt and M. R. Smith, *Tetrahedron Letters*, **1974**, 4141; cf. *Org. Reaction Mech..* **1973**, 275.
[97] A. Aumelas, A. Casadevall, E. Casadevall and C. Largeau, *Tetrahedron*, **30**, 3897 (1974).
[98] T. Suami, S. Ogawa, S. Oki and H. Sato, *Bull. Chem. Soc. Japan*, **47**, 1731, 1737 (1974).
[99] G. Kyriakou and J. Seyden-Penne, *Tetrahedron Letters*, **1974**, 1737; *Org. Reaction Mech.*, **1970**, 72.
[100] P. Coutrot, *Bull. Soc. Chim. France*, **1974**, 1965; P. Coutrot, C. Legris and J. Villieras, *ibid.*, p. 1971.
[101] Z. Chabudzinski, Z. Rykowski and K. Burak, *Rocz. Chem.*, **47**, 2305 (1973); *Chem. Abs.*, **80**, 119815 (1974).
[102] A. S. Gudkova, E. I. Troyanskii and O. A. Reutov, *Dokl. Akad. Nauk S.S.S.R.*, **210**, 855 (1973).
[103] Y. I. Gevaza and V. I. Staninets, *Ukk. Khim. Zh.*, **39**, 1047 (1973); *Chem. Abs.*, **80**, 36454 (1974).
[104] M. Santelli, *Chem. Comm.*, **1974**, 214.
[105] J. M. Hornback, *J. Org. Chem.*, **38**, 4122 (1973); cf. *Org. Reaction Mech.*, **1970**, 73.
[106] P. Wilder and C. van A. Drinnan, *J. Org. Chem.*, **39**, 414 (1974).
[107] G. Illuminati, L. Mandolini and B. Masci, *J. Am. Chem. Soc.*, **96**, 1422 (1974).
[108] D. B. Uliss, R. K. Razdan and H. C. Dalzell, *J. Am. Chem. Soc.*, **96**, 7372 (1974).
[109] R. P. Bell, D. W. Earls and B. A. Timimi, *J.C.S. Perkin II*, **1974**, 811
[109a] J. S. Brimacombe, J. Minshall and L. C. N. Tucker, *Carbohydrate Res.*, **35**, 55 (1974); C. K. De Bruyne, F. Van Wijnendaele and H. Carchon, *ibid.*, **33**, 75 (1974).
[110] I. Tabushi, Y. Tamaru and Z. Yoshida, *Bull. Chem. Soc. Japan*, **47**, 1455 (1974).
[111] S. Ikegami, T. Asai, K. Tsuneoka, S. Matsumura and S. Akaboshi, *Tetrahedron*, **30**, 2087 (1974).
[112] H. Kwart and D. Drayer, *J. Org. Chem.*, **39**, 2157 (1974).

[113] G. A. Olah, P. W. Westerman, E. G. Melby and Y. K. Mo, *J. Am. Chem. Soc.*, **96**, 3565 (1974); cf. *Org. Reaction Mech.*, **1973**, 304.
[114] W. J. Hehre and P. C. Hiberty, *J. Am. Chem. Soc.*, **96**, 2665 (1974).
[115] T. A. Smolina, A. A. Medvedev, L. N. Batyuk, L. Y. Ignatov and O. A. Reutov, *Dokl. Akad. Nauk., S.S.S.R.*, **214**, 1091 (1974); T. A. Smolina, A. A. Medvedev, E. D. Gopius and O. A. Reutov, *ibid.*, **217**, 110 (1974); *Chem. Abs.*, **81**, 104417 (1974).
[116] T. A. Smolina, E. D. Gopius and O. A. Reutov, *Zh. Org. Khim.*, **10**, 1121 (1974); *Chem. Abs.*, **81**, 48971 (1974).
[117] A. S. Gudkova, K. U. Uteniyazov, S. G. Zavgorodnii and O. A. Reutov, *Dokl. Akad. Nauk. S.S.S.R.*, **210**, 595 (1974); A. S. Gudkova, K. Uteniyazov and O. A. Reutov, *ibid.*, **214**, 572 (1974); A. S. Gudkova, K. Uteniyazov, S. G. Zavgorodnii and O. A. Reutov, *ibid.*, p. 815.
[118] B. Giese, *Tetrahedron Letters*, **1974**, 3579.
[119] R. D. Wieting, R. H. Staley and J. L. Beauchamp, *J. Am. Chem. Soc.*, **96**, 7552 (1974).
[120] G. Bellucci, G. Ingrosso, F. Marioni, A. Marsili and I. Morelli, *Gazz. Chim. Ital.*, **104**, 69 (1974).
[121] G. Bellucci, A. Marsi, E. Mastrorilli, I. Morelli and V. Scartoni, *J.C.S. Perkin II*, **1974**, 201.
[122] T. A. Smolina, E. D. Gopius, V. N. Gruzdneva and O. A. Reutov, *Dokl. Akad. Nauk S.S.S.R.*, **209**, 872 (1973).
[123] H. Marschall, K. Tantau and P. Weyerstahl, *Chem. Ber.*, **107**, 887 (1974).
[124] L. A. Dotsenko and B. A. Ershov, *Zh. Org. Khim.*, **10**, 1342 (1974); *Chem. Abs.*, **81**, 90842 (1974).
[125] A. Hassner and S. S. Burke, *Tetrahedron*, **30**, 2613 (1974).
[126] N. Bosworth, P. Magnus and R. Moore, *J.C.S. Perkin I*, **1973**, 2694.
[127] S. Oae and T. Numata, *Tetrahedron*, **30**, 2641 (1974).
[128] H. L. Slates, Z. S. Zelawski, D. Taub and N. L. Wendler, *Tetrahedron* **30**, 819 (1974).
[129] P. L. Durette and H. Paulsen, *Chem. Ber.*, **107**, 937 (1974).
[130] P. L. Durette and H. Paulsen, *Chem. Ber.*, **107**, 951 (1974).
[131] J. M. Coxon, M. P. Hartshorn and W. H. Swallow, *J. Org. Chem.*, **39**, 1142 (1974).
[132] J. H. Coy, A. F. Hegarty, E. J. Flynn and F. L. Scott, *J.C.S. Perkin II*, **1974**, 53.
[133] D. St. C. Black, R. F. C. Brown, B. T. Dunstan and S. Sternhell, *Tetrahedron Letters*, **1974**, 4283.
[134] S. P. McManus and D. W. Ware, *Tetrahedron Letters*, **1974**, 4271.
[135] M. Langin and J. Huet, *Tetrahedron Letters*, **1974**, 3115.
[136] M. Pánková and M. Tichý, *Coll. Czech. Chem. Comm.*, **39**, 1447 (1974).
[137] G. Lamaty and A. Sivade, *Bull. Soc. Chim. France*, **1974**, 2143.
[138] G. Lamaty and A. Sivade, *Bull. Soc. Chim. France*, **1974**, 2149.
[139] C. F. Hammer, S. R. Heller and J. H. Craig, *Tetrahedron*, **28**, 239 (1972).
[140] S. Ikegami, K. Uoji and S. Akaboshi, *Tetrahedron*, **30**, 2077 (1974).
[141] J. K. Crandall and W. W. Conover, *J. Org. Chem.*, **39**, 63 (1974).
[142] A. Picot and X. Lusinchi, *Tetrahedron Letters*, **1974**, 679.
[143] R. Schwesinger, H. Fritz and H. Prinzbach, *Angew. Chem. Internat. Ed.*, **12**, 994 (1973).
[144] A. F. Hegarty, P. Quain, T. A. F. O'Mahony and F. L. Scott, *J.C.S. Perkin II*, **1974**, 997.
[145] S. Sicsic, *Tetrahedron*, **30**, 277 (1974).
[146] R. H. Staley, R. R. Corderman, M. S. Foster and J. L. Beauchamp, *J. Am. Chem. Soc.* **96**, 1260 (1974).
[147] P. K. G. Hodgson, R. Shepherd and S. Warren, *Chem. Comm.*, **1974**, 633.
[148] D. Howells and S. Warren, *J.C.S. Perkin II*, **1974**, 992; cf. *Org. Reaction Mech.*, **1973**, 207; **1971**, 72.
[149] F. Allen, O. Kennard, L. Nassimbeni, R. Shepherd and S. Warren, *J.C.S. Perkin II*, **1974**, 1530; *Nature*, **248**, 670 (1974).
[150] D. M. Chauncey, L. J. Andrews and R. M. Keefer, *J. Am. Chem. Soc.*, **96**, 1850 (1974).
[151] *Synthesis of Carbocyclic Spiro Compounds via Intramolecular Alkylation Routes*, A. P. Krapcho, *Synthesis*, **1974**, 383.
[152] F. G. Bordwell and J. B. O'Dwyer, *J. Org. Chem.*, **39**, 2519 (1974); F. G. Bordwell and M. D. Wolfinger, *ibid.*, p. 2521.
[153] F. G. Bordwell and E. Doomes, *J. Org. Chem.*, **39**, 2526, 2531 (1974).
[154] M. Schlosser and G. Fouquet, *Chem. Ber.*, **107**, 1171 (1974).
[155] M. Schlosser and G. Fouquet, *Chem. Ber.*, **107**, 1162 (1974).
[156] J. Ranfaing, B. Calas, J. M. Fabre and L. Giral, *Tetrahedron Letters*, **1974**, 1439.
[157] G. Stork, L. D. Cama and D. R. Coulson, *J. Am. Chem. Soc.*, **96**, 5268 (1974).
[158] G. Stork and J. F. Cohen, *J. Am. Chem. Soc.*, **96**, 5270 (1974).
[159] A. I. Meyers, E. D. Mihelich and K. Kamata, *Chem. Comm.*, **1974**, 768.
[160] V. I. Sokolov, V. V. Bashilov, P. V. Petrovskii and O. A. Reutov, *Dokl. Akad. Nauk, S.S.S.R.*, **213**, 1103 (1973).

[161] D. D. Davis and H. T. Johnson, *J. Am. Chem. Soc.*, **96**, 7576 (1974).
[162] P. C. Hariharan, L. Radom, J. A. Pople and P. von R. Schleyer, *J. Am. Chem. Soc.*, **96**, 599 (1974); see also Chapter 8, p. 313.
[163] P. H. Blustin and J. W. Linnett, *J.C.S. Faraday II*, **1974**, 297.
[164] *Org. Reaction Mech.*, **1972**, 75.
[165] S. Hiršl-Starčevič, Z. Majerski and D. E. Sunko, *J. Am. Chem. Soc.*, **96**, 3659 (1974).
[166] G. Goto, K. Yoshioka and K. Hiraga, *Tetrahedron*, **30**, 2107 (1974).
[167] T. V. Ilyukhina, A. V. Kamernitzkii and I. I. Voznesenskaya, *Tetrahedron*, **30**, 2239 (1974).
[168] M. Fărcaşiu, D. Făvcaşiu, J. Slutsky and P. von R. Schleyer, *Tetrahedron Letters*, **1974**, 4059.
[169] M. A. McKinney and P. P. Patel, *J. Org. Chem.*, **38**, 4064 (1973).
[170] M. A. McKinney and P. P. Patel, *J. Org. Chem.*, **38**, 4059 (1973).
[171] E. T. J. Bathurst and J. M. Coxon, *Chem. Comm.*, **1974**, 131; E. T. J. Bathurst, J. M. Coxon and M. P. Hartshorn, *Austral. J. Chem.*, **27**, 1505 (1974).
[172] J. Thierry, F. Frappier, M. Païs and F.-X. Jarreau, *Tetrahedron Letters*, **1974**, 2149; *Org. Reaction Mech.*, **1972**, 78.
[173] V. P. Vitullo and E. A. Logue, *Chem. Comm.*, **1974**, 228.
[174] J.-C. Jacquesy, R. Jacquesy and U. H. Ly, *Tetrahedron Letters*, **1974**, 2199.
[175] *Org. Reaction Mech.*, **1970**, 88.
[176] R. A. Moss and P. E. Schueler, *J. Am. Chem. Soc.*, **96**, 5792 (1974).
[177] R. A. Moss and J. Banger, *Tetrahedron Letters*, **1974**, 3549; cf. *Org. Reaction Mech.*, **1973**, 311.
[178] P. I. Meikle and D. Whittaker, *J.C.S. Perkin II*, **1974**, 318; cf, *Org. Reaction Mech.*, **1972**, 79.
[179] R. J. Hill and G. Stedman, *J.C.S. Perkin II*, **1973**, 2153.
[180] E. H. White, R. H. McGirk, C. A. Aufdermarsh, H. P. Tiwari and M. J. Todd, *J. Am. Chem. Soc.*, **95**, 8107 (1973).
[181] J. Sabo, J. Hiti and R. C. Fort, *J. Org. Chem.*, **39**, 250 (1974).
[182] B. A. Kazanskii, E. A. Shokova and S. I. Knopova, *Izv. Akad. Nauk, S.S.S.R.*, **1968**, 2645.
[183] W. Kirmse, W. J. Baron and U. Seipp, *Angew. Chem. Internat. Ed.*, **12**, 924 (1973).
[184] J. R. Mohrig, K. Keegstra, A. Maverick, R. Roberts and S. Wells, *Chem. Comm.*, **1974**, 780.
[185] W. Kirmse and U. Seipp, *Chem. Ber.*, **107**, 745 (1974).
[186] J. Alberti, R. Siegfried and W. Kirmse, *Annalen*, **1974**, 1605.
[187] W. Kirmse, J. Alberti and H.-G. Varbelow, *Chem. Ber.*, **107**, 2788 (1974).
[188] P. Beak, J. T. Adams and J. A. Barron, *J. Am. Chem. Soc.*, **96**, 2494 (1974).
[189] J. F. J. Engbersen and J. B. F. N. Engberts, *Tetrahedron*, **30**, 1215 (1974).
[190] R. Curci, F. DiFuria, J. Ciabattoni and P. W. Concannon, *J. Org. Chem.*, **39**, 3295 (1974).
[191] G. A. Olah and J. Welch, *Synthesis*, **1974**, 896.
[192] J. S. Lomas and J. E. Dubois, *Tetrahedron Letters*, **1974**, 4239.
[193] H. Marschall and F. Vogel, *Chem. Ber.*, **107**, 2176 (1974).
[194] G. Mehta, S. K. Kapoor, T. N. G. Row and K. Venkatesan, *Tetrahedron Letters*, **1974**, 2653.
[195] H. Kropf, M. Ball, H. Schröder and G. Witte, *Tetrahedron*, **30**, 2943 (1974).
[196] J.-C. Jacquesy, R. Jacquesy and J.-F. Patoiseau, *Bull. Soc. Chim. France*, **1974**, 1959.
[197] B. G. McKinnie, N. S. Bhacca, F. K. Cartledge and J. Fayssoux, *J. Am. Chem. Soc.*, **96**, 2637 (1974); cf. *Org. Reaction Mech.*, **1973**, 314.
[198] G. D. Homer and L. H. Sommer, *J. Am. Chem. Soc.*, **95**, 7700 (1973); cf. *Org. Reaction Mech.*, **1973**, 314.
[199] T. J. Pinnavaia and J. A. McClarin, *J. Am. Chem. Soc.*, **96**, 3012 (1974).
[200] H. J. Reich and D. A. Murcia, *J. Am. Chem. Soc.*, **95**, 3418 (1973).
[201] R. Corriu, G. Lanneau and M. Leard, *J. Organometal. Chem.*, **64**, 79 (1974); R. J. P. Corriu and G. F. Lanneau, *ibid.*, **67**, 243 (1974).
[202] R. J. P. Corriu and B. J. L. Henner, *J. Organometal. Chem.* **71**, 393 (1974); cf. *Org. Reaction Mech.*, **1973**, 315, ref. 147.
[203] M. H. Leard and R. J. P. Corriu, *J. Organometal. Chem.*, **65**, C39 (1974); R. Corriu and M. Henner, *Bull. Soc. Chim. France*, **1974**, 1447; *Org. Reaction Mech.*, **1971**, 83.
[204] J. Pijselman and M. Pereyre, *J. Organometal. Chem.*, **63**, 139 (1973); *Org. Reaction Mech.*, **1971**, 85.
[205] T. H. Chan and W. Mychajlowskij, *Tetrahedron Letters*, **1974**, 3479.
[206] C. Eaborn and I. D. Jenkins, *J. Organometal. Chem.*, **69**, 185 (1974).
[207] R. Alexander, W. A. Asomaning, C. Eaborn, I. D. Jenkins and D. R. M. Walton, *J. C.S. Perkin II*, **1974**, 490.
[208] B. Bøe, C. Eaborn and D. R. M. Walton, *J. Organometal. Chem.*, **82**, 327 (1974).
[209] B. Bøe, C. Eaborn and D. R. M. Walton, *J. Organometal Chem.*, **82**, 13 (1974).
[210] M. Ali, C. Eaborn and D. R. M. Walton, *J. Organometal. Chem.*, **78**, 83 (1974).
[211] B. Bøe, *J. Organometal. Chem.*, **73**, 297 (1974).

[212] B. Bøe, C. Eaborn and D. R. M. Walton, *J. Organometal. Chem.*, **82**, 17 (1974).
[213] C. G. Swain, K.-R. Pörschke, W. Ahmed and R. L. Schowen, *J. Am. Chem. Soc.*, **96**, 4700 (1974).
[214] A. Modro and R. L. Schowen, *J. Am. Chem. Soc.*, **96**, 6980 (1974).
[215] J. R. Chipperfield and G. E. Gould, *J.C.S. Perkin II*, **1974**, 1324.
[216] J. Chojnowski and W. Stańczyk, *J. Organometal. Chem.*, **73**, 41 (1974).
[217] V. V. Yastrebov and A. I. Chernyshev, *Kinet. Katal.*, **14**, 933 (1973); *Chem. Abs.*, **79**, 145599 (1973).
[218] A. Radecki, H. Lamparczyk and J. Halkiewicz, *J. Organometal. Chem.*, **77**, 307 (1974).
[219] L. P. Razumovskii, Y. V. Moiseev, N. A. Markova, A. G. Kuznetsova and G. E. Zaikov, *Izv. Akad. Nauk, S.S.S.R., Ser Khim.*, **1973**, 2448; *Chem. Abs.*, **80**, 69992 (1974).
[220] D. Mansuy, J. F. Bartoli and J. C. Chattard, *J. Organometal. Chem.*, **77**, C49 (1974).
[221] I. M. Podgornyi, S. A. Golubtsov, K. A. Andrianov and E. G. Mangalin, *Zh. Obsch. Khim.*, **44**, 767, 774 (1974).
[222] J. R. Chipperfield and G. E. Gould, *J.C.S. Perkin II*, **1974**, 1331.
[223] G. Pirazzini, R. Danielli, A. Ricci and C. A. Boicelli, *J.C.S. Perkin II*, **1974**, 853.
[224] I. N. Nazarova, I. I. Seifullina, E. M. Belousova and D. I. Chubar, *Zh. Obsch. Khim.*, **43**, 1518 (1973); *Chem. Abs.*, **80**, 47113 (1974).
[225] F. Ramirez and I. Ugi, *Bull. Soc. Chim. France*, **1974**, 453.
[226] S. A. Bone, S. Trippett, M. W. White and P. J. Whittle, *Tetrahedron Letters*, **1974**, 1795.
[227] R. R. Holmes, *J. Am. Chem. Soc.*, **96**, 4143 (1974).
[228] M. Eisenhut, H. L. Mitchell, D. D. Traficante, R. J. Kaufman, J. M. Deutch and G. M. Whitesides, *J. Amer. Chem. Soc.*, **96**, 5385 (1974).
[229] G. M. Whitesides, M. Eisenhut and W. M. Bunting, *J. Am. Chem. Soc.*, **96**, 5398 (1974).
[230] F. Ramirez, I. Ugi, F. Lin, S. Pfohl, P. Hoffman and D. Marquarding, *Tetrahedron*, **30**, 371 (1974).
[231] A. Klaebe, A. Carrelhas, J.-F. Brazier and R. Wolf, *J.C.S. Perkin II*, **1974**, 1668.
[232] R. G. Cavell, D. D. Poulin, K. I. The and A. J. Tomlinson, *Chem. Comm.*, **1974**, 19.
[233] F. Ramirez, Y. F. Chaw, J. F. Marecek and I. Ugi, *J. Am. Chem. Soc.*, **96**, 2429 (1974).
[234] D. Hellwinkel and W. Krapp, *Angew. Chem. Internat. Ed.*, **13**, 542 (1974).
[235] F. Ramirez, V. A. V. Prasad and J. F. Marecek, *J. Am. Chem. Soc.*, **96**, 7269 (1974).
[236] T. Koizumi, Y. Watanabe, Y. Yoshida and E. Yoshii, *Tetrahedron Letters*, **1974**, 1075.
[237] D. Bernard and R. Burgada, *Compt. rend., Ser. C.*, **279**, 883 (1974).
[238] D. G. Gorenstein, *J. Am. Chem. Soc.*, **95**, 8060 (1973).
[239] C. R. Hall and D. J. H. Smith, *Tetrahedron Letters*, **1974**, 1693; *Org. Reaction Mech.*, **1973**, 320.
[240] R. B. Wetzel and G. L. Kenyon, *J. Am. Chem. Soc.*, **96**, 5199 (1974).
[241] E. E. Schweizer, T. Minami and S. E. Anderson, *J. Org. Chem.*, **39**, 3038 (1974).
[242] D. W. Allen, B. G. Hutley and M. T. J. Mellor, *J.C.S. Perkin II*, **1974**, 1690; *Org. Reaction Mech.*, **1973**, 318.
[243] F. Y. Khalil and G. Aksnes, *Acta Chem. Scand.*, **27**, 3822 (1973).
[244] K. E. De Bruin and D. M. Johnson, *Phosphorus*, **4**, 17 (1974); *Chem. Abs.*, **81**, 135316 (1974); cf. *Org. Reaction Mech.*, **1973**, 318.
[245] D. A. Tyssee, L. P. Bausher and P. Haake, *J. Am. Chem. Soc.*, **95**, 8066 (1973).
[246] T. Koizumi, Y. Kobayashi and E. Yoshii, *Chem. Comm.*, **1974**, 678.
[247] T. Koizumi and P. Haake, *J. Am. Chem. Soc.*, **95**, 8073 (1973).
[248] R. D. Cook, C. E. Diebert, W. Schwarz, P. C. Turley and P. Haake, *J. Am. Chem. Soc.*, **95**, 8088 (1973).
[249] D. B. Cooper, J. M. Harrison and T. D. Inch, *Tetrahedron Letters*, **1974**, 2697.
[250] W. S. Wadsworth and Y.-G. Tsay, *J. Org. Chem.*, **39**, 984 (1974); cf. *Org. Reaction Mech.*, **1973**, 318.
[251] R. D. Bertrand, H. J. Berwin, G. K. McEwen and J. G. Verkade, *Phosphorus*, **4**, 81 (1974); *Chem. Abs.*, **81**, 151336 (1974).
[252] K. L. Marsi, *J. Org. Chem.*, **39**, 265 (1974).
[253] D. R. Hogg and J. Stewart, *J.C.S. Perkin II*, **1974**, 43, 436.
[254] D. R. Hogg and J. Stewart, *J.C.S. Perkin II*, **1974**, 1040.
[254a] J. Carlsson, M. P. J. Kierstan and K. Brocklehurst, *Biochem. J.*, **139**, 221 (1974).
[255] L. E. Overman, D. Matzinger, E. M. O'Connor, and J. D. Overman, *J. Am. Chem. Soc.*, **96**, 6081 (1974); cf. *Org. Reaction Mech.*, **1973**, 318.
[256] T. Hata and M. Sekine, *J. Am. Chem. Soc.*, **96**, 7363 (1974).
[257] S. Braverman and D. Reisman, *Tetrahedron*, **30**, 3891 (1974).
[258] N. C. Collins and W. K. Glass, *J.C.S. Perkin II*, **1974**, 713.
[259] M. Makosza and M. Fedoryński, *Synthesis*, **1974**, 274.
[260] J. Meinwald and S. Knapp, *J. Am. Chem. Soc.*, **96**, 6532 (1974).
[260a] V. Bertini, A. De Munno, A. Menconi and A. Fissi, *J. Org. Chem.*, **39**, 2294 (1974).

[261] M. Mikolajczyk and J. Drabowicz, *Chem. Comm.*, **1974**, 775.
[262] T. Durst, M. J. Le Belle, R. Van den Elzen and K.-C. Tin, *Can. J. Chem.*, **52**, 761 (1974).
[263] S. Oae, M. Moriyama, T. Numata and N. Kunieda, *Bull. Chem. Soc. Japan*, **47**, 179 (1974); cf. *Org. Reaction Mech.*, **1972**, 90.
[263a] P. Bonvicini, A. Levi and G. Scorrano, *Gazz. Chim. Ital.*, **104**, 1 (1974).
[264] N. Furukawa, T. Masuda, M. Yakushiji and S. Oae, *Bull. Chem. Soc. Japan*, **47**, 2247 (1974).
[265] J. C. Martin and E. F. Perozzi, *J. Am. Chem. Soc.*, **96**, 3155 (1974).
[266] E. F. Perozzi, J. C. Martin and I. C. Paul, *J. Am. Chem. Soc.*, **96**, 6735 (1974).
[267] J. H. Krueger and R. J. Kiyokane, *Inorg. Chem.*, **13**, 2522 (1974).
[268] H. Matsuyama, H. Minato and M. Kobayashi, *Bull. Chem. Soc. Japan*, **46**, 3828 (1973).
[269] R. V. Vizgert, *Zh. Obsch. Khim.*, **32**, 628 (1962); *Chem. Abs.*, **58**, 428 (1963).
[270] L. Senatore, L. Sagramora and E. Ciuffarin, *J.C.S. Perkin II*, **1974**, 722.
[271] E. Ciuffarin and L. Senatore, *Tetrahedron Letters*, **1974**, 1635.
[272] E. Maccarone, G. Musumarra and G. A. Tomaselli, *J. Org. Chem.*, **39**, 3286 (1974).
[273] A. Arcoria, E. Maccarone, G. Musumarra and G. A. Tomaselli, *J. Org. Chem.*, **39**, 1689 (1974).
[274] W. K. Kim and I. Lee, *Daehan Hwahak Hwojee*, **18**, 8 (1974); *Chem. Abs.*, **81**, 77138 (1974).
[275] J. Kim, *J. Korean Nucl. Soc.*, **5**, 321 (1973); I. Lee, and J. E. Yie, *ibid.*, **6**, 23 (1974); *Chem. Abs.*, **80**, 119860; **81**, 104298 (1974).
[276] U. R. Kim, K. Y. Lee, S. H. Bai and I. Lee, *Daehan Hwahak Hwojee*, **18**, 3 (1974); I. Lee, S. H. Bai and U. R. Kim, *ibid.*, p. 171; *Chem. Abs.*, **81**, 119765, 151319 (1974).
[277] I. Lee and S. C. Kim, *Daehan Hwahak Hwojee*, **17**, 406 (1973); *Chem. Abs.*, **80**, 107569 (1974).
[278] R. W. Hay, C. R. Clark and J. A. G. Edmonds, *J.C.S. Dalton*, **1974**, 9.
[279] W. J. Spillane, C. B. Goggin, N. Regan and F. L. Scott, *Internat. J. Sulfur Chem.*, **3**, 281 (1973); *Chem. Abs.*, **79**, 125612 (1973).
[280] M. R. Jones and D. J. Cram, *J. Am. Chem. Soc.*, **96**, 2183 (1974).
[281] B. Lindgren, *Tetrahedron Letters*, **1974**, 4347.
[282] W. E. McEwen and C. T. Lin, *Phosphorus*, **3**, 229 (1974); *Chem. Abs.*, **81**, 90768 (1974).
[283] J. H. Krueger, P. F. Blanchet, A. P. Lee and B. A. Sudbury, *Inorg. Chem.*, **12**, 2714 (1973); P. F. Blanchet and J. H. Krueger, *ibid.*, **13**, 719 (1974); B. A. Sudbury and J. H. Krueger, *ibid.*, p. 1736; J. H. Krueger, B. A. Sudbury and P. F. Blanchet, *J. Am. Chem. Soc.*, **96**, 5733 (1974).
[284] S. Oae, T. Sakurai and S. Kozuka, *Chem. Letters (Tokyo)*, **1974**, 621; *Chem. Abs.*, **81**, 48974 (1974).
[285] P. Sarthou, F. Guibé and G. Bram, *Chem. Comm.*, **1974**, 377.
[286] P. Hrnčiar and F. Szemes, *Coll. Czech. Chem. Comm.*, **39**, 1744 (1974).
[287] A. de Groot and B. J. M. Jansen, *Rec. Trav. chim.*, **93**, 153 (1974).
[288] W. D. Korte, K. Cripe and R. Cooke, *J. Org. Chem.*, **39**, 1168 (1974).
[280] J. A. Bryson and R. B. Gammill, *Tetrahedron Letters*, **1974**, 3963.
[290] P. Houdewind and U. K. Pandit, *Tetrahedron Letters*, **1974**, 2359.
[291] J. A. Katzenellenbogen and A. L. Crumrine, *J. Am. Chem. Soc.*, **96**, 5662 (1974).
[292] D. R. Hogg and A. Robertson, *Tetrahedron Letters*, **1974**, 3783.
[293] P. J. Stang, *Progr. Phys. Org. Chem.*, **10**, 205 (1973); G. Modena and U. Tonellata, *Chim. Ind. (Milan)*, **56**, 207 (1974); *Chem. Abs.*, **81**, 90667 (1974).
[294] P. Beltrame, A. Gavezzotti and M. Simonetta, *J.C.S. Perkin II*, **1974**, 502.
[295] R. F. Cunico and Y.-K. Han, *J. Organometal. Chem.*, **81**, C9 (1974).
[296] H. U. Siehl, J. C. Carnahan, L. Eckes and M. Hanack, *Angew. Chem. Internat. Ed.*, **13**, 675 (1974).
[297] T. S. Abram and W. E. Watts, *Chem. Comm.*, **1974**, 857.
[298] R. H. Summerville and P. von R. Schleyer, *J. Am. Chem. Soc.*, **96**, 1110 (1974); *Org. Reaction Mech.*, **1972**, 93.
[299] Z. Rappoport, J. Kaspi and Y. Apeloig, *J. Am. Chem. Soc.*, **96**, 2612 (1974).
[300] Z. Rappoport and J. Kaspi, *J. Amer. Chem. Soc.*, **96**, 586, 4518 (1974).
[301] K. Yates and J.-J. Périé, *J. Org. Chem.*, **39**, 1902 (1974).
[302] P. J. Stang, R. J. Hargrove and T. E. Dueber, *J.C.S. Perkin II*, **1974**, 843; cf. *Org. Reaction Mech.*, **1973**, 323.
[303] Z. Rappoport and Y. Apeloig, *J. Am. Chem. Soc.*, **96**, 6428 (1974); *Org. Reaction Mech.*, **1973**, 323.
[304] R. H. Summerville, C. A. Senkler, P. von R. Schleyer, T. E. Dueber and P. J. Stang, *J. Am. Chem. Soc.*, **96**, 1100 (1974).
[305] M. Hanack, T. Bässler, W. Eymann, W. E. Heyd and R. Kopp, *J. Am. Chem. Soc.*, **96**, 6686 (1974); *Org. Reaction Mech.*, **1973**, 322.
[306] H. Stutz and M. Hanack, *Tetrahedron Letters*, **1974**, 2457.
[307] G. Hammen, T. Bäßler and M. Hanack, *Chem. Ber.*, **107**, 1676 (1974).
[308] I. L. Reich and H. J. Reich, *J. Am. Chem. Soc.*, **96**, 2654 (1974).

[309] P. Bassi and U. Tonellato, *J.C.S. Perkin II*, **1974**, 1283; *Org. Reaction Mech.*, **1972**, 94.
[310] K.-P. Jäckel and M. Hanack, *Tetrahedron Letters*, **1974**, 1637.
[311] C. C. Lee, A. J. Cessna, B. A. Davis and M. Oka, *Can. J. Chem.*, **52**, 2679 (1974); cf. *Org. Reaction Mech.*, **1973**, 324.
[312] Y. Riad and E. M. E. Mansour, *U.A.R. J. Chem.*, **14**, 537 (1971); *Chem. Abs.*, **79**, 136331 (1973).
[313] G. A. Olah, R. J. Spear, P. W. Westerman and J.-M. Denis, *J. Am. Chem. Soc.*, **96**, 5855 (1974).
[314] D. Plouin, C. Coeur and R. Glénat, *Bull. Soc. Chim. France*, **1974**, 244.
[315] M. V. Mavrov, A. P. Rodionov and V. F. Kucherov, *Izv. Akad. Nauk S.S.S.R., Ser. Khim*, **23**, 369 (1974).
[316] S. Holand and R. Epstein, *Tetrahedron Letters*, **1974**, 1897.
[317] J.-C. Chalchat and F. Theron, *Bull. Soc. Chim. France*, **1974**, 953.
[318] J.-C. Chalchat and F. Theron, *Bull. Soc. Chim. France*, **1974**, 1543.
[319] J. Chanet-Ray and R. Vessière, *Bull. Soc. Chim. France*, **1974**, 1661.
[320] B. Cheminat and B. Mege, *Bull. Soc. Chim. France*, **1974**, 2233.
[321] L. M. Litvinenko, A. F. Popov, L. J. Kostenko and I. I. Tormosin, *Dokl. Akad. Nauk S.S.S.R.*, **211**, 353 (1973).
[322] G. H. Posner and D. J. Brunelle, *Chem. Comm.*, **1973**, 907.
[323] J. J. Brunet, B. Fixari and P. Caubère, *Tetrahedron*, **30**, 1237, 1245 (1974); *Org. Reaction Mech.*, **1972**, 95.
[324] J. J. Brunet, B. Fixari and P. Caubère, *Tetrahedron*, **30**, 2931 (1974); *Org. Reaction Mech.*, **1973**, 325.
[325] C. J. Devlin and B. J. Walker, *J.C.S. Perkin I*, **1974**, 453.
[326] J. Klein and R. Levene, *Tetrahedron Letters*, **1974**, 2935.
[327] M. Charpentier-Morize, *Bull. Soc. Chim. France*, **1974**, 343.
[328] M. Charpentier-Morize, J. M. Lefour and N. T. Anh, *Tetrahedron Letters*, **1974**, 1729.
[329] D. Baudry, M. Charpentier-Morize, D. Lefort and J. Sorba, *Tetrahedron Letters*, **1974**, 2499.
[330] J.-P. Bégué and D. Bonnet, *Tetrahedron*, **30**, 141 (1974).
[331] A. M. Bianchi, V. Rosnati, F. Sannicolò, F. Soccolini and G. Zecchi, *Gazz. Chim. Ital.*, **103**, 801 (1973); See *Org. Reaction Mech.*, **1970**, 110.
[332] A. M. Bianchi, V. Rosnati and G. Zecchi, *Gazz. Chim. Ital.*, **104**, 445 (1974); see *Org. Reaction Mech.*, **1969**, 120.
[333] W. T. Brady and A. D. Patel, *J. Org. Chem.*, **39**, 1949 (1974).
[334] D. B. Denney and F. A. Wagner, *Phosphorus*, **3**, 27 (1973); *Chem. Abs.*, **81**, 119786 (1974).
[335] N. N. Magdesieva and I. V. Leont'eva, *Khim. Geterotsikl. Soedin.*, **1973**, 910; *Chem. Abs.*, **79**, 125441 (1973).
[336] T. Wakui, Y. Otsuji and E. Imoto, *Bull. Chem. Soc. Japan*, **47**, 1522, 2267 (1974).
[337] J. H. Smith and E. T. Kaiser, *J. Org. Chem.*, **39**, 728 (1974).
[338] W.-D. Stohrer, *Chem. Ber.*, **107**, 1795 (1974); cf. *Org. Reaction Mech.*, **1971**, 103.
[339] L. Bobkiewicz and B. Zurawski, *Bull. Acad. Pol. Sci., Ser. Sci. Chim.*, **22**, 457 (1974); *Chem. Abs.*, **81**, 90912 (1974).
[340] D. K. Bohme, G. I. Mackay and J. D. Payzant, *J. Am. Chem. Soc.*, **96**, 4027 (1974); *Org. Reaction Mech.*, **1973**, 328; **1972**, 97.
[341] J. I. Brauman, W. N. Olmstead and C. A. Lieder, *J. Am. Chem. Soc.*, **96**, 4030 (1974); C. A. Lieder and J. I. Brauman, *ibid.*, p. 4028.
[342] R. C. Dougherty, J. Dalton and D. J. Roberts, *Org. Mass. Spectrom.*, **8**, 77 (1974); R. C. Dougherty and J. D. Roberts, *ibid.*, p. 81; R. C. Dougherty, *ibid.*, p. 85; *Chem. Abs.*, **81**, 12641, 12774, 12719 (1974).
[343] P. H. Anderson, B. Stephenson and H. S. Mosher, *J. Am. Chem. Soc.*, **96**, 3171 (1974); *Org. Reaction Mech.*, **1972**, 98.
[344] O. E. Edwards and C. Grieco, *Can. J. Chem.*, **52**, 3561 (1974).
[345] K. Brocklehurst, *Tetrahedron*, **30**, 2397 (1974).
[346] See *Org. Reaction Mech.*, **1973**, 328; **1972**, 59.
[347] P. O. I. Virtanen and R. Korhonen, *Acta Chem. Scand.*, **27**, 2650 (1973).
[348] M. Arbelot, J. Metzger, M. Chanon, C. Guimon and G. Pfister-Guillouzo, *J. Am. Chem. Soc.*, **96**, 6217 (1974).
[349] G. Sturtz and M. Rio, *Bull. Soc. Chim. France*, **1974**, 2180, 2187.
[350] P. G. Le Grâs, R. L. Dyer, P. J. Clifford and C. D. Hall, *J.C.S. Perkin II*, **1973**, 2064.
[351] T. Austad, S. Esperas and J. Songstad, *Acta Chem. Scand.*, **27**, 3594 (1973).
[352] D. M. E. Reuben and T. C. Bruice, *Chem. Comm.*, **1974**, 113.
[353] W. H. Mandeville and G. M. Whitesides, *J. Org. Chem.*, **39**, 400 (1974).

[354] N. A. Radomsky and M. J. Gibian, *J. Am. Chem. Soc.*, **95**, 8713 (1973).
[355] J. B. Hendrickson, A. Giga and J. Wareing, *J. Am. Chem. Soc.*, **96**, 2275 (1974).
[356] P. Müller and B. Siegfried, *Helv. Chim. Acta*, **57**, 987 (1974).
[357] A. I. Meyers and K. Kamata, *J. Org. Chem.*, **39**, 1603 (1974).
[358] J. D. Morrison and W. F. Masler, *J. Org. Chem.*, **39**, 270 (1974).
[359] P. L. Bock and G. M. Whitesides, *J. Am. Chem. Soc.*, **96**, 2826 (1974); P. L. Bock, D. J. Boschetto, J. R. Rasmussen, J. P. Demers and G. M. Whitesides, *ibid.*, p. 2814.
[360] G. M. Whitesides, J. S. Sadowski and J. Lilburn, *J. Am. Chem. Soc.*, **96**, 2829 (1974).
[361] G. H. Posner and J.-S. Ting, *Tetrahedron Letters*, **1974**, 683.
[362] S. Masamune, G. S. Bates and P. E. Georghiou, *J. Am. Chem. Soc.*, **96**, 3686 (1974).
[363] A. McKillop and M. E. Ford, *Tetrahedron*, **30**, 2467 (1974).
[364] M. Kluba and N. Zwierzak, *Zesz. Nauk. Politech. Lodz., Chem.*, **27**, 159 (1973); *Chem. Abs.*, **80**, 132389 (1974).
[365] B. D. Chernokal'skii and L. A. Vorob'eva, *Zh. Obshch. Khim.*, **43**, 1939 (1973); *Chem. Abs.*, **80**, 2819 (1974).
[366] P. E. Iversen and H. Lund, *Acta Chem. Scand.*, *B*, **28**, 827 (1974).
[367] S. Bank and D. A. Noyd, *J. Am. Chem. Soc.*, **95**, 8203 (1973).
[368] M.-C. Wu, L. Anderson, C. W. Slife and L. J. Jensen, *J. Org. Chem.*, **39**, 3014 (1974).
[369] A. Knöchel, G. Rudolph and J. Thiem, *Tetrahedron Letters*, **1974**, 551.
[370] A. G. Pinkus and W. H. Lin, *Synthesis*, **1974**, 279.
[371] A. G. Anderson and D. H. Kono, *Tetrahedron Letters*, **1973**, 5121.
[372] H. Vorbrüggen, *Annalen*, **1974**, 821.
[373] S. Oae, T. Aida, M. Nakajima and N. Furukawa, *Tetrahedron*, **30**, 947 (1974).
[374] P. Wilder, L. A. Feliu-Otero and G. A. Diegnan, *J. Org. Chem.*, **39**, 2153 (1974).
[375] H. R. Hudson and J. C. Roberts, *J.C.S. Perkin II*, **1974**, 1575.
[376] J. G. Duboudin and J. Valade, *Bull. Soc. Chim. France*, **1974**, 272.
[377] J. Bachelet, P. Demerseman and R. Royer, *Bull. Soc. Chim. France*, **1974**, 2631.
[378] L. René, J.-P. Buisson and R. Royer, *Bull. Soc. Chim. France*, **1974**, 475.
[379] O. Exner, *Coll. Czech. Chem. Comm.*, **39**, 515 (1974).
[380] M. J. Duffy, G. Pass and G. O. Phillips, *J.C.S. Perkin II*, **1974**, 1466.
[381] T. Painter. *Acta Chem. Scand.*, **27**, 2463 (1973).
[382] H. K. Hall, L. J. Carr, R. Kellman and F. De Blauwe, *J. Am. Chem. Soc.*, **96**, 7265 (1974).
[383] A. McKillop, J.-C. Fiaud and R. P. Hug, *Tetrahedron*, **30**, 1379 (1974).
[384] S. Osterman-Golkar, L. Ehrenburg and F. Solymosy, *Acta Chem. Scand.*, *B*, **28**, 215 (1974).
[385] J. C. Tou, L. B. Westover and L. F. Sonnabend, *J. Phys. Chem.*, **78**, 1096 (1974); cf. *Org. Reaction Mech.*, **1967**, 63.
[386] S. Arakawa, S. Hariya, H. Itsuki and S. Terasawa, *Nippon Kagaku Kaishi*, **1974**, 1170; *Chem. Abs.* **81**, 119325 (1974).
[387] L. V. Aleksanyan and V. F. Shvets, *Arm. Khim. Zh.*, **25**, 986 (1972); *Chem. Abs.*, **79**, 145614 (1973).
[388] F. B. Ghoneim, *Egypt. J. Chem.*, **15**, 371 (1972); *Chem. Abs.*, **80**, 69975 (1974).
[389] D. N. Kozachuk, Y. A. Serguchev, Y. A. Fialkov and L. M. Yagupol'skii, *Zh. Org. Khim.*, **9**, 1918 (1973); D. N. Kozachuk, Y. A. Serguchev, M. M. Kremlev, Y. A. Fialkov and L. M. Yagupol'skii, *ibid.*, **10**, 1230 (1974); *Chem. Abs.* **79**, 145610 (1973); **81**, 90746 (1974).
[390] A. M. Azzam and E.-H. M. Diefallah, *Z. Phys. Chem. (Frankfurt)*, **91**, 44 (1974).
[391] S. M. Gabrielyan, Y. A. Treger and R. M. Flid, *Azerb. Khim. Zh.*, **1973**, 50; *Chem. Abs.*, **81**, 90726 (1974).
[392] S. Y.-K. Tam and B. Fraser-Reid, *Tetrahedron Letters*, **1973**, 4897.
[393] C. Agami and J. Levisalles, *J. Organometal. Chem.*, **64**, 281 (1974).
[394] E. Meléndez and M. del Carmen Pardo, *Bull. Soc. Chim. France*, **1974**, 632.
[395] H. Hauptmann, *Tetrahedron Letters*, **1974**, 3587, 3589, 3593.
[396] S. Senda and K. Hirota, *Chem. Comm.*, **1974**, 483.
[397] C. A. Grob and M. G. Schlageter, *Helv. Chim. Acta*, **57**, 509 (1974); V. J. Baker, I. D. Blackburne, A. R. Katritzky, R. A. Kolinski and Y. Takeuchi, *J.C.S. Perkin II*, **1974**, 1563; T. Yoshino, S. Inaba, H. Komura and Y. Ishido, *Bull. Chem. Soc. Japan*, **47**, 405 (1974); C. D. Mengler, *Annalen*, **1974**, 1543; M. Davis, L. W. Deady and E. Homfeld, *Austral. J. Chem.*, **27**, 1221, 1917 (1974); E. Bălă, *Bull. Soc. Chim. France*, **1974**, 577; C. Roussel, R. Gallo, M. Chanon and J. Metzger, *J.C.S. Perkin II*, **1974**, 1304; J. Daunis, *Bull. Soc. Chim. France*, **1974**, 999; R. Jacquier, J. L. Olive, C. Petrus and F. Petrus, *ibid.*, p. 1651; M. Pánková, J. Krupička and J. Závada, *Coll. Czech. Chem. Comm.*, **39**, 167 (1974); G. L. Szendey and E. Mutschler, *Arch. Pharm. (Weinheim, Ger.)*, **307**, 647 (1974); *Chem. Abs.*, **81**, 151346 (1974); G. B. Sergeev, C. A. Batýuk, M. B. Stepanov and T. N. Lukina, *Kinet.*

Katal., **15**, 333 (1974); *Chem. Abs.*, **81**, 12765 (1974); R. D. Otzenberger, K. B. Lipkowitz and B. P. Mundy, *J. Org. Chem.*, **39**, 319 (1974); J. E. Oliver and A. B. De Milo, *ibid.*, p. 2225; J. E. Oliver, *ibid.*, p. 2235; V. J. Baker, I. D. Blackburne and A. R. Katritzky, *J.C.S. Perkin II*, **1974**, 1557; L. Cariello, S. Crescenzi, G. Prota and L. Zanetti, *Tetrahedron*, **30**, 3611 (1974).

398 M. Adinolfi, G. Laonigro and L. Mangoni, *Gazzetta*, **104**, 309 (1974); S. Kumazawa, T. Sakakibara, R. Sudoh and T. Nakagawa, *Angew. Chem. Internat. Ed.*, **12**, 921 (1973); R. Schwesinger, H. Fritz and H. Prinzbach, *ibid.*, p. 993; E. Bengsch, M. Corval, R. Viallard and A. Brunissen, *Bull. Soc. Chim. France*, **1974**, 877; C. Sabaté-Alduy, J. Bastide, P. Berçot and J. Lematre, *ibid.*, p. 1942; J. Eichler and I. Dobáš, *Coll. Czech. Chem. Comm.*, **38**, 3461 (1973); M. I. Farberov, A. V. Bondarenko, V. M. Obukhov, T. V. Tsilyurik and I. P. Stepanova, *Dokl. Akad. Nauk, S.S.S.R.*, **214**, 358 (1974); L. M. Kozlov, E. B. Filippov and R. S. Khatypova, *Khim. Tekhnol. Elementoorg. Soedin. Polim.*, **1972**, 108; *Chem. Abs.*, **81**, 151218 (1974); K. A. Arutyunyan, A. O. Tonoyan, S. P. Davtyan, B. A. Rosenberg and N. S. Enikolopyan, *Dokl. Akad. Nauk S.S.S.R.*, **214**, 832 (1974); B. L. Vorob'ev and A. L. Shapiro, *Organic Reactivity* (*Tartu*), **10**, 1111 (1973); S. M. Danov, G. V. Dvoeglazov and L. L. Sibiryakova, *Zh. Prikl. Chim.* (*Leningrad*), **47**, 1129 (1974); *Chem. Abs.*, **81**, 36962 (1974); M. F. Sorokin, L. D. Shode, L. A. Dobrovinskii and S. I. Nogteva, *Zh. Org. Khim.*, **10**, 1149 (1974); *Chem. Abs.*, **81**, 77170 (1974); J. P. Girard, J. P. Vidal, R. Granger, J. C. Rossi and J. P. Chapat, *Tetrahedron Letters*, **1974**, 943; H. Paulsen, K. Eberstein and W. Koebernick, *ibid.*, p. 4377; S. Krishnamurthy, R. M. Schubert and H. C. Brown, *J. Am. Chem. Soc.*, **95**, 8486 (1974); G. Teutsch and R. Bucourt, *Chem. Comm.*, **1974**, 763; R. A. Bekker, G. V. Asratyan, B. L. Dyatkin and I. L. Knunyants, *Tetrahedron*, **30**, 3539 (1974); M. Naruse, K. Utimoto and H. Nozaki, *ibid.*, p. 3037; C.B. Rose and S. K. Taylor, *J. Org. Chem.*, **39**, 578 (1974); C. J. Chang, R. F. Kiesel and T. E. Hogen-Esch, *J. Am. Chem. Soc.*, **95**, 8446 (1973).

399 P. Vermeer, J. Meijer, C. de Graaf and H. Schreurs, *Rec. Trav. chim.*, **93**, 46 (1974); J. M. Coxon, M. P. Hartshorn and B. L. S. Sutherland, *Austral. J. Chem.*, **27**, 679 (1974); P. Schiess and M. Wisson, *Helv. Chim. Acta*, **57**, 981 (1974); L. Červený, J. Bartoň and V. Ružička, *Coll. Czech. Chem. Comm.*, **39**, 2470 (1974); B. A. Arbuzov, Z. G. Isaeva and V. A. Shaikhutdinov, *Dokl. Akad. Nauk S.S.S.R.*, **210**, 837 (1973); V. F. Shvets, A. V. Romashkin and V. V. Yudina, *Kinet. Katal.*, **14**, 928 (1973); *Chem. Abs.*, **79**, 145602 (1973); P. O. I. Virtanen and T. Kuokkanen, *Suomen Kem.*, *B*, **46**, 267 (1973); J. Novak, *Int. J. Chem. Kinet.*, **5**, 919 (1973); *Chem. Abs.*, **80**, 69994 (1974); R. Durand, P. Geneste, G. Lamaty and J. P. Roque, *Compt. Rend.*, *Ser. C.*, **277**, 1395 (1973); M.-L. Leriverend, *ibid.*, **279**, 755 (1974); A. Balsamo, P. Crotti, B. Macchia and F. Macchia, *J. Org. Chem.*, **39**, 874 (1974); G. L. Batten and H. B. Miller, *ibid.*, p. 3058; R. S. Dhillon, B. R. Chabra, M. S. Wadia and P. S. Kalsi, *Tetrahedron Letters*, **1974**, 401; T. H. Chan and J. R. Finkenbine, *ibid.*, p. 2091; P. Y. Bruice, T. C. Bruice, H. G. Selander, H. Yagi and D. M. Jerina, *J. Am. Chem. Soc.*, **96**, 6814 (1974); A. M. Jeffrey, H. J. C. Yeh, D. M. Jerina, R. M. De Marinis, C. H. Foster, D. E. Piccolo and G. A. Berchtold, *ibid.*, p. 6929; D. Baldwin and J. R. Hanson, *Chem. Comm.*, **1974**, 211; C. Battistini, A. Balsamo, G. Berti, P. Crotti, B. Macchia and F. Macchia, *ibid.*, p. 712; G. Berti, G. Catelani, M. Ferretti and L. Monti, *Tetrahedron*, **30**, 4013 (1974); R. D. H. Murray, R. W. Mills, A. J. McAlees and R. McCrindle, *ibid.*, p. 3399.

400 D. M. Jerina and J. W. Daly, *Science*, **185**, 573 (1974).

401 H. J. Nestler and H. Bestian, *Annalen*, **1974**, 460; H. Stamm and L. Schneider, *Chem. Ber.*, **107**, 2870 (1974); W. G. Taylor and W. A. Remers, *Tetrahedron Letters*, **1974**, 3483; J. F. Bunnett, R. L. McDonald and F. P. Olsen, *J. Am. Chem. Soc.*, **96**, 2855 (1974); P. A. Capps and A. R. Jones, *Chem. Comm.*, **1974**, 320; A. T. Bottini, L. R. Sousa and B. F. Dowden, *J. Org. Chem.*, **39**, 355 (1974).

402 P. Raynolds, S. Zonnebelt, S. Bakker and R. M. Kellogg, *J. Am. Chem. Soc.*, **96**, 3146 (1974).

403 M. Dessolin, M. Laloi-Diard and M. Vilkas, *Tetrahedron Letters*, **1974**, 2405.

404 P. Frøyen and R. F. Hudson, *Acta Chem. Scand.*, **27**, 4001 (1974).

405 S. F. Nelsen and J. C. Calabrese, *J. Am. Chem. Soc.*, **95**, 8385 (1973); cf. *Org. Reaction Mech.*, **1973**, 331; **1970**, 117.

406 R. Aneja, A. P. Davies and J. A. Knaggs, *Tetrahedron Letters*, **1974**, 67.

407 R. Aneja and A. P. Davies, *J.C.S. Perkin I*, **1974**, 141; see *Org. Reaction Mech.*, **1972**, 103.

408 R. U. Lemieux and T. Kondo, *Carbohydrate Res.*, **35**, C4 (1974).

CHAPTER 10

Carbanions and Electrophilic Aliphatic Substitution

R. B. Boar

Department of Chemistry, Chelsea College, University of London

The appearance this year of books and review articles which cover virtually all aspects of carbanion chemistry has done much to redress recent shortcomings in this area.[1]

Carbanion Structure and Stability

There has been a predictable increase in the application of ^{13}C-NMR (CMR) spectroscopy to problems of carbanion structure. A compilation of ^{13}C chemical shifts and ^{13}C–H coupling constants in organolithium compounds has appeared.[2] The temperature-dependence of the proton-decoupled CMR spectrum of the allylic carbanion (**1**) in liquid ammonia allowed the barrier to phenyl rotation to be estimated at 10.9 kcal mol^{-1}. Solvent-separated ion pairs were evidently present since the rotational barrier was effectively independent of the counterion.[3] PMR spectroscopy indicated that crotyl-lithium existed predominantly as the *Z*-anion (**2**), a conclusion that was confirmed by quenching studies[4] and agrees with CNDO/2 calculations on the optimum geometry of allyl anions.[5] Proton-decoupled CMR spectroscopy allowed direct observation of the

Na+
Ph Ph

(**1**)

H
H H
H Me

(**2**)

conformers of acyclic pentadienyl anions. The predominant conformation of pentadienyl-lithiums was W with substantial amounts of sickle present in some cases.[6] Addition of

18-crown-6 to a pyridine solution of sodium acetylacetonate caused a progressive shift from the U-conformer (**3**) to the W-conformer (**4**).[7] Like the parent bicyclo[3.2.2]-

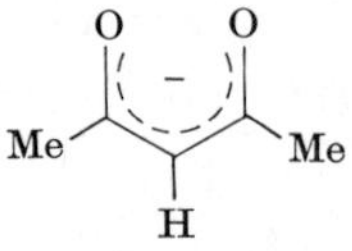

(**3**) 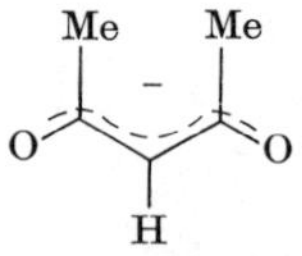 (**4**)

nonatrienyl anion,[8] the 6,7-benzo-derivative (**5**) readily underwent degenerate rearrangement.[9] Failure to detect the anion (**6**) by NMR spectroscopy or by quenching studies was interpreted as evidence that an allylic carbanion is more stable than its benzylic counterpart by at least 2 kcal mol^{-1}. Other compounds whose anions have

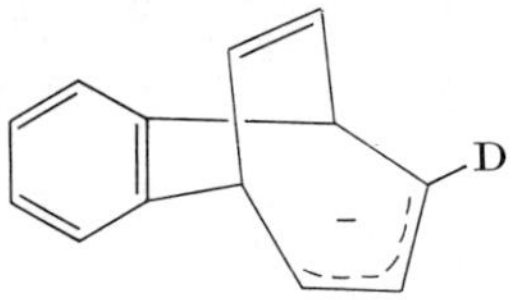

(**5**) 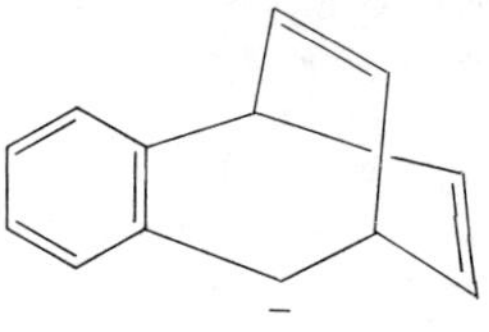 (**6**)

been examined spectroscopically include 6*H*-benzo[*cd*]pyrene,[10] *trans*-α-cyano-β-methoxystilbenes,[11] and *para*-substituted acetophenones.[12]

The effect of the cation on the structural and spectroscopic properties of fluorenyl ion pairs has been assessed,[13] and ^{7}Li chemical shifts of various organolithium compounds have been determined.[14] The pure organolithiums required for the latter study were conveniently prepared by cleavage of diorganomercury compounds with lithium metal.[15] Transient generation of a free benzyl carbanion by electron attachment to dibenzylmercury by a pulse radiolysis method allowed its absorption spectrum (λ_{max} 362 nm) to be compared with that of its sodium ion-paired counterpart (λ_{max} 355 nm). Proton-transfer from various alcohols was more rapid to the ion-paired species, as found for other reactions in which a charge-delocalized reactant was converted into a charge-localized product.[16] Electron-transfer reactions of carbanions are sufficiently rapid to compete with proton capture.[17] Whilst free triphenylmethide ions or crown ether-complexed ion pairs were best trapped by oxygen in *tert*-butyl alcohol, tight potassium triphenylmethide ion pairs were more efficiently trapped by nitrobenzene in *tert*-butyl alcohol.[18] These criteria were used to evaluate the nature of triphenylmethide ions produced in the base-induced cleavage of benzopinacolone and benzoylazotriphenylmethane.[19] Addition of ethanol to a diethyl ether solution of the lithium salt of 10-*tert*-butyl-9,10-dihydro-9-methylanthracene afforded solely the *cis*-hydrocarbon (**7**).

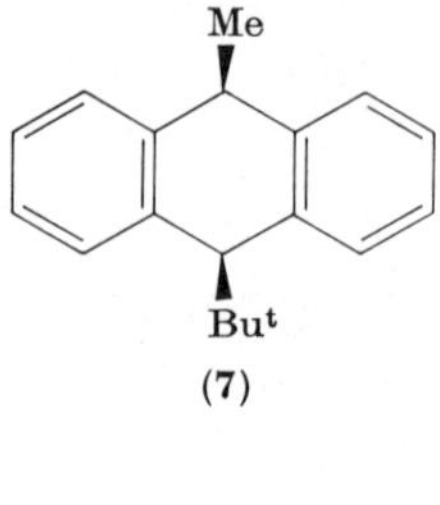

(**7**)

Prior addition of hexamethylphosphoramide (6 equivalents) transformed the nature of the ions present, since protonation then gave only the *trans*-isomer of (**7**).[20]

Aromaticity

PMR spectroscopy showed that the benzocyclononatetraenyl anion (**8**) was strongly diatropic,[21] but the isoelectronic nitrogen derivative (**9**) showed no indication of a diamagnetic ring current.[22] Although the parent cyclononatrienyl anion undergoes *trans*-to-*cis* isomerization on warming, the aromatic *cis*-anion (**8**) rearranged at ambient temperature to the *trans*-isomer (**10**). Relief of the severe *peri*-H–H interactions which are present in (**8**) must motivate this rearrangement.[23] Nitrogen heterocycles that did

(**8**) (**9**) (**10**)

exhibit diamagnetic ring currents included the aza[13]annulene (**11**), the aza[17]-annulene[24] and the bridged didehydroaza[17]annulenes (**12**).[25] In each case the corresponding anion was considerably more diatropic than the parent compound. The

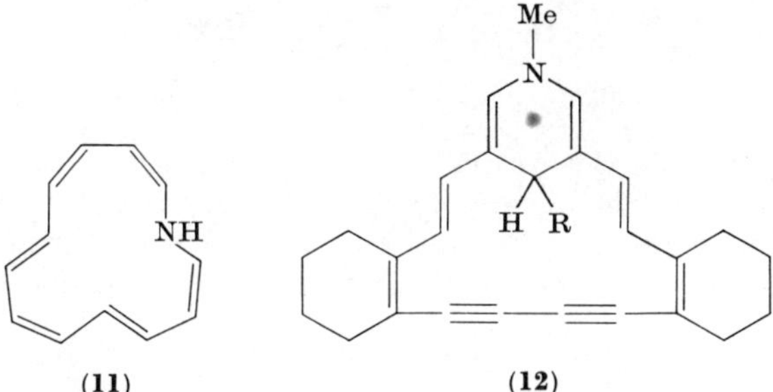

(**11**) (**12**)

bridged 1,6- and 1,7-methano[12]annulenes were reduced by lithium to dianions [1,7-isomer (**13**)] which were completely stable at room temperature and satisfied the normal criteria for aromatic systems.[26] Methylation of the dianion[27] of the bridged

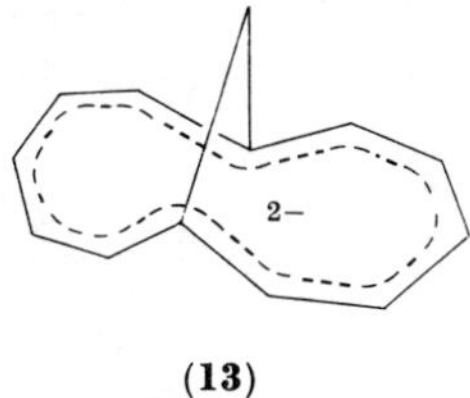

(**13**)

sulphone (**14**) followed by photochemical extrusion of sulphur dioxide completed an attractive synthesis of 1,4-dimethylcyclo-octatetraene. This and the 1,2-, 1,3- and

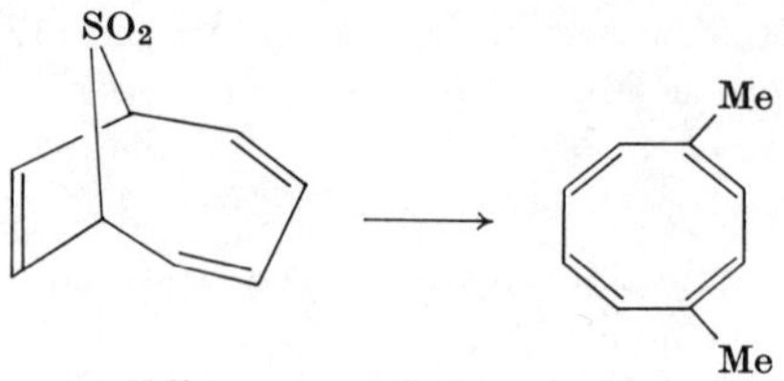

SO$_2$ Me Me

(14)

1,5-isomers were each reduced by potassium in liquid trideuterioammonia to dianions which were diatropic and essentially planar (PMR spectroscopy).[28] X-Ray analysis of 1,3,5,7-tetramethylcyclo-octatetraene dianion showed this to be planar with average C—C distances of 1.407 Å.[29] Cyclo-octatetraene dianions were also formed by deprotonation of cyclo-octatrienes in strongly basic media.[30]

Reduction of *cis*-bicyclo[6.1.0]nona-2,4,6-triene (**15**) and various methyl derivatives by potassium in liquid trideuterioammonia afforded homoaromatic dianions, but their basicity was such that only in favourable cases could they even be detected by NMR spectroscopy. In the case of the parent system the NMR spectrum of the reaction mixture showed only the monodeuteriated anion (**16**). Under the same conditions cyclo-octatetraene dianions were quite stable.[31] Since stabilization of a benzylic

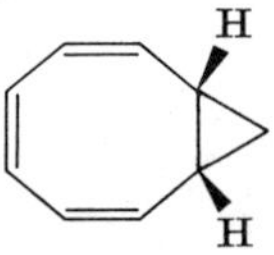

(15) 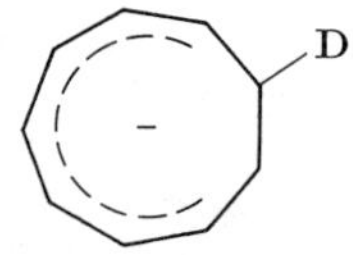 **(16)**

carbanion by an α-cyclopropyl substituent was minimal (compare carbonium ions and radicals), kinetic results on the base-catalysed rearrangement of tricyclo[7.1.0^{2,7}]deca-2(7),3,5-triene (**17**) were interpreted by invoking participation of a homoindenyl anion (**18**) which possessed some homoaromatic character.[32] Participation of the bishomo-

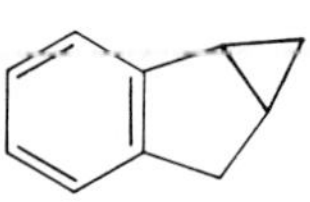

(17) 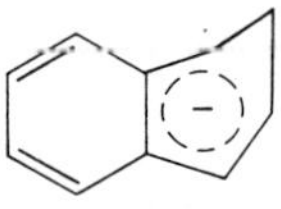 **(18)**

aromatic anion (**20**) was used to explain the unexpectedly rapid cleavage of homocuneone (**19**) by base.[33] The red-brown solution which resulted when cycloheptatriene was

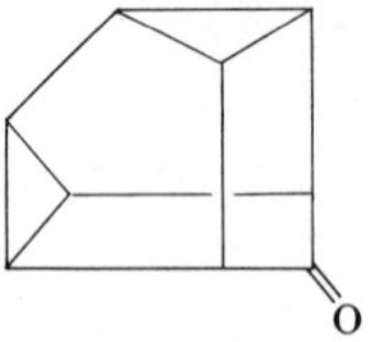

(19) 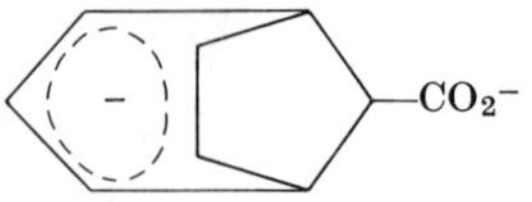 **(20)**

added to a solution of potassium amide in liquid ammonia contained the anion (**21**); deuterium-labelling studies indicated that (**21**) arose by rapid stepwise cycloaddition of the antiaromatic cycloheptatrienyl anion to cycloheptatriene.[34] Treatment of

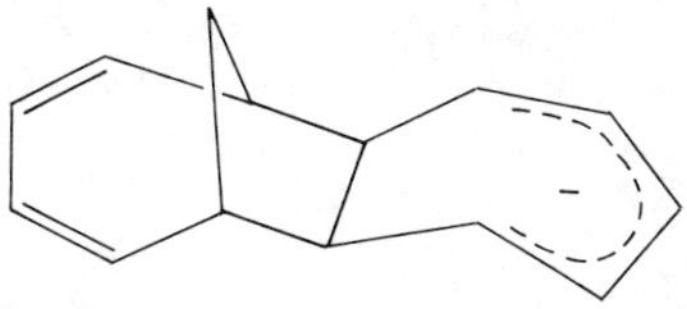

(**21**)

1,2,3-tribenzoyl-1-halocyclopropanes with various bases failed to produce the anti-aromatic tribenzoylcyclopropenyl anion, although in some cases the expected intermediate tribenzoylcyclopropene could be detected.[35] Treatment of bullvalene (**22**)

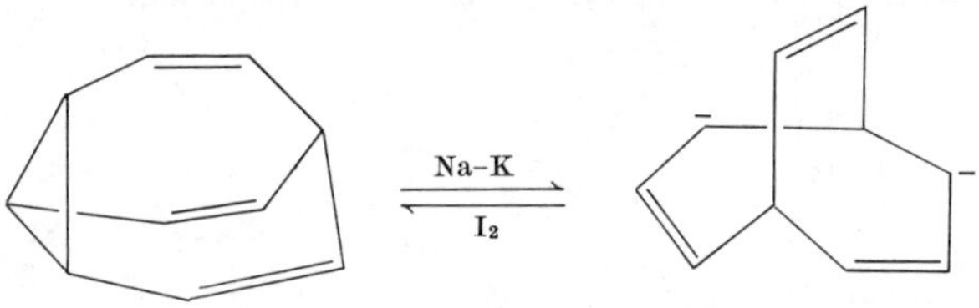

Na–K

I_2

(**22**) (**23**)

with sodium–potassium alloy at room temperature yielded the longicyclic bicyclo[3.3.2]-decatrienyl dianion (**23**).[36] Dianions of various simple non-conjugated alkenes have been prepared, even though the products have no form of aromatic stabilization.[37]

Reactions of Carbanions

Some two-phase reactions for which crown ethers act as catalysts have been described.[38] *N*,*N*-Dialkylamides afforded synthetically useful α-anions when treated with lithium di-alkylamide in hexamethylphosphoramide,[39] whilst di-isopropylformamide reacted with the same reagent at –78° to form an aminocarbonyl anion; quenching with deuterium oxide gave ^{2}H-CONPr$^i_2$ in 70% yield.[40] Similar evidence established the formation of the anion $^-$COOEt when ethyl formate was added to a solution containing a carboxylic acid dianion.[41] The preference of the cyanohydrin anion (an acyl anion equivalent) for conjugate addition to enone systems has been explained in terms of the transition state (**24**).[42] Formation of the bis(trimethylsilyl) derivative (**25**) when the

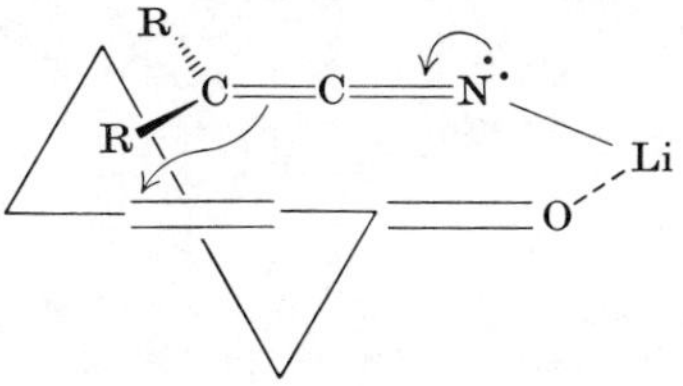

R, R, C=C=N, Li, O

(**24**)

reaction of allyl cyanide with an excess of *n*-butyl-lithium was quenched with trimethylsilyl chloride occurred via the sequence monoanion → mono-derivative → monoanion → (**25**), and not via a dianion.[43] Allylic ethers were readily metallated by alkyl-lithiums at

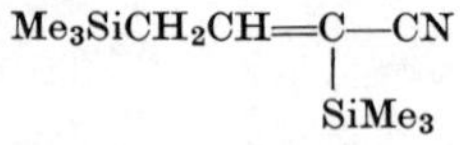

$$Me_3SiCH_2CH{=}C(SiMe_3){-}CN$$

(**25**)

–78°; Wittig rearrangement was not observed, but electrophiles reacted to give α- and γ-substituted products, the former affording a route to β-substituted aldehydes.[44] Details have appeared of the thermal ring-opening of cyclopropyl anions reported last year, and the same reactions have been achieved photochemically for the first time.[45] The racemization of 1-cyano-2,2-diphenylcyclopropane by sodium methoxide in methanol/dimethyl sulphoxide mixtures gave a linear plot of $\log k_0$ vs. H_-; rates varied by a factor of more than 10^8.[46] The isocyanide anion (**26**) retained its stereochemistry at temperatures where the corresponding cyanide anion racemized completely.[47]

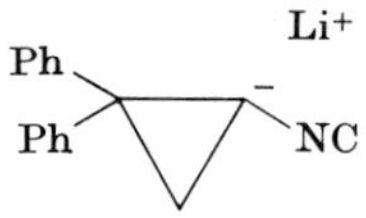

Ph Ph Li+ NC

(**26**)

1,5-Diphenylpentadienyl anion, generated at 225° from *trans,trans*-dibenzylideneacetone semicarbazone and alkoxide, underwent *in situ* disrotatary electrocyclic closure to afford primarily *cis*- and *trans*-1,2-diphenylcyclopentenyl anions in a ratio which varied with the nature of the base (Scheme 1).[48] The cycloaddition of pentadienyl-

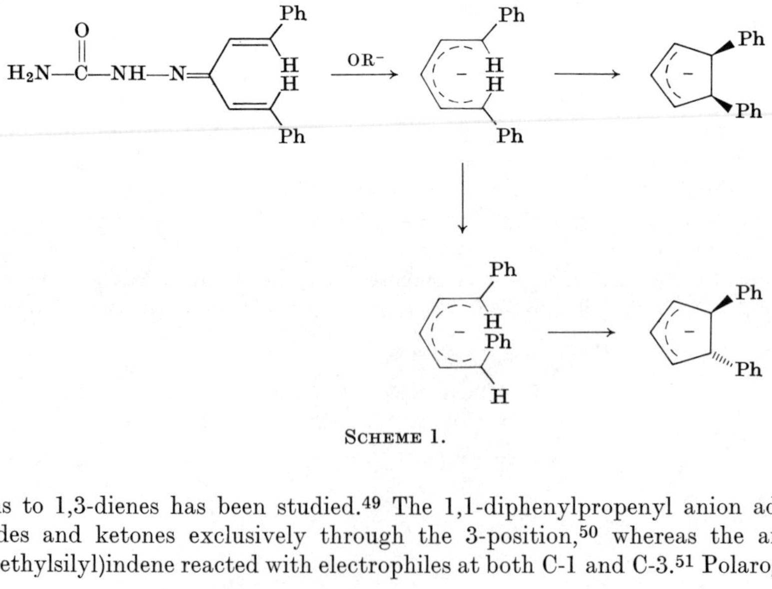

O H2N—C—NH—N= Ph H H Ph OR− Ph H H Ph Ph Ph

Ph H Ph H Ph Ph

SCHEME 1.

lithiums to 1,3-dienes has been studied.[49] The 1,1-diphenylpropenyl anion added to aldehydes and ketones exclusively through the 3-position,[50] whereas the anion of 1-(trimethylsilyl)indene reacted with electrophiles at both C-1 and C-3.[51] Polarographic reduction potentials of acetylenes were found to be inconsistent with the often assumed intermediacy of radical anions or dianions in the reduction of acetylenes by sodium in liquid ammonia or hexamethylphosphoramide. Reduction by sodium in hexamethylphosphoramide–tetrahydrofuran containing *tert*-butyl alcohol (to prevent isomerization[52]) was best explained by the mechanism of Scheme 2. The product was normally >95% *trans*, but increased temperature or reduction of the sodium concentration favoured equilibration of the radical (**27**) and led to 10–15% *cis*-alkene[53] (see also Chapter 4). Applications of α-silylcarbanions to alkene synthesis have been detailed.[54]

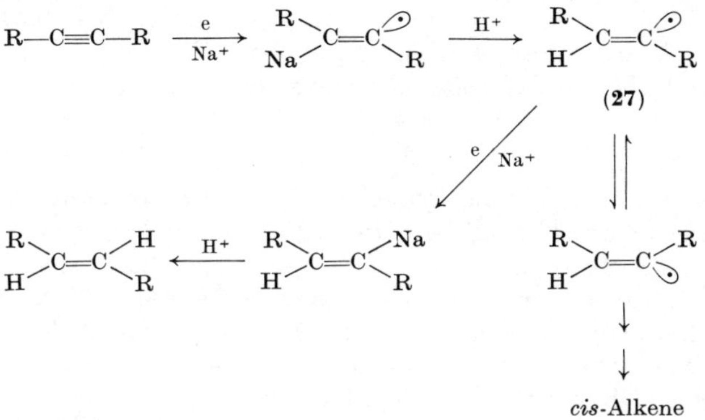

(**27**)

cis-Alkene

SCHEME 2.

Perfluoro-*tert*-butyl anion reacted with perfluoronitrosoalkanes to afford oxime ethers $[(CF_3)_3C{-}O{-}N{=}C(R_F)_2]$, and not nitrones as previously claimed.[55] When 2,3-bis-(*p*-methoxyphenyl)butyronitrile was reduced by lithium aluminium hydride in tetrahydrofuran without the exclusion of air, the alcohol (**28**) was the unexpected major product.[56] Dimer formation in the reduction of benzylic halides by sodium naphthalene

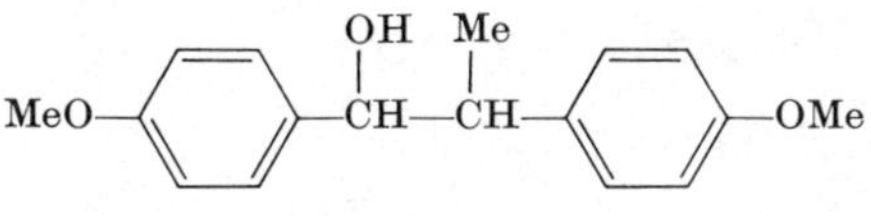

(**28**) 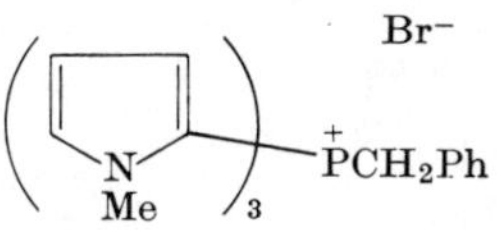 (**29**)

involved carbanionic intermediates since, in the presence of [*O*-^{2}H]-*tert*-butyl alcohol, Ph_2CH-^{2}H, but no dimer, was formed.[57] Hydrolysis of the phosphonium salt (**29**) afforded 1-methylpyrrole and toluene (2:3), in contrast to the corresponding furan and thiophene systems which gave only the heterocyclic component.[58]

Equilibration of *erythro*- and *threo*-products has been observed in the Reformatsky reaction of ethyl 2-bromopropionate with *para*-substituted acetophenones.[59] The Cu(I)-catalysed reaction of *o*-bromobenzoic acids with carbanions (Hurtley reaction) has been studied.[60] Amongst base-catalysed processes to have received mechanistic attention are the Dieckmann reaction of $MeOOCCH_2SCH_2CH_2COOMe$,[61] the decomposition of diacetone alcohol,[62] the epimerization of the alkaloid pilocarpine,[63] and the isomerization of δ-keto-epoxides to dihydropyrans.[64]

Enolates

When generated in the presence of acetone enolate anion, phenyl radicals were trapped as phenylacetone and related products, whereas phenyl anions simply formed benzene; this distinction was used to determine the mode of cleavage of various PhX compounds by potassium in liquid ammonia.[65] The intermediacy of cyclohepta-1,2-diene has been proposed to explain the predominant formation of bicyclo[5.2.0]non-1-en-8-ols when enolates of di-isopropyl and methyl phenyl ketones were condensed with 1-chlorocycloheptene.[66] Further applications of preformed lithium enolates to aldol condensations have been described.[67] Interest continues in the factors affecting the ratio of *C*- to *O*-alkylation of ethyl acetoacetate by ethyl toluene-*p*-sulphonate.[68] In keeping with the theory of stereoelectronic control, the axial protons at C-2 of 4-*tert*-butylcyclohexanone were exchanged by NaOD/D_2O some 5.5 times faster than the equatorial ones.[69]

Dianions

Reactions of radical anions and dianions of aromatic hydrocarbons have been reviewed,[70] and details have appeared of the dehydrogenation of dihydroaromatic compounds via dianion formation.[71] The isomerization of *cis*- to *trans*-stilbene by radical anions is well known, and it has now been shown that the pseudo-first-order rate constant for the isomerization of *cis*-stilbene catalysed by sodium anthracenide ($A^{\overset{-}{\cdot}}Na^+$) is proportional to $([A^{\overset{-}{\cdot}}Na^+]/[A])^2$. Thus, isomerization of *cis*-stilbene dianion must be the rate-determining step.[72] Under the appropriate conditions *N*-methylpyrrole reacted with *n*-butyl-lithium to afford predominantly the 2,4-dianion, contradicting previous beliefs that such heterocycles are metallated almost exclusively at the α-positions.[73] β-Keto-esters[74] and β-keto-sulphoxides[75] are further[76] classes of compounds whose dianions (formed by treatment with sodium hydride and then *n*-butyl-lithium) react with electrophiles exclusively at the γ-position. A useful synthesis of vinyl ketones is thus available (Scheme 3).[77]

$$CH_3\text{—}\underset{}{\overset{SOPh}{CH}}\text{—}CO\text{—}CH_2R \longrightarrow CH_3\text{—}\overset{SOPh}{CH}\text{—}CO\text{—}\overset{E}{CHR} \xrightarrow{\Delta} CH_2\text{=}CH\text{—}CO\text{—}\overset{E}{CHR}$$

SCHEME 3.

Anions α to Sulphur

Reaction of prop-2-enethiol with *n*-butyl-lithium in TMEDA gave a doubly metallated species which reacted with electrophiles predominantly at the γ-position to give products (**30**) of exclusively *cis*-geometry. Hydrolysis of (**30**) by the novel method indicated completed a route to β-substituted aldehydes.[78] Regioselectivity in reactions of the

2− S 2Li+ —(1) E^1, (2) E^2→ E^1…SE^2 (**30**)

↓ RSH

E^1…C(H)(SE^2)(RS) ⟶ E^1…CHO (H, O)

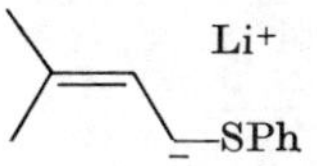

Li+

SPh

(31)

carbanion (**31**) could be controlled by the use of additives which affected the solvation of the counterion, and hence the nature of the ion pairs present.[79] Anions of allylic dithioacetals (**32**) underwent rapid [2.3]sigmatropic rearrangement at low temperatures; hydrolysis of the products then afforded β,γ-unsaturated aldehydes.[80] Lithium

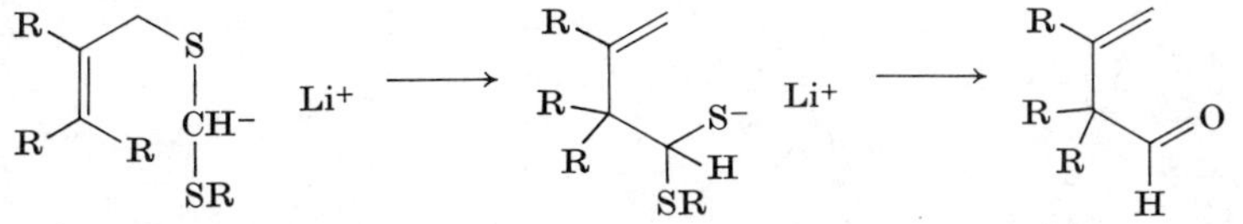

(32)

derivatives of *cis*-4,6-dimethyl-1,3-dithiane reacted with electrophiles to give exclusively equatorial 2-substituted products. Thus, 2-*tert*-butyl-*cis*-4,*cis*-6-dimethyl-1,3-dithiane (**33**) was converted by treatment with *n*-butyl-lithium in TMEDA, then acid, into the *trans*-isomer (**34**) with greater than 99.7% stereoselectivity; the stereoselectivity, which was shown to be largely the result of thermodynamic rather than kinetic factors, was attributed to a preference (by at least 6 kcal mol^{-1}) for the intermediate 2-dithianyl-lithium to have the metal in an equatorial position. A possible rationale for this is embodied in structure (**35**).[81]

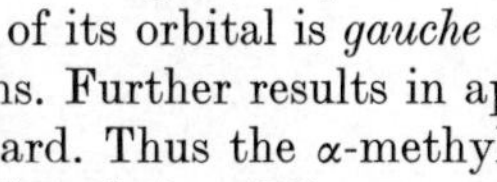

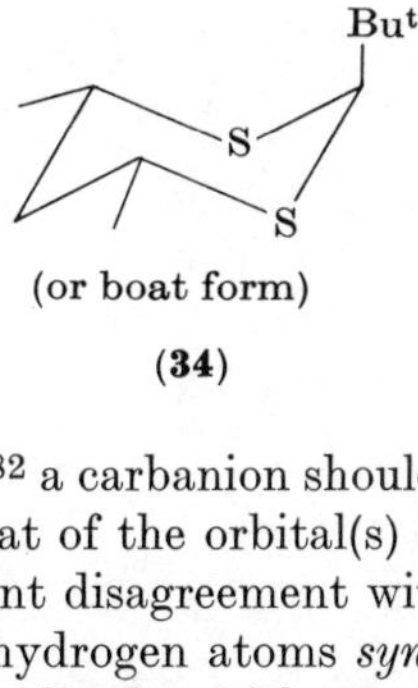

(or boat form)

(33) **(34)** **(35)**

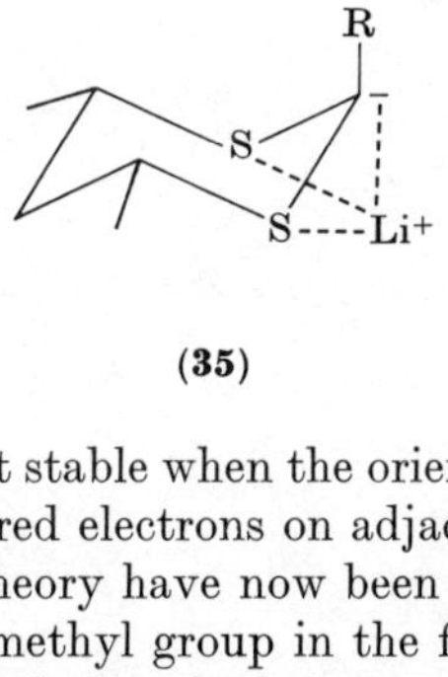

According to the *gauche* effect,[82] a carbanion should be most stable when the orientation of its orbital is *gauche* to that of the orbital(s) or unshared electrons on adjacent atoms. Further results in apparent disagreement with this theory have now been put forward. Thus the α-methylene-hydrogen atoms *syn* to the methyl group in the five-membered ring (**36**) were more acidic than either pair of α-methylene-hydrogen atoms in the six-membered ring (**37**), and the rate ratio for diastereotopic α-methylene-hydrogen atoms was considerably greater in the five- than the six-membered ring.[83] For (**38**) the relative rates of exchange ($H^1:H^2:H^3:H^4 \approx 1:200:3:3$) were in agreement

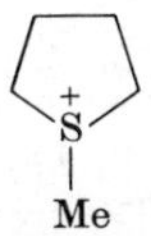

(36) 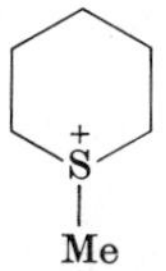 **(37)** 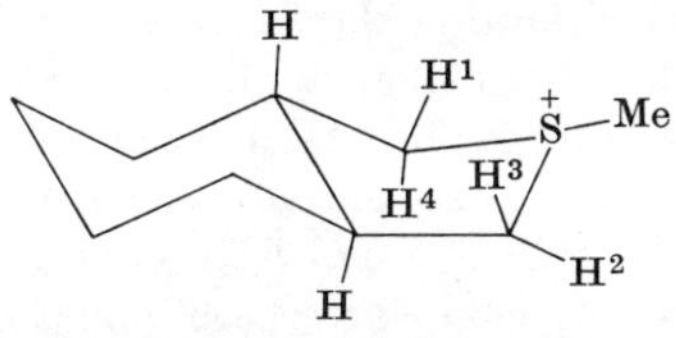 **(38)**

with the *gauche* effect for the pair of pseudoequatorial hydrogens H^1 and H^2, but not for the pseudoaxial hydrogens H^3 and H^4.[84] Factors of probable relevance to these discrepancies are the relationship of kinetic to thermodynamic acidity and, more particularly, solvation effects. In the latter context, the micelle-forming allyl-lauryl-methylsulphonium salt (**39**) underwent base-catalysed hydrogen–deuterium exchange much more rapidly than (**40**), and the rate enhancement was greater for the allyl- than for the methyl-hydrogen atoms. Also, for (**40**), the relative rates of exchange of the allyl- and methyl-hydrogens varied in a complex manner with solvent.[85] Hydrogen exchange in organic sulphides has been studied by others.[86]

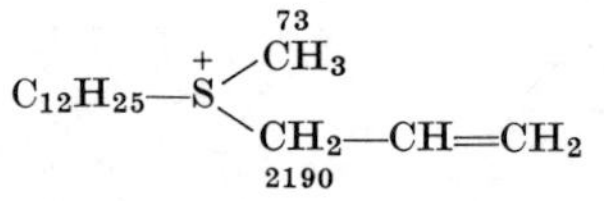

$C_{12}H_{25}$—$\overset{+}{S}$(CH_3, 73)(CH_2—CH=CH_2, 2190) (**39**)

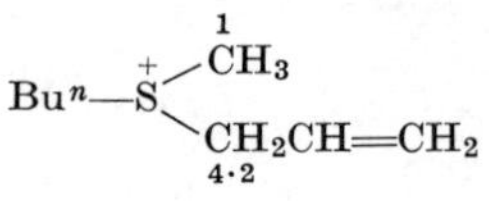

Bu^n—$\overset{+}{S}$(CH_3, 1)(CH_2CH=CH_2, 4·2) (**40**)

Numbers are relative rates of H–D exchange in D_2O–NaOD at 42°.

Further results in support of the mechanism discussed last year[87] for the halogenation of thiane 1-oxides have appeared,[88] and the formation of 4-chlorothiane 1,1-dioxides in the chlorination of *cis*-4-hydroxythiane 1-oxides has been explained by the mechanism of Scheme 4.[89] Bromination of benzyl methyl sulphoxide by bromine in pyridine

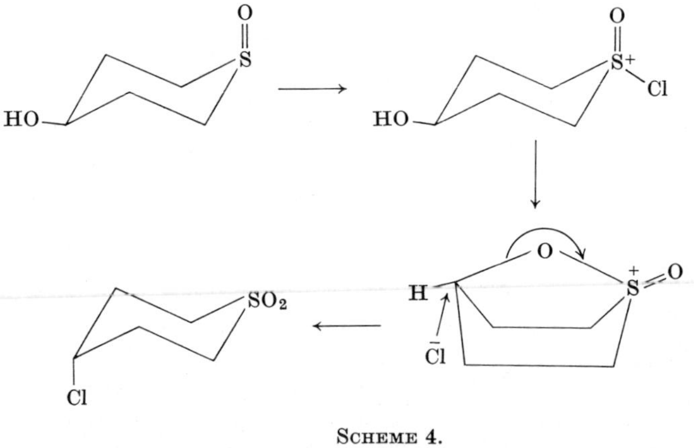

SCHEME 4.

proceeded with retention of configuration at sulphur for substitution of the methyl group, but inversion for substitution at the benzylic position.[90] The effect of solvent (ion pairing) on the stereoselectivity of anions of benzyl methyl sulphoxide has been studied.[91] Further work on hydrogen–deuterium exchange by NaOMe/NaOD of the rigid sulphoxides (**41**) has confirmed that the hydrogen eclipsed by the S—O bond is exchanged preferentially, but the magnitude of this preference was not as dramatic as previously suggested.[92] In contrast, in 2*H*-naphtho[1,8-*bc*]thiophene 1-oxide (**42**) the hydrogen eclipsed to the sulphur lone pair had the greater kinetic acidity.[93] Differences in the preferred geometries of the anions could explain this variation.

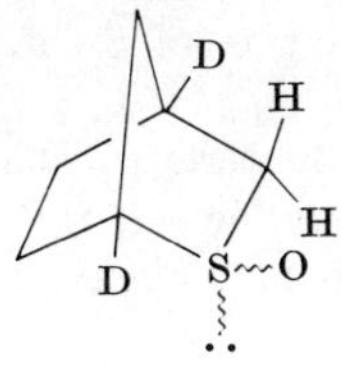

(41)

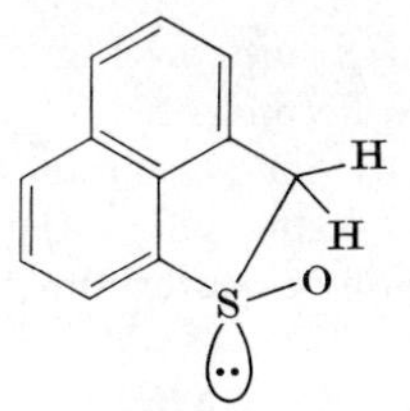

(42)

Further studies on the mechanism of the Ramberg–Backlund reaction have appeared.[94,95] The general mechanism is illustrated for the "*erythro*"-bromo-sulphone (**43**). For the "*threo*"-isomer of (**43**) the rate was independent of methoxide concentration, indicating that decomposition of the thiirane 1,1-dioxide had become rate-determining. Two novel reactions of related interest have been described this year:

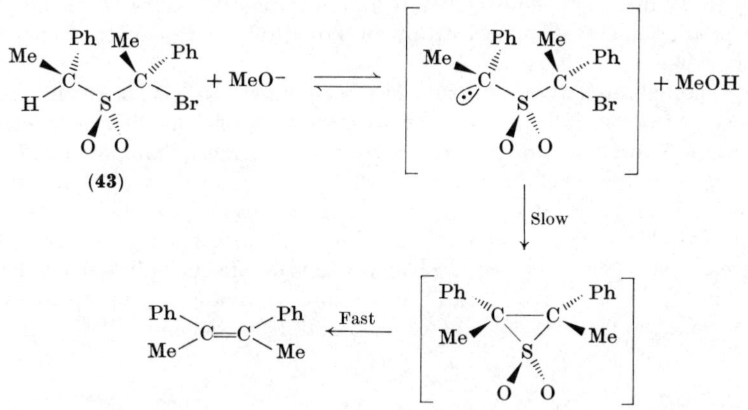

(43) + MeO⁻ ⇌ [carbanion] + MeOH

Slow

Fast

anions of 2,5-dialkyltetrahydrothiophene dioxides (as **44**) have been shown to undergo sulphur extrusion when heated with lithium aluminium hydride in refluxing dioxan,[96] and α,α'-dianions of sulphones were converted by cupric chloride into alkenes;[97] thus $EtOOC\bar{C}HSO_2\bar{C}HCOOEt$ afforded diethyl fumarate (54%) plus diethyl α-chloro-fumarate (16%).

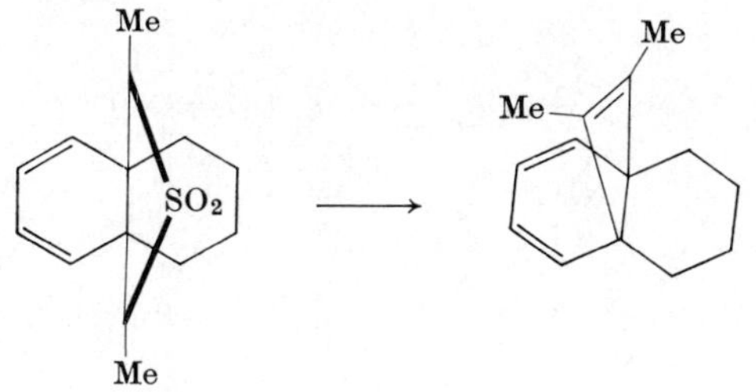

(44)

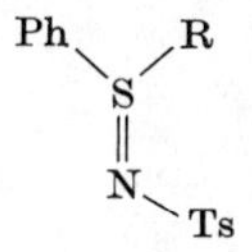

(45)

Other reactions to have been studied include the isomerization and hydrogen–deuterium exchange of 2-halo-3-morpholinothietan 1,1-dioxides,[98] the haloform reaction of β-sulphonyl-methyl ketones,[99] and the condensation of β-keto-sulphones with α-halo-ketones.[100]

At pH > 7 the sulphinimines (**45**) underwent hydrogen–deuterium exchange some ten times faster than the corresponding sulphoximines. For (**45**; $R = PhCH_2$), incorporation of deuterium was seven times slower than for benzyl phenyl ketone.[101]

Proton-transfer and Hydrogen Isotope Effects

(–)-1-Methylindene is rearranged by quinidine about 70% more rapidly than is the (+)-isomer. Thus, when a 0.4 M-solution of racemic 1-methylindene in DMSO was treated with quinidine (0.04 M) at –27° for 48 h, 37% of (+)-1-methylindene with an optical purity of 28% could be recovered. This represents a very appreciable kinetic resolution. Interestingly, the isomerization proceeded considerably faster in frozen DMSO than in supercooled solutions at the same temperature.[102] Kinetic solvent isotope effects (k_{ROH}/k_{ROD}) for the racemization of (+)-1-methylindene and (+)-1,3-dimethylindene by RO^- in ROH and ROD were in the range 0.49–0.54,[103] in keeping with previous predictions.[104] The prototropy of 2-methyl-1,3-diphenylpropenes has also been studied.[105]

Tetraethylammonium thiophenoxide has been shown to have a strong, but as yet unexplained, catalytic influence on the racemization of 1-nitro-1-phenylethane and 1,2-diphenylpropan-1-one under conditions where tetraethylammonium 2,4-dinitrophenoxide is quite inactive. Such catalysis of proton removal by thiolate anions could have important implications as regards the mechanism by which certain enzyme systems operate.[106] Various problems regarding proton-transfer in relation to enzyme mechanisms have been discussed.[107] Kinetic isotope effects indicated that the rate-determining step in the conversion of L-serine into pyruvate by tryptophan synthetase was proton-removal from an initially formed Schiff base (Scheme 5).[108]

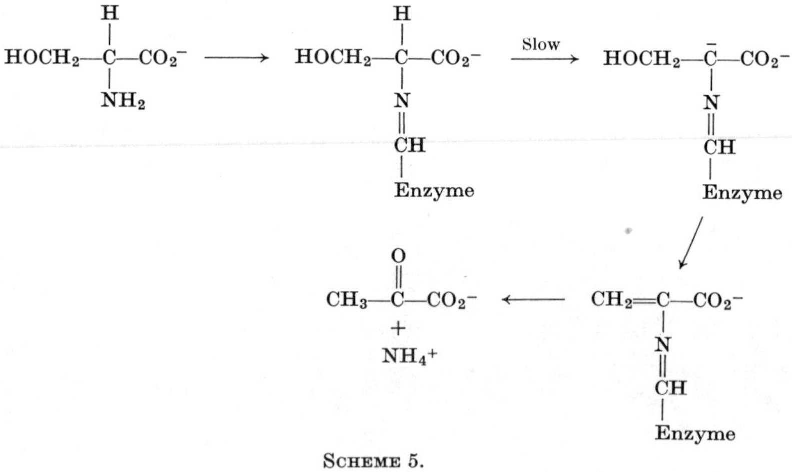

$$HOCH_2{-}CH(NH_2){-}CO_2^- \longrightarrow HOCH_2{-}CH(N{=}CH{-}\text{Enzyme}){-}CO_2^- \xrightarrow{\text{Slow}} HOCH_2{-}\bar{C}(N{=}CH{-}\text{Enzyme}){-}CO_2^-$$

$$\longrightarrow CH_2{=}C(N{=}CH{-}\text{Enzyme}){-}CO_2^- \longrightarrow CH_3{-}C({=}O){-}CO_2^- + NH_4^+$$

SCHEME 5.

Proton-transfer from di-(*p*-nitrophenyl)methane to *tert*-butoxide ion in 10% v/v toluene in *tert*-butyl alcohol showed none of the characteristics that would be expected if tunnelling were important.[109, 110] On empirical grounds, it was suggested that tunnelling is most likely to be observed for proton-transfers with activation energies of about 10 kcal mol^{-1} or more.[109]

The nitroalkanes are a group of compounds that have received widespread attention as a result of their anomalous behaviour in various aspects of proton-transfer reactions. There has been considerable discussion this year of the occurrence of Brønsted coefficients that lie outside the normal range of 0–1, and of the interpretation of such values in terms of transition-state structure. Particular studies have concerned the ionization of nitroethane,[111] 2-nitropropane,[112,113] phenylnitromethane,[114] 1,1-dinitropropane,[115] α-halonitroalkanes,[116] α,α-dinitroacetamides,[117] and nitroalkanes in general.[118,119] It now seems likely that in the transition state the partial bonds to the proton undergoing transfer are of approximately equal strength, even for reactions having ΔG^0 rather different from zero. Generalizations about even closely related systems are, however, clearly tenuous.

Absolute and Relative Acidities

Apparent equilibrium acidities of weak carbon acids are frequently much greater in solvents where ion pairing is important than in DMSO where salts are highly dissociated. Thus the apparent acidity of phenylacetylene is approximately 15.8 (diethyl ether), 20.5 (cyclohexylamine), or 26.5 (DMSO).[120] DMSO has the advantage that it allows an accurate comparison of equilibrium acidities in a single solvent over a wide pK range (0–30), and with a minimum of interference from counterion effects. Many of the basic measurements necessary for setting up an extended McEwan-type pK scale with this solvent have now been reported.[121]

A new method of determining the acidities of carbon acids has been described. The rate of detritiation (k_1^{T}) of a standard carbon acid (e.g. [9-^{3}H]-9-*tert*-butylfluorene) in a series of DMSO/water solutions containing a fixed concentration of hydroxide ($[OH^-]_1$) is determined. The procedure is then repeated in the presence of the test acid ($[A_2H]_{init.}$), which must be appreciably ionized under these conditions. The resulting decrease in hydroxide ion concentration ($[OH^-]_2$) (and hence H_-) is reflected in a decreased rate of detritiation (k_2^{T}) of the standard acid. The relevant equations are:

$$[OH^-]_2 = [OH^-]_1 \cdot k_2^{T}/k_1^{T} \qquad [A_2^-] = [OH^-]_1 - [OH^-]_2$$

$$[A_2H] = [A_2H]_{init.} - [A_2^-] \qquad H_- = pK_{A_2H} - \log[A_2^-]/[A_2H]$$

Fluorene and 2-nitro-4-chloroaniline were thus assigned pK_a 21.0 and 16.9, respectively, values in good agreement with the literature.[122]

Gas-phase acidities, which are valuable in that they allow intrinsic and solvent-induced substituent effects to be separated, have been determined for a range of carbon acids from proton-transfer equilibrium measurements. Substituents increased the acidity of methane in the following order: phenyl ≈ vinyl < acetyl < cyano < benzoyl. Apart from a transposition of acetyl and cyano, this is the same order as that which applies for aqueous solutions.[123] Relative acidities deduced from *ab initio* molecular-orbital calculations on various alcohols, alkylacetylenes, phenols, and xylenes agreed well with gas-phase data from ion-cyclotron resonance studies.[124]

Ab initio SCF calculations have been carried out in order to evaluate the effects that angle distortions in methanes,[125] ethanes, and ethylenes[126] have on the acidity of adjacent C—H bonds. The results of these calculations supported the previous view that the enhanced kinetic acidity of a C—H bond adjacent to a strained ring is a reflection of increased *s*-character. It is generally supposed that there is a close parallel between the *s*-character in the hydrocarbon and in the derived carbanion. Cubane (**46**) is a case where this assumption might well not be true; a rehybridization of the anionic

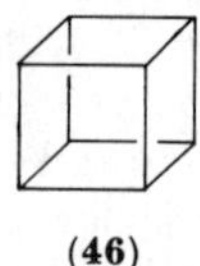

(**46**)

centre leading to endocyclic orbitals with greater *p*-character would be very favourable in terms of strain relief and would concurrently produce an exocyclic orbital of much increased *s*-character. The kinetic acidity of cubane towards lithium cyclohexylamide in cyclohexylamine has now been determined,[127] and in line with these arguments was found to be some 10^3 times that of cyclopropane.

Attempts to determine the acidities of nitroalkanes in ammonia by NMR spectroscopy failed. The relatively high concentrations needed for this method led to various complications believed to be associated with temperature- and concentration-dependent ion-aggregation effects.[128,129] Regiospecific metallation of 2,4-dimethylpyridines was not as simple as previous results suggested;[130] synthetically useful selectivity could be obtained, however, provided that due attention was paid to the choice of solvent, metallating agent and reaction time.[131] The p*K*'s of some very acidic oxocarbon acids (e.g. phenylhydroxycyclobutenedione p*K* –0.22) have been determined spectrophotometrically, and the possible effect of electron delocalization in the anions of these species has been discussed.[132] Acidities of some 3,5-diaryl-1,3,4-thiadiazolium salts have been measured,[133] and the effect of the perfluorophenyl group on the acidity of substituted methanes has been assessed.[134]

Organometallics

Intramolecular addition of a silicon—lithium bond to a carbon—carbon double bond has been observed. Thus, treatment of bis[dimethyl(pent-4-en-1-yl)silyl]mercury with lithium metal afforded, after hydrolysis, 1,1,2-trimethylsilacyclopentane (**47**).[135] The

Si Me
Me Me

(**47**)

rate of formation of ethyl-lithium in tetrahydrofuran was proportional to the ethyl bromide concentration and the surface area of the lithium metal.[136] ^{2}H-NMR spectroscopy has been used to monitor the exchange of methyl groups between tetramethyltitanium and trimethylaluminium: rates were extremely dependent on any diethyl ether solvent remaining in the system.[137] Rate constants (k) for the Zerewitinoff reaction of various Grignard reagents with diethylamine obeyed the relationship $\log k/k_0 = -7.16\sigma^* + 1.33E_s$, where k_0 was the rate constant for methylmagnesium bromide.[138] Rate constants have been reported for the cleavage of $(PhMe_2Si)_2$ by bromine[139] and of siletanes and disiletanes by mercuric acetate.[140]

Insertion of sulphur dioxide into the iron—methyl bond of the complex (**48**) proceeded with high stereoselectivity at iron.[141] For a similar complex it was shown that the

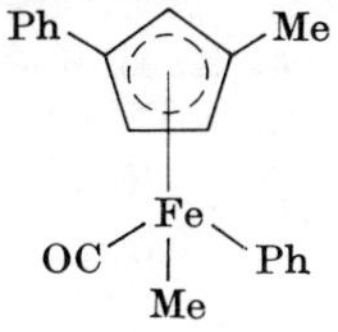

(48)

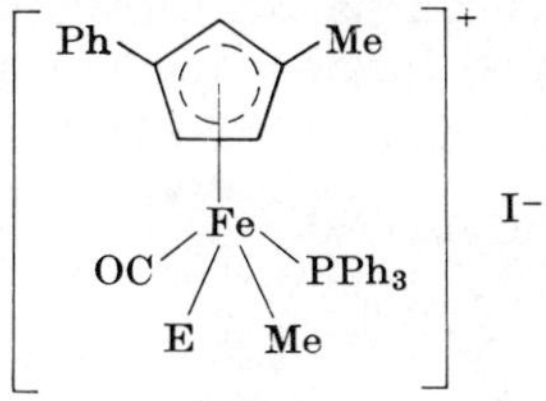

(49)

reaction involved retention of configuration at iron. [142] The observed stereochemistry for the electrophilic cleavage of (**48**) by hydrogen iodide, iodine or mercuric iodide, including the partial epimerization of recovered starting material, could be explained by invoking the intermediate (**49**; E = H, I or HgI).[141] Evidence has been presented that insertion of sulphur dioxide into tetraorganostannanes in liquid sulphur dioxide proceeds by an S_E2(open) mechanism.[143] Reaction of iodine–iodide with Me_3SnCN in DMSO was unimolecular, but with $Me_3SnCH{=}CH_2$ under the same conditions an S_E2 mechanism was evident.[144]

Differentiation between S_E2(open) and S_E2(cyclic) mechanisms remains difficult. Details have appeared of the criteria[145] on which the S_E2 substitution of tetraethyltin by Hg(II) carboxylates in methanol is believed to proceed via an open transition state (**50**);[146] in particular, rate acceleration by the presence of electron-attracting substi-

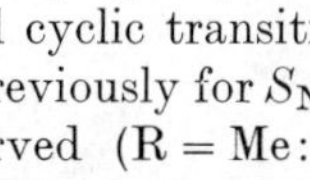

(50)

tuents in R is more consistent with an open transition state. Calculations have been carried out on the steric effects of alkyl groups in the reaction $RHgX + {}^*HgX_2 \rightarrow R^*HgX + HgX_2$ with both open and cyclic transition-state models.[147] Methods were basically similar to those employed previously for S_N2 reactions; for an open transition state the agreement between observed (R = Me:Et:Bu^tCH_2:*sec*-Bu = 1:0.42:0.33:0.06) and calculated (1:0.50:0.38:0.05) relative rate constants was excellent, but for the cyclic transition state there was no concord.

Mercurideboronation of tris-(*erythro*-1,2-dideuterio-3,3-dimethylbutyl)borane by mercuric acetate in tetrahydrofuran occurred with inversion of configuration,[148] in contrast to the similar reaction of tris-(*exo*-norbornyl)borane with mercuric benzoate for which retention of configuration was established.[149] Electrophilic reactions of Hg(I) are normally masked by the presence of Hg(II) but it has now been shown that the former is some 10^3 times more reactive than Tl(III) towards various transition-metal alkyls.[150] Further details have appeared of the exchange of organomercury compounds with mercury metal,[151] and CIDNP has been observed for the methylene-protons in Ph_3CCH_2Ar derived from the reaction of Ph_3CBr with $ArCH_2HgBr$.[152]

The chromium–carbene complex (**51**), with an equilibrium acidity approximately the same as that of *p*-cyanophenol, becomes one of the most acidic carbon acids known; reaction with *n*-butyl-lithium at –78° gave an anion that formed a stable bis(triphenylphosphine)iminium salt; the PMR spectrum of the latter indicated structure (**52**) for the anion.[153] Incorporation of deuterium into some bisarene complexes of chromium

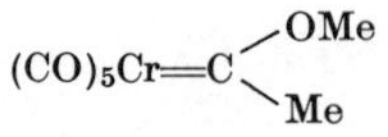

$(CO)_5Cr{=}C(OMe)Me$

(**51**)

$(CO)_5\bar{C}r{-}C(OMe){=}CH_2$

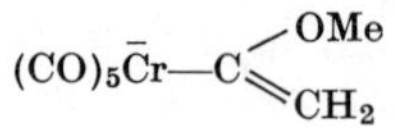

(**52**)

has been studied.[154] Hydrogen–deuterium exchange of [β-^{14}C]ethylbenzene with platinum salt catalysis involved no scrambling of the label, which is inconsistent with occurrence of reversible dealkylation;[155] with longer sidechains, predominantly α- and ω-exchanges were observed.[156] It was proposed that α-exchange occurred via a π-benzyl species (**53**), and ω-exchange by an independent, reversible, ligand-to-metal hydrogen-transfer from a localized π-complex.

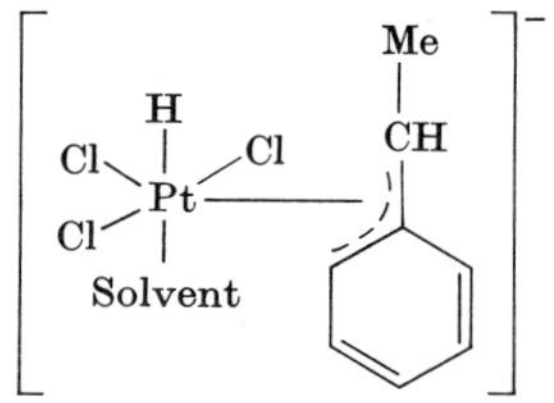

(**53**)

Other Reactions

A comprehensive review entitled "Stereochemistry and Reactivity in Cyclopropane Ring Cleavage by Electrophiles" has appeared.[157] The stereochemistry of electrophilic cleavage of a particular cyclopropane is evidently determined by a delicate balance. 7,7-Dimethylnorcarane (**54**) was cleaved by mercuric acetate in methanol with predominant inversion of configuration at the site of attack by the electrophile [formation of (**55**)], and hydroxymercuration of norcarane proceeded with complete inversion at

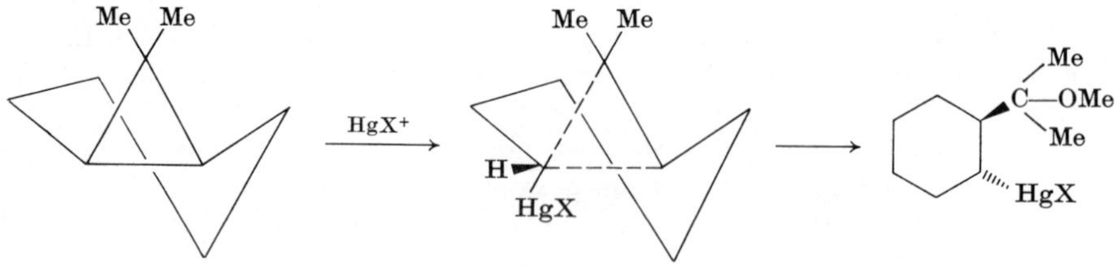

(**54**) $\xrightarrow{HgX^+}$ → (**55**)

the site of attack by the nucleophile.[158] The reactivity of tetracyclo[4.2.1.1^{2,5}.0^{1,6}]decane (**56**) towards electrophiles has been studied: the novel products (**57**) and (**58**) were rapidly formed when (**56**) was treated with dimethyl acetylenedicarboxylate at 25°;

the ratio (**57**):(**58**) increased with increasing solvent polarity but was insensitive to changes in dimethyl acetylenedicarboxylate concentration.[159]

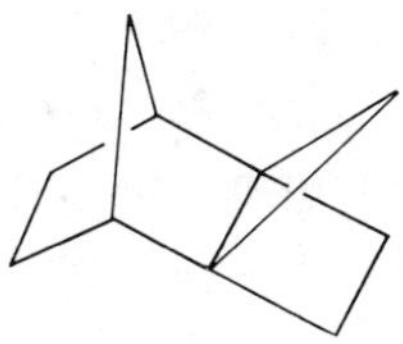

MeOOC COOMe COOMe COOMe MeOOC COOMe

(**56**) 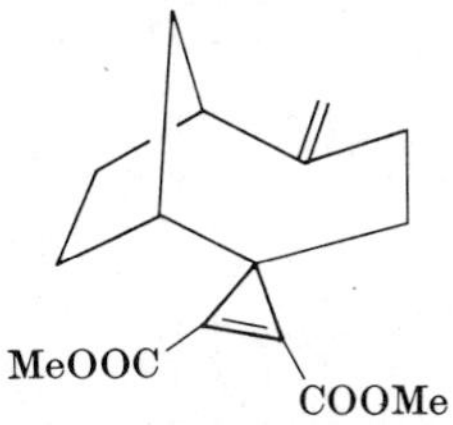 (**57**) 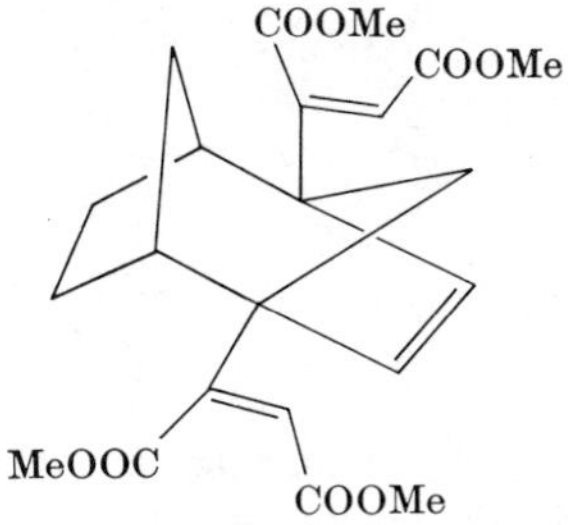 (**58**)

The bicyclopropene (**59**; R = Ph) reacted with bromine to give exclusively the stable triphenylcyclopropenium cation, but for the related (**59**; R = H) ring opening was favoured and a mixture of 1,6-dibromotetraphenylhexatrienes resulted.[160] Adamantane

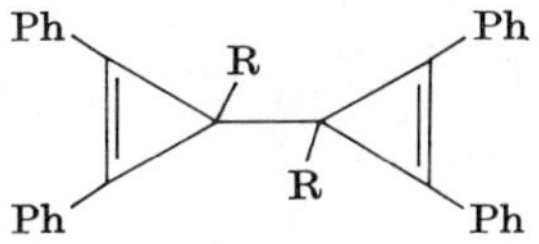

Ph R Ph Ph R Ph H Me Me B

(**59**) 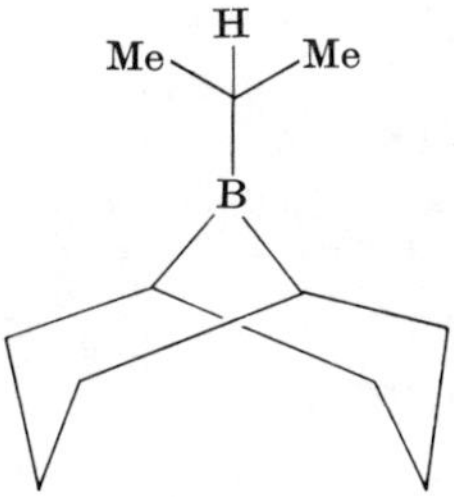 (**60**)

and related compounds were found to be much more susceptible to bromination by liquid bromine than previously believed.[161] Competitive bromination of *B*-isopropyl-9-borabicyclo[3.3.1]nonane (**60**) revealed that the tertiary hydrogen of the isopropyl group was 5.5 times more reactive than that of cumene; a free-radical mechanism was proposed, the high reactivity being attributed to stabilization of the odd electron by interaction with the vacant *p*-orbital of the boron atom.[162]

Other topics to have received attention are the reaction of alkyl sulphones[163] and disulphides[164] with sulphuryl chloride, bromination of 1,3-dioxan,[165] and cleavage of α-polynitroalkyl suphides by electrophiles.[166]

In the absence of solvent, or in chlorinated solvents, dinitrogen tetroxide converted 1-(acetoxymethyl)perhydro-3,5-dinitro-1,3,5-triazine (**61**) into perhydro-1,3-dinitro-5-nitroso-1,3,5-triazine (**62**); the reaction was of the first order in (**61**) and dinitrogen tetroxide, and the mechanism of Scheme 6 was suggested. In acetonitrile solution, the major product of this reaction was perhydro-1,3,5-trinitro-1,3,5-triazine; this solvent would reduce the concentration of free nitrosonium ions by complex formation, and the observed product would then be formed by reaction of (**61**) with nitrate ion.[167]

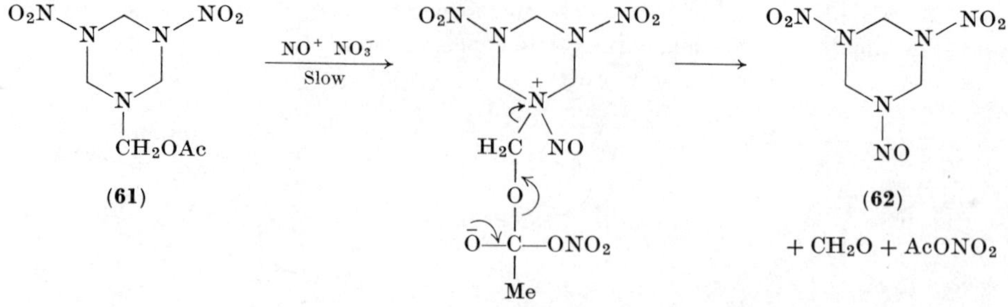

(61) $\xrightarrow[\text{Slow}]{NO^+\ NO_3^-}$ → (62) + CH_2O + $AcONO_2$

SCHEME 6.

The effect of substituents in positions 2 and 6 of piperidine on the rates of *N*-nitrosation has been determined.[168] The *N*-amidino-derivatives (**63**) constitute another class[169] of hydrazones for which the rate-determining step in bromination is *E*-to-*Z* isomerization of the starting material.[170]

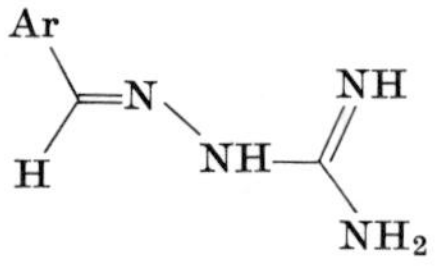

(**63**)

The reaction of 3-alkyl-1-aryltriazenes with benzoic acids involved a mechanism in which proton transfer ($k_H/k_D = 2.47$) and departure of the alkyl cation (relative rates for $R_3C = Me:Et:Pr^i$ were 1:1.7:7.4) were synchronous and rate-determining (Scheme 7).[171] Large solvent effects were observed for these reactions, as they were for

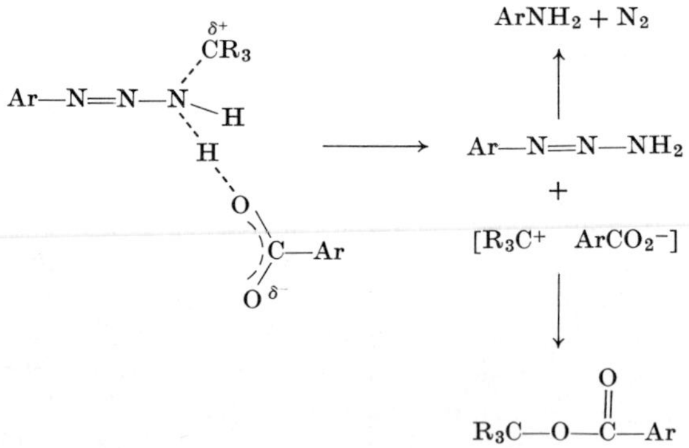

$ArNH_2 + N_2$

Ar—N=N—NH_2 + [R_3C^+ $ArCO_2^-$]

R_3C—O—C(=O)—Ar

SCHEME 7.

the similar acid-induced decomposition of diphenyldiazomethane.[172] The kinetics of the reactions of diazonium salts with dimethyl phosphonate[173] and with secondary amines[174] have been investigated.

The kinetics of the reaction of phenylacetylene with iodine have been determined,[175] and the interaction of haloacetylenes and thiolate anion has been probed by extended Hückel MO theory.[176] Salts of 1,1,3,3-tetracyano-2-azapropenide[177] and imido-betaines[178] have been prepared, and reactions of nitromalonaldehyde have been studied.[179]

References

1 E. Buncel, *Carbanions: Mechanistic and Isotopic Aspects*, Elsevier, Amsterdam, 1975; J. R. Jones, *The Ionisation of Carbon Acids*, Academic Press, London, 1973; M. Szwarc (Ed.), *Ions and Ion Pairs in Organic Reactions*, Wiley, New York, 1974, Vol. 2; E. M. Kaiser and D. W. Slocum, "Carbanions", in S. P. McManus (Ed.), *Organic Reactive Intermediates*, Academic Press, New York, 1973; G. Wittig, "From Diyls over Ylides to my Idyll", *Accounts Chem. Res.*, **7**, 6 (1974); J. W. Henderson, "Chirality in Carbonium Ions, Carbanions, and Radicals", *Chem. Soc. Rev.*, **2**, 397 (1973); D. J. Cram and J. M. Cram, "Carbanions and Carbonium Ions. Stereochemical Analogs", *Intra-Sci. Chem. Rep.*, **7**, 1 (1973); R. D. Guthrie, "Carbanions. Electron Transfer vs. Proton Capture", *Intra-Sci. Chem. Rep.*, **7**, 27 (1973); M. Szwarc, "Effect of Ionic Aggregation on the Equilibrium and Rates of Reaction of Carbanions and Radical Anions", *Intra-Sci. Chem. Rep.*, **7**, 35 (1973); J. R. Murdoch and A. Streitwieser, "Stabilisation of Ion Pairs of Carbanions", *Intra-Sci. Chem. Rep.*, **7**, 45 (1973); T. Durst and R. Viau, "α-Sulphinyl Carbanions", *Intra-Sci. Chem. Rep.*, **7**, 63 (1973); J. Smid, "Ion Pair Solvation in Carbanion Salt Solutions", *Intra-Sci. Chem. Rep.*, **7**, 75 (1973); H. H. Szmant, "Chemistry of DMSO", *Dimethyl Sulphoxide*, **1**, 1 (1971); R. P. Bell, "Recent Advances in the Study of Kinetic Hydrogen Isotope Effects", *Chem. Soc. Rev.*, **3**, 513 (1974).

2 J. P. C. M. van Dongen, H. W. D. van Dijkman and M. J. A. de Bie, *Rec. Trav. chim.*, **93**, 29 (1974).

3 R. J. Bushby and G. J. Ferber, *Tetrahedron Letters*, **1974**, 3701; *Org. Reaction Mech.*, **1972**, 105.

4 R. B. Bates and W. A. Beavers, *J. Am. Chem. Soc.*, **96**, 5001 (1974).

5 A. Atkinson, A. C. Hopkinson and E. Lee-Ruff, *Tetrahedron*, **30**, 2023 (1974).

6 W. T. Ford and M. Newcomb, *J. Am. Chem. Soc.*, **96**, 309 (1974).

7 E. A. Noe and M. Raban, *J. Am. Chem. Soc.*, **96**, 6184 (1974); see also A. A. Petrov, S. M. Esakov and B. A. Ershov, *Zh. Org. Khim.*, **10**, 336 (1974); *Chem. Abs.*, **81**, 77350 (1974); S. M. Esakov, A. A. Petrov, V. A. Lovchikov and B. A. Ershov, *Zh. Org. Khim.*, **10**, 890 (1974); *Chem. Abs.*, **81**, 12999 (1974).

8 *Org. Reaction Mech.*, **1972**, 106.

9 M. V. Moncur, J. B. Grutzner and A. Eisenstadt, *J. Org. Chem.*, **39**, 1604 (1974).

10 I. Murata, K. Yamamoto and O. Hara, *Tetrahedron Letters*, **1974**, 2047.

11 I. N. Juchnovski and I. G. Binev, *Tetrahedron Letters*, **1974**, 3645.

12 V. A. Lovchikov, S. M. Esakov and B. A. Ershov, *Zh. Org. Khim.*, **10**, 885 (1974); *Chem. Abs.*, **81**, 12711 (1974).

13 R. Zerger, W. Rhine and G. D. Stucky, *J. Am. Chem. Soc.*, **96**, 5441 (1974).

14 P. A. Scherr, R. J. Hogan and J. P. Oliver, *J. Am. Chem. Soc.*, **96**, 6055 (1974).

15 J. B. Smart, R. J. Hogan, P. A. Scherr, M. T. Emerson and J. P. Oliver, *J. Organometal. Chem.*, **64**, 1 (1974).

16 B. Bockrath and L. M. Dorfman, *J. Am. Chem. Soc.*, **96**, 5708 (1974).

17 R. D. Guthrie, *Intra-Sci. Chem. Rep.*, **7**, 27 (1973).

18 R. D. Guthrie, G. R. Weisman and L. G. Burdon, *J. Am. Chem. Soc.*, **96**, 6955 (1974).

19 R. D. Guthrie and G. R. Weisman, *J. Am. Chem. Soc.*, **96**, 6962 (1974).

20 E. J. Panek and T. J. Rodgers, *J. Am. Chem. Soc.*, **96**, 6921 (1974).

21 A. G. Anastassiou and R. C. Griffith, *J. Am. Chem. Soc.*, **96**, 611 (1974).

22 A. G. Anastassiou and E. Reichmanis, *Angew. Chem. Int. Ed.*, **13**, 404 (1974).

23 A. G. Anastassiou and E. Reichmanis, *Angew. Chem. Int. Ed.*, **13**, 728 (1974).

24 G. Schroder, G. Frank, H. Rottele and J. F. M. Oth, *Angew. Chem. Int. Ed.*, **13**, 205 (1974).

25 P. J. Beeby, J. M. Brown, P. J. Garratt and F. Sondheimer, *Tetrahedron Letters*, **1974**, 599.

26 J. F. M. Oth, K. Mullen, H. Konigshofen, M. Mann, Y. Sakata and E. Vogel, *Angew. Chem. Int. Ed.*, **13**, 284 (1974).

27 *Org. Reaction Mech.*, **1973**, 344.

28 L. A. Paquette, S. V. Ley, R. H. Meisinger, R. K. Russell and M. Oku, *J. Am. Chem. Soc.*, **96**, 5806 (1974).

29 S. Z. Goldberg, K. N. Raymond, C. A. Harmon and D. H. Templeton, *J. Am. Chem. Soc.*, **96**, 1348 (1974).

30 S. W. Staley, G. M. Cramer and A. W. Orvedal, *J. Am. Chem. Soc.*, **96**, 7433 (1974).

31 S. V. Ley and L. A. Paquette, *J. Am. Chem. Soc.*, **96**, 6670 (1974).

32 M. J. Perkins and P. Ward, *J.C.S. Perkin I*, **1974**, 667.

33 W. G. Dauben and R. J. Twieg, *Tetrahedron Letters*, **1974**, 531.

34 S. W. Staley and A. W. Orvedal, *J. Am. Chem. Soc.*, **96**, 1618 (1974).

35 R. Breslow, K. Ehrlich, T. Higgs, J. Pecoraro and F. Zanker, *Tetrahedron Letters*, **1974**, 1123.

[36] M. J. Goldstein, S. Tomoda and G. Whittaker, *J. Am. Chem. Soc.*, **96**, 3676 (1974).
[37] R. B. Bates, W. A. Beavers, M. G. Greene and J. H. Klein, *J. Am. Chem. Soc.*, **96**, 5640 (1974).
[38] M. Makosza and M. Ludwikow, *Angew. Chem. Int. Ed.*, **13**, 665 (1974).
[39] T. Cuvigny, P. Hullot, M. Larcheveque and H. Normant, *Compt. rend., Ser. C*, **278**, 1105 (1974).
[40] R. R. Fraser and P. R. Hubert, *Can. J. Chem.*, **52**, 185 (1974).
[41] G. K. Koch and J. M. M. Kop, *Tetrahedron Letters*, **1974**, 603.
[42] G. Stork and L. Maldonado, *J. Am. Chem. Soc.*, **96**, 5272 (1974).
[43] S. Brenner and M. Bovete, *Tetrahedron Letters*, **1974**, 1377.
[44] D. A. Evans, G. C. Andrews and B. Buckwalter, *J. Am. Chem. Soc.*, **96**, 5560 (1974); W. C. Still and T. L. Macdonald, *J. Am. Chem. Soc.*, **96**, 5561 (1974).
[45] M. Newcomb and W. T. Ford, *J. Am. Chem. Soc.*, **96**, 2968 (1974); *Org. Reaction Mech.*, **1973**, 342.
[46] A. A. Youssef and S. M. Sharaf, *J. Org. Chem.*, **39**, 1705 (1974).
[47] H. M. Walborsky and M. P. Periasamy, *J. Am. Chem. Soc.*, **96**, 3711 (1974).
[48] C. W. Shoppee and G. N. Henderson, *Chem. Comm.*, **1974**, 561.
[49] M. Newcomb and W. T. Ford, *J. Org. Chem.*, **39**, 232 (1974).
[50] R. Boyce, W. S. Murphy and E. A. O'Riordan, *J.C.S. Perkin I*, **1974**, 792.
[51] G. A. Taylor and P. E. Rakita, *J. Organometal. Chem.*, **78**, 281 (1974).
[52] *Org. Reaction Mech.*, **1973**, 342.
[53] H. O. House and E. F. Kinloch, *J. Org. Chem.*, **39**, 747 (1974).
[54] T. H. Chan and E. Chang, *J. Org. Chem.*, **39**, 3264 (1974); *Org. Reaction Mech.*, **1973**, 341.
[55] B. L. Dyatkin, L. G. Martynova, B. I. Martynov and S. R. Sterlin, *Tetrahedron Letters*, **1974**, 273.
[56] D. J. Collins and J. J. Hobbs, *Austral. J. Chem.*, **27**, 1731 (1974).
[57] Y.-J. Lee and W. D. Closson, *Tetrahedron Letters*, **1974**, 1395.
[58] D. W. Allen, B. G. Hutley and M. T. J. Mellor, *J.C.S. Perkin II*, **1974**, 1690; *Org. Reaction Mech.*, **1973**, 346.
[59] A. Balsamo, P. L. Barili, P. Crotti, M. Ferretti, B. Macchia and F. Macchia, *Tetrahedron Letters*, **1974**, 1005.
[60] K. A. Cirigottis, E. Ritchie and W. C. Taylor, *Austral. J. Chem.*, **27**, 2209 (1974).
[61] O. Hromatka, D. Binder and K. Eichinger, *Monatsh. Chem.*, **104**, 1520 (1973).
[62] C. Kalidas and N. Chattanathan, *J. Indian Chem. Soc.*, **51**, 479 (1974).
[63] M. A. Nunes and E. Brochmann-Hanssen, *J. Pharm. Sci.*, **63**, 716 (1974).
[64] L. A. Dotsenko and B. A. Ershov, *Zh. Org. Khim.*, **10**, 1342 (1974); *Chem. Abs.*, **81**, 90842 (1974).
[65] R. A. Rossi and J. F. Bunnett, *J. Am. Chem. Soc.*, **96**, 112 (1974).
[66] J. J. Brunet, B. Fixari and P. Caubere, *Tetrahedron*, **30**, 1237, 1245 (1974).
[67] H. O. House, W. C. Liang and P. D. Weeks, *J. Org. Chem.*, **39**, 3102 (1974); *Org. Reaction Mech.*, **1973**, 343.
[68] A. L. Kurts, S. M. Sakembaeva, I. P. Beletskaya and O. A. Reutov, *Vestn. Mosk. Univ., Khim.*, **14**, 213 (1973); *Chem. Abs.*, **79**, 114912 (1973); *Zh. Org. Khim.*, **9**, 1553 (1973); *Chem. Abs.*, **79**, 125476 (1973); *Dokl. Akad. Nauk SSSR*, **211**, 602 (1973); *Chem. Abs.*, **79**, 125494 (1973); cf. *Org. Reaction Mech.*, **1973**, 343.
[69] G. B. Trimitsis and E. M. van Dam, *Chem. Comm.*, **1974**, 610.
[70] N. L. Holy, *Chem. Rev.*, **74**, 243 (1974).
[71] R. G. Harvey and H. Cho, *J. Am. Chem. Soc.*, **96**, 2434 (1974); *Org. Reaction Mech.*, **1973**, 346.
[72] G. Levin, T. A. Ward and M. Szwarc, *J. Am. Chem. Soc.*, **96**, 270 (1974).
[73] D. J. Chadwick, *Chem. Comm.*, **1974**, 790.
[74] S. N. Huckin and L. Weiler, *Can. J. Chem.*, **52**, 1343, 2157 (1974); *J. Am. Chem. Soc.*, **96**, 1082 (1974).
[75] I. Kuwajima and H. Iwasawa, *Tetrahedron Letters*, **1974**, 107; P. A. Grieco and C. S. Pogonowski, *J. Org. Chem.*, **39**, 732 (1974).
[76] *Org. Reaction Mech.*, **1973**, 346.
[77] P. A. Grieco, D. Boxler and C. S. Pogonowski, *Chem. Comm.*, **1974**, 497.
[78] K. Geiss, B. Seuring, R. Pieter and D. Seebach, *Angew. Chem. Int. Ed.*, **13**, 479 (1974).
[79] P. M. Atlani, J. F. Biellmann, S. Dube and J. J. Vicens, *Tetrahedron Letters*, **1974**, 2665.
[80] S. Julia, V. Ratovelomanana and C. Huynh, *Compt. rend., Ser. C*, **278**, 371 (1974).
[81] E. L. Eliel, A. A. Hartmann and A. G. Abatjoglou, *J. Am. Chem. Soc.*, **96**, 1807 (1974).
[82] S. Wolfe, *Accounts Chem. Res.*, **5**, 102 (1972).
[83] O. Hofer and E. L. Eliel, *J. Am. Chem. Soc.*, **95**, 8045 (1973).
[84] G. Barbarella, A. Garbesi, A. Boicelli and A. Fava, *J. Am. Chem. Soc.*, **95**, 8051 (1973).
[85] T. Okonogi, T. Umezawa and W. Tagaki, *Chem. Comm.*, **1974**, 363.

[86] A. I. Shatenshtein and E. A. Gvozdeva, *Khim. Seraorg. Soedin., Soderzhashchikhsya Neft. Nefteprod.*, **9**, 129 (1972); *Chem. Abs.*, **80**, 95077 (1974).
[87] *Org. Reaction Mech.*, **1973**, 345.
[88] M. Cinquini, S. Colonna and F. Montanari, *J.C.S. Perkin I*, **1974**, 1723.
[89] H. Stollar and J. Klein, *J.C.S. Perkin. I*, **1974**, 1763.
[90] M. Cinquini, S. Colonna and F. Montanari, *J.C.S. Perkin I*, **1974**, 1719.
[91] J. F. Biellmann and J. J. Vicens, *Tetrahedron Letters*, **1974**, 2915.
[92] U. Folli, D. Iarossi, I. Moretti, F. Taddei and G. Torre, *J.C.S. Perkin II*, **1974**, 1655.
[93] U. Folli, D. Iarossi and F. Taddei, *J.C.S. Perkin II*, **1974**, 1658.
[94] F. G. Bordwell and J. B. O'Dwyer, *J. Org. Chem.*, **39**, 2519 (1974); F. G. Bordwell and M. D. Wolfinger, *ibid.*, p. 2521; F. G. Bordwell and E. Doomes, *ibid.*, p. 2526.
[95] J. Kattenberg, E. R. de Waard and H. O. Huisman, *Tetrahedron*, **30**, 463, 3177 (1974).
[96] J. M. Photis and L. A. Paquette, *J. Am. Chem. Soc.*, **96**, 4715 (1974).
[97] J. S. Grossert, J. Buter, E. W. H. Asweld and R. M. Kellogg, *Tetrahedron Letters*, **1974**, 2805.
[98] S. Bradamante, P. D. Buttero, D. Landini and S. Maiorana, *J.C.S. Perkin II*, **1974**, 1676.
[99] P. D. Buttero and S. Maiorana, *Gazz. Chim. Ital.*, **103**, 809 (1973).
[100] B. Koutek, L. Pavlickova and M. Soucek, *Coll. Czech. Chem. Comm.*, **39**, 192 (1974).
[101] M. Kobayashi, A. Mori and H. Minato, *Bull. Chem. Soc. Japan*, **47**, 891 (1974).
[102] L. Meurling, *Chem. Scripta*, **6**, 92 (1974).
[103] J.-O. Levin, *Chem. Scripta*, **6**, 89 (1974).
[104] *Org. Reaction Mech.*, **1973**, 347.
[105] J. M. Gamboa, C. Saa and J. M. Figuera, *J.C.S. Perkin II*, **1973**, 2025.
[106] D. F. DeTar and D. M. Coates, *J. Am. Chem. Soc.*, **96**, 942 (1974).
[107] K. Brocklehurst, *Tetrahedron*, **30**, 2397 (1974).
[108] E. W. Miles and P. McPhie, *J. Biol. Chem.*, **249**, 2852 (1974).
[109] J.-H. Kim and K. T. Leffek, *Can. J. Chem.*, **52**, 592 (1974).
[110] See also A. Jarczewski, *Uniw. Poznaniu, Wydz. Mat., Fiz. Chem.* (*Pr.*), *Ser. Chem.*, **1972**, 74; *Chem. Abs.*, **80**, 46999 (1974).
[111] T. Matsui, L. G. Hepler and E. M. Woolley, *Can. J. Chem.*, **52**, 1910 (1974).
[112] M. H. Davies, *J.C.S. Perkin II*, **1974**, 1018.
[113] A. J. Kresge, D. A. Drake and Y. Chiang, *Can. J. Chem.*, **52**, 1889 (1974).
[114] J. R. Keefe and N. H. Munderloh, *Chem. Comm.*, **1974**, 17.
[115] R. P. Bell and R. L. Tranter, *Proc. Roy. Soc., A*, **337**, 517 (1974).
[116] A. G. Bazanov, M. V. Chistyakova, M. F. Kozlova, I. V. Tselinsky and B. V. Ghidaspov, *Organic Reactivity* (*Tartu*), **10**, 816 (1973); A. G. Bazanov, I. V. Tselinsky, N. B. Nikol'skaya, M. F. Kozlova and B. V. Ghidaspov, *ibid.*, p. 824.
[117] I. V. Tselinsky and V. K. Krylov, *Organic Reactivity* (*Tartu*), **10**, 795 (1973).
[118] I. V. Tselinsky and V. N. Dronov, *Organic Reactivity* (*Tartu*), **10**, 806 (1973).
[119] A. J. Kresge, *Can. J. Chem.*, **52**, 1897 (1974).
[120] F. G. Bordwell and W. S. Matthews, *J. Am. Chem. Soc.*, **96**, 1214 (1974).
[121] F. G. Bordwell and W. S. Matthews, *J. Am. Chem. Soc.*, **96**, 1216 (1974).
[122] A. F. Cockerill, D. W. Earls, J. R. Jones and T. G. Rumney, *J. Am. Chem. Soc.*, **96**, 575 (1974).
[123] T. B. McMahon and P. Kebarle, *J. Am. Chem. Soc.*, **96**, 5940 (1974).
[124] L. Radom, *Chem. Comm.*, **1974**, 403.
[125] A. Streitwieser and P. H. Owens, *Tetrahedron Letters*, **1973**, 5221.
[126] A. Streitwieser, P. H. Owens, R. A. Wolf and J. E. Williams, *J. Am. Chem. Soc.*, **96**, 5448 (1974).
[127] T.-Y. Luh and L. M. Stock, *J. Am. Chem. Soc.*, **96**, 3712 (1974).
[128] J. A. Zoltewicz and J. K. O'Halloran, *J. Org. Chem.*, **39**, 89 (1974).
[129] *Org. Reaction Mech.*, **1973**, 350.
[130] *Org. Reaction Mech.*, **1973**, 350.
[131] E. M. Kaiser and W. R. Thomas, *J. Org. Chem.*, **39**, 2659 (1974).
[132] E. Patton and R. West, *J. Am. Chem. Soc.*, **95**, 8703 (1973).
[133] G. Scherowsky, *Chem. Ber.*, **107**, 1092 (1974).
[134] V. M. Vlasov and G. G. Yakobson, *Zh. Org. Khim.*, **10**, 573 (1974); *Chem. Abs.*, **80**, 132620 (1974).
[135] T. W. Dolzine, A. K. Hovland and J. P. Oliver, *J. Organometal. Chem.*, **65**, Cl (1974).
[136] A. N. Plyusnin, Yu. G. Kryazhev, N. A. Loginova and T. I. Kochmareva, *Kinet. Katal.*, **15**, 799 (1974); *Chem. Abs.*, **81**, 104312 (1974).
[137] L. S. Bresler, A. S. Khachaturov and I. Ya. Poddubnyi, *J. Organometal. Chem.*, **64**, 335 (1974).
[138] C. Tuzun and E. Erdik, *Commun. Fac. Sci. Univ. Ankara, Ser. B*, **20**, 41 (1973); *Chem. Abs.*, **81**, 62990 (1974).

[139] J. Reffy, J. Nagy and I. Lazanyi, *Period. Polytech., Chem. Eng.*, **18**, 105 (1974); *Chem. Abs.*, **81**, 90744 (1974).
[140] N. S. Nametkin, V. M. Vdovin, M. V. Pozdnyakova, E. D. Babich and I. V. Silkina, *Izv. Akad. Nauk SSSR, Ser. Khim.*, **1973**, 1681; *Chem. Abs.*, **79**, 136112 (1973).
[141] T. G. Attig and A. Wojcicki, *J. Am. Chem. Soc.*, **96**, 262 (1974).
[142] T. C. Flood and D. L. Miles, *J. Am. Chem. Soc.*, **95**, 6460 (1973).
[143] U. Kunze and J. D. Koola, *J. Organometal. Chem.*, **80**, 281 (1974).
[144] I. P. Beletskaya, A. N. Kashin, A. Ts. Malkhasyan, A. A. Solov'yanov, E. Yu. Bekhli and O. A. Reutov, *Zh. Org. Khim.*, **10**, 678 (1974); *Chem. Abs.*, **81**, 12738 (1974).
[145] *Org. Reaction Mech.*, **1973**, 354.
[146] M. H. Abraham and D. F. Dadjour, *J.C.S. Perkin II*, **1974**, 233.
[147] M. H. Abraham, P. L. Grellier and M. J. Hogarth, *J.C.S. Perkin II*, **1974**, 1613.
[148] M. Gielen and R. Fosty, *Bull. Soc. Chim. belges*, **83**, 333 (1974).
[149] R. C. Larock and H. C. Brown, *J. Organometal. Chem.*, **26**, 35 (1971).
[150] D. Dodd and M. D. Johnson, *J.C.S. Perkin II*, **1974**, 219.
[151] K. P. Butin, A. B. Ershler, V. V. Strelets, A. N. Kashin, I. P. Beletskaya, O. A. Reutov and K. Marcushova, *J. Organometal. Chem.*, **64**, 171 (1974); K. P. Butin, V. V. Strelets, A. N. Kashin, I. P. Beletskaya and O. A. Reutov, *ibid.*, p. 181; K. P. Butin, V. V. Strelets, I. P. Beletskaya and O. A. Reutov, *ibid.*, p. 189.
[152] I. P. Beletskaya, S. V. Rykov and A. L. Buchachenko, *Organic Mag. Res.*, **5**, 595 (1973).
[153] C. P. Casey and R. L. Anderson, *J. Am. Chem. Soc.*, **96**, 1230 (1974).
[154] D. N. Kursanov, V. N. Setkina, B. G. Gribov and E. V. Bykova, *Izv. Akad. Nauk SSSR, Ser. Khim.*, **1974**, 751; *Chem. Abs.*, **81**, 37090 (1974).
[155] J. L. Garnett and R. S. Kenyon, *Austral. J. Chem.*, **27**, 1023 (1974).
[156] J. L. Garnett and R. S. Kenyon, *Austral. J. Chem.*, **27**, 1033 (1974).
[157] C. H. DePuy, *Fortschritte Chem. Forsch.*, **40**, 73 (1973).
[158] F. R. Jensen, D. B. Patterson and S. E. Dinizo, *Tetrahedron Letters*, **1974**, 1315.
[159] D. H. Aue and R. N. Reynolds, *J. Org. Chem.*, **39**, 2315 (1974).
[160] R. Weiss and H. P. Kempcke, *Tetrahedron Letters*, **1974**, 155.
[161] E. Osawa, *Tetrahedron Letters*, **1974**, 115.
[162] H. C. Brown and N. R. DeLue, *J. Am. Chem. Soc.*, **96**, 311 (1974).
[163] I. Tabushi, Y. Tamaru and Z. Yoshida, *Tetrahedron*, **30**, 1457 (1974).
[164] T. P. Vasil'eva, M. G. Lin'kova, O. V. Kil'disheva and I. L. Knunyants, *Izv. Akad. Nauk SSSR, Ser. Khim.*, **1974**, 643; *Chem. Abs.*, **81**, 3049 (1974).
[165] D. L. Rakhmankulov, V. S. Martem'yanov, Z. L. Ayupova, S. S. Zlotskii and V. I. Isagulyants, *Zh. Prikl. Khim.* (*Leningrad*), **47**, 1435 (1974); *Chem. Abs.*, **81**, 90747 (1974).
[166] V. I. Erashko, A. V. Sultanov, S. A. Shevelev and A. A. Fainzil'berg, *Izv. Akad. Nauk SSSR, Ser. Khim.*, **1974**, 1350; *Chem. Abs.*, **81**, 90769 (1974).
[167] T. G. Bonner, R. A. Hancock and J. C. Roberts, *J.C.S. Perkin II*, **1974**, 653.
[168] A. R. Jones, W. Lijinsky and G. M. Singer, *Cancer Res.*, **34**, 1079 (1974).
[169] *Org. Reaction Mech.*, **1973**, 355.
[170] A. F. Hegarty, T. A. F. O'Mahony, P. Quain and F. L. Scott, *J.C.S. Perkin II*, **1973**, 2047.
[171] N. S. Isaacs and E. Rannala, *J.C.S. Perkin II*, **1974**, 899.
[172] N. S. Isaacs and E. Rannala, *J.C.S. Perkin II*, **1974**, 902.
[173] E. S. Lewis and E. C. Nieh, *J. Org. Chem.*, **38**, 4402 (1973).
[174] M. Remes, J. Divis, V. Zverina and M. Matrka, *Chem. Prum.*, **24**, 138 (1974); *Chem. Abs.*, **81**, 24681 (1974).
[175] Yu. P. Filinov, I. M. Vasil'kevich and G. F. Dvorko, *Zh. Obshch. Khim.*, **44**, 467 (1974); *Chem. Abs.*, **80**, 119828 (1974).
[176] P. Beltrame, A. Gavezzotti and M. Simonetta, *J.C.S. Perkin II*, **1974**, 502.
[177] J. Perchais and J.-P. Fleury, *Tetrahedron*, **30**, 999 (1974).
[178] D. Leguern, G. Morel and A. Foucaud, *Tetrahedron Letters*, **1974**, 955.
[179] S. M. Kvitko, V. V. Perekalin and Yu. V. Maksimov, *Zh. Org. Khim.*, **9**, 2228 (1973); *Chem. Abs.*, **80**, 145229 (1974).

CHAPTER 11

Elimination Reactions

A. C. KNIPE

Chemistry Department, The New University of Ulster

Stereochemistry and Orientation in *E*2 Reactions

It has been confirmed[1] that (*erythro*-1,2-diphenylpropyl)trimethylammonium salts undergo *syn*-elimination when treated with Bu^tOK–Bu^tOH; mechanisms involving product isomerization, α-epimerization via an ylide, or a pre-equilibrium *E*1*cB* reaction have been rejected.

The role of ion pairing in phenoxide-ion-promoted elimination reactions of 3-hexyl toluene-*p*-sulphonate, fluoride and trimethylammonium ion has been investigated.[2,3] Results[3] for reactions of stereospecifically deuteriated (4-$^2H_1$)-3-hexyl fluorides with RO^-/ROH reveal that a large leaving group (cf. R_3N^+–) is not a necessary prerequisite for *syn*-elimination (68 and 14% where R = Bu^t and Bu^n, respectively) which must therefore be correlated with poor leaving-group ability. While strongly basic reagents (more reactant-like transition state) may further help to promote *syn*-elimination from fluorides, ion pairing is apparently a decisive factor since only 20% of *syn*-elimination occurs in Bu^tO^-/DMSO. This interpretation is supported by the variation of elimination stereochemistry for reaction of 3-hexyl toluene-*p*-sulphonate in 95% DMSO/5% Bu^tOH induced by LiOPh (16.5% *syn*), KOPh (9.3% *syn*) and $KOC_6H_4 \cdot NO_2$-*p* (3.3% *syn*). In the presence of tetramethylammonium iodide *syn*-elimination by lithium phenoxide is, however, completely suppressed and this has been attributed to a decrease in proportion of ion-paired PhO^- through establishment of the following equilibrium (eqn. 1):

$$RO^-M^+ + RNMe_3^+ + X^- \rightleftarrows M^+X^- + RNMe_3^+ + RO^- \qquad (1)$$

Thus, the results indicate that a poor leaving group and an ion-paired base are conducive to *syn*-*E*2-reactions of neutral substrates.

In marked contrast, the free base is more effective than ion-paired base in promoting such reactions of onium salts.[2] This has previously been attributed to electrostatic attraction between the positive leaving group and the negative base in a configuration conducive to attack on the *syn*-β-hydrogen atom. Thus, for reaction of 3-hexyltrimethylammonium iodide with potassium phenoxides in 95% DMSO/5% Bu^tOH the percentage of *syn*-elimination is greatest (69%) for *p*-nitro and least (26%) for the less dissociated *p*-methoxy anion.[2] For reactions with KOPh in 0, 20, and 95% DMSO the contributions of *syn*-elimination to formation of *trans*-olefin are 14, 34 and 38%, respectively; the main variation occurs between 0 and 20% DMSO and is therefore believed to correlate with dissociation of PhOK rather than with the solvent-dependence of its basicity. An apparent inconsistency is the greater degree of *syn* → *trans* elimination that occurs in 20% DMSO with lithium (74% *syn*) and sodium (56% *syn*) co-cations, even though they would be expected to form ion pair more effectively than potassium does (34% *syn*). With reference to equation (1) it has therefore been suggested that the stability of the ion pair M^+X^- may control the position of equilibrium so that the concentration of free base increases in the order K < Na < Li. This explanation requires, however, that PhOK is not fully dissociated in 20% DMSO. It has also been found that dicyclohexyl-18-crown-6 has no significant effect on the proportion of *syn*-elimination by Bu^tOK–Bu^tOH or Bu^nOK–Bu^nOH and it has been argued that control of the concentration of free base by equation (1) dominates any such influence. A change from potassium to tetra-*n*-butylammonium *n*-butoxide is, however, also without effect. It is clear that many aspects of the role of ion pairing in this system have still to be resolved.

For reactions of *trans*-2-arylcyclopentyl toluene-*p*-sulphonates (**1**), the transition-state differences for *syn*-eliminations, induced by (i) associated and dissociated metal alkoxides, and (ii) dissociated alkoxide ion bases of varying strength have been probed.[4] It has already been established that aryl-activated *syn*-elimination and unactivated *anti*-eliminations give (**2**) and (**3**), respectively, while formation of (**3**) by unactivated *syn*-elimination is unimportant. The ratio of (**2a**):(**3a**) obtained upon reaction of (**1a**) with Bu^tOK–Bu^tOH is decreased from 90:10 to 30:70 in the presence of dicyclohexyl-18-crown-6 and it has been suggested that these ratios are characteristic of reactions of associated and dissociated Bu^tOK, respectively. From the distribution of products for (**1a–g**) it has been established that the ρ-value for formation of (**2**) by *syn*-elimination is increased from +2.2 to +3.1 in the presence of crown ether. Enhanced carbanion character for reaction of the dissociated base is probably a consequence of poorer leaving-group ability, since crown ether has little effect on the corresponding β-deuterium isotope effect ($k_H/k_D = 5.1$–5.3). For reactions of (**1**) with dissociated alkoxide (ROK–

H Y Z OTs → H Y + Y Z

(**1**) (**2**) (**3**)

	a	b	c	d	e	f	g
Y =	H	H	*m*-Cl	*p*-Cl	*m*-Me	*p*-Me	*p*-OMe
Z =	H	D	H	H	H	H	H

ROH in presence of crown ether) a marked influence of the base–solvent system on the transition-state character for *syn*-elimination is shown by variations in k_H/k_D (1.2, 1.4, 5.1) and ρ (0.3, –, 3.1) with R (Bu^n, Bu^s and Bu^t, respectively). A change in mechanism seems unlikely and the trends have been explained within the framework of the variable *E*2 transition-state theory. These results are in agreement with the decrease in carbanionic character which accompanies decrease in base strength for *anti*-eliminations of phenethyl bromide in a common solvent but are in sharp contrast to the large decrease in k_H/k_D but moderately enhanced ρ, observed for *anti*-eliminations from (phenethyl)-trimethylammonium bromide when the base–solvent system is changed from Bu^tONa–Bu^tOH to EtONa–EtOH. This contrasting behaviour has still to be reconciled.

The importance of base association for the stereochemistry of base-promoted β-eliminations from norbornyl derivatives has also been demonstrated.[5] Thus, the almost exclusive *exo-syn*-elimination reaction of *exo*-2-norbornyl *exo*-[$^2H_1$]-toluene-*p*-sulphonate with the sodium salt of 2-cyclohexylcyclohexanol in triglyme is modified in the presence of 18-crown-6 such that 2-[$^2H_1$]-norbornene (27.2%), norbornene (70%) and nortricyclene (2.8%) are obtained by *anti*-, *syn*- and α-eliminations, respectively. An isotope effect, $k_H/k_D = 5$–7, has been determined for the *syn-exo*-elimination promoted by the dissociated base, and a relative propensity for *syn-exo*- and *anti-endo*-hydrogen elimination of ca. 15:1 has been estimated. The latter ratio contrasts with >100:1 for reaction of the associated base which is believed to account for previous reports of predominant *syn-exo*-eliminations of substituted norbornanes.

The complex base $NaNH_2$–Bu^{t}ONa promotes *syn*-elimination of substituted cyclohexanes while *E2H* mechanisms seem to be more probable for analogous reactions of cyclopentanes and cycloheptanes.[6] Dehydrobromination of 3-*endo*-bromobornan-2-*exo*-ol and 2-*endo*-bromobornan-3-*exo*-ol by RO^-–ROH (R = Me, Et, Bu^t or $EtCMe_2$) occurs by a *syn-E*2 mechanism to form bornan 2- and -3-one, respectively.[7]

The degree of complexity of anionic oxygen-, nitrogen- and carbon-bases required to cause steric control of positional orientation in eliminations from 2-iodobutane has been probed, with DMSO as the solvent in order to minimize base association with the alkali-metal cations.[8] It was established that fundamental control of positional orientation is by base strength and that the dependence decreases in the order O > N > C. The stereochemistry of each base for which steric control could be discerned was such that a plane passed through the basic atom is intersected by other atomic centres, even when the remaining portion of the anion is arranged so as to relieve such contact.

By application of ion-cyclotron resonance techniques, elimination of HF from fluorinated ethanes, promoted by methoxide ion in the gas phase, has been studied.[9,10] A chemically activated intermediate, formed upon collision, decomposes either by *syn*-eliminative formation of alkene and CH_3OHF^- or by proton transfer to give the fluorinated carbanion from which F^- is abstracted by neutral fluoroethane. This technique affords a valuable means of study of ionic *E*2 processes in the absence of solvent effects.

Close agreement has been found between experimental observations and calculated stereochemical paths involving the least motion of atoms during acetylene and allene formation and 1,2-, 1,3- or 1,4-elimination reactions.[11] The stereochemistries of enolization, homoenolization, concerted 1,2-hydride shifts and epoxidation are also explicable by the PLM approach.

The corresponding *trans*-cinnamic acid is the product of *E*2 debromination of *erythro*-$RC_6H_4CHBrCHBrCOOH$ by KI in 80% aqueous ethanol, for which $\rho = -0.84$. The rate increases slightly with solvent polarity, but NaCl and LiBr exert negative salt effects.[12]

The *E1cB* Mechanism

The *E*1*cB* reaction has been reviewed with respect to stereochemistry and the influence of the reaction medium.[13]

An intramolecular reaction of the conjugate α-sulphonyl carbanion is believed[14] to account for the formation of methyl phenyl sulphone, styrene and CO_2 upon reaction of (**4a**) (and related substrates) in Bu^{t}O$^-$/Bu^{t}OH. Quenching with DCl revealed that (**4a**) is substantially ionized in the reaction mixture. Although such ionization should decrease the rate of elimination of $ArSO_2CR^1{}_2COO^-$ by an $E2$ process, the rate constant for reaction of (**4e**) is actually smaller than that for (**4a**). Activation parameters for reactions of (**4e**) are quite different from those for (**4a–d**) for which an alternative intramolecular *syn*-elimination mechanism has been proposed, since (**4f**) and (**4g**) form [α-$^2H_1$]-*trans*-stilbene- and *trans*-stilbene, respectively. *syn*-Elimination is favoured by strong base in a solvent of low dielectric constant as shown[15] by the increase and decrease in deuterium content of *trans*-stilbene formed from (**4f**) and (**4g**), respectively, with the base–solvent systems KOH–H_2O–EtOH, MeOK–MeOH, EtOK–EtOH and Bu^{t}OK–Bu^{t}OH. In the presence of an equimolar amount of dicyclohexyl-18-crown-6, however, (**4f**) and (**4g**) react by predominant *anti*-elimination in Bu^{t}OK–Bu^{t}OH. At the high base concentrations employed, this may be attributed to dissociation of Bu^{t}OK, with concomitant acceleration of the *anti*-$E2$ process, combined with deceleration of the *syn*-elimination mechanism by conformational interference through co-ordination to the cation within the carbanion–potassium ion pair.

X–C_6H_4–SO_2–CR^1_2–C(=O)–O–CH(R^2)–C(H)(R^3)–C_6H_4–Y $\xrightarrow[\text{Bu}^t\text{OH}]{\text{Bu}^t\text{OK}}$ X–C_6H_4–$SO_2CHR^1_2$ + Y–C_6H_4–CR^3=CHR^2 + CO_2

(**4**)

(**4a**) X = Y = all R = H

(**4b**) X = H; Y = Me: all R = H

(**4c**) X = H; Y = Cl; all R = H

(**4d**) X = Me; Y = H; all R = H

(**4e**) X = Y = H; R^1 = Me; R^2 = R^3 = H

(**4f**) *erythro*- (X = Y = H; R^1 = H; R^2 = Ph; R^3 = D)

(**4g**) *threo*- (X = Y = H; R^1 = H; R^2 = Ph; R^3 = D)

(**4h**) X = Y = H; R^1 = R^3 = H; R^2 = Ph

It has been established that *trans*-2-ethoxyvinyl *p*-tolyl sulphone is formed upon reaction of *cis*- or *trans*-2-phenoxyvinyl *p*-tolyl sulphone with EtO^-/EtOH by an anionic addition–$(E1cB)_R$ mechanism and not by alternative S_N2 or elimination–addition routes.[16]

For elimination of PhO^- from 1-nitro-2-phenoxyethane in EtONa–EtOH, NaOH–H_2O or Et_3N–H_2O the ($E1$)–anion mechanism has been assigned, since the substrate is almost completely in the form of its conjugate base from which loss of PhO^- is rate-determining.[17] The $E2$ mechanism has, however, been tentatively assigned to account for reaction of the corresponding (phenylthio)nitroethane for which both the elimination rate and primary β-deuterium isotope effect (7.4) are larger than predicted (from results for model substrates[18]) for rate-determining deprotonation. Rate constants for reaction of 1-aryl-2-nitro-1-(phenylthio)ethane are, however, in close agreement with such predictions and ρ = ca. 0.8 has been estimated, as required of the $(E1cB)_I$ mechanism. It is interesting that, contrary to expectation, the elimination is retarded by an α-phenyl substituent. Similar rate depressions, upon α-phenyl-substitution, have been reported for substrates which react by the $(E1cB)_R$ mechanism and it has now been established that this is also the effect on deprotonation of related non-eliminating compounds.[18]

Thus, the slower ionization of 2-nitro-1,1-diphenylethane than of 2-nitro-1-phenylethane has been attributed to steric interaction between substrate and base. For a series of aliphatic nitro-compounds $YCH_2CH_2NO_2$, rates of ionization in $NaOH$–H_2O ($\rho^* = 2.07$) and in EtONa–EtOH ($\rho^* = 1.88$) have been measured[18] in order to facilitate elucidation of the mechanisms of these and related β-elimination reactions.

The effect[19] of replacing Cl in p-$NO_2C_6H_4CH_2CH_2Cl$ by SCH_2COOH or $SCPh_2COOH$ is to change the β-elimination mechanism from $E2$ to $(E1cB)_R$. An $E1cB$ mechanism has also been proposed to account for the formation of (E)-ethyl β-nitro-α,β-unsaturated carboxylates from a 50:50 mixture of *threo*- and *erythro*-ethyl α-acetoxy-β-nitro-carboxylates.[20] In contrast, AcOH is eliminated from ethyl β-acetoxy-α-nitro-carboxylates by an $E2$ process.

Hammett ρ-values[21] for bimolecular elimination of HF from series of 1-fluoro- ($\rho = 3.24$), 1,1-difluoro- ($\rho = 3.56$) and 1,1,1-trifluoro- ($\rho = 4.04$) -2-arylethanes in Bu^tOK–Bu^tOH indicate that in each case the transition state is highly carbanionic in character. This is also suggested by the relatively low value of k_H/k_D (4.50 and 2.77 for monofluoro- and difluoro-phenylethane, respectively) for these reactions which proceed without exchange of the β-hydrogen of unchanged substrate. It is not clear whether carbanion intermediates are actually formed.

Further investigation of dehydrochlorination of DDT has led to the firm conclusion that the $(E1cB)_I$ mechanism applies.[22,23] For eliminations induced by PhS^-, p-$NO_2C_6H_4O^-$, PhO^- and EtO^- in EtOH, and by MeO^- in MeOH, $k_{\beta\text{-H}}/k_{\beta\text{-D}}$ passes through a maximum as the base strength increases.[22] As expected of a process involving *only* proton transfer, the maximum arises when the pK_a of the proton-acceptor equals the pK_a (ca. 17) estimated for the proton-donor (DDT). In contrast, for reactions of Ar_2CHCCl_3 with MeO^-/MeOH and Bu^tO^-/Bu^tOH the isotope effects are comparable, being ca. 4.7 and (3.4–4.0), respectively, throughout a range of substrate acidity of over 11 pK_a units. This anomaly has not been adequately explained; however, an excellent fit of the rate constants, for dehydrochlorination in MeO^-/MeOH to the Brønsted equation for deprotonation of hydrocarbons in the same medium provides support for an $(E1cB)_I$ mechanism.[23] The close agreement between the dependence of DDT-elimination and fluorene-deprotonation rates on basicity functions, recently defined[24] for MeONa–MeOH, further supports this mechanism and implies that internal return of hydrogen is insignificant. ρ-Values of 2.54, 2.99 and 2.11, respectively, for reactions in MeO^-/MeOH, Bu^tO^-/Bu^tOH and PhS^-/EtOH at 30° suggest that the transition state is carbanion-like and that the degree of proton-transfer is greater for the stronger base. There is, however, no evidence of direct conjugation with a p-nitro-substituent, and the Brønsted α (= 0.309) for deprotonation of hydrocarbons, including Ar_2CHCCl_3, is relatively low. Dehydrochlorinations of 1,1-diaryl-2,2-dichloroethanes (DDD) in MeO^-/MeOH have also been further investigated;[25] the ρ-value (1.93) is much less than for DDT and the reaction rate, which exceeds that predicted for deprotonation, is less sensitive to the MeONa basicity function; it has been suggested that proton tunnelling is a feature of the proton-transfer since $k_H/k_D = 10.8$ at 30°, the activation energy difference $(E_a{}^D - E_a{}^H) = 5.3 \pm 1.2$ kcal mol^{-1}) is considerably greater than the zero-point energy-difference (1.15 kcal mol^{-1}), and $\log(A^D/A^H) = 2.8 \pm 1.4$; it has been concluded that DDD eliminates predominantly by an $E2$ mechanism.

Rate-determining loss of cyanide ion during $E1cB$ dehydrocyanation of certain polycyanoethanes is indicated by heavy-atom (^{14}C) kinetic isotope effects.[26] A full account[27,28a] of an investigation[28b] of competitive elimination and H–D exchange reactions of 1-methoxyacenaphthenes has emphasized the stereochemical importance of ion pairing of Bu^tOM in Bu^tOH. Reactions of BuLi with alkoxy-β,β-dichloroethanes involve competitive α- and β-elimination from the intermediate alkoxydichloroethyllithium.[29]

Pre-equilibrium *E*1c*B* mechanisms have been established for the hydrolysis of aryl phenylthionocarbamates with formation of phenyl isothiocyanate[30] and for the elimination of bisulphite ion from a series of uracil–bisulfite adducts.[31] Irreversible mechanisms have been suggested for the formation[32] of 1,2,3-triazoles by dehydration of 5-hydroxy-Δ^2-1,2,3-triazolines in hot KOH–MeOH, and to account for formation[33] of *trans*- and *cis*-1-ethylideneindenes, in differing proportions, from diastereomeric 1-(1-acetoxyethyl)indenes and 3-(1-acetoxyethyl)indene in MeO^-/MeOH.

Although alkaline solvolyses of phosphoramidic halides, acyl carbamates and aryl (*N*-methylamino)sulphonates[34] have been reported to occur by the *E*1c*B* process, this mechanism does not contribute significantly to reaction of the corresponding aryl methyl phosphoramidates and phosphoramidothioates.[35]

Pyrolytic Elimination Reactions

Increase in rate of fragmentation of azo-compounds (**5a** < **5b** < **5c** < **5d**) with increase in dihedral angle between the cyclopropyl plane and the rest of the structure has been attributed to concomitant destabilization of the cyclopropyl e_S and stabilization of π(N=N) and n_+ molecular orbitals.[36]

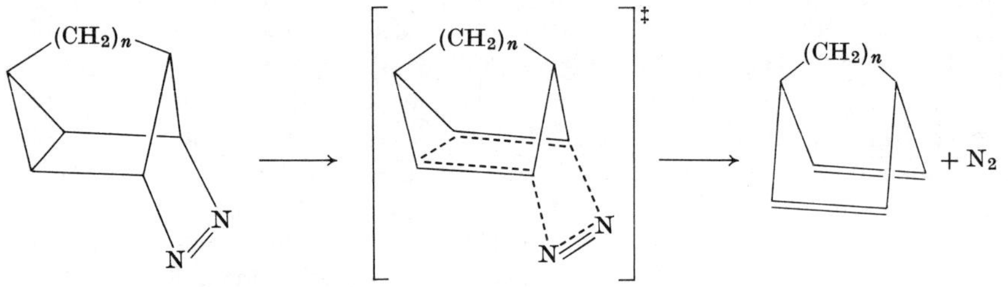

$(CH_2)_n$ ⟶ [$(CH_2)_n$]‡ ⟶ $(CH_2)_n$ + N_2

(**5a**) $n = 0$; (**5b**) $n = 1$

(**5c**) $n = 2$; (**5d**) $n = 4$

The rate of thermolysis of pyrazoline (**6a**) is successively increased by 10^5- and 10^3-fold, respectively, upon introduction of a first (**6b**) and a second (**6c** or **6d**) vinyl group.[37] This is in contrast to rate enhancements of ca. 10^7 and <6 effected when the first and the second C—N bond become allylic in linear azoalkanes, for which a stepwise biradical mechanism has already been proposed. It has therefore been argued that, in keeping with secondary deuterium isotope effects previously reported for (**6e–g**), decomposition of alkylpyrazolines occurs by a concerted process.

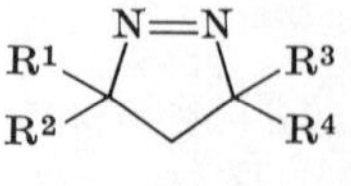

(**6a**) all R = H

(**6b**) R^1 = Et; R^3 = vinyl; $R^{2,4}$ = H

(**6c**) $R^{1,3}$ = vinyl; $R^{2,4}$ = H

(**6d**) $R^{1,4}$ = vinyl; $R^{2,3}$ = H

(**6e**) $R^{1,2}$ = D; $R^{3,4}$ = H

(**6f**) all R = D

(**6g**) R^1 = vinyl; R^2 = H; $R^{3,4}$ = D

Decomposition of 1,1-azobisformamide in DMSO occurs by competitive denitrogenation (to form formamoyl radicals) and cyclization to 1,2,4-triazoline-3,5-dione which subsequently fragments yielding N_2, CO and urea;[38] this constitutes only the second investigation of thermolysis of an azoformamide in solution. Formation of azirines from vinyl azides is believed[39] to involve neighbouring-group participation rather than a recently favoured[40] nitrene mechanism. Upon flash photolysis 2,3-diazidonaphthalene forms *trans*-1,2-dicyano-1,2-dihydrobenzocyclobutene as expected from orbital-symmetry-controlled conrotatory closure of an intermediate *cis,cis*-dicyanoquinodimethane.[41] Kinetic results for elimination of nitrous acid from mononitroalkanes[42] are consistent with a polar transition state.

Dehydrations of alcohols on solid acidic or basic catalysts have been studied.[43–49] The phosphoric-acid-catalysed dehydration sequence **(7)** → **(8)** → **(9a, b)** has been confirmed[43] by deuterium labelling; the product **(9)** retained 91.3% of the label, which was distributed between **(9a)** (38%) and **(9b)** (52%). Kinetic isotope effects[44] for dehydration of deuteriopropan-2-ols at 300° on oxide catalysts reveal that the mechanism may vary from $E2$-like (on Al_2O_3) to $E1$-like (on SiO_2). An alkoxonium–carbonium ion mechanism has been proposed to account for the Brønsted-acid-catalysed dehydration of 4-methylpentan-2-ol and butan-2-ol.[45] Procedures for pyrolytic identification of alcohols have been developed.[48, 49]

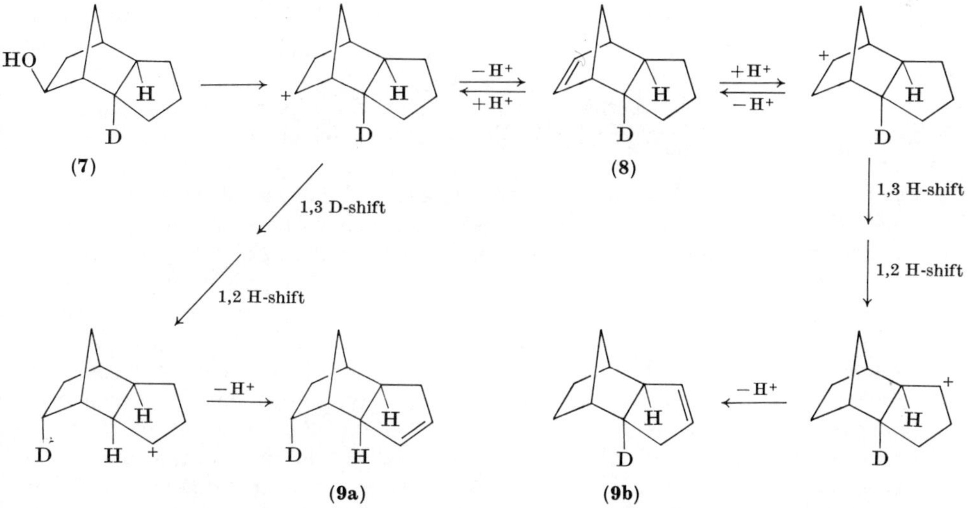

Extended Hückel and iterative extended Hückel calculations[50] suggest that the transition state for β-*cis*-dehydrochlorination of ethyl chloride is more similar to kinetic models of Setser and Hasler than to the electrostatic models of Benson and Haugen. By application of RRKM (Rice–Ramsperger–Kassel–Marcus) theory, to four- and three-centered transition-state models, it has been estimated that for dehydrochlorination of chemically activated 1,1,2-trichloroethane the threshold energies E_0 for the $\beta\alpha$, $\alpha\beta$ and $\alpha\alpha$ processes are 59, 57 and 60 kcal mol^{-1}, respectively.[51] Dehydrohalogenations of alkyl halides in the presence of metal salts[52–55] and on a range of metal zeolites[56] have been reported.

From results of ^{14}C- and ^{3}H-labelling experiments it has been concluded that flash thermolysis of phenol occurs through formation of cyclopentadiene (or an isomeric hydrocarbon) by elimination of the carbonyl group of the corresponding cyclohexadienone tautomer.[57] The decarbonylative formation of butadiene from cyclopent-3-enone has also been investigated.[58]

A study of the mechanism of pyrolysis of ethyl cyanoformate[59] has prompted the suggestion that pyrolyses of compounds of general formula XCOOR ($X = H$, C_nH_m, C_nH_mO, Cl or CN; $R = C_nH_m$) can be explained by a common mechanism, whereby the efficacy of competing substitution (to form XR and CO_2) and elimination (to form XCOOH and alkene) mechanisms is dependent upon the nature of R and X. Where X is such that –NH– is linked to the carbonyl group, isocyanates may also be formed. Activation energies for elimination reactions, where R = Et and X = CN, Cl, MeO, Me, EtO or Et, have been correlated by the equation: $E_a/\text{kcal mol}^{-1} = 48.19 - 1.548\sigma^*$, such that E_a decreases as the electron-withdrawing ability of X increases. Eliminative formation of ethylene from ethyl cyanoformate is believed to involve formation of cyanoformic acid (which subsequently decomposes to CO_2 and HCN) via a four-centre transition state, rather than by the six-centre fragmentation previously proposed. Cyanic acid and bromoacetylurea are the primary products of thermolysis of 1-bromoacetylbiuret.[60]

There have been several studies of decarboxylation of carboxylic acids.[61–65] For reaction in *m*-nitrotoluene, large values of ΔG^{+} for decarboxylation of 3-methoxy- and 5-methoxy-, relative to 4-methoxy-pyridine-2-carboxylic acid have been attributed to steric interactions between adjacent groups.[61] Deuterium kinetic isotope effects have been determined for oxidation–decarboxylation of several aliphatic carboxylic acids,[62] and the effect of phthalocyanine derivatives of transition metals on reactions of oxalic acid has been reported.[63] Decarboxylations of 1-naphthaleneacrylic acid in quinoline containing a copper catalyst[64] and of oxalic acid in resorcinol or pyrocatechol[65] have been studied.

A pattern of pyrolysis behaviour has been identified for amino-acids.[66] Thus, while α-amino-acids generally undergo decarboxylation and condensation reactions, β-amino-acids form α,β-unsaturated acids by deaminative elimination. An α-alkyl substituent is sufficient to promote a deaminative S_Ni reaction of α-amino-acids whereby a ketone is formed upon decarboxylation of the intermediate α-lactone. Lactams are the major products from γ, δ and ϵ-amino-acids.

Investigation of the mechanism of decomposition of alkyl vinyl ethers has been extended to include the Pr^n, Bu^i and 2-methoxyethyl compounds.[67] In each case acetaldehyde and the corresponding alkene are obtained. The effect of electron-withdrawing and -releasing β-substituents is, respectively, to increase and to decrease E_a. These effects are small when compared with those of α-substituents and this is consistent with a six-centred transition state primarily featuring alkyl—O bond-polarization rather than bond-making between a β-H and the terminal vinyl carbon.

Unimolecular decomposition of 2,3-dihydro-*p*-dioxin occurs at 335–400° to form ethylene and glyoxal;[68] although the reaction is believed to be concerted, it is probable that the activated complex is asymmetric, with biradical character. Benzocyclobutene and 3-oxabicyclo[3.2.0]hepta-1,4-diene are formed upon gas-phase fragmentation of 2,3-dimethylene-7-oxabicyclo[2.2.1]heptane by competitive loss of water or ethylene, respectively.[69] Kinetics of pyrolysis of acetone dimethyl acetal have also been studied.[70]

The kinetics of gas phase retro-ene reactions of diallyl ether,[71] allylmethylamine[72] and hepta-1,6-diene[73] have been determined. It has already been established that the activation energies for *N*-alkylallylamine decompositions are very sensitive to substituents at positions X and Y (equation 2; X = N). In an investigation of allyl-

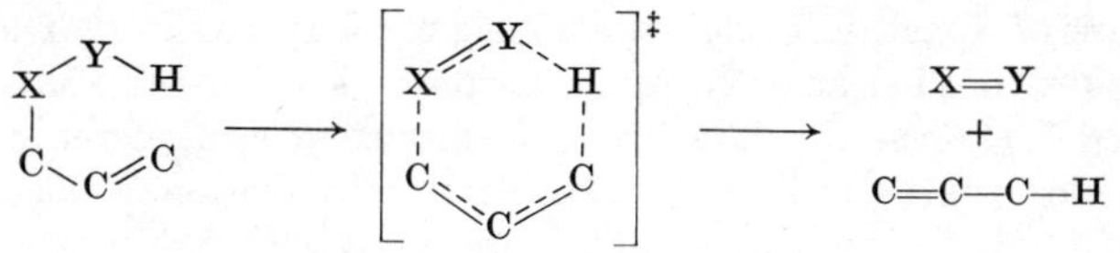

(2)

methylamine (AMA) it has been estimated[72] that for concerted decomposition of di- and tri-allylamine the activation energies are respectively 5.4 and 5.1 kcal mol^{-1} less than for AMA. The vinyl group is known to stabilize a carbonium ion by 13 kcal mol^{-1} and it has therefore been concluded that, for decomposition of alkylallylamines, bond-breaking and bond-formation are non-synchronous processes. The degree of polarization is, however, less than for dehydrohalogenations, for which the corresponding stabilizing effect of the vinyl group is even greater (8 kcal mol^{-1}). In keeping with an earlier suggestion, that there is little charge separation in the transition state for decomposition of alkyl allyl ethers (equation 2; X = O), results obtained for diallyl ether[71] reveal a much smaller effect (2.7 kcal mol^{-1}) of vinyl substitution on atom Y in such systems. Predominant formation of propylene and butadiene upon thermal decomposition of hepta-1,6-diene is also best described as a concerted retro-ene reaction and the kinetic results[73] are in line with those for systems where X = NR, O, CO or CR_2 and Y = CR_2 or O.

While thermolysis of (**10a**) gives (**11**) and (**12**) by a reverse Diels–Alder reaction, it has also been shown that (**10b**) undergoes predominant desulphonative rearrangement

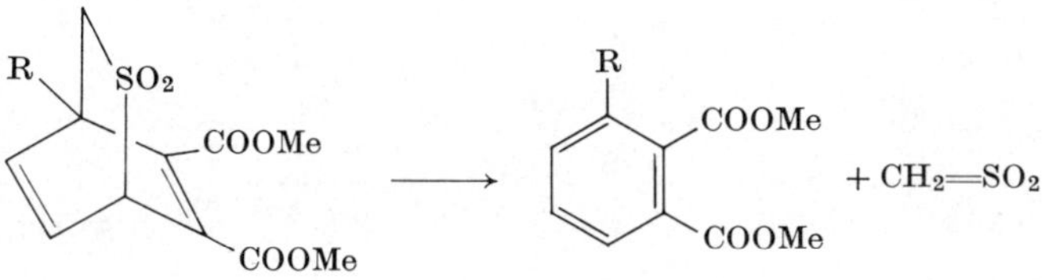

(**10**) (**11**) (**12**)

(**a**) R = Ph

(**b**) R = H

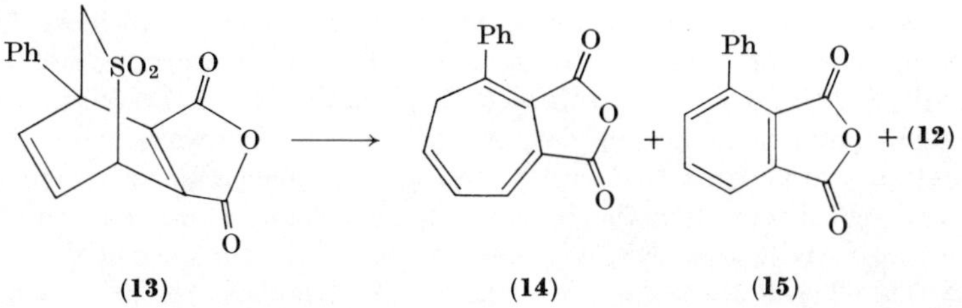

(**13**) (**14**) (**15**)

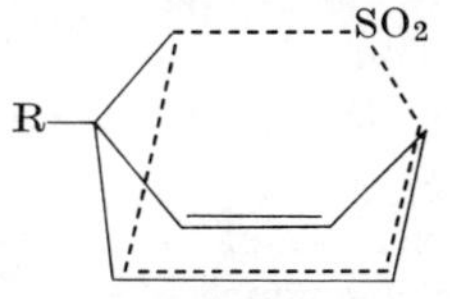

(**16**) 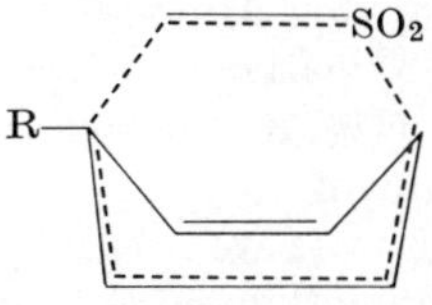 (**17**)

to form a mixture of dimethyl cycloheptatrienedicarboxylates.[74a, b] The formation of comparable amounts of (**14**) and (**15**) upon decomposition of the anhydride (**13**) has now been reported[74]: and these results have been discussed with reference to substituent effects on transition states (**16**) and (**17**). Sulphur extrusion is also a feature of decomposition of 5,5-diphenyldithiohydantoin derivatives.[75]

Pyrolysis of phenyl 2,2,2-trideuterioacetate in a nitrogen stream at 635° gives phenol, which is free from deuterium at the *o*-position, along with ketene which can be trapped as the corresponding 2,2-dideuterioacetanilide.[76] This observation is inconsistent with a concerted sigmatropic rearrangement and an alternative radical chain mechanism is favoured:

$$PhO\cdot + MeCOOPh \rightarrow PhOH + \cdot CH_2COOPh \rightarrow CH_2{=}C{=}O + PhO\cdot$$

Homolytic decompositions of *tert*-butyl *trans*-2-substituted cyclopropylperoxyacetates have been studied.[77]

The Hammett ρ-value (–3.44) for decomposition of sterically hindered 1-aryl-1-phenylpropyl hydrogen phthalates in DMSO clearly indicates a heterolytic mechanism for formation of phthalic acid and the corresponding 1-aryl-2-methylstyrene.[78] This precludes an alternative free-radical mechanism, it already having been established that a six-centred transition state is not a feature of reactions of *tert*-alkyl hydrogen phthalate or related esters.

A change in elimination mechanism from *cis*-concerted to nearly *E*1-type is suggested by the 10^3-fold rate acceleration effected[79a] by 1-phenyl substitution (**18b**) of ethylphenylsulphilimine (**18a**). However, for derivatives ($X \neq H$) of (**18b**) ρ_X = ca. 0.2

$XC_6H_4CH^{\alpha}CH_3{}^{\beta}$ / $YC_6H_4S{\rightarrow}NTs$ (**18**) ⟶ [S, C, C, H, N, Ts five-centred transition state]‡ (**19**) ⟶ $XC_6H_4CH{=}CH_2$ + YC_6H_4SNHTs

(**a**) $(XC_6H_4) = Y = H$ (**b**) $X = Y = H$

(**c**) $(XC_6H_4) = Y = H$ (**d**) $X = Y = H;\ H^{\beta} = {}^2H_1$

for reaction in benzene while in nitrobenzene the rates are less than five-fold greater. These observations combined with estimates of the deuterium isotope effect, $k(\mathbf{18b})/k(\mathbf{18d}) = 3.9$, and $\rho_Y = 0.9$ rule out *E*1 or radical mechanisms in which C—S bond cleavage is alone rate-determining. A carbanion mechanism has also been ruled out since only a two-fold rate increase occurs upon β-phenyl substitution of (**18a**). The results are consistent with the *cis*-concerted mechanism previously proposed.[79b] It is interesting to note that $k(\mathbf{18a})/k(\mathbf{18c}) = 3.03$ while, for derivatives ($Y \neq H$) of (**18a**), $\rho_Y = 0.9$. These values are almost identical with those for (**18b**) which reacts 10^3-fold more rapidly. Although the accelerative effect of the α-phenyl group has been attributed to conjugative stabilization of the incipient double bond in the five-centred transition state (**19**), it is not clear to the reviewer why this is not also the effect of a β-phenyl substituent.

In marked contrast are results[80] of a Hammett investigation of the Cope elimination of (1-arylethyl)dimethylamine oxides in aqueous media, for which $\rho = -6.5$. The large negative ρ-value combined with the large positive ΔS^{+} (+32 eu) determined for the *p*-methoxy-derivative suggest that the transition state is not a quasi-five-membered ring and that styrenes are formed by heterolysis. The generation and subsequent trapping of 3-homoadamantene from pyrolysis of *N*,*N*-dimethyl-3-aminohomoadamantane *N*-oxide constitutes the first preparation of a highly strained bridgehead alkene by this procedure.[81]

To facilitate interpretation of kinetics of eliminative decomposition of vibrationally "hot" β-trifluoroethylsilanes $CF_3CH_2SiF_xMe_{3-x}$ ($x = 0$–3) it has been necessary to determine[82] Arrhenius parameters for the corresponding thermal reactions, for which four-centre transition states have been proposed. As x increases, E_a decreases in direct relation to the bond index of the newly formed Si—F bond. For $x = 0$ some results have been obtained for decomposition of the "hot" molecule.

The mechanisms of thermolysis of *N*-silyl-*N*-(silylalkyl)urethanes[83] and of *trans*-2-benzylidene-3-cyclohexylamino-4,4-dimethyl-1-tetralone[84] have been investigated. Dehydrogenations of ethylene,[85] cyclopentene,[86] butene[87] and formamide[88] have been studied.

Other Topics

A book[89] entitled *Mechanisms of Elimination Reactions* has been published and elimination reactions of carbohydrates[90a] and steroids[90b] have been briefly reviewed.

The arguments in support of *E2C*, *E2H* or ion-pair mechanisms for eliminations promoted by weak bases have been reviewed[91] and the effect of a β-fluorine substituent on cyclopentyl, cyclohexyl and hexyl bromides has been interpreted[92] with respect to the rate and orientation of reaction with Et_4NCl and KI.

Primary deuterium isotope effects have been determined for bimolecular elimination reactions of cyclohexyl toluene-*p*-sulphonates and bromides throughout the spectrum of transition-state structures from *E2C*- to *E2H*-like, attained by increasing the acidity of the substrate, increasing the H-basicity of the base or by changing the leaving group from toluene-*p*-sulphonate to bromide.[93a] The k_H/k_D values pass through a maximum either as the basicity of the base increases ($Br^- < Cl^- < AcO^- < EtO^-$) or (for reactions of PhO^-, AcO^-, EtO^- and Bu^tO^-) as the acidity of the substrate is increased. This is in accord with conventional ideas about proton-transfer between acids and bases. Of particular interest, however, is the very low sensitivity of k_H/k_D to change in base strength on the *E2C* side of the spectrum, combined with a much greater sensitivity on the *E2H* side. This is in agreement with recent estimates[93b] of the isotope effects to be expected of bent relative to linear transition states. Recent unpublished results have led Parker to reconsider the structure of the *E2C* transition state. Structure (**20**) is still favoured although (**21**) cannot be ruled out. The very small α-substituent effects previously reported need not be inconsistent with positive charge development at C_α since the presence of two loosely bound anions may reduce the substituent dependence of carbonium ion stability. With respect to alternative *E2C* transition-state structures

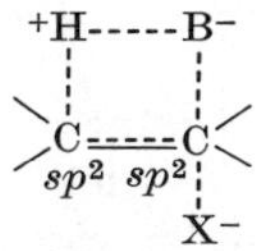

(**20**)

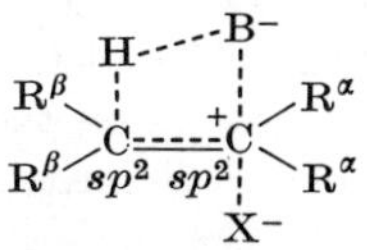

(**21**)

(**20**) and (**21**), the increase in k_H/k_D as the reaction becomes more *E2H*-like can be attributed to increased linearity of C—H—B accompanied either by increased transfer of the proton from substrate to base or by tightening of the C_β—H bond, respectively.

The effect of a π-complexed $Cr(CO)_3$ group on *E*2 reactions of phenethyl and α-methylphenethyl bromide and toluene-*p*-sulphonate in RO^-/ROH ($R = Bu^t$ or Et) is to increase the rate (by 11–495-fold) and percentage of alkene formation, without altering the distribution of isomers obtained from the secondary substrates.[94] β-Deuterium isotope effects do not vary appreciably upon complexation or *p*-nitro-substitution of the phenethyl derivatives and, in contrast to $Cr(CO)_3$, the nitro-group exhibits resonance stabilization of the carbanion-like transition state. Trends observed indicate that reactions of tricarbonylchromium substrates fit the *E*2 spectrum of transition states proposed for uncomplexed substrates. Since complexation of an aryl ring with $Cr(CO)_3$ causes strong acceleration of reactions involving development of pronounced positive or negative charge at the α-carbon atom of a side chain (by 10^{4-5} and 10^2-fold, respectively), its effects on eliminations under *E2C* conditions have been studied.[95] The rate of *E*1 reaction of 1-phenyl-1-chloropropane in acetone containing *n*-Bu_4NClO_4 is increased by >400-fold upon complexation, whereas elimination induced by *n*-Bu_4NCl is essentially unaffected. The latter result is inconsistent with an *E2H* transition state lying between the central and *E*1-like extreme but supports an *E2C* structure (**20**) with well-developed double-bond character. The effect of complexation is to decrease the percentage of elimination (F_E) of α-methylphenethyl toluene-*p*-sulphonate from 8.7 to 0.5% upon reaction in acetone–Cl^-, while in ROH–RO^- F_E is increased from 90 to 100%. $Cr(CO)_3$-stabilization of the *E*1*cB*-like transition state causes acceleration of elimination in the latter case, whereas 5-fold deceleration of the former *E2C* reaction (of the corresponding bromide, for which $F_E = 100\%$) has been ascribed to decreased conjugation with the incipient double bond.

E2C-Like transition states have been implicated[96, 97] in interpretation of reactions of dialkyl tetramethylammonium phosphates with 2-bromopentane in DMF and CH_3CN.

It has been found that the ^{14}N-isotope effect for *E*2 reaction of 2-arylethylammonium ions with $EtO^-/EtOH$ decreases with electron-withdrawing ability of *p*-substituents while the corresponding β-deuterium isotope effect increases.[98a] Thus, for the transition state of this system, in which hydrogen is known to be more than half transferred, electron-withdrawing substituents cause both C—H and C—N^+ bond rupture to decrease. This is in accord with the Thornton model. The values of $\rho = 3.67$ and $k_H/k_D = 2.98$ (Ar = Ph), however, suggest that the transition state is carbanion-like, in which case the More O'Ferrall model can also accommodate the results (provided that substituent effects are greater along the reaction co-ordinate than perpendicular to it). Such considerations may also explain the clear discrepancy between these results and the proposal that, for reactions of alkyl halides, etc., there is a continuum of *E*2 transition states extending from carbanion-like to carbonium-ion-like extremes. The degree of C—H and C—N bond rupture is less for the phenethyl than for the ethyl 'onium substrate and a contrary conclusion[98b] is believed to be erroneous. The nitrogen isotope effect decreases with base strength.

Secondary deuterium isotope effects of multiple deuteriation of trimethylammonium leaving groups have been determined for *E*2 reactions of axial steroidal and phenethyl substrates.[99] The effects on the former *E*1-like reaction are large in comparison with those on the latter *E*1*cB*-like process and are more comparable with those for *E*1 reaction of a related system. Secondary isotope effects may therefore provide another criterion of transition-state structure for *E*2 reactions of 'onium compounds.

Radical anions are believed to be intermediates in the reductive cleavage of aliphatic and aromatic trimethylammonium iodides with lithium naphthalenide in THF,[100] and a cyclic transition state has been proposed to account for a base-induced β-elimination of 1,1,1-trisubstituted hydrazinium salts, $[NRMe_2(NH_2)]^+X^-$.[101]

There have been several studies of amino acid decarboxylation.[102–109] As expected of a mechanism involving an intermediate azlactone, Dakin–West reaction of *N*-methyl-5-pyrrolidone-2-carboxylic acid in Ac_2O–pyridine is precluded by Bredt's rule.[102] Stereoselective formation of β-branched aspartic acids has been achieved by decarboxylation of $AcNHC(COOEt)_2CHRCOOEt$ in aqueous alkali,[103] and kinetics of decarboxylation of Schiff bases of α-amino-acids have been reported.[104] For the first time[105] a small molecule has been found to display the absolute chiral recognition of a prochiral centre (C_{aabd}) that is a common feature of enzymic reactions. Thus, considerable asymmetric induction is achieved upon decarboxylation of aminoalkylmalonic acids mediated by a dissymmetric cobalt complex. The mechanisms of decarboxylation of 1,3-dimethylorotic acid[106] and oxaloacetate[107,108] (catalysed by Cu(II) and Zn(II)) have been studied.

The mechanism and stereochemistry of decarbonylation of acid chlorides by chlorotris(triphenylphosphine)rhodium(I) (**22**) has been investigated.[110–112] Intermediate complexes (**24**) and (**25**) have been isolated for reactions of *p*-substituted-benzoyl (**23a**) and -phenylacetyl (**23b**) chlorides.[110] Linear free-energy relationships suggest that the reactions are homolytic since there is little charge developed during conversion of (**24**) into (**25**) [$\rho = 0.265$ and -0.612 for (**23a**) and (**23b**), respectively] or during the

$(Ph_3P)_3RhCl$ (**22**) ⇌ [$-PPh_3$, $+RCOCl$ (**23**)] $Cl-Rh(PPh_3)_2(Cl)(COR)$ (**24**) ⇌ [$-CO$] $Rh(Cl)_2(CO)(COR)(PPh_3)$

$Rh(Cl)_2(CO)(COR)(PPh_3)$ ⇌ [$+(23)$] $Ph_3P-Rh(Cl)(CO)-PPh_3$ (**26**)

(**24**) ⇌ $Rh(Cl)_2(R)(CO)(PPh_3)_2$ (**25**) ⇌ [$-(Olefin + HCl)$ or $-RCl$ (**27**)] (**26**)

(**a**) $R = p\text{-}XC_6H_4$—
(**b**) $R = p\text{-}XC_6H_4CH_2$—
(**c**) $R = C_6H_5CH(CF_3)$—
(**d**) $R = CH_3CHPhCHPh$—
(**e**) $R = CH_3(CH_2)_2CHCH_3$—
(**f**) $R = CH_3(CH_2)_2C(CH_3)_2$—
(**g**) $R = (CH_3)_2CHCH(C_2H_5)$—
(**h**) $R = (CH_3)_2CHCH(CH_3)$—

subsequent formation of (**26**) along with a benzyl or aryl chloride. Decarbonylation of (*S*)-(–)-(**23c**) with (**22**) gave racemic (**27c**) (71% yield); while it was shown that optically active (**24c**) also gave racemic (**27c**), the intermediate (**25c**) could not be isolated or prepared independently.[111] It is probable, however, that racemization occurs during the acyl-to-(alkyl + carbonyl) rearrangement (**24**) → (**25**) since (*S*)-α-(trifluoromethyl)-benzyl chlorosulphite loses SO upon reaction with (chlorocarbonyl)bis(diethylphenylphosphine)Rh(I) to form an optically active intermediate [analogous to (**25**)] from which (*R*)-(**27**) is obtained with 62% net inversion in the two steps. Decarbonylation of *erythro*- and *threo*-(**23d**) gives exclusively *trans*- and *cis*-methylstilbene, respectively.[112] Saytzeff orientation governs alkene formation upon decarbonylation of (**23e–h**) with *trans*-(chlorocarbonyl)bis(triphenylphosphine)rhodium(I).

The high degree of stereospecificity in the photochemical decarbonylation of (+)-(*R*)-(h^5-C_5H_5)Fe(CO)(Ph_3P)C(O)Et to optically pure (−)-(*S*)-(h^5-C_5H_5)Fe(CO)-(Ph_3P)Et has been attributed[113] to ethyl migration into the site vacated by CO.

o-Phenylenediamine catalyses the decarboxylation of MeCOCHR(COOH), where R = H, Me, Et or Bu, through the intermediacy of an arylcarbamic acid.[114] Decarboxylation of $HOOCCOCH_2COOH$ is not catalysed since the active site is apparently blocked by the additional carboxyl group. The alkylative decarboxylation of *N*-carbalkoxypyrazoles has been shown to require a polar aprotic solvent and to be catalysed by halide ions.[115] Nucleophilic attack on the α-carbon of the alkoxy-group yields alkyl halide, CO_2 and the pyrazolyl anion which is subsequently *N*-alkylated by reaction with either the alkyl halide or the parent ester.

Decarboxylative ring contraction of dichlorotetrahydroquinolinediones to dioxindole derivatives,[116] and acid-catalysed decarboxylation of 2,4,6-trihydroxybenzoic, anthranilic and *p*-aminobenzoic acids,[117] have been studied.

Lithium-ion-catalysed decompositions of octane-2-diazotate in phosphoric hexamethyl triamide containing nucleophilic salts have been discussed in terms of nitrogen-separated and intimate ion pairs.[118] Secondary deuterium isotope effects for solvolysis of cyclopentyl *p*-bromobenzenesulphonate have indicated that, in EtOH–H_2O mixtures, products of substitution and stereospecific *trans*-elimination are formed by rate-determining reactions of the reversibly formed ion pair. In CF_3COOH–H_2O formation of the solvent-separated ion pair is rate-determining and the subsequent fast elimination is non-stereospecific.[119]

The rate of *E*1 reaction of 2-phenyl-2-chloropropane varies[120] with solvent in the order DMSO > CH_3NO_2 > CH_3CN > DMF. Both the Hammett constant ρ (ca. −4.6) and the secondary β-$^2H_1$ isotope effect (1.06–1.15) (estimated by comparison with rates of reactions of the *p*-bromo- and 1,1,1,3,3,3-hexadeuterio-derivative, respectively) are little affected by a change from protic to aprotic solvents. Conformational constraints in certain *syn*- and *anti*-toluene-*p*-sulphonyloxy- and bromo-2,4-bridged 5,6-benzopyridinophanes render these compounds resistant to S_N2 and *E*2 reactions; however, they undergo highly stereospecific ionization reactions under forcing conditions.[121] Product distributions suggest that reactions of *cis*- and *trans*-4-*tert*-butylcyclohexyl and *erythro*- and *threo*-1-methyl-2-phenylpropyl chloroformate with $AgSbF_6$ in AcOH resemble deaminations while, in contrast, those of the corresponding chlorides resemble solvolyses.[122]

In the presence of $AgSbF_6$ as catalyst, benzoylcyclohexene is the product of dehydrobromination of 1-benzoyl-1-bromocyclohexane in CH_2Cl_2, while in MeOH debrominative rearrangement gives methyl 1-phenylcyclohexanecarboxylate.[123]

Dehydration of etheral solutions of ferrocenylcyclopropylcarbinols on silica gel[124] and of 1-arylbut-3-ene-1,2-diols and 1-aryl-2-methylbut-3-ene-1,2-diols[125] in H_2SO_4 are also complicated by carbonium ion rearrangements. Alkene intermediates have been detected[126] during Koch–Haaf carboxylation of cyclohexanols in H_2SO_4.

Results for solvolyses of tertiary α-silyl bromides RMe_2SiCMe_2Br in EtOH—H_2O have been contrasted with those for their carbon analogues.[127] When R = Me, solvolysis occurs 38,000-fold slower than for $(CH_3)_3CC(CH_3)_2Br$, and the sole product is isopropenyltrimethylsilane. While it is probable that Me_3Si is less able than Me_3C to stabilize an incipient carbonium ion, the relative rates may also reflect unusual ground-state stability resulting from interaction between non-bonding electron pairs on bromine and the empty 3*d*-orbitals of Si. Surprisingly, no substitution product is obtained even though the α-silylcarbonium ion would be more open to nucleophilic attack. It is concluded that the solvolysis mechanism is not "limiting" and this is supported by the Grunwald–Winston m-value (0.66). The degree of positive charge development is, however, substantial since $\rho_X = -1.1$ where $R = XC_6H_4$. In contrast with results for $XC_6H_4(CH_3)_2CC(CH_3)_2Cl$, there is no evidence of anchimeric assistance by the aryl group.

The mechanism of pH-dependent hydrolysis of tertiary α-(*N*-substituted amino)-carbonitriles is believed to involve intermediate ketiminium ions formed by ejection of CN^- upon nucleophilic participation by the amine lone pair.[128] An *E*1 reaction of the conjugate acid of *cis*-4-*tert*-butyl-*N*-methyl-1-phenylcyclohexylamine in formic acid has been proposed.[129] Eliminative deamination, of related compounds, does not occur if the amino-group is in the equatorial position or if the α-phenyl substituent is replaced by Me or H.

Solvolyses of *cis*- and *trans*-2-halovinyl 2,4,6-trinitrobenzenesulphonates ($HalCR^2{=}C(O_3SAr)R^1$) in AcOH and CH_3NO_2–MeOH have been studied;[130] the *trans*-iodo-compounds are more reactive than the *cis*-isomers, give substitution products of retained *trans*-stereochemistry, react ca. 10^4-fold faster than the corresponding 2-chlorovinyl derivatives and are believed to feature anchimeric assistance by iodine. Nucleophilic attack of the solvent on the intermediate bridged cation may occur either at the unsaturated carbon atoms or at the iodine atom to give substitution and alkyne-elimination products, respectively.

Benzyltrimethylammonium hydroxide[131] (TRITON B) and tetraethylammonium fluoride[132] have been used as effective bases for formation of alkynes from halogeno-ethylenes. Such reactions of Et_4NF in CH_3CN with (**28**) are uncomplicated by competing substitution or addition and are believed to involve *anti*-elimination since (**29a**) is unreactive while (**29b**) gives the alternative *p*-nitrophenylallene.[132] Dehydrohalogenations of (**28**) and (**29**) are also effected by KF in dipolar aprotic solvents and the rate is enhanced by dicyclohexyl-18-crown-6.

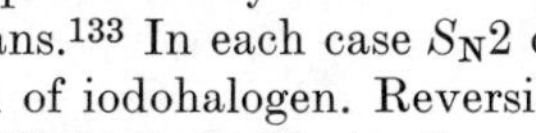

$p\text{-}NO_2C_6H_4(H)C{=}C(X)(R)$ | $p\text{-}NO_2C_6H_4(H)C{=}C(R)(X)$

(**28a**) X = Br or Cl; R = H

(**28b**) X = Br or Cl; R = Me

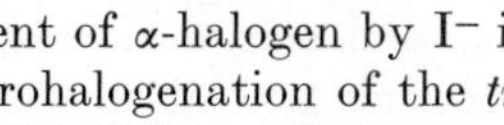

(**29a**) X = Br; R = H

(**29b**) X = Br or Cl; R = Me

Investigation of dehalogenation of vicinal halides (in which α-halogen is activated by ether-oxygen) promoted by NaI–acetone has been extended to include *trans*-2,3-dihalo-tetrahydrofurans.[133] In each case S_N2 displacement of α-halogen by I^- is followed by *cis*-elimination of iodohalogen. Reversible dehydrohalogenation of the *trans*-dihalides also occurs. Vicinal dechlorination of diaryl-1,2-dichloroethylene, catalysed by $NaOH–Et_2O–DMSO$, is believed to involve an intermediate radical anion.[134] A mechanism has been formulated for epoxide-induced dehydrohalogenation of phosphonium halides.[135] Assistance (carbonyl formation) and participation (epoxide formation) feature in dehydrohalogenation reactions of *trans*-fused bicyclo(4.*n*.0)alkane halohydrins with aqueous alkali;[136] dehydrochlorination of $ClCH_2CH_2(OH)OPh$ is also effected under such conditions.[137]

7,7-Dichlorobicyclo[4.1.0]hept-3-ene (**30**) labelled at C-7 with ^{12}C probably undergoes base-catalysed double dehydrochlorination according to equation 3 and not by a skeletal rearrangement;[138] this is believed to be the first application of the technique of ^{12}C-labelling (which has many advantages over ^{13}C-labelling) to elucidation of a reaction

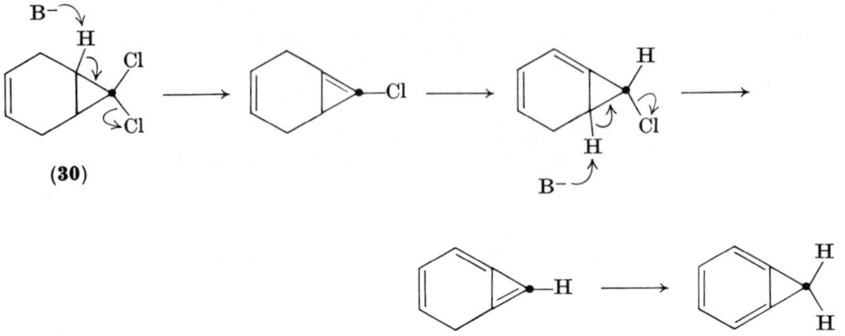

(**30**) (3)

mechanism. Multiple dehydrohalogenations of sodium *threo,threo*-9,10,12,13-tetrachloro-octadecanoate[139] and solid *meso*-β,β'-dihaloadipate[140] have been studied.

Dehydrations of 3-methylbutane-1,3-diol,[141] polyhydroxy-bi- and -ter-phenyls[142] (catalysed by acid) and of 2,2′-biphenol[143] (catalysed by $ZnCl_2$) have received attention, while reaction between bromine and 1,1-diphenylalkan-1-ols has been shown to proceed via an alkene.[144] Eliminations of hydroxylic fragments from molecular ions have been studied.[145–147] Electron-impact-induced eliminations of AcOH from 1,2-disubstituted ethyl acetates, or of H_2O from acyclic secondary alcohols, are stereochemically comparable to corresponding thermal reactions.[145] By deuterium labelling it has likewise been shown that *cis*-1,2-elimination occurs upon electron impact of epimeric androstanyl acetates,[147] and of bornyl alcohols or acetates.[146] By mass spectrometric examination of metastable peaks it has been established that concerted 1,2-elimination of H_2 from $CH_3CH_3^{\cdot +}$, $CH_2{=}OH^+$, $CH_2{=}NH_2^+$, $CH_2{=}SH^+$ and $CH_3NH_2^{\cdot +}$ proceed by symmetry-forbidden routes in which intended crossings are avoided by release of kinetic energy;[148] the experimental results suggest that concerted eliminations of hydrogen that proceed without significant release of kinetic energy are not 1,2-eliminations and on this basis it has been concluded[149] that dehydrogenations of $C_6H_7^+$, $C_2H_5^+$ and $C_2H_4^{\cdot +}$ are not concerted while those of $C_3H_7^+$ and $C_7H_9^+$ are.

Oxidative dimerization of phenylacetylene,[150] oxidative transfer dehydrogenation of α,β-unsaturated alcohols,[151] dehydrogenation of 9,10-dihydroanthracenes by substituted quinones[152] and of *p*-menthadienes by strong bases[153] have been reported.

CNDO/2 Calculations have been used to account for the observations that Et_3N-induced formation of 2-chloro-2-methylketene from 2-chloropropionyl chloride proceeds through an enolate salt whereas an acyl-ammonium salt is the intermediate of dehydrochlorination of isobutyryl chloride.[154] Allenes are formed upon reaction of ethyl 2-methyl-3-halo (or -3-cyano)but-3-enoates (*E* and *Z* isomers) with Bu^tO^-, whereas with EtO^- or Bu^tS^- subsequent addition reactions give a mixture of substitution products.[155] In contrast, EtS^- and PhS^- react by direct substitution.

Further evidence has been obtained for the intermediacy of sulphenes in reactions of alkanesulphonyl chlorides with amines. It has been shown,[156] by deuterium exchange and kinetic experiments, that kinetic resolution previously effected upon reaction of (+)-camphor-10-sulphonyl chloride with amines must occur in a sulphene-trapping step.

This report necessitates that kinetic results for many reactions of amines with sulphonyl chlorides be reinterpreted. An elimination–addition mechanism has also been proposed[157] to account for formation of trichloromethanesulphinate, $[NHEt_3]^+[CCl_3SO_2]^-$, on reaction of dichloromethanesulphonyl chloride with triethylamine in Et_2O or C_6H_6.

It has been shown[158] that the formation of SO_2, Cl^- and alkenes from 2-chloroalkanesulphinates in aqueous alkali proceeds by a *trans*-coplanar eliminative fragmentation and does not involve an intermediate thiirane dioxide, 1,2-oxathietane 2-oxide or ethylene sulphinate. It has been pointed out that 2-alkoxyalkanesulphinate ions may behave similarly and that their intermediacy in the Ramberg–Bäcklund reaction of α-halo-sulphones cannot yet be confidently ruled out.

Despite the fact that the W-configuration of the α-halo-sulphone (**31**) should be conducive to a concerted 1,3-elimination it has been shown[159a] that a carbanion mechanism accounts for thiirane dioxide formation in the Ramberg–Bäcklund reaction of this compound with MeO^-/MeOH. The results support the view[159b] that concerted 1,3-eliminations are rare, if they exist at all.

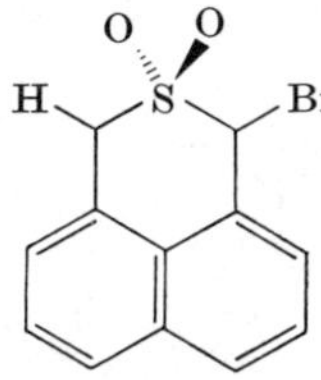

(**31**)

The synthetic utility of 1-bromovinyl sulphones has prompted a study[160] of the stereochemistry of Et_3N-induced dehydrobromination of (**32**) and (**33**). The *threo*-isomers [(**32a**) and (**33a**)] give only (*Z*)-(**34a**) by *anti*-elimination, in a wide variety of solvents, whereas the erythro-dibromides [(**32b**) and (**33b**)] form a mixture of (*Z*)- and (*E*)-(**34b**) which can be isomerized to (*Z*)-(**34b**) alone by a catalytic amount of bromine.

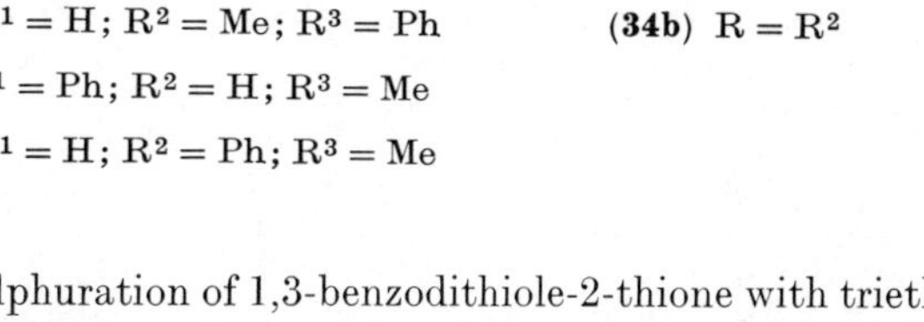

$$\xrightarrow{Et_3N} RCH{=}C(Br)SO_2R^3$$

(**32a**) $R^1 = Me$; $R^2 = H$; $R^3 = Ph$ (**34a**) $R = R^1$

(**32b**) $R^1 = H$; $R^2 = Me$; $R^3 = Ph$ (**34b**) $R = R^2$

(**33a**) $R^1 = Ph$; $R^2 = H$; $R^3 = Me$

(**33b**) $R^1 = H$; $R^2 = Ph$; $R^3 = Me$

The products of desulphuration of 1,3-benzodithiole-2-thione with triethyl phosphite, in the presence of alcohols, enamines, carbonyl compounds or piperidine, have been attributed to reactions of the 1,3-dipolar intermediate of thiophilic attack on the phosphite.[161] A fast reversible α-elimination of arenesulphinate ion, to give an intermediate aroylaminonitrene, followed by tautomerism with loss of N_2, is believed[162] to constitute the mechanism of the McFadyen–Stevens reaction of $XC_6H_4CONHNHSO_2$-

C_6H_4Y (as its conjugate base in aqueous Na_2CO_3). α-Elimination reactions of anil-idoximes in basic media,[163] and of *N,N*-dichloromethanesulphonamide catalysed by copper,[164] have been reported.

Two mechanisms have been proposed for the Chichibabin reaction of pyridine and phenanthridine.[165] An addition–ring-opening–ring-closure sequence is believed[166] to explain the reaction of 2-bromo-4-phenylpyrimidine with KNH_2 in liquid NH_3. Elimination reactions have featured in the formation of cyclohexenethiocarboxamides[167] and decomposition of *S,S'*-dimethyl-1,4-dithanium bisfluoroborates.[168] The rates of removal of ethanol from EtOCHMeCH(COMe)COOEt in AcOH can be correlated with the crystal lattice energy of general-salt catalysts.[169]

A radical chain process is believed[170] to account for the formation of an equilibrium mixture of the two 6-bromocholest-4-en-3-ones upon dehydrobromination of 5α,6β- or 5β,6α-dibromocholestan-3-one in $CDCl_3$.

Base-catalysed β-eliminations from $(Me_3SiOSiMe_2CH_2)_3B$ and $(C_6H_5Me_2SiOSiMe_2CH_2)_3B$ have been used to generate 1-silaethylene intermediates.[171]

Elimination reactions of isopropylamine[172] (catalysed by HBr in the gas phase), bis-(*p*-nitrophenyl)methyl chloride,[173] pectin,[174] dibromochalcone,[175] *N*-acyl derivatives of α-chloroglycine,[176] 1,2-dihydroisoquinoline,[177] pyridine-2- and -4-diazotate,[178] 5-fluorouracil,[179] 3,3-disubstituted 1,2-dioxetanes,[180] and tetracarbonyl-*cis*-dihydrido-osmium[181] have also been studied. Hexamethylphosphoramide-induced dehydration of alcohols has been further investigated,[182] and rate constants for elimination reactions of 2-arylethyl bromides and sulphonium salts have been correlated with corresponding NMR shifts of the α- and β-methylene protons.[183]

References

1 J. K. Borchardt and W. H. Saunders, Jr., *J. Org. Chem.*, **39**, 99 (1974).
2 J. K. Borchardt and W. H. Saunders, Jr., *J. Am. Chem. Soc.*, **96**, 3912 (1974).
3 J. K. Borchardt, J. C. Swanson and W. H. Saunders, Jr., *J. Am. Chem. Soc.*, **96**, 3918 (1974).
4 R. A. Bartsch, E. A. Mintz and R. M. Parlman, *J. Am. Chem. Soc.*, **96**, 4249 (1974).
5 R. A. Bartsch and R. H. Kayser, *J. Am. Chem. Soc.*, **96**, 4346 (1974).
6 G. Guillaumet, V. Lemmel, G. Coudert and P. Caubere, *Tetrahedron*, **30**, 1289 (1974).
7 K. Marks, M. Szkoda, *Rocz. Chem.*, **47**, 2295 (1973); *Chem. Abs.*, **81**, 3076 (1974).
8 R. A. Bartsch, K. E. Wiegers and D. M. Guritz, *J. Am. Chem. Soc.*, **96**, 430 (1974).
9 D. P. Ridge and J. L. Beauchamp, *J. Am. Chem. Soc.*, **96**, 637 (1974).
10 D. P. Ridge and J. L. Beauchamp, *J. Am. Chem. Soc.*, **96**, 3595 (1974).
11 O. S. Tee, J. A. Altmann and K. Yates, *J. Am. Chem. Soc.*, **96**, 3141 (1974).
12 C. Tuzun and K. H. Malik, *Commun. Fac. Sci. Univ. Ankara, Scr. B*, **21**, 11 (1974); *Chem. Abs.*, **81**, 49124 (1974).
13 D. H. Hunter, *Intra-Sci. Chem. Rep.*, **7**, 19 (1973); *Chem. Abs.*, **81**, 90658 (1974).
14 A. Sera, H. Mano and K. Maruyama, *Bull. Chem. Soc. Japan*, **47**, 1754 (1974).
15 H. Mano, A. Sera and K. Maruyama, *Bull. Chem. Soc. Japan*, **47**, 1758 (1974).
16 M. J. van der Sluijs and C. J. M. Stirling, *J.C.S. Perkin II*, **1974**, 1268.
17 P. F. Cann and C. J. M. Stirling, *J.C.S. Perkin II*, **1974**, 820.
18 P. F. Cann and C. J. M. Stirling, *J.C.S. Perkin II*, **1974**, 817.
19 Y. Riad and H. M. Attia, *U.A.R. J. Chem.*, **14**, 447 (1971); *Chem. Abs.*, **79**, 136287 (1973).
20 C.-G. Shin, Y. Yonezawa and J. Yoshimura, *Nippon Kagaku Kaishi*, **1974**, 718; *Chem. Abs.*, **81**, 3040 (1974).
21 C. H. DePuy and A. L. Schultz, *J. Org. Chem.*, **39**, 878 (1974).
22 D. J. McLennan and R. J. Wong, *J.C.S. Perkin II*, **1974**, 526.
23 D. J. McLennan and R. J. Wong, *J.C.S. Perkin II*, **1974**, 1373.
24 A. Streitwieser, C. J. Chang and A. T. Young, *J. Am. Chem. Soc.*, **94**, 4888 (1972).
25 A. B. N. Gray and D. J. McLennan, *J.C.S. Perkin II*, **1974**, 1377.

[26] F. M. Fovad, P. G. Farrell and A. G. Abdel-Rehiem, *Tetrahedron Letters*, **1974**, 3355.
[27] D. H. Hunter, Y-t. Lin, A. L. McIntyre, D. J. Shearing and M. Zvagulis, *J. Am. Chem. Soc.*, **95,** 8327 (1973).
[28a] D. H. Hunter and D. J. Shearing, *J. Am. Chem. Soc.*, **95**, 8333 (1973).
[28b] See *Org. Reaction Mech.*, **1971**, 139.
[29] J. Villiéras, C. Bacquet and J. F. Normant, *Bull. Soc. Chim. France*, **1974**, 1731.
[30] G. Sartore, *Tetrahedron Letters*, **1974**, 3133.
[31] G. S. Rork and I. H. Pitman, *J. Am. Chem. Soc.*, **96**, 4654 (1974).
[32] C. E. Olsen, *Acta Chem. Scand.*, **27**, 2983 (1973).
[33] A. Thibblin and P. Ahlberg, *Acta Chem. Scand.*, *B*, **28**, 818 (1974).
[34] A. Williams and K. T. Douglas, *J.C.S. Perkin II*, **1974**, 1727.
[35] N. K. Hamer and R. D. Tack, *J.C.S. Perkin II*, **1974**, 1184.
[36] H. Schmidt, A. Schweig, B. M. Trost, H. B. Neubold and P. H. Scudder, *J. Am. Chem. Soc.*, **96,** 622 (1974).
[37] R. J. Crawford and M. Ohno, *Can. J. Chem.*, **52**, 3134 (1974).
[38] J. E. Herweh and R. M. Fantazier, *J. Org. Chem.*, **39**, 786 (1974).
[39] G. Làbbé and G. Mathys, *J. Org. Chem.*, **39**, 1778 (1974).
[40] M. Komatsu, S. Ichijima, Y. Ohshiro and T. Ogawa, *J. Org. Chem.*, **38**, 4341 (1973).
[41] M. E. Peek, C. W. Rees and R. C. Storr, *J.C.S. Perkin I*, **1974**, 1260.
[42] V. V. Dubikhin and G. M. Nazin, *Izv. Akad. Nauk SSSR, Ser. Khim.*, **1974**, 1345; *Chem. Abs.*, **81,** 90900 (1974).
[43] M. Gates and J. L. Zabriskie, Jr., *J. Org. Chem.*, **39**, 222 (1974).
[44] K. Kochloefl and H. Knoezinger, *Catal., Proc. Int. Congr., 5th*, **2**, 1171 (1973); *Chem. Abs.*, **80,** 119998 (1974).
[45] T. Yamaguchi and K. Tanabe, *Bull. Chem. Soc. Japan*, **47**, 424 (1974).
[46] P. Canesson, F. N. Gnonlounfoun and M. Blanchard, *Bull. Soc. Chim. France*, **1973**, 3056.
[47] L. P. Loseva, Ya. M. Paushkin, D. I. Metelitsa, Yu. P. Loseva and V. N. Isakovich, *Dokl. Akad. Nauk SSSR*, **215**, 134 (1974); *Chem. Abs.*, **80**, 132365 (1974).
[48] G. Ingram and S. M. H. Rizvi, *Mocrochem. J.*, **19**, 253 (1974); *Chem. Abs.*, **81**, 77263 (1974).
[49] B. C. Capelin, G. Ingram and J. Kokolis, *Microchem. J.*, **19**, 229 (1974); *Chem. Abs.*, **81,** 77264 (1974).
[50] I. Tvaroska, V. Klimo and L. Valko, *Tetrahedron*, **30**, 3275 (1974).
[51] K. C. Kim and D. W. Setser, *J. Phys. Chem.*, **78**, 2166 (1974).
[52] M. Misono, Y. Aoki and Y. Yoneda, *Chem. Letters* (*Tokyo*), **1974**, 535.
[53] P. Andeu and J. Madrid, *Acta Cient. Venez., Supl.*, **24**, 169 (1973); *Chem. Abs.*, **81**, 62721 (1974).
[54] S. Villalba and H. Noller, *Acta Cient. Venez, Supl.*, **24**, 165 (1973); *Chem. Abs.*, **81**, 62720 (1974).
[55] R. M. Rodova, L. A. Shevtsova, S. V. Levanova, A. M. Rozhnov and I. K. Garkushin, *Izv. Vyssh. Ucheb. Zaved., Khim. Tekhnol.*, **17**, 379 (1974); *Chem. Abs.*, **81**, 12862 (1974).
[56] W. Kladnig and H. Noller, *Acta Cient. Venez., Supl.*, **24**, 142 (1973); *Chem. Abs.*, **81**, 90756 (1974).
[57] R. Cypres and B. Bettens, *Tetrahedron*, **30**, 1253 (1974).
[58] W. R. Dolbier, Jr., and H. M. Frey, *J.C.S. Perkin II*, **1974**, 1674.
[59] N. Barroeta, V. DeSantis and M. Rincón, *J.C.S. Perkin II*, **1974**, 911.
[60] E. Catalina, M. Nutiu and G. Ostrogovich, *Bul. Stiint. Teh. Inst. Politeh. Timisoara, Ser. Chim.*, **18**, 167 (1973); *Chem. Abs.*, **81**, 119616 (1974).
[61] E. V. Brown, R. J. Moser and M. B. Shambu, *Trans. Ky. Acad. Sci.*, **34**, 22 (1973); *Chem. Abs.*, **80**, 26492 (1974).
[62] V. E. Agabekov, N. I. Mitskevich, V. A. Azarko and N. L. Budeiko, *Dokl. Akad. Nauk Beloruss. SSR*, **17**, 826 (1973); *Chem. Abs.*, **79**, 145746 (1973).
[63] S. A. Borisenkova, L. M. Il'ina, E. V. Leonova and A. P. Rudenko, *Zh. Org. Khim.*, **9**, 1827 (1973); *Chem. Abs.*, **79**, 145693 (1973).
[64] A. M. Shur and R. I. Ishchenko, *Izv. Vyssh. Ucheb. Zaved., Khim. Khim. Tekhnol.*, **17**, 844 (1974); *Chem. Abs.*, **81**, 90859 (1974).
[65] M. A. Haleem and M. Azeem, *Pak. J. Sci. Ind. Res.*, **16**, 18 (1973); *Chem. Abs.*, **80**, 36542 (1974).
[66] M. A. Ratcliff, Jr., E. E. Medley and P. G. Simmonds, *J. Org. Chem.*, **39**, 1481 (1974).
[67] T. O. Bamkole, *J.C.S. Perkin II*, **1974**, 801.
[68] H. M. Frey and R. A. Smith, *J.C.S. Perkin II*, **1974**, 1407.
[69] P. Vogel and M. Hardy, *Helv. Chem. Acta*, **57**, 196 (1974).
[70] J. A. Garcia Dominguez and M. J. Molera, *An. Quím.*, **70**, 186 (1974); *Chem. Abs.*, **81**, 77266 (1974).
[71] P. Vitins and K. W. Egger, *J.C.S. Perkin II*, **1974**, 1292.
[72] P. Vitins and K. W. Egger, *J.C.S. Perkin II*, **1974**, 1289.

[73] K. W. Egger and P. Vitins, *J. Am. Chem. Soc.*, **96**, 2714 (1974).
[74a] J. F. King and E. G. Lewars, *Can. J. Chem.*, **51**, 3044 (1973).
[74b] N. H. Fischer and H. N. Lin, *J. Org. Chem.*, **38**, 3073 (1973).
[74c] J. F. King, R. M. Enanoza and E. G. Lewars, *Can. J. Chem.*, **52**, 2409 (1974).
[75] J. Fetter, J. Nyitrai and K. Lempert, *Acta Chim.* (*Budapest*), **79**, 197 (1973); *Chem. Abs.*, **80**, 26594 (1974).
[76] A. C. Barefoot and F. A. Carroll, *Chem. Comm.*, **1974**, 357.
[77] T. Shono and I. Nishiguchi, *Tetrahedron*, **30**, 2173 (1974).
[78] R. M. Ottenbrite and J. W. Brockington, *J. Org. Chem.*, **39**, 2463 (1974).
[79a] S. Oae, K. Harada, K. Tsujihara and N. Furukawa, *Bull. Chem. Soc. Japan*, **46**, 3482 (1973).
[79b] See *Org. Reaction Mechs.*, **1970**, 150.
[80] G. G. Smith and E. Silber, *J. Org. Chem.*, **38**, 4172 (1973).
[81] B. L. Adams and P. Kovacic, *J. Am. Chem. Soc.*, **96**, 7014 (1974).
[82] T. N. Bell, R. Berkley, A. E. Platt and A. G. Sherwood, *Can. J. Chem.*, **52**, 3158 (1974).
[83] V. D. Sheludyakov, E. S. Rodionov and V. F. Mironov, *Zh. Obshch. Khim.*, **44**, 1044 (1974); *Chem. Abs.*, **81**, 49068 (1974).
[84] G. Glaros and N. H. Cromwell, *J. Org. Chem.*, **38**, 4226 (1973).
[85] Yu. P. Yampol'skii, *Izv. Akad. Nauk SSSR, Ser. Khim.*, **1974**, 564; *Chem. Abs.*, **81**, 12811 (1974).
[86] D. A. Knecht, *J. Am. Chem. Soc.*, **95**, 7933 (1973).
[87] G. Andrei, G. Şerban and C. Floriţa, *Rev. Roumaine Chim.*, **19**, 727 (1974).
[88] K. Schwetlick and F. Kretzschmar, *Z. Chem.*, **14**, 192 (1974).
[89] W. H. Saunders and A. F. Cockerill, *Mechanism of Elimination Reactions*, Wiley-Interscience, New York, 1973.
[90a] J. G. Buchan, *MTP* (*Med. Tech. Publ. Co.*) *Int. Rev. Sci.: Org. Chem. Ser. 1*, **7**, 31 (1973).
[90b] W. Nagata and S. Uyeo, *MTP* (*Med. Tech. Publ. Co.*) *Int. Rev. Sci.: Org. Chem. Ser. 1*, **8**, 40 (1973).
[91] W. T. Ford, *Accounts Chem. Res.*, **6**, 410 (1973).
[92] Y. Gounelle and D. Solgadi, *Bull. Soc. Chim. France*, **1973**, 3019.
[93a] D. Cook, R. E. J. Hutchinson, J. K. MacLeod and A. J. Parker, *J. Org. Chem.*, **39**, 534 (1974).
[93b] See *Org. Reaction Mechs.*, **1972**, 146.
[94] A. Ceccon and G. Catelani, *J. Organometal. Chem.*, **72**, 179 (1974).
[95] A. Ceccon, *J. Organometal. Chem.*, **72**, 189 (1974).
[96] G. Sturtz and M. Rio, *Bull. Soc. Chim. France*, **1974**, 2180.
[97] G. Sturtz and M. Rio, *Bull. Soc. Chim. France*, **1974**, 2187.
[98a] P. J. Smith and A. N. Bourns, *Can. J. Chem.*, **52**, 749 (1974).
[98b] A. Fry, *Chem. Soc. Rev.*, **1**, 163 (1972).
[99] G. H. Cooper, J. S. Bartlett, A. M. Farid, S. Jones, D. J. Mabbott, J. McKenna, J. M. McKenna and D. G. Orchard, *Chem. Comm.*, **1974**, 950.
[100] I. Angres and H. E. Zieger, *J. Org. Chem.*, **39**, 1013 (1974).
[101] Y. Tamura, J. Minamikawa, O. Nishikawa and M. Ikeda, *Chem. Pharm. Bull.*, **22**, 1202 (1974); *Chem. Abs.*, **81**, 48965 (1974).
[102] N. L. Allinger, G. L. Wang and B. B. Dewhurst, *J. Org. Chem.*, **39**, 1730 (1974).
[103] M. Bochenska and J. F. Biernat, *Rocz. Chem.*, **48**, 445 (1974); *Chem. Abs.*, **81**, 49164 (1974).
[104] M. Malherbe and G. Chatelus, *Compt. rend.*, *Ser. C*, **278**, 1205 (1974).
[105] R. C. Job and T. C. Bruice, *J. Am. Chem. Soc.*, **96**, 809 (1974).
[106] P. Beak and B. Siegel, *J. Am. Chem. Soc.*, **95**, 7920 (1973).
[107] W. D. Covey and D. L. Leussing, *J. Am. Chem. Soc.*, **96**, 3860 (1974).
[108] N. V. Raghavan and D. L. Leussing, *J. Am. Chem. Soc.*, **96**, 7147 (1974).
[109] T. C. Owen and P. R. Young, Jr., *FEBS Letters*, **43**, 308 (1974).
[110] J. K. Stille and M. T. Regan, *J. Am. Chem. Soc.*, **96**, 1508 (1974).
[111] J. K. Stille and R. W. Fries, *J. Am. Chem. Soc.*, **96**, 1514 (1974).
[112] J. K. Stille, F. Huang and M. T. Regan, *J. Am. Chem. Soc.*, **96**, 1518 (1974).
[113] A. Davison and N. Martinez, *J. Organometal. Chem.*, **74**, C17 (1974).
[114] V. A. Latenko, T. S. Boiko and A. A. Yasnikov, *Dopov. Akad. Nauk Ukr. RSR, Ser. B*, **35**, 636 (1973); *Chem. Abs.*, **79**, 145745 (1973).
[115] J. L. Wilczynski and H. W. Johnson, Jr., *J. Org. Chem.*, **39**, 1909 (1974).
[116] G. Kollenz and T. Kappe. *Annalen*, **1974**, 1634.
[117] A. V. Willi, M. H. Cho and W. Chen, *Z. Phys. Chem.* (*Frankfurt*), **91**, 193 (1974).
[118] R. A. Moss and P. E. Schueler, *J. Am. Chem. Soc.*, **96**, 5792 (1974).
[119] K. Humski, V. Sendijarević and V. J. Shiner, Jr., *J. Am. Chem. Soc.*, **95**, 7722 (1973).
[120] T. Uchida, S. Marui, Y. Miyagi and K. Maruyama, *Bull. Chem. Soc. Japan*, **47**, 1549 (1974).

[121] W. E. Parham, P. E. Olson, K. R. Reddy and K. B. Sloan, *J. Org. Chem.*, **39**, 172 (1974).
[122] P. Beak, J. T. Adams and J. A. Barron, *J. Am. Chem. Soc.*, **96**, 2494 (1974).
[123] J.-P. Bégué and D. Bonnet, *Tetrahedron*, **30**, 141 (1974).
[124] W. M. Horspool and B. J. Thomson, *Tetrahedron Letters*, **1974**, 3529.
[125] G. Dana, Sa Le Thi Thuan and J. Gharbi-Benarous, *Bull. Soc. Chim. France*, **1974**, 2089.
[126] J. A. Peters, J. Rog and N. van Bekkum, *Rec. Trav. chim.*, **93**, 243 (1974).
[127] F. K. Cartledge and J. P. Jones, *J. Organometal. Chem.*, **67**, 379 (1974).
[128] J. Taillades and A. Commeyras, *Tetrahedron*, **30**, 127 (1974).
[129] S. Sicsic and Z. Welvart, *Bull. Soc. Chim. France*, **1974**, 1477.
[130] P. Bassi and U. Tonellato, *J.C.S. Perkin II*, **1974**, 1283.
[131] A. Gorgues, *Compt. rend., Ser. C*, **278**, 287 (1974).
[132] F. Naso and L. Ronzini, *J.C.S. Perkin I*, **1974**, 340.
[133] J. Pichler and I. Borkovcová, *Coll. Czech. Chem. Comm.*, **39**, 779 (1974).
[134] M. Ballester, J. Castañer and A. Ibáñez, *Tetrahedron Letters*, **1974**, 2147.
[135] J. Buddrus and W. Kimpenhaus, *Chem. Ber.*, **107**, 2062 (1974).
[136] A. Aumelas, A. Casadevall, E. Casadevall and C. Lacgeau, *Tetrahedron*, **30**, 3897 (1974).
[137] L. N. Finyakin, N. V. Kafarov, M. F. Sorokin and L. G. Shode, *Izv. Vyssh. Ucheb. Zaved., Khim. Khim. Tekhnol.*, **17**, 774 (1974); *Chem. Abs.*, **81**, 62732 (1974).
[138] J. Prestien and H. Günther, *Angew. Chem. Int. Edn.*, **13**, 276 (1974).
[139] M. Ketola, *Tetrahedron*, **30**, 2717 (1974).
[140] G. Friedman, M. Lahav and G. M. J. Schmidt, *J.C.S. Perkin II*, **1974**, 428.
[141] V. Z. Sharf, L. Kh. Friedlin, V. I. Kheifets, V. V. Yakubenok and E. A. Shefer, *Neftekhimiya*, **13**, 832 (1973); *Chem. Abs.*, **80**, 81644 (1974).
[142] B. G. Pring, *Chem. Commun., Univ. Stockholm*, **1973**, 14; *Chem. Abs.*, **81**, 48892 (1974).
[143] P. D. Pistrova and G. D. Kharlampovich, *Khim. Tverd. Topl.*, **1974**, 108; *Chem. Abs.*, **80**, 107565 (1974).
[144] D. W. Grant and R. Shilton, *J.C.S. Perkin I*, **1974**, 135.
[145] M. M. Green, J. M. Moldowan and J. G. McGrew, II, *J. Org. Chem.*, **39**, 2166 (1974).
[146] R. Robbiani and J. Seibl, *Org. Mass Spectrom.*, **7**, 1153 (1973); *Chem. Abs.*, **80**, 26495 (1974).
[147] R. Robbiani and J. Seibl, *Helv. Chem. Acta*, **57**, 674 (1974).
[148] D. H. Williams and G. Hvistendahl, *J. Am. Chem. Soc.*, **96**, 6753 (1974).
[149] D. H. Williams and G. Hvistendahl, *J. Am. Chem. Soc.*, **96**, 6755 (1974).
[150] L. G. Fedenok, V. M. Berdnikov and M. S. Shvartsberg, *Zh. Org. Khim.*, **9**, 1781 (1973); *Chem. Abs.*, **79**, 145624 (1973).
[151] Y. Sasson and G. L. Rempel, *Can. J. Chem.*, **52**, 3825 (1974).
[152] R. C. Obach and W. Y. Lim, *Philipp. J. Sci.*, **100**, 251 (1971); *Chem. Abs.*, **80**, 81661 (1974).
[153] A. Ferro and Y.-R. Naves, *Helv. Chem. Acta*, **57**, 1141 (1974).
[154] W. T. Brady and G. A. Scherubel, *J. Am. Chem. Soc.*, **95**, 7447 (1973).
[155] J. C. Chalchat and F. Theron, *Bull. Soc. Chim. France*, **1974**, 953.
[156] J. F. King, S.-K. Sim and S.-K. L. Li, *Can. J. Chem.*, **51**, 3914 (1973).
[157] T. Kempe and T. Norin, *Acta Chem. Scand., B*, **28**, 609 (1974).
[158] T. Kempe and T. Norin, *Acta Chem. Scand., B*, **28**, 613 (1974).
[159a] F. G. Bordwell and E. Doomes, *J. Org. Chem.*, **39**, 2531 (1974).
[159b] See also *Org. Reaction Mech.*, **1973**, 377; **1971**, 144.
[160] J. C. Philips, M. Aregullin, M. Oku and A. Sierra, *Tetrahedron Letters*, **1974**, 4157.
[161] G. Scherowsky and J. Weiland, *Chem. Ber.*, **107**, 3155 (1974).
[162] S. B. Matin, J. Cymerman-Craig and R. P. K. Chan, *J. Org. Chem.*, **39**, 2285 (1974).
[163] J. Garapon and B. Sillion, *Tetrahedron Letters*, **1974**, 41.
[164] T. Shingaki, N. Torimoto, M. Inagaki and T. Nagai, *Chem. Letters (Tokyo)*, **1973**, 1243.
[165] S. V. Kessar, U. K. Nadir and M. Singh, *Indian J. Chem.*, **11**, 825 (1973); *Chem. Abs.*, **79**, 145633 (1973).
[166] A. P. Kroon and H. C. van der Plas, *Rec. Trav. chim.*, **92**, 1020 (1973); *Chem. Abs.*, **80**, 26476 (1974).
[167] H. Singh and S. Singh, *Indian J. Chem.*, **11**, 1055 (1973); *Chem. Abs.*, **80**, 81656 (1974).
[168] K.-D. Gundermann, W. Hönig, M. Berrada, H. Giesecke and H.-G. Paul, *Annalen*, **1974**, 809.
[169] S. S. Yufit and I. A. Esikova, *Dokl. Akad. Nauk SSSR*, **217**, 394 (1974); *Chem. Abs.*, **81**, 90733 (1974).
[170] P. B. D. de la Mare and R. D. Wilson, *Tetrahedron Letters*, **1974**, 3777.
[171] S. P. Hooper and J. S. Fine, *J. Organometal. Chem.*, **80**, C21 (1974).
[172] A. Maccoll and S. S. Nagra, *J.C.S. Perkin II*, **1974**, 1099.

[173] R. Tenfik, F. M. Fouad and P. G. Farrell, *J.C.S. Perkin II*, **1974**, 31.
[174] M. J. H. Keijbets and W. Pilnik, *Carbohydrate Res.*, **33**, 359 (1974).
[175] F. G. Weber and C. Bandlow, *Z. Chem.*, **13**, 467 (1973).
[176] D. Matthies, *Arch. Pharmazie*, **307**, 801 (1974).
[177] J. Knabe and A. Ecker, *Archiv Pharmazie*, **307**, 727 (1974).
[178] C. A. Bunton M. J. Minch and B. B. Wolfe, *J. Am. Chem. Soc.*, **96**, 3267 (1974).
[179] F. A. Sedor, D. G. Jacobson and E. G. Sander, *Bioorg. Chem.*, **3**, 221 (1974).
[180] W. H. Richardson, F. C. Montgomery, M. B. Yelvington and H. E. O'Neal, *J. Am. Chem. Soc.*, **96**, 7525 (1974).
[181] J. Evans and J. R. Norton, *J. Am. Chem. Soc.*, **96**, 7577 (1974).
[182] S. Arimatsu, R. Yamaguchi and M. Kawanisi, *Bull. Chem. Soc. Japan*, **47**, 1693 (1974).
[183] L. F. Blackwell, P. D. Buckley and K. W. Jolley, *Austral. J. Chem.*, **27**, 2283 (1974).

CHAPTER 12-I

Addition Reactions. I. Polar Addition

A. C. Knipe

Chemistry Department, The New University of Ulster

Electrophilic Additions

New addition reactions of carbon—carbon multiple bonds,[1] and catalysis of olefin and CO insertion reactions,[2] have been reviewed.

The effect of a cyclopropyl group on the rate of electrophilic addition to alkenes has been investigated.[3] Thus, it has been found that, relative to a phenyl group, a cyclopropyl substituent may be ca. 10^3-fold better able to accelerate any such reaction that leads to development of a "relatively open" positive charge on the adjacent carbon. The two substituents have comparable influence when their resonance stabilization of the charge is precluded either by bridged-ion formation or by presence of an even more strongly stabilizing substituent. The magnitude of the effect therefore provides a useful indication of transition-state structure. Reactions of vinylcyclopropene that have thus been assigned to the former category include bromination (Br_2/AcOH) and hydration (H_2O/H_2SO_4), which are in contrast to its reaction with sulphenyl halide and to the hydration of α-cyclopropylvinyl phosphate in aqueous HCl. This criterion correlates well with ρ-values (−4.7, −3.6, −2.4 and −2.1, respectively) for the corresponding reactions of styrene.

NMR techniques have been used[4] to distinguish between (*a*) σ-bonded alkenium or halonium complexes and (*b*) molecularly bound π-complexes (containing an oriented two-electron, three-centre bond between the double bond and the electrophile) formed upon reaction of a variety of electrophiles with ethylene, 2,3-dimethylbut-2-ene or adamantylideneadamantane (**1**). The π-complexes (alkonium ions) (e.g. **2a–c**) are formed preferentially by (**1**), while β-substituted carbonium ions (alkenium ions) or onium ions containing a three-membered ring are generally obtained upon reaction of the less sterically hindered alkenes. Alkene halonium ion complexes have been prepared directly for the first time (by use of "positive halogen" complexes such as $ICN^{\delta+}$–$SbF_5^{\delta-}$ in SO_2), and the first direct observation of a carbonium ion (**3**) formed by addition of hydrogen halide to an alkene was recorded during reaction of (**1**) with HF–SO_2. Of particular interest is (**2a**) which can be obtained by reaction of (**1**) with Br_2, NBS (*N*-bromosuccinimide) or cyanogen bromide in SO_2–HSO_3F or, alternatively, by reaction with Br_2 in CCl_4. Spectra of (**2**) correspond to a symmetrical structure, and the bridging has been attributed to π-bonding since σ-bond formation would be sterically hindered. This view is supported by the atypical behaviour of (**2a**) which is readily converted into (**1**) upon reaction with cyclohexene or other nucleophiles.

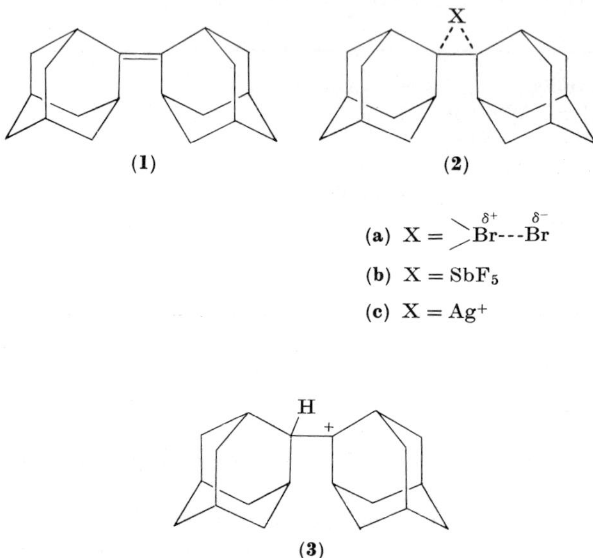

By CNDO/2 study of the ethylene–chlorine reaction it has been estimated that addition by direct attack of Cl^- on the ethylene–chlorine charge-transfer complex is energetically more favourable than a route involving prior heterolytic dissociation of the complex.[5] Electron-diffraction patterns determined during gas-phase reaction of ethylene with atomic bromine require that the bromine atom be symmetrically bound to each of the two carbon atoms in the intermediate.[6]

Linear correlations between the ionization potentials of alkenes and their reactivity ($\log k$) in fourteen addition reactions (and in eight reactions involving formation of atmospheric pollutants) have been reported.[7]

Halogen and Related Addition

Earlier, reactant-like, transition states of π-complex nature have been proposed to account for the smaller range of relative bromination rates found for reactions of alkenes and alkynes in 1,1,2-trichlorotrifluoroethane than in more polar solvents.[8] The exclusive *trans*-stereochemistry is that expected of Br^- attack on the three-membered ring bromonium ion (σ-complex) formed by participation of the non-bonded electron pairs of bromine in the molecularly bonded π-complex. It has been found that the effect of structure on rate of bromination of alkenes in MeOH–H_2O is slightly less than for the 10^5-fold slower reactions of alkynes, although the solvent-dependence ($m = 0.8$–0.9)

suggests that identical bromonium ion mechanisms apply;[9] phenylacetylene is atypical in that it reacts only 10^2-fold slower than styrene, probably via an intermediate vinyl cation. The conversion of phenylacetylene into PhC≡CI upon iodination in DMF ($E_a = 10.3$ kcal mol^{-1}) and to PhCI=CHI in MeOH ($E_a = 7.7$ kcal mol^{-1}), respectively, has also been studied.[10]

Halogenation of phenylallene (**4**) takes place primarily on the internal double bond to form a cyclic halonium ion (**5**). Formation of 1,2- or 2,3-addition products (**6** + **7**) (equation 1) is dependent upon the lifetime of the intermediate and upon the reactivity of the nucleophile.[11] In MeOH the fraction of 1,2-adduct obtained varies in the order I > Br > Cl and is reduced by electron-donating aryl substituents and by isomerization to the 1,3-adduct at temperatures above 0°.

$$C_6H_5CH{=}C{=}CH_2\ (\mathbf{4}) \xrightarrow{X^+} C_6H_5CH\text{—}C{=}CH_2\ (\text{bridged by } X^+)\ (\mathbf{5}) \underset{}{\overset{Nu^-}{\rightleftharpoons}} C_6H_5CH(Nu)\text{—}C(X){=}CH_2\ (\mathbf{6})$$

$$(\mathbf{5}) \rightleftharpoons C_6H_5CH\text{---}C(X)\text{---}CH_2\ (\text{allyl cation, }+) \overset{Nu^-}{\rightleftharpoons} C_6H_5CH{=}C(X)CH_2Nu\ (\mathbf{7}); \quad (\mathbf{6}) \xrightarrow{-Nu^-} \text{allyl cation} \qquad (1)$$

An intermediate delocalized allyl radical is believed to be formed during selective chlorination of allenes by the dichloride of iodobenzene.[12]

Yields of the kinetic products of polar addition of bromine to cyclopentadiene, cyclohexa-1,3-diene and three hexa-2,4-dienes (*E,E*, *Z,Z* and *E,Z*), in solvents of widely different polarity, have been estimated.[13] The initially formed bromonium bromide ion pair is believed to form predominantly *cis*-1,4-adduct by an S_N2' reaction of its anion. Alternatively it may undergo reorientation to an ion pair from which *trans*-1,2- and 1,4-adducts are competitively obtained. The non-stereospecific 1,2-addition to hexa-2,4-diene suggests that the charge of the vinylic bromonium ion is extensively dispersed in this case. In contrast with this[13] and with the conclusions from earlier studies, it has been claimed[14] that cyclopentadiene reacts with chlorine in CH_2Cl_2, CCl_4, and C_5H_{12} to give the *cis*-3,4-dichloro-adduct in 38, 27 and 13% yield, respectively, and that the ratios of 1,2- to 1,4-addition are 1:0.3, 1:1 and 1:1.3.

Kinetically controlled dichlorination of buta-1,3-diene by antimony pentachloride in chlorinated hydrocarbon solvents is stereoselective towards formation of *cis*-1,4-dichlorobut-2-ene[15] (23–41% yield; cf. < 1% obtained upon reaction with Cl_2). No comparable reaction occurs with *trans-trans*-hexa-2,4-diene, and a mechanism of symmetry-allowed antarafacial transfer of two chlorine atoms of the trigonal bipyramidal $SbCl_5$ to the cisoid diene has therefore been proposed.

Several investigations of cyclohexene halogenation have been reported.[16–21] Distribution of the four *trans*-bromide-chlorides (**11**–**14**) obtained upon reaction of 3-*tert*-butylcyclohexene (**8**) with BrCl and with monopyridine bromine(I) chloride is consistent with an ionic two-step mechanism.[16] Only with the former reagent is formation of (**10**) stereoselectively favoured (64–79%) relative to that of (**9**). Subsequent nucleophilic attack on both intermediates occurs preferentially at C-1, to form (**11**) and (**14**), respectively. It has been suggested that electrophilic addition constitutes the rate-determining step for reaction with BrCl. In the case of PyBrCl rate-determining nucleophilic attack on reversibly formed bromonium ion may account for the change in product distribution obtained.

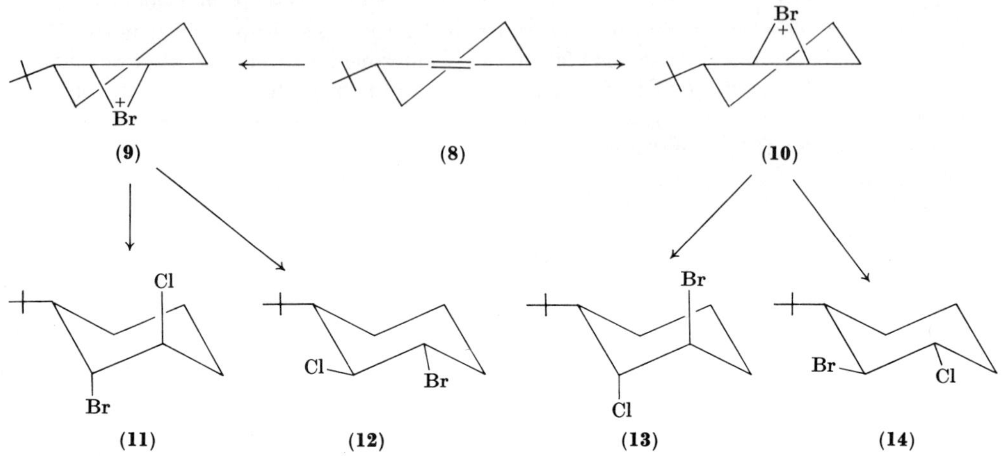

It has been argued (on the basis of the kinetic influence of 1- and 2-alkyl substituents) that transition states for bromination of cyclopentene and cyclohexene in MeOH are bromonium-ion-like, whereas for reaction of cyclo-octene the charge distribution resembles that of an α-bromo-carbonium ion;[17] the order of reactivity ($C_5 > C_6 \gg C_8$-*cis*) does not therefore reflect the ring-strain energy ($C_8 > C_5 > C_6$). An Ad_EC_1 mechanism, subject to primary and secondary transannular steric effects, has been invoked to account for rates of bromination of substituted cyclohexenes.[18] Iodination of cyclohexene has also been studied.[20, 21]

Bromination of *trans*-, *endo-cis*- and *exo-cis*-5,6-dichloronorbornene in acetic acid gives, stereospecifically, the corresponding *trans*-, *exo-cis*- and *trans*-adducts, respectively.[22] The amount of *exo-cis*-addition to the *trans*-dichloride increases as the solvent polarity is decreased. This has been explained in terms of a mechanism whereby, in poor solvents, the bromonium ion intermediate is disfavoured relative to a bromonium– or carbonium–tribromide ion pair. The ion pairs form the *cis*-adduct by intramolecular collapse but are apparently susceptible to attack by added bromide ion which thereby enhances *trans*-addition. The results complement those of an earlier study of dicyanonorbornenes. Additions of halogen to 2,3-homotropone,[23] Dewar-benzene,[24] and 18-norandrost-13-enes[25] have also been reported.

Kinetic results for bromination of stilbenes ($XC_6H_4C_xH{=}C_yHC_6H_4Y$) in MeOH have been interpreted with reference to alternative carbonium- and bromonium-ion-like transition states having different charge distributions.[26] Bromination occurs predominantly via carbonium ions C_x^+ and C_y^+ (without significant participation by the bromine atom) when either aryl substituent is electron-donating. Bromination rates calculated according to the corresponding LFE equations:

$$\log(k_x/k_0) = \rho_\alpha \sigma_x^+ + \rho_\beta \sigma_y \quad \text{and} \quad \log(k_y/k_0) = \rho_\alpha \sigma_y^+ + \rho_\beta \sigma_x$$

(where ρ_α and ρ_β are -5.07 and -1.41, respectively) are, however, much less than those observed when both X and Y are strongly deactivating. The apparent attenuation of substituent effects has been ascribed to emergence of a bromonium ion mechanism, whereby the charge is delocalized from C_x and C_y on to the bromine atom of the symmetrical intermediate; this mechanism is governed by the expression

$$\log k = -1.0(\sigma_x + \sigma_y) + 1.3.$$

Hammett studies of bromination of 3-arylpropenes in chloroform[27,28] and in acetic acid[29] have also been reported. It has been argued[27,28] that 2-aryl-1,3-dibromopropane is formed (particularly when Ar bears an electron-donating substituent) via a phenonium ion which is the product of an anchimerically assisted bromination rather than a bromonium ion rearrangement. Formation of di- and tetra-hydrofurans upon iodination of hex-5-en-2-one and 2-allylcyclohexanone in aqueous KI/I_2 has been attributed[30] to participation by carbonyl-oxygen. Kinetics of iodination of $CH_2{=}CHCH_2CMe(OH)$-CH_2OH, $CH{=}CHCMeClCH_2OH$, 4,5-epoxy-4-methylpent-1-ene and 6,7-epoxy-6-methylhept-1-ene have also been interpreted.[31]

Quantitative *trans*-bromination of alkenes is apparently[32] initiated by abstraction of a brominium cation from an orange complex of $MgBr_2$ and benzoyl peroxide in THF; the dependence on aryl-substitution and on alkene structure has been investigated. The use of MeOH as a solvent for bromine in kinetic studies has also been discussed.[33]

It has been demonstrated that antimony pentachloride, in admixture with an equimolar amount of bromine or LiBr, forms a chlorobrominating agent that is superior to those based on Lewis acids[34] (e.g. $FeCl_3$, $SnCl_4$ and $TiCl_4$). Antimony pentachloride has also been used for direct chlorination of alkenes in chlorinated hydrocarbons.[35] A remarkable feature of the reaction with simple alkenes is the kinetically controlled predominance of *cis*-addition; this is particularly favoured by increase in temperature or solvent polarity; the multiple products of reaction with cyclopentene and norbornene do not, however, include *cis*-1,2-adducts and have been interpreted with reference to reactions of chlorocyclopentyl and chloronorbornyl cations.

Bromine in $CHCl_3$–CH_2Cl_2 is believed to react with phenylcyclopropane by electrophilic attack on the cyclopropane ring to give a benzylic carbonium ion.[36] This may either trap Br^- or react by an elimination–addition sequence, affording 1,3-dibromo- and 1,2,3-tribromo-1-phenylpropane, respectively. 1,3-Dibromo-1,3-diphenylpropane is similarly obtained upon reaction of *trans*-1,2-diphenylcyclopropane, and it has been argued that the *dl*:*meso* ratio of unity is best attributed to the intermediacy of a cation that is not stabilized by 1,3-bromine participation.

Markovnikov additions of BrCl to alkenes have been performed in CCl_4 containing ethylene oxide and cyclohexene oxide.[37] The products of bromination of methyl *trans*-cinnamate in hydroxylic solvents have been tabulated.[38] No interpretation has been offered for the contrasting results for addition of bromine to dibenzo[*b,f*]thiepin (in CH_2Cl_2) and its 5,5-dioxide (in CCl_4); under conditions of kinetic control the thiepin gives *trans*-addition, whilst the dioxide gives *cis*-addition; *trans*-addition to the dioxide occurs in acetic acid.[39]

Application of XeF_2 in CH_2Cl_2 as a halogenating agent for alkenes, in the presence of HF or CF_3COOH, has been extended to *cis*- and *trans*-stilbenes.[40] Vicinal difluorides are the only products of each HF-catalysed reaction, whereas CF_3COOH promotes competitive trifluoroacetolysis (ca. 50%). The reactions are non-stereospecific, with some preference for *anti*-addition to *trans*-stilbene, and are therefore in contrast with the general preference for *syn*-addition of molecular fluorine. It has also been found[41] that, upon reaction with XeF_2 in CH_2Cl_2 containing HF, acetylenes $PhC{\equiv}CR$ (R = Ph, Me or *n*-Pr) undergo ready tetrafluorination to $PhCF_2CF_2R$. The intermediate difluoroalkene, which could not be isolated, is also more reactive than the parent acetylene towards addition of molecular fluorine. 9,9,10,10-Tetrafluoro-9,10-dihydrophenanthrene has likewise been obtained from 9,10-difluorophenanthrene.

The mechanism of addition of iodonium nitrate to alkenes, in the presence of pyridine, has been further investigated;[42] *trans*-stereospecific and regiospecific formation of iodoalkyl nitrate esters and iodoalkylpyridinium nitrates occurs upon reaction with a series of *Z–E* pairs of alkenes and cannot be attributed to restricted rotation of a carbonium ion intermediate. The yield of the latter compound decreases with steric hindrance to addition. The proposed bridged iodonium ion intermediate has (unlike that formed in other pseudo-halogen additions) sufficient carbonium ion character to facilitate attack of the nucleophile at the more stable centre, despite the greater steric hindrance encountered.

The chlorination of α,β-unsaturated ketones p-$YC_6H_4CH{=}CHCOX$ (Y = H, MeO or NO_2; X = Me or Ph) in MeOH has been found to result in predominant *trans*-addition of the elements of methyl hypochlorite with little formation of dichlorides.[43] The results are consistent with formation of an open or partially bridged carbocation which undergoes only limited rotation about the C—C bond before capture by solvent. Formation of intermediate acetals may account for HCl-catalysis (detected for Y = H; X = Me). The expected dichlorides are obtained upon reaction with chlorine in CF_3COOH. A radical mechanism has been proposed[44] to account for the formation of vicinal dihalides upon reaction of alkyl halides with $SbCl_5$, and the stereochemistry of bromination of vinyl sulphones has been investigated.[45]

Addition of Hydrogen Halides

The relative reactivities of alkynes (**15**)–(**20**), towards HCl in acetic acid, by competing Ad_E2 (via carbonium–chloride ion pairs) and $Ad3$ (reverse $E2$) mechanisms are indicated, respectively, by the figures above and below the position of proton addition.[46] The ion

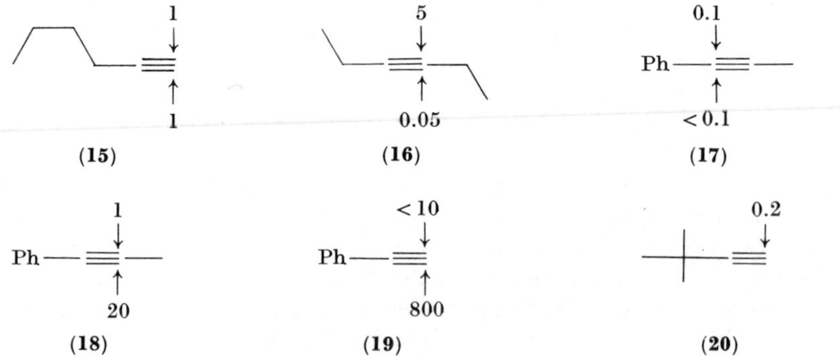

pair formed in the slow protonation step of the Ad_E2 mechanism may collapse to give an acetate (which rapidly hydrolyses to form a ketone) or a chloride. The effect of added Cl^- (0.2 M) is to decrease the chloride: ketone product ratio for Ad_E2 reaction of (**19**), from 12:1 to 7:1, and this has been attributed to a medium effect on ion-pair stability. The converse trend applies to $Ad3$ reactions. *syn*-Stereochemistry, attributed[46] to the Ad_E2 mechanism, has been found to predominate upon addition of HCl to 2-$^2H_1$-(**19**) in CH_2Cl_2, AcOH, CH_3NO_2 and sulpholane.[47]

The hydrogen dichloride anion has been implicated in *trans*-addition of HCl to α-acetylenic tertiary amines to form α- and β-chloro-α-amino-alkenes.[48] Hydrochlorinations of acetylene[49] and chloroacetylene[50] catalysed by HCl–CuCl, have been studied. The ethynyl ketone (**21**) forms (**22**) and (**23**) upon reaction with HCl in benzene at high and ambient temperatures, respectively.[51]

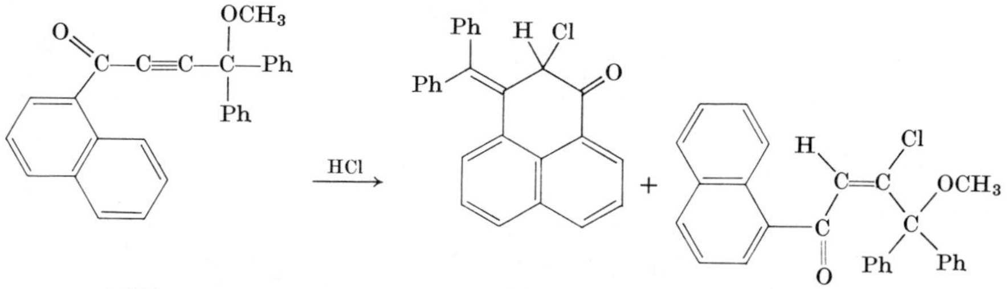

(**21**) (**22**) (**23**)

Acetyl iodide is rapidly formed upon addition of HI to ketene (at ca. 500°) and is subsequently converted into methane and acetaldehyde by a homolytic mechanism.[52] Propyne, 2-chloropropane and 2,2-dichloropropane, which are obtained[53] upon gas-phase hydrochlorination of allene, are the products expected of a vinyl cation intermediate. Equilibria between 2-chloropropene and 1:1 complexes of HCl with allene and propyne have been implicated in the mechanism. Alkene–HCl complex formation may also facilitate hydrochlorination of propene and isobutene since at room temperature the latter reaction has been found to occur ca. 10^9-fold faster than predicted by consideration of the equilibrium established at higher temperature.[54]

Products of hydrochlorination of cyclohexene and isomeric methylcyclohexenes in concentrated hydrochloric acid have been characterized.[55] The rates and products of reactions of hexene and cyclohexene with HBr–AcOH are consistent with *Ad*3 mechanisms whereby AcOH and HBr additions are *anti*-stereospecific and subject to catalysis by HBr and Br^-, respectively.[56] Both *syn*- and *anti*-Ad_E3 mechanisms have been proposed to account for electrophilic additions of *cis*- and *trans*-but-2-ene, hex-3-ene, and cyclopentene under the same conditions.[57] The asymmetrically bridged cation route is believed[58] to account for formation of 52% of 2-*endo*- and 48% of 2-*exo*-methoxy-5-*exo*-chloronorbornane upon hydrochlorination of 2-*exo*-methoxy-5-norbornene in CH_2Cl_2.

It has been necessary to propose[59] a common intimate ion-pair intermediate in order to correlate polarimetric and titrimetric rate constants and products of solvolysis of optically active 1-phenylethyl chloride in AcOH, with the distribution of products formed upon hydrochlorination of styrene.[60] Relative rates of hydrochlorination of *cis*- and *trans*-1-arylpropenes in AcOH have also been determined.[61]

Hydration and Related Reactions of Alkenes

It has been established that many esterifications of cycloalkenes by carboxylic acids involve intermediate carbocations which are formed by general-acid-catalysed slow proton-transfer to the alkene. Such reactions have been catalysed by Lewis acids,[62,63] lithium perchlorate[64] and methanesulphonates[65,66] and in each case it has been found that the rate is governed by an expression of the form: $\log k_{\text{exp}} = -\alpha H_0 + \beta$. Carboxylic acid addition to butoxyethene has also been studied.[67]

Reaction of (**24**) in 80% methanol–water has been found[68] to proceed by an A-S_E2 mechanism, whereby rate-limiting general-acid-catalysed protonation is followed by rapid attack of methanol on the intermediate carbonium ion (**25**) to give (**26**). The system is unique in that the three species (**24**)–(**26**) coexist at measurable concentrations and it has therefore been possible to determine the four rate constants for their interconversion. Rate constants for protonation of (**24**) by lyonium ion and for deprotonation

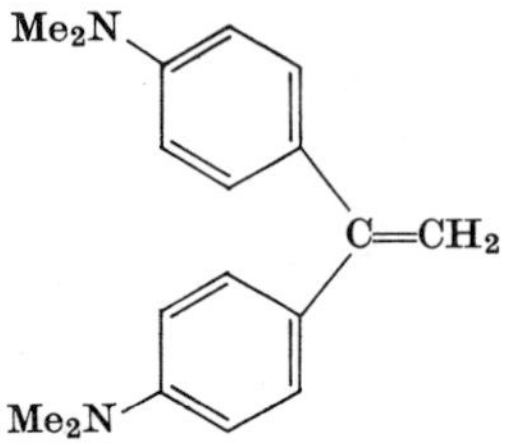

(**24**) 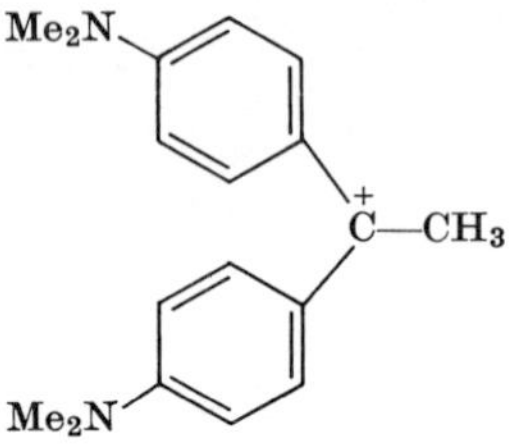 (**25**) 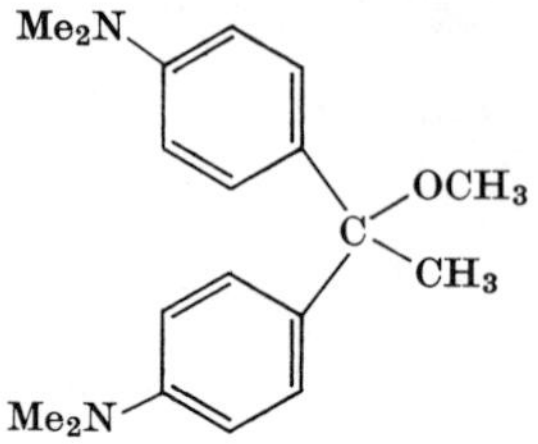 (**26**)

of (**25**) by solvent are lower than expected (23 M^{-1} sec^{-1} and 4.8×10^{-2} sec^{-1} at 25°, respectively), as is the rate ratio, $k^{MeO^-}/^{MeOH}$ = ca. 2×10^4, determined for nucleophilic attack on (**25**); the study was complicated by involvement of the diethylamino-groups in acid–base equilibria.

The catalytic reactivity and selectivity of rhodium carbonyls in olefin hydroformylation reactions has been reviewed[69] and the mechanisms of hydroformylation of alkynes[70] and styrenes[71] have been investigated; direct hydroformylation of the triple bond has been substantiated by isolation of the intermediate α,β-unsaturated aldehyde, (*E*)-2-methylbut-2-enal, during conversion of but-2-yne into 2-methylbutanal on a rhodium catalyst.[70] It has been found[71] that the $[\alpha]_D$ value and configuration of the product (hydratropaldehyde) of hydroformylation of styrene with phosphine–rhodium catalysts is markedly dependent upon the phosphine:rhodium ratio, the reaction temperature (20–110°), and the partial pressure of CO and H_2 (10–100 atm.). The results have been explained on the basis of an equilibrium between active species [(**27**), (**28**) and (**29**)] of

$$\underset{(\mathbf{27})}{HRh(CO)_3} \underset{CO,\ -P^*}{\overset{P^*,\ -CO}{\rightleftarrows}} \underset{(\mathbf{28})}{HRh(CO)_2P^*} \underset{CO,\ -P^*}{\overset{P^*,\ -CO}{\rightleftarrows}} \underset{(\mathbf{29})}{HRh(CO)P^*_2}$$

comparable reactivity; where P* = (*R*)-benzylmethylphenylphosphine, formation of racemic, *S*-, and *R*-aldehyde has been attributed to predominant catalysis by (**27**), (**28**) and (**29**), respectively. Styrene (in contrast to simple olefins) has a marked tendency to be formylated on the inner carbon and this has reasonably been ascribed to the effect of π-(benzyl)rhodium complexation.

The marked influence of *p*-methoxybenzonitrile on the reaction rate and product distribution observed upon hydroformylation of cyclopentene in the presence of $HCo(CO)_4$ has been related to the concentration of $HCo(CO)_3$ available.[72] With an excess of catalyst, under nitrogen, the effect of the nitrile is to increase the yield of cyclopentane (a by-product) from 3% to 52%.

Mechanisms of reactions between $HCo(CO)_4$ and hex-1-ene or hex-2-ene[73,74] and of related hydroformylations[75] have also been discussed.

Labelling experiments[76] have established that $HOCH_2CH_2CMePhOH$ is an intermediate of the Prins reaction of CH_2O with $PhCMe{=}CH_2$ in aqueous acid, to form 4-methyl-4-phenyl-1,3-dioxan. It has also been found that the second-order rate constants for Prins reaction of $RCH{=}CH_2$ (R = Me, Et, Pr, Bu or Pr^i) and $RCMe{=}CH_2$ (R = Me, Et or Pr) correlate with σ^* constants and steric-effect parameters.[77] The rate-determining step apparently involves isomerization of the CH_2O–olefin π-complex to the hydroxylated carbonium ion.[77]

Kinetics of hydrosilylation of non-1-ene and 2-methylpent-1-ene by $MeSiHCl_2$, Et_2SiHCl and Et_2SiHCl (with H_2PtCl_6 as catalyst) have been interpreted with reference to complexes (containing olefin, platinum and silane) formed during an induction period.[78] The effects of ligands on binuclear platinum–olefin catalysts, which participate in the mechanism of hydrosilylation of styrene by $MeSiHCl_2$, have also been investigated kinetically.[79,80]

Asymmetric hydrogenation of α-ethylstyrene and methyl atropate has been catalytically promoted by 2,3-dihydroxy-2,3-*O*-isopropylidene-1,4-bis(diphenylphosphino)butanerhodiumRh(I) in benzene solution.[81] A much lower catalytic efficiency was achieved when a suspension of an analogous rhodium complex, supported on a Merrifield resin, was used. Asymmetric hydrosilylation of ketones was, however, promoted effectively by either soluble or insoluble catalyst, the optical yield (up to 58%) being greater for dihydro- than for monohydro-silanes.

A convenient synthetic procedure has now been developed for the *anti*-Markovnikov esterification of alkenes whereby monosubstituted olefins can be directly converted into the corresponding primary esters, under mild conditions, by a hydroboration–mercuration–iodination sequence;[82] primary alkylmercuric acetates can be prepared from terminal alkenes, by hydroboration and subsequent mercuration, and are known to form the corresponding *anti*-Markovnikov hydrobromination products upon bromination *in situ* but, in contrast, the primary alkyl acetate is obtained in high yield upon reaction with iodine:

$$\begin{aligned}
3RCH{=}CH_2 + BH_3 &\rightarrow (RCH_2CH_2)_3B \\
(RCH_2CH_2)_3B + 3Hg(OOCR')_2 &\rightarrow 3RCH_2CH_2HgOOCR' + B(OOCR')_3 \\
RCH_2CH_2HgOOCR' + I_2 &\rightarrow RCH_2CH_2I + IHgOOCR' \\
IHgOOCR' + B(OOCR')_3 &\rightarrow IHgB(OOCR')_4 \\
IHgB(OOCR')_4 + RCH_2CH_2I &\rightarrow RCH_2CH_2OOCR' + HgI_2 + B(OOCR')_3
\end{aligned}$$

The procedure is limited to synthesis of esters derived from weaker carboxylic acids and monosubstituted olefins. Hydroboration of α,β-unsaturated epoxides has also been studied.[83]

A sequence of acetoxymercuration, halodemercuration and displacement is believed[84] to account for formation of cyclohexane-1,2-diol derivatives upon reaction of cyclohexene with $Hg(OOCCH_3)_2/I_2$. It has been established[85] that dimethylaminomercuration of but-2-enes proceeds by *trans*-addition (> 97%).

Evidence has been presented in support of an earlier suggestion that reaction of HgO/I_2 with olefins in aprotic solvents proceeds through electrophilic attack by iodine oxide.[86] The reaction products (rearranged carbonyl compounds, vinyl iodides and 2,2′-di-iodo-ethers) are derived from iodohypoiodites and their iodonium ion precursors.

The role of mercurinium ion intermediates in oxymercuration of alkenes has been challenged.[87] *cis-exo*-Mercurial adducts are the exclusive ($\geqslant 98\%$) products of reaction of norbornene and its 1-methyl, 2-methyl, 7,7-dimethyl and 1,7,7-trimethyl derivatives with mercuric acetate in aqueous THF. It has therefore been concluded that the reaction proceeds through a two-stage addition process that does not involve a cyclic mercurinium ion intermediate [e.g. (**30**), which would be sterically hindered by a *syn*-methyl group]. Nor is there evidence to suggest that *exo*-attack of the nucleophile can be attributed to reaction with a σ-bridged intermediate [e.g. (**31**) or (**32**)] since no rearranged by-product is obtained and 1-methylnorbornene gives essentially equal amounts of 1- and 4-methyl-*exo*-norbornanol (upon *in situ* demercuration of the corresponding oxymercurials by $NaBH_4$) even though (**31**) would be much more stable than (**32**). It was therefore suggested that electrophilic attack by XHg^+ forms a mercury-substituted carbonium ion (in which much of the charge should be retained on mercury) which reacts rapidly with solvent from the *trans*-direction, if that is unhindered, but from the *cis*-direction in norbornyl derivatives where *endo*-approach is strongly hindered. The degree of positive charge developed on carbon is apparently sufficient to induce Markovnikov addition but insufficient to cause rapid carbonium ion rearrangements.

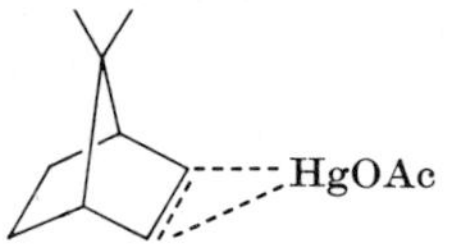

HgOAc
(**30**)

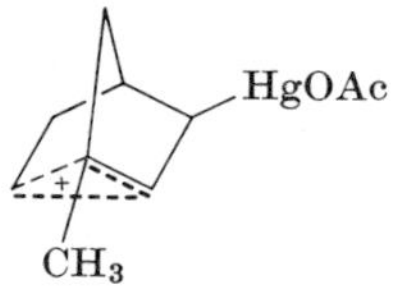

HgOAc
CH_3
(**31**)

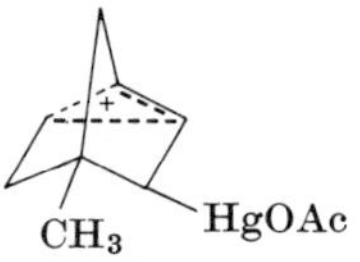

CH_3 HgOAc
(**32**)

Arrhenius parameters have been determined for the addition of unassociated Et_3Al to styrene and 2-methylhept-1-ene.[88] High values of A and E for sterically encumbered alkenes have been attributed to rate-determining formation of an initial π-complex and are in contrast with the low values obtained (e.g. for reactions of *n*-alk-1-enes and styrene) when subsequent four-centre decomposition of the complex apparently becomes rate-limiting. There was no evidence for formation of an intermediate aromatic π-complex during reaction of styrene, which differs from other alkenes in that ethylation occurs at the least hindered carbon.

To account for the kinetics of insertion of methyl acrylate (Ol) into the metal—H bond of *trans*-bis(triethylphosphine)hydridoplatinum(II) nitrate ($HPtL_2X$), an intermediate four-co-ordinated hydrido-olefinic complex ($HPtL_2Ol^+$) has been proposed.[89]

Miscellaneous Additions to Alkynes and Dienes

Evidence has been presented in support of concerted formation of two rings of a steroid system upon Lewis acid-catalysed cationic cyclization of a polyolefin.[90,91] Novel caged compounds have been obtained by electrophile-induced transannular cross-bonding of π-bonds of tricyclo[4.2.2.$0^{2,5}$]deca-3,7-diene derivatives,[92a] and the formation of benzene by trimeric cyclization of acetylene over Ni(0) complexes has been studied.[92b]

Additions of alkyl,[93,94] silyl,[95,96] acyl[97] and sulphenyl chlorides[98–101] to alkynes have been investigated. 1:1-Addition products (**37a–c**) of the *E*-configuration are obtained by predominant *anti*-addition of (**33a–c**) to phenylacetylene (**34**), catalysed by Lewis acids in CH_2Cl_2. The same products (**37a–c**) are obtained by predominant *syn*-addition of HCl to (**35a–c**). It has therefore been suggested that in each case a linear vinyl cation (**36**), preferentially attacked by Cl^- at the least hindered side, is a common intermediate of the two mechanisms. 1:1-Addition products of reaction of (**33a**) and (**33b**) with diphenylacetylene are consistent with this interpretation.[94] Indene derivatives are also obtained from (**33b**) and are the predominant products of reaction of (**33c**).

$$RCl + HC{\equiv}CPh \; (\mathbf{33}) \; (\mathbf{34}), \quad HCl + RC{\equiv}CPh \; (\mathbf{35}) \longrightarrow \underset{(\mathbf{36})}{RHC{=}\overset{+}{C}{-}Ph \;\; Cl^-} \longrightarrow \underset{(\mathbf{37})}{RHC{=}C(Ph)Cl}$$

R = (**a**) Bu^t; (**b**) $PhCH_2$; (**c**) Ph_2CH

Stereospecific *cis*-addition of dichlorosilane to alkynes has been catalysed by chloroplatinic acid, by Pt on carbon, and by benzoyl peroxide.[95] Internal and external adducts (*trans*-alkenyldichlorosilanes) are obtained (ca. 20:80) from terminal alkynes. The latter adducts are more reactive than SiH_2Cl_2 and form novel bis-(*trans*-dialkenyl)-dichlorosilanes upon reaction with a further molecule of alkyne. Vinyltrichlorosilane has also been prepared by platinum-catalysed addition of trichlorosilane to acetylene.[96]

Equilibration of X^- with $Ni(CO)_4$ to form $[Ni(CO)_3X^-]$ has been implicated in a halide-ion-catalysed ($I^- > Br^- > Cl^-$) cyclization of RCOCl (R = Ph or Me) with acetylene in the presence of water or acetone, to give β,γ-unsaturated butenolides.[97]

For reaction of 2,4-dinitrobenzenesulphenyl chloride with 1-phenylacetylene in AcOH, β-deuteriation gives rise to a small inverse isotope effect (0.95) while an elevenfold rate increase obtains upon β-methylation.[98] These trends are consistent with the formation of an intermediate episulphonium ion. For the β-methyl-derivative, however, the reaction rate is relatively insensitive to aryl substitution ($\rho = -1.46$), in contrast to the corresponding reactions of 1-arylpropenes ($\rho = -2.64$) and styrenes ($\rho = -2.41$); it has therefore been argued that addition is to the acetylenic π-bond which is orthogonal to the $p\pi$-orbitals of the aromatic ring. Results[99] for both orientation and kinetic order of addition of PhSCl to ethyl but-3-ynoate suggest that reaction is by an Ad_E2 process, whereas competitive Ad_E3 and Ad_E4 mechanisms account for reactions of ICl.

Disulphur dichloride forms vinyl sulphides by Markovnikov and anti-Markovnikov *trans*-addition to $PhC{\equiv}CR$ in DMF where R exerts a –I or +I effect, respectively.[100] Divinyl disulphides and benzothiophenes may also be obtained.[101]

In a basic aqueous medium[102] the addition of H_2O to $PhC{\equiv}CNR_2$ is general-acid catalysed, with a Brønsted $\alpha = 0.74$ and solvent deuterium isotope effect of ca. 4. Proton-transfer occurs much faster to this ynamine than to corresponding alkynyl ethers or enamines and is essentially an ionization process (when water acts as the general acid) as evidenced by $\Delta S^{+} = -39$ eu. Kinetics of hydration have been determined for $HC{\equiv}CCH{=}CHOEt$ (catalysed[103] by H_2SO_4 in H_2O/EtOH), and for C_2H_2 (catalysed in solution by Pd(II) and Fe(III) salts,[104] and in the gas phase by metal ions on molecular sieves).[105]

Results of a kinetic investigation of the addition of di-isobutylaluminium hydride to oct-4-yne are consistent with a mechanism of electrophilic attack by the monomeric hydride on the alkyne, via a π-complex-like transition state.[106]

A cyclic transition state has been proposed to account for the low pre-exponential factor determined for reaction of NO_2 with propyne or but-2-yne.[107]

It has been established,[108] by investigation of alkenylidenecyclopropanes (–)(*R*)-(**38a**) and (**38b** or **c**), that the bulky Ph and R^1 substituents promote antifacial attack by chlorosulphonyl isocyanate at the perpendicular (relative to the ring plane) *p*-orbital on C-4. Disrotatory ring-opening of the cyclopropyl cation gives the dipolar ion (**39**) which may then collapse upon intramolecular nucleophilic attack by nitrogen or oxygen on either end of the allyl cation. The products (**40**)–(**43**) derived from (–)(*R*)-(**38a**)

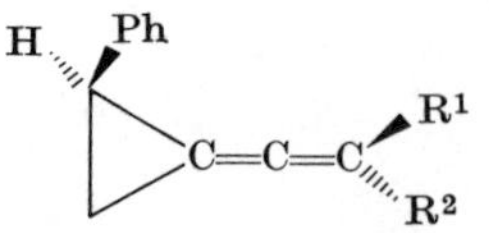

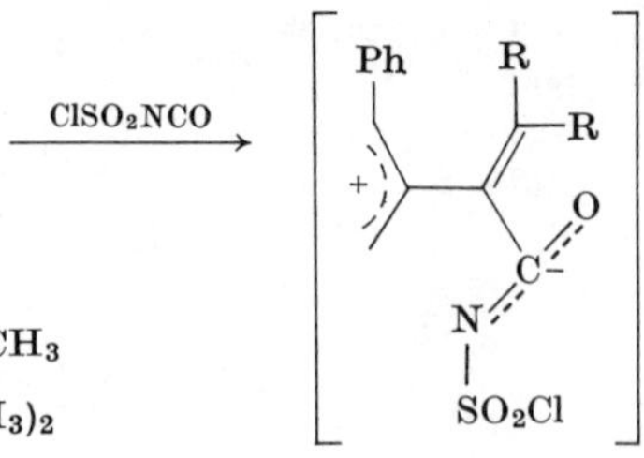

(**38a**) $R^1 = R^2 = CH_3$

(**38b**) $R^1 = CH_2CH(CH_3)_2$; $R^2 = CH_3$

(**38c**) $R^1 = CH_3$; $R^2 = CH_2CH(CH_3)_2$

$ClSO_2NCO$

(**39**)

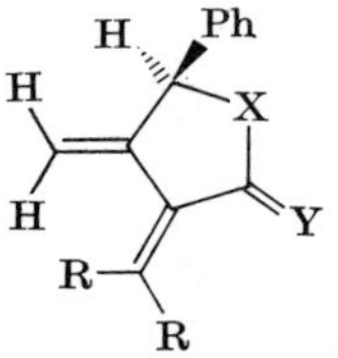

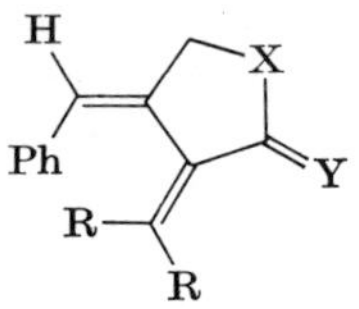

(**40**) $X = NSO_2Cl$; $Y = O$

(**41**) $X = O$; $Y = NSO_2Cl$

(**42**) $X = NSO_2Cl$: $Y = O$

(**43**) $X = O$; $Y = NSO_2Cl$

indicate that bond rotation within (**39a**), with concomitant loss of optical activity, is much slower than its cyclization at 0°, whereas at 60° optically inactive products are obtained.

At 0° dialkylallenes have been found to react with iodine isocyanate, to give a kinetically controlled 1:1 mixture of isomeric adducts, corresponding to reaction at either of the double bonds, whereas at –78° predominant addition to the substituted double bond occurs.[109] Likewise, addition of 2,4-dinitrobenzenesulphenyl chloride to monosubstituted aliphatic allenes,[110] RCH=C=CH_2, gives mainly RCHClC(SAr)=CH_2. Terminal addition does not occur and this is in contrast with earlier results for phenylallene which forms *trans*-PhCH=C(SAr)CH_2Cl as the sole product. Three dipolar-ion intermediates have been implicated in the mechanism of formation of five rearranged products upon addition of bis(trifluoromethyl)ketene to 1,3,3-trimethylcyclopropene.[111]

Complexes (**44**) and (**45**), which feature C—Pd σ-bonds have been obtained by addition of arylpalladium chloride to 1,5-cyclo-octadiene and *endo*-dicyclopentadiene, respectively.[112]

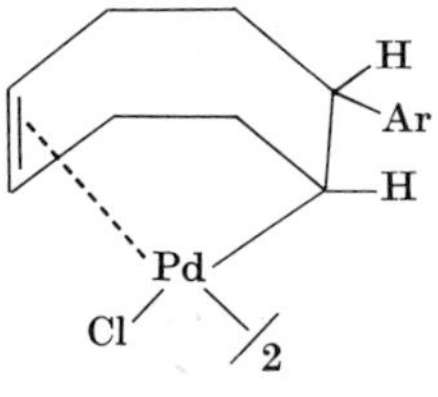

(**44**)

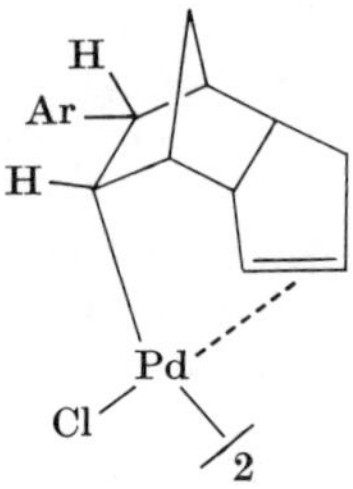

(**45**)

A mechanism has been proposed for the co-dimerization of butadiene and ethylene, promoted by a σ-arylnickel(II) compound;[113] the catalytically active species is apparently a nickel hydride complex formed upon dissociative addition of olefin to the catalyst. π-Allylnickel complexes have been implicated in the formation of 1:1 and 2:1 adducts upon reaction of amines[114] and active methylene compounds[115] with 1,3-dienes, catalysed by nickel salts. Dimerization and telomerization of butadienes have also been catalysed by palladium complexes.[116,117]

The sites of protonation (positions 4 and 1, respectively) of methyl-substituted 1- and 2-ethoxybuta-1,3-dienes during their acid-catalysed hydrolysis to corresponding conjugated carbonyl derivatives have been determined by D_3O^+-labelling experiments.[118] The results have been compared with predictions based on CNDO/2 calculations.

Miscellaneous Electrophilic Additions

The stereochemistry of electrophilic ring-opening of cyclopropanes by D^+ has been determined under conditions uncomplicated by regioselectivity.[119] Thus (**46**) in AcOD (containing a trace of D_2SO_4) undergoes attack by D^+ with 68% retention and 32% inversion of configuration while the nucleophile reacts with at least 95% inversion to form diastereoisomeric 3-deuterio-1,2-dimethylbutyl acetates; behaviour with other acid systems was very similar. Assuming that ring-opening of (**47**) also occurs with nearly complete inversion at the site of attachment of the nucleophile, it was possible to show that for this isomer each ring-carbon is equally reactive towards D^+, which attacks C^t with 68% retention and 32% inversion, and C^c with a 3:1 preference for approach from the *cis*- rather than the *trans*-substituted side. The results are best accounted for by in-plane-approach of D^+ to give a configurationally stable corner-protonated cyclopropane intermediate (**48**) having trigonal-bipyramidal geometry about the site of attack.

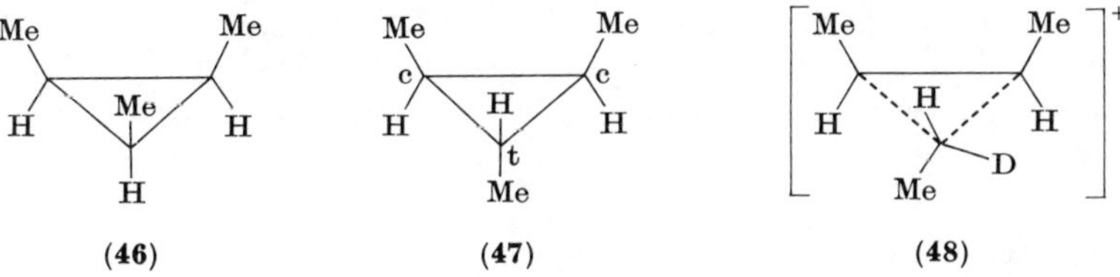

(**46**) (**47**) (**48**)

These results are not inconsistent with those of extended Hückel calculations[120] on five reaction co-ordinates for electrophilic ring-opening of hydroxy- and methyl-substituted cyclopropane. Thus, although it has been established that in-plane C-1—C-2 edge approach of H^+, with retention, is the preferred reaction pathway ($E_a = 0$), it has also been predicted that formation of corner- or edge-protonated intermediates should occur through initial C-1—C-2 edge attack.

It has been found that (diphenylmethylene)cyclopropane, in contrast to methylenecyclopropane, undergoes ring-opening upon reaction with several electrophiles.[121] This is believed to be a consequence of the longer lifetime of the less readily trapped intermediate cyclopropylcarbinyl cation (cf. ref. 108).

Hydroxy-sulphonium salts $R_2C(OH)CR_2OS^+(CH_3)_2$ have been prepared[122] by stereospecific (inversion) ring-opening of epoxides with DMSO, in the presence of 2,4,6-trinitrobenzenesulphonic acid. Kinetics of acid-catalysed ring-opening of epoxides have also been reported.[123]

It has been concluded that chlorinations of alkenes and arenes, promoted by simple chloramines in polar solvents at low pH, do not involve intermediate nitrenium ions but probably occur by direct electrophilic transfer of positive chlorine from protonated chloramine.[124]

It has been found that, in common with ethylene, α- and β-methyl substitution of styrene causes a cumulative increase in rate of reaction with *p*-chlorobenzenesulphenyl chloride (in $Cl_2CHCHCl_2$) provided that a *cis*-β-methyl substituent is not included.[125] In contrast, steric interaction between the phenyl group and a β-methyl substituent causes a decrease in rate, relative to that for styrene, there being little additional effect of further methyl-substitution. These results do not support a recent claim that an alkene rate ratio $k_{cis}:k_{trans} > 1$ is diagnostic of a bridged transition state. An independent study has also established that *trans*-1-arylpropenes are more reactive than their *cis*-isomers towards 2,4-dinitrobenzenesulphenyl chloride in AcOH (the corresponding Hammett ρ-values being −2.64 and −2.3), whereas the converse is true of reactions of isomers of hept-2-ene and 4-methylpent-2-ene.[126] With respect to transition-state structure the former trend has been ascribed to a difference in charge-transfer interaction between substrate and electrophile, while the latter trend is apparently determined mainly by Coulombic interactions. A plot of the log reactivities of the alkenes against their ionization potentials (as expected for a charge-transfer-controlled reaction) has been found to afford two lines corresponding to aromatic and aliphatic alkenes severally.

Steric hindrance to formation of an intermediate episulphonium ion during stereospecific *anti*-addition of *p*-chlorobenzenesulphenyl chloride to ethylene and its *tert*-butyl, 1,1-di-*tert*-butyl, *cis*- and *trans*-1,2-di-*tert*-butyl and 1,1,2-tri-*tert*-butyl derivatives in $Cl_2CHCHCl_2$ is believed to account for the relative rates observed[127] ($1:1.5:4.9 \times 10^{-4}:13:8.2 \times 10^{-5}:1.3 \times 10^{-4}$). The maximum hindrance is achieved when Bu^t groups are in the 1,1- or *trans*-1,2-positions whereupon a further Bu^t substituent has little effect.

The orientation of Ad_E addition of PhSCl to alkenes in AcOH has been found to be sensitive to the concentration of perchlorate ions.[128]

β-Chloro-*tert*-alkyl ethers have been prepared by BF_3-catalysed reaction of *tert*-butyl hypochlorite with alkenes[129] and the kinetics of reaction of $(EtO)_2PS_2H$ with substituted styryl phosphonates[130] and of the hydration of CO_2 catalysed by carbonic anhydrase[131] have been studied.

Nucleophilic Additions

Deuterium-labelling experiments have established that, in common with the platinum-catalysed process, palladium-promoted amination of but-2-enes is *trans*-stereospecific.[132] A study of the oxidative hydrolysis of olefins in the presence of $[Pd(dien)H_2O]^{2+}$ or $[Pd(en)(H_2O)_2]^{2+}$ has indicated that in the former circumstance attack (probably *trans*) is by OH^- on $[Pd(en)(H_2O)_2]^{2+}$ whereas in the latter case prior co-ordination of OH^- results in a faster *cis*-attack.[133] Formation of methoxy-π-allyl complexes upon reaction of MeOH with isoprene in the presence of $PdCl_2$ has been shown,[134] by kinetic investigation, to involve attack of unco-ordinated MeOH on two π-complexes of isoprene in which one or both double bonds are co-ordinated to Pd.

Certain base-catalysed carbon–carbon additions have been reviewed.[135] Addition of *sec*- or *tert*-BuLi to *trans*-cyclo-octene occurs readily, as theoretically predicted, and is followed by formation of 1- or 3-butyl-*cis*-cyclo-octene upon elimination of LiH, with concomitant release of transannular strain.[136] Kinetics of reaction of butadiene with Bu^sLi in iso-octane have also been measured.[137]

The stereochemistry of addition of organometallic and hydride reagents to anthraquinone and 9-alkyl-9-hydroxyanthrones,[138] and the role of alkali-metal cations in the fluoradenyl-carbanion-initiated epoxide ring-opening of ethylene oxide, have been studied.[139]

By kinetic analysis it has been established that cyclization of hex-5-enyl (R) to cyclopentylmethyl Grignard reagents[140] is of only first order in R_2Mg or RMgX, and that upon 1:1 or 2:1 complexation with added salts (LiX or R'_4NX; X = halide or perchlorate) mono- and di-organomagnesium compounds react more rapidly with benzonitrile.[141]

A first-order dependence[142] on [ZC_6H_4OCN] or [ROCN] and on [R''_2Mg] or [$R''MgX$] (R'' = Bu) is consistent with proposed[143] alternative four-centre mechanisms for their reaction to form $R''CN$; for reaction of aryl cyanates with $R''MgBr$, $\rho_{m-Z} = 0.97$ (R'' = Bu) and $\rho_{p-Y} = -0.80$ ($R'' = YC_6H_4$, Z = H), while for corresponding reactions of isobutyl cyanates[144] ρ_Y = ca. -1.5 has been determined. The reactivity order ArMgBr > Bu_2Mg > BuMgBr has been established[144] for reaction of PhOCN. Electrochemical evidence supports the view that the transition state for reaction of RMgBr with benzophenone has significant radical character.[145]

Multiple products have been obtained upon Darzens reaction of dibromo- or dichloroacetonitrile with carbonyl compounds in alcoholic alkali.[146,147] Rates of addition of alcohols to phenyl isocyanates[148] and of phenol and amines to aryl cyanide have been determined.[149]

Michael additions of triazolides to alkynes,[150] of ethylene glycol and water to acrylonitrile[151,152] (and acetylene[153]) and of nitromethane to *p*-halobenzylidenemalonates[154] in MeONa–MeOH have been reported. A BF_3-catalysed Michael addition of acetophenone to substituted chalcones has been found to be accelerated and decelerated by electron-donating *p*-substituents on the aryl and aroyl groups, respectively.[155] The kinetics of formation of cyclopropyl ketones upon reaction between sulphoxonium ylide and α,β-unsaturated ketones[156] and of addition of pyridine to acrylic acid[157] have been studied.

N-Hydroxyenamines are believed to be intermediates of reactions of *N-tert*-butyl- and *N*-cyclohexyl-hydroxylamine with *p*-toluenesulphonylacetylene to form the isomeric nitrones, which may prove to be useful spin traps.[158] In contrast, *N-tert*-butylhydroxylamine reacts as an oxygen nucleophile with the corresponding propyne.

While piperidine-catalysed addition of thiophenols to benzoylphenylacetylenes in C_6H_6 gives a mixture (ca. 1:3) of (*E*)- and (*Z*)-1-aryl-3-arylthio-3-phenylprop-2-en-1-ones, only the *Z*-isomers are obtained in MeOH or aqueous dioxan.[159] Rates of addition of substituted thiophenoxides to phenyl vinyl sulphone have been found to correlate better with their equilibrium constants for formation of Meisenheimer adducts, with 1,3,5-trinitrobenzene, than with their protonation equilibrium constants.[160] 3-Alkoxyallyl sulphides have been conveniently prepared by reaction of trimethylsilyl sulphides with α,β-unsaturated acetals.[161] *cis-trans*-Isomerization of oleic acid has been catalysed by thiols,[162] and additions of thioureas to ethyl (2-furyl)- and (5-nitro-2-furyl)-propiolate,[163] have been studied.

The pH-rate profiles have been determined for addition of thiophenol to 2-cyano-3-piperonylacrylic acid[164] and for reactions of glycine, thioglycollic acid and OH^- ions with isocyanates bearing neighbouring sulphide or sulphonyl groups.[165]

Kinetic study of deamination of cytidine, deoxycitidine and dihydro-1-methylcytosine by bisulphite has further clarified the addition–deamination–elimination mechanism of formation of corresponding uracil derivatives;[166] optimum rates of deamination occur at pH 5 in a high bisulphite concentration. Additions of a variety of nucleophiles (including HSO_3^-, SO_3^{2-}, $HOCH_2CH_2S^-$, $HONH^-$ and OH^-) to the C(5)=C(6) and C(4)=N^+(3) groups of 1,3-dimethyl-5-nitrouracil[167] and quinazoline cation,[168] respectively, have been studied; linear relationships between functions of the equilibrium constants for addition and/or protonation of the nucleophile and a measure (γ) of the ability of its conjugate acid to add to carbonyl functions have been discerned.

While 1:1 adducts could not be isolated, stable 2:1 adducts have been obtained by reaction of strong nucleophiles (including HSO_3^-, barbituric acid and its thio-analogue) across the 5,6- and 7,8-double bonds of pteridin-4-one and related compounds.[169] Barbituric acid also adds across the 1,6-double bond of purine and its derivatives (including the 2-oxo-, 2-thioxo- and 8-methylsulphonyl compounds) which are, however, much less reactive[170] than pteridines (as expected, in view of the respective lengths of the pteridine 3,4- and purine 1,6-bond). Additions of HSO_3^- to 5-fluorouracil,[171] of cyanide ion to nicotinamide-adenine nucleotides,[172] and of benzenesulphinate to acrylic monomers[173] have also been studied. Competition experiments have established[174] that the reactive intermediate $Me_2Si{=}CH_2$ behaves as a dipolar silicon electrophile, as shown by its order of reactivity towards a variety of substrates: $Ph_2CO > ROH$, $ArOH \gg m\text{-}ClC_6H_4NH_2 > CH_3CN$. This interpretation is consistent with analogous behaviour of $R_3P{=}CH_2$, and with Hückel calculations reported[175] for $H_2Si{=}CH_2$.

For formation of *N*,*N*-diethylphenethylamine by lithium diethylamide-catalysed addition of Et_2NH to styrene, the Hammett ρ-value (+5.0) is consistent with an anionic nucleophilic mechanism.[176] Addition of primary and secondary amines to the ester and amide of phenyl- and *p*-nitrophenyl-propiolic acid has been studied.[177] Mechanisms of base-promoted amidoalkylation of *N*-acyl-α-chloroglycyl derivatives[178] and of formation of 1-arylpyrazoles upon reaction of β-chloro-epoxides with aryl hydrazines[179] have been suggested.

Kinetic resolution of amines achieved through reaction with camphor-10-sulphenyl chloride has been ascribed to a sulphene-trapping step.[180] Hydrolysis of 2-cyano-3-piperonylacrylic acid,[181] and the addition of Bu^sLi and Bu^tLi to *trans*-cyclo-octene,[136] have also been studied.

References

[1] R. Oda, *Kagaku* (*Kyoto*), **29**, 606 (1974); *Chem. Abs.*, **81**, 119298 (1974).
[2] G. P. Chiusoli, *Accounts Chem. Res.*, **6**, 422 (1973).
[3] D. G. Garratt, A. Modro, K. Oyama, G. H. Schmid, T. T. Tidwell and K. Yates, *J. Am. Chem. Soc.*, **96**, 5295 (1974).
[4] G. A. Olah, P. Schilling, P. W. Westerman and H. C. Lin, *J. Am. Chem. Soc.*, **96**, 3581 (1974).
[5] B. Nelander, *Tetrahedron*, **30**, 1337 (1974).
[6] T. L. Leggett, R. E. Kennerly and D. A. Kohl, *J. Chem. Phys.*, **60**, 3264 (1974); *Chem. Abs.*, **81**, 36949 (1974).
[7] D. Grosjean, P. Masclet and G. Mouvier, *Bull. Soc. Chim. France*, **1974**, 573.
[8] G. A. Olah and T. R. Hockswender, Jr., *J. Am. Chem. Soc.*, **96**, 3574 (1974).
[9] J. M. Kornprobst and J. E. Dubois, *Tetrahedron Letters*, **1974**, 2203.
[10] Yu. P. Filinov, I. M. Vasil'kevich and G. F. Dvorko, *Dopov. Akad. Nauk Ukr. RSR*, *Ser. B*, **36**, 153 (1974); *Chem. Abs.*, **80**, 132359 (1974).
[11] T. Okuyama, K. Ohashi, K. Izawa and T. Fueno, *J. Org. Chem.*, **39**, 2255 (1974).
[12] M. C. Lasne and A. Thullier, *Bull. Soc. Chim. France*, **1974**, 249.
[13] G. E. Heasley, V. L. Heasley, S. L. Manatt, H. A. Day, R. V. Hodges, P. A. Kroon, D. A. Redfield, T. L. Rold and D. E. Williamson, *J. Org. Chem.*, **38**, 4109 (1973).
[14] V. L. Heasley, P. D. Davis, D. M. Ingle, K. D. Rold and G. E. Heasley, *J. Org. Chem.*, **39**, 736 (1974).
[15] R. P. Vignes and J. Hamer, *J. Org. Chem.*, **39**, 849 (1974).
[16] G. Bellucci, G. Ingrosso, F. Marioni, E. Mastrorilli and I. Morelli, *J. Org. Chem.*, **39**, 2562 (1974).
[17] J. E. Dubois and P. Fresnet, *Tetrahedron Letters*, **1974**, 2195.
[18] J. E. Dubois and P. Fresnet, *Tetrahedron*, **29**, 3407 (1973).
[19] P. L. Barili, G. Bellucci, G. Berti, M. Golfarini, F. Marioni and V. Scartoni, *Gazz. Chim. Ital.*, **104**, 107 (1974); *Chem. Abs.*, **81**, 62753 (1974).

[20] E. A. Lissi and E. Sanheuza, *Rev. Latinoamer. Quim.*, **5**, 26 (1974); *Chem. Abs.*, **81**, 24635 (1974).
[21] M. Abufhele, C. Andersen, A. E. Lissi and E. Sanhueza, *Rev. Latinoamer. Quim.*, **5**, 67 (1974); *Chem. Abs.*, **81**, 37116 (1974).
[22] Y. Yanagida, H. Shigesato, M. Nomura and S. Kikkawa, *Bull. Chem. Soc. Japan*, **47**, 677 (1974).
[23] N. Iskikawa, A. Nagashima and A. Sekiya, *Chem. Letters (Tokyo)*, **1974**, 1225.
[24] G. Mehta and N. Pattnaik, *Indian J. Chem.*, **11**, 1313 (1973); *Chem. Abs.*, **80**, 132368 (1974).
[25] C. L. Hewett, I. M. Gilbert, J. Redpath, D. S. Savage, J. Strachan, T. Sleigh and R. Taylor, *J.C.S. Perkin I*, **1974**, 897.
[26] M.-F. Ruasse and J.-E. Dubois, *J. Org. Chem.*, **39**, 2441 (1974).
[27] D. Fain, J. Toullec and J.-E. Dubois, *Tetrahedron Letters*, **1974**, 1725.
[28] J.-E. Dubois, J. Toullec and D. Fain, *Tetrahedron Letters*, **1973**, 4859.
[29] V. R. Kartashov, I. V. Bodrikov, K. N. Tishkov and T. I. Zaikina, *Dokl. Akad. Nauk SSSR*, **216**, 325 (1974); *Chem. Abs.*, **81**, 77181 (1974).
[30] V. I. Staninets, T. A. Degurko and Yu. V. Melika, *Ukr. Khim. Zh.* (*Russian Ed.*), **39**, 1043 (1973); *Chem. Abs.*, **80**, 36459 (1974).
[31] V. I. Staninets and Yu. I. Gevaza, *Ukr. Khim. Zh.* (*Russian Ed.*), **39**, 589 (1973); *Chem. Abs.*, **79**, 145620 (1973).
[32] M. Ōkubo, H. Kusakabe, T. Hiwatashi and K. Kishida, *Bull. Chem. Soc. Japan*, **47**, 860 (1974).
[33] M. J. Nanjan and R. Ganesan, *J. Indian Chem. Soc.*, **50**, 563 (1973).
[34] S. Uemura, A. Onoe and M. Okano, *Bull. Chem. Soc. Japan*, **47**, 143 (1974).
[35] S. Uemura, A. Onoe and M. Okano, *Bull. Chem. Soc. Japan*, **47**, 692 (1974).
[36] R. T. LaLonde and A. D. Debboli, Jr., *J. Org. Chem.*, **38**, 4228 (1973).
[37] A. L. Shabanov, M. M. Movsum-Zade, R. A. Babakhanov and R. G. Movsum-Zade, *Azerb. Khim. Zh.*, **4**, 87 (1973); *Chem. Abs.*, **80**, 145011 (1974).
[38] J. M. Agoff, M. C. Cabaleiro and J. C. Podesta, *Chem. Ind. (London)*, **1974**, 305.
[39] M. Nógrádi, W. D. Ollis and I. O. Sutherland, *J.C.S. Perkin I*, **1974**, 621.
[40] M. Zupan and A. Pollak, *Tetrahedron Letters*, **1974**, 1015.
[41] M. Zupan and A. Pollak, *J. Org. Chem.*, **39**, 2646 (1974).
[42] J. W. Lown and A. V. Joshua, *J.C.S. Perkin I*, **1973**, 2680.
[43] M. C. Cabaleiro and A. B. Chopa, *J.C.S. Perkin II*, **1974**, 452.
[44] J. L. Luche, J. Bertin and H. B. Kagan, *Tetrahedron Letters*, **1974**, 759; J. Bertin, J. L. Luche and H. B. Kagan, *ibid.*, p. 763.
[45] J. C. Philips, M. Aregullin, M. Oku and A. Sierra, *Tetrahedron Letters*, **1974**, 4157.
[46] R. C. Fahey, M. T. Payne and Do-J. Lee, *J. Org. Chem.*, **39**, 1124 (1974).
[47] F. Marcuzzi, G. Melloni and G. Modena, *Tetrahedron Letters*, **1974**, 413.
[48] J. Cousseau and L. Gouin, *Tetrahedron Letters*, **1974**, 2889.
[49] L. A. Gasparyan, *Sb. Nauch. Tr., Ivanov. Energ. Inst.*, **14**, 241 (1972); *Chem. Abs.*, **80**, 132430 (1974).
[50] L. G. Bruk, S. M. Brailovskii, O. N. Temkin and R. M. Flid, *Kinet. Katal.*, **15**, 516 (1974); *Chem. Abs.*, **81**, 24727 (1974).
[51] W. Ried and A. Marhold, *Annalen*, **1974**, 1010.
[52] L. Szirovicza, R. Walsh, *J.C.S. Faraday I*, **1974**, 33.
[53] F. Amar, D. R. Dalton, G. Eisman and M. J. Haugh, *Tetrahedron Letters*, **1974**, 3037.
[54] F. Amar, D. R. Dalton, G. Eisman and M. J. Haugh, *Tetrahedron Letters*, **1974**, 3033.
[55] U. Kh. Agaeva, A. T. Alyev and S. I. Sadykhzade, *Azerb. Khim. Zh.*, **1974**, 112; *Chem. Abs.*, **81**, 104315 (1974).
[56] R. C. Fahey, C. A. McPherson and R. A. Smith, *J. Am. Chem. Soc.*, **96**, 4534 (1974).
[57] D. J. Pasto, G. R. Meyer and B. Lepesak, *J. Am. Chem. Soc.*, **96**, 1858 (1974).
[58] E. Kantolahti, *Acta Chem. Scand.*, **27**, 2667 (1973).
[59] A. F. Diaz and M. Blanco, *J. Org. Chem.*, **39**, 1313 (1974).
[60] R. C. Fahey and C. A. McPherson, *J. Am. Chem. Soc.*, **91**, 3863 (1969).
[61] K. Izawa, T. Okuyama and T. Fueno, *Bull. Chem. Soc. Japan*, **47**, 1477 (1974).
[62] J. Guenzet and M. Camps, *Tetrahedron*, **30**, 849 (1974).
[63] J. Guenzet and M. Camps, *Bull. Soc. Chim. France*, **1973**, 3167.
[64] J. Guenzet, M. Toumi and A. Toumi, *Tetrahedron*, **30**, 159 (1974).
[65] R. Corriu and C. Reye, *Bull. Soc. Chim. France*, **1974**, 1327.
[66] R. Corriu and C. Reye, *Tetrahedron*, **30**, 3935 (1974).
[67] B. A. Trofimov, S. E. Korostova, N. A. Nedolya, T. K. Pogodayeva and M. G. Voronkov, *Organic Reactivity (Tartu)*, **10**, 979 (1973).
[68] C. F. Bernasconi and W. J. Boyle, Jr., *J. Am. Chem. Soc.*, **96**, 6070 (1974).

[69] L. Marko, *Acta Cient. Venez., Supl.*, **24**, 49 (1973); *Chem. Abs.*, **81**, 36836 (1974).
[70] C. Botteghi and C. Salomon, *Tetrahedron Letters*, **1974**, 4285.
[71] M. Tanaka, Y. Watanabe, T. Mitsudo and Y. Takegami, *Bull. Chem. Soc. Japan*, **47**, 1698 (1974).
[72] A. C. Clark and M. Orchin, *J. Org. Chem.*, **38**, 4004 (1973).
[73] V. Yu. Gankin, *Catal., Proc. Int. Congr. 5th*, **1**, 421 (1973); *Chem. Abs.*, **80**, 107688 (1974).
[74] V. Yu. Gankin, V. A. Dvinin and V. A. Rybakov, *Kinet. Katal.*, **15**, 60 (1974); *Chem. Abs.*, **80**, 119864 (1974).
[75] T. E. Zhesko and A. A. Polyakov, *Organic Reactivity* (*Tartu*), **11**, 257 (1974).
[76] V. I. Isagulyants, G. V. Isagulyants, I. R. Khairudinov and D. L. Rakhmankulov, *Izv. Akad. Nauk SSSR, Ser. Khim.*, **1973**, 1870; *Chem. Abs.*, **79**, 145639 (1973).
[77] M. G. Safarov, V. I. Isagulyants and N. G. Nigmatullin, *Zh. Org. Khim.*, **10**, 1365 (1974); *Chem. Abs.*, **81**, 90818 (1974).
[78] L. A. Efremova, Z. V. Belyakova, K. K. Popkov and S. A. Golubtsov, *Zh. Prikl. Spektrosk.*, **19**, 367 (1973); *Chem. Abs.*, **79**, 114805 (1973).
[79] V. O. Reikhsfel'd and M. I. Astrakhanov, *Zh. Obshch. Khim.*, **43**, 2431 (1973); *Chem. Abs.*, **80**, 107597 (1974).
[80] M. I. Astrakhanov and V. O. Reikhsfel'd, *Zh. Obshch. Khim.*, **43**, 2439 (1973); *Chem. Abs.*, **80**, 107599 (1974).
[81] W. Dumont, J.-C. Poulin, T.-P. Dang and H. B. Kagan, *J. Am. Chem. Soc.*, **95**, 8295 (1973).
[82] R. C. Larock, *J. Org. Chem.*, **39**, 834 (1974).
[83] M. Zaidlewicz and A. Uzarewicz, *Rocz. Chem.*, **48**, 467 (1974); *Chem. Abs.*, **81**, 49155 (1974).
[84] C. Georgulis and J.-M. Valery, *Bull. Soc. Chim. France*, **1974**, 178.
[85] J.-E. Bäckvall and B. Åkermark, *J. Organometal. Chem.*, **78**, 177 (1974).
[86] C. P. Forbes, A. Goosen and H. A. H. Lave, *J.C.S. Perkin I*, **1974**, 2346.
[87] H. C. Brown and J. H. Kawakami, *J. Am. Chem. Soc.*, **95**, 8665 (1973).
[88] P. E. M. Allen and R. M. Lough, *J.C.S. Faraday I*, **1973**, 2087.
[89] H. C. Clark and C. S. Wong, *J. Am. Chem. Soc.*, **96**, 7213 (1974).
[90] P. A. Bartlett and W. S. Johnson, *J. Am. Chem. Soc.*, **95**, 7501 (1973).
[91] P. A. Bartlett, J. I. Brauman, W. S. Johnson and R. A. Volkmann, *J. Am. Chem. Soc.*, **95**, 7504 (1973).
[92a] T. Sasaki, K. Kanematsu and A. Kondo, *J. Org. Chem.*, **39**, 2246 (1974).
[92b] G. A. Chukhadzhyan, Zh. I. Abramyan and G. A. Gevorkyan, *Zh. Obshch. Khim.*, **43**, 2012 (1973); *Chem. Abs.*, **80**, 36401 (1974).
[93] R. Maroni, G. Melloni and G. Modena, *J.C.S. Perkin I*, **1973**, 2491.
[94] R. Maroni, G. Melloni and G. Modena, *J.C.S. Perkin I*, **1974**, 353.
[95] R. A. Benkeser and D. F. Ehler, *J. Organometal. Chem.*, **69**, 193 (1974).
[96] M. Kraus, *Coll. Czech. Chem. Comm.*, **39**, 1318 (1974).
[97] M. Foa and L. Cassar, *Gazz. Chim. Ital.*, **103**, 805 (1973); *Chem. Abs.*, **81**, 62803 (1974).
[98] T. Okuyama, K. Izawa and T. Fueno, *J. Org. Chem.*, **39**, 351 (1974).
[99] J. Tendil, M. Verney and R. Vessiere, *Tetrahedron*, **30**, 579 (1974).
[100] W. Ried and W. Ochs, *Chem. Ber.*, **107**, 1334 (1974).
[101] W. Ried and W. Ochs, *Annalen*, **1974**, 1248.
[102] W. F. Verhelst and W. Drenth, *J. Am. Chem. Soc.*, **96**, 6692 (1974).
[103] A. E. Tsil'ko and I. A. Maretina, *Zh. Org. Khim.*, **10**, 929 (1974); *Chem. Abs.*, **81**, 48986 (1974).
[104] D. V. Sokol'skii, S. S. Segizbaeva and Ya. A. Dorfman, *Khim. Atsetilena Tekhnol. Karbida Kal'tsiya*, **1972**, 384; *Chem. Abs.*, **79**, 125487 (1973).
[105] G. Gut and K. Aufdereggen, *Helv. Chem. Acta*, **57**, 441 (1974).
[106] J. J. Eisch and S.-G. Rhee, *J. Am. Chem. Soc.*, **96**, 7276 (1974).
[107] B. V. Ashmead and J. H. Thomas, *Symp.*, (*Int.*) *Combust.*, [*Proc.*], **14**, 593 (1973); *Chem. Abs.*, **80**, 132431 (1974).
[108] D. J. Pasto and J. K. Borchardt, *J. Am. Chem. Soc.*, **96**, 6937 (1974).
[109] T. Greibrokk, *Acta Chem. Scand.*, **27**, 3368 (1973).
[110] T. L. Jacobs and R. C. Kammerer, *J. Am. Chem. Soc.*, **96**, 6213 (1974).
[111] D. H. Ave and G. S. Helwig, *Chem. Comm.*, **1974**, 925.
[112] A. Kasahara, T. Izumi, K. Endo, T. Takeda and M. Ookita, *Bull. Chem. Soc. Japan*, **47**, 1967 (1974).
[113] N. Kawata, K. Maruya, T. Mizoroki and A. Ozaki, *Bull. Chem. Soc. Japan*, **47**, 2003 (1974).
[114] R. Baker, A. H. Cook, D. E. Halliday and T. N. Smith, *J.C.S. Perkin II*, **1974**, 1511.
[115] R. Baker, A. H. Cook and T. N. Smith, *J.C.S. Perkin II*, **1974**, 1517.
[116] J. Beger and H. Reichel, *J. prakt. Chem.*, **315**, 1067 (1973).
[117] J. Beger, C. Duschek and H. Reichel, *J. prakt. Chem.*, **315**, 1077 (1973).

[118] J.-P. Gouesnard and M. Blain, *Bull. Soc. Chim. France*, **1974**, 338.
[119] C. H. DePuy, A. H. Andrist and P. C. Fünfschilling, *J. Am. Chem. Soc.*, **96**, 948 (1974).
[120] A. H. Andrist, *J. Am. Chem. Soc.*, **95**, 7531 (1973).
[121] E. Dunkelblum, *Israel J. Chem.*, **11**, 557 (1973).
[122] M. A. Khuddus and D. Swern, *J. Am. Chem. Soc.*, **95**, 8393 (1973).
[123] R. Durand, P. Geneste, G. Lamaty and J.-P. Roque, *Compt. rend. Ser. C.*, **277**, 1395 (1973).
[124] O. E. Edwards, G. Bernath, J. Dixon, J. M. Paton and D. Volelle, *Can. J. Chem.*, **52**, 2123 (1974).
[125] G. H. Schmid and D. G. Garratt, *Can. J. Chem.*, **52**, 1807 (1974).
[126] K. Izawa, T. Okuyama and T. Fueno, *Bull. Chem. Soc. Japan*, **47**, 1480 (1974).
[127] C. L. Dean, D. G. Garratt, T. T. Tidwell and G. H. Schmid, *J. Am. Chem. Soc.*, **96**, 4958 (1974).
[128] I. V. Bodrikov, L. G. Gurvich, N. S. Zefirov, V. R. Kartashov and A. L. Kurts, *Zh. Org. Khim.*, **10**, 1545 (1974); *Chem. Abs.*, **81**, 104296 (1974).
[129] C. Walling and R. T. Clark, *J. Org. Chem.*, **39**, 1962 (1974).
[130] A. N. Pudovik, R. A. Cherkasov and G. A. Kutyrev, *Zh. Obshch. Khim.*, **43**, 1466 (1973); *Chem. Abs.*, **80**, 59172 (1974).
[131] Y. Pocker and J. S. Y. Ng, *Biochemistry*, **12**. 5127 (1973).
[132] B. Åkermark, J. E. Bäckvall, K. Siirala-Hanseń, K. Sjoberg and K. Zetterberg, *Tetrahedron Letters*, **1974**, 1363.
[133] C. Burgess, F. R. Hartley and G. W. Searle, *J. Organometal. Chem.*, **76**, 247 (1974).
[134] R. Pietropaolo, F. Faraone, D. Pietropaolo and P. Piraino, *J. Organometal. Chem.*, **64**, 403 (1974).
[135] H. Pines, *Accounts Chem. Res.*, **7**, 155 (1974).
[136] R. D. Bach, K. W. Bair and C. L. Willis, *J. Organometal. Chem.*, **77**, 31 (1974).
[137] V. N. Zgonnik, E. Yu. Shadrina, K. Kalnins and B. L. Erusalimskii, *Makromol. Chem.*, **174**, 81 (1973); *Chem. Abs.*, **80**, 81646 (1974).
[138] M. Montebruno, F. Fournier, J.-P. Battioni and W. Chodkiewicz, *Bull. Soc. Chim. France*, **1974**, 283.
[139] C. J. Chang, R. F. Kiesel and T. E. Hogan-Esch, *J. Am. Chem. Soc.*, **95**, 8446 (1973).
[140] H. G. Richey, Jr., and H. S. Veale, *J. Am. Chem. Soc.*, **96**, 2641 (1974).
[141] M. Chastrette, R. Amouroux and M. Subit, *J. Organometal. Chem.*, **78**, 303 (1974).
[142] E. Huge-Jensen and A. Holm, *Acta Chem. Scand.*, *B*, **28**, 757 (1974).
[143] A. Holm and E. Huge-Jensen, *Acta Chem. Scand.*, *B*, **28**, 705 (1974).
[144] A. Holm, T. Holm and E. Huge-Jensen, *Acta Chem. Scand.*, *B*, **28**, 781 (1974).
[145] T. Holm, *Acta Chem. Scand.*, *B*, **28**, 809 (1974).
[146] P. Coutrot, *Bull. Soc. Chim. France*, **1974**, 1965.
[147] P. Coutrot, C. Legris and J. Villieras, *Bull. Soc. Chim. France*, **1974**, 1971.
[148] R. P. Tiger, L. S. Bekhli, S. P. Bondarenko and S. G. Entelis, *Zh. Org. Khim.*, **9**, 1563 (1973); *Chem. Abs.*, **79**, 114795 (1973).
[149] L. Bacaloglu, R. Bacaloglu, D. Martin and K. Nadolski, *J. Prakt. Chem.*, **316**, 529 (1974).
[150] Y. Tanaka and S. I. Miller, *Tetrahedron*, **29**, 3286 (1973).
[151] M. Polievka, V. Macho and L. Uhlar, *Petrochemia*, **13**, 127 (1973); *Chem. Abs.*, **81**, 3036 (1974).
[152] F. Wolf, K.-H. Bergk and H. Wagenschein, *Z. Chem.*, **14**, 272 (1974).
[153] G. A. Shitov, R. D. Yakubov and I. N. Azerbaev, *Khim. Atsetilena Tekhnol. Karbida Kal'tsiya*, **1972**, 274; *Chem. Abs.*, **79**, 125489 (1973).
[154] J. Lange, J. Piechaczek and K. Rudzinski, *Rocz. Chem.*, **48**, 861 (1974); *Chem. Abs.*, **81**, 62673 (1974).
[155] Z. Csuros, P. Sallay and G. Deak, *Acta Chim.* (*Budapest*), **79**, 341 (1973); *Chem. Abs.*, **80**, 59142 (1974).
[156] F. Rocquet and A. Sevin, *Bull. Soc. Chim. France*, **1974**, 881.
[157] A. Le Berre and A. Delacroix, *Bull. Soc. Chim. France*, **1974**, 1896.
[158] J. A. Sanders, K. Hovius and J. B. F. N. Engberts, *J. Org. Chem.*, **39**, 2641 (1974).
[159] M. T. Omar and M. N. Basyouni, *Bull. Chem. Soc. Japan*, **47**, 2325 (1974).
[160] P. De Maria, A. Fini and F. M. Hall, *Chim. Ind.* (*Milan*), **55**, 808 (1973); *Chem. Abs.*, **80**, 36466 (1974).
[161] Y. Shirota, T. Nogami, Y. Hasegawa and H. Mikawa, *Chem. Letters* (*Tokyo*), **1974**, 1009.
[162] W. Cr. Niehaus, *Bioorganic Chemistry*, **3**, 302 (1974).
[163] E. Åkerblom, *Chem. Scripta*, **6**, 35 (1974).
[164] T.-R. Kim and T.-S. Huh, *Daehan Hwahak Hwoejee*, **17**, 363 (1973); *Chem. Abs.*, **80**, 69980 (1974).
[165] V. Knoppová and M. Uher, *Coll. Czech. Chem. Comm.*, **38**, 3852 (1973).
[166] R. Shapiro, V. DiFate and M. Welcher, *J. Am. Chem. Soc.*, **96**, 906 (1974).
[167] I. H. Pitman, M. J. Cho and G. S. Rock, *J. Am. Chem. Soc.*, **96**, 1840 (1974).

[168] M. J. Cho and I. H. Pitman, *J. Am. Chem. Soc.*, **96**, 1843 (1974).
[169] A. Albert and J. J. McCormack, *J.C.S. Perkin I*, **1973**, 2630.
[170] W. Pendergast, *J.C.S. Perkin I*, **1973**, 2759.
[171] F. A. Sedor, D. G. Jacobson and E. G. Sander, *Bioorganic Chemistry*, **3**, 221 (1974).
[172] T. Okubo and N. Ise, *J. Biol. Chem.*, **249**, 3563 (1974).
[173] J. D. Margerum, R. G. Brault, A. M. Lackner and L. J. Miller, *J. Phys. Chem.*, **77**, 2720 (1973).
[174] R. D. Bush, C. M. Golino, G. D. Homer and L. H. Sommer, *J. Organometal. Chem.*, **80**, 37 (1974).
[175] M. D. Curtis, *J. Organometal. Chem.*, **60**, 63 (1973).
[176] T. Narita, T. Yamaguchi and T. Tsuruta, *Bull. Chem. Soc. Japan*, **46**, 3825 (1973).
[177] E. Åkerblom, *Chem. Scripta*, **3**, 232 (1973).
[178] D. Matthies, *Archiv Pharmazie*, **307**, 801 (1974).
[179] P. Bouchet and C. Coquelet, *Bull. Soc. Chim. France*, **1973**, 3153, 3159.
[180] J. F. King, S.-K. Sim and S.-K. L. Li, *Can. J. Chem.*, **51**, 3914 (1973).
[181] T.-R. Kim and K.-I. Lee, *Daehan Hwahak Hwoejee*, **17**, 269 (1973); *Chem. Abs.*, **79**, 145621 (1973).

CHAPTER 12-II

Addition Reactions. II. Cycloaddition

W. E. WATTS

School of Physical Sciences, The New University of Ulster

A review article by Epiotis[1] has appeared dealing with the effect of configuration interaction (see 1973 Report[2]) upon the preferred reaction pathways, reaction rates, and stereochemical outcome of pericyclic reactions including cycloadditions. Hendrickson[3] has reviewed thermal 6-electron pericyclic reactions and has proposed a new nomenclature system to aid classification of such processes. The removal of orbital-symmetry restrictions on pericyclic reactions through involvement of transition-metal orbitals has been discussed by Mango.[4]

The second-order perturbation MO treatment, restricted to HOMO–LUMO interactions, has been developed to account satisfactorily for the periselectivity of a wide range of known cycloadditions and to allow prediction of the favoured route for thermal two-component reactions having available several symmetry-allowed pathways.[5] Theoretical treatments[6,7] of 1,3-dipolar cycloadditions have also appeared which successfully account for the regioselectivity of such reactions from consideration of frontier orbital interactions.

2 + 2-Cycloaddition

The various types of thermal addition ($\pi^2 + \pi^2$; $\pi^2 + \sigma^2$; $\sigma^2 + \sigma^2$) can be differentiated mechanistically in terms of the phase continuity *and* the triplet stability of the restricted Hartree–Fock MO's relevant to the entire reaction system.[8] It can then be shown that the former criterion allows discrimination between symmetry-allowed and symmetry-forbidden processes while the latter permits distinction between diradical and non-radical reactions. MINDO/3 calculations show that the lowest energy of the available pathways for the dimerization of ethylene to cyclobutane involves a diradical intermediate.[9] It has been established that $PhCF{=}CF_2$ dimerizes in a head-to-head manner giving the *cis*- and *trans*-isomers of hexafluoro-1,2-diphenylcyclobutane,[10] contrary to an earlier conclusion.[11]

Huisgen *et al.* have shown that the addition of TCNE (tetracyanoethylene) to enol ethers proceeds *via* a zwitterion which, before closure to a cyclobutane, may be intercepted by alcohols giving acyclic acetals.[12] Equilibration of the stereoisomeric forms of the zwitterion by internal rotation is slower than capture by alcohol which occurs stereospecifically.[13] Thermal cycloaddition reactions of cyclopropylalkenes with TCNE and with $ArSO_2NCO$ give only the usual four-membered ring products;[14] a two-step mechanism involving an intermediate zwitterion seems less likely in this case since products of rearrangement of an intermediate cyclopropylcarbenium species were not detected.

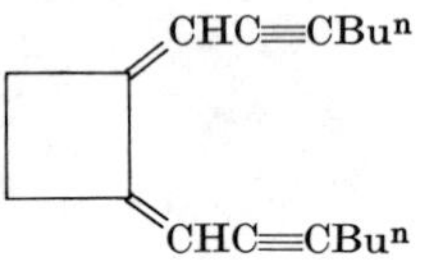

(1) 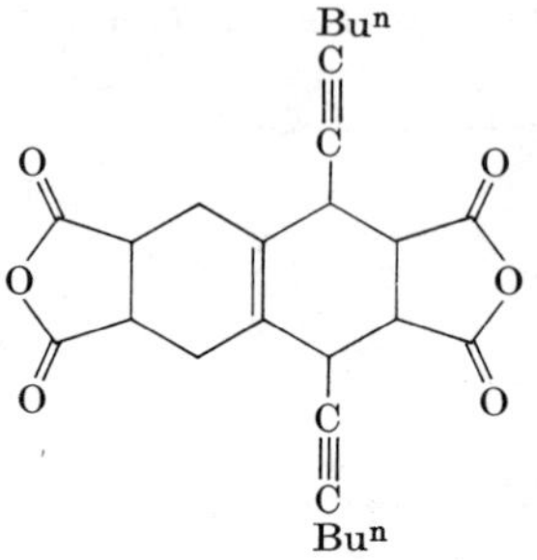 (2) 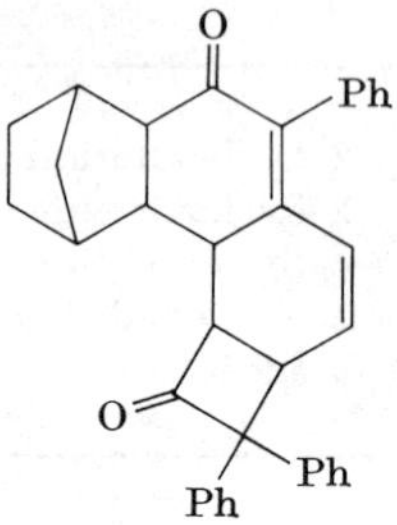 (3)

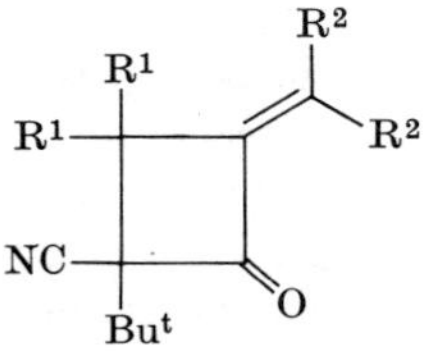

(4) (5) (6)

(**a**) $R^1 = Me$; $R^2 = H$

(**b**) $R^1 = H$; $R^2 = Me$

(**c**) $R^1 = Me$; $R^2 = CD_3$

(**d**) $R^1 = CD_3$; $R^2 = Me$

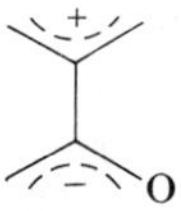

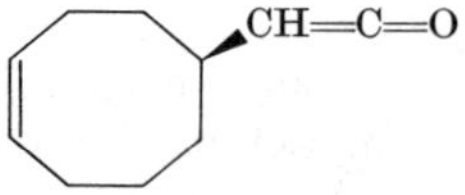

Dimerization of nona-1,2-dien-4-yne gives a highly reactive, non-isolable dimethylenecyclobutane, thought to have structure (**1**), which forms a bisadduct (**2**) with maleic anhydride.[15] A study of the secondary deuterium-isotope effect in the addition of dimethylketene to [α-^{2}H]-styrene gives k_H/k_D = ca. 0.8, in accord with a concerted rather than a two-step mechanism.[16] The thermal reaction of norbornene with diphenylketene gives both the $\pi^2 + \pi^2$ adduct and a 1:2 adduct for which the structure (**3**) is proposed.[17] Cycloaddition of *tert*-butylcyanoketene to allenes gives cyclobutanones in good yields;[18] with 1,1-dimethylallene, products (**4a**) and (**4b**) of addition to each double bond are formed. The activation energies for the thermal isomerization of (**4a**) to (**4b**) and for the equilibration of (**4c**) and (**4d**) have been measured (ca. 39 kcal mol^{-1}), and involvement of zwitterions of the type (**5**) both in these isomerizations and in the original cycloadditions has been suggested.[19] Treatment of (cyclo-oct-4-enyl)acetyl chloride with triethylamine gives the two stereoisomeric tricyclic ketones resulting from intramolecular $\pi^2s + \pi^2a$ cycloaddition of the intermediate ketene (**6**).[20]

Cycloaddition reactions of diphenylketene with azobenzenes[21] (giving diazetidinones) and with nitrosobenzenes[22] (giving oxazetidin-3- and -4-ones) are relatively insensitive to solvent and to substitution in the benzene rings, indicating a concerted rather than a two-step mechanism. In the latter reactions, the oxazetidin-4-ones spontaneously eliminate CO_2 and the resulting Schiff bases add diphenylketene, giving azetidinones.[22] Vinyl sulphene adds to the carbon—carbon double bond of enamines with complete lack of stereospecificity and mixtures of *cis*- and *trans*-substituted thietane dioxides are formed.[23] From perturbation calculations, a three-centre model has been proposed[24] for the thermal 2 + 2 addition of an electron-donor carbon—carbon double bond to an electron-acceptor carbonyl group (e.g. in CO_2). The advantage of this particular configuration, which is free of the transition-state strain associated with the $\pi^2s + \pi^2a$ mode, stems mainly from the significant stabilization energy provided by the HOMO(C=C)–LUMO(C=O) charge-transfer interaction.

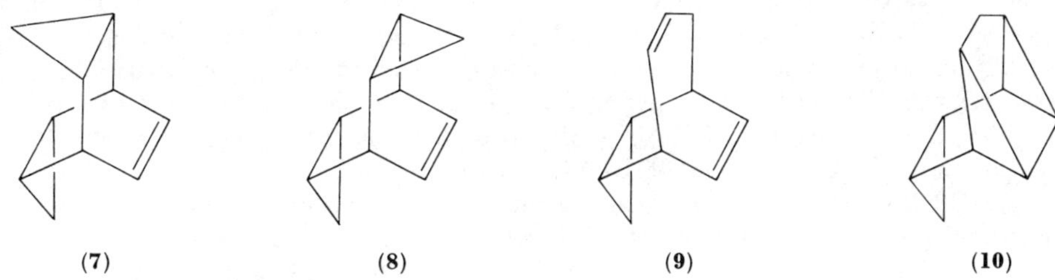

(7) **(8)** **(9)** **(10)**

Several intramolecular $\pi^2 + \sigma^2$ additions involving opening of a cyclopropane bond have been studied. Thermolysis (600°) of the *endo,exo*-tetracyclodecene (**7**) gives a complex mixture of products including the *exo,exo*-isomer (**8**), the tricyclic diene (**9**), and the "trishomobenzene" (**10**).[25] Treatment of (**7**) with a rhodium(I) catalyst gives (**8**), while photolysis in pentane at 20° affords (**10**).[26] Thermolysis of various derivatives of *endo*-tricyclo[3.2.1.$0^{2,4}$]oct-6-ene has likewise provided new synthetic routes to tetracyclo[3.2.1.$0^{2,7}$.$0^{4,6}$]octanes.[27] It has been shown that the Ni(0)-catalysed $\pi^2 + \sigma^2$ additions of dimethyl maleate and fumarate to bicyclo[2.1.0]pentane, giving norbornanes, proceed with predominant retention of alkene configuration in contrast to the corresponding thermal reactions which are non-stereospecific.[28] Evidence, as yet inconclusive, has been obtained to suggest that certain intramolecular $\pi^2 + \pi^2$ cycloadditions may involve two steps, viz. a di-π-methane rearrangement followed by a $\sigma^2 + \pi^2$ addition.[29]

2 + 3-Cycloaddition

The variation in the Hammett ρ-constants for the 1,3-dipolar addition of aryl azides to a variety of C=C systems (e.g. maleic anhydride, $\rho = -1.1$; *N,N*-tetramethylenecyclohexenamine, $\rho = +2.54$) has been discussed from the viewpoint of FMO theory.[30] Study of the relative rates of addition of phenyl azide to alkenes has shown that strained double bonds react faster with these 1,3-dipoles.[31] Methyl diazoacetate adds to 3,3-dimethylcyclopropene to give the dihydropyridazine (**11**) and the 2:1 adduct (**12**), both resulting from ring-opening of the first-formed product of 2 + 3-cycloaddition.[31] Both 1:1 and 2:1 adducts can be obtained from the cycloaddition of diazoalkanes to 7-substituted norbornadienes;[32] the preferential formation of the *endo,anti*-monoadducts from the 7-halonorbornadienes has been attributed to a direct interaction between the HOMO(C=C) and σ^*-orbital of the carbon—halogen bond.

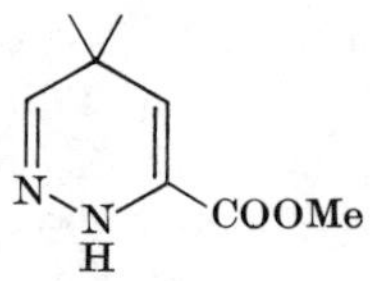

(11)

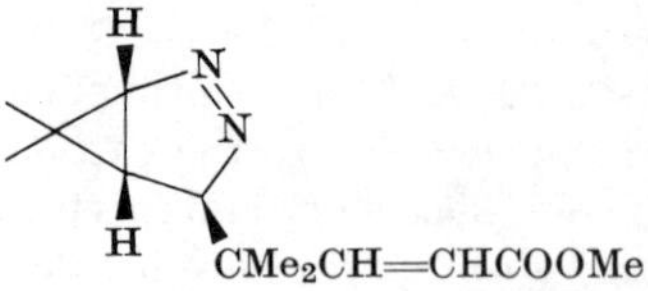

(12)

The regioselectivity of the 2 + 3-additions of diazoalkanes to carbonyl-activated alkenes and alkynes is sensitive to the nature of substituents attached to the dipolarophile and the carbon atom of the dipole.[33] The spiropyrazolenines formed in additions of α-diazocycloalkanones to activated acetylenes undergo a spontaneous [1,5]-shift (C → N) of the carbonyl group,[34] and analogous rearrangements have been encountered in additions of diazofluorene to alkynes.[35] Steric and electronic factors controlling the ratio of stereoisomeric *exo*-Δ^1-pyrazolines formed in the reactions of diazoalkyl phosphonates, $RC(N_2)P(O)(OMe)_2$, with norbornadiene have been discussed.[36] A kinetic study of the additions of Ph_2CN_2 to unsaturated phosphonates of the type $(RO)_2P(O)CR{=}CH_2$ suggests a cyclic transition state;[37] depending on the conditions, the first-formed Δ^1-pyrazolines either rearrange to the Δ^2-isomers or eliminate N_2 giving cyclopropanes. The Δ^1-pyrazolines formed by addition of silyldiazoalkanes [e.g. $Me_3SiC(Ph)N_2$] to diethyl maleate and fumarate, on the other hand, undergo a thermal [1,3]-shift (C → N) of the silyl group, giving Δ^2-pyrazolines.[38] The relative proportions of Δ^2-1,2,3- and Δ^3-1,3,4-thiadiazolines resulting from addition of diazomethane to adamantanethione vary considerably with solvent [cf. $\Delta^3/\Delta^2 = 6.7$ (petrol), 0.28 (methanol)].[39]

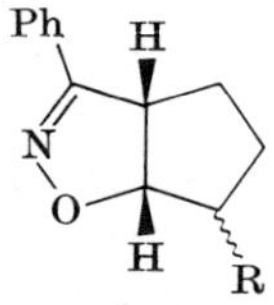

(13)

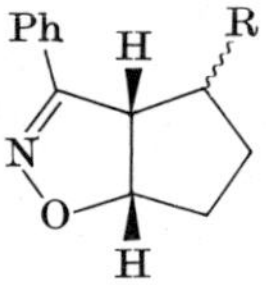

(14)

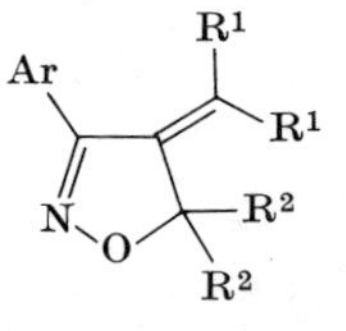

(15)

(**a**) $R^1 = H$; $R^2 = Ph$

(**b**) $R^1 = Ph$; $R^2 = H$

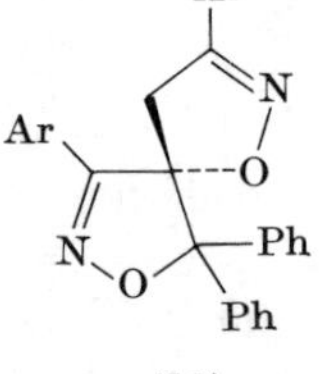

(16)

The importance of steric and polar effects in the addition of benzonitrile oxide to 3-substituted cyclopentenes has been evaluated through analysis of the distributions of the four possible isomeric products, namely *syn*- and *anti*-R-(**13**) and *syn*- and *anti*-R-(**14**), formed in each case.[40] Steric factors favour the formation of the *syn*-isomer of a given epimeric pair, while the more nucleophilic carbon atom of the cyclopentene

C=C bond interacts preferentially with the carbon atom of the CNO group in the transition state. Variation in the rates of formation of 2 + 3-cycloadducts from the additions of diazomethane and benzonitrile oxide to arylacetylenes can be rationalized by consideration of FMO interactions.[41] In a related study of the kinetics of addition of ArCNO to phenylacetylene, a reaction constant $\rho = +1.04$ has been evaluated for the 2 + 3-cycloaddition.[42] A reinvestigation of the reaction of 3,5-dichloro-2,4,6-trimethylbenzonitrile oxide with 1,1-diphenylallene has established[43] the correct structures of the resulting monoadducts (**15a**) and (**15b**) and bisadduct (**16**) which had previously been incorrectly assigned. Valence tautomerism of the 2 + 3-cycloadducts derived from bullvalene and a variety of 1,3-dipolar addends (e.g. RCNO) has been studied by NMR spectroscopy.[44] Other 1,3-dipolar cycloadditions which have been investigated include additions of azomethine ylides to activated alkenes,[45] and of mesoionic imidazoles[46] and Reissert salts[47] to alkenes and alkynes. Intramolecular 2 + 3-cycloadditions between nitrilimine and alkyne groups have been reported.[48]

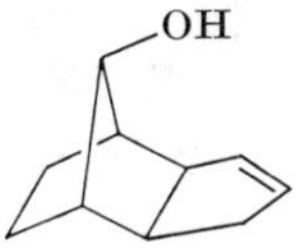

OH

(**17**)

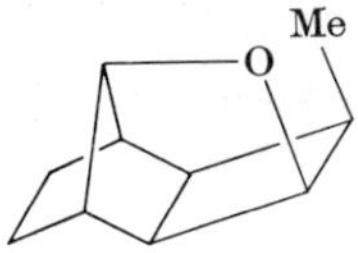

Me
O

(**18**)

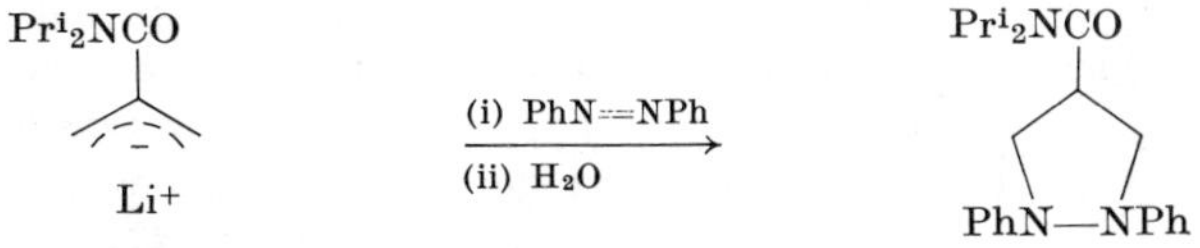

Pr^i_2NCO

Li^+

(**19**)

(i) PhN=NPh
(ii) H_2O

Pr^i_2NCO

PhN—NPh

Treatment of *syn*-7-allyloxynorbornene with methyl-lithium followed by hydrolysis gives a mixture of the alcohol (**17**) and cyclic ether (**18**), but it has not been established whether a concerted intramolecular cycloaddition (ene + allyl anion) or a two-step mechanism operates.[49] Analogous products can be obtained by similar treatment of *syn*-7-benzyloxynorbornene. A two-step mechanism has been established, however, for the addition of the 2-amidoallyl-lithium (**19**) to azobenzene which affords the pyrazolidine shown.[50] Pyrrolidine derivatives have been obtained in good yields from 2 + 3-additions of 2-aza-1,3-diphenylallyl-lithium to a series of vinyl- and allyl-silanes.[51]

2 + 4-Cycloaddition

Several investigations have been reported[52–55] concerning the effect of pressure and temperature upon Diels–Alder reactions and the topic has been reviewed.[56] A correlation has been found between the free energy of activation and activation volume for a range of $\pi^2 + \pi^4$ additions, and a contribution to volume contraction resulting from secondary orbital interactions has been suggested.[52] The activation volume appears to be a good diagnostic probe for the transition state in such cycloadditions. In the reaction of methyl acrylate with penta-1,3-diene, increase in pressure leads to a significant increase in the *cis*:*trans* ratios for both the 4- and 5-methoxycarbonyl-3-methylcyclohexene products. Increase in temperature, on the other hand, produces a decrease in these *cis*:*trans* ratios and a more marked decrease in the 4-COOMe/5-COOMe product ratios for both *cis*- and *trans*-stereoisomers.[55] The rate of cycloaddition of maleic anhydride to isoprene in the vicinity of the upper critical solution points of the two liquid–liquid mixtures shows an anomalous increase, possibly related to the abrupt changes of the solution activities of dilute solutes near a critical point.[57] The enthalpy of addition of maleic anhydride to cyclopentadiene has been determined by flow calorimetry.[58]

The regioselectivity observed in Diels–Alder reactions of fulvenes with unsymmetrically substituted cyclic 1,3-dienes can be rationalized by an FMO model in which the HOMO(fulvene)–LUMO(diene) interaction exerts a controlling effect;[59] in each reaction studied, the fulvene acts as dienophile and *endo*-periselectivity invariably predominates. The results of a kinetic study of Diels–Alder reactions with inverse electron demand (e.g. norbornene + tropone) have also been interpreted on the basis of FMO theory, photoelectron-spectroscopic data being used as a basis for a qualitative perturbational treatment.[60] Stereoselectivity and stereospecificity of 2 + 4-additions of perchlorocyclopentadienes to norbornadienes have been investigated[61,62] and the low reactivity of perchlorodienophiles in Diels–Alder reactions has been noted.[63] Octachlorofulvalene functions as a dienophile in reactions with conjugated dienes (e.g. furan, cyclopentadiene) but as a diene in addition to norbornene.[64] Relative reactivity, selectivity and stereochemical studies of Diels–Alder reactions of cyclopentadiene[65] (with activated alkenes), tetracyclones (with alkenes[66] and styrenes[67]), anthracenes (with maleic anhydride[68] and *N*-arylmaleimides[69]), and naphthalene[70] (with *N*-amidomaleimides) have been reported.

The valence-tautomers (**20a**) and (**20b**) of 2,3,4,5,6,7-[$^2H_6$]-1-substituted cyclooctatetraenes are the only ones which give cyclo-adducts with a series of dienophilic alkenes; with the more reactive dienophiles, k_1 was effectively rate-determining and was found to be subject to a small secondary deuterium isotope effect ($k_H/k_D \simeq 1.08$). However, with less reactive dienophiles, when equilibration of the valence tautomers preceded cycloaddition a very much greater isotope effect was found (1.5). This must principally be attributable to a large equilibrium isotope effect, since hybridization at the site of isotopic discrimination does not alter in the cycloaddition step.[71]

R   H   D6 k_1 k_{-1} D   R   H   D4   D   +   D   D   R   D4   H

(**20a**)   (**20b**)

R = Me; COOEt;

The influence of Lewis acids upon the course of Diels–Alder reactions of furan with ethyl propiolate[72] and dimethyl acetylenedicarboxylate,[73] and of β-nitrostyrene with 1,2-[74] and 1,4-diphenylbuta-1,3-diene[75] has been studied. Bradsher *et al.* have studied the polar Diels–Alder additions of alkenes to acridizinium cations and have found that the rates of reaction are very sensitive to substitution in the cation and to the nature of the alkene.[76] A linear relationship between HOMO(diene) energy and relative reaction rates has been found for additions of maleic anhydride to 1-arylbuta-1,3-dienes;[77] regioselectivity in these reactions is apparently controlled by the terminal electron densities of the diene frontier orbital. Product analysis of the 2 + 4-additions of substituted cyclopentadienes to activated alkenes has underlined the importance of steric effects in determining *endo*-selectivity.[78] The stereochemistry of the 2:1 adduct from cyclopentadiene and norbornene has been elucidated.[79]

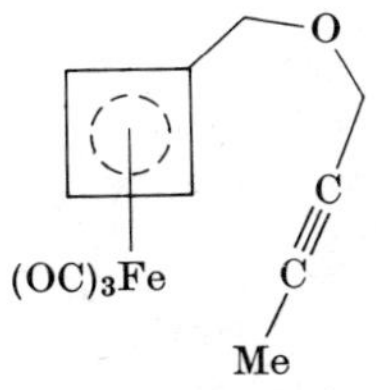

(21) 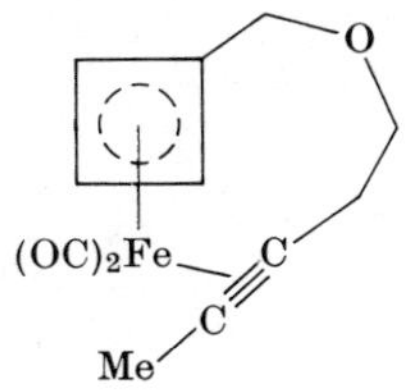 **(22)**

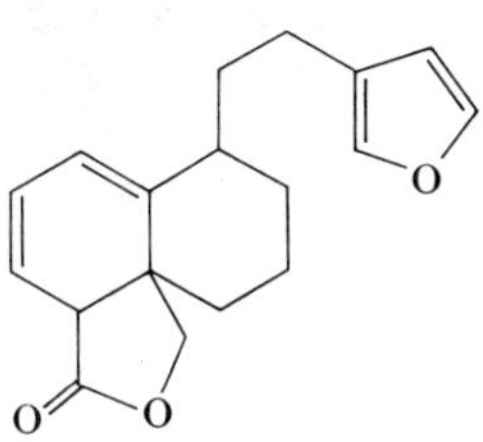

(23) 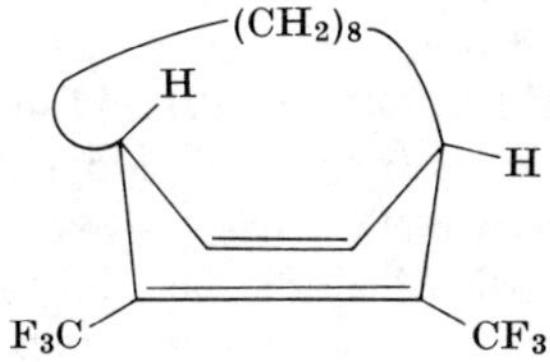 **(24)**

Treatment of the cyclobutadiene-iron complexes (**21**) and (**22**) with a cerium(IV) salt gives 4-methylphthalan and 5-methylisochroman, respectively.[80] These products probably arise through intramolecular capture of the cyclobutadiene group of the liberated ligand by the alkyne and aromatization of the resulting Dewar-benzene derivatives. Intramolecular Diels–Alder reactions of 6-allylpentachlorocyclohexa-2,4-dienone,[81] *N*-alkenyl-4-cyano-1,2-dihydropyridines,[82] and the furan derivative[83] (**23**) have been described. The "inside-outside" bicyclo-diene (**24**) can be obtained in 78% yield by addition of hexafluorobut-2-yne to *cis,trans*-cyclododeca-1,3-diene.[84] Predominant *endo*-selectivity has been found for additions of 1,2-dicyanocyclobutene to cycloalka-1,3-dienes whereas, with furan, the *exo*-adduct is the major product.[85] Thermolysis of this cyclobutene affords 2,3-dicyanobuta-1,3-diene which undergoes normal Diels–Alder additions to alkenes.[85]

Holmberg *et al.* have elucidated the structure and stereochemistry of the products of 2 + 4-dimerization of a number of *o*-quinols and related cyclohexa-2,4-dienones.[86] Coumalic acid and its derivatives behave as dienes in 2 + 4-additions to acenaphthylene.[87] The resulting cycloadducts readily eliminate CO_2 to give cyclohexa-1,3-dienes which either aromatize, giving fluoranthenes or give products of 2 + 4-cycloaddition of a second acenaphthylene molecule. In a similar reaction, acenaphthylene and *sym*-tetrazines give diazafluoranthenes after elimination of nitrogen.[87] Methyl coumalate, on the other hand, reacts both as a diene and as a dienophile in cycloadditions to acyclic 1,3-dienes, but exclusively as a diene in reactions with cyclic 1,3-dienes.[88]

Differentiation between concerted and two-step mechanisms for the cycloaddition of amidomethylium ions, $RCON^+H{=}CH_2$, to 2-arylpropenes is possible through investigation of the stereochemistry of the oxazine products.[89] The tetramethylketene immonium ion, $Me_2C{=}C{=}N^+Me_2$, reacts with cyclic 1,3-dienes to give products of 2 + 4-cycloaddition of the C=N bond whereas, with most acyclic 1,3-dienes, products of 2 + 2-cycloaddition of the C=C bond result. With 2,3-dimethylbuta-1,3-diene, products of both 2 + 2- and 2 + 4-cycloaddition are formed.[90]

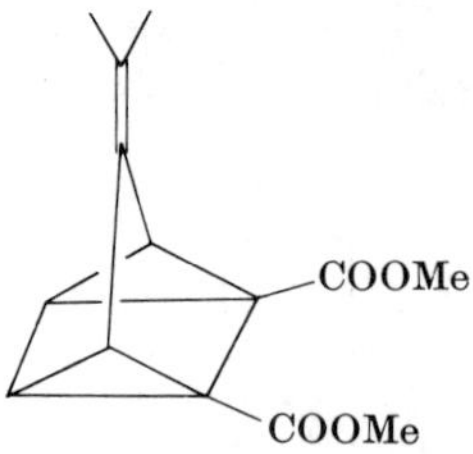

COOMe
COOMe

(25)

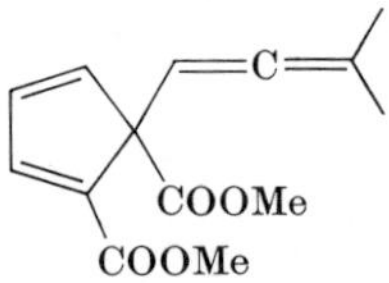

COOMe
COOMe

(26)

The allenylcyclopentadiene (**26**), obtained by treatment of the quadricyclane (**25**) with $AgBF_4$, reacts with dienophiles to give products of 2 + 4-cycloaddition to the 1,3-diene system.[91] With acrolein, 1- and 2-ethoxy-1,3-dienes give preferentially *cis*-substituted cyclohexene adducts which epimerize in the presence of base to the more stable *trans*-substituted isomers.[92] It has also been found that 1-alkoxy-1,3-dienes add to aldehydes and alkyl glyoxylates giving dihydropyrans, and the stereochemistry of the reaction has been investigated.[93] Other 2 + 4-cycloadditions involving heteroatoms studied include the addition of 1,3-dienes to thiofluorenone,[94] and the reactions of aroyl isothiocyanates with Schiff bases and with iminodithiocarbonates.[95]

Other Cycloadditions

Reactions of the strained hydrocarbon (**27**) with a range of dienophilic reagents have been studied,[96] both under thermal conditions and in the presence of $AgClO_4/Na_2CO_3$. Evidence has been found that the first-formed products are those (**28**) of normal Diels–Alder addition of the dienophiles (X═X) to the 1,3-diene system, but that these rearrange thermally or, more efficiently, in the presence of silver(I) to give the arenes (**29**).

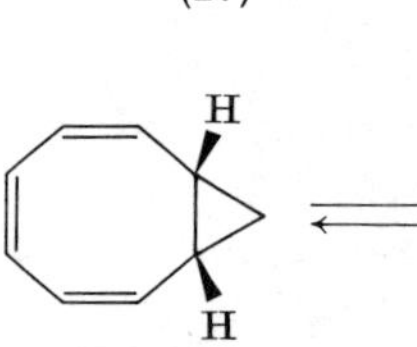

(27) 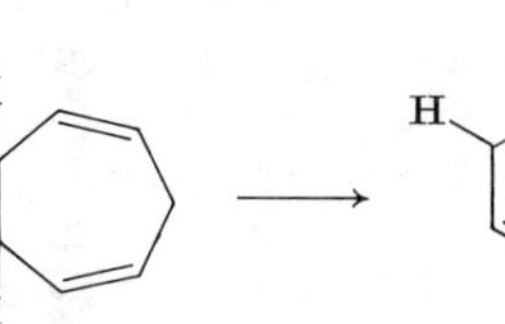 (28) 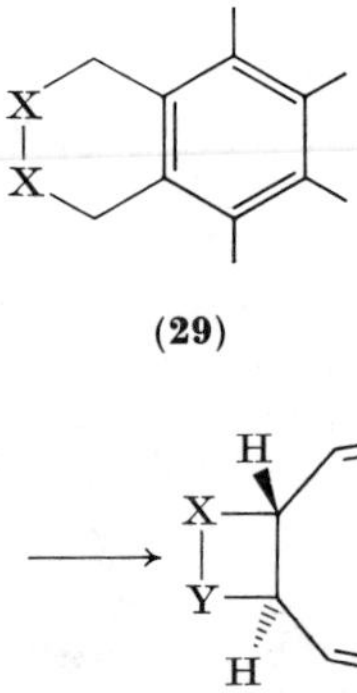 (29)

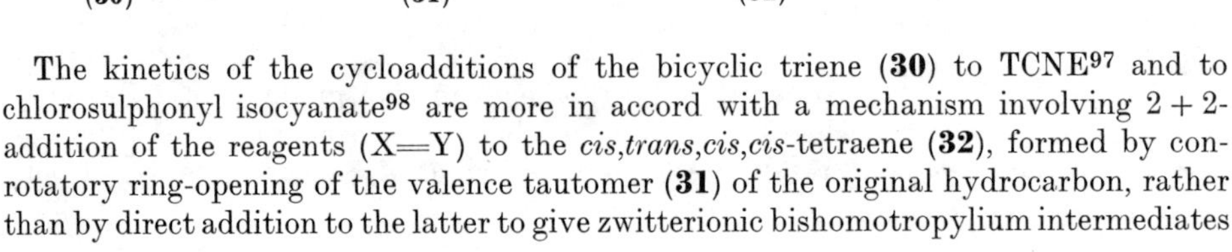

(30) (31) (32)

The kinetics of the cycloadditions of the bicyclic triene (**30**) to TCNE[97] and to chlorosulphonyl isocyanate[98] are more in accord with a mechanism involving 2 + 2-addition of the reagents (X═Y) to the *cis,trans,cis,cis*-tetraene (**32**), formed by conrotatory ring-opening of the valence tautomer (**31**) of the original hydrocarbon, rather than by direct addition to the latter to give zwitterionic bishomotropylium intermediates as previously proposed.[99] A 2 + 8-cycloadduct obtained from the reaction of (**30**) with 1,3-diphenylisobenzofuran is also thought to be derived from the tetraene (**32**).[100] An ionic two-step mechanism is suggested for the 2 + 6-cycloadditions of nitrosobenzenes to cycloheptatriene.[101]

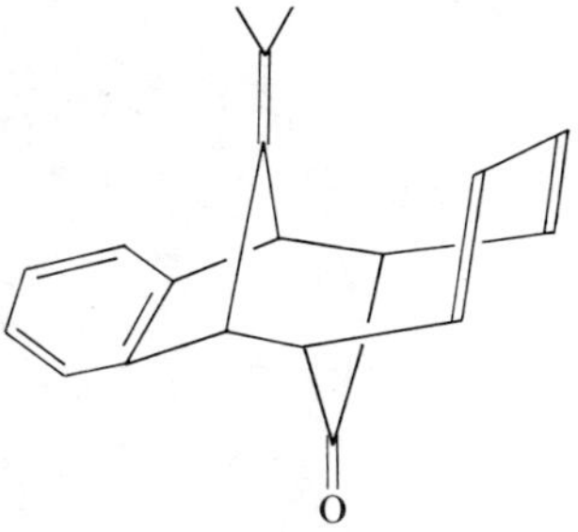

(**33**) 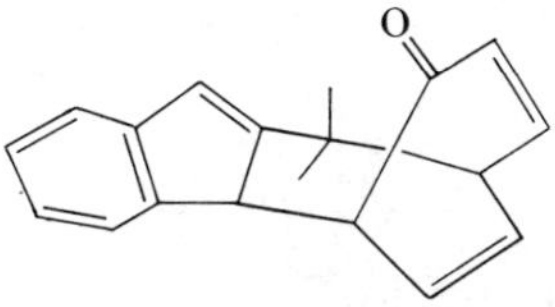 (**34**)

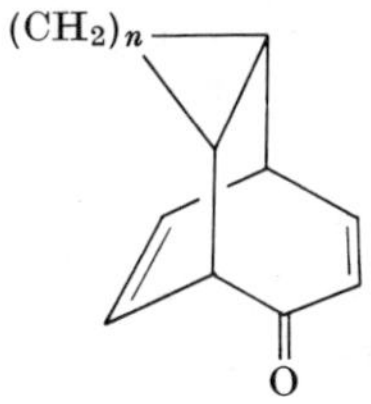

(**35**) 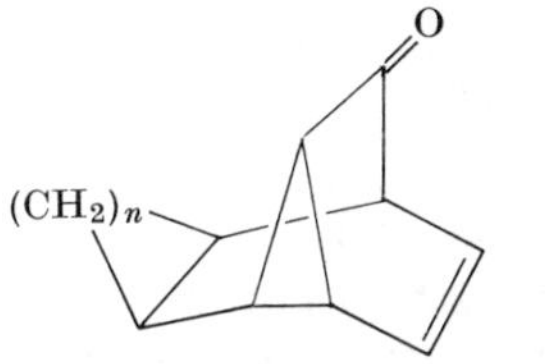 (**36**)

In agreement with theoretical prediction,[5] 8,8-dimethylisobenzofulvene undergoes π^8s + π^6s cycloaddition to tropone, giving the adduct (**33**) which is converted into the isomer (**34**) at 120° by Cope rearrangement.[102] Reactions of tropone with benzonitrile oxides give complex mixtures of products formed both by 1,3-dipolar cycloaddition to the carbonyl group and by cycloadditions to the conjugated alkene system.[103] Simple cycloalkenes react with tropylium perchlorate in aqueous dioxan to give alcohol mixtures which afford ketones of the types (**35**) and (**36**) upon CrO_3-oxidation.[104]

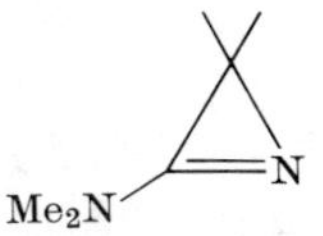

(**37**) 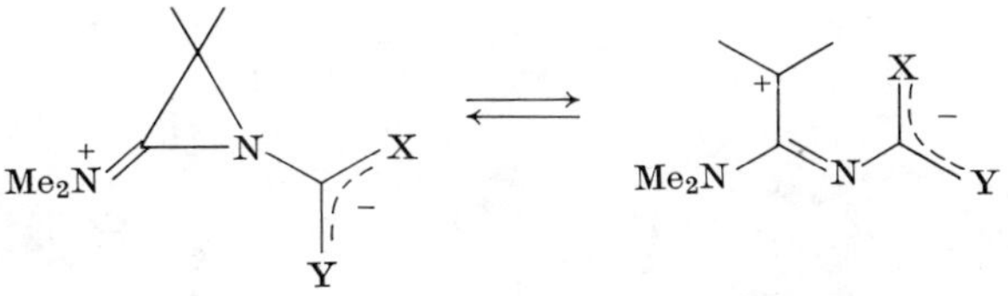 (**38**)

Thermal reactions of the azirine (**37**) with heterocumulenes (RNCO, RNCS, CS_2, ketenes, thioketenes) give cycloadducts derived from intermediate dipolar species of the type (**38**).[105] Analogous zwitterions are formed in reactions of dimethylketene *O,N*-acetals with isocyanates and isothiocyanates.[106] Diphenylketene undergoes both 2 + 2- and 4 + 2-cycloaddition reactions with 1,3-dienes, giving cyclobutanones and dihydropyrans, respectively, by different modes of closure of common dipolar intermediates.[107] Acyclic 2:1 adducts, on the other hand, result from thermal reactions of diphenylketene with diaziridines, whereas the latter give triazolidinones when heated with benzoyl isocyanate.[108] Photochemical reactions of diketene with azobenzenes give hexahydropyridazine-3,5-diones.[109]

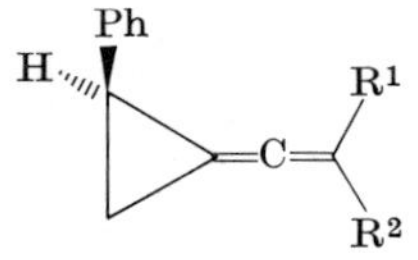

(39)

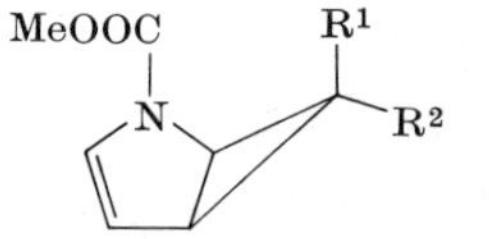

(40)

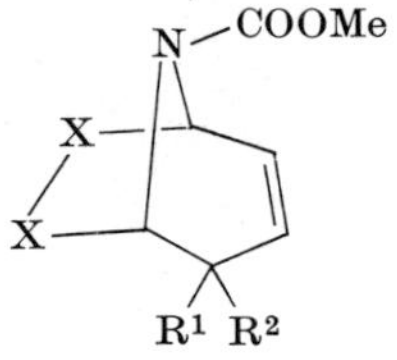

(41)

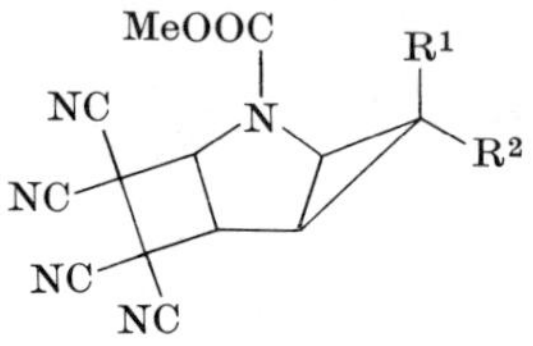

(42)

A detailed study of the stereochemical outcome of reactions of chiral allenes of the type (**39**) with 4-phenyl-1,2,4-triazoline-3,5-dione leads to the conclusion that the process represents a concerted $(\pi^2 + \pi^2 + \sigma^2) + \pi^2$ cycloaddition involving *both* double bonds *and* the cyclopropane C-2—C-3 bond of the allene in an 8-electron Möbius-type transition state.[110] The kinetics and stereochemistry of reactions of azabicyclohexenes (**40**) with electrophilic addends have been studied;[111] with dimethyl acetylenedicarboxylate and *N*-phenylmaleimide, products of the type (**41**) result whereas, with TCNE, azatricyclo-octanes (**42**) are formed via an intermediate dipolar species. The stereochemistry of the ene reactions of deuterium-labelled β-pinenes with benzyne and with methyl phenylglyoxylate has established a stereoselective migration of the allylic methylene H(D) atom which is located *anti* to the CMe_2 bridge.[112] Full details of the $\pi^2 + \pi^4$ cycloadditions to furan of allylic cations derived from α,α-dibromocycloalkanones have appeared,[113] and the formation of cyclic adducts by stepwise addition of pentadienyl-lithium to 1,3-dienes has been described.[114]

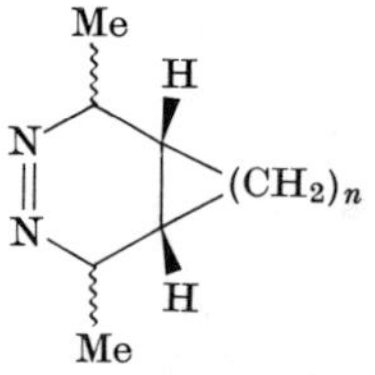

(43)

Berson *et al.* have reported details of an elegant study of 2 + 2 + 2-cyclorevision processes.[115] Diazabicycloalkanes of the type (**43**; $n = 1$) lose N_2 when heated, giving hepta-2,5-dienes. The concerted nature of these reactions is shown by the enhanced rates, the lack of ring-closure products, and the generation of the dienes with high (>99.5%) stereospecificity by an "inward disrotatory" process. The cyclobutane analogues (**43**; $n = 2$) decompose to octa-2,6-dienes by a similar concerted pathway; although much slower reaction rates are observed, stereospecificity remains high.[115]

From the close quantitative correlation found between the log rates of alkene epoxidation and alkyne peroxidation, it has been concluded that the reactions of alkynes with per-acids may involve oxirene intermediates.[116] Lead tetra-acetate oxidation of *N*-aminoimides in the presence of alkenes gives aziridines by stereospecific capture of an intermediate nitrene.[117]

References

1 N. D. Epiotis, *Angew. Chem. Internat. Ed.*, **13**, 751 (1974).
2 *Org. Reaction Mech.*, **1973**, 397.
3 J. B. Hendrickson, *Angew. Chem. Internat. Ed.*, **13**, 47 (1974).
4 F. D. Mango, *Fortschr. Chem. Forsch.*, **45**, 39 (1974).
5 M. N. Paddon-Row, *Austral. J. Chem.*, **27**, 299 (1974).
6 J. Bastide and O. Henri-Rousseau, *Bull. Soc. Chim. France*, **1974**, 1037.
7 R. Grée, F. Tonnard and R. Carrié, *Tetrahedron Letters*, **1974**, 135.
8 T. Okada, K. Yamaguchi and T. Fueno, *Tetrahedron*, **30**, 2293 (1974).
9 M. J. S. Dewar and S. Kirschner, *J. Am. Chem. Soc.*, **96**, 5246 (1974).
10 P. D. Bartlett and G. M. Cohen, *J. Am. Chem. Soc.*, **95**, 7923 (1973).
11 M. P. Votinov, V. A. Kosobutskii and A. F. Dokukina, *Vysokomol. Soedin., Ser. A*, **10**, 1137 (1968); *Polymer Sci. USSR*, **10**, 1318 (1968).
12 R. Huisgen, R. Schug and G. Steiner, *Angew. Chem. Internat. Ed.*, **13**, 80 (1974).
13 R. Huisgen, R. Schug and G. Steiner, *Angew. Chem. Internat. Ed.*, **13**, 81 (1974).
14 F. Effenberger and O. Gerlach, *Chem. Ber.*, **107**, 278 (1974).
15 W. J. Gensler and A. Whitehead, *J. Org. Chem.*, **38**, 3843 (1973).
16 N. S. Isaacs and B. G. Hatcher, *Chem. Comm.*, **1974**, 593.
17 L. A. Feiler, R. Huisgen and P. Koppitz, *J. Am. Chem. Soc.*, **96**, 2270 (1974).
18 H. A. Bampfield and P. R. Brook, *Chem. Comm.*, **1974**, 171.
19 H. A. Bampfield and P. R. Brook, *Chem. Comm.*, **1974**, 172.
20 S. Moon and T. F. Kolesar, *J. Org. Chem.*, **39**, 995 (1974).
21 R. C. Kerber, T. J. Ryan and S. D. Hsu, *J. Org. Chem.*, **39**, 1215 (1974).
22 R. C. Kerber and M. C. Cann, *J. Org. Chem.*, **39**, 2552 (1974).
23 V. N. Drozd, V. V. Sergeichuk and N. D. Antonova, *Zh. Org. Khim.*, **10**, 1498 (1974); *Chem. Abs.*, **81**, 104605 (1974).
24 S. Inagaki, T. Minato, S. Yamabe, H. Fujimoto and K. Fukui, *Tetrahedron*, **30**, 2165 (1974).
25 D. Kaufmann and A. de Meijere, *Tetrahedron Letters*, **1974**, 3831.
26 A. de Meijere, D. Kaufmann and O. Schallner, *Tetrahedron Letters*, **1974**, 3835.
27 H.-D. Martin, *Chem. Ber.*, **107**, 477 (1974); H.-D. Martin and H. L. Grafetstätter, *ibid.*, p. 680.
28 R. Noyori, Y. Kumagai and H. Takaya, *J. Am. Chem. Soc.*, **96**, 634 (1974).
29 H. Hart and M. Kuzuya, *J. Am. Chem. Soc.*, **96**, 3709 (1974).
30 R. Sustmann, *Tetrahedron Letters*, **1974**, 963; see also V. V. Mel'nikov, I. V. Tselinskii, S. L. Churakov and B. V. Gidaspov, *Zh. Org. Khim.*, **10**, 1360 (1974); *Chem. Abs.*, **81**, 90727 (1974).
31 D. H. Aue and G. S. Helwig, *Tetrahedron Letters*, **1974**, 721.
32 M. Franck-Neumann and M. Sedrati, *Angew. Chem. Internat. Ed.*, **13**, 606 (1974).
33 J. Bastide, O. Henri-Rousseau and L. Aspart-Pascot, *Tetrahedron*, **30**, 3355 (1974).
34 M. Martin and M. Regitz, *Annalen*, **1974**, 1702.
35 H. Dürr and W. Schmidt, *Annalen*, **1974**, 1140.
36 H. Cohen and C. Benezra, *Can. J. Chem.*, **52**, 66 (1974).
37 A. N. Pudovik, R. D. Gareev, L. A. Stabrovskaya, G. I. Evstaf'ev and A. B. Remizov, *Zh. Obshch. Khim.*, **43**, 1674 (1973); *Chem. Abs.*, **79**, 145603 (1973).
38 A. R. Bassindale and A. G. Brook, *Can. J. Chem.*, **52**, 3474 (1974).
39 A. P. Krapcho, M. P. Silvon, I. Goldberg and E. G. E. Jahngen, *J. Org. Chem.*, **39**, 860 (1974).
40 P. Caramella and G. Cellerino, *Tetrahedron Letters*, **1974**, 229.
41 J. Bastide, O. Henri-Rousseau and E. Stephan, *Compt. rend., Ser. C*, **278**, 195 (1974).
42 A. Dondoni and G. Barbaro, *J.C.S. Perkin II*, **1974**, 1591.
43 P. Beltrame, P. L. Beltrame, M. G. Cattania and G. Zecchi, *J.C.S. Perkin II*, **1974**, 1301.

[44] A. Gamba Invernizzi, R. Gandolfi and M. Strigazzi, *Tetrahedron*, **30**, 3717 (1974).
[45] F. Texier and R. Carrié, *Bull. Soc. Chim. France*, **1974**, 310.
[46] G. Singh and P. S. Pande, *Tetrahedron Letters*, **1974**, 2169.
[47] W. E. McEwan, P. E. Stott and C. M. Zepp, *J. Am. Chem. Soc.*, **95**, 8452 (1973).
[48] R. Fusco, L. Garanti and G. Zecchi, *Tetrahedron Letters*, **1974**, 269.
[49] G. W. Klumpp and R. F. Schmitz, *Tetrahedron Letters*, **1974**, 2911.
[50] W. Bannwarth, R. Eidenschink and T. Kauffmann, *Angew. Chem. Internat. Ed.*, **13**, 468 (1974).
[51] E. Popowski, *Z. Chem.*, **14**, 360 (1974).
[52] K. Seguchi, A. Sera and K. Maruyama, *Bull. Chem. Soc. Japan*, **47**, 2242 (1974).
[53] J. Rimmelin and G. Jenner, *Tetrahedron*, **30**, 3081 (1974).
[54] J. R. McCabe and C. A. Eckert, *Ind. Eng. Chem. Fundam.*, **13**, 168 (1974); *Chem. Abs.*, **81**, 77291 (1974).
[55] S. K. Shakhova and B. S. El'yanov, *Izv. Akad. Nauk SSSR, Ser. Khim.*, **1973**, 1504; *Chem. Abs.*, **79**, 114806 (1973).
[56] J. R. McCabe and C. A. Eckert, *Accounts Chem. Res.*, **7**, 251 (1974).
[57] R. B. Snyder and C. A. Eckert, *A. I. Ch. E. J.*, **19**, 1126 (1973); *Chem. Abs.*, **80**, 59129 (1974).
[58] K. J. Breslauer and D. S. Kabakoff, *J. Org. Chem.*, **39**, 721 (1974).
[59] K. N. Houk and L. J. Luskus, *J. Org. Chem.*, **38**, 3836 (1973).
[60] H. R. Pfaendler, H. Tanida and E. Haselbach, *Helv. Chim. Acta*, **57**, 383 (1974).
[61] L. T. Byrne, A. R. Rye and D. Wege, *Austral. J. Chem.*, **27**, 1961 (1974).
[62] K. MacKenzie, *Tetrahedron Letters*, **1974**, 1203.
[63] A. I. Naimushin, V. V. Simonov and A. F. Annshchenko, *Khim. Vysokomol. Soedin. Neftekhim.*, **1973**, 17; *Chem. Abs.*, **80**, 132579 (1974).
[64] V. Mark, *Chem. Comm.*, **1973**, 910.
[65] A. I. Konovalov, G. I. Kamasheva and M. P. Loskutov, *Zh. Org. Khim.*, **9**, 2048 (1973); *Chem. Abs.*, **80**, 26682 (1974).
[66] A. I. Konovalov, Ya. D. Samuilov and V. I. Uba, *Zh. Org. Khim.*, **9**, 2084 (1973); *Chem. Abs.*, **80**, 26633 (1974).
[67] A. I. Konovalov, Ya. D. Samuilov, L. F. Slepova and V. A. Breus, *Zh. Org. Khim.*, **9**, 2086 (1973); *Chem. Abs.*, **80**, 26659 (1974); A. I. Konovalov and B. N. Solomonov, *Dokl. Akad. Nauk SSSR*, **211**, 1115 (1973); *Chem. Abs.*, **79**, 125446 (1973).
[68] V. D. Kiselev and A. I. Konovalov, *Zh. Org. Khim.*, **10**, 6 (1974); *Chem. Abs.*, **80**, 107814 (1974).
[69] A. I. Konovalov, B. N. Solomonov and A. N. Ustyugov, *Dokl. Akad. Nauk SSSR*, **213**, 349 (1973); *Chem. Abs.*, **80**, 69955 (1974).
[70] S. M. Verma and O. Subba Rao, *Tetrahedron*, **30**, 2371 (1974).
[71] L. A. Paquette, W. Kitching, W. E. Heyd and R. H. Meisinger, *J. Am. Chem. Soc.*, **96**, 7371 (1974).
[72] A. W. McCulloch, D. G. Smith and A. G. McInnes, *Can. J. Chem.*, **52**, 1013 (1974).
[73] A. W. McCulloch, D. G. Smith and A. G. McInnes, *Can. J. Chem.*, **51**, 4125 (1973).
[74] Y. N. Mukerjee, P. C. Jain and N. Anand, *Indian J. Chem.*, **12**, 331 (1974); *Chem. Abs.*, **81**, 119468 (1974).
[75] P. C. Jain, Y. N. Mukerjee and N. Anand, *J. Am. Chem. Soc.*, **96**, 2996 (1974).
[76] C. K. Bradsher and F. H. Day, *J. Heterocyclic Chem.*, **10**, 1031 (1973); C. K. Bradsher, N. A. Porter and T. G. Wallis, *J. Org. Chem.*, **39**, 1172 (1974); N. A. Porter, I. J. Westerman, T. G. Wallis and C. K. Bradsher, *J. Am. Chem. Soc.*, **96**, 5104 (1974).
[77] P. V. Alston and R. M. Ottenbrite, *J. Org. Chem.*, **39**, 1584 (1974).
[78] J. M. Mellor and C. F. Webb, *J.C.S. Perkin II*, **1974**, 17; B. C. C. Cantello, J. M. Mellor and C. F. Webb, *ibid.*, p. 22; J. M. Mellor and C. F. Webb, *ibid.*, p. 26.
[79] J. Mantzaris and E. Weissberger, *J. Org. Chem.*, **39**, 726 (1974).
[80] R. H. Grubbs, T. A. Pancoast and R. A. Grey, *Tetrahedron Letters*, **1974**, 2425; cf. J. Rebek and F. Gaviña, *J. Am. Chem. Soc.*, **96**, 7112 (1974).
[81] G. M. Brooke, *J.C.S. Perkin I*, **1974**, 233.
[82] H. Greuter and H. Schmid, *Helv. Chim. Acta*, **57**, 1204 (1974).
[83] E. L. Ghisalberti, P. R. Jefferies and T. G. Payne, *Tetrahedron*, **30**, 3099 (1974).
[84] P. G. Gassman, S. R. Korn and R. P. Thummel, *J. Am. Chem. Soc.*, **96**, 6948 (1974).
[85] D. Belluš, K. von Bredow, H. Sauter and C. Weis, *Helv. Chim. Acta*, **56**, 3004 (1973).
[86] E. Adler and K. Holmberg, *Acta Chem. Scand.*, *B*, **28**, 465, 549 (1974); K. Holmberg, *ibid.*, p. 857; K. Holmberg, H. Kirudd and G. Westin, *ibid.*, p. 913.
[87] T. Sasaki, K. Kanematsu and T. Hiramatsu, *J.C.S. Perkin I*, **1974**, 1213.
[88] T. Imagawa, N. Sueda and M. Kawanisi, *Tetrahedron*, **30**, 2227 (1974).
[89] R. R. Schmidt and A. R. Hoffmann, *Chem. Ber.*, **107**, 78 (1974).

[90] J. Marchand-Brynaert and L. Ghosez, *Tetrahedron Letters*, **1974**, 377.
[91] S. F. Nelsen, J. P. Gillespie, P. J. Hintz and E. D. Seppanen, *J. Am. Chem. Soc.*, **95**, 8380 (1973).
[92] J. P. Gouesnard, G. J. Martin and M. Blain, *Tetrahedron*, **30**, 151 (1974); cf. J. Bertran, V. Forero, F. Mora and J. I. Fernandez-Alonso, *An. Quím.*, **70**, 195 (1974); *Chem. Abs.*, **81**, 24849 (1974).
[93] S. D. Yablonovskaya, N. M. Shekhtman, S. V. Bogatkov, S. M. Makin and N. S.,Zefirov, *Mater. Vses. Konf. Din. Stereokhim. Konformatsionnomu Anal.*, 1st, **1971**, 3; *Chem. Abs.*, **80**, 81866 (1974).
[94] B. König, J. Martens, K. Praefcke, A. Schönberg, H. Schwarz and R. Zeisberg, *Chem. Ber.*, **107**, 2931 (1974).
[95] K. Milzner and K. Seckinger, *Helv. Chim. Acta*, **57**, 1615 (1974).
[96] H. Hogeveen and W. F. J. Huurdeman, *Tetrahedron Letters*, **1974**, 1255; H. Hogeveen, W. F. J. Hurrdeman and E. P. Schudde, *ibid.*, p. 4211.
[97] G. Boche, H. Weber and J. Benz, *Angew. Chem. Internat. Ed.*, **13**, 207 (1974).
[98] J. E. Baldwin and D. B. Bryan, *J. Am. Chem. Soc.*, **96**, 319 (1974).
[99] *Org. Reaction Mech.*, **1973**, 410.
[100] T. Sasaki, K. Kanematsu and Y. Yukimoto, *Heterocycles*, **2**, 1 (1974); *Chem. Abs.*, **80**, 119827 (1974).
[101] S. Itô, S. Narita and K. Endo, *Bull. Chem. Soc. Japan*, **46**, 3517 (1973).
[102] M. N. Paddon-Row and R. N. Warrener, *Tetrahedron Letters*, **1974**, 3797.
[103] C. De Micheli, R. Gandolfi and P. Grünanger, *Tetrahedron*, **30**, 3765 (1974).
[104] S. Itô, I. Itoh, I. Saito and A. Mori, *Tetrahedron Letters*, **1974**, 3887.
[105] E. Schaumann, E. Kausch and W. Walter, *Chem. Ber.*, **107**, 3574 (1974).
[106] E. Schaumann, S. Sieveking and W. Walter, *Chem. Ber.*, **107**, 3589 (1974).
[107] J. P. Gouesnard, *Tetrahedron*, **30**, 3113 (1974).
[108] M. Komatsu, N. Nishikaze, M. Sakamoto, Y. Ohshiro and T. Agawa, *J. Org. Chem.*, **39**, 3198 (1974).
[109] T. Kato, M. Sato and K. Tabei, *J. Org. Chem.*, **39**, 3205 (1974).
[110] D. J. Pasto and J. K. Borchardt, *J. Am. Chem. Soc.*, **96**, 6944 (1974).
[111] S. R. Tanny and F. W. Fowler, *J. Org. Chem.*, **39**, 2715 (1974).
[112] V. Garsky, D. F. Koster and R. T. Arnold, *J. Am. Chem. Soc.*, **96**, 4207 (1974).
[113] J. G. Vinter and H. M. R. Hoffmann, *J. Am. Chem. Soc.*, **96**, 5466 (1974).
[114] M. Newcomb and W. T. Ford, *J. Org. Chem.*, **39**, 232 (1974).
[115] J. A. Berson, E. W. Petrillo and P. Bickart, *J. Am. Chem. Soc.*, **96**, 636 (1974); J. A. Berson, S. S. Olin, E. W. Petrillo and P. Bickart, *Tetrahedron*, **30**, 1639 (1974).
[116] K. M. Ibne-Rasa, R. H. Pater, J. Ciabattoni and J. O. Edwards, *J. Am. Chem. Soc.*, **95**, 7894 (1973).
[117] H. Person, C. Fayat, F. Tonnard and A. Foucaud, *Bull. Soc. Chim. France*, **1974**, 635.

CHAPTER 13

Molecular Rearrangements

A. R. Butler

Department of Chemistry, University of St. Andrews

G. V. Meehan

Department of Chemistry and Biochemistry, James Cook University of North Queensland, Australia

and M. J. Perkins

Department of Chemistry, Chelsea College, University of London

Aromatic Rearrangements

Benzene Derivatives

Competition between $AlCl_3$-catalysed intramolecular alkyl-group migration and intermolecular alkyl transfer in mixtures of propylbenzene and benzene has been studied by a ^{14}C-labelling procedure.[1]

A reaction has been found whereby *o*-acylphenols (**1**; R = H or alkyl) may be converted into *o*-hydroxyanilides;[2] the proposed mechanism, shown, allies this to the Dakin reaction.

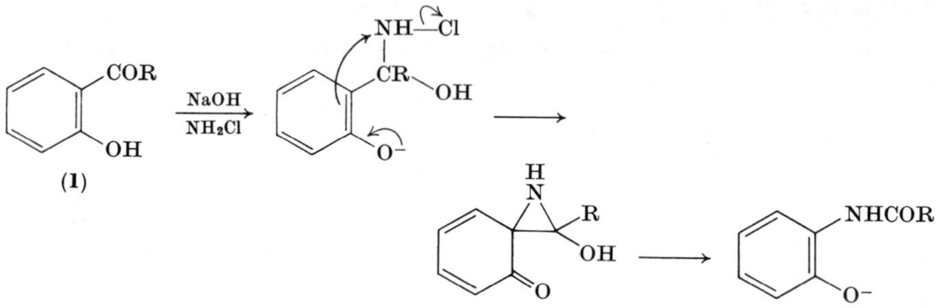

(**1**)

In new work on strong acids,[3] the concentrations or activities of various individual species in sulphuric acid have been summarized as a function of $-H_0$; it was suggested that HSO_4^- rather than H_2O should act as a nucleophile or base in some reactions, especially above 80% H_2SO_4; the Wallach rearrangement of azoxybenzene was selected to exemplify general acid-catalysed reactions in concentrated sulphuric acid and was found to be catalysed by H_2SO_4 and $H_3SO_4^+$, but not by H_3O^+; the results were consistent with the intermediacy of $Ph\overset{+}{N}\equiv\overset{+}{N}Ph$ as proposed by Buncel. The substitution patterns in the products of Wallach rearrangement of azoxynaphthalenes[4] and substituted azoxybenzenes[5] have been examined; in the latter studies, sulphuric acid and chlorosulphonic acid gave different results.

The photochemical rearrangement of *N,N'*-dimethylhydrazoarenes has been examined for mixtures of unlabelled and symmetrically di-labelled compounds.[6] The reaction yielded scission products (*N*-methylarylamines) and crossed starting materials, Ar(Me)NN(Me)Ar'; both of these could be consistent with radical processes, although the crossed hydrazobenzenes were suggested to arise in a bimolecular four-centre reaction. However, *o*-semidines (**2**) were also obtained, and these showed no evidence of crossed reaction; it was suggested that they might result from an intramolecular rearrangement.

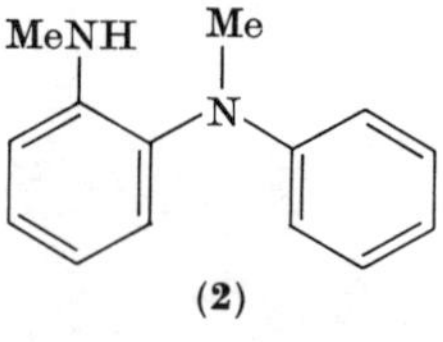

(**2**)

The Fischer–Hepp rearrangement of *N*-methyl-*N*-nitrosoaniline in the presence of urea and other nucleophiles has been studied. Concurrent rearrangement and denitrosation were found.[7] In HCl, NOCl was formed which reacts rapidly with the added nucleophiles; no evidence for *direct* transnitrosation was found.

The mechanism of the Fries rearrangement was scrutinized,[8–10] and factors, especially the nature of the Lewis acid,[10] influencing product distribution were discussed. Examples of alkyl-group migration from a blocked *ortho*-position were examined,[9] and a fundamental difference was found between the behaviour of *o*-Me and *o*-Bu^t substituents; in the case of *tert*-butyl groups, migration usually precedes Fries rearrangement and is intermolecular; with methyl groups, intramolecular 1,2-shifts follow the Fries rearrangement.

The interesting observation has been made that, whilst H^+-catalysed or photochemical rearrangement of *N*-benzylanilines give *o*- and *p*-benzylanilines, if the photochemical reaction is effected in the presence of $EtAlCl_2$ the major product is the *meta*-isomer; no explanation was offered. Amongst other Lewis acids examined, Et_2AlCl gave no *meta*-product.[11]

The production of (**4**) from *p*-toluenesulphinic acid and *N*,*N*-dimethylaniline in

2 Me–C_6H_4–SO_2H ⟶ Me–C_6H_4–$SOSO_2$–C_6H_4–Me

↓ $PhNMe_2$

Me_2N–C_6H_4–CH_2–SO_2–C_6H_4–Me (**5**) ⟵ C_6H_5–N(Me)–CH_2–SO_2–C_6H_4–Me ⟶ MeNH–C_6H_4–CH_2–SO_2–C_6H_4–Me (**4**)

\+ Me–C_6H_4–SOH (**3**)

ethanol has been attributed to the acid-catalysed rearrangement of (**3**), itself considered to be formed as shown.[12] Formation of (**5**) is attributed to intermolecular trapping of $^+CH_2SO_2Ar$.

The arylhydroxylamine derivatives (**6**) rearrange thermally to *N*-(*o*-acyloxyaryl)-*p*-toluenesulphonamides (**7**) at ca. 120°, apparently by a concerted [3,3]-sigmatropic mechanism since carbonyl-^{18}O label in (**6**) is found exclusively in the phenoxy-oxygen of the product.[13] Intramolecular concerted mechanisms are also proposed for the

TsN(OCOPh)–C_6H_4–R ⟶ TsNH–C_6H_3(OCOPh)–R

(**6**; R = H, Me, Cl) (**7**)

rearrangement of the sulphonamide (**8**) to the amino-sulphone (**9**), irrespective of whether the reaction is acid- or base-catalysed, or occurs thermally in the absence of catalyst.

The base-promoted reaction, which is effected by alkyl-lithium, is considered to proceed as indicated.[14] Another sigmatropic rearrangement is involved as a key step in the

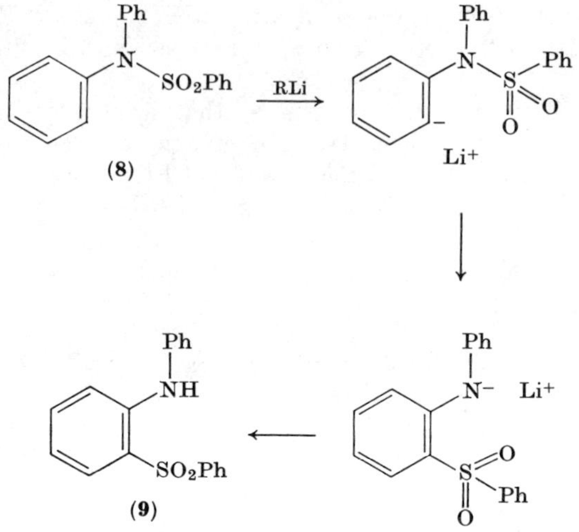

(8)

(9)

conversion of the cyanohydrazine (**10**) into the benzimidazole derivative (**11**) by BrCN. Labelling experiments with ^{14}C and ^{15}N identified the fate of different carbon and nitrogen atoms, and the mechanism shown accommodates the results.[15] An alternative

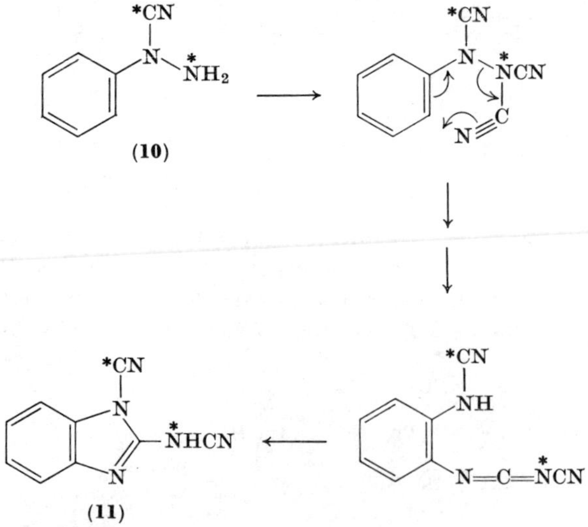

(10)

(11)

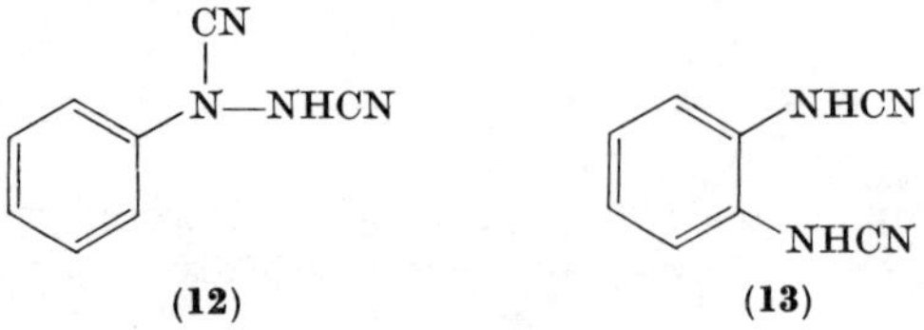

(12)

(13)

scheme, in which rearrangement occurs in the dicyanohydrazine (**12**), was considered to be inapplicable since this should pass through the symmetrical intermediate (**13**).

Heterocyclic Derivatives

The results of investigations of the Cornforth rearrangement of oxazoles (**14**)[16] have now been discussed in detail;[17] substituent and solvent effects reveal surprisingly little development of polar character in the transition state, despite the zwitterionic formulation of the intermediate. Rearrangements of the oxadiazoles (**15**) parallel the Lander

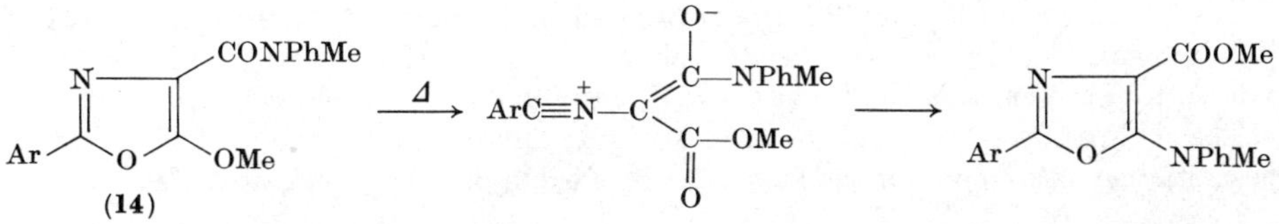

(**14**)

rearrangement of acyclic imino-ethers;[18] the product incorporates the alkyl group of an external alkyl halide, and a mechanism involving the intermediate (**16**) was preferred. Both esters (**17a**) and thiol-esters (**17b**) rearrange as shown, although competing reactions are dominant in the case of (**17b**).[19] A further oxadiazole rearrangement has been invoked to explain the formation of (**19**) from (**18**) and PhNCS.[20] This is the first sulphur

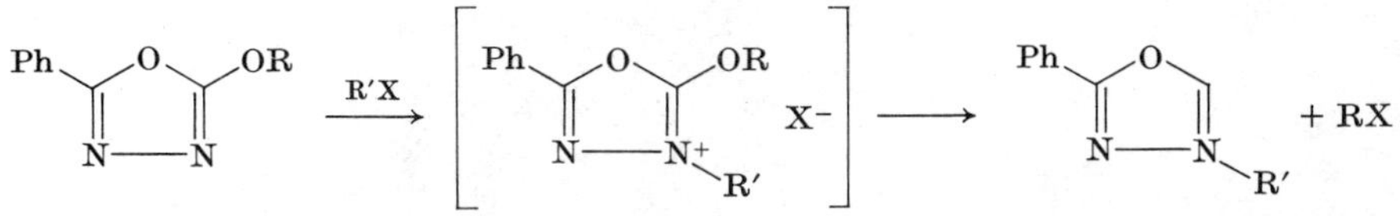

(**15**) (**16**)

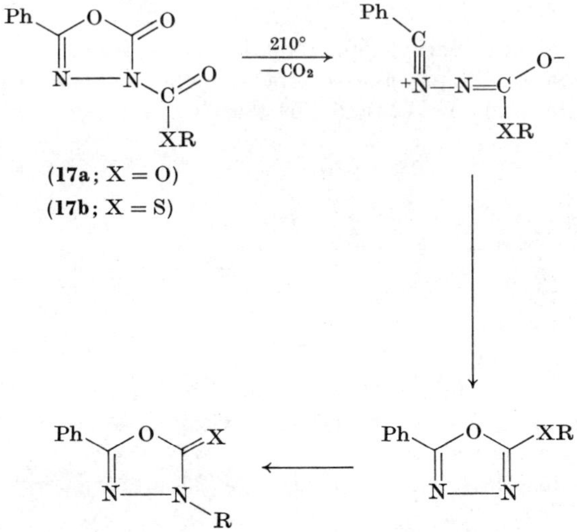

(**17a**; X = O)
(**17b**; X = S)

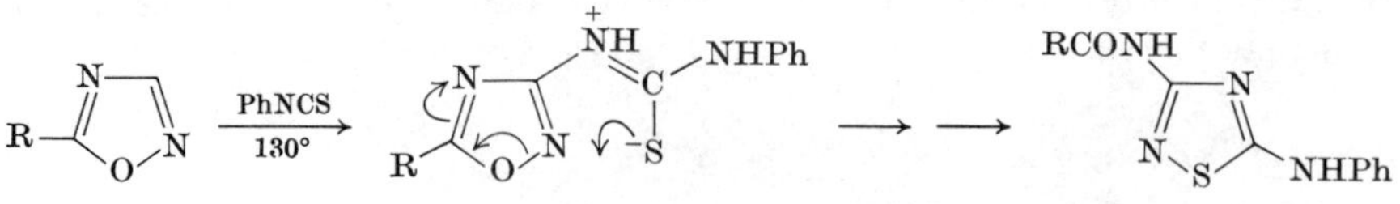

(18)

(19)

example of a well-known rearrangement type. Details of rearrangements of oxadiazolium thiolates and related heterocyclic betaines have been given,[21] and further studies of the Dimroth rearrangement have been reported.[22]

The formation of (**21**) and (**22**) from (**20**), and the subsequent isomerization of (**21**) to (**22**), are considered to involve the tricyclic intermediate (**23**).[23] The reaction was considered to represent a model for the role of biotin in enzymic carboxylation.

The acid-catalysed rearrangement of 3-alkyl-1-allylindoles to give 2-allyl isomers has been studied as a model for the biosynthesis of echinulin-type compounds.[23a]

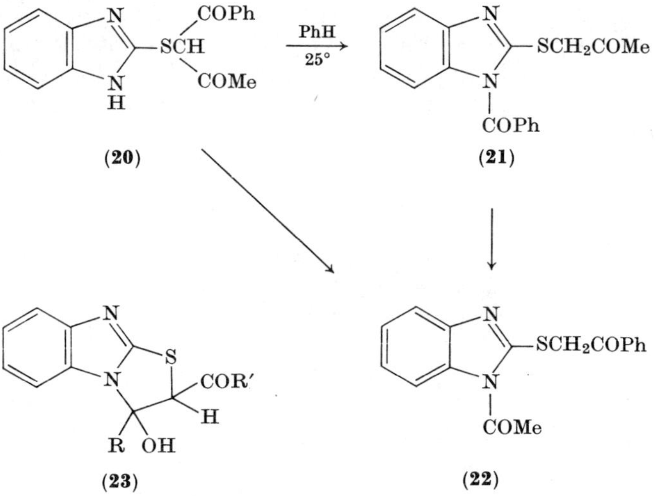

(20) (21)

(23) (22)

The rate of rearrangement of 5-(carboxymethylene)hydantoin (**24**) to orotic acid by M^+OH^- increases with ionic radius of M^+.[24] This was attributed to more rapid rearrangement in the ion pair formed from the dianion of (**24**) and M^+ than in the free dianion.

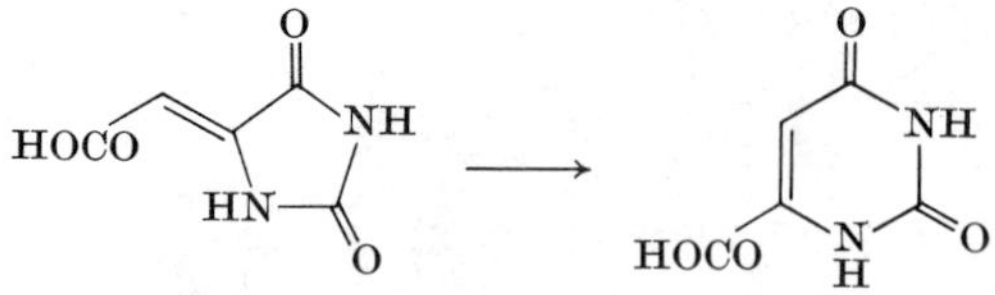

(24)

Thermolysis of derivatives of 5,5-diphenyldithiohydantoin and related compounds,[24a] and the intramolecular aryl migrations that occur when 5-alkylthio-2,4,4-triaryl-4*H*-imidazoles are heated[24b] have been examined.

A possible mechanism for the transformation of the diazetidine (**25**) into the pyridone (**26**) in TFA is indicated in the scheme.[25]

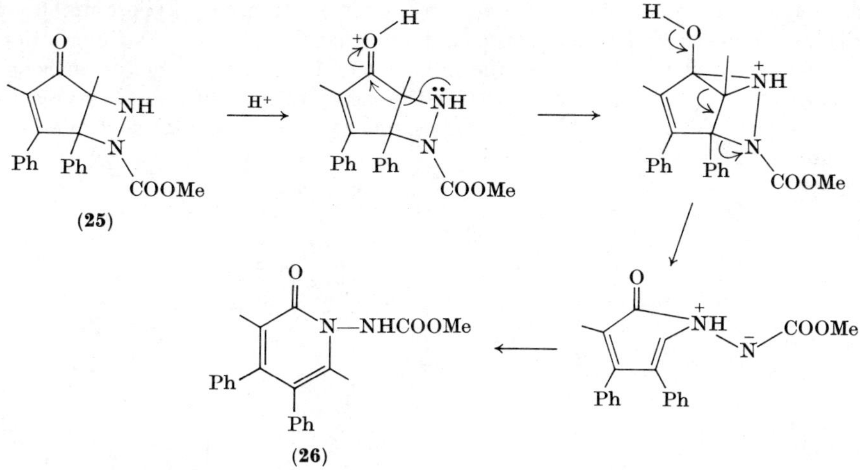

(25)

(26)

In extension of work on the reactions between acetylenes and pyridine *N*-oxides,[26] it has been found that benzyne gives *o*-hydroxyphenylpyridines.[27] The α-derivative (**29**) is produced in significant yield only when positions 3 and 5 are both blocked. The initial cycloadduct (**27**), which would normally rearrange rapidly via (**28**), may slowly open to (**29**). Alternatively (**29**) may arise via (**30**) or (**31**), both of which were isolated from reactions with 3,5-lutidine *N*-oxide carried out below 45°.

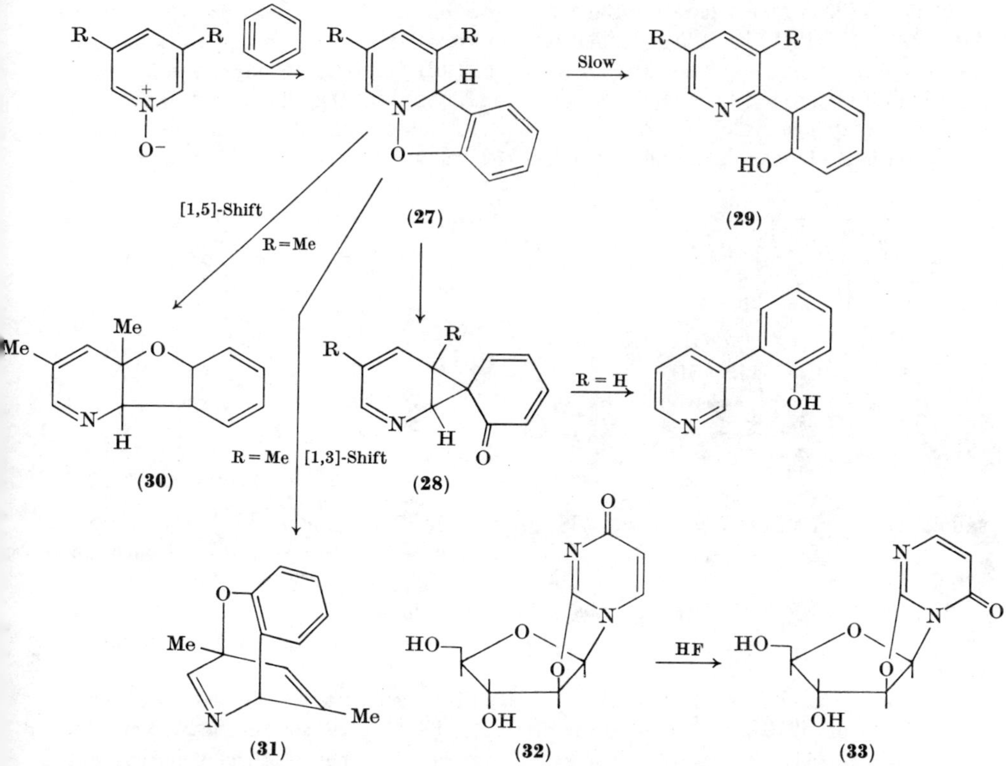

(27)

(29)

(30)

(28)

(31)

(32)

(33)

The anhydronucleoside (**32**) rearranges in liquid HF to give (**33**), presumably by ring-opening followed by recyclization,[28] and 4-acetoxythiocoumarin rearranges to the 3-acetyl-4-hydroxy-derivative under the influence of PPA or pyridine, or on heating alone; cross-over experiments showed that in the case of the pyridine-catalysed reaction the process was largely intermolecular, proceeding via (**34**).[29] Similar acyl migrations were found with *O*-acetyltetronic acids.[29]

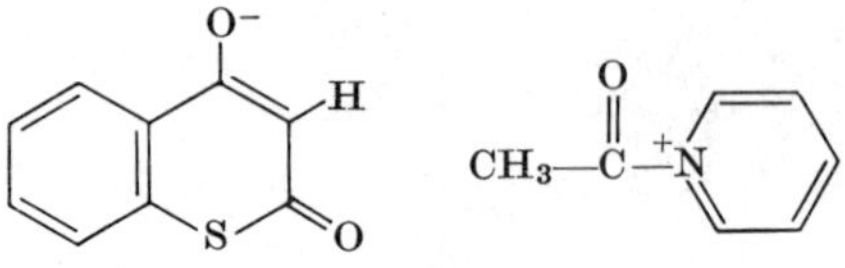

(**34**)

Cyclohexadiene Derivatives

The dienone–phenol rearrangement affords a continuing source of interest and frequently brings forth the unexpected.

Comparison of the rates of migration of CH_3 and CD_3 in 4,4-dimethylcyclohexa-2,5-dienone gives $k_{CH_3}/k_{CD_3} = 1.144$, confirming that the methyl migration occurs in the rate-determining step; the magnitude of the isotope effect was discussed in terms of the transition-state geometry.[30]

The migration of ethoxycarbonyl groups in acid-catalysed rearrangement of a series of dienones (**35**) roughly parallels the carbonium-ion stabilizing abilities of the R groups (R = Me, 1; Et, 0.45; Ph, 135).[31] When R is isopropyl or benzyl, a fragmentation process leads to ethyl *p*-hydroxybenzoate. In the series (**36**), the relative rates of R-group migration were found to be R = Me, 1; COOEt, 14; Et, 55; Ph, 200. The successful migration of the ethoxycarbonyl group was suggested to depend on back-donation from the carbonyl π-electrons which could stabilize the transition state (**37**). The migration of alkoxycarbonyl occurs when dienes (**38**) are treated with triphenylmethyl fluoroborate in MeCN.[32] Competing alkyl migration or fragmentation were again found with some R-groups.

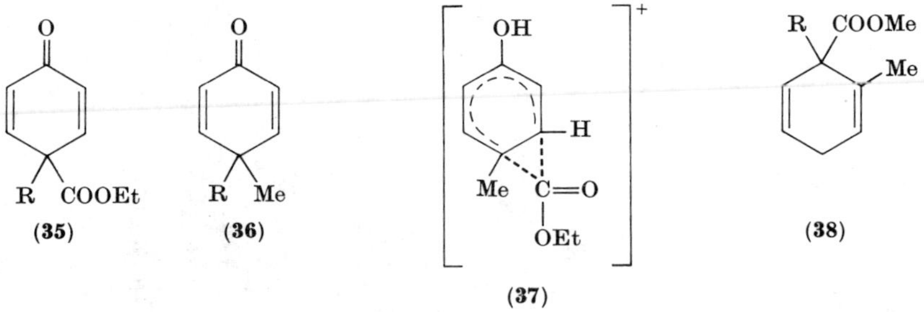

(**35**) (**36**) (**37**) (**38**)

In a further report[33] of the kinetics of dienone–phenol rearrangements, the unusual phenol–phenol rearrangement (**39**) → (**40**) was disclosed, and a mechanism was suggested.[34]

Systematic exclusion of all reasonable alternatives has led Miller to interpret the appearance of (**42**) and (**43**) among the products of acid-catalysed rearrangement of the *ortho*-dienone (**41**) in terms of concerted [1,3]-sigmatropic shifts of the 4-methylbenzyl

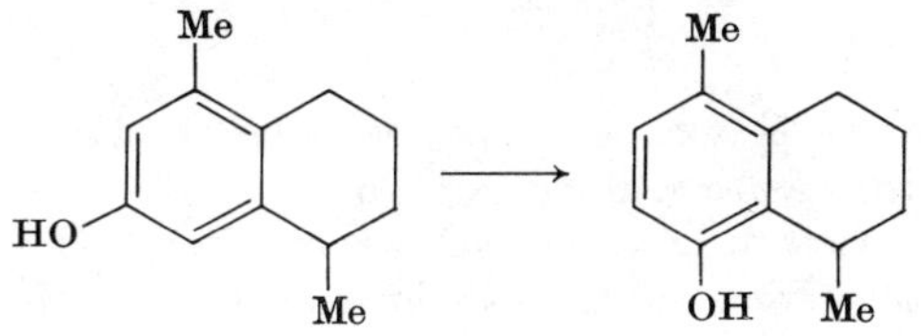

(39) (40)

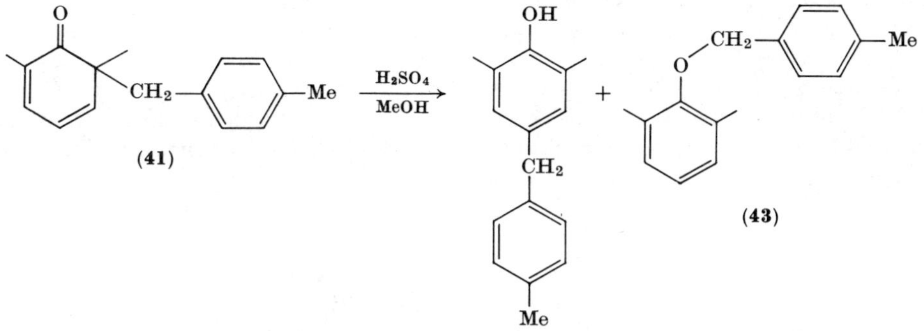

(41) $\xrightarrow[MeOH]{H_2SO_4}$ (42) + (43)

group.[35] The spiro-dienone (**44**) is opened by HCl–MeOH with nucleophilic addition at one of the double bonds to give, *inter alia*, (**45**).[36]

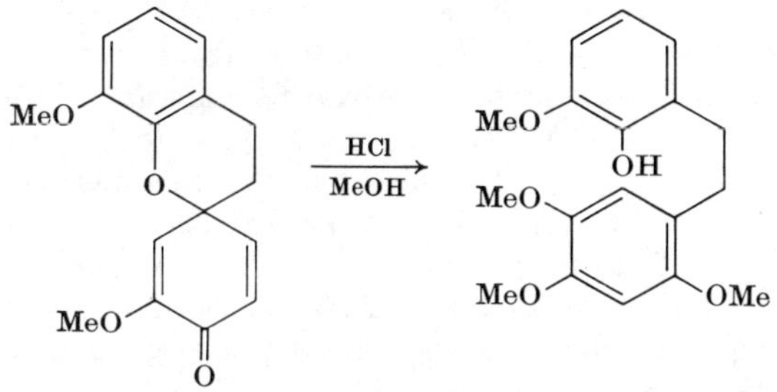

(44) $\xrightarrow[MeOH]{HCl}$ (45)

Rearrangements involving cyclohexadienyl cations which occur in the course of electrophilic aromatic substitution reactions have been reviewed.[37] A recent example is the intramolecular [1,3]-shift of a nitro-group observed when the adduct (**46**) formed on nitration of 3-cyano-*o*-xylene in Ac_2O is treated with acetic acid[38] (see also Chapter 7).

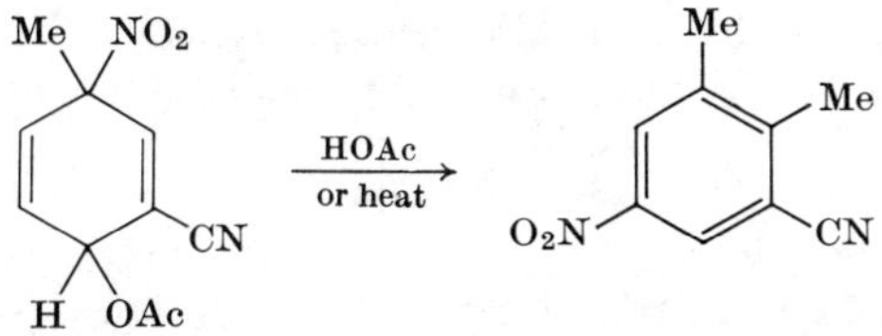

(46) $\xrightarrow[\text{or heat}]{HOAc}$

Sigmatropic Rearrangements

A systematic classification of six-electron pericyclic reactions has been presented.[39] The principle of orbital isomerism has been discussed[40] and applied[41, 42] to the analysis of various pericyclic processes, particularly those of the Woodward–Hoffmann "forbidden" class. Other specific theoretical calculations include those on the degenerate rearrangements of cyclopentadiene,[43] cyclopentadienylsilane,[43, 44] bicyclo[2.2.0]butyl cations[45] (*via* the Woodward–Hoffmann "forbidden" *bisected* cyclopropenylcarbinyl cation, as a result of subjacent orbital control,[46] and bicyclo[3.1.0]hexenyl and homotropylium cations.[47] The stereochemistry of [3,3]- and [5,5]-sigmatropic rearrangements has been reviewed.[48]

The thermal (160–200°) rearrangement of allyl (trimethylsilyl)methyl ethers (**47** → **48**)[49] is not a concerted dyotropic[50] reaction; various lines of evidence indicate that trimethylsilyl migration (intramolecular) is in advance of allyl migration (partly intermolecular) at the transition state.

R R O $SiMe_3$ → R R O $SiMe_3$

(**47**) (**48**)

[*3,3*]-*Migrations*

Claisen and Related Rearrangements. A detailed mechanistic study[51] (e.g., substituent, salt and solvent isotope effects) of the *ortho*-Claisen rearrangements of allyl aryl ethers in trifluoroacetic acid confirms that these rearrangements are of the charge-induced [3,3]-type.[52] Thermal (200°) Claisen rearrangement of (–)-(*S*)-*trans*-4-(1-methylbut-2-enyloxy)toluene affords mainly (+)-(*R*)-*trans*-4-methyl-2-(1-methylbut-2-enyl)phenol with high stereoselectivity. The $ZnCl_2$-catalysed (120°, decalin) rearrangement of the same substrate involves both intra- and inter-molecular processes.[53] Controlled (–35°) Claisen rearrangement of allyl trihalovinyl ethers has been reported; the rate of rearrangement is strongly dependent on the halogen substituents, fluorine producing greater acceleration than chlorine.[54] Formation of 8-(3-methylbut-2-enyl)guanine by the reaction of 2-amino-6-chloropurine with sodium 3-methyl-but-2-enyl oxide has been shown to result from a double [3,3]-sigmatropic (Claisen–Cope) rearrangement of the expected substitution product, 6-(3-methylbut-2-enyloxy)guanine.[55] Thermal rearrangements of 2-allyloxy and 2-(allylthio)benzothiazoles to the corresponding *N*-allylbenzothiazolin-2-ones or -2-thiones involve concerted Claisen and thio-Claisen processes.[56, 57]

Base-catalysed (NaOEt–EtOH, 78°) rearrangement of the catechol monoallyl ethers (**49**) and (**50**) involves successive sigmatropic rearrangements to give products that differ from those obtained under normal Claisen conditions ($PhNMe_2$, 180°). Specifically, both (**49b**) and (**50b**) give a 1 : 1 mixture of (**51**) and (**52**) probably by rapid interconversion of intermediates (**53**) and (**54**) formed by initial [2,3]-sigmatropic rearrangement of the respective phenoxides from (**49b**) and (**50b**).[58]

Formation of (**55**) and (**56**) in the thermolysis of 7-propargyloxycycloheptatriene is believed to involve [3,3]-sigmatropic rearrangement of the 3- and 1-isomers formed by well-precedented [1,5]-sigmatropic shifts (Scheme 1). Under similar conditions, 2-

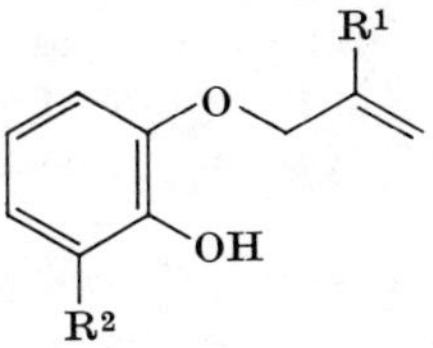

R1 O OH R2 (49) OH O R2 R1 (50) OH OH Me (51)

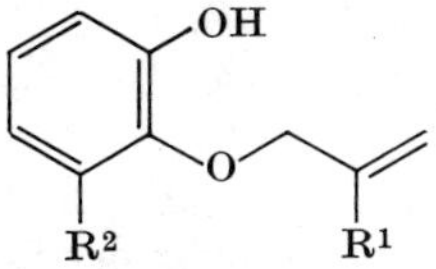

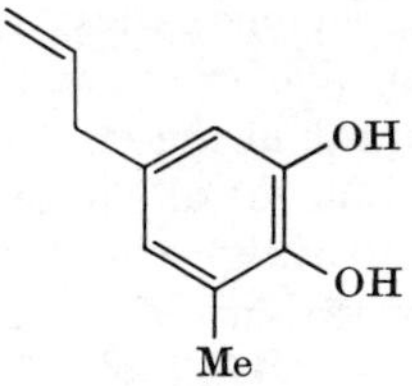

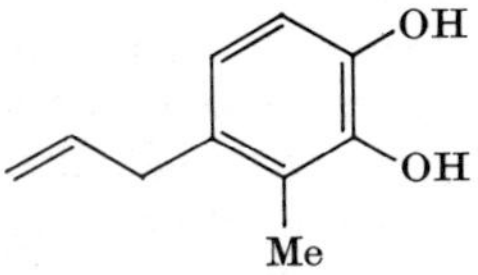

OH OH Me (52) O⁻ O Me (53) O O⁻ Me (54)

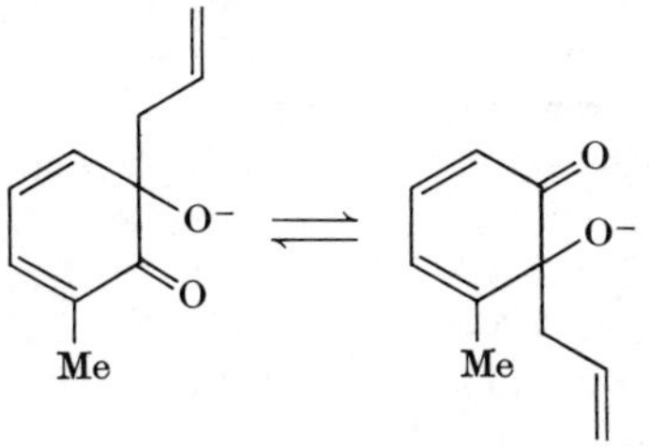

a. $R^1 = R^2 = H$
b. $R^1 = H$, $R^2 = Me$
c. $R^1 = R^2 = Me$

propargyloxycycloheptatrienone undergoes initial [3,3]-sigmatropic rearrangement followed by cyclization to give 2-methyl-8*H*-cyclohepta[*b*]furan-8-one.[59]

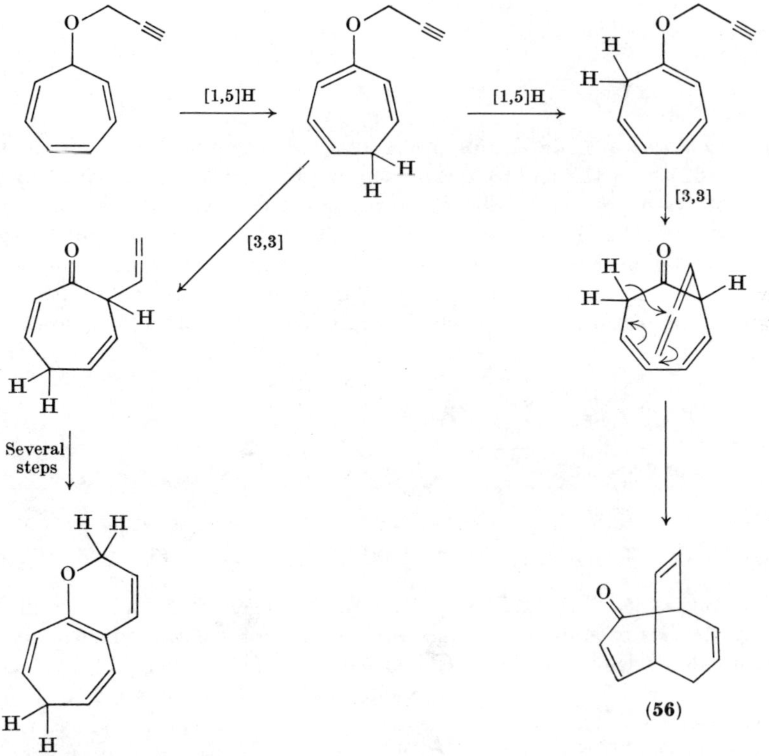

[1,5]H [1,5]H [3,3] [3,3] Several steps (55) (56)

SCHEME 1.

Nucleophilic catalysis of the thio-Claisen rearrangement is thought to be initiated by nucleophilic attack at the allylic carbon of the substrate. This results in a small displacement in the electron density of the carbon–sulphur bond and formation of a *p*-orbital on the allylic carbon to accommodate the orbital requirements and geometry of the [3,3]-sigmatropic transition state.[60]

New variants of the thio-Claisen rearrangement continue to be reported. Thermolysis (DMF or DMA, 120°) of allyl-2-naphthyl sulphoxide affords (**58**) as a result of an intramolecular *cis*-addition reaction of the sulphenic acid (**57**) formed in a preceding [3,3]-sigmatropic rearrangement.[61] The products obtained by heating alk-1-ynyl alk-2-ynyl sulphides with dialkylamines or dialkylphosphines can be rationalized on the basis of an initial [3,3]-sigmatropic rearrangement to thioketenes.[62] A new route to α,β-unsaturated aldehydes involves as a key step, the [3,3]-sigmatropic rearrangement of *S*-allyl dithiocarbamates.[63] Other related studies include the rearrangement of allylketene mercaptals to α-allyldithioesters,[64] of 1-(allylthio)-1-aminoalkenes to thioamides,[65] and of certain ketene *N*-allyl-*O*,*N*-acetals to dihydro-oxazines.[66] 2-Alkoxy-substituents promote the thio-Claisen rearrangement of allyl 1,1-dicyanovinyl sulphides and their propargyl analogues.[67]

S OH Me S+ O−

(**57**) (**58**)

Allylic trichloroacetimidates (**59**) are converted into the corresponding trichloroacetamides (**61**) by a thermal [3,3]-sigmatropic process that is catalysed by mercuric ion. The catalysed process is believed to occur *via* the cyclic intermediate (**60**).[68] A so-called "amide"-Claisen rearrangement of 4-(*N*-allyl-*N*-tolueneamido)-1-naphthaldehyde *N*-methylimine has been reported.[69] That thermal rearrangement of 1-aryl-2-cyano-3,3-diphenyldiazetidin-4-ones to imidazo[1,2-*a*]benzimidazoles involves initial [3,3]-sigmatropic rearrangement has been confirmed by ^{15}N-labelling studies.[70]

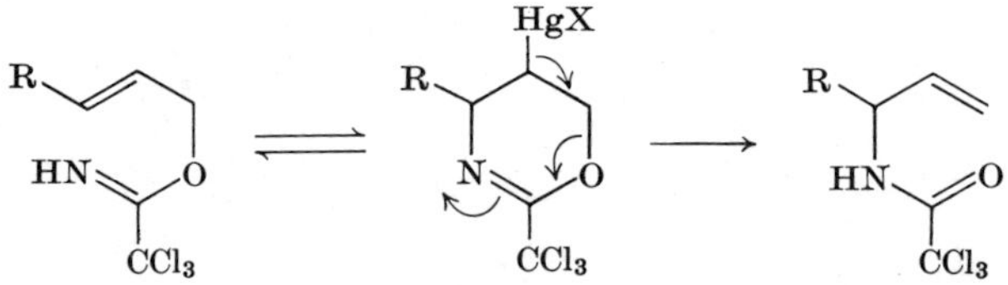

R HN O CCl_3 HgX R N O CCl_3 R HN O CCl_3

(**59**) (**60**) (**61**)

Cope Rearrangements. The fluxional barrier to the degenerate Cope rearrangement of semibullvalene has been determined experimentally for the first time; analysis of the temperature-dependent 251 MHz ^{1}H NMR spectra affords the following activation parameters, $\Delta G^\ddagger = 5.5 \pm 0.1$ kcal mol^{-1} (at −143°), $\Delta H^\ddagger = 4.8 \pm 0.2$ kcal mol^{-1} and $\Delta S^\ddagger = -5.4 \pm 3$ e.u.[71] The analogous rearrangement of bullvalene has been re-investigated by using ^{13}C-NMR techniques.[72,73] In one study,[72] the activation energy (13.10 kcal mol^{-1}) is in accord with that (13.13 kcal mol^{-1}) determined in a parallel ^{1}H-NMR investigation; the other[73] reports a value (14.5 kcal mol^{-1}) considerably larger than any previously recorded. Substitution of the semibullvalene nucleus leads to a unidirectional thermodynamic imbalance in its Cope rearrangement, a wide variety of substituents preferring attachment to olefinic > cyclopropyl > aliphatic carbon atoms.[74]

Unlike 8-oxabicyclo[5.1.0]octa-2,5-diene, the 4,8-dioxa-analogue does not undergo a thermal Cope rearrangement below 90°. It is, however, prone to acid-catalysed rearrangement, giving 4*H*-pyran-4-carbaldehyde, probably *via* a dioxahomotropylium ion intermediate.[75]

Although constrained to a boat-like transition state, the degenerate Cope rearrangement of 1,4-dimethylenecyclohexane shows a $\Delta G^{\ddagger}$ only 6 kcal mol^{-1} greater than that for rearrangement of 1,5-hexadiene *via* the favoured chair transition state.[76] Chair-like transition states are also implicated in the Cope rearrangement of germacrene-type sesquiterpenes.[77]

The minimal variation in the ratio of oxy-Cope rearrangement: cleavage observed for a series of phenyl-1,5-hexadien-3-ols rules out biradical intermediates for these processes,[78] in contrast to their reported implication in the Cope rearrangement of the corresponding phenyl-1,5-hexadienes.[79]

Stereochemical aspects of the unusually easy[88] diaza-Cope rearrangement of 1,2-diaryl-*N*,*N'*-diarylideneethylenediamines (**62**) have been investigated. This process equilibrates the (±)- and *meso*-isomers at temperatures above 120°. At higher tempera-

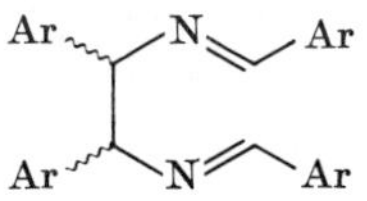

(**62**) 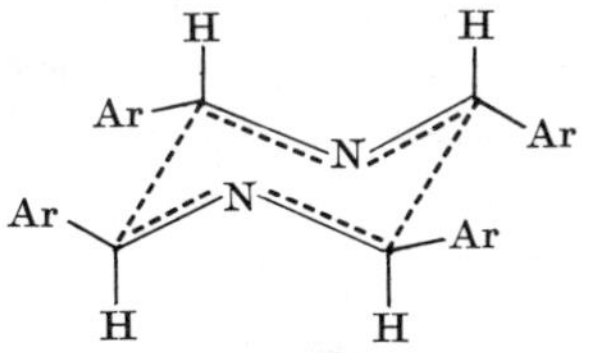 (**63**) 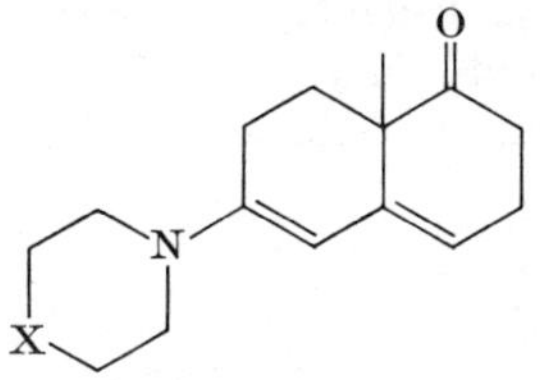 (**64**)

tures, the *meso*-isomer remained unchanged, but the NMR spectrum of the (±)-isomer shows further broadening due to a fluxional process with $\Delta G^{\ddagger} = 23$–24 kcal mol^{-1} (calculated from the extrapolated coalescence temperature, $T_c = 220$–$230°$). This value agrees well with that (23.9 kcal mol^{-1}), determined polarographically for the racemization of either the (+)- or the (−)-isomer. Chair-like transition states, e.g. (**63**), are implicated for these interconversions.[81] Multiple aza-Cope rearrangements have been postulated to account for the products isolated from the reactions of certain dienamines (**64**; X = CH_2 or O) with allylic halides.[82]

[2,3]-*Migrations*

The formation of substituted nitrones by thermolysis of oxime *O*-allyl ethers, previously characterized as a [2,3]-sigmatropic rearrangement,[83] has been shown to proceed, at least in part, by a radical-pair mechanism.[84] Rearrangement of 4-methoxybenzyl trichloromethanesulphenate to 4-methoxybenzyl trichloromethyl sulphoxide is believed to occur by an ion-pair mechanism rather than by [2,3]-sigmatropic rearrangement.[85]

2-Naphthyl prop-2-ynyl sulphoxides rearrange thermally to give condensed naphthothiophens as a result of consecutive [2,3]- and [3,3]-sigmatropic migrations.[86] Synthetic applications of the [2,3]-sigmatropic rearrangement of acetylenic and allenic sulphonium ylides have been reported.[87,88] For example, the transformation (**65**) → (**66**) is a key step in a new synthesis of artemesia ketone.[88] β,γ-Unsaturated neopentyl ketones of type (**68**) are produced by [2,3]-sigmatropic rearrangement of the carbanions (**67**) derived from α-allyloxyacetonitriles.[89]

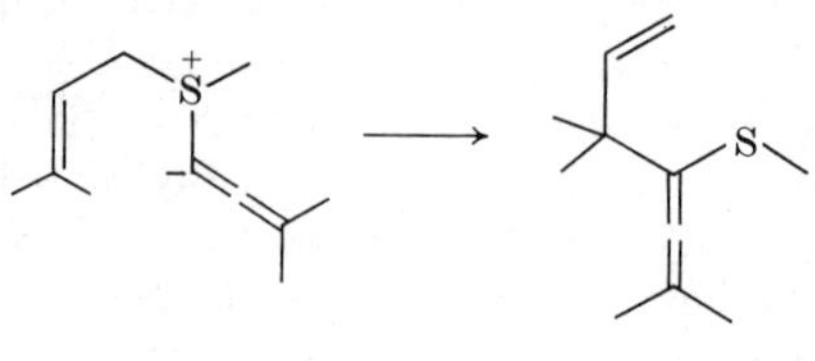

(**65**) (**66**)

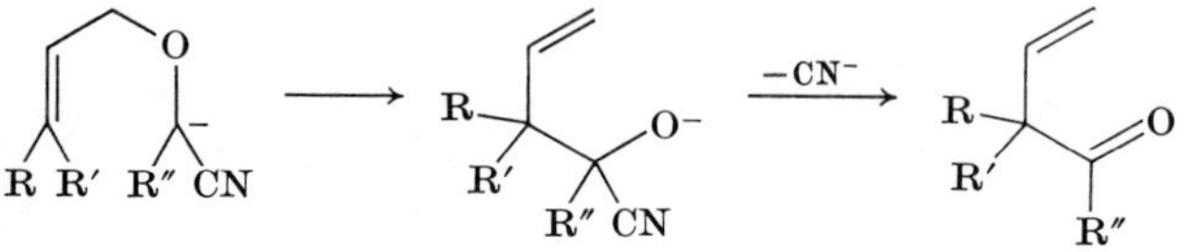

$R^{\prime\prime}$ CN — $-CN^-$ — R, R′, R″

(**67**) (**68**)

A new [2,3]-sigmatropic reaction has been reported which allows the base-catalysed conversion of acetonyl allyl ethers into 3-hydroxyalk-5-en-2-ones. This is a genuine suprafacial [2,3]-sigmatropic rearrangement since both (−)-tetrahydro-*cis*-5-methyl-5-vinyl-2-furyl methyl ketone (**69a**) and its (+)-*trans*-isomer (**69b**) yield the same (+)-hydroxy-ketone (**70**).[90] The cupric sulphate-catalysed vinylogous Wolff rearrangement

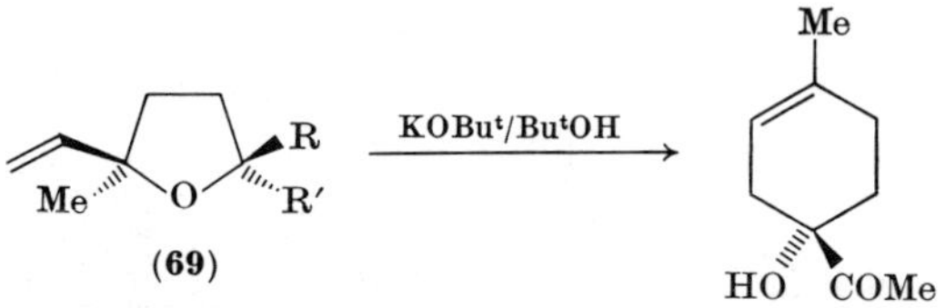

KOBu^t/Bu^{t}OH

(**69**)

a; R = COMe; R′ = H

b; R = H; R′ = COMe

(**70**)

MeOH, $CuSO_4$

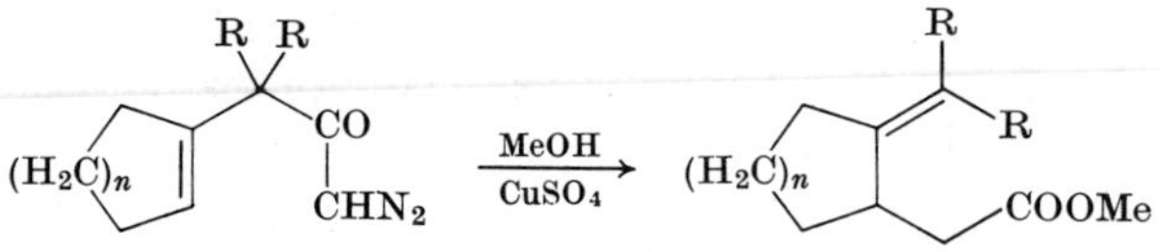

(**71**) (**72**)

(**71**) → (**72**) could formally proceed by [2,3]-sigmatropic rearrangement of a carbene (or carbenoid) intermediate.[91]

[1,*n*]-*Migrations*

A concerted [1,2]-trimethylsilyl shift in an intermediate silylsulphonium ylide could account for the [1,2]-rearranged product observed in the reaction of ethyl trimethylsilyl sulphide with substituted carbenes. Equally probable is a mechanism involving a geminate radical pair.[92]

Thermolysis (*ca.* 300°) of either (**73**) or (**74**) affords (**75**) with greater than 99% stereospecificity. Since retention of configuration would be forced on the migrating group (C-1), these rearrangements probably do not proceed by direct [1,3]-sigmatropic shifts. Rather, the favoured mechanism involves *endo*- or *exo*-7-propenylbicyclo[4.1.0]-hept-2-ene intermediates as outlined in Scheme 2.[93]

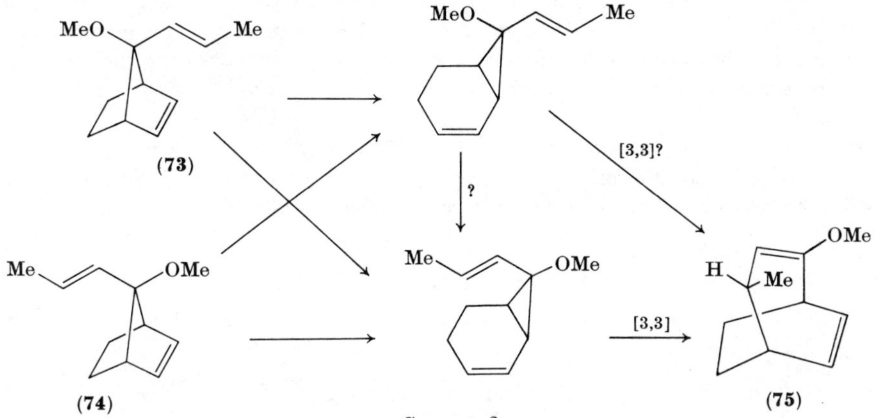

(73)

(74)

(75)

SCHEME 2.

A degenerate thermal rearrangement (formally a [1,3]-sigmatropic shift) of norpinene has been detected.[94] The thermal rearrangement of 1-acyloxy-1-alkoxy-λ^5-phosphorins to 6-acyl-1-alkoxy-1-oxo-1-λ^5-phospha-2,4-cyclohexadienes, probably involves suprafacial [1,3]- or [1,7]-sigmatropic acyl shifts.[95]

Further details have been reported[96] for the suprafacial [1,3]-sigmatropic alkyl shifts from nitrogen to carbon in stable 1,4-dialkyl-1,4-dihydropyrazines.[97] Contrary to a previous report,[98] the thermal rearrangement of certain unsaturated quaternary quinuclidine-3-carboxylic esters does not involve two consecutive suprafacial [1,3]-sigmatropic shifts. Instead, the rearrangement is initiated by nucleophilic ring-opening by the counterion.[99]

Quantitative CIDNP measurements indicate that the rearrangement of 2-alkoxy-quinoline *N*-oxide to *N*-alkoxy-2-quinolone, formally a [1,4]-sigmatropic rearrangement, involves a significant contribution (up to *ca.* 75%) from a radical-pair mechanism.[100]

The products isolated from the gas-phase pyrolysis of *cis*- and *trans*-1-ethynyl-2-methylcyclopropane can be rationalized by the intermediacy of hexa-1,2,5-triene, formed by an initial homo-[1,5]-sigmatropic hydrogen shift in the *cis*-isomer.[101] An analogous homo-[1,5]-hydrogen shift accounts for the thermal lability of *endo*- but not *exo*-7-methylbicyclo[4.1.0]hex-2-ene.[102] The occurrence of [1,5]-sigmatropic shifts during the thermal rearrangement of **(76)** to **(77)**, is supported by trapping of the product of a single ester migration, **(78)**, as an adduct with dimethyl acetylenedicarboxylate.[103]

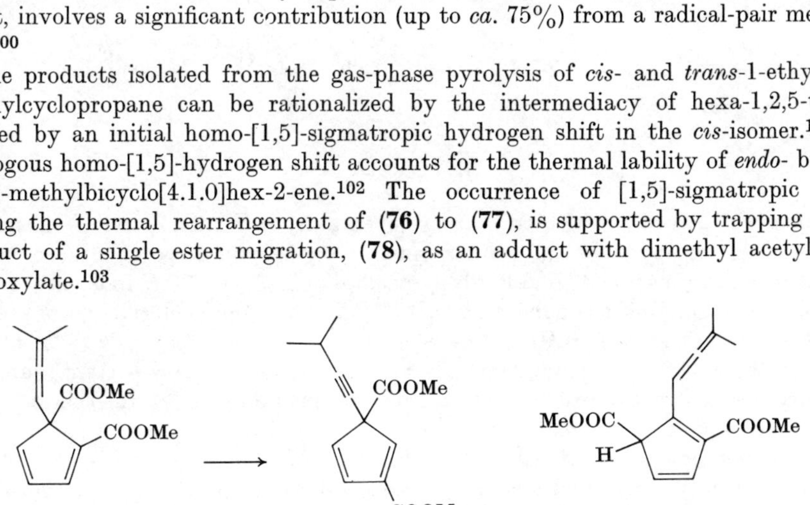

(76) (77) (78)

In the reversible thermal isomerization of silylindenes, the equilibrium distribution of isomers is sensitive to the nature of the silicon substituents, but not in an easily explicable way.[104] The preference for 4-substituted-phenyl over phenyl migration is greater in the photochemical than in the thermal sigmatropic rearrangement of 1,1-diarylindenes; this leads to the suggestion of a radical pathway for the thermal process and a transition state with a significant degree of charge transfer for the photochemical process.[105]

Pyrolysis (510–550°) of unsaturated acetal (**79**) affords (**81**) in a reaction with the characteristics (intramolecular, $k_H/k_D = 3 \pm 1$) of a concerted [1,5]-hydrogen shift; possibly by way of the strained bicyclic transition state (**80**).[106]

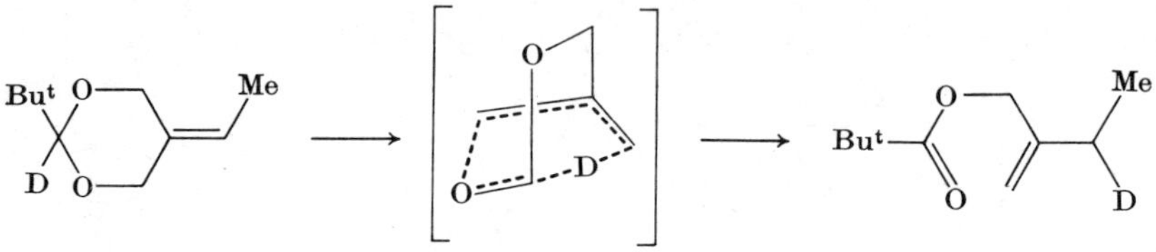

(**79**) (**80**) (**81**)

At low temperatures, 2-acylindazoles can be isolated from the reaction of benzyne with diazobenzils. Their subsequent rearrangement to 1-acylindazoles (the normal end-products from the benzyne reaction) involves a heterolytic dissociation–recombination process rather than an intramolecular [1,5]-sigmatropic acyl shift.[107] [1,5]-Sigmatropic acyl shifts do account for the products formed by reaction of 2-diazocycloalkanones with dimethyl acetylenedicarboxylate.[108] Thermal rearrangement (100–190°) of pyrazole (**82**) affords[109] (**83**) and (**84**), the products of [1,5]-phenyl migration to carbon or nitrogen, rather than (**84**) and (**85**) [the product of multiple ester migration] as previously reported.

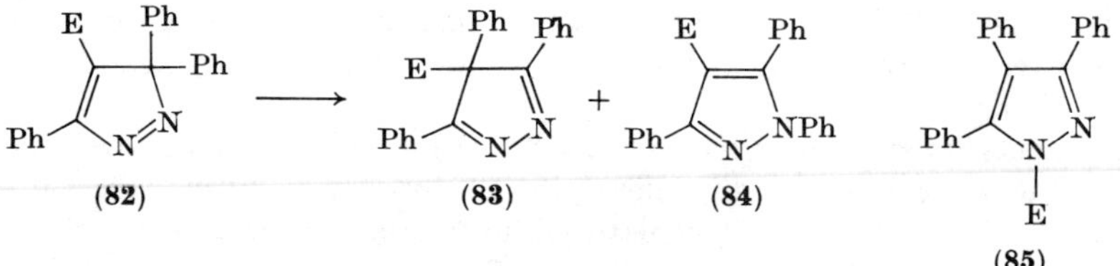

(**82**) (**83**) (**84**) (**85**)

E = COOMe

Dipolar addition of substituted diazocyclopentadienes to mono- and di-substituted alkynes gives spiropyrazoles as primary products; these undergo spontaneous [1,5]-sigmatropic rearrangement to afford pyrazolopyridines and 3a*H*-indazoles or 1*H*-indazoles depending on the substituents.[110] [1,5]-Sigmatropic chlorine shifts occur in certain 2*H*-pyran derivatives.[111]

The observed activation parameters ($E_a = 43.4$ kcal mol^{-1}; $\Delta S^{\ddagger} = -10$ e.u.) support a concerted mechanism *via* a rigid eight-membered transition state for the thermal (350°) fragmentation of ethyl homoallyl ethers (**86**). These and the analogous fragmentations of ethyl homopropargyl ethers provide the first examples of [1,7]-hydrogen transfer for systems with only a terminal olefinic (or acetylenic) receptor site in an otherwise σ-bonded framework.[112] The thermal isomerization of (**87**) to (**88**) is claimed as the first example of an antarafacial [1,7]-sigmatropic hydrogen shift in a rigid system.[113]

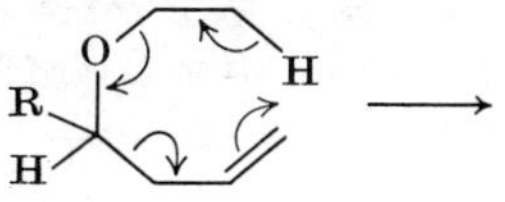

$RCHO + CH_2{=}CH_2 + CH_2{=}CHMe$

(86)

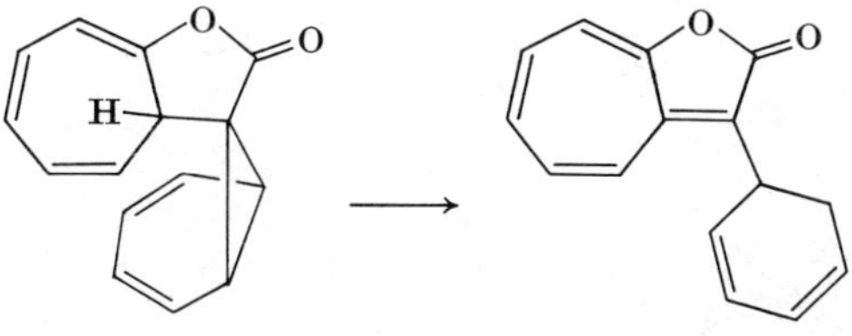

(87) **(88)**

Electrocyclic Reactions

The preferred modes of concerted electrocyclic ring openings can be predicted theoretically from a symmetry-independent index T, describing the energy difference $\Delta E = E_{con} - E_{dis}$ as a function of the reaction co-ordinate for small values of the rotational angle.[114] This index is complementary to the bond-order rule previously developed for the corresponding ring closures.[115] An extensive MINDO/3 study of Woodward–Hoffmann "forbidden" electrocyclic reactions leads to the general conclusion (*inter alia*) that the transition states for such reactions should be unsymmetrical unless symmetry is enforced by severe geometrical constraints.[92]

Other theoretical studies of specific electrocyclic reactions include those on the hexa-1,3,5-triene to 1,3-cyclohexadiene cyclization,[116] cyclopropyl to allyl transformations,[117] and the cyclobutene to buta-1,3-diene ring opening.[118, 119]

Pyrolysis of the *trans*-propenylbenzocyclobutene (**89**) affords 1,2-dihydro-2-methylnaphthalene, whereas the corresponding *cis*-isomer (**90**) gives 1-(*o*-tolyl)butadiene. These products most reasonably arise *via* electrocyclization and a [1,7]-hydrogen shift in the respective *o*-xylylene intermediates (**91**) and (**92**), formed by initial electrocyclic ring opening.[120]

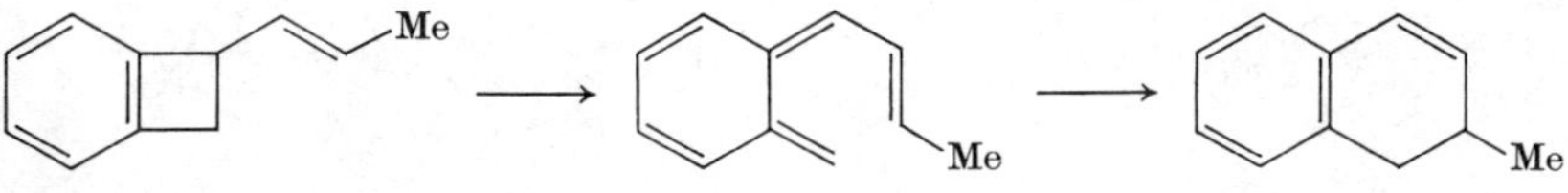

(89) **(91)**

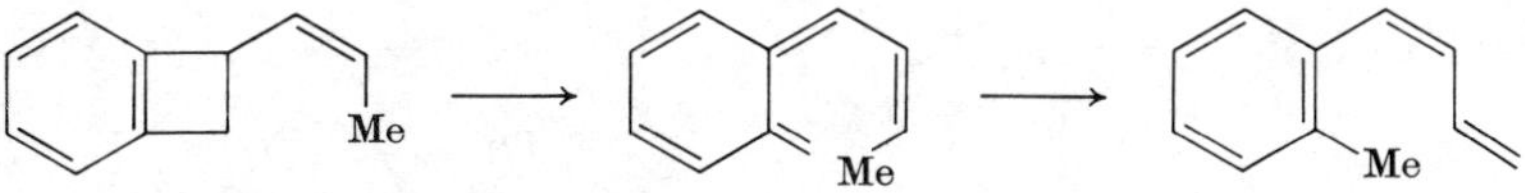

(90) **(92)**

2,3,4,4-Tetraphenylcyclobut-2-en-1-one rearranges thermally to 2,3,4-triphenyl-1-naphthol through initial electrocyclic ring opening, followed by electrocyclization of the vinylketene intermediate.[121] The quantitative thermal isomerization ($CDCl_3$, 100°, 5 min) of (**93**) to tiglaldehyde is thought to involve the sterically favoured conrotatory mode of ring opening to the dienol (**94**), which undergoes bond rotation before tautomerization to the observed product. The alternative conrotation gives (**95**) which by an intramolecular [1,5]-hydrogen shift would lead to angelaldehyde.[122]

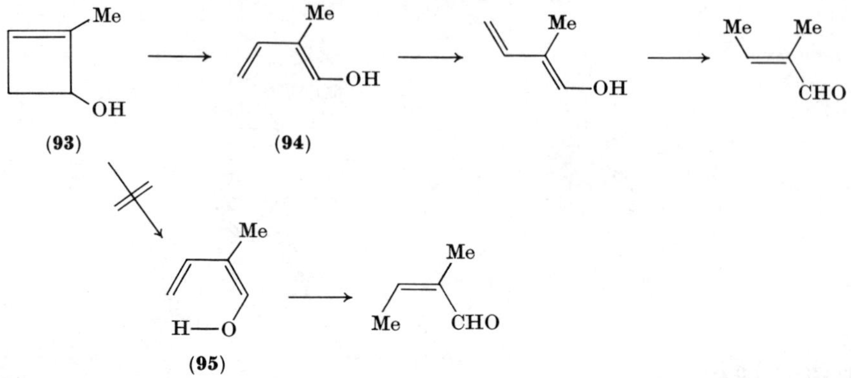

(**93**) (**94**) (**95**)

Thermolysis (50–62°) of Dewar acetophenone affords acetophenone in the triplet excited state but in very low yield.[123] A further study of the isomerization of *gem*-dihalocyclopropanes to 2,3-dihaloalkenes has been reported.[124] Stereoselective electrocyclization of a 1,5-diphenylpenta-1,4-dienyl anion has been observed.[125] Thiocarbonyl ylides are generated by photochemical conrotatory electrocyclization of 2-naphthyl vinyl sulphides.[126]

Thermal (180°) isomerization of (**96**) to (**98**) is believed to occur *via* initial disrotatory ring closure to (**97**), followed by a concerted [$_{\pi}4_{a}$ + $_{\sigma}2_{s}$ + $_{\sigma}2_{s}$]-pericyclic process involving conrotatory opening of the newly formed σ-bond, with concomitant suprafacial [1,7]-shift of the hydrogen *anti* to the 10-phenyl group.[127] The transformation (**97** → **98**) is formally the result of two orbital symmetry-"forbidden" processes, namely conrotatory ring opening (**97** → **99**) and suprafacial [1,3]-hydrogen migration (**99** → **98**).

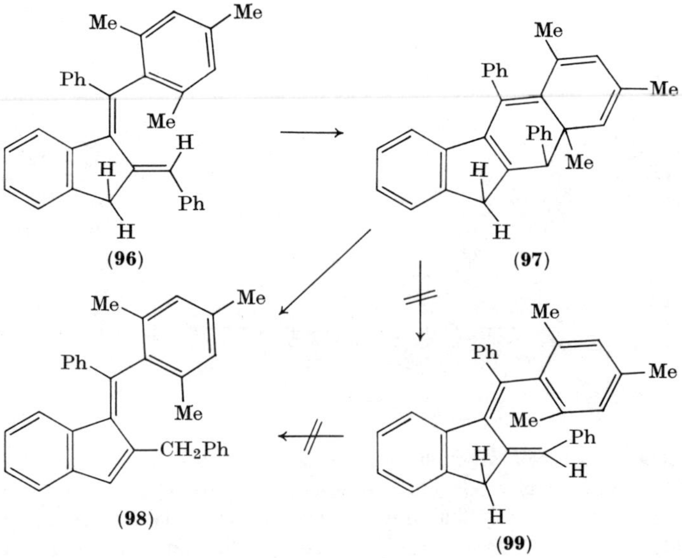

(**96**) (**97**) (**98**) (**99**)

Anionic Rearrangements*

Dewar and Ramsden[128] have reported the results of MINDO/3 calculations on a concerted (but Woodward–Hoffmann "forbidden") pathway for the Stevens rearrangement of trimethylammonium methylide (**100**). For this strongly exothermic reaction ($\Delta H =$ -87 kcal mol^{-1}) the transition state was found to be reactant-like, and the activation energy for the process was only 4 kcal mol^{-1}. It was argued that although CIDNP had in some instances indicated radical-pair formation, a predominantly concerted mechanism may obtain in many instances and would accommodate the high degree of retention

$$Me_3\overset{+}{N}-\overset{-}{C}H_2 \longrightarrow Me_2N-CH_2CH_3$$

(**100**)

of configuration of the migrating group which has been encountered in experimental work.

Pyrolyses of 1- and 2-adamantyltrimethylammonium hydroxide give the corresponding dimethylaminoadamantane as the major products (S_N2), together with adamantane and the Stevens products, adamantylmethyldimethylamines.[129] No evidence for adamantene formation was obtained. These reactions were considered to be the first examples of migration of secondary or tertiary aliphatic alkyl groups. A radical pair (**101**) mechanism was preferred, which would also accommodate adamantane formation. With 3-homoadamantyltrimethylammonium hydroxide (**102**), similar

$$1(2\text{-})Ad-\overset{+}{N}Me_2-\overset{-}{C}H_2 \longrightarrow [1(2\text{-})Ad\cdot + Me_2\overset{+}{N}-CH_2\cdot]\ (\mathbf{101})$$

$$\mathbf{101} \xrightarrow{RH} AdH \qquad \mathbf{101} \longrightarrow 1(2\text{-})AdCH_2NMe_2$$

rearrangement was found, but in this case the formation of C_{22}-hydrocarbon indicated parallel formation of homoadamant-3-ene, which dimerized.[130] A parallel Meisenheimer rearrangement accompanied homoadamantene dimer production in the pyrolysis of (**103**).[131]

(**102**; X = $^{+}NMe_3\ OH^{-}$)

(**103**; X = $^{+}NMe_2\ O^{-}$)

Both Stevens and Sommelet products were formed when (**104**) was allowed to react with a KOH melt.[132] More conventional conditions gave little or no rearranged amines.

* See also Chapter 10.

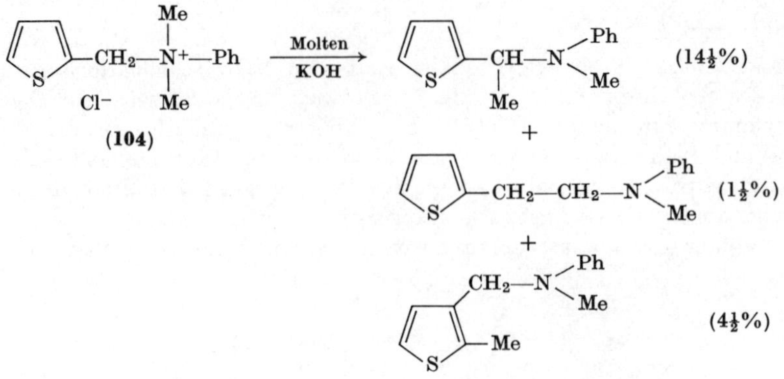

(104) $\xrightarrow{\text{Molten KOH}}$ (14½%) + (1½%) + (4½%)

Carbamoylaminimides (**105**) undergo a Stevens-like rearrangement to semicarbazides on heating when R = benzyl or allyl, but not when R = propargyl; a homolytic dissociation–recombination mechanism was proposed,[133] since when R = butenyl products

$$\text{PhNHCO}\overset{-}{\text{N}}\overset{+}{\text{N}}\text{RMe}_2 \longrightarrow \text{PhNHCONRNMe}_2$$

(105)

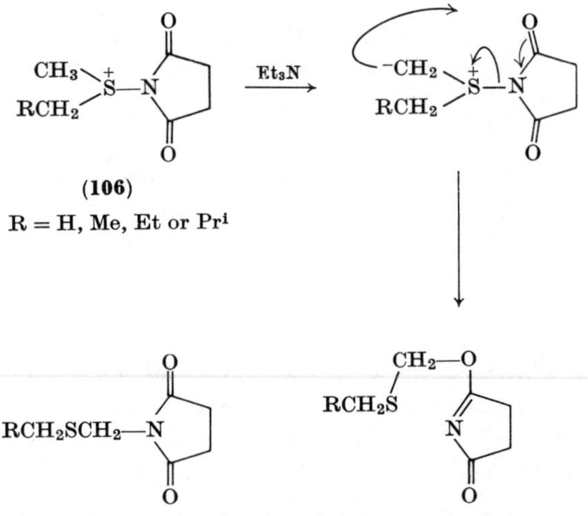

(106) $\xrightarrow{Et_3N}$ → (107) + (108)

R = H, Me, Et or Pr[i]

of both [1,2]- and [3,2]-migration were found. An unusual concerted Stevens-like rearrangement was encountered when the succinimidosulphonium salts (**106**) were treated with Et_3N;[134] the major product (⩾90%) is the [2,3]-rearrangement product (**107**), accompanied by small quantities of (**108**). The sulphonium salt (**106**; R = H) transfers SMe_2 to phenols to give (**109**); these undergo a Sommelet-type rearrangement under the influence of triethylamine to give (**110**).[135]

The ratio of Sommelet product to Stevens product of 0.36 found in rearrangement of (**111**) by aqueous hydroxide ions increases to a limit of 1.5 on addition of cycloheptaamylose;[136] the cause of this effect was not identified. A Stevens-like rearrangement of

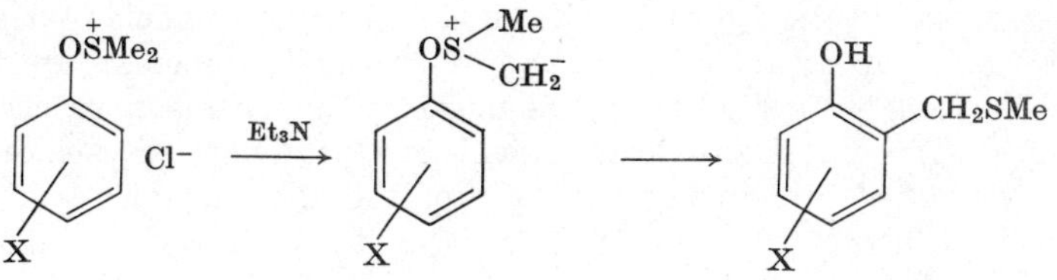

(109) (110)

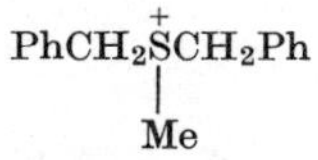

(111)

the sulphimides (**112**) is catalysed by Cl^-.[137] The reaction involves rate-determining nucleophilic attack of chloride at carbon, followed by reaction of the sulphenamide anion with RCl; nucleophilic attack at sulphur followed by rearrangement was discounted. The kinetics of rearrangement of a range of phosphonates into phosphates have been determined.[138]

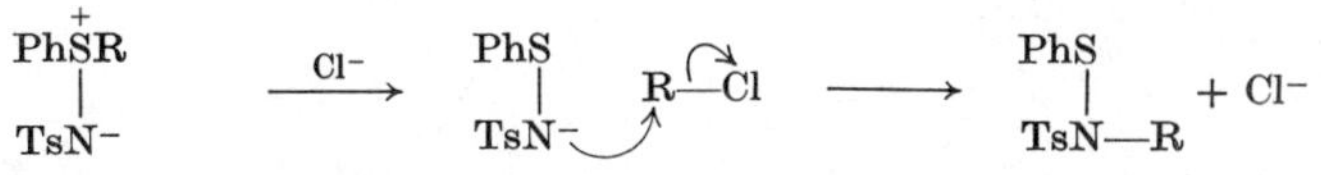

(112)

R = Me or CH_2Ph

[1,2]-Migrations of vinyl groups in the Wittig rearrangements of the conjugate bases of benzyl vinyl ethers have been examined;[139] with alkyl-lithium the corresponding benzyl vinyl sulphides form carbanions that do not rearrange. The mechanisms of [1,2]-aryl shifts in Wittig rearrangements have been further discussed.[140]

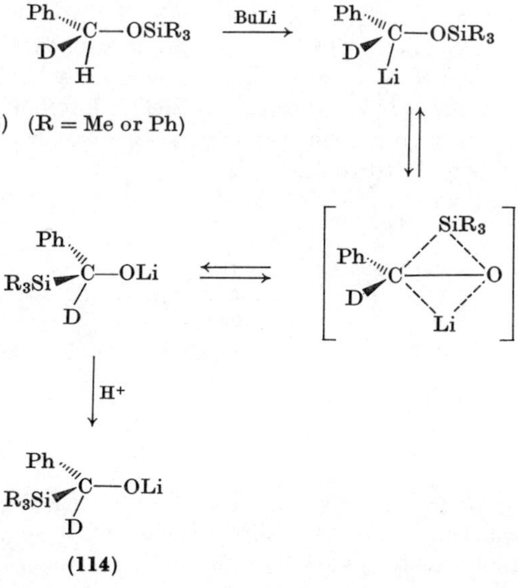

(113) (R = Me or Ph)

(114)

Rearrangements involving a [1,2]-shift of silyl groups to anionic centres have been reviewed,[141] and new examples uncovered;[142–144] similar processes are found in germanium chemistry.[143,144] Whilst alkyl migration in similar reactions may often have some diradical character, Si and Ge are capable of expanding their valence shell so that reaction may proceed without bond dissociation through a pentaco-ordinate Si or Ge species; this is consistent with the >99% inversion of configuration found on conversion of (**113**) into (**114**) (p. 475),[145] and on rearrangement of the corresponding germanium compound with R = Me.

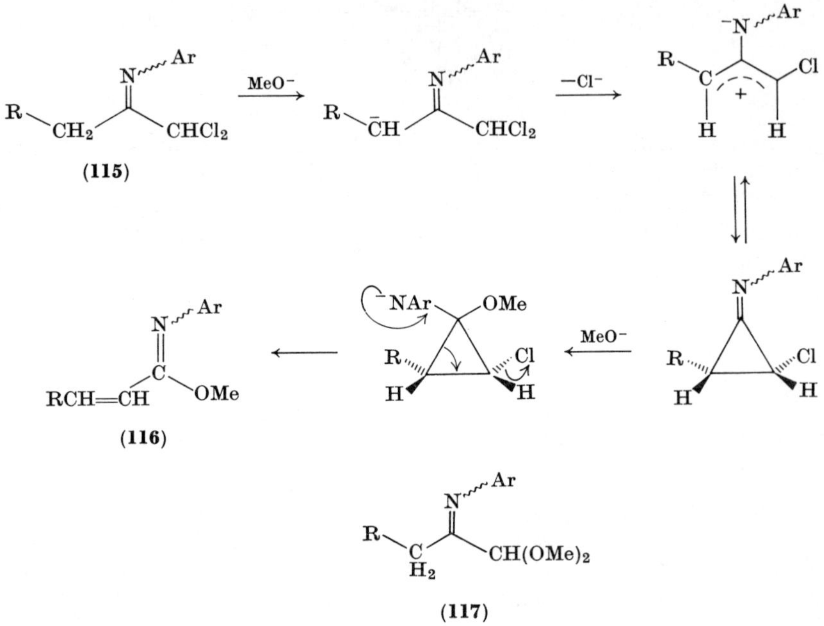

A novel Favorskii-like reaction has been found in the reactions of *N*-[(dichloromethyl)alkylidene]anilines (**115**) with sodium methoxide which lead, *inter alia*, to the unsaturated imidates (**116**).[146] The proposed mechanism is indicated in the formulae. The by-product (**117**) is formed in MeOH but not in Et_2O. A quasi-Favorskii reaction affords a route to 1-phenylcycloalkanecarboxylic acids (**118**) accompanied by some direct substitution product.[147]

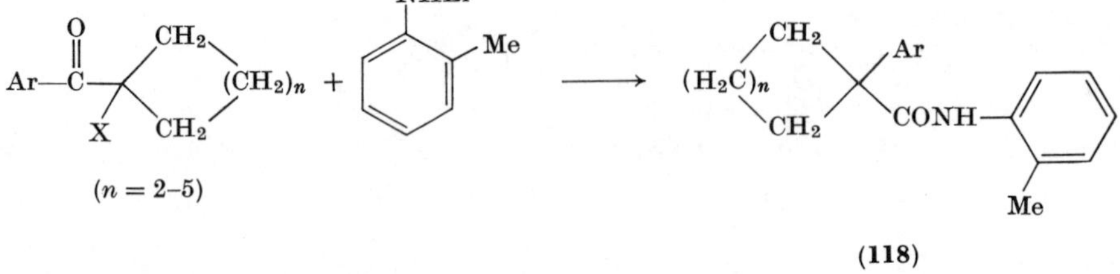

Homocuneone (**119**), prepared by a new route, undergoes rapid base-promoted cleavage to (**121**);[148] that this reaction is faster by at least 10^3 than cleavage of homocubanone (**122**) was attributed to the bis-homoaromatic character of the intermediate

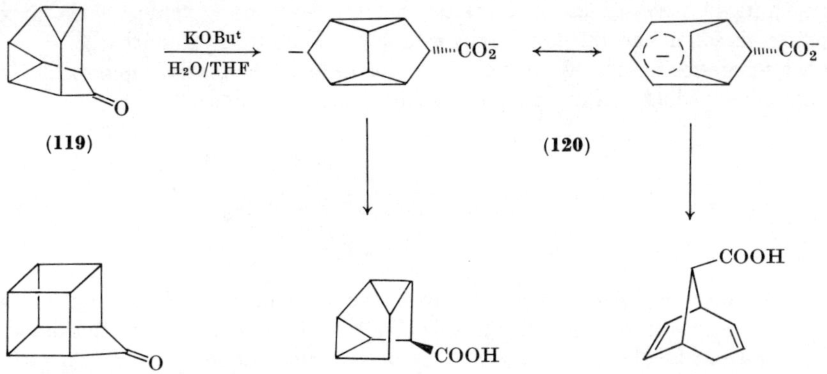

KOBu^t / H_2O/THF

CO_2^- $\longleftrightarrow$ CO_2^-

(119) (120)

COOH

COOH

(122) (2%) (121) (98%)

dianion (**120**). The rearrangement of anion (**123**), which parallels that of the parent bicyclo[3.2.2]nonatrienyl anion,[149] has been employed as a means of comparing benzyl and allyl anion stabilities;[150] the result is that the allylic structure (**123**) is more stable

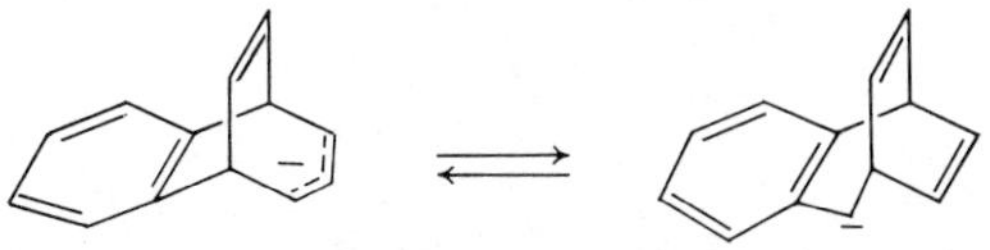

(123) (124)

by at least 2 kcal mol^{-1} than the benzylic ion (**124**) which was not directly observable. However, the extent to which this can be related to simple benzylic and allylic stabilities is questionable in view of the laticyclic interactions in (**123**) and (**124**). That the benzylic ion (**124**) must be formed was shown by scrambling of a deuterium label initially at C-2; rearrangement was found to occur both by vinyl migration (→ **124**) and by aryl migration.

Rearrangements Involving Electron-deficient Heteroatoms

Several studies of the Beckmann rearrangement have been reported. It has been found, for example, that the dichlorotin(IV) salt of the sulphonic acid derivative (**125**) is an especially effective catalyst.[151] With the $SnCl_4$ complex of cyclohexanone oxime, for example, exchange takes place to give an oxime sulphate complex which undergoes

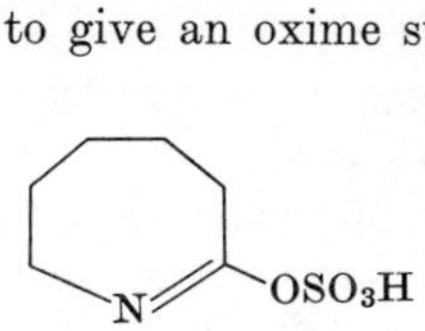

N OSO_3H

(125)

rearrangement to regenerate the catalyst. Efficient conversion into the $SnCl_4$ complex of caprolactam results. A zirconium salt of (**125**) was also effective. The ease of migration of various alkyl groups has been investigated for a series of alkan-2-one oximes catalysed by polyphosphoric and related acids in several solvents;[152] the extent of competing fission of the *anti*-alkyl oximes to give acetonitrile was noted.

In the ferrocene series, the oximes (**126**) are reduced to iminium ions with $PhSO_2Cl$ in acetone;[153] when R = Ph some phenyl migration was found.

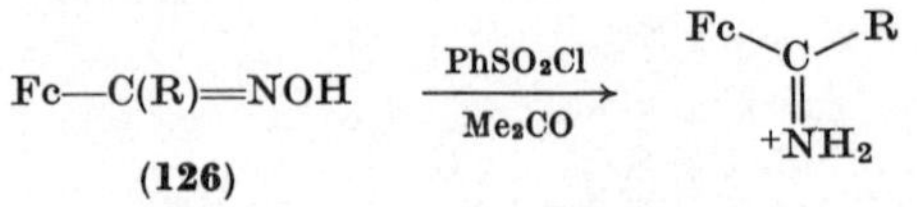

Fc—C(R)=NOH $\xrightarrow[Me_2CO]{PhSO_2Cl}$ Fc(R)C=$\overset{+}{N}H_2$

(**126**)

Abnormal rearrangements of oximes have been encountered in attempts to acetylate the hydroxyiminocholestane derivative (**127**),[154] and on treatment of the tetrahydrocinnoline derivative (**128**) with acid.[155] In the latter case a similar reaction was found under Schmidt conditions; in the Beckmann case the intermediacy of (**129**) was proposed.

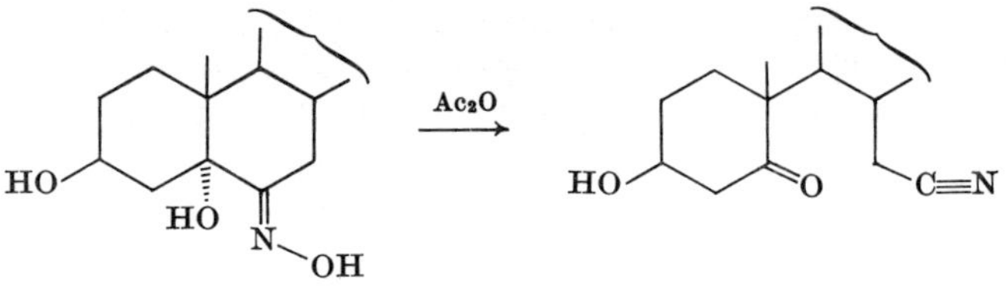

(**127**)

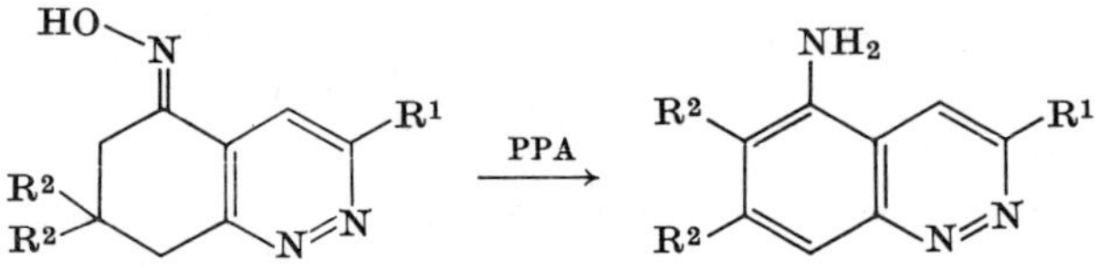

(**128**)

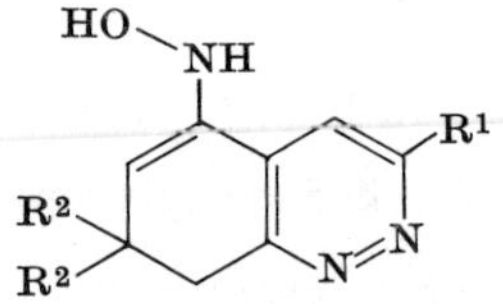

(**129**)

Two competing anomalous pathways have been encountered on acid-catalysed rearrangement of (**130**) or its *O*-benzoyl derivative, differing according to the site of protonation which initiates the reaction.[156]

A theoretical model for the rearrangement step of the Baeyer–Villiger reaction has been discussed, and CNDO/2 and *ab initio* calculations appear to agree well with experiment.[157] For the optimized transition-state geometry (**131**), there is little reorganization of the migrating group; this is consistent with the small α-deuterium isotope effects determined experimentally.[158] The experimental β-deuterium isotope effect (determined[158] for benzyl migration in 3,3,3-trideuterio-1-phenylpropan-2-one) is consistent with the calculated development of carbonyl double-bond character at the migration transition state.

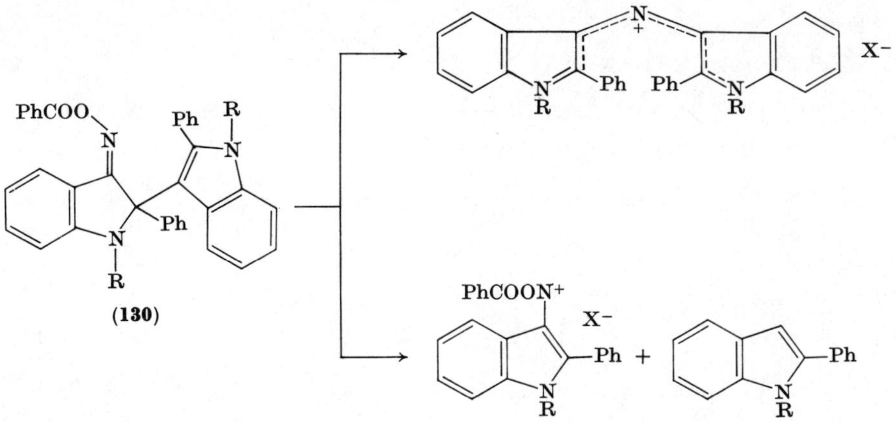

(130)

A study of Lossen and Hofmann rearrangements of dialkylhomophthalimide (isoquinoline-1,3-dione) derivatives (**132**) showed that the bulk of the alkyl group determined the direction of the rearrangement.[159] In neither case was ^{18}O incorporated from labelled water.

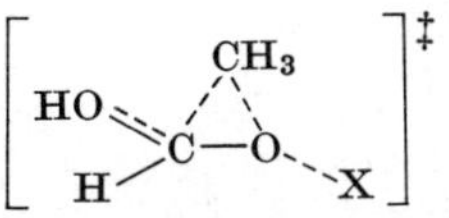

(131)

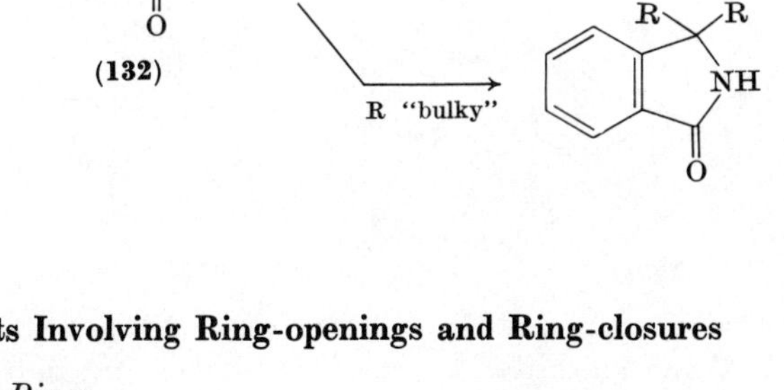

R "small"

R "bulky"

(132)

Rearrangements Involving Ring-openings and Ring-closures

Three-membered Rings

Several reports have appeared of rearrangements involving derivatives of cyclopropyl ketones or aldehydes. For example, the *endo*-pyrrolidineaminal (**133**) isomerizes to the *exo*-isomer in a mildly acid-catalysed process at 80°, probably by way of the cyclopropylidene enamine (**134**);[160] at 140°, further rearrangement to the homofulvenes (**135**) occurs, also by way of (**134**). Heating, followed by acidic work-up, transforms the aminal

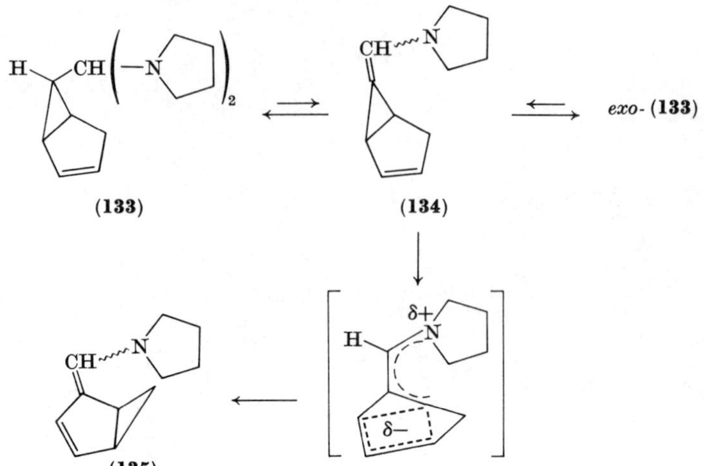

H CH(—N pyrrolidine)$_2$ ⇌ CH〰N ⇌ *exo*-(**133**)

(**133**) (**134**)

δ+ H N δ−

CH〰N

(**135**)

(**136**) into mixtures of (**137**) and (**138**), presumably by a related mechanism.[161] Bamford–Stevens reactions of several cyclopropyl ketone tosylhydrazones are described on p. 234.

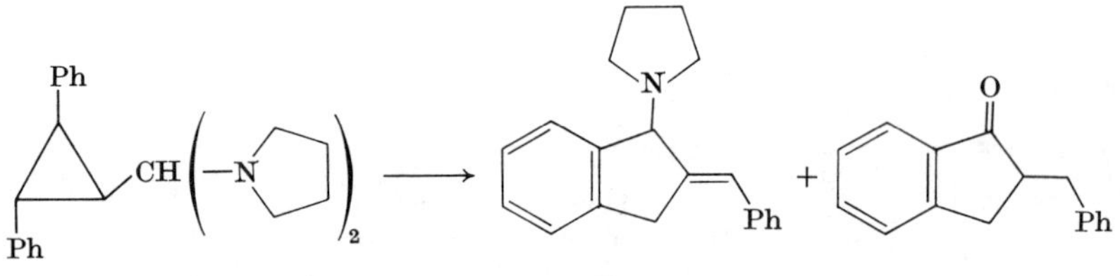

Ph CH(—N)$_2$ Ph → N Ph + O Ph

(**136**) (**137**) (**138**)

Further work on ring-expansion reactions of cyclopropanols to cyclobutanones includes a study of the pyrolytic (100°) rearrangement of 1-vinylcyclopropanols; since (**139**) gave (**140**), it was argued that the process involved intramolecular *cis*-addition to the double bond rather than a pericyclic process which, it was stated, should have led to the opposite stereochemistry.[162]

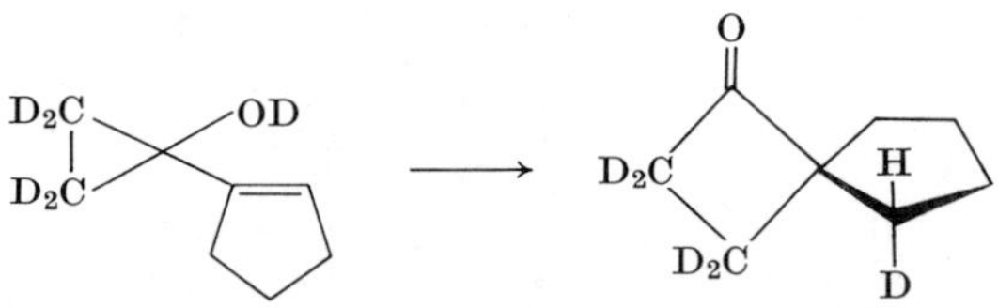

D_2C OD D_2C → O D_2C H D_2C D

(**139**) (**140**)

Ring-opening of the dicyclopropylamine (**141a**) gives (**142**), probably through the azomethine ylide (**143**) and 1,5-shift of hydrogen. Orbital overlap for rearrangement to (**143**) is much poorer in the *anti*-isomer (**141b**), and higher temperatures are required for reaction to occur.[163]

The importance of stereochemical factors on the thermal transformation of 1-phenyl-2-vinylaziridines into dihydrobenz[*b*]azepines, imines, or 3-pyrrolines has been examined.[164]

Kinetic parameters have been obtained for the formation of cyclopentanone and cyclopent-2-en-1-ol by gas-phase pyrolysis of 6-oxabicyclo[3.1.0]hexane;[165] the results

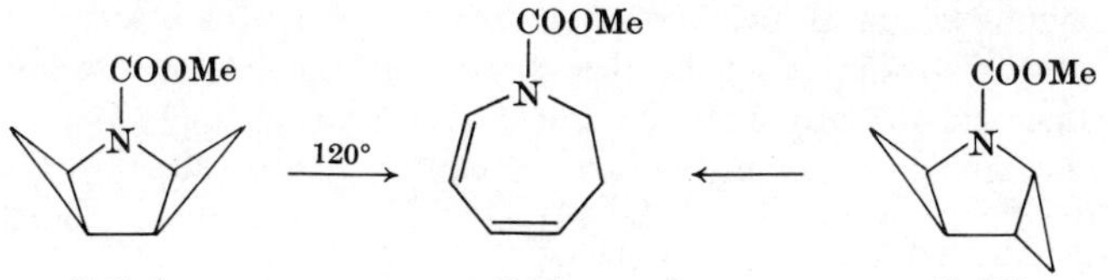

COOMe N 120° (141a) COOMe N (142) COOMe N (141b)

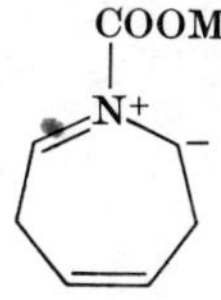

COOMe N+ (143)

were inconsistent with the formation of the two products through a common biradical intermediate (**144**), although this may have been the cyclopentanone precursor, with a

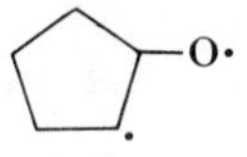

O· (144)

concerted process leading to the alcohol. It was not possible to differentiate between a biradical and a zwitterion intermediate in the formation of phenylacetaldehyde on

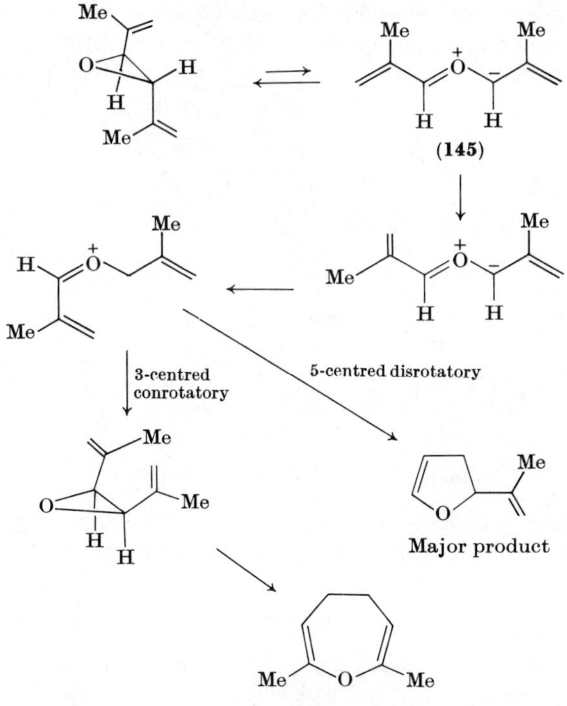

Me O H H Me ⇌ Me Me Ȯ H H (145)

H Ȯ Me Me ← Me Ȯ Me H H

3-centred conrotatory

5-centred disrotatory

Me O Me H H

Me O

Major product

Me O Me

Minor product

pyrolysis of styrene oxide at 250°;[166] at higher temperatures toluene is formed. The mechanism proposed for thermal racemization and rearrangement of *trans*-di-isopropenyloxirane does involve a zwitterionic intermediate (**145**) as indicated.[167] A dipolar species may also be involved in the conversion of (**146**) into (**148**) [probably through (**147**)];[168] similar reactions of [(**146**); O replaced by NMe or CH_2] give analogues of (**147**), and there is evidence that the rearrangement is concerted.

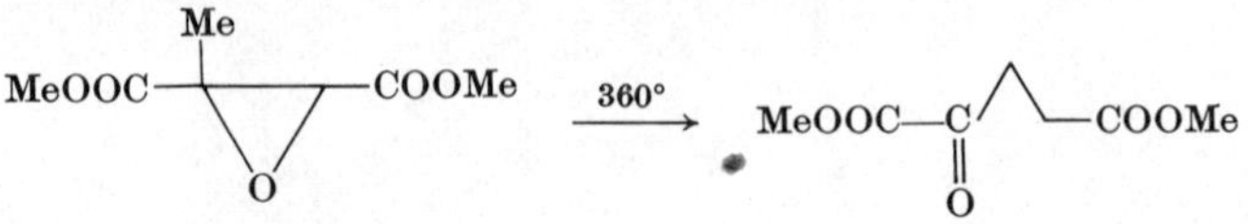

MeOOC–C(Me)(O)–COOMe $\xrightarrow{360°}$ MeOOC–C(=O)–CH=CH–COOMe

(**146**) (**147**)

MeOOC–C(=CH_2)–O–CH_2COOMe

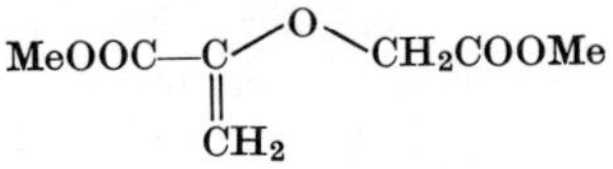

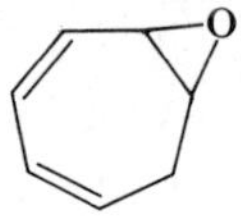

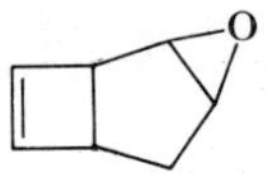

(**148**) (**149**) (**150**)

Pyrolysis (250°) of 8-oxabicyclo[5.1.0]octa-2,4-diene (**149**) gives heptatrienal. At higher temperatures *o*-cresol and benzaldehyde are obtained; these are formed by way of bicyclo[3.2.0]hept-2-en-7-one and 2,3- and 2,5-dihydrobenzaldehyde, by thermally allowed pericyclic rearrangements.[169] Photochemical isomerization of (**149**) gives the aldehyde, together with cyclohepta-3,5-dienone and the tricyclic compound (**150**). Heterolytic reactions of (**149**) have also been thoroughly investigated.[170]

The thermal and photochemical rearrangements of the epoxy-ketone (**151**) give a spirodiketone (**152**), most reasonably by way of (**153**).[171] In contrast, the BF_3-catalysed rearrangement gives (**154**).

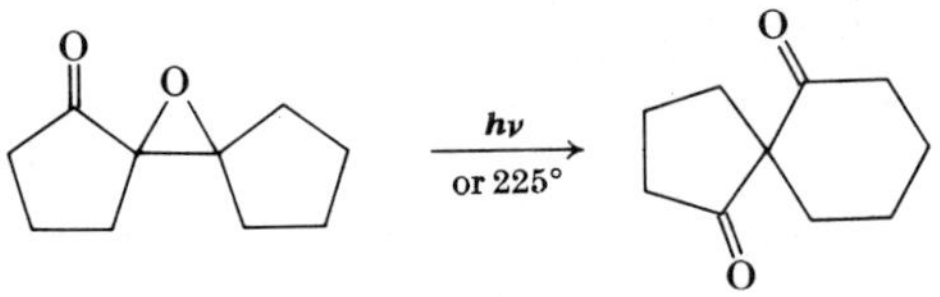

$\xrightarrow[\text{or } 225°]{h\nu}$

(**151**) (**152**)

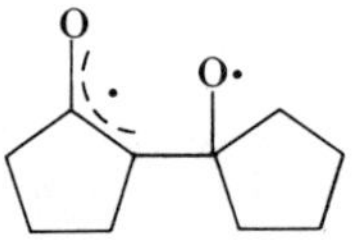

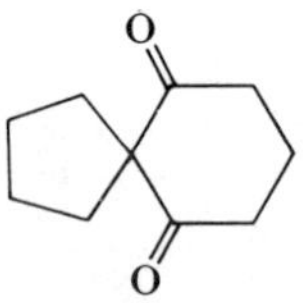

(**153**) (**154**)

The oxabicyclohexene (**155**) reacts with water at pH 8.2 by a pH-independent mechanism to give *cis*-2,4-pentadienal (32%), cyclopent-3-enone (33%), and a mixture

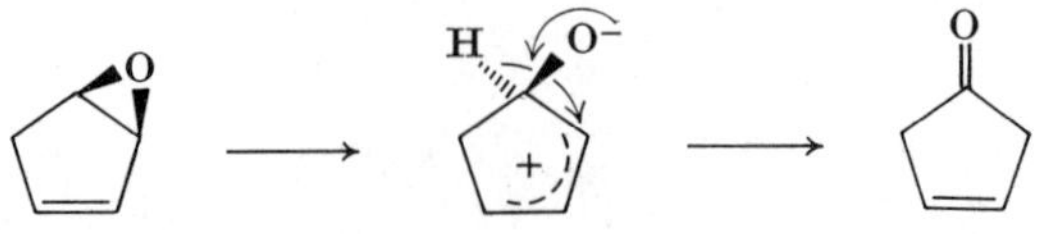

(**155**)

of dihydroxycyclopentenes (35%);[172] when reaction was effected in D_2O the cyclopentenone did not incorporate deuterium, indicating that it was not formed through its enol but rather by an NIH-shift, presumably in the zwitterionic intermediate illustrated.

Other epoxide reactions are discussed in Chapter 9.

Four-membered Rings

Several 2-alkylidenecyclobutanones have been prepared and their ring-contraction to cyclopropyl ketones or aldehydes on heating or exposure to acid has been studied;[173] 2-hydroxycyclobutanone ketals, e.g. (**156**), were also shown to give cyclopropane derivatives on heating. Cyclobutanone itself undergoes ring-contraction to cyclopropane-

OH OMe OMe →(220°) CHO OMe

(**156**)

carboxylic acid under a variety of oxidative conditions [Tl(III), SeO_2–H_2O_2–H^+].[174] The bicyclic chloro-ketones (**157**) rearrange readily with base, but rearrangement of the epimeric compounds (**158**) is more difficult, particularly when R is bulky, in which case 2-alkyltropones accompany the bicyclo[3.1.0]hexenes.[175]

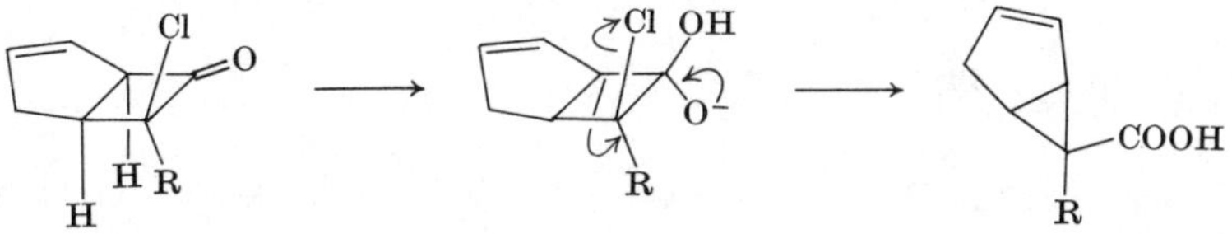

(**157**) → → COOH R

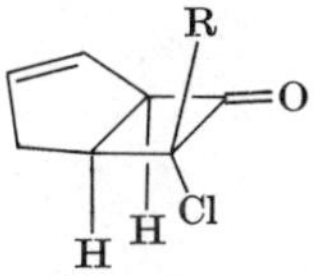

(**158**)

The formation of (**160**) by dimerization of (**159**), and the kinetics of its thermal isomerization into (**161**) have been studied;[176] these reactions were discussed in terms of a general analysis of factors governing reactions which proceed through bisallyl diradicals.

$2(Me_3CC{\equiv}C)_2C{=}C{=}CClCMe_3$ →

(**159**)

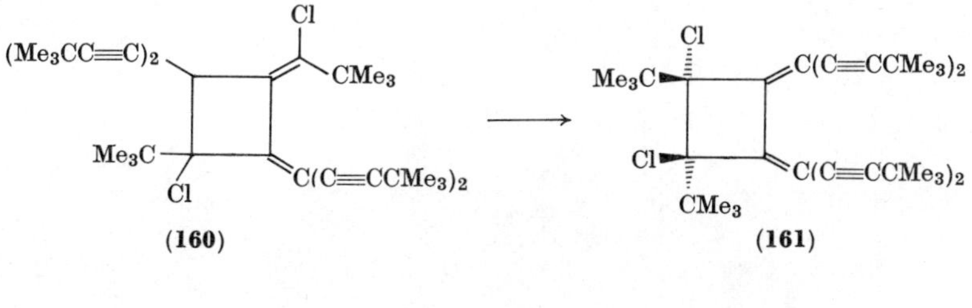

(**160**) → (**161**)

The mechanisms of thermal rearrangement of 4-phenyl-2-oxa-3-azabicyclo[3.2.0]hepta-3,6-diene into pyrrole derivatives[177] and of rearrangements of an imino-oxetane[178] an iminoazetine,[179] and a thietane 1,1-dioxide[180] have been discussed.

Five-membered Rings

When *N*-acyl-lactams are heated with calcium oxide, cyclic imines are formed with loss of the elements of carbon dioxide; it has now been shown by means of isotopic labelling that the acyl carbon is retained, perhaps by the sequence shown for (**162**).[181]

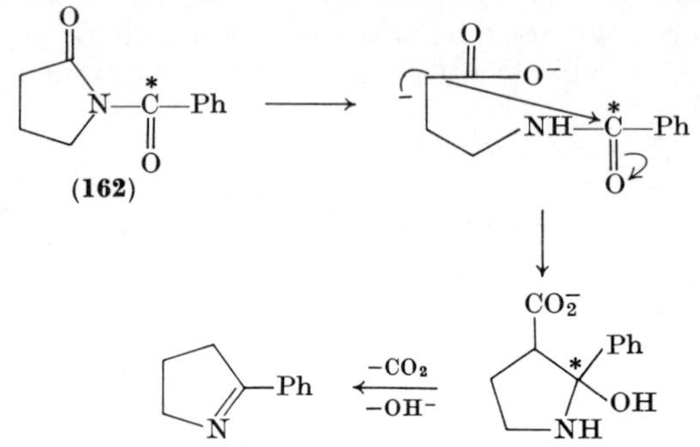

(**162**)

$-CO_2$
$-OH^-$

It has been suggested that the isomerization of the 4-acyl-5-alkyl-3-phenylisoxazoles (**163**) to oxazoles (**165**) proceeds with initial N—O bond fission and recyclization to

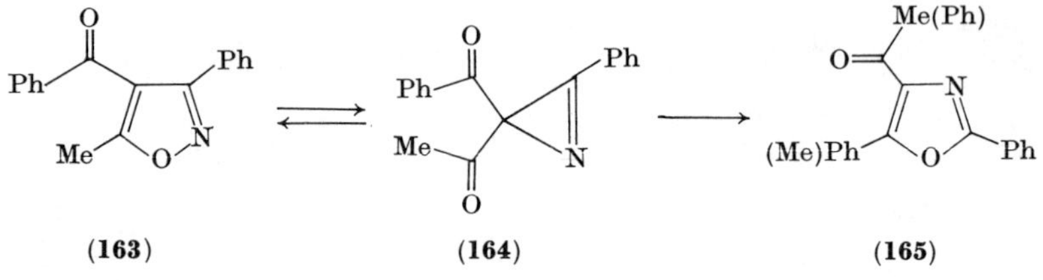

(**163**) (**164**) (**165**)

intermediate 2*H*-azirines (**164**).[182] The labelled furan derivative (**166**), on pyrolysis (700°), gives the ring-contraction product (**167**); retention of both deuterium atoms

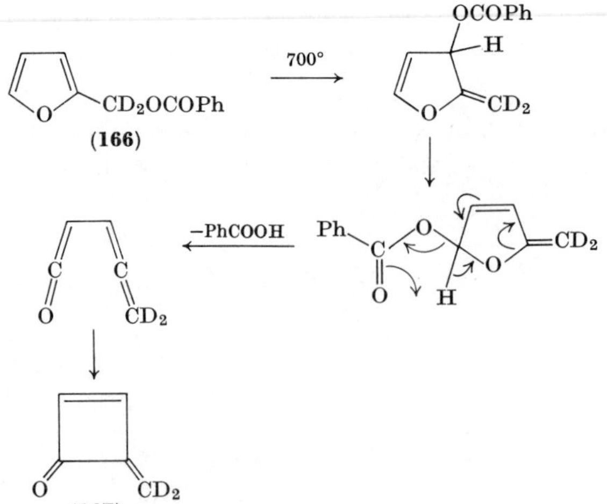

700°

(**166**)

−PhCOOH

(**167**)

excludes the possibility of an α-elimination mechanism but is consistent with the sequence of events indicated.[183] This was considered to be supported by the fact that the 5-methyl analogue of (**166**) gave 2,5-dihydro-2,5-dimethylenefuran under comparable conditions.

The kinetics of rearrangement of several 3-arylidenephthalides into 2-arylindane-1,3-diones have been studied.[184]

Six-membered and Seven-membered Rings

Pyrolysis of the labelled cyclohexadiene (**168**) (560°) gives complete scrambling of the deuterium label, indicating the occurrence of ring-opening and [1,7]-hydrogen shifts, rather than [1,5]-methyl shifts.[185] In a related reaction, the dienones (**169**) are converted into acid chlorides (**170**) by way of pyran derivatives; these presumably undergo

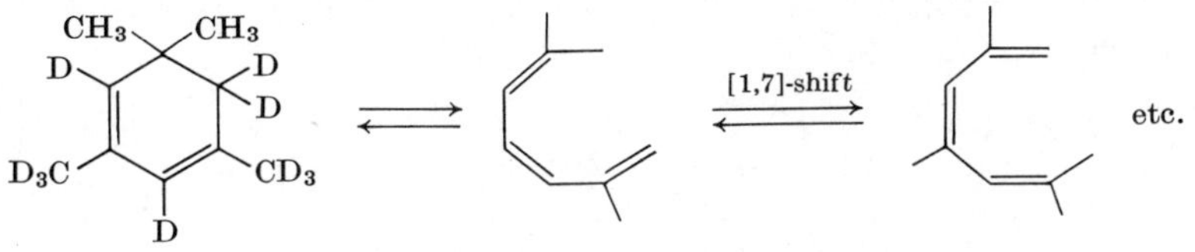

CH_3 CH_3 D D D D_3C CD_3 D [1,7]-shift etc.

(**168**)

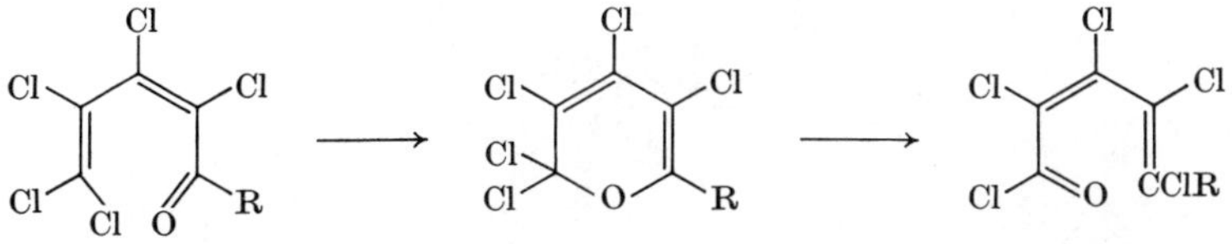

(**169**)
R = alkyl or aryl

(**170**)

"[1,5]-chlorine shifts" involving pyrylium chloride intermediates.[186] Other work with related compounds has been discussed,[187] as have factors influencing the position of equilibrium between alkylated pentadienones and the corresponding 2*H*-pyrans.[188]

The triazine (**171**) is converted above its melting point into the triphenylmelamine (**172**).[189] The amino-acid (**173**) gives a methylene-lactam when heated with acetic anhydride by way of the ring-opened mixed anhydride shown.[190]

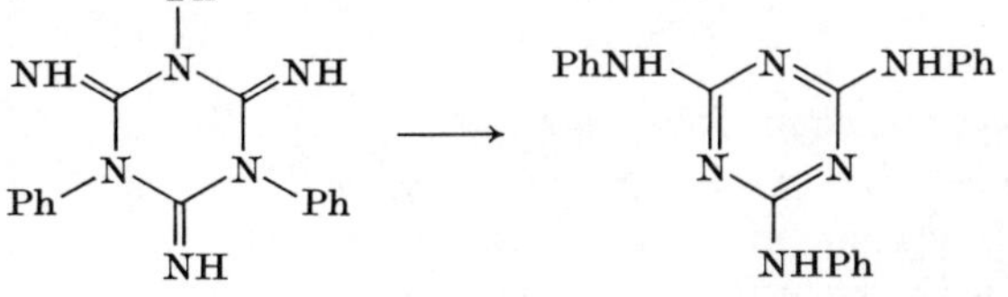

(**171**) (**172**)

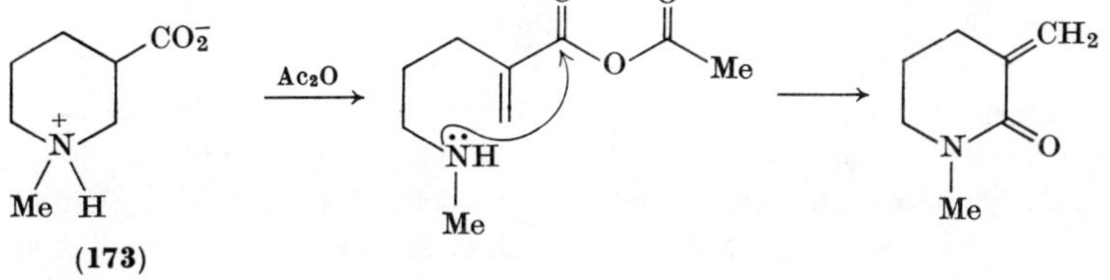

Ac_2O

(**173**)

The rearrangement of the bridged heterocycle (**174**),[191] catalysed by TFA, parallels the thermal reaction reported last year;[192] this probably involves the intermediate (**175**). Gas-phase pyrolysis of (**176**) gives the products indicated, for the formation of which biradical mechanisms are discussed;[193] the oxygen affords only weak activation to the reactions in comparison with its effect on the decomposition of oxaquadricyclanes.

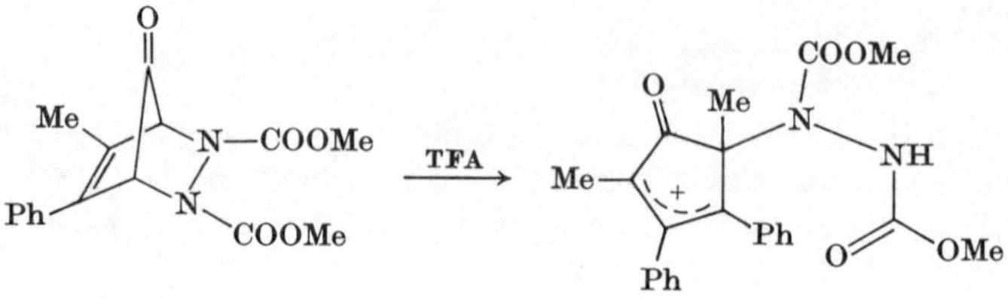

(**174**) (**175**)

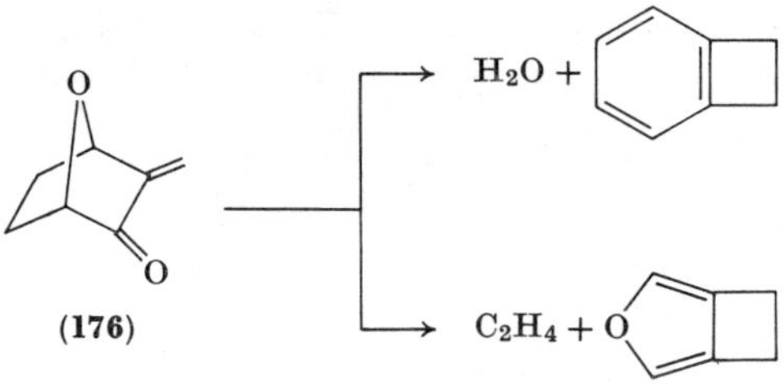

(**176**)

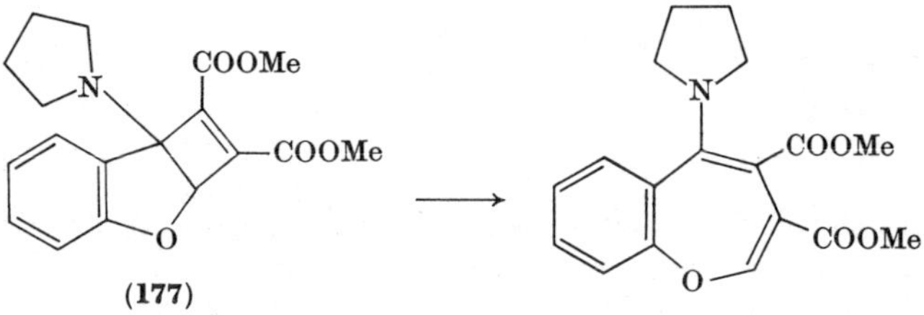

(**177**)

Ring-contraction reactions of 1,2-diazepines[194] and of a benz[*b*]oxepin[195] have been discussed. A benzoxepin was formed by the thermal ring expansion of the [2 + 2]-cycloadduct (**177**) of 3-pyrrolidinylbenzofuran and dimethyl acetylenedicarboxylate.

Isomerizations

Cyclopropane and Cyclobutane Reorganizations

Dynamic trajectories on the potential energy surface for the geometric and optical isomerization of cyclopropanes have been calculated; only for the ring-opening and -closure stages of the reaction do these lie close to the static minimum-energy path as calculated by *ab initio* molecular orbital methods.[196] The isomerization of cyclopropane to propene may become an "allowed" concerted reaction as a result of configuration interaction (mixing of ground and doubly excited configurations).[197]

An extensive *ab initio* study of the potential-energy surface for the methylenecyclopropane rearrangement shows the orthogonal methylene-allylic trimethylenemethane diradical to be the most stable reaction midpoint by almost 20 kcal mol^{-1}. The calculated transition state corresponds to a methylene twist angle of *ca.* 45°, giving an activation energy of 44.7 kcal mol^{-1} in good agreement with experiment.[198] CNDO/2 calculations suggest that the allene oxide–cyclopropanone isomerization involves transition states in which significant bending around the sp^2 centre has preceded bond-breaking, rather than the planar oxyallyl intermediate (or transition state) previously assumed.[199]

Formation of (**181**) and (**182**) in the thermolysis (40–65°) of the *syn*- and *anti*-isomers of 1-allylidene-2-vinylcyclopropane (**178**), is most satisfactorily explained by initial methylenecyclopropane rearrangement of (**178**) to *cis*- and *trans*-divinylmethylenecyclopropanes, (**179**) and (**180**) (for a discussion of the positional selectivity in the degenerate rearrangement of vinylmethylenecyclopropanes see ref. 200). Under similar conditions, 1-(1-buta-1,3-dienyl)-2-methylenecyclopropane affords 1-methylene-4-vinylcyclopent-2-ene, probably by a concerted [1,3]-sigmatropic shift or its diradical equivalent.[201]

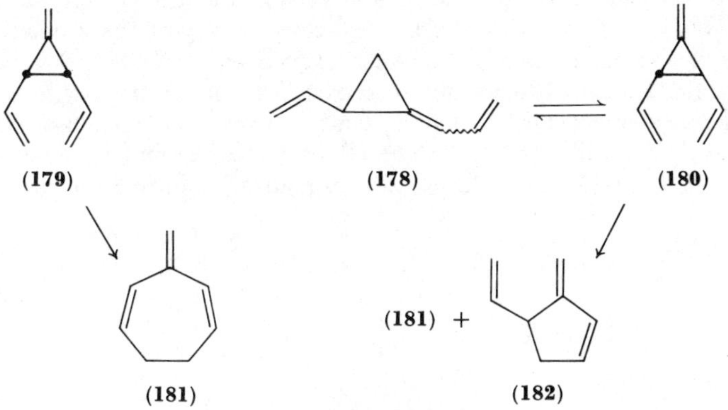

(**179**) (**178**) (**180**)

(**181**) (**181**) + (**182**)

The stereoselectivity observed in the thermolysis (178°) of dialkenylcyclopropanes (**183**) and (**184**) has been rationalized on the basis of a one-centre (C-1 or C-2) epimerization to a *cis*-dialkenylcyclopropane intermediate, followed by rapid Cope rearrangement to the cycloheptadiene product. Severe methyl–methyl interactions in the boat-like transition state for Cope rearrangement of the *cis,cis,cis*-isomer (**185**), make the rate of its isomerization to the *cis,trans,cis*-isomer (**186**) comparable with those for the ring expansions of (**183**) and (**184**).[202]

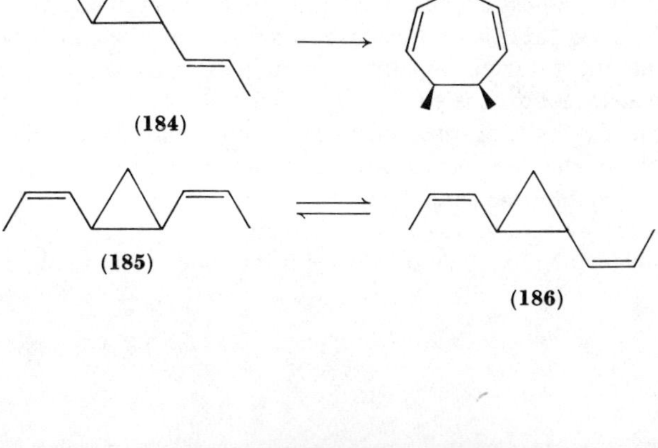

(**183**)

(**184**)

(**185**) (**186**)

In the solid state, 1,2,5,6-tetracyano-*anti*-tricyclo[4.2.0.0^{2,5}]octane rearranges thermally to 1,2,5,6-tetracyano-(Z,E)-cyclo-octa-1,5-diene, the product of orbital symmetry-"allowed" ($_{\sigma}2_s + {}_{\sigma}2_a$) cycloreversion. This observation which contrasts with those from thermolysis of related compounds, where (Z,Z)-cyclo-octa-1,5-dienes are the only products isolated, may be the consequence of kinetic factors related to minimal lattice displacements during reaction.[203] The ratios of cracking (to styrene): epimerization for *cis*-1,2-diphenylcyclobutane are 2.0, 7.1 and 2.6 for reaction initiated thermally (190–210°), photochemically, and photochemically with acetone-sensitization, respectively. These observations lead to the suggestion that there may be a significant concerted component to the cycloreversion in the first singlet excited state.[204]

At 160–165° the alkylidenecyclobutanones (**187a**) and (**189a**) are readily equilibrated, and (**187b**) isomerizes to (**189b**), probably by way of the zwitterionic intermediate (**188**) or its singlet diradical equivalent. However, pyrolysis (180°) of the related ethylidenecyclobutanones (**190**) results in ready E/Z-isomerization with virtually no change in the ring stereochemistry. This E/Z-interconversion (which also occurs during base-catalysed deuterium exchange at the allylic methyl group) is believed to occur through the dienol (**191**).[205] In contrast, Z/E-isomerization during the pyrolysis (330°)

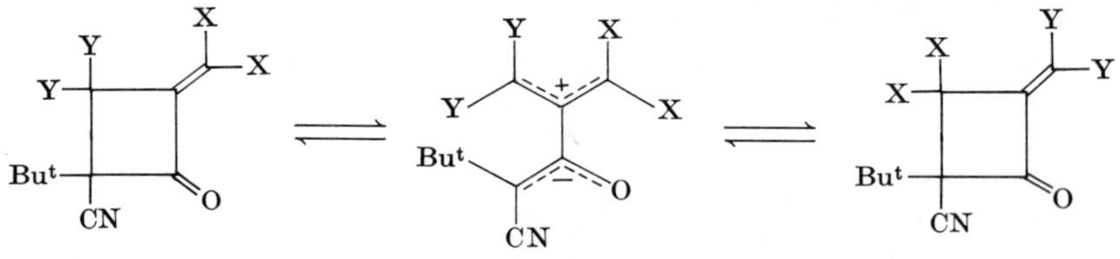

(**187**) (**188**) (**189**)

a: X = CD_3; Y = Me
b: X = H; Y = Me

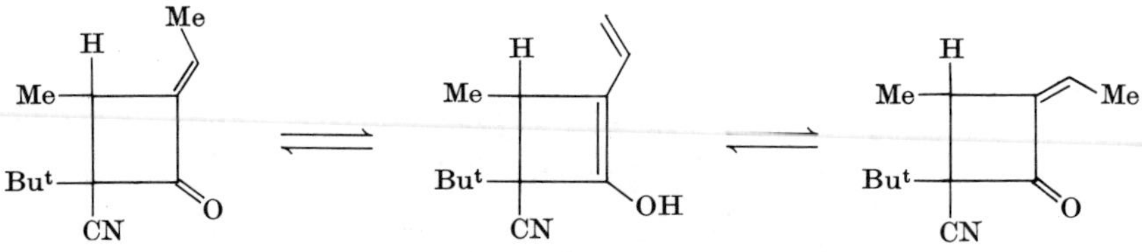

(**190**) (**191**)

of optically active (Z)-2-ethylidene-3,4,4-trimethylcyclobutanone, is accompanied by inversion of configuration at C-3,[206] and this result is in accord with 1,3-carbon migration involving predominant antarafacial allylic participation, as observed in the automerization of methylenecyclobutanes.[207]

Gas-phase pyrolysis (500°) of *trans*-1,2-diethynylcyclobutane affords vinylacetylene, 1,2-dihydropentalene and the previously unknown C_8H_8 isomer, bicyclo[4.2.0]octa-1,5,7-triene. A reasonable mechanistic route to this latter compound involves initial conversion of the *trans*-cyclobutane into its *cis*-isomer by a diradical process, followed by Cope rearrangement to cyclo-octa-1,2,4,5-tetraene and final disrotatory ring closure.[208]

Polycyclic Rearrangements

Substituent effects on the norcaradiene–cycloheptatriene equilibrium have been studied. The norcaradiene isomer is stabilized by 1-, 2- or 3-substituents in the relative order 1-Me < H < 3-Me < 2-Me ≈ 2-Br < 2-Ph, while for 7-substituents the relative stabilizing effect is CN < COOMe < COOH.[209] The degenerate rearrangement of norcaradienes has been shown to proceed with inversion of configuration at the migrating carbon, whereas orbital-symmetry control would lead to retention. The observed process corresponds to the least-motion pathway and may be subject to subjacent orbital control.[210]

Full details of the thermally induced degenerate skeletal rearrangement and isomerization reactions of cyclo-octatetraenes have been reported.[211] In all cases the valence isomeric bicyclo[4.2.0]octatrienes are implicated. Further rearrangements are believed to occur by formation (through intramolecular Diels–Alder cyclization) and subsequent ($_{\sigma}2_s + _{\sigma}2_s + _{\pi}2_s$) opening of intermediate tetracyclo[$4.2.0.0^{2,8}.0^{5,7}$]octenes and/or [1,5]-sigmatropic shifts in the same trienes. The former pathway predominates for structurally unconstrained cyclo-octatetraenes (e.g. the dimethyl derivatives)[212] and is followed exclusively for annulated bicyclo[4.2.0]octatrienes (propellenes).[213]

The relative ease of thermal epimerization of *syn*-9-substituted-*cis*-bicyclo[6.1.0]-nonatrienes (**192**; Y = F > OMe > NMe_2 > CN) is not consistent with initial cleavage of the 1,9-bond to give a diradical species. Rather, the epimerization is believed to occur *via* tricyclo[$4.3.0.0^{7,9}$]nona-2,4-diene intermediates. Further pyrolysis of the *anti*-isomers (**194**; Y = F or NMe_2) affords, *inter alia*, bicyclo[4.2.1]nonatriene derivatives (**195**; Y = F or NMe_2), probably as a result of [1,5]-sigmatropic carbon migration from the folded conformation (**193**).[214]

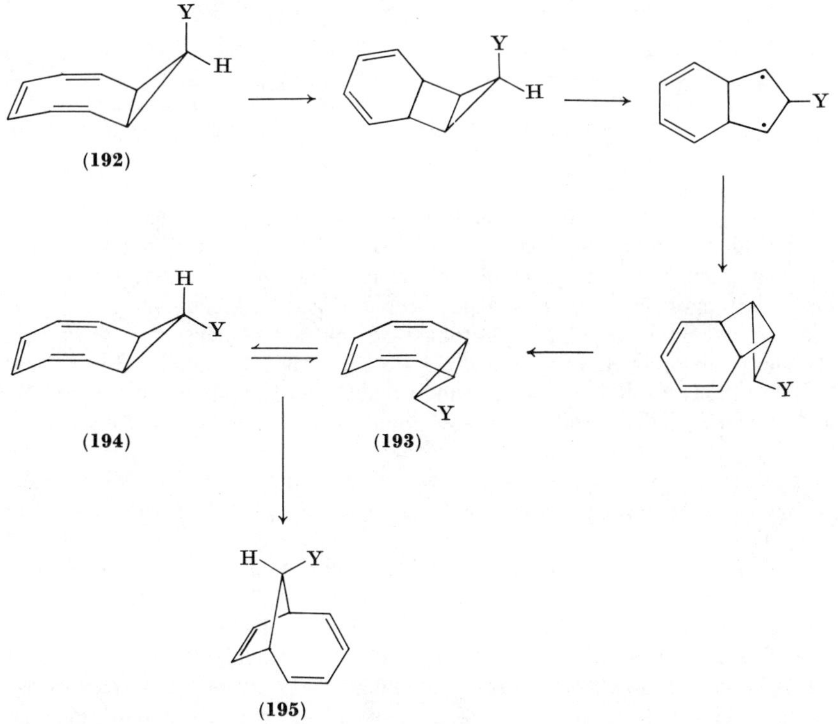

(**192**)

(**194**) (**193**)

(**195**)

A related kinetic study, confirms theoretical predictions that thermal rearrangement of 9-substituted *cis*-bicyclo[6.1.0]nonatrienes should be accelerated by π-donor and retarded by π-acceptor substituents.[215] Reaction of *cis*-bicyclo[6.1.0]nonatriene with benzylidenetricarbonyliron affords, amongst other products, the tricarbonyliron complex of tricyclo[4.3.0.0^{7,9}]nona-2,4-diene. The free hydrocarbon can be isolated by oxidation of the complex with Ce(IV) at −20° and its reversion to *cis*-bicyclo[6.1.0]nonatriene has been examined ($\Delta G^{\ddagger} = 22.9$ kcal mol^{-1}). In a similar fashion, tricyclo[4.4.0.0^{2,5}]-deca-7,9-diene can be obtained from bicyclo[6.2.0]deca-2,4,6-triene.[216] Thermal (300°) rearrangement of 3,4-bis(trimethylsilyloxy)-*endo*-tricyclo[4.2.1.0^{2,5}]nona-3,7-diene affords a different bis(trimethylsilyloxy)-*cis*-8,9-dihydroindene isomer than would have been predicted from studies with the parent hydrocarbon.[217]

Substituent effects on the rearrangement of 1- and 2-substituted pentafluorobicyclo-[2.2.9]hexa-2,5-dienes to pentafluorobenzenes have been investigated.[218] The reaction pathways for the extensive skeletal rearrangements accompanying dehydrogenation of thujopsene and widdrol have been established by the isolation and subsequent dehydrogenation of stable reaction intermediates.[219] A further kinetic study of the gas- and liquid-phase pyrolysis of α-pinene has been reported.[220] On the basis of deuterium-labelling experiments, the automerisation of bicyclo[3.1.0]hex-2-ene involves cleavage of the internal cyclopropane bond; the observed distribution of label can be explained either with a diradical formalism involving a complex scheme of interconverting intermediates, or with the competition of at least three concerted pathways.[221]

Acid-catalysed isomerization of 7-methylenebicyclo[4.1.0]hept-2-ene gives 3-methylcyclohepta-1,3,5-triene.[222] Thermal equilibration of bicyclic methylenecyclopropanes (**196**) and (**197**) is entropy-controlled near the boundaries of Bredt's rule.[223]

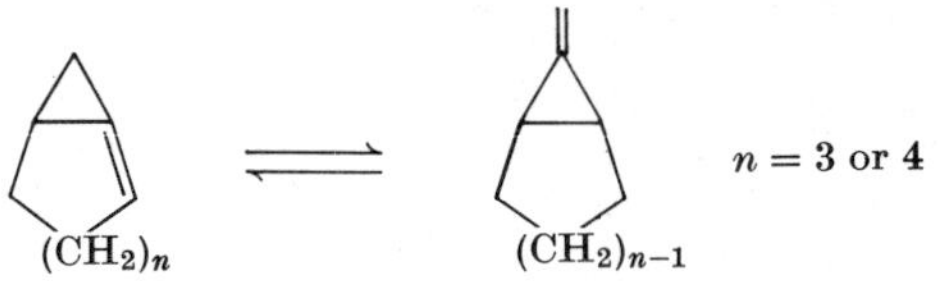

$(CH_2)_n$ ⇌ $(CH_2)_{n-1}$ $n = 3$ or 4

(**196**) (**197**)

Formation of azulene in the gas-phase pyrolysis (600°) of triquinacene is thought to involve initial concerted loss of molecular hydrogen. This removes the molecule from the $(CH)_{10}$ energy surface and precludes rearrangement to *cis*-9,10-dihydronapthalene, the probable precursor to other pyrolysis products observed at higher temperatures.[224] Contrary to a previous report,[225] photolysis of (**198**) gives (**199**), not (**201**). However, (**199**) is thermally labile and rearranges (71–83°; $E_a = 29.2$ kcal mol^{-1}) to a 6:5 mixture of (**198**) and (**201**), probably through the formal Cope rearrangement product (**200**).[226]

Further mechanistic details of the thermal rearrangements of 7-alkoxy- and 7-phenyl-bicyclo[2.2.1]heptadienes to cycloheptatrienes have been reported.[227] Photochemical rearrangement of 2-cyanobarrelene involves initial vinyl–cyanovinyl bridging[228] and thus is not controlled by the polar AX mechanism for $2\pi + 2\pi$ pericyclic reactions.[229] For the degenerate thermal rearrangement of 6-methylenebicyclo[3.2.1]oct-2-ene, results of deuterium-labelling studies are equally consistent with mechanisms involving pure Cope rearrangement and with the intermediacy of completely randomized, or symmetrical but incompletely equilibrated diradical species.[230]

The cyclohexylidenecyclopropane (**202**) is thermally stable below *ca.* 185°, whereas the related compound (**203**) rearranges to (**204**) at 150°. The latter reaction is interpreted as involving intramolecular trapping of an intermediate trimethylenemethane

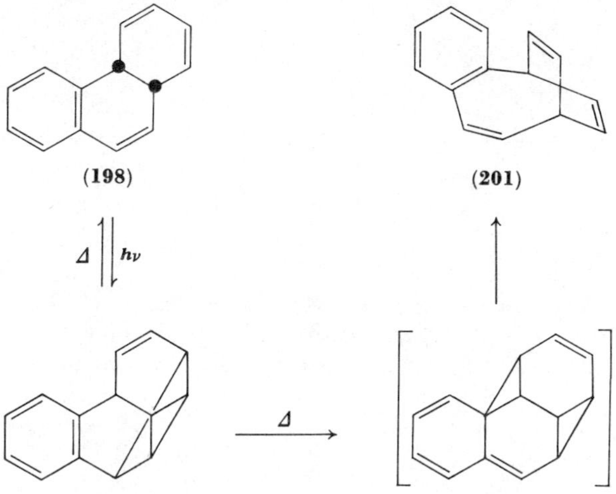

(198) (201)

(199) (200)

diradical by the suitably oriented double bond of **(203)**. In **(202)** the double bond is less favourably oriented for the homoallylic participation needed for rearrangement.[231]

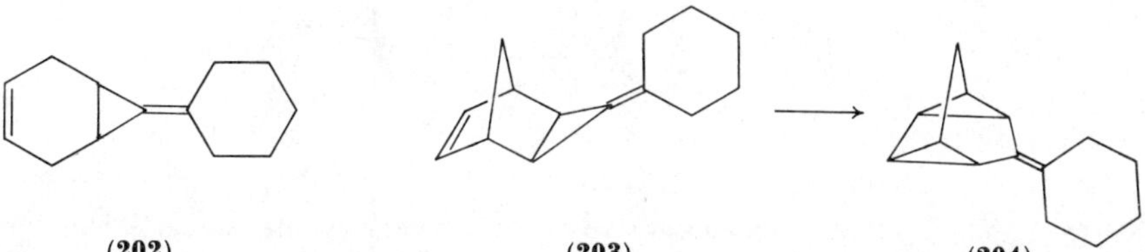

(202) (203) (204)

Thermal cycloreversions of the anthracene–cyclopentadiene adducts **(205)** and **(207)** provide examples of formally symmetry-"forbidden" processes that occur more rapidly than "allowed" processes. Thus, both fragmentation of **(205)** to anthracene and cyclopentadiene, and its isomerization to **(207)**, are kinetically favoured over fragmentation of **(207)** to anthracene and cyclopentadiene. These observations are explicable on the basis of a common diradical intermediate **(206)**, formed more rapidly from **(205)** owing to the relief of H–H eclipsing strain.[232]

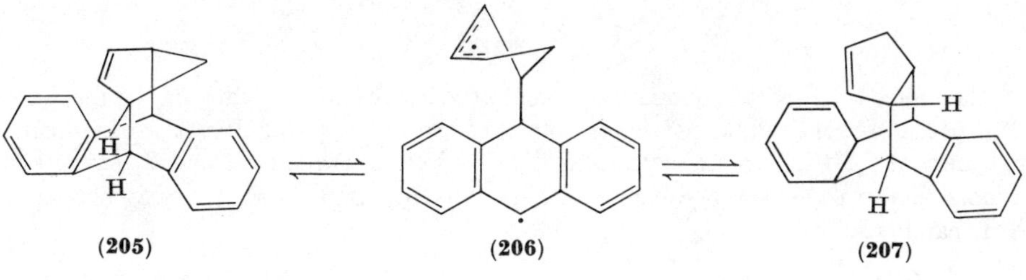

(205) (206) (207)

Pyrolysis (300°) of **(209)** affords **(210)**, the product of [1,5]-sigmatropic hydrogen migration. Isomer **(208)**, for which such migration is precluded, is stable at 300°, but at 320° affords products that most reasonably arise by initial vinylcyclopropane rearrangement of **(208)** to **(209)**. Deuterium-labelling studies support these conclusions.[233]

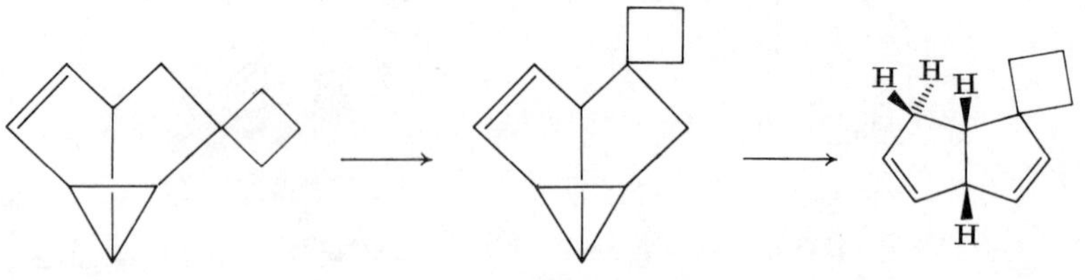

(208) (209) (210)

Zero-bridged *cis*-trishomobenzenes (**211**; X = CH_2, O or COOMe; R = COOMe, CF_3 or Me) undergo four-electron ($2\sigma \rightarrow 2\pi$) thermal isomerization to tricyclic dienes (**212**).[234] In contrast, the analogous methylene-bridged compounds (**213**; X = O, NH or NAc) isomerize to bicyclic trienes (**214**) by a six-electron ($3\sigma \rightarrow 3\pi$) process.[235]

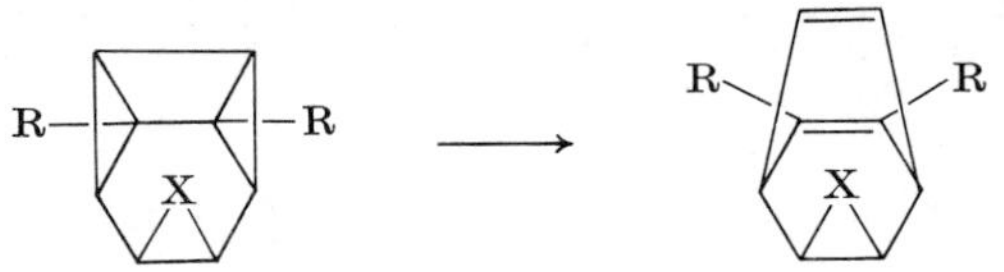

(211) (212)

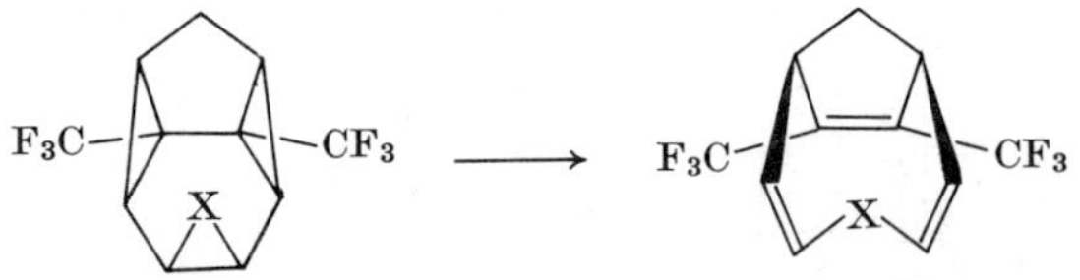

(213) (214)

Two new branches have been elaborated in the network of interconvertible polycyclic $C_{11}H_{10}O$ ketones. Thus, tricyclic homobullvalenone (**215**) is isomerized by mercuric bromide to the bicyclic ketone (**216**), and the latter undergoes thermal (100°) isomerization to the tetracyclic ketone (**217**).[236] The same transformations previously had been affected with BF_3.[237]

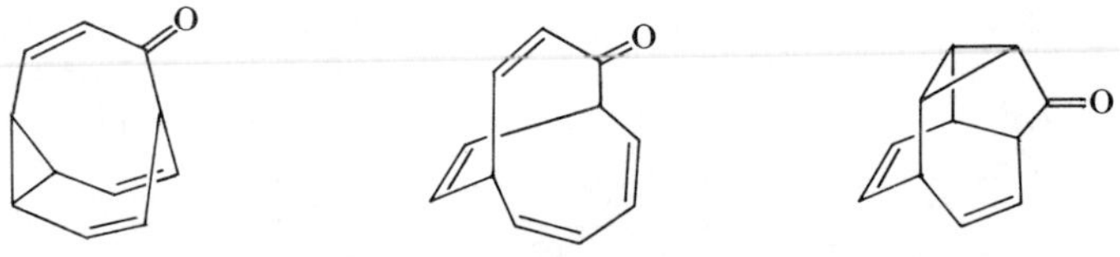

(215) (216) (217)

Other relevant reports include those dealing with the reversibility of the rearrangement of the bicyclo[4.2.2]decatrienyl radical to the bicyclo[3.3.2]decatrienyl radical,[238] the thermal ($_\sigma2 + _\sigma2$) isomerizations of 2,4-didehydroadamantane and 2,4-didehydrohomoadamantane,[239] and the isomerization of $C_{12}H_{20}$ alkyladamantanes by aluminium oxide catalysts.[240]

*Prototropic Rearrangements**

In the base-catalysed isomerization of 1-(arylalkyl)indenes to the 3-(arylalkyl) isomers, the 1-trityl-derivative reacts very slowly; this was attributed to a conformationally

* See also Chapter 10.

unfavourable transition state.[241] Equilibrium studies show that 1-methylindene is less stable than the 3-methyl compound. For the *tert*-butyl analogue, interference between the *peri*-hydrogen and *tert*-butyl reduces the relative stability of the 3-*tert*-butyl-isomer; surprisingly, this effect is less marked for the trityl-derivatives.[241] Further work on isomerization of substituted indenes[242, 243] includes a study of the acid-catalysed isomerization of 1-methyl-3-(1-piperidyl)indene.[243]

New results have been reported on the base-promoted migration of unsaturation in terpenoid cycloalkenes,[244] in hex-1-en-4-yne,[245] and in strained cycloalkenes,[246] and the acid-catalysed isomerization of alkenes has been discussed in terms of π-complex formation and donor–acceptor properties of the catalyst.[247]

The reactions of isopentenyl isomerase with artificial substrates [3-ethylbut-3-enyl (**218**) and *trans*-3-methylpent-3-enyl pyrophosphate (**219**)] have been examined.[248] Compound (**219**) is isomerized reversibly into *trans*-3-methylpent-2-enyl pyrophosphate (**220**) in a process similar to that between isopentenyl and dimethylallyl pyrophosphate,

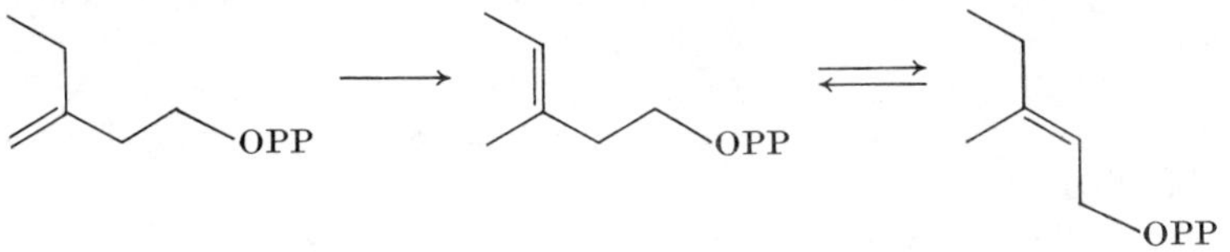

OPP OPP OPP

(**218**) (**219**) (**220**)

but the predominant, and unusual, formation of (**219**) from (**218**) is irreversible. These results were discussed in terms of the existence of at least two catalytic sites for proton abstraction and addition on the enzyme, and of distortion of (**218**) in the enzyme–substrate complex due to the presence of the bulky ethyl group.

*Racemizations**

The racemization of cystine in strong acid conditions suitable for protein hydrolysis has been attributed to the intermediacy of an acid enol which is stabilized by a partially or fully charged β-hetero-atom.[249]

First-order rate constants have been measured for the thermal racemization of *trans*-1,2-dibromocycloalkanes and related compounds; activation parameters were consistent with a 1,2-switching mechanism by way (at least in polar solvents) of a bromonium–bromide ion pair (**221**).[250]

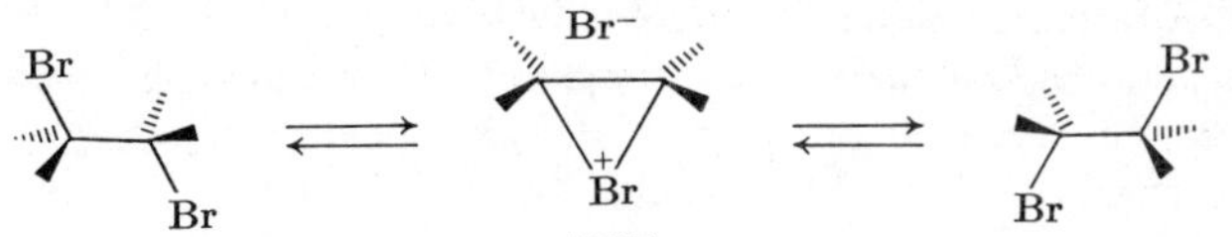

Br Br Br^- Br Br Br

(**221**)

It has been demonstrated that thermal racemization of dihydro-1,4-thiazine sulphoxides, e.g. (**222**), occurs through sulphenic acids, e.g. (**223**);[251] the *syn*-axial arrangement of the sulphinyl group and the tertiary hydrogen is a pre-requisite for racemization, suggesting an intramolecular process as indicated. Racemization of the sulphilimine (**224**) and of related *N*-methylated sulphonium salts has been examined and interpreted

* See also Chapter 10.

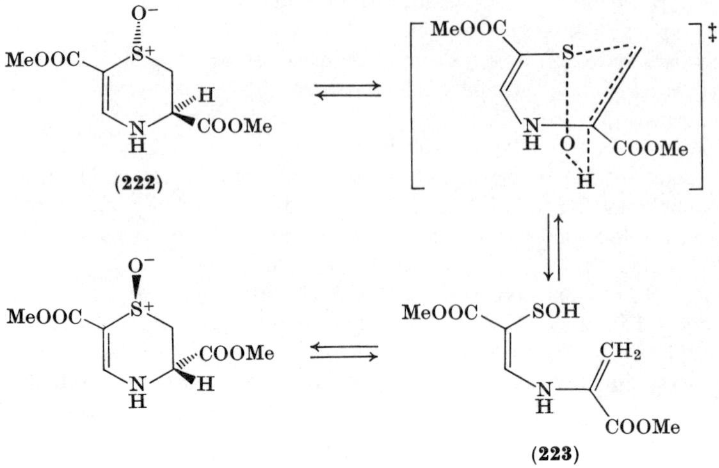

(222)

(223)

in terms of pyramidal inversion at sulphur;[252] in the case of (**224**), dissociation to a nitrene was discounted.

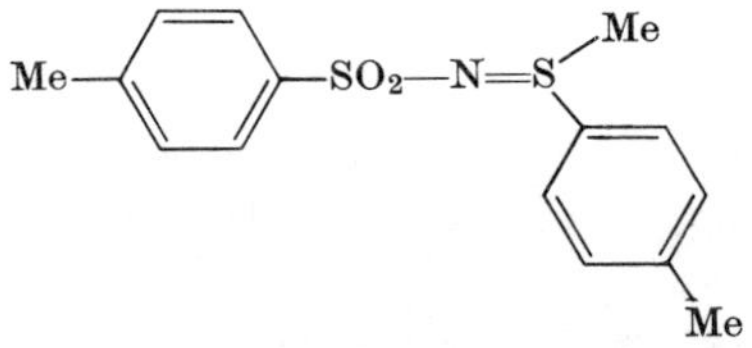

(224)

Solvent-induced racemization of halosilanes may involve a hexaco-ordinate octahedral intermediate or a pentaco-ordinate bipyramidal siliconium ion; pseudorotation of a pentaco-ordinate intermediate was excluded.[253]

Activation parameters for racemization of hepta-, octa- and nona-helicene are not in accord with bond-breaking to a diradical; a purely conformational process was preferred.[254]

Geometrical Isomerizations

The thermal Z/E-isomerization of olefins has been reviewed.[255]

The kinetics and mechanism of isomerization of maleic acid, both alone[256] and promoted by thiourea,[257] have been reported, and both acid- and base-catalysed isomerization of its dimethyl ester have been studied.[258]

Overcrowding leads to very low rotational barriers about the central double bond of tetrabenzofulvalenes; dynamic NMR studies on the symmetrical dibenzodi-(2,3-naphtho)-compound gave $\Delta G^{\ddagger} = 23.5$ kcal mol^{-1} at 180°.[259]

An interesting phenomenon has been encountered in the photo-isomerization of thioindigo dyes: when appropriately substituted, these compounds form monolayers, irradiation of which causes exclusive *cis* → *trans* rearrangement.[260] This contrasts with solution behaviour which can proceed in either direction and is wavelength-dependent. The result was discussed in terms of molecular packing within the monolayer.

Numerous studies of isomerization about the C–N double bond relate to azomethine dyes,[261] imidates,[262] isothiuronium salts[263] and isothioureas,[264] azole-carbaldoximes,[265] isoimides,[266] and both aldimines[267] and ketimines.[268]

Kinetics of *syn–anti*-isomerization of *p*-nitrobenzenediazotate have been studied,[269] and it has been concluded that *cis–trans*-isomerization of substituted 4-(dimethylamino)-azobenzenes proceeds by inversion at one of the nitrogen atoms of the azo-group.[270]

Miscellaneous

The rate of thermal rearrangement of the oxime thiocarbamates (**225**) is insensitive to changes in solvent polarity and substituent effects, suggesting that little charge separation is present in the transition state;[271] the formation of crossed products, and detection by ESR of iminyl radicals is consistent with a dissociation–recombination mechanism.

$$R^1R^2C{=}N{-}O{-}C({=}S){-}NR^3_2 \longrightarrow R^1R^2C{=}N{-}S{-}C({=}O){-}NR^3_2$$

(225)

Further studies of the alkyl isocyanide → cyanide rearrangement have been carried out employing optically active alkyl groups.[272] The extent of racemization was discussed in terms of radical stability and (in diphenyl ether) asymmetric solvation.

Isomerization of thiourea to ammonium thiocyanate, and the exchange of sulphur between thiourea and diphenyl tetrasulphide, have been studied;[273] in the latter reaction only one sulphur atom exchanges at 100°, presumably in a bimolecular process, whilst at 120° two sulphur atoms exchange by initial homolysis of the tetrasulphide.

Metal-catalysed Rearrangements*

Acyclic Systems

The isomerization of 3-methylpent-1-ene to *cis*- and *trans*-3-methylpent-2-enes, catalysed by $Ni[P(OEt)_2Ph]_4$, is a stereoselective process; in the presence of an added optically active phosphine, [*R*-(–)-P(Me)(Pr^n)(Ph)], unreacted starting material is enriched in the *S*-(+)-isomer; the optical yield reaches a limiting value at high phosphine concentrations, a fact which can be related to saturation in the exchange of neutral ligand with the added phosphine.[274] Isomerization of 3-chlorobut-1-ene to 1-chlorobut-2-ene, formally a 1,3-chlorine shift, is catalysed by various iridium, rhodium and copper complexes.[275] Photochemical activation enhances the catalytic activity of the copper complexes in this isomerization.[276]

cis–trans-Isomerization and double-bond migrations encountered in the hydrogenation of octadecenoates over silica-supported nickel catalysts, have been shown to occur at different sites on the catalyst surface, of the NiH_2- and NiH-type, respectively.[277] Rhodium(III)-catalysed thermal isomerization of 2-methylideneglutaric esters to *cis*- and *trans*-2-methylglutaconic esters has been investigated.[278]

The oxoruthenium acetate complex, $Ru_3O(OOCMe)_7$ is a selective catalyst for the homogeneous rearrangement of alk-1-en-3-ol substrates to saturated ketones.[279] A study of this general rearrangement, using deuterium-labelled substrates and heterogeneous copper metal and cupric oxide catalysts, leads to the following conclusions: the rearrangement is intermolecular, the hydroxyl-hydrogen and the hydrogen at C-3 do not

* See also Chapter 8.

become equivalent during rearrangement, and the rate-determining step involves cleavage of the allylic carbon–hydrogen bond ($k_H/k_D \simeq 4$). A radical mechanism is proposed (Scheme 3).[280]

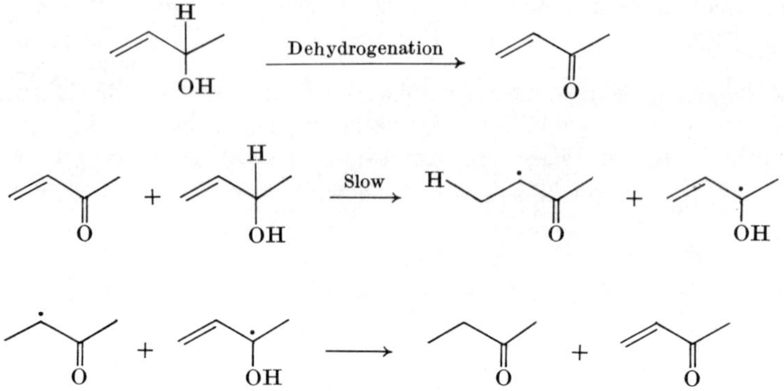

SCHEME 3.

Full details of the nickel-catalysed rearrangements of 1,4-dienes have been reported.[281] The major rearrangement pathways, exemplified by the isomerization of hexa-1,4-diene to 2-methylpenta-1,3-diene and of 3-methylpenta-1,4-diene to hexa-1,4-diene, are believed to involve cyclopropylcarbinylnickel and alkenylnickel intermediates.[282] Support for these conclusions is provided by studies of the nickel-promoted rearrangement of methylvinylcyclopropanes[283] and by the nickel- and rhodium-catalysed additions of alkenes to substituted 1,3-dienes.[284]

By analogy with the known mechanism of the Ag(I)-catalysed propargyl ester ⇌ allenyl ester rearrangement,[285] Ag(I)-catalysed rearrangement of aryl propargyl ethers, e.g. (**226**) → (**227**), is believed to proceed by [3,3]-sigmatropic rearrangement of a pre-equilibrium Ag(I)-alkyne π-complex. With 2,6-dimethylphenyl propargyl ethers (**228**), the initially formed allenyldienones (**229**) undergo a novel Ag(I)-catalysed dienone–phenol rearrangement to give 3-allenylphenols (**230**).[286]

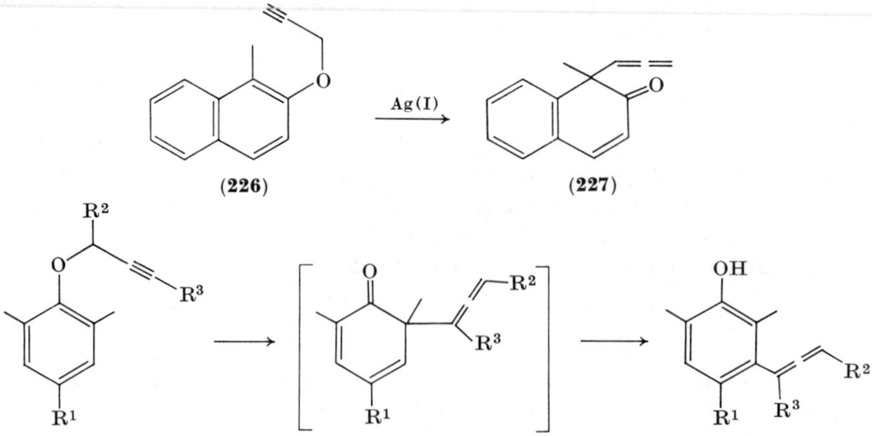

(**226**) (**227**)

(**228**) (**229**) (**230**)

Cyclic Systems

A stepwise mechanism, involving a co-ordinated cyclopropylcarbinyl cation, most reasonably accounts for the steric and conformational effects observed in the Rh(I)-catalysed rearrangement of vinylcyclopropanes to *trans*-pentadiene derivatives.[287] Tetrahydro-2,5-divinylfuran is isomerized to 3,4-dihydro-6-ethyl-3-vinyl-2*H*-pyran and 3,4-dihydro-6-ethyl-3-ethylidene-2*H*-pyran, by a ruthenium trichloride–triphenylphosphine catalyst.[288] On the basis of ^{13}C-NMR studies, a symmetrical intermediate, with a flat (or nearly flat), octagonally symmetrical C_8H_8 ring, is proposed for the fluxional rearrangement of (cyclo-octatetraene)tricarbonylmolybdenum.[287]

Zinc(II)-catalysed rearrangment of 1-phenylbicyclo[2.1.0]pentane-5,5-d_2 affords 3-phenylcyclopentene-2,3-d_2. This distribution of label is consistent with mechanisms involving a metallocyclic intermediate formed by cleavage of the 1,4-bond, or with a metal-assisted [$_\sigma 2_s + _\sigma 2_s$] pericyclic reaction.[290] Photochemical or Ag(I)-catalysed rearrangement of (**231**) affords (**232**), whereas thermal (150°) rearrangement gives (**233**). For the thermal reaction the labelling pattern is consistent with a concerted pathway, [$_\sigma 2_a + _\sigma 2_s$] ring-opening and subsequent disrotatory closure, while the formation of (**232**) in the photochemical reaction suggests a stepwise diradical process. These experiments cannot, however, distinguish between two reasonable alternatives (A and B) for the Ag(I)-catalysed rearrangement.[291]

A kinetic study has revealed mechanistic differences between the Rh(I)-, Pd(II)- and Ag(I)-catalysed isomerizations of quadricyclanes. The Pd(II)- and Ag(I)-catalysed processes are second-order reactions but Rh(I)-catalysis involves a pre-equilibrium step, probably formation of a co-ordinatively unsaturated Rh(I) complex.[292] Further study of the Rh(I)- and Pt(II)-catalysed rearrangements of oxaquadricyclanes[293] has shown that the products observed depend strongly on the presence of traces of protic solvents and the relative amount of catalyst used. For example, rearrangement of (**234**) in $CDCl_3$

with 15 mole-% of $[Rh(CO)_2Cl]_2$ affords 60–70% of the new oxepine derivative (**235**), along with smaller amounts of hydroxyfulvene (**236**) and *o*-hydroxyphthalate (**237**). In methanol as solvent, (**236**) is the major product (>90%). These results are discussed in terms of competition between acid-catalysed symmetry-"allowed" pathways, leading predominantly to (**236**), and metal-assisted symmetry-"forbidden" pathways leading to (**235**).[294]

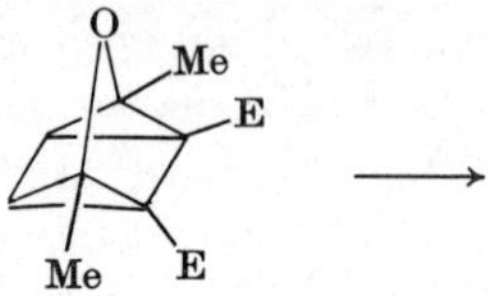

(**234**)

E = COOMe

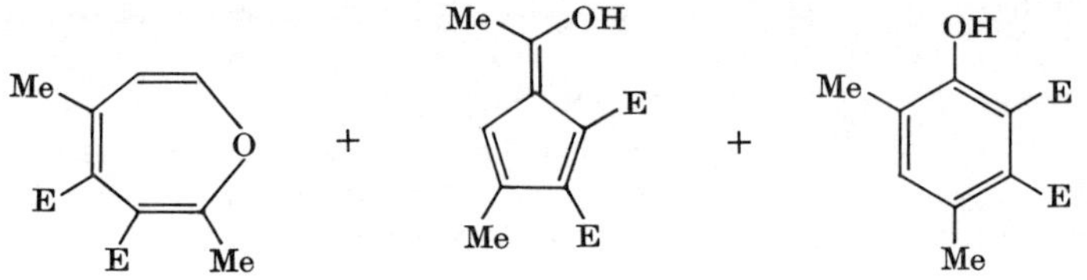

(**235**) + (**236**) + (**237**)

Hexamethyl-Dewar-benzene–palladium(II) chloride, prepared at room temperature, shows autocatalytic decomposition to hexamethyl-Dewar-benzene and $PdCl_2$. Samples of the complex prepared and stored at low temperatures (–28°) exhibit first-order kinetics. The latter may be a *bona fide* example of Pd(II)-catalysis of a symmetry-"forbidden" process, but it is equally consistent with a two-step mechanism involving rate-determining formation of a dipolar intermediate.[295] Tetracarbonyliron complexes (**238**) and (**241**) rearrange in refluxing hexane to (**240**) and (**242**), with half-lives of 2.5 hr. and 2 hr., respectively (for the free hydrocarbons the corresponding half-lives are *ca.* 16 hr. and *ca.* 3 weeks); rearrangement is inhibited by added carbon monoxide and alkenes and is believed to involve disrotatory ring-opening (symmetry-"forbidden" in the free ligand) of intermediate tricarbonyliron complexes, e.g. (**239**) formed by initial loss of carbon monoxide.[296]

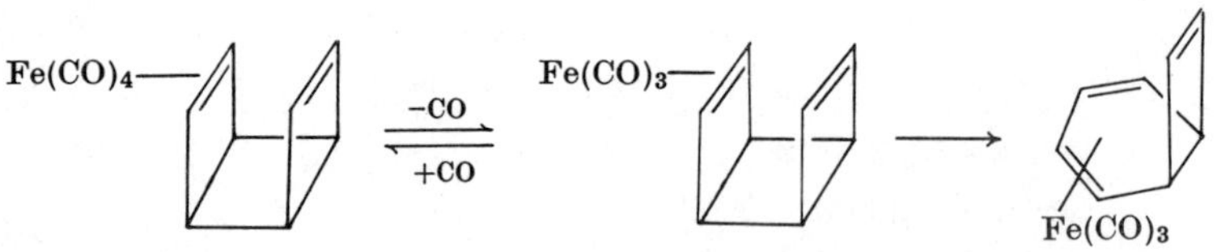

$Fe(CO)_4$ (**238**) $\underset{+CO}{\overset{-CO}{\rightleftharpoons}}$ $Fe(CO)_3$ (**239**) ⟶ $Fe(CO)_3$ (**240**)

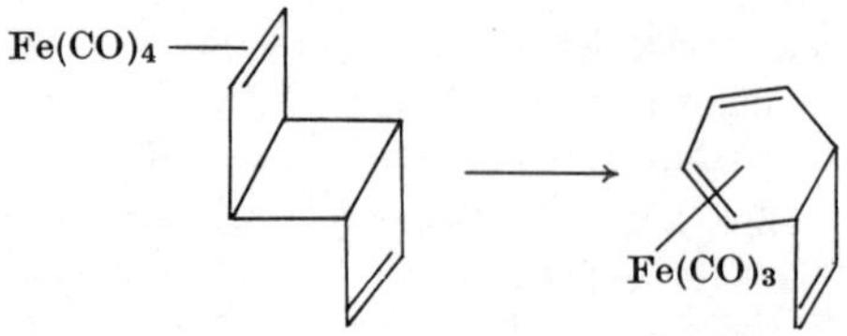

$Fe(CO)_4$ (**241**) ⟶ $Fe(CO)_3$ (**242**)

Isomerization of unsaturated propellanes (**243**; X = H or D) to annulated cyclo-octatetraenes (**244**; X = H or D) by $Mo(CO)_6$, formally involves a 1,5-cyclobutene shift. Bridged (cyclo-octatetraene)tetracarbonylmolybdenum complexes are probably intermediates in these rearrangements, since reaction of (**243**; $n = 2$, X = H) with $Mo(CO)_3$-(diglyme)$_3$ allows the isolation of (**245**) which in turn rearranges thermally to (**244**; $n = 2$, X = H).[297] These transformations are in contrast to the purely thermal rearrangements of the same propellanes, which proceed by way of tetracyclo[$4.2.0.0^{2,8}.0^{5,7}$]octene intermediates.[213]

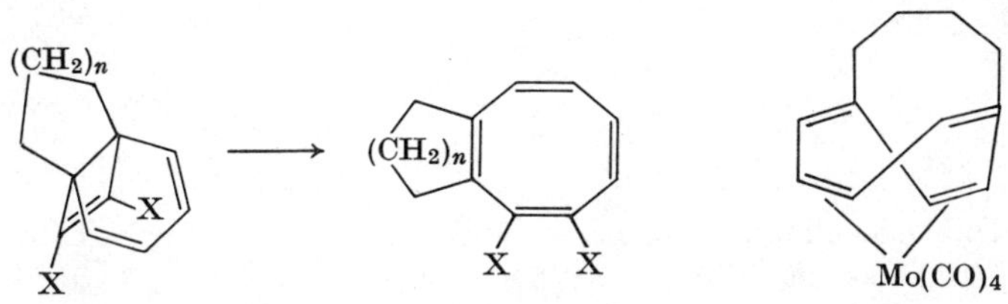

(**243**) (**244**) (**245**)

Silver(I) and Rh(I)-catalysed rearrangements of diademane (**246**) involve different sites of attack by the transition metal. For Ag(I)-catalysis, such attack takes place from the side of the basal six-membered ring, leading ultimately to triquinacene (**247**). In contrast, oxidative addition by Rh(I) occurs from the side of one of the apical five-membered rings to give snoutene (**248**).[298] The latter is itself subject to further rearrangement (e.g. by $[Rh(CO)_2Cl]_2$) to (**249**). Similar Rh(I)-catalysed isomerizations are observed for other systems that incorporate a bicyclopropyl unit fixed in the s-*cis*-conformation.[299]

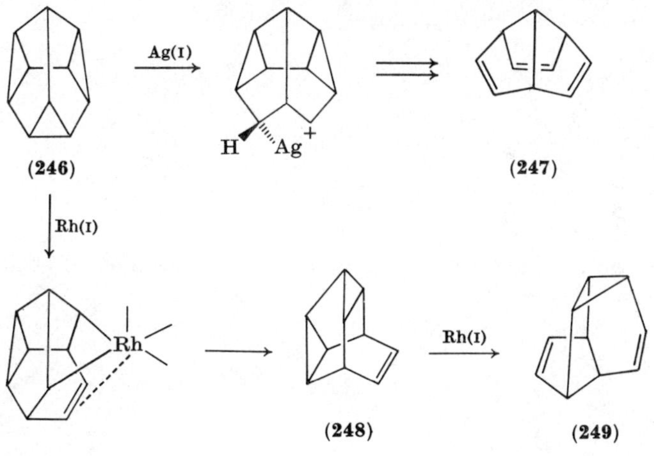

(**246**) (**247**)

(**248**) (**249**)

Miscellaneous

*Cationic Processes**

Acid-catalysed rearrangements of β,γ-unsaturated ketones have been surveyed.[300] The unsaturated ketone (**250**) rearranges to the spiro-ketone (**251**) or the octalin derivative (**252**) on treatment with acid, in processes that are initiated by protonation at carbon rather than at the (more usual) oxygen.[301]

* See also Chapters 8 and 9.

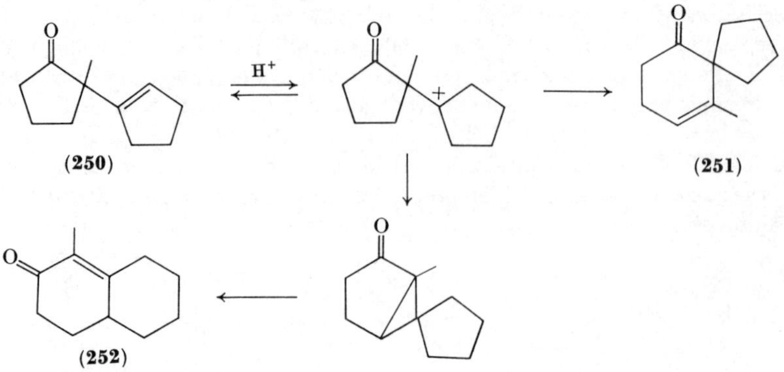

(250) (251)

(252)

Hart's group have conducted extensive investigations into the acid-catalysed rearrangement of epoxy-ketones (**253**) and (**254**).[302] Intramolecular 1,5-hydride shifts are involved in the cyclization reactions of, e.g., (**255**) which are effected by BF_3/Et_2O.[303]

(253) (254)

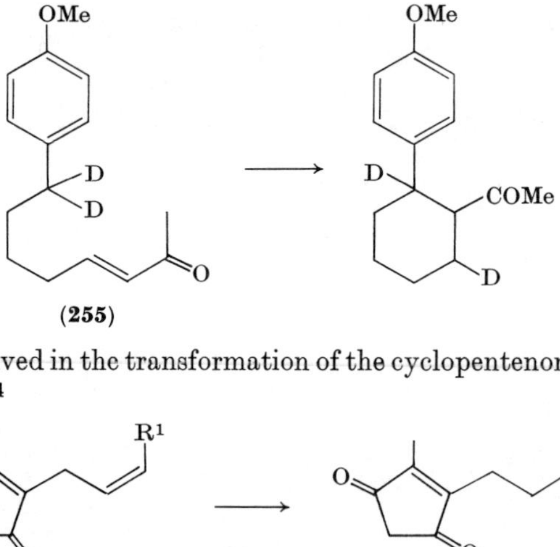

(255)

Prototropic shifts involved in the transformation of the cyclopentenones (**256**) into (**257**) have been discussed.[304]

(**256**) R^1 = H, Me or Et
R^2 = H or Me

(**257**)

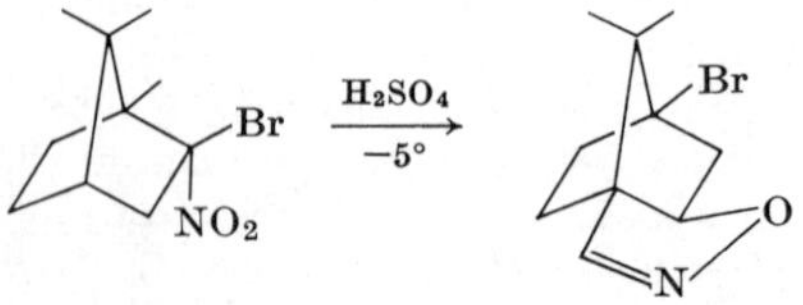

H_2SO_4, −5°

(**258**) (**259**)

The transformation of bromonitrocamphane (**258**) into the "anhydro"-compound (**259**) in sulphuric acid occurs by a complex sequence of reactions initiated by protonation of the nitro-group.[305] The interconversion of the homocamphene isomers (**260**)–(**262**) has also been studied.[306]

CHMe CH$_2$ CH$_2$

(**260**) (**261**) (**262**)

Deuterium-labelling experiments have implicated the bridged ion (**264**) in the epimerization of (**263**),[307] and the fact that tetrahydroabietic acid gives racemic product on rearrangement in concentrated sulphuric acid implicates the ring-opened intermediate (**265**).[308]

AcO D D OAc

(**263**) (**264**)

COOH H H H $+H^+$ $-H_2O$ $-CO$ $-H^+$ H COOH H H H H

(**265**)

1,2-Shifts of alkynyl groups have been encountered in the formation of (**266**) from α-halo-aldehydes and PhC≡CMgX,[309] and an ion-pair mechanism has been discussed for the interconversion of various α,ω-derivatives typified by (**267**).[310]

$R^1R^2C(X)$—CHO $\xrightarrow{PhC\equiv CMgX}$ $R^1R^2C(X)$—CH(OMgX)—C≡CPh

$R^1R^2C(CHO)$—C≡CPh R^1R^2CH—C(=O)—C≡CPh

(**266**) (*ca.* 15%) (*ca.* 85%)

S ‖ OCPh Y O_2N Y SCPh ‖ O O_2N

(**267**) Y = Cl or I

Other Reactions

"Non-classical" Arbuzov reactions of $\alpha\beta$-unsaturated carbonyl compounds with trialkyl phosphites have been surveyed.[311] Results on the $AlCl_3$-catalysed epimerization of ferrocenyl ketones (**268**) and (**269**) have been elaborated.[312] 1,2-Aryl migration (giving biaryls) in the oxidation of tetra-arylborate anions occurs both in the $Ar_4B\cdot$ radical and in electrophilic adducts of the anion, depending on the oxidizing agent employed.[313]

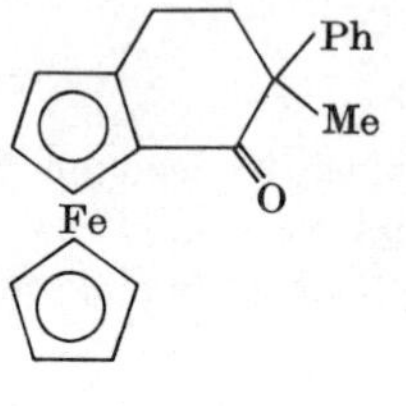

(**268**) 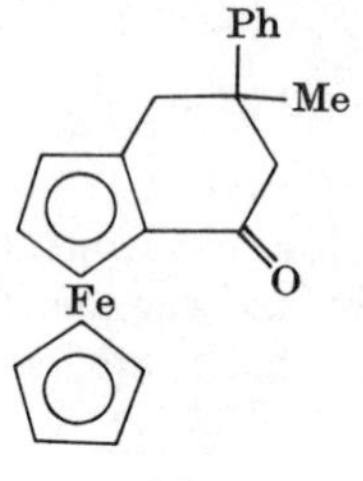 (**269**)

The rates of decomposition of *N*-benzyhydryl-*N*-nitrosobenzamides are insensitive to changes in solvent or to the substituent in the benzhydryl group, indicating a non-polar rearrangement to diazo-ester in the slow step, prior to ionization and loss of N_2.[314] The α-halogenated epoxide (**270**) is rearranged in refluxing Et_3N to give the products indicated; phenyl migration in an initial α-keto-cation was considered responsible.[315]

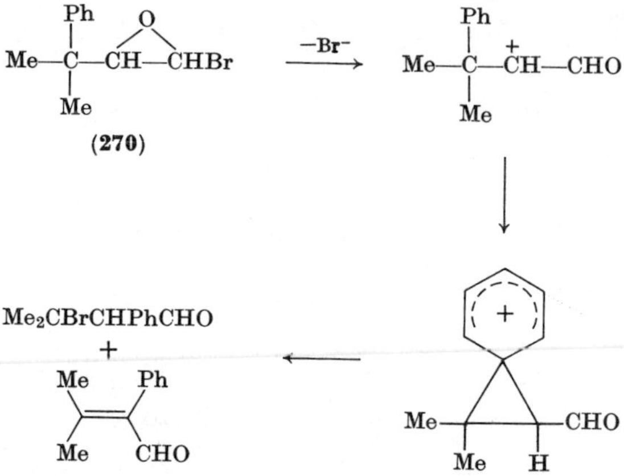

(**270**)

$Me_2CBrCHPhCHO$ +

References

[1] V. G. Lipovich, L. E. Latysheva and V. V. Chenets, *Khim. Aromat. Nepredal'nykh Soedin.*, **1971**, 110; *Chem. Abs.*, **80**, 94838 (1974).
[2] R. A. Crochet, F. R. Sullivan and P. Kovacic, *J. Org. Chem.*, **39**, 3094 (1974).
[3] R. A. Cox, *J. Am. Chem. Soc.*, **96**, 1059 (1974).
[4] A. Dolenko and E. Buncel, *Canad. J. Chem.*, **52**, 623 (1974).
[5] J. Yamamoto, *Yuki Grosei Kagaku Kyokai Shi*, **31**, 605 (1973); *Chem. Abs.*, **79**, 145722 (1973).
[6] J. D. Cheng and H. J. Shine, *J. Org. Chem.* **39**, 2835 (1974).
[7] D. L. Williams and J. A. Wilson, *J.C.S. Perkin II*, **1974**, 13.
[8] R. Martin, *Bull. Soc. Chim. France*, **1974**, 983.
[9] R. Martin, *Bull. Soc. Chim. France*, **1974**, 1519.
[10] R. Martin, *Bull. Soc. Chim. France*, **1974**, 1523.

[11] J. Furukawa, K. Omura and S. Sawada, *J.C.S. Chem. Comm.*, **1974**, 78.
[12] S. Oae, O. Yamada and T. Maeda, *Bull. Chem. Soc. Japan*, **47**, 166 (1974).
[13] S. Oae, T. Sakurai, H. Kimura and S. Kozuka, *Chem. Letters* (*Tokyo*), **1974**, 671.
[14] D. Hellwinkel and M. Supp, *Angew. Chem. Internat. Ed.*, **13**, 270 (1974).
[15] C. W. Bird and C. K. Wong, *Tetrahedron Letters*, **1974**, (1251).
[16] See *Org. Reaction Mech.*, **1973**, 478.
[17] M. J. S. Dewar and I. J. Turchi, *J. Am. Chem. Soc.*, **96**, 6148 (1974).
[18] M. Golfier and R. Milcent, *Tetrahedron Letters*, **1974**, 3871.
[19] M. Golfier, M.-G. Guillerez and R. Milcent, *Tetrahedron Letters*, **1974**, 3875.
[20] M. Raccia, N. Vivona and G. Cusmeno, *J.C.S. Chem. Comm.*, **1974**, 358.
[21] A. R. McCarthy, W. D. Ollis and C. A. Ramsden, *J.C.S. Perkin I*, **1974**, 627; see also *Org. Reaction Mech.*, **1971**, 268.
[22] R. Jacquier, H. Lopez and G. Maury, *J. Heterocyclic Chem.*, **10**, 755 (1973); M. S. S. Siddiqui and M. F. G. Stevens, *J.C.S. Perkin I*, **1974**, 609; D. J. Brown, and K. Ienega, *ibid.*, p. 372.
[23] Y. Alkasaki and A. Ohno, *J. Am. Chem. Soc.*, **96**, 1957 (1974).
[23a] G. Casnati, R. Marchelli, and A. Pochini, *J.C.S. Perkin I*, **1974**, 754.
[24] B. A. Ivin, G. V. Rutkovskii, S. A. Andreev and E. G. Sochilin, *Zh. Org. Khim.*, **9**, 2194 (1973); *Chem. Abs.*, **80**, 36540 (1974).
[24a] J. Fetter, J. Nyitrai and K. Lempert, *Acta Chim.* (*Budapest*), **79**, 197 (1973); *Chem. Abs.* **80**, 26594 (1974).
[24b] J. Nyitrai and K. Lempert, *Chem. Ber.*, **107**, 1637 (1974); J. Nyitrai, K. Lempert and T. Cserfalvi, *ibid.*, p. 1645; E. Koltai, J. Nyitrai, K. Lempert and L. Bursics, *ibid.*, p. 1649.
[25] D. MacKay and L. I. Wong, *J.C.S. Chem. Comm.*, **1974**, 621.
[26] See *Org. Reaction Mech.*, **1973**, 423.
[27] R. A. Abramovitch and I. Shinkai, *J. Am. Chem. Soc.*, **96**, 5265 (1974).
[28] J. O. Palazzi, D. L. Leland and M. P. Katick, *J. Org. Chem.*, **39**, 3114 (1974).
[29] J. Lehmann and H. Wamhoff, *Annalen*, **1974**, 1287.
[30] V. P. Vitullo and E. A. Logue, *J.C.S. Chem. Comm.*, **1974**, 228.
[31] J. N. Marx, J. C. Argyle and L. R. Norman, *J. Am. Chem. Soc.*, **96**, 2121 (1974).
[32] R. M. Acheson and R. F. Flowerday, *J.C.S. Perkin I*, **1974**, 2339.
[33] See *Org. Reaction Mech.*, **1973**, 429.
[34] M. J. Hughes and A. J. Waring, *J.C.S. Perkin II*, **1974**, 1043.
[35] B. Miller, *J. Am. Chem. Soc.*, **96**, 7155 (1974).
[36] A. M. Choudhury, *J.C.S. Perkin I*, **1974**, 132.
[37] S. R. Hartshorn, *Chem. Soc. Rev.*, **3**, 167 (1974).
[38] A. Fischer and C. C. Greig, *Canad. J. Chem.*, **52**, 1231 (1974).
[39] J. B. Hendrickson, *Angew. Chem. Internat. Ed.*, **13**, 47 (1974).
[40] M. J. S. Dewar, S. Kirschner and H. W. Kollmar, *J. Am. Chem. Soc.*, **96**, 5240 (1974).
[41] M. J. S. Dewar, S. Kirschner, H. W. Kollmar and L. E. Wade, *J. Am. Chem. Soc.*, **96**, 5242 (1974).
[42] M. J. S. Dewar and S. Kirschner, *J. Am. Chem. Soc.*, **96**, 5244 (1974).
[43] G. A. Shchembelov and Y. A. Ustynyuk, *J. Am. Chem. Soc.*, **96**, 4189 (1974).
[44] G. A. Shchembelov and Y. A. Ustynyuk, *J. Organometal. Chem.*, **70**, 343 (1974).
[45] A. J. P. Devaquet and W. J. Hehre, *J. Am. Chem. Soc.*, **96**, 3644 (1974).
[46] See *Org. Reaction Mech.*, **1973**, 431.
[47] W. J. Hehre, *J. Am. Chem. Soc.*, **96**, 5207 (1974).
[48] H.-J. Hansen and H. Schmid, *Tetrahedron*, **30**, 1959 (1974).
[49] M. T. Reetz, *Angew. Chem. Internat. Ed.*, **13**, 402 (1974).
[50] M. T. Reetz, *Angew. Chem. Internat. Ed.*, **11**, 129 (1972).
[51] U. Svanholm and V. D. Parker, *J.C.S. Perkin II*, **1974**, 169.
[52] See *Org. Reaction Mech.*, **1973**, 432.
[53] E. K. Aleksandrova, L. I. Bunina-Krivorukova, A. V. Moshinskaya and Kh. V. Bal'yan, *Zh. Org. Khim.*, **10**, 1039 (1974); *Chem. Abs.*, **81**, 49151 (1974).
[54] J.-F. Normant, O. Reboul, R. Sauvêtre, H. Deshayes, D. Masure and J. Villieras, *Bull. Soc. Chim. France*, **1974**, 2072.
[55] N. J. Leonard and C. R. Frihart, *J. Am. Chem. Soc.*, **96**, 5894 (1974).
[56] T. Takahashi, Y. Okaue, A. Kaji and J. Hayami, *Bull. Inst. Chem. Res., Kyoto Univ.*, **51**, 173 (1973); *Chem. Abs.*, **79**, 125583 (1973).
[57] T. Takahashi, A. Kaji and J. Hayami, *Bull. Inst. Chem. Res., Kyoto Univ.*, **51**, 163 (1973); *Chem. Abs.*, **79**, 125590 (1973).
[58] W. D. Ollis, R. Somanathan and I. O. Sutherland, *Chem. Comm.*, **1974**, 494.

[59] A. Pryde, J. Zsindely and H. Schmid, *Helv. Chem. Acta*, **57**, 1598 (1974).
[60] H. Kwart and J. L. Schwartz, *J. Org. Chem.*, **39**, 1575 (1974).
[61] Y. Makisumi, S. Takada and Y. Matsukura, *Chem. Comm.*, **1974**, 850.
[62] J. Meijer, P. Vermeer, H. J. T. Bos and L. Brandsma, *Rec. Trav. chim.*, **93**, 26 (1974).
[63] T. Nakai, H. Shiono and M. Okawara, *Tetrahedron Letters*, **1974**, 3625.
[64] L. Dalgaard, L. Jensen and S.-O. Lawesson, *Tetrahedron*, **30**, 93 (1974).
[65] F. C. V. Larsson and S.-O. Lawesson, *Tetrahedron*, **30**, 1283 (1974).
[66] R. E. Ireland and A. K. Willard, *J. Org. Chem.*, **39**, 421 (1974).
[67] K. Hartke and G. Gölz, *Chem. Ber.*, **107**, 566 (1974).
[68] L. E. Overman, *J. Am. Chem. Soc.*, **96**, 597 (1974).
[69] S. Inada, A. Arikawa and M. Okazaki, *Yuki Gosei Kagaku Kyokai Shi*, **31**, 597 (1973); *Chem. Abs.*, **79**, 145718 (1973).
[70] C. W. Bird, M. W. Kaczmar and C. K. Wong, *Tetrahedron*, **30**, 2549 (1974).
[71] A. K. Cheng, F. A. L. Anet, J. Mioduski and J. Meinwald, *J. Am. Chem. Soc.*, **96**, 2887 (1974).
[72] J. F. M. Oth, K. Müllen, J.-M. Gilles and G. Schröder, *Helv. Chim. Acta*, **57**, 1415 (1974).
[73] H. Nakanishi and O. Yamamoto, *Tetrahedron Letters*, **1974**, 1803.
[74] L. A. Paquette, D. R. James and O. H. Birnberg, *Chem. Comm.*, **1974**, 722.
[75] H. Klein and W. Grimme, *Angew. Chem. Internat. Ed.*, **13**, 672 (1974).
[76] J. J. Gajewski, L. K. Hoffman and Chung Nan Sih, *J. Am. Chem. Soc.*, **96**, 3705 (1974).
[77] K. Takeda, *Tetrahedron*, **30**, 1525 (1974).
[78] A. Viola, A. J. Padilla, D. M. Lennox, A. Hecht and R. J. Proverb, *Chem. Comm.*, **1974**, 491.
[79] See *Org. Reaction Mech.*, **1973**, 434.
[80] See *Org. Reaction Mech.*, **1973**, 436.
[81] F. Vögtle and E. Goldschmitt, *Angew. Chem. Internat. Ed.*, **13**, 480 (1974).
[82] P. Handewind and U. K. Pandit, *Tetrahedron Letters*, **1974**, 2359.
[83] See *Org. Reaction Mech.*, **1973**, 436.
[84] A. E. Eckersley and N. A. J. Rogers, *Tetrahedron Letters*, **1974**, 1661.
[85] S. Braverman and B. Drendi, *Tetrahedron*, **30**, 2379 (1974).
[86] Y. Makisumi and S. Takada, *Chem. Comm.*, **1974**, 848.
[87] P. A. Grieco, M. Meyers and R. S. Finkelhor, *J. Org. Chem.*, **39**, 119 (1974).
[88] D. Michelot, G. Linstrumelle and S. Julia, *Chem. Comm.*, **1974**, 10.
[89] B. Cazes and S. Julia, *Tetrahedron Letters*, **1974**, 2077.
[90] A. F. Thomas and R. Dublini, *Helv. Chim. Acta*, **57**, 2084 (1974).
[91] A. B. Smith, *Chem. Comm.*, **1974**, 695.
[92] W. Ando, K. Konishi, T. Hagiwara and T. Migita, *J. Am. Chem. Soc.*, **96**, 1601 (1974).
[93] J. A. Berson, T. Tsutomu and G. Jones, III, *J. Am. Chem. Soc.*, **96**, 3468 (1974).
[94] K. Dietrich and H. Musso, *Chem. Ber.*, **107**, 731 (1974).
[95] M. Constenla and K. Dimroth, *Chem. Ber.*, **107**, 3501 (1974).
[96] J. W. Lown, M. H. Akhtar and R. S. McDaniel, *J. Org. Chem.*, **39**, 1998 (1974).
[97] See *Org. Reaction Mech.*, **1973**, 438.
[98] K.-H. Hasselgren, J. Dolby, J. L. G. Nilsson and M. Elander, *Tetrahedron Letters*, **1971**, 2917.
[99] K.-H. Hasselgren, J. Dolby, M. M. AlHolly, M. Elander and J. L. G. Nilsson, *J. Org. Chem.*, **39**, 1355 (1974).
[100] F. Gerhart and L. Wilde, *Tetrahedron Letters*, **1974**, 475.
[101] V. Dalacker and H. Hopf, *Tetrahedron Letters*, **1974**, 15.
[102] M. Schneider and A. Erben, *Z. Naturforsch.*, **29B**, 288 (1974).
[103] S. F. Nelson, J. P. Gillespie, P. J. Hintz and E. D. Sephanen, *J. Am. Chem. Soc.*, **95**, 8380 (1973).
[104] P. E. Rakita and G. A. Taylor, *J. Organometal. Chem.*, **61**, 71 (1973).
[105] J. J. McCullough and M. R. McClory, *J. Am. Chem. Soc.*, **96**, 1962 (1974).
[106] F. Mutterer and J.-P. Fleury, *J. Org. Chem.*, **39**, 640 (1974).
[107] T. Yamazaki, G. Baum and H. Schechter, *Tetrahedron Letters*, **1974**, 4421.
[108] M. Martin and M. Regitz, *Annalen*, **1974**, 1702.
[109] P. J. Abbott, R. M. Acheson, R. F. Flowerday and G. W. Brown, *J.C.S. Perkin I*, **1974**, 1177.
[110] H. Dürr and R. Sergio, *Chem. Ber.*, **107**, 2027 (1974).
[111] A. Roedig, H.-A. Renk, V. Schaal and D. Scheutzow, *Chem. Ber.*, **107**, 1136 (1974).
[112] A. Viola, S. Madhavan, R. J. Proverb, B. L. Yates and J. Larrahondo, *Chem. Comm.*, **1974**, 842.
[113] N. Morita, T. Asao and Y. Kitahara, *Chem. Letters (Tokyo)*, **1974**, 747.
[114] E. E. Weltin, *J. Am. Chem. Soc.*, **96**, 3409 (1974).
[115] See *Org. Reaction Mech.*, **1973**, 487.
[116] A. Komornicki and J. W. McIver, Jr., *J. Am. Chem. Soc.*, **96**, 5798 (1974).

[117] P. Merlet, S. D. Peyerimhoff, R. J. Buenker and S. Shih, *J. Am. Chem. Soc.*, **96**, 959 (1974).
[118] J. P. Daudey, J. Langlet and J. P. Malrieu, *J. Am. Chem. Soc.*, **96**, 3393 (1974).
[119] M. J. S. Dewar and S. Kirschner, *J. Am. Chem. Soc.*, **96**, 6809 (1974).
[120] M. R. De Camp, R. H. Levin and M. Jones, Jr., *Tetrahedron Letters*, **1974**, 3575.
[121] E. W. Neuse and B. R. Green, *Annalen*, **1974**, 1534.
[122] C. W. Jefford, A. F. Boschung and C. G. Rimbault, *Tetrahedron Letters*, **1974**, 3387.
[123] N. J. Turro, G. Schuster, J. Pauliquen, R. Pettit and C. Maudlin, *J. Am. Chem. Soc.*, **96**, 6797 (1974).
[124] A. I. Ioffe and O. M. Nefedov, *Izv. Akad. Nauk SSSR, Ser. Khim.*, **1974**, 1536; *Chem. Abs.*, **81**, 104431 (1974).
[125] C. W. Shoppee and G. H. Henderson, *Chem. Comm.*, **1974**, 561.
[126] A. G. Schultz and M. B. De Tar, *J. Am. Chem. Soc.*, **96**, 296 (1974).
[127] J. S. Hastings, H. G. Heller, H. Tucker and K. Smith, *Chem. Comm.*, **1974**, 348.
[128] M. J. S. Dewar and C. A. Ramsden, *J.C.S. Perkin I*, **1974**, 1839.
[129] J. L. Fry, M. G. Adlington, R. C. Badger and S. K. McCullough, *Tetrahedron Letters*, **1974**, 429; see also B. L. Adams, J.-H. Liu and P. Kovacic, *ibid.*, p. 427.
[130] B. L. Adams and P. Kovacic, *J. Org. Chem.*, **39**, 3090 (1974).
[131] B. L. Adams and P. Kovacic, *J. Am. Chem. Soc.*, **96**, 7014 (1974).
[132] A. Griumanini and G. Lercker, *Gazz. Chim. Ital.*, **104**, 415 (1974).
[133] R. F. Smith, R. D. Blondell, R. A. Abgott, K. B. Lipkowitz, J. A. Richmond and K. A. Fountain, *J. Org. Chem.*, **39**, 2036 (1974).
[134] E. Vilsmaier, K. H. Dittrich and W. Sprungel, *Tetrahedron Letters*, **1974**, 3061.
[135] P. G. Gassman and D. R. Amick, *Tetrahedron Letters*, **1974**, 889.
[136] M. Mitani, R. Tsuchida and K. Koyama, *Tetrahedron Letters*, **1974**, 869.
[137] S. Oae, T. Aida and N. Furukawa, *J.C.S. Perkin II*, **1974**, 1231.
[138] N. Yuksekisik, *Commun. Fac. Sci. Univ. Ankara, Ser. B.*, **20**, 49 (1973); *Chem. Abs.*, **81**, 62748 (1974); G. V. Romanov, M. Sh. Yafarov, A. I. Konovalov, A. N. Pirdovik, I. V. Konovalova and T. N. Yusupova, *Zh. Obsch. Khim.*, **43**, 2378 (1973); *Chem. Abs.*, **80**, 70073 (1974).
[139] V. Rautenstrauch, G. Buchi and H. Wnest, *J. Am. Chem. Soc.*, **96**, 2576 (1974).
[140] J. J. Eisch, C. A. Kovacs and S.-G. Rhee, *J. Organometal. Chem.*, **65**, 289 (1974).
[141] A. G. Brook, *Accounts Chem. Res.*, **7**, 77 (1974).
[142] A. G. Brook and J. M. Duff, *J. Am. Chem. Soc.*, **96**, 4692 (1974).
[143] A. Wright and R. West, *J. Am. Chem. Soc.*, **96**, 3214 (1974).
[144] A. Wright and R. West, *J. Am. Chem. Soc.*, **96**, 3222 (1974).
[145] A. Wright and R. West, *J. Am. Chem. Soc.*, **96**, 3227 (1974).
[146] N. DeKimpe and N. Schamp, *Tetrahedron Letters*, **1974**, 3779.
[147] C. L. Stevens, P. M. Pillai and K. G. Taylor, *J. Org. Chem.*, **39**, 3158 (1974).
[148] W. G. Dauben and R. J. Twigg, *Tetrahedron Letters*, **1974**, 531.
[149] See *Org. Reaction Mech.*, **1972**, 106.
[150] M. V. Moncur, J. B. Grutzner and A. Leisenstadt, *J. Org. Chem.*, **39**, 1604 (1974).
[151] M. Masaki, J. Kita, K. Fukui and S. Matsunami, *Tetrahedron Letters*, **1974**, 191.
[152] T. Ogata, H. Sato, H. Yoshida and S. Inokawa, *Nippon Kagaku Kaishi*, **1974**, 382; *Chem. Abs.*, **80**, 132518 (1974).
[153] H. Patin, *Tetrahedron Letters*, **1974**, 2893.
[154] C. R. Narayanan and M. S. Parkar, *Chem. & Ind. (London)*, **1974**, 163.
[155] R. Nayarayan and R. K. Shah, *Chem. Comm.*, **1973**, 926.
[156] M. Collona, L. Greci and L. Marchetti, *Gazz. Chim. Ital.*, **104**, 395 (1974).
[157] V. A. Stoute, M. A. Winnik and I. G. Csizmadia, *J. Am. Chem. Soc.*, **96**, 6388 (1974).
[158] M. A. Winnik, V. Stoute and P. Fitzgerald, *J. Am. Chem. Soc.*, **96**, 1977 (1974).
[159] N. A. Johnsson and P. Moses, *Acta Chem. Scand., B.*, **28**, 441 (1974).
[160] M. Rey and A. S. Dreiding, *Helv. Chim. Acta*, **57**, 734 (1974).
[161] M. K. Huber and A. S. Dreiding, *Helv. Chim. Acta*, **57**, 748 (1974).
[162] J. Salaun, B. Garnier and J. M. Conia, *Tetrahedron*, **30**, 1413 (1974).
[163] S. R. Tanny and F. W. Fowler, *J. Am. Chem. Soc.*, **95**, 7320 (1973).
[164] A. Sauleau, J. Sauleau, H. Bourget and J. Huet, *Compt. Rend., C*, **279**, 473 (1974).
[165] M. C. Flowers and D. E. Penny, *J.C.S. Faraday I*, **1974**, 355.
[166] J. M. Watson and B. L. Young, *J. Org. Chem.*, **39**, 117 (1974).
[167] J.-C. Paladini and R. J. Crawford, *Can. J. Chem.*, **52**, 2098 (1974); see also *Org. Reaction Mech.*, **1973**, 479.
[168] P. Dowd and K. Kang, *J.C.S. Chem. Comm.*, **1974**, 258.

[169] P. Schiess and M. Wesson, *Helv. Chim. Acta*, **57**, 1692 (1974).
[170] P. Schiess and M. Wesson, *Helv. Chim. Acta*, **57**, 980 (1974).
[171] J. R. Williams, G. M. Sarkisian, J. Quigley, A. Hasiuk and R. Vander Vennen, *J. Org. Chem.*, **39**, 1028 (1974).
[172] D. L. Whalen and A. M. Ross, *J. Am. Chem. Soc.*, **96**, 3678 (1974).
[173] J. P. Barnier, J. M. Denis, J. Salaun and J. M. Conia, *Tetrahedron*, **30**, 1405 (1974).
[174] J. Salaun, B. Garnier and J. M. Conia, *Tetrahedron*, **30**, 1423 (1974).
[175] P. R. Brook, A. J. Duke, J. M. Harrison and K. Hunt, *J.C.S. Perkin I*, **1974**, 927.
[176] K. G. Miglionese and S. I. Miller, *Tetrahedron*, **30**, 385 (1974).
[177] T. Tezuka and T. Mukai, *Chem. Comm.*, **1974**, 696; see also A. Padwa, J. Smalanoff and A. Tremper, *Tetrahedron Letters*, **1974**, 29.
[178] J. A. Green and L. A. Singer, *Tetrahedron Letters*, 4153 (1974).
[179] L. DeVries, *J. Org. Chem.*, **39**, 1707 (1974).
[180] P. Del Buttero, S. Maiorana, D. Pocar, G. D. Andreetti, G. Bocelli and P. Sgarabotto, *J.C.S. Perkin II*, **1974**, 1483; P. Del Buttero, S. Maiorana and M. Trautluft, *J.C.S. Perkin I*, **1974**, 1411.
[181] B. P. Mundy, K. B. Lipkowitz, M. Lee and B. R. Larsen, *J. Org. Chem.*, **39**, 1963 (1974).
[182] A. Padwa and E. Chen, *J. Org. Chem.*, **39**, 1976 (1974).
[183] W. S. Trahanovsky and M. G. Park, *J. Org. Chem.*, **39**, 1448 (1974).
[184] M. Livar, P. Hrnciar and A. Kopecka, *Chem. Zvesti*, **28**, 402 (1974); *Chem. Abs.*, **81**, 104418 (1974).
[185] B. C. Baumann and A. S. Dreiding, *Helv. Chim. Acta*, **57**, 1872 (1974).
[186] A. Roedig, F. Frank and G. Röbke, *Annalen*, **1974**, 630.
[187] A. Roedig, H.-A. Renk, M. Schlosser and T. Neukam, *Annalen*, **1974**, 1206; A. Roedig and H.-A. Renk, *ibid.*, p. 1214.
[188] T. A. Gosink, *J. Org. Chem.*, **39**, 1942 (1974).
[189] V. V. Korshak, D. F. Kutepov, V. A. Pantatov, N. P. Antsiferova and S. V. Vinogradova, *Isv. Akad. Nauk SSSR., Ser. Khim.*, **1973**, 1408; *Chem. Abs.*, **79**, 145715 (1973).
[190] D. L. Lee, C. J. Marrow and H. Rapoport, *J. Org. Chem.*, **39**, 983 (1974).
[191] D. MacKay and C. W. Pelger, *Can. J. Chem.*, **52**, 1114 (1974).
[192] See *Org. Reaction Mech.*, **1973**, 435.
[193] P. Vogel and M. Hardy, *Helv. Chim. Acta*, **57**, 196 (1974).
[194] D. J. Harris, M. T. Thomas, V. Snieckus and E. Klingsberg, *Can. J. Chem.*, **52**, 2805 (1974).
[195] D. N. Reinhoudt and C. G. Koumenhoven, *Rec. Trav. chim.*, **93**, 129 (1974).
[196] Y. Jean and X. Chapuisat, *J. Am. Chem. Soc.*, **96**, 6911 (1974).
[197] J. E. Baldwin and M. W. Grayston, *J. Am. Chem. Soc.*, **96**, 1630 (1974).
[198] W. J. Hehre, L. Salem and M. R. Willcott, *J. Am. Chem. Soc.*, **96**, 4328 (1974).
[199] M. E. Zandler, C. E. Choc and C. K. Johnson, *J. Am. Chem. Soc.*, **96**, 3317 (1974).
[200] See *Org. Reaction Mech.*, **1973**, 461.
[201] W. E. Billups, W. Y. Chow, K. H. Leavell and E. S. Lewis, *J. Org. Chem.*, **39**, 274 (1974).
[202] J. E. Baldwin and C. Ullenius, *J. Am. Chem. Soc.*, **96**, 1542 (1974).
[203] D. Bellus, H.-C. Mez, G. Rihs and H. Sauter, *J. Am. Chem. Soc.*, **96**, 5007 (1974).
[204] G. Jones II, and V. L. Chow, *J. Org. Chem.*, **39**, 1447 (1974).
[205] H. A. Bampfield and P. R. Brook, *Chem. Comm.*, **1974**, 173.
[206] M. Bertrand, J.-L. Gras and G. Gil, *Tetrahedron Letters*, **1974**, 37.
[207] See *Org. Reaction Mech.*, **1973**, 464.
[208] L. Eisenhuth and H. Hopf, *J. Am. Chem. Soc.*, **96**, 5667 (1974).
[209] F.-G. Klarner, *Tetrahedron Letters*, **1974**, 19.
[210] F.-G. Klarner, *Angew. Chem. Internat. Ed.*, **13**, 268 (1974).
[211] L. A. Paquette, M. Oku, W. E. Heyd and R. H. Meisinger, *J. Am. Chem. Soc.*, **96**, 5815 (1974).
[212] L. A. Paquette and M. Oku, *J. Am. Chem. Soc.*, **96**, 1219 (1974).
[213] L. A. Paquette, R. E. Wingard, Jr., and J. M. Photis, *J. Am. Chem. Soc.*, **96**, 5801 (1974).
[214] J. M. Brown and M. M. Ogilvy, *J. Am. Chem. Soc.*, **96**, 292 (1974).
[215] G. Boche and G. Schneider, *Tetrahedron Letters*, **1974**, 2449.
[216] G. Scholes, G. R. Graham and M. Brookhart, *J. Am. Chem. Soc.*, **96**, 5665 (1974).
[217] R. C. DeSelms, *J. Am. Chem. Soc.*, **96**, 1967 (1974).
[218] E. Ratajczak, *Bull. Acad. Pol. Sci., Ser. Sci. Chim.*, **21**, 691 (1973); *Chem. Abs.*, **79**, 136300 (1973).
[219] S. Itô, M. Yatagai, K. Endo and Y. Ueda, *Chem. Letters (Tokyo)*, **1974**, 117.
[220] K. Riistama and O. Harva, *Finnish Chem. Letters*, **1974**, 132.
[221] R. S. Cooke and U. H. Andrews, *J. Am. Chem. Soc.*, **96**, 2974 (1974).
[222] A. V. Tarakanova, N. P. Vinnikova, E. M. Mil'vitskaya and A. F. Plate, *Zh. Org. Khim.*, **10**, 669 (1974); *Chem. Abs.*, **80**, 132530 (1974).

[223] A. S. Kende and E. E. Riecke, *Chem. Comm.*, **1974**, 383.
[224] L. T. Scott and G. K. Agopian, *J. Am. Chem. Soc.*, **96**, 4325 (1974).
[225] L. A. Paquette, M. J. Kukla and J. C. Stowell, *J. Am. Chem. Soc.*, **94**, 4902 (1972).
[226] E. Vedejs, R. P. Steiner and E. S. C. Wu, *J. Am. Chem. Soc.*, **96**, 4040 (1974).
[227] R. K. Lustgarten and H. G. Richey, Jr., *J. Am. Chem. Soc.*, **96**, 6393 (1974).
[228] C. O. Bender and S. S. Shugarman, *Chem. Comm.*, **1974**, 934.
[229] N. D. Epiotis, *J. Am. Chem. Soc.*, **94**, 1941 (1972).
[230] J. A. Berson and J. M. Janusz, *J. Am. Chem. Soc.*, **96**, 5939 (1974).
[231] M. S. Newman and M. C. Van der Zwan, *J. Org. Chem.*, **39**, 761 (1974).
[232] G. Kaupp and R. Dyllick-Brenzinger, *Angew. Chem. Internat. Ed.*, **13**, 478 (1974).
[233] R. Srinivasan, *Tetrahedron Letters*, **1974**, 2725.
[234] H. Prinzbach, H. Fritz, H. Hagemann, D. Hunkler, S. Kagabu and G. Philippossian, *Chem. Ber.*, **107**, 1971 (1974).
[235] H. Prinzbach, S. Kagabu and H. Fritz, *Angew. Chem. Internat. Ed.*, **13**, 482 (1974).
[236] M. J. Goldstein and S.-H. Dai, *Tetrahedron Letters*, **1974**, 535.
[237] See *Org. Reaction Mech.*, **1973**, 48.
[238] H.-P. Loffler, *Chem. Ber.*, **107**, 2691 (1974).
[239] J. E. Baldwin and M. W. Grayston, *J. Am. Chem. Soc.*, **96**, 1629 (1974).
[240] E. I. Bagrii, T. N. Doglopolova and P. I. Sanin, *Neftekhimiya*, **13**, 482 (1973); *Chem. Abs.*, **79**, 125491 (1973).
[241] L. Meurling, *Acta Chem. Scand.*, *B*, **28**, 369 (1974).
[242] L. Meurling, *Acta Chem. Scand.*, *B.*, **28**, 399 (1974).
[243] U. Edlund and G. Bergson, *Acta Chem. Scand.*, *B.*, **28**, 157 (1974).
[244] A. Ferro and Y.-R. Naves, *Helv. Chim. Acta*, **57**, 1141, 1152 (1974).
[245] Y. A. Slobodin and I. Z. Egenburg, *Zh. Org. Khim.*, **9**, 1791 (1973); *Chem. Abs.*, **79**, 145719 (1973).
[246] Y. A. Ranneva, G. A. Kudryatseva, A. V. Dolidze and A. I. Shatenshtein, *Zh. Obsch. Khim.*, **44**, 671 (1974); *Chem. Abs.*, **80**, 145126 (1974).
[247] S. I. Faingold and T. Lesment, *Sin. Primen. Novykh Poverkh. Vesch.*, **1973**, 8; *Chem. Abs.*, **80**, 132520 (1974).
[248] T. Koyama, K. Ogura and S. Seto, *J. Biol. Chem.*, **248**, 8043 (1973).
[249] S. J. Jacobson, C. G. Willson and H. Rapoport, *J. Org. Chem.*, **39**, 1074 (1974).
[250] G. Belluci, A. Marsili, E. Mastronilli, I. Morelli and V. Scartoni, *J.C.S. Perkin II*, **1974**, 201.
[251] R. J. Stoodley and R. B. Wilkins, *J.C.S. Perkin I*, **1974**, 1572.
[252] D. Darwesh and S. K. Datta, *Tetrahedron*, **30**, 1155 (1974).
[253] R. Corriu and M. Henner-Leard, *J. Organometal. Chem.*, **64**, 351 (1974); **65**, C39 (1974).
[254] R. H. Martin and M. J. Marchant, *Tetrahedron*, **30**, 347 (1974).
[255] H. Kessler, *Tetrahedron*, **30**, 1861 (1974).
[256] M. F. Hughes and R. T. Adams, *Am. Chem. Soc., Div. Petrol. Chem. Prepr.*, **17**, B109 (1972); *Chem. Abs.*, **79**, 136250 (1973).
[257] A. I. Konovalov, L. K. Konovalova and E. G. Kateev, *Zh. Org. Khim.*, **9**, 1830 (1973); *Chem. Abs.*, **79**, 145721 (1973).
[258] F. Kasper and G. Bachmann, *Z. Chem.*, **14**, 241 (1974).
[259] M. Rabinovitz, I. Agranat and A. Weitzer-Dagen, *Tetrahedron Letters*, **1974**, 1241.
[260] D. G. Whitten, *J. Am. Chem. Soc.*, **96**, 594 (1974).
[261] W. G. Herkstroeter, *J. Am. Chem. Soc.*, **95**, 8686 (1973).
[262] C. O. Meese, W. Walter and M. Berger, *J. Am. Chem. Soc.*, **96**, 2259 (1974); A. C. Satterthwaite and W. P. Jencks, *ibid.*, p. 7045.
[263] G. Toth, L. Toldy, I. Toth and B. Rezessy, *Tetrahedron*, **30**, 1219 (1974).
[264] G. Toth and L. Toldy, *Acta Chim. (Budapest)*, **80**, 349 (1974); *Chem. Abs.*, **81**, 24759 (1974).
[265] I. N. Samin, N. I. Shapranova and S. G. Kusnetsov, *Khim. Geterotsikl. Soedin.*, **1973**, 1176; *Chem. Abs.*, **80**, 14463 (1974).
[266] C. K. Sauers and H. M. Relles, *J. Am. Chem. Soc.*, **95**, 7731 (1973).
[267] J. Bjørgo, D. R. Boyd, C. G. Watson, W. B. Jennings and D. M. Jerina, *J.C.S. Perkin II*, **1974**, 1081.
[268] J. Bjørgo, D. R. Boyd, C. G. Watson and W. B. Jennings, *J.C.S. Perkin II*, **1974**, 757.
[269] V. A. Ketlinskii and I. L. Bagal, *Zh. Org. Khim.*, **9**, 1915 (1973); *Chem. Abs.*, **79**, 145720 (1973).
[270] T. Sueyoshi, N. Nishimura, S. Yamamoto and S. Hasegawa, *Chem. Letters (Tokyo)*, **1974**, 1131.
[271] R. F. Hudson, A. J. Lawson and K. A. F. Record, *J.C.S. Perkin II*, **1974**, 869.
[272] M. Shibasaki, T. Sato, N. Ohashi, S. Terashima and S. Yamada, *Chem. Pharm. Bull.*, **21**, 1868 (1973); *Chem. Abs.*, **80**, 14462 (1974).
[273] G. Peyronel, R. Battistuzzi and G. Marcotrigiano, *J.C.S. Perkin II*, **1974**, 896.

[274] G. Strukul, M. Bonivento, R. Ros and M. Graziani, *Tetrahedron Letters*, **1974**, 1791.
[275] W. Strohmeier and E. Eder, *Z. Naturforsch.*, **29**, 280 (1974).
[276] W. Strohmeier, *Z. Naturforsch.*, **29**, 282 (1974).
[277] I. Heertje, G. K. Koch and W. J. Wösten, *J. Catal.*, **32**, 337 (1974).
[278] C. D. Weis and T. Winkler, *Helv. Chim. Acta*, **57**, 856 (1974).
[279] Y. Sasson and G. Rempel, *Tetrahedron Letters*, **1974**, 4133.
[280] G. Eadon and M. Y. Shiekh, *J. Am. Chem. Soc.*, **96**, 2288 (1974).
[281] R. G. Miller, P. A. Pinke, R. D. Stauffer, H. J. Golden and D. J. Baker, *J. Am. Chem. Soc.*, **96**, 4211 (1974).
[282] P. A. Pinke and R. G. Miller, *J. Am. Chem. Soc.*, **96**, 4221 (1974).
[283] P. A. Pinke, R. D. Stauffer and R. G. Miller, *J. Am. Chem. Soc.*, **96**, 4229 (1974).
[284] H. J. Golden, D. J. Baker, and R. G. Miller, *J. Am. Chem. Soc.*, **96**, 4235 (1974).
[285] F. A. Cotton, D. L. Hunter and P. Lahuerta, *J. Am. Chem. Soc.*, **96**, 4723 (1974).
[286] See *Org. Reaction Mech.*, **1973**, 450.
[287] H. W. Voigt and J. A. Roth, *J. Catal.*, **33**, 91 (1974).
[288] T. D. J. D'Silva, W. E. Walker and R. M. Manyik, *Tetrahedron*, **30**, 1015 (1974).
[289] V. Koch-Pomeranz, H.-J. Hansen and H. Schmid, *Helv. Chim. Acta*, **56**, 2981 (1973).
[290] M. A. McKinney and S. K. Chan, *Tetrahedron Letters*, **1974**, 1145.
[291] I. Murata, T. Tatsuoka and Y. Sugihara, *Tetrahedron Letters*, **1974**, 199.
[292] H. Hogeveen and B. J. Nusse, *Tetrahedron Letters*, **1974**, 159.
[293] See *Org. Reaction Mech.*, **1973**, 454.
[294] R. Roulet, J. Wenger, M. Hardy and P. Vogel, *Tetrahedron Letters*, **1974**, 1479.
[295] G. F. Koser and D. R. St. Cyr, *Tetrahedron Letters*, **1974**, 3015.
[296] W. Slegeir, R. Case, J. S. McKennis and R. Pettit, *J. Am. Chem. Soc.*, **96**, 287 (1974).
[297] L. A. Paquette, J. M. Photis, J. Fayos and J. Clardy, *J. Am. Chem. Soc.*, **96**, 1217 (1974).
[298] A. de Meijere, *Tetrahedron Letters*, **1974**, 1845.
[299] A. de Meijere and L.-U. Meyer, *Tetrahedron Letters*, **1974**, 1849.
[300] R. L. Cargill, T. E. Jackson, N. P. Peet, and D. M. Pond, *Accounts Chem. Res.*, **7**, 106 (1974).
[301] P. Bakuzis, G. C. Magalhaes, H. Martius and M. L. F. Bakusis, *J. Org. Chem.*, **39**, 2427 (1974).
[302] H. Hart, I. Huang and P. Lavrik, *J. Org. Chem.*, **39**, 999 (1974); H. Hart and I. Huang, *ibid.*, p. 1005.
[303] R. S. Atkinson and R. H. Green, *J.C.S. Perkin I*, **1974**, 394.
[304] G. Pattenden and R. Storer, *J.C.S. Perkin I*, **1974**, 1606.
[305] S. Ranganathan and H. H. Raman, *Tetrahedron*, **30**, 63 (1974).
[306] W. R. Vaughan and D. M. Teegarden, *J. Am. Chem. Soc.*, **96**, 4902 (1974).
[307] S. J. Cristol and R. J. Bopp, *J. Org. Chem.*, **39**, 1336 (1974).
[308] B. E. Cross, M. R. Firth and R. E. Markwell, *Chem. Comm.*, **1974**, 930.
[309] J. J. Riehl, A. Smolikiewcz and L. Thil, *Tetrahedron Letters*, **1974**, 1451.
[310] D. H. R. Barton and S. Prabhakar, *J.C.S. Perkin I*, **1974**, 781.
[311] B. A. Arbuzov, *Z. Chem.*, **14**, 41 (1974).
[312] H. Des Abbayes and R. Badard, *J. Organometal. Chem.*, **61**, C51 (1973).
[313] J. J. Eisch and R. J. Wiksok, *J. Organometal. Chem.*, **71**, C21, (1974).
[314] E. H. White and P. M. Dzadzic, *J. Org. Chem.*, **39**, 1517 (1974).
[315] L. Duhamel and J. Gralak, *Bull. Soc. Chim. France*, **1974**, 2124.

Author Index 1974

In this index each page number is followed, in parentheses, by the relevant reference number(s) listed on that page.

Subject Index 1970–1974

Errata for Organic Reaction Mechanisms 1969

P. 346, 9 lines from the bottom: *For* Mn^{III} *read* Mn^{II}.

Errata for Organic Reaction Mechanisms 1970

P. 150: The solvent for the reaction should be triglyme, not THF.

Errata for Organic Reaction Mechanisms 1973

P. 345: In formula (**37**) the bond from phosphorus should be to C-2 of thiophen, and not to C-3 as shown.

U.C.W. ABERYSTWYTH
LIBRARY